49th American Association of Swine Veterinarians Annual Meeting 2018

GLOBAL KNOWLEDGE: Individual Applications

San Diego, California, USA
3 – 6 March 2018

ISBN: 978-1-5108-5980-7

Printed from e-media with permission by:

Curran Associates, Inc.
57 Morehouse Lane
Red Hook, NY 12571

Some format issues inherent in the e-media version may also appear in this print version.

Printed by Curran Associates, Inc. (2018)

For permission requests, please contact American Association of Swine Veterinarians
at the address below.

American Association of Swine Veterinarians
830 26th Street
Perry, IA 50220-2328
USA

Phone: 515-465-5255
Fax: 515-465-3832

aasv@aasv.org

Additional copies of this publication are available from:

Curran Associates, Inc.
57 Morehouse Lane
Red Hook, NY 12571 USA
Phone: 845-758-0400
Fax: 845-758-2633
Email: curran@proceedings.com
Web: www.proceedings.com

GLOBAL KNOWLEDGE:
Individual Application

49th AASV Annual Meeting

March 3-6, 2018

San Diego, California

Table of Contents

Howard W. Dunne Memorial Lecture

How culture, geography, and socioeconomic status affect global animal source food (ASF) consumption: Applications for swine veterinarians 15
William R. DuBois, DVM, DABVP

Alex Hogg Memorial Lecture

This is our time, the choices are yours 29
Rodger Main, DVM

Sunday: Research Topics

Effect of antimicrobial administration on the prevalence and abundance of selected antibiotic resistance genes in perinatal piglets 39
J. Lowe; B. Blair; M. Zeineldin; et al.

Microbial killing capacity of aqueous and gaseous ozone on different surfaces contaminated with manure 43
Ameer A. Megahed; Brian Aldridge; James F. Lowe

Evaluation of the survival of viral pathogens in contaminated feed ingredients using transboundary shipment models 44
Scott Dee; Fernando V. Bauermann; Megan C. Niederwerder; et al.

Spatial autocorrelation and what it means for swine surveillance 45
M. Rotolo; C. Wang; M Haddad; et al.

Pathogenesis of a porcine parainfluenza virus-1 isolate (USA/MN25890NS/2016) in conventional and CDCD piglets 46
M. Welch, BS; J. Park, PhD; K. Harmon, PhD; et al.

Breeding herd factors associated with influenza in piglets at weaning 51
F. Chamba; A. Schelkopf; M. Allerson; et al.

Is influenza vaccination a key driver of influenza genetic diversity in piglets? 52
J. Nirmala; A. Perez; M. Culhane; et al.

Novel approaches for influenza surveillance in swine breeding herds 53
Jorge Garrido Mantilla; Julio Alvarez; Marie Culhane; et al.

Case report: Identification of *Mycoplasma hyopneumoniae* infection in a breeding herd through tracheo-bronchial swab monitoring 54
F. Vangroenweghe; E. Willems; O. Thas; et al.

Genotypic and phenotypic differences between LA-MRSA ST5 and clinical MRSA ST5 from humans 57
Samantha J. Hau; Anna Allué-Guardia; Brigida Rusconi; et al.

Processing fluids for PRRSV monitoring and surveillance systems 61
Will A. López, DVM; Jose Angulo, DVM; Clayton Johnson, DVM; et al.

PRRS outbreak investigation pilot program: Lessons learned 63
Derald Holtkamp, DVM, MS; Kimberlee Baker; Rita Neat; et al.

Evaluating natural planned exposure protocols on rotavirus shedding patterns in gilts and the impact on their suckling pigs 64
Amanda V. Anderson; Jamie Madigan, BS; Derald Holtkamp, DVM, MS; et al.

Evaluation of the persistence of Senecavirus A during elimination program in a sow farm 67
Deb Murray; Levi Johnson; Matthew Sturos; et al.

Shedding and persistence of Senecavirus A in boars: Natural exposure and experimental infection with an historical and a contemporary strain 68
Matthew Sturos; Deborah Murray; Darwin Reicks; et al.

Sunday Concurrent Session #1: Student Seminar

Developing a useful model to evaluate the judicious use of a ceftiofur crystalline free acid to control umbilical infections and hernias in a commercial production system 70
Cassandra Fitzgerald, BS; Pete Thomas, DVM; Richard Swalla, et al.

The effect of increased feed intake during late gestation on litter characteristics of commercial sows 71
K. O'Brien, BS; A.Ornelas; F. Zarate, MS; et al.

Evaluation of antibiotics at weaning on swine respiratory disease in the early wean-finish phase 73
Joel Steckelberg; Ethan Spronk, DVM; Noel Garbes, DVM; et al.

Comparison of oral fluids, environmental swabs, and pooled feces for detection of dysentery-associated *Brachyspira* spp. in pigs 75
W. Webb, BS; H. Warneke, MS; J. Seate, DVM; et al.

Comparison of biosecurity aspects between swine breeding herds with low and high incidence of porcine reproductive and respiratory syndrome virus outbreaks 80
Kimberlee L. Baker, BS; Gustavo Silva, DVM, MS; Robert Morrison, DVM, MBA, PhD; et al.

Putting to test a diagnostic protocol for confirmation of *Mycoplasma hyopneumoniae* elimination in sow farms 81
B. Chapman, BS; R. Evelsizer, DVM; M. Pieters, DVM

Evaluation of washing-disinfectant-TADD conditions required to produce a negative PCR result when monitoring the presence of PRRSV on metal surfaces 83
Rita Anne Neat; Jianqiang Zhang; Hai Hoang; et al.

Litter porcine reproductive and respiratory syndrome (PRRS) status at processing and relationship to PRRS status at weaning using different pools of sampling tissues 84
Mikalah Smith, BS; John Waddell, DVM, MBA; Greg Hartsook, BS; et al.

Vaccination shedding of a porcine reproductive and respiratory syndrome (PRRS) modified live virus in nursery pigs 87
E. Koep, BS; B. Leuwerke, DVM, MS; S. Holst, DVM

Evaluation of Senecavirus A transmission in laboratory and wild type mice strains under experimental conditions 90
E. R. Houston; G. Temeeyasen; L. G. Gimenez-Lirola; et al.

Utilizing piglet-processing fluids to detect PRRSV by PCR in a low-prevalence population 95
Jacob Baker; Deb Murray; Amanda Sponheim; et al.

Geographic distribution and genetic diversity of porcine circovirus type 3 from clinical samples in the US swine farms 96
Zhen Yang; Todd P. Knutson, PhD; Fangzhou Chen, MS; et al.

Use of a hurricane fogger for *Mycoplasma hyopneumoniae* inoculation in nursery age gilts 97
M. Nickel, BS; E. Toohill, DVM; J. Lehman, DVM, MS

A comparison of the use of tails versus testicles in the production of processing fluids for detection of porcine reproductive and respiratory syndrome virus 99
M. Hood, BS; S. Hough, DVM; J. Angulo, et al.

Evaluating the effects of an isotonic protein solution in nursery pigs 101
Stephanie Betbeze, BS; Ava Firth, BS, DVM, MVS, MANZCVS, DACVECC, DECVECC; Chris Sievers, DVM, MS; et al.

Sunday Concurrent Session #2: Industrial Partners

Effect of increasing GleptoForte dosage in newborn pigs on sow and litter performance 103
H. Williams, MS; J. DeRouchey, MS, PhD; J. Woodworth, MS, PhD; et al.

Effect of ButiPEARL™ Z level and feed form on nursery pig growth performance for the first 6 weeks after weaning 106
Ana Lucia de Souza; B. Kremer; T. Marsteller

Is it possible that an antibiotic may actually reduce antibiotic resistance? Flavomycin review 111
D. Nolan, DVM; C. Sparks, PhD

Mucosal IgA production in response to a killed, Carbigen-adjuvanted vaccine administered intranasally 113
G. Weaver; D. Rudrappa; J. Kula; D. Guggenbiller; et al.

Evaluation of a killed virus vaccine (MJPRRS®) as part of a PRRS virus mitigation program 116
K Kinsley, DVM; D Guggenbiller, DVM, MBA; J Newberry, DVM, MBA, MS; et al.

Field studies to evaluate the efficacy of BOAR BETTER®, a new pheromone spray for sows 121
Michael Kuhn, DVM, MBA; John J. McGlone, PhD; Garrett Thompson, MS

Impact of sire-specific mortality on commercial swine profitability 124
D. W. Newcom; C. Witte; M. Brubaker; et al.

Effect of Tonisity Px™ administration to sows and their piglets 128
Ava M. Firth, BS, DVM, MVS, MANZCVS, DACVECC, DECVECC; Guillermo López Cano, DVM MSc; Ahmad R. Seradj, PhD; et al.

Production performance evaluation of Porcilis™ Ileitis in a commercial finisher using a *Lawsonia intracellularis* seeder pig challenge model 131
Brett O'Brien, VMD, MS; Scott Dee, DVM, PhD, Dipl ACVM; Joel Nerem, DVM; et al.

Efficacy of swine influenza virus replicon particle vaccination including cellular and humoral responses after challenge 134
Pravina Kitikoon, DVM, PhD; Channing Sebo, DVM; Mark Mogler, PhD; et al.

Efficacy and field safety of Prime Pac PRRS™ RS, a new tool against porcine reproductive and respiratory syndrome virus 138
E. Strait, DVM, PhD; F. Roerink, MS; S. Knetter, PhD; et al.

Swine models: An integrated approach to providing management and nutritional solutions 142
N. S. Ferguson; M. Swantek; D. Woods

Trends of mycotoxin contamination in US corn 144
E. G. Hendel, VMD, PhD; S. Masching, DVM; T. Jenkins, PhD; et al.

Polymeric formulations are efficient and versatile adjuvants for swine vaccines 147
S. B. Xu; J. Ben Arous; H. Imbault

Increasing intestinal mucin production using algae extracts 149
M. Gallissot, MSc; L. Miel, MSc; C. Rousseaux, PhD; et al.

Seneca Valley virus VI positive flies *(Musca domestica)*: What questions does this pose for the swine industry? 153
Mark Engesser, DVM; Phil Gauger, DVM, MS, PhD; Jianqiang Zhang, MD, MS, PhD; et al.

Use estimates of in-feed antimicrobials in swine production in the United States, 2012 156
E. Bush, DVM, MS; K. Bjork, DVM, MSPH, PhD; C. Kopral, PhD

Development of multiplex PCRs for a full swine respiratory panel 159
E. van Esch, DVM

How heat placement in farrowing affects nutrient intake and pre-wean mortality (PWM) 162
B. Strobel, PE, MS, MBA

Sunday Concurrent Session #3: Industrial Partners

Improving the speed of health-based decision making 165
E. J. Lowe, DVM; J. T. Holck, DVM, MS, MBA; D. D. Polson, DVM, MS, PhD; et al.

The impact of zinc amino acid complex supplementation on the porcine response to subclinical *Lawsonia intracellularis* infection 168
F. Leite, DVM, MS, PhD Candidate; E. Vasquez, DVM; F. Vannucci, DVM, MSc, PhD; et al.

Effect of avilamycin (Kavault®) on incidence and severity of diarrhea in presence of pathogenic *Escherichia coli* in nursery pigs 172
D. C. Beckler, DVM, MS; R. Evelsizer, DVM; C. L. Puls, PhD; et al.

Use of tilmicosin (Pulmotil AC®) water medication for *Mycoplasma hyopneumoniae* elimination 176
P. Yeske, DVM, MS; R. Evelsizer, DVM; M. Pieters, DVM, PhD; et al.

Controlling bacterial pathogens in the growing pig: New technology 177
Keith Wilson, DVM; Eric Bumgardner, MS; Paulraj Lawrence, PhD

Streptococcus suis: Research overview and field experiences 180
Paulraj Lawrence, PhD; Eric Bumgardner, MS

Molecules produced by probiotics prevent enteric colibacillosis in pigs 183
R. Nordeste; A. Tessema; S. Sharma; et al.

Genetic selection for health and welfare using meat inspection data 187
P. K. Mathur, MVSc, PhD; J. M. Eggert, MS, MBA, PHD, Dipl. ACAG

Evaluation of two serological assays for *Lawsonia intracellularis* antibodies 191
Erika Vasquez; Lacey Marshall Lund; Fabio Vannucci; et al.

Comparison of DLI® controlled immunity, Enterisol® and Porcilis® Ileitis vaccines in grower pigs challenged with a *Lawsonia intracellularis* mucosal homogenate 195
Nathan Winkelman, DVM; Adam Mueller, DVM; Rickie Domangue, PhD

Fat-soluble vitamins: Beyond nutritional needs 200
Robert L. Stuart, PhD

Vaccination and challenge study for an H3N2 swine influenza recombinant DNA vaccine 204
Tim Miller; Laura Trygstad; Dustin Petrik; et al.

Fast Genetics: The use of technology to increase stability and profitability in the swine production chain 208
R. Leach, PhD; B. Belstra, PhD; D. Grossi, DVM, PhD; et al.

Lactobacillus acidophilus fermentation product as an alternative to therapeutic zinc oxide in weaned pig diets on performance and response to *E coli* challenge 211
X. Guan; B. E. Bass; P. J. van der Aar; et al.

Use of statistical quality control and metrics in PRRS control programs 213
Jose Angulo, DVM

Evaluation of a novel PCV + *M hyopneumoniae* vaccine in swine experimentally challenged with PCV2d and *M hyopneumoniae* under commercial conditions 218
L. Galina, DVM, PhD; B. Cowles, DVM, MS, MBA; D. Amodie, MS; et al.

Safety comparative study of two reproductive vaccines administered during lactation 222
I. Rodriguez-Ballara; G. Ibanez

The use of LucraFit® Feed and OutPace® Feed Additive to support the reduction of antibiotics and animal protein in young pig diets 224
Dan McManus, DVM; T. Weeden, PhD; B. de Rodas, PhD; et al.

Management considerations and tactics to maximize sow lactation feed intake 227
J. C. Pinilla, DVM, MS; S. Casiro, DVM. MS; S. Canavate, DVM; et al.

Sunday Concurrent Session #4: Industrial Partners

The chemistry of biosecurity 231
Jeff Odle

Efficacy of Circo/MycoGard™ to reduce clinical disease related to *Mycoplasma hyopneumoniae* and porcine circovirus type 2 234
D. Rosener, DVM; R. Kaptur, DVM; S. Ma, PhD; et al.

Determination of the effectiveness of Aivlosin® Water Soluble Granules (62.5% w/w tylvalosin) in drinking water of pigs for the control of swine respiratory disease (SRD)....239
R. Kaptur, DVM; D. Rosener, DVM; E. Abbott, BVMS, PhD, MRCVS; et al.

Maximus Sow™ next-generation sow ESF and management information system....242
Thomas Stein, DVM, MS, PhD

Development of Ingelvac Provenza™: A new tool against IAV-S 243
Marc A. Eichmeyer; Wesley Johnson; Eric Vaughn; et al.

Laboratory efficacy evaluation of intranasal vaccination with Ingelvac Provenza™ in newborn piglets 245
R. A. Smiley, MS; W. Johnson, MS; B. Fergen, PhD; et al.

The effect of intranasal vaccination with Ingelvac Provenza™ in newborn piglets on IAV-S shedding in nasal swabs 248
Marika Genzow; Christa Goodell; Troy J. Kaiser; et al.

Understanding factors influencing pig quality measured using pig weaning weight....249
C. Bruns, PhD; R. Noel; B. McNeil; et al.

eYeGrow - weight monitor for finisher....250
Guido Klement, Dr. agr; Simon Lague; Patrick van Dijk, B.Eng., B.ICT

Use of the swine bioassay test to evaluate the infectivity of UV-irradiated commercial liquid porcine plasma 253
E. Blázquez; J. Ródenas; C. Rodríguez; et al.

High quality semen program – HQS....255
P. Quadros, DVM, MS; K. Calderam, DVM, MS; J. Rocha, DVM; et al.

Cleaning and disinfecting critical surfaces: Farrowing crates and trailers....260
Ricardo Muñoz

Protecting animal health and profitability via air filtration solutions, a front line defense against PRRSV, PEDV, and influenza....262
J. McGuire; M. Osborne, PhD

Field safety study of a vaginal mucosal vaccination for porcine reproductive and respiratory syndrome virus (PRRSV) in sows....265
Glen Almond; Cary Sexton; Juliana Bonin Ferreira; et al.

The effects of Defusion® products on the performance of nursery pigs fed diets contaminated with vomitoxin....267
Kari Saddoris-Clemons, PhD; Sara Ebarb, MS; Colleen Fowler, MS; et al.

Effects of an antioxidant support technology on sows and piglets 270
K. Price, PhD; C. Fowler, MS; D. Giesting, PhD; et al.

Impact of two training implementation plans on post-training competency expression of personnel on farm....274
S. Probst Miller, DVM

The impact of Swine Awake®, a new, non-antimicrobial protein, in natural disease conditions in newly weaned pigs....276
J. Tyler Holck, DVM, MS, MBA; Rob Musser, PhD; Ran Song, PhD; et al.

Next generation sequencing for veterinary diagnostics and vaccinology....278
B. Hause, MS, PhD

Poster Session: Veterinary Students

Pooling sensitivity of oropharyngeal swabs when sampling for influenza A and porcine reproductive and respiratory syndrome viruses 279
Rachel Schulte; Thomas Wetzell, DVM; Ethan Schmaling; et al

Medium chain fatty acids improve growth and alter fecal microbial populations in nursery pigs 280
Jordan Gebhardt, BS; Steve Dritz, DVM, PhD; Jason Woodworth, PhD; et al.

An assessment of hemoglobin concentrations in relation to reproductive stage and parity in sows 282
Kayla Castevens; Glen Almond, DVM; Juliana Bonin Ferreira, DVM

Pre-challenge microbiome composition is associated with improved weight gain in pigs after vaccination with a PRRS MLV vaccine followed by challenge with PRRSV and PCV2b 284
L.A. Constance, J.B. Thissen, C.J. Jaing; et al.

Investigations of prevalence and causes of lameness in various age groups of gilts 285
Shelby Perkins, BS; Emily Severt, DVM; Darin Madson, DVM; et al.

The cost of porcine reproductive and respiratory syndrome virus and porcine epidemic diarrhea virus in finishing pigs 286
K. Henness, BS; K. Dion, DVM; D. Holtkamp, DVM, MS; et al.

Immuno-modulatory influence of antibiotic therapy on the humoral immune response to vaccination in weaned pigs 287
J. Scott, BS; G. Almond, DVM, PhD; J. Fogle, DVM, PhD, DACVIM

Comparison of serological responses in traditional and parity segregated flows for *Mycoplasma hyopneumoniae* and *Lawsonia intracellularis* 289
Elizabeth Noblett, MAS; Joshua Duff, DVM; Samuel Copeland, DVM; et al.

Utilizing serum antibody levels in post vaccinated piglets to evaluate maternal antibody interference and vaccination compliance 291
J. Tubbs, MS; E. Byers, DVM

Assessment of the association between respiratory pathogen burden and the productivity of growing pigs 293
R. Stika; D. Holtkamp, MS, DVM; C. Goodell, MS, DVM, PhD; et al.

A novel approach to water medication calculations 295
Brent Sexton, BS; Alejandro Ramirez, DVM, MPH, PhD

Comparison of individual oral fluids, pooled oral fluid samples and Swiffer™ environmental samples of drinkers for the detection of IAV-S and PRRSV by PCR 296
Taylor Homann; Chris Sievers; Jana Morgan; et al.

Evaluation of serological responses to Circumvent PCV M G2 in sows vaccinated pre-farrowing and in pigs with maternally-derived antibodies at time of first vaccination 297
Jamie Madigan, BS; Amanda Anderson; Jeremy S Pittman, DVM, MS, DABVP

Validation of a urine dipstick as a tool to assess the potential impact of urinary tract health on reproductive performance and sow mortality 299
J. Bremer, BS; B. Peters, DVM; M. Jansen, DVM; et al.

Evaluation of antibody titer of the weaned pig following vaccination with a killed influenza A virus vaccine in the presence of maternal antibodies both pre and post heterologous influenza A virus challenge 301
Abigail Ruane, BS; Mark Wagner, DVM; Abigail Redalen, DVM; et al.

Characterization of skin variations that may influence transdermal absorption of topical drugs 303
S. Anthony, BS; J. Brown, DVM; D. Madson, DVM, PhD, DACVP; et al.

Monitoring mature sows for *Mycoplasma hyopneumoniae* (Mhp) pre-and post-Mhp inoculated gilt entry 305
Megan E. Bloemer, BS; Thomas Painter, DVM; Adam Schelkopf, DVM

Testing of the efficacy of various methods of urine collection for detection of flunixin meglumine (Banamine) in gilts and barrows 306
C. Brooks, BS; T. Mays, MS; B. Dominguez, DVM, MS

Effects of Improvest®, a gonadotropin releasing factor analog-diphtheria toxoid conjugate, on boar behavior during trailer transport 308
Alexandria D. Bufford; Noel Garbes, DVM; Wesley Lyons, DVM

Assessing the impact of perinatal antibiotics on performance and antibiotic resistant bacteria in pre-weaned pigs 309
B. Burton; M. Zeineldin; A. Megahed; et al.

The impact of observer experience and training methods on behavioral scores in piglets 310
E. DeCaluwe-Tulk, HBSc; H. Golightly, HBSc; K. Reynolds, BSc, MSc, DVM; et al.

Validation and application of accelerometers to sow behavior and management 312
Mary Jane Drake, PhD; Thomas D. Parsons, VMD, PhD, DACAW

Multiplexed phylogenetic comparison of PRRSV sequences using the MinION nanopore sequencer 314
T. Engle, BS, MS; S. M. Todd, BS, MS; K. K. Lahmers, BS, DVM, PhD, DACVP

Evaluation of the impact of iron dosage on post-weaning weight gain, and mortality 315
L. Fredericks, BS; C. Olsen, DVM; A. Maschhoff, DVM; et al.

Assessment of intra- and interrater agreement of pain behavior scores in castrated piglets 316
H. Golightly, HBSc; E. DeCaluwe-Tulk, HBSc; K. Richardson; et al.

In vitro culture of porcine B cells 317
Kevin L. Gustafson; Michael P. Murtaugh

Understanding the impact of farrowing induction on pigs born alive 319
Olivia Hardebeck, BS; Luis Ochoa, PhD; Laura Greiner, PhD; et al.

Physical and behavioral indicators of animal welfare on farms using electronic sow feeders 321
J. Hurst; T. D. Parsons, VMD, PhD., DACAW; M. K. Pierdon, VMD, DACAW

Efficacy of interferon-inducing PRRSV vaccine candidate A2P90 against challenge with atypical pathogenic strain MN184 322
R. Kass, BS; Y. Zhang, PhD; Z. Ma, PhD; et al.

Determining the necessity of peri-processing antimicrobial use in piglets 323
K. Klemesrud; A. Forseth, DVM; E. Akin, BS; et al.

Dynamics of porcine parainfluenza virus shedding in sows and their piglets 324
Allison Knox, BS; Amber Stricker, DVM, MS

Comparison of pathogenic *Salmonella* spp. detection from porcine oral fluids, over-the-shoe booties, and fecal samples 326
S. Naberhaus; A. Krull, DVM, PhD; B. Arruda, DVM, PhD; et al.

Worker satisfaction and animal empathy on sow farms using electronic sow feeders 327
A. Palmieri; T. Parsons, VMD, PhD, DACAW; M. Pierdon, VMD, DACAW

Evaluation of analgesics for castration and tail docking in 3-5 day old piglets 328
Andrea Patterson, BScH; Karine Talbot, DVM

Using estrone sulfate urine test to determine pregnancy in sows and gilts 329
Meagan Rennie, BS; Brandon J. Dominguez, DVM MS

Evaluation of BoarBetter® pheromone in optimizing swine reproductive efficiencies in two commercial breeding farms 330
M. Rohde, BS; J. Brown, DVM; S. Anthony, BS; et al.

Evaluation of farrowing stall sanitation protocols with ATP bioluminescence 332
Marjorie Schleper, BS; Laura Dalquist, DVM; Tyler Te Grotenhuis, DVM; et al.

Effects of mechanical lead system for boar control during heat detection on libido and salivary androstenone 334
N. Sugai; S. Probst Miller, DVM; R. V. Knox, PhD

Comparison of oral fluids versus tonsil swabs to diagnose *Mycoplasma hyosynoviae* infection prior to onset of clinical signs.............336
Grzegorz Tarasiuk; Luis Ochoa, PhD; Micah Jansen, DVM; et al.

Medicated starter feed in weaned piglets and ensuing effects on overall health.............337
Bex Weaver, BS; Owen Rae, DVM; Brian Payne, DVM

Poster Session: Research Topics

Comparative evaluation of family oral fluids and piglet sera to detect PRRSV RNA by PCR.............339
Marcelo N. de Almeida; Jeffrey J. Zimmerman; Derald J. Holtkamp; et al.

Assessment of slaughter surveillance based on oral fluid samples.............340
Marcelo N. de Almeida; Jeffrey J. Zimmerman; Fabio Vannucci; et al.

Human-like H3N2 and H3N1 swine influenza A viruses are resistant to neutralization due to antigenic variation.............341
Srivishnupriya Anbalagan, M.Sc., Ph.D; Joshua D Elston; Jessica L Peterson

Minimal *Mycoplasma hyopneumoniae* genetic variability within production flows.............342
A. Anderson Betlach, BS; E. Fano, DVM, MS, PhD; A. Sponheim, DVM; et al.

New duplex real time RT-PCR to detect porcine Teschovirus and Sapelovirus.............344
M. Barrera; F. Chamba; T. Knutson; et al.

Potential use of outer membrane proteins as subunit vaccines against *Streptococcus suis* and *Haemophilus parasuis*.............345
Susan L. Brockmeier, DVM, PhD; Kirsten Eberle, PhD; Samantha Hau, DVM, PhD; et al.

Comparison of historic and contemporary strains of Senecavirus A.............347
Alexandra Buckley; Baoqing Guo; Vikas Kulshreshtha; et al.

Serological evaluation of influenza A virus vaccine antibody responses in replacement gilts on twelve breeding farms in the United States.............348
J. Carlson, MS; A. L. Vincent, DVM, PhD; J. Zhang, PhD; et al.

Evaluation of chlorine dioxide virucidal efficacy against porcine epidemic diarrhea virus, porcine reproductive and respiratory syndrome virus, Senecavirus A and influenza A virus.............349
Qi Chen; Julian N. Rosenberg; Douwe F. Mason; et al.

Identification and genetic characterization of a novel deltacoronavirus.............351
Qi Chen; Chenghuai Yang; Ying Zheng; et al.

Effect of chemical clarification of oral fluids on the detection of PRRSV IgG.............353
Y. A. Henao-Díaz, DVM; L. Giménez-Lirola, PhD; K. Poonsuk, DVM, PhD; et al.

Novel behavioral analysis of piglets treated with Metacam® thirty minutes before and at time of castration.............354
J. Law, DVM; G. Cunningham, BSc, DVM, MVetSc; N.J. Cook, PhD

Porcine parvovirus type 2 associated with lymphocytic perivascular lesions in nursery and finishing pigs.............356
Lacey Marshall Lund; Stephanie A Rossow; Fabio A Vannucci

Age susceptibility of pigs to *Mycoplasma hyorhinis* challenge.............357
Brian Martinson; F. C. Minio; Jeremy Kroll; et al.

Fecal microbiota transplantation improves outcome in pigs following co-infection with porcine reproductive and respiratory syndrome virus and porcine circovirus type 2d.............358
M. C. Niederwerder, DVM, PhD; R. R. R. Rowland, PhD; L. A. Constance, MS; et al.

Increased microbiome diversity at the time of infection is associated with improved growth rates of pigs after co-infection with porcine reproductive and respiratory syndrome virus (PRRSV) and porcine circovirus type 2 (PCV2).............359
Rebecca A. Ober; James B. Thissen; Crystal J. Jaing; et al.

Short summary on the low prevalence of PCV2 on pigs at birth.............361
L. Ochoa, PhD; L. Greiner, PhD; J. Connor, DVM; et al.

Effects of PCV2-viremia level of piglets at birth on the efficacy of PCV2 vaccine against exposure to PCV2-b determined by viremia levels and growth rate of pigs....363
L. Ochoa, PhD; L. Greiner, PhD; J. Connor, DVM; et al.

Measuring air filter efficiencies in the discrete PRRS virus range....365
M. Osborne, PhD; D. Sharpley

Analysis of the importation of feed ingredients, live animals and genetic material into the United States to reduce the risk of foreign swine disease and health hazard introduction....366
Gilbert Patterson, VMD, MPH, DACVPM

Oral fluid specimens can be clarified ("cleaned up") without affecting porcine epidemic diarrhea virus (PEDV) isotype-specific (IgG, IgA) ELISA responses....368
K. Poonsuk; L. Giménez-Lirola; R. Magtoto; et al.

Work in progress: Development of an ELISA for the detection of FMDV 3ABC antibody in swine oral fluid specimens....369
K. Poonsuk; L. Giménez-Lirola; J. Chiotti; et al.

General trends of *Mycoplasma hyopneumoniae* detection by PCR in cases submitted to the Iowa State University Veterinary Diagnostic Laboratory from 2004 - 2016....370
G. Rawal DVM; P. Arruda MS, PhD; C. J. Rademacher DVM, BS; et al.

Development of a duplex assay for simultaneous detection of PCV2 and PCV3 in association with histological lesions....372
TP Resende, DVM, MS, PhD student; Benjamin Miller, BS; Todd Knutson, BS, PhD; et al.

Factors that impact PRRS time to stability....374
Juan Sanhueza, DVM, MS, PhD; Carles Vilalta, DVM, PhD; Andreia G. Arruda, DVM, MS, PhD

Phenotypic relationship and amino acid variability of the VP7 gene in porcine rotavirus B and C....376
F.K. Shepherd, BS; F. Chen, BS; T.P. Knutson, PhD; et al.

Benchmarking pen-side sequencing for infectious disease diagnosis....378
Shaoyuan Tan; Michael P. Murtaugh

Associations between PRRS farm status and productive parameters in Spanish sow farms....380
D. Torrents; J. Miranda; A. Puigredon; et al.

Live virus immunization (LVI) with a recent 1-7-4 PRRSV isolate elicits broad protection against PRRSV challenge in finishing age swine....382
Albert G. M. van Geelen; Alexandra Buckley; Vikas Kulshreshtha; et al.

Good Vaccination Practice: It all starts with a good vaccine storage temperature....383
F. Vangroenweghe; K. Kelderman; F. Meulders; et al.

Cup waterer:pig ratio on growth performance of growing and finishing pigs....384
C. M. Vier, DVM; S. S. Dritz, DVM, PhD; M. D. Tokach, PhD; et al.

Model to estimate PRRSV introduction in filtered farms with negative pressure....386
C. Vilalta, DVM, PhD; K. Janni, PhD; C. Alonso, DVM, MSc, PhD; et al.

Use of tylvalosin (Aivlosin®) water medication for *Mycoplasma hyopneumoniae* elimination....387
P. Yeske, DVM, MS; R Kaptur, DVM; D. Rosener, DVM; et al.

Poster Session: Industrial Partners

Preventing PRRS outbreaks in swine environments: Monitoring air filtration for airborne diseases....388
N. Woolard; M. Osborne, PhD

Pursuing American Board of Veterinary Practitioners (ABVP) certification in Swine Health Management....389
Jeffrey A. Husa, DVM, Diplomate ABVP

Identification of the complete swine reproductive pheromone....390
John J. McGlone, PhD; Garrett Thompson, MS; Michael Kuhn, DVM, MBA

Exploring profitability differences when average wean-finish production metrics show minor numerical differences between groups 392
Dennis DiPietre, PhD; Lance Mulberry, BS; Eduardo Fano, DVM MC PhD; et al.

PEDV, PRRSV, and Seneca Valley virus PCR positive flies (*Musca domestica*). What now? 394
Grant Allison, DVM; Phil Gauger, DVM, MS, PhD; Jianqiang Zhang, MD, MS, PhD; et al.

A *Brachyspira* challenge study to evaluate medication options 396
Chris Puls, PhD; Tom Weber, PhD; Matt Ritter, et al.

Key VFD learnings from 2017 398
B. Cowan, MS; J. T. Holck, DVM, MS, MBA; C. J. Rademacher, DVM

Efficacy of Porcilis™ Ileitis against *Lawsonia intracelluaris* challenge and post-vaccination seroresponse kinetics 400
Brad Thacker, DVM, PhD; Greg Armbruster, DVM; Brett O'Brien, VMD, MS; et al.

Streptoccus suis: Research overview 402
Brittanny Wiener, BS; Eric Bumgardner, MS; Paulraj Lawrence, PhD

Mineral source matters conception to market 404
B. Lawrence, PhD; T. Loeffler, DVM; Y. Ma, PhD; et al.

How accurate and precise are current options for testing deoxynivalenol? 406
Sarah Weiland

Comparing the immunogenicity of Circo/MycoGard™ with other combination vaccines used in the field 407
A. Hintz, DVM; R. Kaptur, DVM; D. Rosener, DVM; et al.

Effects of level of Defusion® Prime on performance of growing pigs fed diets containing vomitoxin 409
C. Fowler, MS; S. Ebarb, MS; P. Xue, PhD; et al.

Effect of diet and sex on feed efficiency and the fecal microbiome at slaughter weight 411
L. M. G. Verschuren, MSc; M. P. L. Calus, PhD; A. J. M. Jansman, PhD; et al.

Organic acids modes of action and impacts on intestinal microbiota and performance in swine 412
J. A. Mesonero Escuredo, DVM, ECPHM Resident; P. Bruinenberg, PhD; N. de Groot, DVM; et al.

A dose confirmation study evaluating 100 g/ton of LINCOMIX® (lincomycin hydrochloride) for the reduction of the severity of effects of respiratory disease associated with *Mycoplasma hyopneumoniae* 414
Michael Senn, DVM, MS; Raksha Tiwari, BVSc & AH, MS, PhD; Tom Painter, DVM

Monday General Session: Global Knowledge: Individual Application

Evaluating data: What do we really know? 415
Eric R. Burrough, DVM, PhD, DACVP

Portraying the industry in a positive light 418
Erin Brenneman

Agriculture 2025: Global, local and high tech 420
Lowell B. Catlett, PhD

Monday Concurrent Session #1: Not Your Father's Sow Farm

Experiences with new sow farm technologies 423
N. H. Williams, MS, PhD; H. Williams, MS

Managing highly prolific sows 425
Deborah Murray, DVM

Leading change to improve piglet survival 427
Larry L. Coleman, DVM

Update on feeding strategies for the highly prolific sow 429
M. B. Menegat, DVM, MS; S. S. Dritz, DVM, PhD; C. M. Vier, DVM; et al.

Sow mortality: Impact on performance and root causes 437
Clayton L. Johnson, DVM

Monday Concurrent Session #2: Emerging Diseases

Next generation sequencing and metagenomics: Interpretation for the practitioner 441
D. Marthaler, PhD; T. Knutson, PhD

Porcine circovirus type 3: Practitioner perspective 445
Emily B. Byers, DVM

Porcine circovirus type 3: A pathologist's perspective 448
S. E. Detmer, DVM, PhD; F. Vannucci, DVM, MSc, PhD

Porcine parainfluenza virus – Practitioner perspective 450
Aaron Lower

Porcine parainfluenza virus type 1 (PPIV-1): Diagnostic laboratory perspective 452
P. Gauger, DVM, PhD; M. Welch, BS; J.Y. Park, PhD, PhD; et al.

Porcine sapelovirus: Practitioner perspective 458
Brian Payne, DVM; Abby Patterson, DVM, PhD; Paulo Arruda, DVM, PhD; et al.

Porcine sapelovirus: Pathologist/research perspective 462
B. Arruda, DVM, PhD; K. Schwartz, DVM, MS; P. Arruda, DVM, MS, PhD

Porcine deltacoronavirus: Researcher's and practitioner's perspectives 464
Dick Hesse, MS, PhD; Katherine Wedel, DVM

Seneca Valley virus A, a practitioner's perspective 467
Dr. Laura Bruner

Senecavirus A: Pathologist/researcher perspective 469
Fabio A. Vannucci, DVM, MSc, PhD

Monday Concurrent Session #3: Managing Endemic Disease

Regional IAV-S vaccination strategies for breeding herds 475
Clayton Johnson, DVM

Systems approach to PRRSV management 478
Hans Rotto, DVM, MS; Peter Schneider, DVM, MS; Jessica Arnold, DVM; et al.

Perennial herd closure 480
E. Spronk, DVM

Tech tools for disease management 481
Maryn Ptaschinski

Mycoplasma hyopneumoniae: Lateral transmission and gilt exposure methods 482
P. Yeske

Experiences managing wean-to-finish PEDV in a production system 485
Lynn Pavlovic, VMD

Subclinical ileitis: Diagnostic monitoring, R^2, and economics 488
Nathan Winkelman, DVM

Tuesday General Session: Antibiotics

Crucial to criminal: The range of perspectives on antimicrobial use in pork production....494
L. Karriker, DVM, MS, DACVPM

Antimicrobial use: Current EU perspective....497
MEC White, BVSc, LLB, DPM, FRCVS

What to expect when you're not expecting: A veterinarian, producer and feed mill perspective of an educational FDA audit....505
Peter Schneider, DVM, MS

Measuring antibiotic use in pork production: Why, how, and for whom?....507
Peter Davies

Antibiotic use monitoring in cattle....510
Michael D. Apley, DVM, PhD, DACVCP

Swine Health Information Center 2017 progress....513
P. Sundberg, DVM, PhD, DACVPM

Howard Dunne Memorial Lecture

How culture, geography, and socioeconomic status affect global animal source food (ASF) consumption: Applications for swine veterinarians

William R. DuBois, DVM, DABVP

In remembrance and honor of
Howard W. Dunne

Dr. Howard W. Dunne received his veterinary degree from Iowa State University in 1941. He was in engaged in private practice for one year and industry for four years before entering graduate school at Michigan State University. After receiving his PhD in 1951, he spent two years as deputy chief of veterinary microbiology at the Chemical Corps Biological Laboratories in Camp Detrick, MD. In 1953, he joined the faculty of Pennsylvania State University, where he headed the veterinary science research program and the large animal diagnostic laboratory. Dr. Dunne was internationally known for his work on swine diseases, particularly hog cholera and the swine reproductive viruses then referred to as the SMEDI syndrome. Dr. Dunne authored more than 90 scientific publications, and is perhaps best known for initiating and editing the textbook *Diseases of Swine*, which is still widely considered the "bible" of swine disease.

In 1971, the American Veterinary Medical Association presented Dr. Dunne with the XII International Veterinary Congress Prize to recognize his contributions to the international understanding of veterinary medicine. At the 1974 Congress of the International Pig Veterinary Society (IPVS), Dr. Dunne was elected president of the 1976 IPVS Congress, to be held in Ames, Iowa. His sudden death at the age of 61 in September of 1974 saddened the swine veterinary community, and the 1976 IPVS Congress was dedicated to his memory.

At the suggestion of Dr. Allen Leman, a financial gift made by the 1976 IPVS Congress organizers to the fledgling American Association of Swine Practitioners was combined with additional association funds to endow this prestigious annual lecture in remembrance and honor of Dr. Howard W. Dunne.

HOWARD DUNNE MEMORIAL LECTURERS

1978	Michael Muirhead	1988	Thomas Alexander	1998	Mike Daniel	2008	Timothy J. Loula
1979	W. J. (Bill) Smith	1989	Steven Henry	1999	Larry Rueff	2009	John U. Thomson
1980	N.C. Nielsen	1990	Stanley Curtis	2000	David Reeves	2010	James F. Lowe
1981	Allen Leman	1991	Roy Schultz	2001	Max Rodibaugh	2011	Locke A. Karriker
1982	Michael Wilson	1992	Joseph Connor	2002	Tim Blackwell	2012	R. Dean Boyd
1983	Edward Bohl	1993	Camille Moore	2003	Lonnie King	2013	Craig Rowles
1984	Richard Ross	1994	Robert Morrison	2004	Robert Desrosiers	2014	Daryl Olsen
1985	Kent Kislingbury	1995	Rodney Johnson	2005	Scott Dee	2015	Gregory Stevenson
1986	Ralph Vinson	1996	Paul Yeske	2006	Gregg BeVier	2016	John Harding
1987	Gary Dial	1997	Barbara Straw	2007	Tom Burkgren	2017	Jeff Zimmerman

How culture, geography, and socioeconomic status affect global animal source food (ASF) consumption: Applications for swine veterinarians

William R. DuBois, DVM, DABVP

Introduction

Culture, geography (both physical geography and human geography), and socioeconomic status are all large and complex topics that shape the lives of every person, past, present, and future. This presentation will focus on how they affect general food choices and specifically, animal source food (ASF) choices. While each of these things can be defined separately, they are intimately interrelated with each one affecting the others. Similarly, while each of them affect people's food choices, the human need to acquire and consume food has played a major role in shaping culture, geography, and socioeconomic status.

Each of these factors is sufficiently complex for whole books to have been written on them individually. This presentation will only be able to provide a superficial treatment of each one, and some important aspects will be left out. There are other important factors that also affect people's ASF choices. Things such as government policy, education, and global economics are extremely influential in different ways. However, geography, culture, and socioeconomic status are arguably the most fundamental to individual food choices.

Much of this presentation will be focused on less developed countries. This is partly because much of my experience is in less developed countries, and it is what I am passionate about. But, it is also because 83% of the world's current population lives in less developed countries, and virtually all future population growth will occur there.[1] More developed countries (MDCs) and less developed countries (LDCs) are defined in part by their per capita gross national income (GNI). All MDCs have a high GNI and all low GNI countries are classified as less developed. But, many lower-middle and upper-middle GNI countries are also considered less developed. There are other factors besides income that are used to assess development. These include average education level, average life expectancy, and gender equality. When discussing things on a national or regional level, gross generalizations will be made about millions of people. While these generalizations are helpful to identify large trends, it is important to realize that they will mischaracterize many individuals and minority groups in those countries.

The original title of this presentation used the term "animal protein consumption" rather than "animal source food consumption." While writing the proceedings article, I realized that the term "animal protein" didn't communicate the full picture. It is true that ASFs are a great source of high quality protein, and they contain the full spectrum of amino acids required in the human diet. But nutritionally, they provide much more than just protein. They provide energy and fats and are great sources of some important micronutrients like iron, zinc, calcium, Vitamin A, and Vitamin B_{12}.[2] But, most people do not eat ASFs because they are seeking to add essential amino acids and micronutrients to their daily caloric intake. They eat ASFs because they like them. The sensory features of flavor and texture provide the diner with a pleasurable experience.

Individual food choices

While much of this presentation will discuss trends at a national level, it is important to remember that all food choices are made at the individual or household level. National, regional or even global trends are all caused by choices being made by millions of individuals. Therefore, it is important to understand how individuals across the world make choices about what food they eat.

In the book *More Than Good Intentions*, an imaginary type of person called an "Econ" is described. Econs make all decisions by looking at the economics.[3] For food choices, an Econ would calculate a ratio of nutrition to price and would choose the least-cost diet that would still supply all his needs. He would be sure to weigh each portion out to ensure that he consumed all the needed nutrients without overconsuming and damaging his health or wasting his food budget. But, people are not Econs and do not make decisions in this way. In fact, most of the big decisions in our lives including those about marriage and family, religious faith, and even career choices are mainly based on our values and emotions rather than on data, evidence or economics.

Although people's food choices aren't based on science or economics, it would be incorrect to say that those choices are random. The book *Food: The Key Concepts* defines how individuals make food choices: "Food choices are the result of a complex negotiation among three competing considerations: the consumer's identity (social and personal), matters of convenience (price, skill, availability), and a sense of responsibility (an awareness of the consequences of what we eat)." The "consumer's identity" is influenced by culture, by personal taste and by social status. It is what the consumer wants to eat based upon how they view themselves as an individual. The consideration of "convenience" includes local or seasonal availability, preparation time, preparation skills and equipment needed. The "sense of responsibility" includes health benefits or harm. It also includes concerns about how the food was produced, including the environmental impact of production, worker conditions, and animal welfare.[4] This food equation is a very helpful starting point in thinking about individual food choices. However, it doesn't incorporate the concept of food security.

Food security

The Food and Agriculture Organization of the United Nations (FAO) describes food security as being based on four "pillars." The first pillar is availability which means that sufficient quantities of adequate quality foods are available for consumers. The second pillar is access which means that each consumer can economically and physically obtain the food that is available. The third pillar is stability which means that individual consumers always have stable, uninterrupted access to food. The fourth pillar is utilization which means that the food supply is both nutritious and safe. Food insecurity is when one or more of these pillars cannot be met by an individual, a household, or an entire geographic area. Food insecurity can be transitory. For example, availability is disrupted by a natural disaster. Or, it can be chronic such as an extremely poor individual who chronically lacks economic access to food. Armed conflict is the most common cause for transitory food insecurity and the second most common cause is weather events such as floods or droughts.[2]

For the purposes of this presentation, I have modified the terminology of the food choice equation above to reflect the food security concepts of availability and access. I have also noted where geography, culture, and socioeconomic status exert influence. This is illustrated in Figure 1. The process in this equation is repeated multiple times per day often on the subconscious level where the consumer is not aware that it is occurring.

Food insecurity affects individual food choices at the point in the equation of availability and access. However, an important distinction should be made. Food security is about consistently and fully satisfying nutritional needs and is not about satisfying personal preferences. For example, certain preferred cultural dishes may not be available to an immigrant living in a different country. However, if he or she has stable availability of and access to nutritious and wholesome food that provide complete nutrition, then he or she is not food insecure.

We will now look deeper into how physical geography shaped agriculture and the availability of food, how culture plays an important role in the way people interact with food, how human geography affects food availability, and finally, how socioeconomic status affects access to food for people throughout the world.

Physical geography and the history of agriculture

Physical geography is defined as the study of the earth's features and processes such as climate, topography and landforms, water sources, and soil categorization. This would have greatly influenced the types of plants and animals available in different regions for early humans to eat. As regional groups of people began working together in their shared quest to attain food, these groups developed a shared identity, shared values, and a shared history which led to the formation of societies and cultures.[5]

These groups likely began as nomadic, hunters and gatherers which over long periods of time developed agricultural practices. Agriculture created surpluses of food that had not previously existed. Surpluses made it possible for some people to accumulate wealth. These surpluses also meant that not everyone needed to farm. A portion of the population could focus on non-agricultural pursuits such as education, art, and business. These societal and economic changes created different social and economic classes.[6]

Different geographic regions had different plants and animals that humans could domesticate for food, for fiber, and for labor. All our current crops and livestock species were domesticated thousands of years ago. Corn (maize) was domesticated in Central America; wheat was domesticated in the Middle East near Syria; and rice was domesticated in northern China. Similarly, the Middle East is where domestication of sheep, goats, pigs and cattle took place; turkeys were domesticated in Central America; pigs were also domesticated independently in China along with the water buffalo; and chickens were domesticated in Southeast Asia.[6]

From these different centers, farmers migrated to other areas with their crops and livestock. There they adapted their agricultural systems to the geographic features of the new areas. In semi-arid grasslands, such as those in East Africa, or mountainous areas,

Figure 1: The equation of how an individual makes a food choice

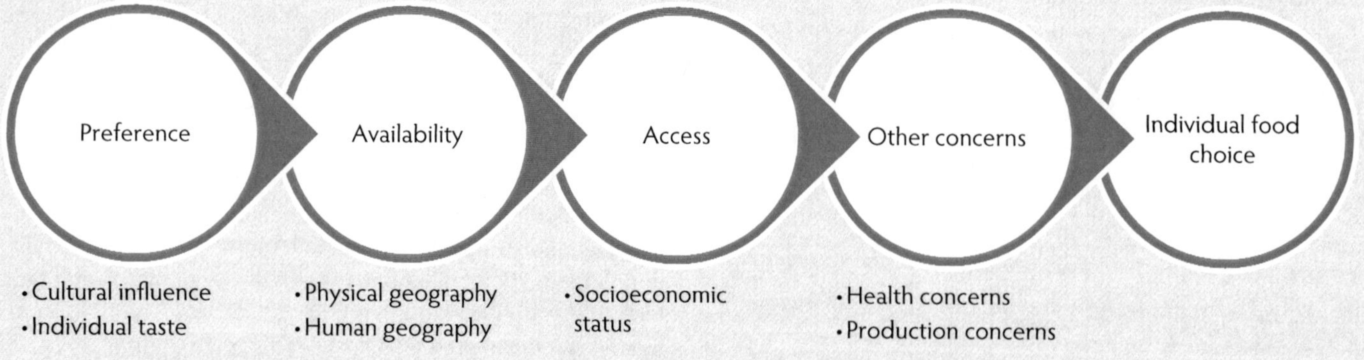

such as Central Asia, pastoralist systems developed. The semi-nomadic, livestock-owning herders in these regions still move their herds of cattle, sheep, and goats through different grazing areas based on season and forage availability. In temperate and tropical regions, forests were cleared, and mixed agricultural systems of crops and livestock were established.[5] Although technologies have changed and advanced through the centuries, the mixed crop and livestock systems is still the predominant agriculture system in the world today. Additionally, fish consumption is very important to societies located near the ocean or bodies of freshwater. Although, in temperate and tropical areas, these societies also raise domestic livestock.

Finally, global trade had a large impact on the spread of domestic plants and livestock through the world. We think of globalization as a modern concept, but China, India, Arabia, and the Roman Empire were trading spices and other goods throughout the known world in the first few centuries A.D. In the 15th century, European explorers were seeking a western sea route to access the Asian spice trade when they discovered the Americas. This discovery led to a massive movement of domestic species known as the Columbian Exchange. Plant species from the New World such as potatoes, tomatoes, corn, chocolate, and chili peppers were taken back to the Old World. Plant species such as wheat, rice, sugar, bananas, and coffee came to the New World along with multiple livestock species and European settlers.[6] So, by the 16th century A.D., all the crops and livestock species that we have today were generally available throughout the world.

Culture and food choices

Culture is defined as a collection of values, ideas, attitudes, and views that are commonly held by a group of people with a common identity. This could be a common national identity, ethnic identity, company identity or even a family identity. These commonly held values, ideas, attitudes and views will lead to behaviors, activities, and production of cultural artifacts (e.g. art, music, clothing, cuisine, buildings, etc.). Although any group culture can influence the members' choices and actions, I will focus mainly on culture as held by ethnic groups or by nationalities. Many of the world's nations have a history of independence that dates back less than 100 years. Yet ethnic groups within those countries have shared histories that date back for many centuries. For this reason, the cultures of ethnic groups are often more developed and more strongly held than cultures based on national identity.

Within cultures, there can be further differentiation along the lines of gender, education level, and social class. Certain cultural values will be expressed or emphasized within those subgroups. There are some important distinctions between the behavior of rural people and urban people within the same culture. Generally, cultural traditions are more pronounced and more strictly adhered to in rural areas and less so in urban areas. Urban dwellers tend to have increased exposure to other ethnicities with different cultures. There is also an increased level of anonymity in urban areas that leads to less social pressure about adhering to cultural norms. Therefore, urban people are less likely to strictly follow their own cultural traditions, and they may adopt certain practices from other cultures.

There are certain cultural dimensions that shape behavior within cultures. There are several different cultural dimensions related to things such as authority, communication, and time. All the dimensions will have some influence on food and eating within that culture. However, the cultural dimension of individualism versus collectivism has the most bearing on people's food choices. Individualist cultures value independence and individual choice. Collectivist cultures value group membership (ethnicity, clan, family, etc.) and maintaining harmony within the group.[7] So, in individualist cultures, individual taste is very important in food choices. Whereas, in collectivist cultures, foods that are considered normative within the group or society will be more important than individual preference.

From commodity to culture

All foodstuffs at their basic level are commodities and do not have cultural value. The same basic raw ingredients can be used in different cultures with dramatically different results. There is a process whereby basic commodities are endowed with cultural meaning and significance. There are cultural influences in every step of the process from purchase, to preparation, to serving, and finally to eating. In many cultures, this process begins in the supermarket. For meat purchases, supermarket customers expect to buy meat that has been slaughtered, chilled, and disassembled into primal cuts at a different facility and out of the customer's sight. Most meat at supermarkets has been further cut, weighed, and pre-packaged for display. Customers only have to select the package with the amount and type of meat they wish to purchase. In other cultures, foodstuffs are sold at traditional open markets sometimes called wet markets. Meat sold at the wet market is from animals that are usually slaughtered on the same day and often on site. In some markets, the animal may be alive at the time of purchase, particularly for poultry and fish. The carcass is cut up in full view of the customer based on the amount and type of meat ordered. The cultural preference for shopping in the traditional market is independent from geographical and socioeconomic influence. There are supermarkets in urban areas throughout the world even in LDCs. However, in many cultures, urban dwellers, even affluent ones, still prefer to buy their food from the traditional open market.[2]

Consumers evaluate quality and desirability of ASF products by using different sensory factors such as visual appearance, smell, texture, and taste. For meat, choices are mainly based on the visual characteristics of color, fat content, and visual moisture level. But, between cultures, there are widely different perceptions of what is desirable. A study conducted on pork color in different nations found that some cultures preferred light colored meat while others preferred dark. The same is true for fat content. Some cultures, particularly in Asia, prefer a higher amount of external and intramuscular fat in pork.[8] Preferences can even extend to the breed of animal. The local breeds of chicken are preferred over commercially raised broilers in many cultures due to differences in flavor and in texture. Finally, certain products such as offal, fish heads, chicken feet, and blood are highly desirable in some cultures. In other cultures, the same products are viewed as something to be discarded.

Preparation and cooking are the steps where the real magic happens. This is where basic commodities become ingredients, and then ingredients are transformed chemically and thermally into a dish that has cultural meaning.[5] For example, wheat flour, beef, and vegetables can be transformed into a hamburger served at a 4th of July picnic in Des Moines, or the same basic commodities could be transformed into a bowl of Chinese noodles served at roadside stall in Hong Kong. The information on how this occurs is passed down from generation to generation. This can be done informally and orally where children and teenagers learn by helping their parents or older members of the community. It can also be formalized in a written recipe or cookbook.[5]

The most important type of eating in any culture is the regular day-to-day eating. This is when most food consumption, including ASF consumption, occurs. This is when the routines and rituals associated with eating are learned, and when the patterns of mealtimes are established. This is when people learn what is considered a proper meal in their culture versus a snack or an incomplete meal. In western cultures, meat is the center of the meal, and other dishes are considered sides. In other cultures, the cereal grain or starch is considered the center of the meal, and meat and vegetables are considered the sides.[5] For instance, in many Asian cultures rice is the center of the meal, and people don't feel like they have eaten a meal unless rice is included.

While most food consumption is of the everyday variety, food plays an important and prominent role in the celebrations of holidays and in the commemoration of major life events such as weddings, funerals, births, and other celebrations. In individualist cultures, the menu for these events is based on personal tastes or the budget of the host. Although, there are some widely followed traditions such as serving turkey on Thanksgiving Day in the US. In many collectivist cultures, there are very specific guidelines for the types of foods, ways food is prepared, amounts prepared, and the order that guests are served. If the rules are not followed, the guests may be insulted, and the host will be shamed. Because of this, families spend large amounts of money to ensure that they meet all the cultural requirements of these events. These cultural feasts always involve ASFs. In the case of holidays, this creates very predictable patterns of high market demand for specific ASFs.

Food taboos are prohibitions on eating certain foods. They may have religious or folk underpinnings, but not always. They can apply to all members of the culture or may be specific for certain age or gender groups. Sometimes, there are perceived health reasons for taboos, but other taboos may have no explainable logic behind them.[9] Taboos often involve the consumption of ASFs. A well-known example is the Hindu prohibition on eating cattle meat. Islam and Judaism both have strict dietary laws, Halal and Kosher, that restrict the types of foods that can be consumed including a prohibition on pork consumption. These laws also regulate how foods are produced and prepared. Food producers serving Muslim or Jewish customers must comply with these laws and be certified.

Human geography effects on ASF availability

Human geography is defined as the study of how human activity affects the earth. This topic has a vast impact on the availability of food. This paper will specifically discuss: landlocked nations; urbanization; infrastructure development; and the structure of agricultural production.

Landlocked nations

The configuration of geopolitical borders was much different 75 years ago. Many new nations were created by the decolonization of Asia and Africa between the 1940's and 1970's and the dissolution of the former Soviet Union in the early 1990's. Generally, this has been very positive with people experiencing more individual freedom. But, it has also caused a geographic difficulty by creating many landlocked nations. Landlocked nations have no coastline thus no in-country access to ports. Today, there are 49 landlocked nations. Of these, 32 are LDCs, and 17 are listed as least-developed countries by the UN. Less developed landlocked countries have a per capita GDP that is only 57% of their non-landlocked neighbors. They also score significantly lower on measurements of education and average life expectancy. This stems from their dependence on neighbors for all trade access.[10] From a food perspective, these countries must rely heavily on domestic food production. All imported foodstuffs will have increased freight costs and are vulnerable to abrupt disruptions from conflict or political disagreement in neighboring countries.[10]

Urbanization

Increasing urbanization has been a global trend for the past several decades. In 1950, only 30% of the world's population lived in urban areas. Currently, that percentage has risen to 54% and is expected to reach 66% by 2050.[11] However, there is a strong socioeconomic influence in urbanization rate. In MDCs, 78% of the population is urban compared to only 49% in LDCs. In the countries considered least developed, only 32% of the population is urban.[1]

Regardless of national development status, urban dwellers have more food choices than rural dwellers. Urban areas have supermarkets as well as traditional open markets. In rural areas, there are only traditional markets or small shops. For pre-prepared food, rural areas and smaller towns have travelling vendors and some small restaurants serving local dishes. In larger cities, there will also be medium and high-end restaurants, and a wider variety of products and food types will be offered.

There are different patterns of food consumption between urban and rural people that are independent of income status. Urban dwellers eat more ASFs than rural dwellers at the same income levels. This pattern is also seen in the consumption of highly-processed foods including processed meat products. Urban dwellers also eat more meals outside of the home than rural dwellers. These differences are likely due to the higher availability of these options in urban areas.[11]

Rural roads

Norman Borlaug was a famous US agronomist whose research in the 1950's and 1960's developed high-yielding Mexican wheat varieties and helped usher in what is termed the "green revolution." In 1970, before awarding him the Nobel Peace Prize, the Nobel Committee proclaimed, "More than any other single person of this age, he has helped to provide bread for a hungry world."[6] Yet he is quoted as saying, "Roads are essential for any type of agricultural development." This is true since roads provide farmers access to supplies and expertise as well as access to markets for their crops and livestock. So, while technologic advances, like those developed by Norman Borlaug, can have a large impact on human well-being, road conditions will limit both the impact and the scope.

There are around 33.8 million km of roads on the earth. About 57% of these are paved. The United States alone contains almost 1/5 of the total roads in the world at 6.5 million km. This is almost double the amounts in both China and India with 3.7 and 3.3 million km respectively.[12] The road infrastructure in the US makes almost any type of ASF available to all US consumers regardless of where it is produced. Allen, South Dakota is the point furthest from the ocean in North America. But, if a person in Allen wants to eat shrimp, all he or she needs to do is drive less than 2 hours to Rapid City where he or she can purchase frozen shrimp at either of 2 Wal-Mart locations or can eat any of several different shrimp dishes at the Rapid City location of the Red Lobster restaurant.

There is a marked contrast in LDCs. The entirety of Sub-Saharan Africa only has 2.5 million km of roads and 30% of them are in one country, South Africa. 85% of these roads are unpaved and many of them become impassable for weeks or months during the rainy season. While Sub-Saharan Africa is the most extreme, rural road conditions in LDCs in other regions are also inadequate.[12] This decreases food security and food choices by limiting the consumers' access to food markets. Equally important is the constraint on rural incomes by limited access to markets for farmers selling their produce and limited access to other employment opportunities.

Energy infrastructure

1.1 billion people did not have access to electricity in 2017. Sub-Saharan Africa accounts for 53% of those without access to electricity, 25% are in India, 19% are in other Asian countries, and 3% are in the rest of the world. This is a significant improvement from the year 2000 when over 1.6 billion people did not have electricity. However, most of the progress occurred in Asia. In Sub-Saharan Africa, slow progress and population growth have caused the number of people without access to increase by 70 million since 2000.[13] Many households do not have access to a home refrigerator due to poor electrical access and low household incomes. In MDCs most households have home refrigerators with many countries approaching ownership rates of 100%.[14] In China, refrigerator ownership increased from 25% in 1994 to 88% in 2014. However, other countries with higher rural populations such as Peru and India have much lower refrigerator ownership rates of 45% and 25% of households respectively.[15] Lack of access to refrigeration limits storage time for ASFs which means that items such as meat and fish must be cooked or consumed on the day of purchase to prevent spoiling. This makes food purchasing much less efficient requiring frequent if not daily trips to the market for any food items that are not shelf-stable.

There are not adequate cold chains in many LDCs due to this limited electrical infrastructure. Many LDCs are in the warm tropics and sub-tropics where spoilage of perishable foods happens quickly without access to chilling and cold storage. This leads to post-harvest food loss and wastage of perishable foodstuffs, including meat, milk and seafood. It also increases the risks of foodborne illness. Around 36% of all food harvested in Sub-Saharan Africa is lost before it can be consumed, though not all of this is because of cold chains. However, 25% of the total food wastage in LDCs could be eliminated by establishing reliable cold chains.[16]

Structure of agriculture production

There are 570 million farms in the world today of which 500 million (88%) would be classified as "family farms." Most farms in the world are very small with 410 million (72%) being less than 1 ha (2.5 ac).[17] There is a significant socioeconomic effect on farm-size at the country level. In low-income and middle-income countries, 95% of all farms are less than 5 ha (7.5 ac). In high income countries, there is a significant shift to bigger farms with 34% being over 5 ha (7.5 ac). However, even in high income countries 66% of all farms are less than 5 ha (7.5 ac).[18] Over the past 50 years, the average farm size has increased in high-income countries and some middle-income countries. In low-income countries and many middle-income countries, the average farm size has decreased during the same time.[17]

The contribution to total agricultural production by different sized farms varies regionally. In North America, South America, Australia and New Zealand large (50-200 ha, 125-500 ac) and very large (> 200 ha, > 500 ac) farms account for 75-100% of grain and livestock production. Conversely in Asia and Sub-Saharan Africa, small farms (< 20 ha, < 50 ac) make up make up over 75% of the production of all commodities. In these same regions, very small farms (≤ 2 ha, ≤ 5 ac) account for around 30% of all commodity production. In China specifically, very small farms account for more than 50% of all food production and for 63% of livestock production.[19]

Socioeconomic status effects on ASF access

The primary driver of socioeconomic status is income. The social status of a person or family rises or falls based upon their economic status. Ultimately, a person's access to money equates to the power that person has over circumstances. Increased access to money increases a person's choices for education, health care, housing, and food. There are many different socioeconomic levels, but the biggest impacts on ASF consumption happen at income levels where people are classified as poor or where they have crossed the threshold to no longer be considered poor. Higher socioeconomic statuses such as upper middle-class or rich still affect people's ASF choices. However, defining these statuses based on income level is much less relevant than defining the level where people move from being poor and into middle class status.

Global definition and measurement of poor versus non-poor

Since food choices happen at the individual or household level, socioeconomic factors at that level will be the most important for ASF consumption. National income classifications based on per capita GNI are useful for looking at national or regional trends. But, there is a high level of variation in household incomes within countries with extremely poor households in high-income countries and vice versa.

Poverty is a complex, multi-dimensional state that often spans multiple generations. It consists of multiple factors such as decreased health, lack of educational opportunities, inadequate nutrition, lack of basic services (e.g. electricity and water) and low income. Since the poor lack power, they are often victimized and oppressed by more powerful entities such as governments, criminals, or other ethnicities.[20] It is difficult to tell which of these problems are root causes and which are symptoms as they feed on each other. For example, a person with a low income can't buy adequate amounts of food. Undernourishment causes her to become ill, and she can't work and thereby loses income. Lack of well-being on so many fronts causes people in poverty to feel hopelessness and despair.

Current poverty data is based solely on income measurements and doesn't adequately account for the other dimensions of poverty. However, people do not typically experience multiple dimensions of poverty without one of them being low income. And conversely, when people's income improves, they are more empowered to deal with the other dimensions of poverty such as food and education. Therefore, I will use an income-based definition of poverty through the rest of this paper while acknowledging that the situation for poor people is much more complex.

In 1990, the World Bank established an international poverty line that is based on the national poverty line of the 15 poorest countries. A standardized measure of purchasing power parity in a common currency (the US dollar) was used to eliminate currency and cost variations. In 2015, the level for extreme poverty was set at $1.90 per person per day, and the threshold for moderate poverty was set at $3.10 per person per day.[21]

For global middle class, the lower threshold of $10/person/day has been widely used. Studies in Mexico, Peru, and Chile showed that at the $10/person/day income level the risk of falling back into poverty was less than 10%. Based on surveys, people self-identified as middle class at $10/person/day in Mexico, Colombia, and Peru, at $16/person/day in Brazil, and at $20/person/day in Chile.[22]

Between the thresholds for poverty and middle class, there is a group that the World Bank has termed "vulnerable to poverty." They have attained incomes that would not be classified as poor, but they may still experience episodes of poverty particularly if there are unforeseen shocks like a poor harvest or illness.[23] This illustrates that while thresholds are very useful for quantitative analysis. The reality for people moving out of poverty is much more complicated.

Poverty has markedly decreased over the past 25 years. Since 1990, extreme poverty has declined from 37% (over 2 billion people) to 11% (770 million) in 2013. The 2013 population for moderate poverty was around 1.5 billion people, meaning that there are around 2.3 billion people living on less than $3.10/person/day.[24] The estimates are much more variable for those vulnerable to poverty and for middle class or higher. There are somewhere between 1.5 and 2.9 billion people with incomes between $3.10 and $10/person/day and between 2.1 and 3.5 billion people with incomes more than $10/person/day.[22,25]

There are some strong demographic and geographic trends within socioeconomic groups. The majority of poor people live in Asia and Africa with several African countries having over 40% of the national population classified as extremely poor. 80% of the extreme poor and 76% of the moderate poor live in rural areas and are mainly employed in agriculture (65% and 52% respectively). Only 20% of people with incomes over $3.10/person/day work in agriculture.[24] In the decade between 2001 and 2011, almost 400 million people rose above the $10 middle class threshold, with over 200 million in China alone. The other areas with large numbers of people entering the middle class were South America and Eastern Europe. North America and Western Europe contain most of the people with higher incomes. 63% of people earning between $20 and $50/person/day and 87% of people making greater than $50/person/day live in North America or Western Europe.[22]

Income effects on animal source food consumption

Multiple national surveys across all world regions show that consumption of ASFs rise whenever household incomes increase. Other studies comparing in-country ASF consumption between rich and poor households in Uganda, India, and Peru showed significantly higher ASF consumption in the rich households. ASFs contribute 28% to worldwide protein consumption, but that share rises to 48% in high income countries.[2] Figure 2 illustrates the change in ASF consumption from 1983 to 2013 for selected regions. During this period, ASF consumption increased in all regions. Although ASF consumption only rose slightly in high income countries, it was already at a high level. ASF consumption has risen significantly in regions containing many LDCs. A notable exception would be Africa where ASF consumption has only slightly increased. In China, which has seen a large amount of household income growth, ASF consumption increased by almost 500%.[26]

Increased incomes allow consumers the power to purchase higher value ASFs like premium quality products such as USDA prime beef. This would also include products with perceived health benefits or that are produced in a way that the consumer prefers. Examples of this would include organic, antibiotic-free, or free-range production. Higher income consumers are able pay the premiums for these types of products. This is even true in LDCs among consumers that have reached sufficient income levels.

On the other hand, low incomes are a major constraint for consumption of ASFs. In addition, people with low incomes are often food insecure and sometimes are unable to provide adequate nourishment for themselves and their families. Protein, particularly from ASFs, is the most expensive component of the diet and is typically what is reduced when there is a lack of money.[2] This leads to clinical undernourishment for hundreds of millions of people.

The prevalence of undernourishment, defined as chronic food deprivation, has been declining for decades. But, the decline has slowed significantly since 2010. From 2010 to 2015, the prevalence of undernourishment only declined from 11.5% to 10.6%. In 2016, it increased to 11%, meaning there were 815 million undernourished people worldwide. There are multiple causes for the increase including conflicts in some regions, regional weather events, and regional economic factors. As would be expected, there is a marked regional variation in undernourishment. The regions with the highest prevalence are Sub-Saharan Africa at 22.7% and the Caribbean at 17.7%. The lowest are North America and Europe at less than 2.5%. Asia and North Africa are close to the world average of 11%. Latin America and Oceania were markedly lower at 5.9% and 6.7% respectively.[27]

Protein-energy malnutrition and micronutrient deficiencies are the 2 primary clinical problems resulting from undernourishment. Protein-energy malnutrition affects all ages, but children have the most severe and long-lasting effects.[28] The 2 clinical syndromes in severely undernourished children are stunting and wasting. Stunting is defined as being more than 2 standard deviations below the median height for that age. There are 162 million stunted children under the age of 5. It is caused by inadequate nutrition during gestation and the first 2 years of life. It is irreversible and impairs both cognitive and physical development.[29] Wasting is caused by severe, acute malnutrition in children under 5 years and is defined as having an upper arm circumference of less than 115 mm or having bilateral pitting edema. Wasting affects at least 52 million children.[30]

The most commonly deficient micronutrients are iron, folate, vitamin A, vitamin B_{12}, and zinc. Iron and folate deficiencies both cause anemia. Anemia, from all causes, affects around 1.6 billion

Figure 2: Animal source protein consumption in selected regions between 1983 and 2013

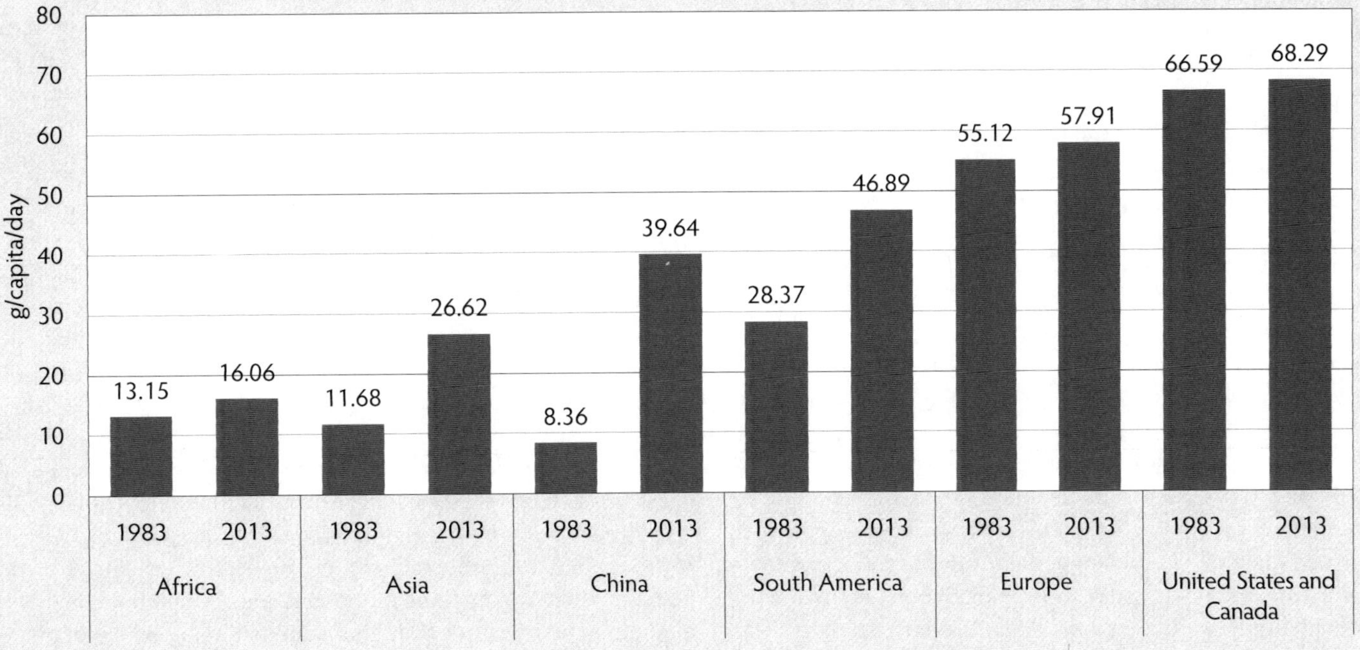

Source: Food and Agriculture Organization of the United Nations. FAOSTAT.

people in the world and impairs the mental development of 40-60% of the children in low income countries. Iron deficiency during pregnancy has been estimated to contribute to 20% of maternal deaths annually. Severe vitamin A deficiency leads to blindness in 250,000 to 500,000 pre-school age children.[31]

Typically, multiple nutritional deficiencies occur concurrently, and even at subclinical levels, protein and micronutrient deficiencies can impair immune function which increases the risk of infectious disease.[28] Undernourishment, both as a primary disease and as a contributing factor to infectious disease, plays a major part in many of the causes of morbidity and mortality in LDCs.

It would not be accurate to say that decreased access to ASFs is the only cause of undernourishment. However, it is widely agreed that regular consumption of ASFs, even in small amounts, can dramatically reduce these conditions. This is especially important for children and for women who are pregnant or lactating. A study in malnourished schoolchildren showed that adding a small amount of meat to their diets over the course of 2 years improved not only height, weight, and muscle area, but also significantly improved their energy levels and cognitive abilities over control children and children receiving a plant-based supplement.[32] ASFs contain high-quality nutrients that are highly digestible and bioavailable. Plant source foods do not contain all the required amino acids, and some plant source foods contain anti-nutritive components, for example phytic acid, that inhibit both protein and mineral absorption.[33]

Applications for swine veterinarians

"Knowing is not enough; we must apply. Willing is not enough; we must do." This quote is from Johann Wolfgang von Goethe who was a German writer that lived between 1749 and 1832. His insightful quote fits very well with the theme of this meeting and with the mindset of many swine veterinarians. Our knowledge and motives are of limited value unless they proceed to action. But, like food choices, application of information is personal and happens at an individual level or at a small group level. So, there are potentially innumerous applications for this information based on individual opinions, circumstances, and values. However, I will offer 3 possible applications. The first is a radical proposal to address a large opportunity seen in the socioeconomic and geography information. The second is a way to apply the cultural information to our current work. The final application would be for our organization, the American Association of Swine Veterinarians.

Livestock services for the poor in less developed countries

Based on what we have learned, much of the growth of ASF consumption will occur in LDCs. This is partly due to population growth which will occur almost exclusively in LDCs. It will also be due to increased demand. As people's incomes increase there, they will begin consuming more ASFs. We have also seen the tragic consequences of the hundreds of millions of people without adequate access to ASFs. Most low-income people with limited access to ASFs live in rural areas and are largely dependent on agriculture for their livelihoods. There are some geographic constraints to availability of ASFs in many of these rural areas due to poor roads and inadequate energy infrastructure. Some nations are further constrained due to being landlocked. So, how do we apply that information to what we do as swine veterinarians?

Over my career as a food animal veterinarian, I have always enjoyed continuing education talks about global economics and about future growth in ASF consumption and livestock production. Many of those talks have carried the message that poor people across the globe are moving to cities, gaining higher incomes, and will want to consume more ASFs. Our production industry only needs to wait for this to happen and prepare to respond to the increased future demand. The question I would pose is: Why wait? We have education, skills, and experience that are valuable to poor farmers. We don't have to passively wait for their low incomes to improve so that they can become our customers. We can begin the process of making them our customers today and be a part of helping them improve their incomes and improve their lives.

I realize that this suggestion is somewhat radical and out of our normal comfort zone. I debated with myself for countless hours about whether to even broach this topic. But, at its core, it's not radical or crazy. As veterinarians, we provide services to livestock owners, and there is a large market that is currently under-served. So, a discussion about expansion of business into that market would be natural and normal. What makes this suggestion radical is that the potential customers are in a markedly different socioeconomic class than our normal customers, and they live in countries where we have little experience or knowledge. It may make us feel uneasy to think about targeting poor people as customers like that is not ethical or not right. However, the goods and services that we would be providing would be beneficial to them and not detrimental or exploitive. Also, the business relationship between a customer and service provider is much more egalitarian than that of a donor and aid recipient. The customer is empowered to choose whether to purchase the goods and services or not.

The theme of the book *The Business Solution to Poverty* is that businesses can greatly improve the incomes and the lives of the billions of people living in poverty by providing them with beneficial goods and services. Some examples are safe drinking water, tools to improve agriculture productivity (irrigation as an example), health care, and low-cost private schools. In 2007, the collective purchasing power of the global poor was estimated at 5 trillion dollars. And, as they move upward out of poverty, their purchasing power will be doubling, tripling or more. As a market, the poor are under-served and largely untargeted while many existing markets are saturated and highly competitive. Businesses have access to the capital that is needed to expand into new regions and into other services, and businesses have the financial incentives to do so. Thus, the potential for businesses to help poor people can often be higher than that of governments or the non-profit sector.[34]

The business opportunities for livestock-related services in LDCs are vast, and the potential impact for the lives and the diets of the poor is even bigger. In LDCs, there are around 1.3 billion poor rural farmers and another 500 million farmers that are vulnerable to poverty.[24] The labor productivity on small-holder farms in LDCs is suboptimal. But, improvements in labor productivity result in lower levels of undernourishment. This happens because of increased amounts of food available for consumption by the farmer's family and also through improved household incomes. Improvement in agriculture productivity is 3 times more effective in reducing extreme poverty in low income countries than growth in other sectors. In Sub-Saharan Africa, it can be up to 11 times more effective.[35] Most smallholder farms are already involved in raising livestock along with their crops, and livestock and crop production are synergistic with animal manure being used as fertilizer and with crop residues being used for livestock feed.[2]

Pre-harvest production losses are higher in LDCs than MDCs. In Sub-Saharan Africa, the largest cause of ASF losses and waste is production losses. These losses would all be attributable to animal health and/or animal nutrition problems.[36] Presently, the animal health services available in LDCs are not sufficient, particularly in rural areas. Related to this is access to animal health products. Basic veterinary drugs such as anthelmintics, antibiotics, vaccines, and vitamins are available in LDCs although choices may be limited. However, there is poor distribution to rural areas and a lack of knowledge and training about product usage.[2]

Animal nutrition is marginal to poor on many smallholder operations. Typical diets consist of low-cost feedstuffs grown by the farmer. They are usually high in carbohydrates and fibrous feedstuffs, and they are often deficient in protein. Commercially produced animal feeds are available in LDCs, but like animal health products, availability is limited in rural areas. Also, the cost of commercial feed is often prohibitive for poor farmers.[2,37]

Many other useful livestock services are not available or only at a limited level. These would include extension/training, housing design, breeding and genetics services (including artificial insemination), diagnostic services, marketing services, and credit. Access to all of these basic services is critical for smallholder producers to be successful and to potentially scale up their operations.[38]

These points are well illustrated in the results of a survey that we recently conducted among smallholder farmers in a local village. These farmers all raise crops for subsistence purposes as well as cash crops. Almost all of them raise either chickens, pigs, or both. When asked where they learned about farming and livestock and where they get new information, 47 of 52 said that they "just know it," 12 said they learned from friends and neighbors, 2 listed the farm store, and 1 said from a book. The feed for chickens consisted of corn, cassava, and rice, and pig feed was made from cassava and sweet potato leaves. None of which would contain adequate levels of protein. No farmers used any animal health products. Many of them described an illness that killed hundreds of chickens within the previous 6 months. The symptoms described would be consistent with Newcastle Disease. Newcastle Disease vaccine is available at a very low cost in the town located about 30 minutes away, but none of the farmers had ever used the vaccine or even heard of it.

I believe that food animal veterinarians and specifically swine veterinarians are highly qualified to provide these types of services due to their wide expertise across all areas of livestock production. Our knowledge and experience go far beyond that of diseases and drugs. We are comfortable evaluating and discussing almost any type of production issue from housing, to nutrition, to genetics, to marketing. And, we are very skilled at training people from various backgrounds and education levels to perform technical health and husbandry tasks.

The service delivery system could be patterned after the community animal health worker model that has been used in several different countries by different organizations.[2,38,39] In that model, local lay people are trained to provide basic animal health services such as vaccination and deworming at the village level. This could be coupled with nutritional services such as ration formulation, small-scale feed manufacturing, and grain drying and storage. Further livestock services such as artificial insemination and marketing could be added to the portfolio in the future. It will be necessary to provide services for livestock other than swine. Initially those services would be very basic, so the impact of knowledge gaps would be minimized. This will allow for learning and adding of new more sophisticated services in the future as demand dictates.

Implementation in different countries would require system modification to accommodate geographic and cultural differences. Relationships and strategic partnerships would have to be built with stakeholder groups such as the government, existing veterinary services, suppliers, and community groups. Legal requirements for foreign-owned businesses and for veterinary care will affect the way things operate in different countries. Varying economic circumstances and availability of products will change the way that pricing and payment are structured. But, the concept of a complete menu of livestock services to smallholder producers could be replicated in all LDCs throughout all the regions of the world.

Apply increased cultural knowledge to our current work

Swine veterinarians are already involved in a global, multi-cultural industry. Many swine production companies have farms in multiple countries. The US pork industry has exported around 2 million metric tons of product every year since 2008.[40] Although exact numbers are not available, a significant percentage of US pork industry workers are immigrants. The first sentence in an April 4, 2017 NPPC position paper on workforce reads: "The US pork industry depends on immigrant labor."[41] Swine genetics companies, pharmaceutical companies, and equipment companies are all international businesses. Many swine veterinarians spend a portion of their time working with pork producers or other entities outside of their home country. International swine veterinary meetings are regularly attended by US swine veterinarians, and many feature speakers from the US. Finally, 47 different nations make up the current AASV membership reflecting the multinational and multicultural nature of our organization. So, whether we realize it or not, each of us is already involved in a cross-cultural profession.

Since we are already providing food and veterinary services for people in other countries, we should seek to better understand the cultural influences of the food choices in those cultures. This will help us to better provide our customers the types of products and services that they want and in the way that they want them. An oversimplified example would be that we don't want to send the "other white meat" to consumers in Asian cultures that prefer high amounts of external and intramuscular fat. There may also be sizable niche markets that desire specific meat qualities, meat cuts, or meat produced in specific ways. Those could be potential growth opportunities for new markets and new products.

Many of us work directly with people from a different culture than our own. Culture not only affects people's food choices, but affects all areas of life. It was previously mentioned that the cultural dimension of individualism versus collectivism influences people's food choices. But, there are many other dimensions of culture that affect the way business is conducted, the way people interact with those in authority (including work supervisors), the way negative information is given and received, the way people view time, and many other things that are relevant to business relationships. Understanding the cultures of the people we work with will be extremely helpful in improving our interactions and communications with our supervisors, employees, trade partners, or colleagues. There are many options for learning about this topic including books, TED talks, podcasts, business seminars and cultural consultants. Much of this information is focused on business topics which would be the most helpful for us.

Application for the AASV

As mentioned earlier, the AASV membership represents 47 different nations. A breakdown of AASV membership by region is listed in Table 1. Outside of North America, Europe is the continent with the highest percentage of members. While the continents of Central/South America, Asia, and Africa only represent a small percentage of the membership.

In the 2007 Howard Dunne Memorial Lecture, Tom Burkgren discussed that any significant expansion of AASV membership would need to occur outside of North America. He also said that the AASV leadership would need to decide if actively seeking members outside North America is desired. If the answer to this is "yes," then the next step would be to assess what member benefits would be attractive to swine veterinarians in those areas.[42] I would propose that we re-visit the question about actively seeking to grow membership outside of North America. In the 11 years since Tom's talk, our industry has become more and more global in nature. This trend does not appear to be stopping soon. Expanding membership would expand our financial base through membership and meeting fees. This would also impact our attractiveness for continued or expanded company sponsorship and advertising. It would expand our knowledge base as well by adding new people with different views, education, and experience. However, this would not be without risk. Human and financial resources would be needed to explore where and how to expand without any assurance that membership will change. We would also need to ensure that expansion did not detract from our mission of increasing the knowledge of the swine veterinarians within our current geographic footprint.

IF the decision to expand is made, the country of China should receive serious consideration as an area to target. China is already the largest producer of pork in the world. We have seen that high numbers of Chinese people are entering the middle class. So, growth in pork consumption will continue. Also, the Chinese pork industry is undergoing a rapid shift towards large, intensive production units which fits with the strengths that we currently have as an organization.[43] Many of our current members have been consulting and working with producers there for several years. And, the University of Minnesota has held the Leman China Swine Conference there annually since 2012. Our current AASV membership from China is 8 people. So, this could be a country with lots of potential for growth. We have many people in the AASV that would be good resources for researching what types of member benefits would be useful to Chinese veterinarians and producers.

Other areas with potential for growth are the rest of Asia (besides China), Africa, and Central/South America. Those areas have a large number of smallholder producers. So, if we want to serve swine veterinarians in those areas, we would need to consider how to expand our education portfolio to include information relevant to smallholder producers. A great example of this type of information would be the article published in the May/June 2017 edition of the *Journal of Swine Health and Production* entitled, "Average

Table 1: 2017 AASV membership by world region

Region	Number of AASV members	Percentage of total membership
United States	916	67.4 %
Europe	145	10.7%
Canada	143	10.5%
Asia	52	3.8%
Mexico	44	3.2%
Central/ South America	39	2.9%
Australia/ New Zealand	18	1.3%
Africa	2	0.1%
Total	1359	100%

daily gain and the impact of starting body weight of individual nursery and finisher Ugandan pigs fed a commercial diet, a forage-based diet, or a silage-based diet." This study provided information that was specific and applicable to small-holder producers in a tropical environment.

References

1. Population Reference Bureau. 2017 World population data sheet with a special focus on youth. Published August 2017. Available at: http://www.prb.org/pdf17/2017_World_Population.pdf. Accessed 11 November 2017.

2. Food and Agriculture Organization of the United Nations. World livestock 2011 – Livestock in food security. Available at: www.fao.org/docrep/014/i2373e/i2373e.pdf. Accessed 1 November 2017.

3. Karlan D, Appel J. *More Than Good Intentions: How a New Economics Is Helping to Solve Global Poverty*. Kindle ed. New York, New York: Dutton Press; 2011.

4. Belasco W. *Food: The Key Concepts*. Kindle ed. New York, New York: Berg Publishers; 2008.

5. Crowther G. *Eating Culture: An Anthropological Guide to Food*. Kindle ed. Toronto, Canada: University of Toronto Press; 2013.

6. Standage T. *An Edible History of Humanity*. Kindle ed. New York, New York: Bloomsbury USA; 2009.

7. Hofstede, G. Dimensionalizing Cultures: The Hofstede Model in Context. *Online Readings in Psychology and Culture*. 2011; 2(1). Available at: https://doi.org/10.9707/2307-0919.1014. Accessed 18 September 2017

8. Font-i-Furnols M, Guerrero L. Consumer preference, behavior and perception about meat and meat products: An overview. *Meat Science*. 2014; 98(3):361-371. Available at: https://doi.org/10.1016/j.meatsci.2014.06.025. Accessed 9 November 2017

9. Meyer-Rochow VB. Food taboos: their origins and purposes. *Journal of Ethnobiology and Ethnomedicine*. 2009; 5:18 Available at: https://doi.org/10.1186/1746-4269-5-18. Accessed 9 November 2017.

10. Faye ML, McArthur JW, Sachs JD, Snow T. The challenges facing landlocked developing countries. *Journal of Human Development*. 2004; 5(4):31-68. Available at: www.tandfonline.com/doi/abs/10.1080/14649880310001660201. Accessed 12 November 2017.

11. Food and Agriculture Organization of the United Nations. 2017 The state of food and agriculture: Leveraging food systems for inclusive rural transformation. Available at: http://www.fao.org/state-of-food-agriculture/en/. Accessed 7 November 2017.

12. Transportation Research Board of the National Academies. The promise of rural roads: review of the role of low-volume roads in rural connectivity, poverty reduction, crisis management, and livability. Published September 2012. Available at: http://www.trb.org/Publications/Blurbs/167787.aspx. Accessed 13 November 2017.

13. International Energy Agency. Energy access outlook 2017: From poverty to prosperity. Published October 2017. Available at: https://www.iea.org/publications/freepublications/publication/weo-2017-special-report-energy-access-outlook.html. Accessed 12 November 2017.

14. USDA – Economic Research Service. Ownership of household amenities among selected countries. Available at: https://www.ers.usda.gov/media/9393/householdamenities.xls. Accessed 8 November 2017

15. James C, Onarinde BA, James SJ. The use and performance of household refrigerators: A review. *Comprehensive Reviews in Food Science and Food Safety*. 2016; 16:160–179. Available at: http://dx.doi.org/10.1111/1541-4337.12242. Accessed 8 November 2017.

16. Food and Agriculture Organization of the United Nations. How access to energy can influence food losses: A brief overview. Available at: http://www.fao.org/3/a-i6626e.pdf. Accessed 10 November 2017

17. Lowder SK, Skoet J, Raney T. The number, size, and distribution of farms, smallholder farms, and family farms worldwide. *World Development*. 2016; 87:16-29. Available at: https://doi.org/10.1016/j.worlddev.2015.10.041. Accessed 16 November 2017

18. Food and Agriculture Organization of the United Nations. 2015 The state of food and agriculture – Social protection and agriculture: breaking the cycle of rural poverty. Available at: http://www.fao.org/state-of-food-agriculture/en/. Accessed 6 September 2017.

19. Herrero M, Thornton PK, Power B, et al. Farming and the geography of nutrient production for human use: a transdisciplinary analysis. *The Lancet Planetary Health*. 2017; 1(1):e33-e42. Available at: https://doi.org/10.1016/S2542-5196(17)30007-4. Accessed 26 October 2017

20. World Bank. Introduction to poverty analysis. Available at: http://documents.worldbank.org/curated/en/775871468331250546/Introduction-to-poverty-analysis. Accessed 2 November 2017.

21. "How is the global poverty line derived? World Bank website. https://datahelpdesk.worldbank.org/knowledgebase/articles/193310-how-is-the-global-poverty-line-derived-how-is-it. Accessed 1 November 2017.

22. Kochhar, R. Pew Research Center. A global middle class is more promise than reality: From 2001 to 2011, nearly 700 million step out of poverty, but most only barely. Available at: http://www.pewglobal.org/2015/07/08/a-global-middle-class-is-more-promise-than-reality/ Accessed 2 November 2017.

23. World Bank. Quantifying vulnerability to poverty: A proposed measure applied to Indonesia. Available at: https://www.openknowledge.worldbank.org/handle/10986/21355 Accessed 2 November 2017.

24. World Bank. Who are the poor in the developing world? Available at: https://openknowledge.worldbank.org/handle/10986/25161 Accessed 2 November 2017.

25. Kharas H. The Brookings Institution. The unprecedented expansion of the global middle class. Available at: https://www.brookings.edu/wp-content/uploads/2017/02/global_20170228_global-middle-class.pdf. Accessed 5 November 2017.

26. FAOSTAT. Available at: http://www.fao.org/faostat/en/#home. Accessed 10 September 2017.

27. Food and Agriculture Organization of the United Nations. 2017 The state of food security and nutrition in the world: Building resilience for peace and food security. Available at: www.fao.org/state-of-food-security-nutrition/en/. Accessed 18 November 2017.

28. Müller O, Krawinkel M. Malnutrition and health in developing Countries. *Canadian Medical Association Journal*. 2005; 173(3): 279–286. Available at: https://doi.org/10.1503/cmaj.050342. Accessed 19 November 2017.

29. World Health Organization. Global nutrition targets 2025: Stunting policy brief. Available at: http://www.who.int/nutrition/publications/globaltargets2025_policybrief_overview/en/. Accessed 2 September 2017.

30. World Health Organization. Global nutrition targets 2025: Wasting policy brief. Available at: http://www.who.int/nutrition/publications/globaltargets2025_policybrief_overview/en/. Accessed 2 September 2017.

31. Demment MW, Young MM, Sensenig RL. Providing micronutrients through food-based solutions: A key to human and national development. *The Journal of Nutrition*. 2003; 133(11): 3879S-3885S. Available at: http://jn.nutrition.org/content/133/11/3879S. Accessed 19 November 2017.

32. Neumann CG, Murphy SP, Gewa C, Grillenberger M, Bwibo NO. Meat supplementation improves growth, cognitive, and behavioral outcomes in Kenyan children. *The Journal of Nutrition*. 2007; 137(4): 1119-1123. Available at: http://jn.nutrition.org/content/137/4/1119. Accessed 27 November 2017.

33. Schönfeldt HC, Hall NG. Dietary protein quality and malnutrition in Africa. *British Journal of Nutrition*. 2012; 108(S2): S69-S76. Available at: https://doi.org/10.1017/S0007114512002553. Accessed 7 September 2017.

34. Polak P, Warwick M. *The Business Solution to Poverty: Designing Products and Services for Three Billion New Customers*. Kindle ed. San Francisco, California: Berrett-Koehler Publishers; 2013.

35. Food and Agriculture Organization of the United Nations. 2015 The state of food security and nutrition in the world: Meeting the 2015 international hunger targets: taking stock of uneven progress. Available at: www.fao.org/3/a-i4646e.pdf. Accessed 14 November 2017.

36. Food and Agriculture Organization of the United Nations. Global food losses and food waste: Extent, causes, and prevention. Available at: http://www.fao.org/docrep/014/mb060e/mb060e00.pdf. Accessed 16 November 2017.

37. Carter NA, Dewey CE, Grace D, Lukuyu B, Smith E, de Lange CFM. Average daily gain and the impact of starting body weight of individual nursery and finisher Ugandan pigs fed a commercial diet, a forage-based diet, or a silage-based diet. *Journal of Swine Health and Production*. 2017; 25(3):121-128. Available at: http://www.aasv.org/shap.html. Accessed 8 May 2017

38. International Fund for Agricultural Development. Smallholder livestock development: Scaling up note. Available at: https://www.ifad.org/what/scaling_up. Accessed 8 September 2017

39. Agronomes et Vétérinaires Sans Frontières. Quality community animal health arrangements. Available at: https://www.avsf.org/en/posts/677/full/quality-community-health-arrangements. Accessed 6 September 2017

40. US Meat Export Federation. Total U.S. pork exports 2007-2016. Available at: https://www.usmef.org/news-statistics/statistics/. Accessed 22 November 2017.

41. National Pork Producers Council. Pork industry needs viable workforce. Available at: http://nppc.org/wp-content/uploads/2017/05/Immigration04-04-17.pdf. Accessed 4 September 2017

42. Burkgren T. 2007. AASV: Good to great. *Proceedings of the 38th Annual Meeting of American Association of Swine Veterinarians*, March 3-6, Orlando, FL, pp.9-25.

43. Patton D. China's pig farmers go north, upending world's top meat, grain market. Reuters website. https://www.reuters.com/article/us-china-farming-pigs/chinas-pig-farmers-go-north-upending-worlds-top-meat-grain-market-idUSKBN1DM12P. Published 22 November 2017. Accessed 23 November 2017.

Alex Hogg Memorial Lecture

This is our time, the choices are yours

Rodger Main, DVM

In remembrance and honor of Alexander Hogg

Alexander Hogg
1920-2006

A native of Scotland, Alex (pronounced "Alec") Hogg immigrated to the United States with his family when he was six years old. After reaching adulthood, he served in the US Navy during World War II before entering veterinary school at Kansas State University. He received his DVM degree in 1950, and was a mixed animal practitioner in Iowa for twenty years. Dr. Hogg returned to college – this time, Iowa State University – to complete a Masters Degree in 1972. He spent the following years serving and educating swine veterinarians and producers in his roles as Nebraska Swine Extension veterinarian and professor at the University of Lincoln-Nebraska. Upon "retirement" from the university in 1990, Alex worked as a technical services consultant for MVP Laboratories. Ever an enthusiastic learner, he graduated from the University of Illinois Executive Veterinary Program in 1995.

An active and longstanding member of the AASV, Dr. Hogg joined what was then the American Association of Swine Practitioners shortly after it was formed. He served as secretary/treasurer of the association from 1974 to 1976, and was elected president in 1979.

The many awards and honors bestowed upon him over the years included several presented by the AASV: The Howard Dunne Memorial Award in 1985, the Meritorious Service Award in 1996, and in 2005, the first-ever AASV Foundation Heritage Award for a lifetime of achievement in veterinary medicine.

ALEX HOGG MEMORIAL LECTURERS

2011 Joseph F. Connor
2012 Lisa Tokach
2013 John T. Waddell
2014 Mark J. Engle
2015 C. Scanlon Daniels
2016 Peggy Anne Hawkins
2017 Matthew Turner

This is our time, the choices are yours

R. Main
Iowa State University, Ames, Iowa

Introduction

As often said, the only thing that is constant is change. Our profession and pork industry that we serve are certainly no exception. As one ages, one begins to have an ever-increasing appreciation for those that have come before us, the very precious and finite amount of time we have, and how our work and contributions of today affect not only the here and now but also those who are to follow. I feel extremely blessed and humbled to have had the opportunity to become a veterinarian and enjoy a variety of work experiences within the US Pork Industry. I am a firm believer that we are all victims of our own experiences and that each experience plays a role in guiding our future decisions, actions, and view of the world. Experiences and position unquestionably influence perspective. Words cannot express my most sincere appreciation for the many mentors, colleagues, co-workers, supervisors, producers, and family members that have allowed me the opportunity to work and learn from the benefit of their insight, skills, knowledge, and experience. The objectives of this communication are simply to provide a personal account of the mega-trends observed in the US Pork Industry and swine veterinary profession (1977 – current), highlight a notable strength and weakness of our current abilities, and provide a few thoughts as to what I see as significant opportunities for swine veterinarians and the US Pork Industry going forward.

I do want to acknowledge to my veterinary colleagues and friends from outside the US that I often refer to observations and opportunities for the "US Pork Industry" in this proceedings paper. This reference is used with acknowledgement and a full appreciation that the US is but part a of the larger North American Pork Industry. Due to the editorial nature of this talk derived from my personal experiences that have been largely limited to the US, I did not want to over generalize in referencing the larger North American Pork Industry. However, I trust many such experiences, observations, and opportunities proposed for consideration would have similar applicability across much of the North American Pork Industry.

Mega-trends (1977 – Current)

Pork production

The US Pork Industry and systems by which pigs are produced have been in a constant and rapid state of change and evolution since the days of rearing swine in the mixed indoor and outside facilities that I grew up with in the mid to late 1970s. Swine farm settings, facilities, and systems in which pigs are produced have become increasingly sophisticated, capital intensive, productive, and efficient. The number of swine housed on individual farm sites has increased, while the number of farm sites producing pigs and number of entities owning pigs has decreased precipitously.[1] Swine farms have evolved from often being a mix of indoor and outside facilities to today's operations where pigs are housed indoors in climate-controlled facilities with slotted flooring throughout all phases of production. Facility design, quality, and overall capabilities in maintaining optimal environments for both pigs and caretakers have continued to make stepwise improvements. Natural and hand-mating systems have long-since been replaced by the now various approaches to artificial insemination. The genetic potential of pigs and the methods by which genetic improvements are delivered to commercial swine farms has improved immensely. Well-managed and coordinated networks of boar studs, genetic nucleus farms, and daughter nucleus (multipliers) herds have created highly repeatable and effective pipelines for delivering genetic improvement and high quality stock across the masses of commercial breeding herds. Nutrition programs have continued to evolve to keep pace with the ever-changing nutritional needs of the pigs being reared and the mix of feedstuffs readily available. Much of the applied nutritional research informing best cost feeding programs has transitioned from smaller-scale university settings to commercial research facilities. Improvements in genetics and nutrition have played a major role in changing the carcass composition and end weight of the pigs being harvested. Slaughter weights in the US have increased by approximately 40 lbs over the past 40 years in the absence of increasing backfat thickness.[1,2] Single-site farrow-to-finish farms have largely been replaced by multi-site production systems. Specifically, multi-site production systems that are often able to more fully capture the benefits of all-in all-out (AIAO) pig flow in the growing pig phase(s) of production.[3] The advent and subsequent wide-spread adoption of multi-site (segregated rearing) swine production that began in the late 1980s and became the norm in the 1990s to present day has unquestionably played a highly significant role in restructuring the US Pork Industry and the overall landscape of how and where pigs are produced in North America and globally. Multi-site production sparked the development and growth of rapidly scalable swine production systems and interdependent pig production networks of highly varied ownership, management, and/or contractual relationship structure. Segregating the growing pigs from breeding herds improved the health of both the sows and the growing pigs and put the US Pork Industry on wheels. Multi-site production created the opportunity for siting breeding herds in less pig dense locations and moving weaned or feeder pigs often significant distances either within or across one or more state lines to be grown. Arguably, the invention and transition to multi-site production is the most significant change in methodology that has influenced the overall structure of the US Pork Industry over the course of the past 40 years. Consolidation, specialization, industrialization, systematization, and sophistication would all be reasonable descriptions of mega-trends observed.

Pork processors and marketing of swine

Market channels, systems for deriving the market or transfer price of top hogs, and the pork processing sector have evolved significantly. Buying stations and related collection points utility have become largely limited to assembling and marketing of non-select market animals and spent breeding stock. Marketing select weight market animals (top hogs) has long since been largely via direct transfer from swine farms to pork processing facilities. As the size and scale of pork producer operations or aligned producer networks became larger, the industry has observed a dramatic reduction in the number of pigs being sold at a negotiated price. Recent estimates suggest that only 2.5 to 3% of top hogs are being sold or transferred at a point in time negotiated price.[4] The vast majority of pigs are being sold or transferred on pre-negotiated pricing agreements that are generally relative to the reported open-market price, wholesale value of primal cuts, or other system-specific transfer price, matrix, or producer-processor agreement. The evolving state of how live hog market values (transfer or selling prices) are being derived has highly significant implications that go beyond the scope of this talk and proceedings paper. While pork producers have historically been integrated on the input side of pork production (e.g., producers of feedgrains), the aforementioned advent of multi-site production was closely followed by a rise in pork-processor ownership of live hog production operations (e.g., often referred to as vertical integrated operations). Albeit such vertically integrated pork operations have largely been established by pork processors either purchasing or building live hog production operations; more recently, there has been a significant growth in pork producers aligning to build and operate pork processing operations. This recent trend has been coupled with a previously unprecedented rate of growth in pork processing capacity in the US. Consolidation within the pork processing sector has been highly significant over the course of the past 40 years. In 2018, the top ten pork processors are expected to account for approximately 87% of the top hog processing capacity in the US.[1] The growth in the size and scale pork producer operations, advent of vertically integrated pork production systems, and consolidation within the pork processing sector have each contributed greatly to the ongoing evolution of the US Pork Industry.

Veterinarians and veterinary services

The model of how veterinary services are delivered and the roles veterinarians play have changed in lockstep with the evolving needs and composition of the industry being served. While mixed animal practitioners continue to play a significant role in the provision of veterinary care to swine operations in the US, swine practice has become an increasingly specialized endeavor. Veterinarians specializing in swine practice and/or swine production management (e.g., swine practitioners) are providing veterinary care for the largest majority of commercial pigs produced in the US. Swine specialty clinics and associated pork production management companies have continued to expand within and across state lines. Swine veterinarians' roles and responsibilities have become increasingly diverse in veterinary practice, pork production, pork business unit management, allied industry, and academia. Such diversity and changing roles are evident even in the name of our professional organization that changed from the American Association of Swine Practitioners to the American Association of Swine Veterinarians in 2001.

Consumers and society

Societal and consumer related interests have unquestionably been having an ever-increasing impact on pork production, swine practice, and how pork is being marketed both domestically and internationally. These range from local or regional facility siting issues for swine buildings (facilities) or pork processing facilities; specified feeding, genetics, housing, or processing requirements to meet predetermined label or market requirements; and antimicrobial use. Each of these issues are having a very clear and direct impact on the US Pork Industry. Such influences are expected to continue going forward.

Exports

The US swine industry has grown markedly dependent on export markets over the past 25 years (Figure 1). While the total amount of pork produced in the US has increased by approximately 50% since the mid-1970s, domestic consumption per capita has remained relatively unchanged.[1,2] A significant portion of the growth in the US Pork Industry since the early 1990s has been due to more pork being purchased by consumers outside of the US. Presently, approximately 23% of US domestic pork production is exported, a 10-fold increase over historical averages. Quality assured and cost competitive products; effective trade agreements; significant international outreach, education, and marketing efforts; improved systems of logistics; and increasing personal wealth and the associated increased appetite for meat-based protein in emerging markets have all been contributing factors to this remarkable growth story. While the advent of multi-site production may likely be the biggest change on the operational side of US pork production, the substantial growth in exports is likely one of the most significant factors (outside of population growth in the US) influencing the overall size and profitability of the US Pork Industry. Current estimates suggest that export markets add approximately $50 of value to each market hog sold in the US.

Notable success (strength) and challenge (weakness)

The swine veterinary profession has provided leadership to a myriad of discoveries, practices, programs, and system changes in swine production that have resulted in the steadily increasing

Figure 1: US pork exports as a percentage of domestic production (carcass weight basis)

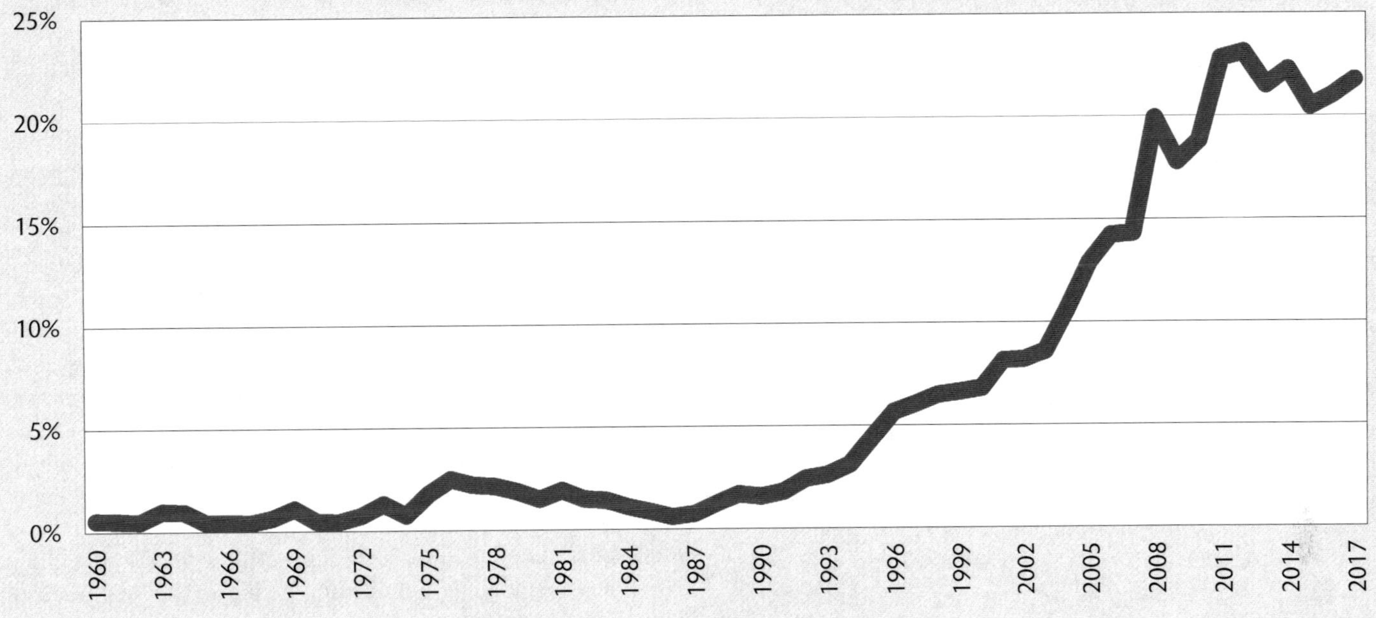

Source: Lee Schulz, Department of Economics, Iowa State University, November 2017.

levels of productivity observed over this course of time. As a gross measure, US pork producers are producing approximately 50% more pork today with 25% fewer sows than was done just 40 years ago.[1,2] While far too many individual contributions to mention, I have attempted to generalize my perception of our (US Pork Industry) singular greatest success (strength), as well as our biggest challenge (weakness), as it relates to the current state of managing the health of pigs in the US.

Success (strength): Ability to control or eliminate diseases at a given farm site or network of related farm sites

Tremendous progress has been made in our ability to confidently, repeatedly, and cost effectively control or eliminate diseases of high consequence from individual farms and networks of related farm sites or premises. Much of this knowledge and capability has its origins in the US Pork Industry's now 30-year long challenge in attempting to mitigate the effects of PRRSV. The more recent lessons learned from the US Pork Industry's PEDV experience have also been extremely insightful. The myriad of highly collaborative public/private research efforts and information sharing amongst veterinarians, pork producers, and researchers has greatly enhanced our understanding of the various routes by which diseases can be transmitted into swine farms. We now much more clearly understand and have an appreciation for how diseases move between farms via something other than the introduction of infected pigs and semen (e.g., transport vehicles, fomites, insects, feed, and aerosol transmission). Such knowledge has prompted a previously unprecedented investment in biosecurity-related infrastructure and procedural-related methodologies. These biosecurity investments have most appreciably been focused on reducing the risk of disease introduction into boar studs, breeding herds, and growing replacement farm sites. Similarly, there have been great advances in our understanding of the shed and spread (circulation and transmission) of pathogens within a given farm site or premises. Such understanding has led to the now common health management practice of coupling breeding herd closure with strict AIAO pig-flow and good sanitary procedures in farrowing facilities to eliminate a growing number of diseases from stand-alone breed-to-wean farms. Diagnostic tools, technologies, and systems used to deliver such services have continued to improve. The advent of the molecular era, user-friendly and more sensitive sample types, over-night shipping, and capable same day/next day diagnostic service providers have enabled practitioners to objectively assess the health status of swine in near real-time and minimize the introduction of high consequence diseases into breeding herds via the more conventional routes of infected breeding stock replacements or semen. We are fortunate to have a network of highly capable and service-centered diagnostic laboratories and diagnosticians in the US. Veterinary biologics and their application in concert with other well-informed management strategies continue to improve and often play an essential role in farm or system level disease control or elimination efforts. In short, our ability to effectively diagnose and eliminate many infectious diseases of high consequence from individual or related farm sites with AIAO pig flow is generally good. Subsequent reinfection (lateral introductions) from something other than the introduction of infected pigs or semen continues to be the primary challenge at hand.

Challenge (weakness): Ability to control or eliminate diseases across a broader area, region, state, or country

With few exceptions, we have not had much sustained success in being able to contain or eliminate emerging or endemic diseases of significance across broad areas or regions of the US with significant domestic pig populations. Since my first day as an undergraduate on the Iowa State University campus in 1987 (which doesn't feel like all that long ago); PRRSV, PCV2b, and PEDV have emerged, spread, and become endemic in the broader context of the US domestic swine herd. While the emergence of each of these pathogens was unique, they share the commonalities of being new pathogens to the US swine industry at the time of their emergence and continue to be pathogens of significance in the US to date. Of these, PRRSV has unquestionably been the most humbling for our profession, the most costly to the US Pork Industry, stimulated the most new knowledge concerning biosecurity and population-level disease control practices, and undeniably has had a profound influence on how and where pigs are raised in the US. The introduction of PCV2b and PEDV share the commonality that they were each diseases that had been causing highly significant losses elsewhere in the world long prior to their transcontinental introduction into North America. The emergence of broadly evident PCVAD (associated with PCV2b) and PEDV associated disease also share the commonality in that they spread across broad and even otherwise isolated regions of the US Pork Industry very quickly. However, one primary difference was that the introduction of PCV2b into farms did not cause the immediate clinical disease as was more visually evident within a few days of PEDV's introduction into previously naïve herds. Outside of that, the timing and distribution across the country following the initial outbreaks were rather similar (e.g., broad segments of the country, industry, and otherwise unrelated herds affected within 12-18 months). The PEDV pandemic more clearly demonstrated how cross-contaminated trucks returning from slaughter plants and contaminated feedstuffs could serve as additional amplifiers of disease spread across the country and borders.[5-8] These two lessons learned (degree of amplification of disease spread via cross-contaminated live-haul transport and contaminated feedstuffs) likely provide highly significant insight as to improved control measures that are needed to better restrain the distribution of diseases of high consequence in the future. Albeit significant progress and sustained success has been observed in keeping large numbers of highly bio-secure and often more remotely located boar studs and breeding herds free of such diseases that are now endemic to the US, such diseases continue to cause significant losses across large segments of the US Pork Industry. The much larger mass of commercial growing pigs serve as an effective reservoir. From a larger industry-wide viewpoint, the shed and spread and continual evolution of high consequence diseases amongst commercial growing pig populations likely represents the primary risk factor and source for subsequent indirect (lateral) disease introductions into breeding herds.

The successful elimination of PRV from US domestic pigs represents the most recent large-scale disease eradication effort and accomplishment in the US Pork Industry. The eradication of PRV had its origins shortly after the US had been officially declared free of CSF in 1976 (CSF elimination program ≈1962-1976).[9,10] Plans and interest in eradicating PRV continued to evolve over the next decade with a more formalized eradication plan being approved by the NPPC in 1987. The national PRV eradication plan moved forward more in earnest in a state-by-state manner in 1989. The US domestic pig population was officially declared free of PRV in 2002. There were many program amendments, controversial and difficult moments, and tremendous examples of industry-led cooperation and leadership provided over the course of the PRV eradication effort. PRV eradication was first achieved in the states/regions with less commercial pig production. Success later worked its way toward the more pig dense areas of the country. The final push and successful elimination of PRV from the more hog dense areas of Iowa was achieved in an environment that closely resembles the industry of today. I feel fortunate for having had the opportunity to be actively engaged as a practitioner and production director over the course of the last five to six years of the PRV eradication program. There were many valuable lessons learned from the PRV eradication experience that are highly relevant for any such large-scale area-regional disease control or elimination efforts. The PRV eradication program was industry led, built upon clearly defined program standards and definitions, executed by veterinarians and their producer clients, and facilitated and monitored by the state and federal veterinary medical officials.

Optimistic outlook for swine veterinarians

The growth and evolution of the US Pork Industry has continued to create additional and increasingly diverse career opportunities for swine veterinarians. While the direct provision of veterinary service and animal health consultation to pork producers remains the core of our profession; the size and scale of the clients being served, the increased complexities of the challenges at hand, and the overall changing dynamics of who and how pigs are owned, managed, served, reared, marketed, and harvested are all factors creating new and diverse roles for swine veterinarians in the US Pork Industry. Veterinarians' core training in the basic sciences, medicine, epidemiology, surgery, and problem solving provides entry level practitioners a solid foundation by which to enter the work force, learn the practical aspects and business of veterinary service and pork production, and subsequently chart their own path. Continuing education and professional development has long been a strength of the veterinary profession and this (AASV) professional organization. The need for such continuing education would seemingly only be on the rise to keep up with the demands of the producers and industry being served. My sense is that the mega-trend toward species-level specialization in production animal medicine and having more veterinarians assuming non-traditional duties or roles will continue for the foreseeable future. Swine interest veterinarians are well positioned to pursue and excel in any number of career paths in the pork industry. Swine veterinarians are often uniquely qualified to provide leadership for efforts, initiatives, or organizations that require the integration of various scientific or technical disciplines; practical (on-farm) knowledge of pig production; and a sound understanding of financial aspects of the business. In closing, I see no shortage in career opportunities (clinical, technical, or managerial) for veterinarians in the US Pork Industry going forward.

Specific opportunities for industry-wide improvement in US swine health and preparedness

Below I have attempted to provide a short list of potential opportunities that I believe this organization's members have the ability to influence both locally in our immediate places of business as well as on a broader industry-wide platform. The topics raised below are simply a few items that I believe warrant our membership's further discussion, consideration, and leadership. As outlined earlier, the US Pork Industry, in which the majority of AASV members serve and actively participate, has experienced transformational change over the course of the past 40 years. Arguably, the transition to multi-site production and the marked growth in, and now dependence on, export markets are two of the most significant industry-impacting changes. These changes have been responsible for a significant portion of the growth in the industry and improvement in biological performance observed. It seems only natural that such significant transitions and changing market risks create and/or expose new opportunities for improvement.

Shed and spread of disease in commercial growing and finishing pigs

Commercial (terminal-cross) growing and finishing pigs represent something close to 80% of the post-weaning age US domestic pig population. On an industry-wide (macro) level, I believe the indirect effects that the shed and spread of disease amongst the commercial growing and finishing pigs has on the health of breeding herds can often be underestimated. Multi-site production has created an environment where the health status of growing pig farm sites and the risk they pose to subsequent disease outbreaks of growing replacement stock, breeding herds, and boar studs can be discounted due to the often significant distance between breeding herds of origin and growing pig farm sites. This disassociation all seems quite natural, as the commercial growing pig sites may be 5, 10, 100, 500, or >1,000 miles away from the sow farm(s) of origin. Additionally, lateral or group-to-group disease introductions into commercial growing pig sites tend to be far less biologically evident and financially less relevant as compared to the same introduction into breeding herds, boar studs, or replacement breeding stock. Thus, the immediate costs, concerns, and impact of, as well as the direct link between, the shed and spread of diseases amongst commercial growing populations and breeding herd health are understandably less evident. However, in a broader context, it seems highly probable that the mass of commercial growing and finishing pigs serve as the primary reservoir for the ongoing evolution of pathogens endemic to the US. Disease transmission amongst and between commercial growing pig farm sites provides a reasonable environment for diseases (most notably viruses) to be maintained and indirectly move both locally and vast distances to infect susceptible breeding stock. In broad-terms, I would suggest that industry-standard biosecurity related practices and efforts to prevent disease introductions into breeding herds are likely unrivaled in commercial livestock production. However, the challenges to prevent such introductions on a 24/7 (365 days/year) basis can be rather immense due to the amount of challenge material that is actively circulating in the greater universe of the US domestic pig population. As mentioned earlier, as a North American Pork Industry, I believe we feel generally confident in our ability to cost effectively and predictably eliminate a growing number of diseases from stand-alone breeding herds without depopulation. However, risks of subsequent reinfection from something other than infected replacement breeding stock and semen continue to be the primary obstacle to larger-scale area-regional or industry-wide elimination efforts. Such very real risks of subsequent reinfection tend to promote extraordinary investments in biosecurity related infrastructure, siting new farms further away from existing swine populations, and/or pro-active acclimation practices of incoming breeding stock in efforts to mitigate economic losses of future unintended introductions. Acclimated, vaccinated, isolated, filtered, or not; breakdowns in breeding herd health status provide the more direct means of large-scale shed and spread through the distribution of infected weaned pigs far and wide. I think opportunities exist to improve upon the status-quo of how we are actively managing the shed and spread of diseases of high consequence among our commercial growing pig populations.

Specifically, I think reassessing our current systems and approach (e.g. standards of practice) towards managing the sanitation of top hog trucks returning from slaughter plants and growing pig immunity for diseases of high consequence warrants thoughtful consideration. The applied research efforts conducted by Lowe, Connor, Yeske, Loula, DuFrense et al, shed great light as to how much cross-contamination of livestock trailers is occurring at the interface of the trailer and unloading dock at slaughter facilities.[5] In essence, this research found that there were approximately 1.4× (3.35× higher risk if immediate previous trailer was positive) more PEDV positive trailers leaving the plant as there were coming into the plant. This work strongly suggests un-sanitized trailers returning from slaughter plants may likely be a significant amplifier of PEDV via the unintended consequences of dragging virus back from the slaughter plants to unsuspecting groups of susceptible finishing pig populations. Such finishing pigs are more than capable of amplifying the virus in great numbers. These late-term finishing breaks of PEDV commonly associated with live-haul transportation infections also revealed the risks that high volumes of virus being shed late in finishing has on the subsequent groups of weaned pigs being placed into the same wean-to-finish facilities, most notably in winter months. It would seem highly probable that this same type of inter-herd transmission associated with finishing pig live-haul is also occurring with other diseases of high consequence; albeit the acute clinical consequence of many other diseases in growing pigs (as compared to PEDV) can often be less visually obvious and/or delayed from the actual point of infection. Improved, high-throughput, robust, and simplistic systems are needed to dramatically reduce the risks of top hog transport recycling and amplifying disease spread amongst commercial growing pigs. I am not so sure the terminal top hog transport vehicles need to be cleaned, disinfected, and rendered non-infectious in the same manner that has become standard of practice as that commonly applied to live-haul transportation vehicles that are backing up to breeding herd sites or hauling pigs to be further grown. However, I do believe broad implementation of systems something akin to the high-throughput "scrape, rinse, and bake" approach that Luc Dufrense described at the 2017 Allen D Leman Conference could provide the industry a step-change in inter-herd disease transmission observed. Similarly, Lowe et al recently completed a study that suggested that simply scraping the unloading area at lairage between loads had a tremendous impact (7-fold reduction in risk of contamination) on reducing the risks of cross-contamination between subsequent trailers.[11]

Vaccination of growing pigs in efforts to reduce clinical signs and shedding has been well demonstrated across any number of vaccines and pathogens. While there are obvious differences as to the degree of efficacy of various vaccines, vaccine strategies, and pathogen combinations; effective vaccines most generally reduce both clinical signs and the magnitude (how much and how long) of shedding. Implementation of a clearly prescribed, aggressive, and compulsory vaccination across all of the growing pigs was clearly the turning point that moved the needle in bringing the final and often controversial push towards eliminating PRV from Iowa. Iowa had long been struggling to make progress. However, the state was cleaned-up (for all practical purposes, with the additional help of depopulating some endemically infected continuous flow sites) in very short order following the decision to proceed with the aforementioned mandatory vaccination of all growing pigs. For benefit of our younger members, prescribed vaccination programs were very aggressive depending on farm-type and location. Pigs in low-density regions were required one full-dose of (gene-deleted) modified-live vaccine as a feeder pig. All growing pigs in designated pig-dense areas with a history of infection received two doses of vaccine, one as a feeder pig and a booster in mid-finishing, in efforts to sustain peak immunity through to slaughter. Pigs from well-vaccinated sows in specified farm-types and locations could receive up to three-vaccinations, once intra-nasally at weaning, once as a feeder pig, and then again in mid-finishing. The profound impact that achieving this heightened and uniform state of immunity had on minimizing the shed and spread across the many pig dense regions across Iowa has always stuck with me. In short, I think the role and impact of growing pig vaccination on larger regional or industry-wide swine health necessitates further applied and innovative study.

Biosecurity of feed supply

The PEDV pandemic (2013-2014) in the US exposed biosecurity risks related to our feed supply that were previously not top of mind. While some concerns over the potential for spray-dried plasma to be a vector for spreading PCV2b were raised during the PCVAD outbreak that spread quickly across the country in 2005-2007, to my knowledge, such evidence for transmission of PCV2 in commercially produced spray-dried pork plasma were never confirmed. However, PEDV clearly demonstrated the potential for its ability to spread via PEDV fecal laden feedstuffs.[6-8] These observations initiated a body of applied research (Pipestone Applied Research, South Dakota State University, Kansas State University, and Iowa State University) assessing the impact of alternative feed additives, feed processing treatments, and the ability for diseases of high consequence to survive on a variety feed ingredients. The observations made over the course of the US PEDV pandemic and the series of applied feed safety related research that has followed seem to clearly suggest managing such risks related to the transmission of disease through contaminated feedstuffs requires our profession's and industry's attention. Feed mill biosecurity, post-processing treatments of higher risk feedstuffs, completed feed processing methods or other treatment, feed additives, and risk-based decision making are all areas of interest.

Traceability and connectivity of information

The digital age of which we now live continues to enable new and ever improving ways of aggregating and summarizing information. Transitioning from conventional paper-based systems of information exchange to more seamless electronic applications creates the opportunity for transformational change in the quality, traceability, value, connectivity, and accessibility of information. The adoption and use of universally recognized data standards is also a critical element that allows data within any number of database applications to be more seamlessly accessed, aggregated, and summarized as needed for a wide variety of specific use cases. Traceability of food products has become increasingly important to consumers, both domestically and internationally. Improving the traceability, quality, and connectivity of livestock movement and animal health related information present has significant opportunities for the current and future needs of the US Pork Industry. Manual transcription of premises or client-level identifiers are not compatible with the capabilities that next generation animal health management applications have to offer. Use of electronic applications for submitting and capturing certificates of veterinary inspection (or swine health production plan equivalent), veterinary diagnostic lab submissions, and inter-premises swine movement information (inclusive of movements to slaughter or other marketing collection points) are universally needed to make such information fit for the digital era. In today's world, the broader US Pork Industry's need for transitioning to electronic systems for capturing these three data elements (certificates of veterinary inspection, veterinary diagnostic lab submissions, and inter-premises swine movement information) seem self-evident. Significant investment, collaboration, and ongoing industry-led leadership will continue to be needed to achieve transformational improvement. All seems very doable with existing technologies and certainly an opportunity for our profession to influence and execute.

Industry-led surveillance, certified health, and continuity of business programs

While I would not profess to have many novel skills, I do feel extremely fortunate to have a somewhat unique set of work experiences dating back to my upbringing. These work experiences have provided me with a reasonable understanding and appreciation for the various aspects of pork production, rural communities and living, pork processing and marketing, veterinary practice, diagnostic medicine, state and federal animal health infrastructure and operations, and the role that state and federal governments or agencies play in animal agriculture. Such experiences suggests to me that perhaps there is the opportunity for the industry (e.g., pork producers, processors, and veterinarians providing counsel or service to such entities) to be playing a larger or more assertive leadership role in providing direction to our state and national animal health related programs and infrastructure. We are fortunate to have many tremendous people and strong leaders across our state and federal agricultural and veterinary related agencies. I can tell you that our AASV members working in such capacities are having a highly significant impact on the US swine industry. These agencies (by and large) clearly understand their roles and mission in serving animal agriculture and most certainly want to do things that are in the best interest of the industries and stakeholders that they serve to protect. However, my sense is that industry can at times put too much responsibility on state and federal agencies for future direction and/or permission to move forward with new or alternative methods, programs, or approaches that are needed to adapt to the changing landscape of industry needs. Additionally, I think it can be easy for industry stakeholders to underestimate their influence and can often over assume the resources and capabilities that exist within our state and federal veterinary medical agencies. The US Pork Industry is fortunate to benefit from the services of some truly committed and capable state and federal veterinary medical officials. However, human resources within such agencies are generally limited and have responsibilities across all species. Two specific opportunities for improvement include: i) addressing the need for an updated (modernized) approach to program-disease surveillance of commercial US swine (e.g. comprehensive swine surveillance) and ii) furthering the development and implementation of the Secure Pork Plan in concert with other industry-led swine health monitoring, compartmentalization, and/or regionalization initiatives.

Swine Health Improvement Plan

The US Poultry Industry established the National Poultry Improvement Plan (NPIP) in the 1930s to eliminate Pullorum Disease. Pullorum Disease (caused by Salmonella typhoid) is a vertically transmitted pathogen that was causing significant losses in the US Poultry Industry.[12] NPIP is a cooperative industry, state, and federal program. NPIP provides a common-set of definitions, guidelines, and requirements to be met for flocks or premises to receive various health status certifications (e.g. individual flocks or states being designated as Monitored or Clean of specific pathogens). Program standards, definitions, and requirements are updated every two years by a committee of industry representatives and veterinary medical officials. Such standards and definitions are subsequently incorporated into the US Code of Federal Regulations (9 CFR, Chapter I, Subchapter G – Livestock Improvement). Biosecurity standards were recently added to the NPIP program. Specific guidance for regionalization or compartmentalization efforts consistent with OIE requirements are also provided. Each state operates their own NPIP Official State Agency. Producer participation in NPIP is voluntary. Each producer and state NPIP Official State Agency determines the NPIP status or requirements they choose to pursue. The national NPIP standards are set-forth as the minimum requirement for such NPIP health status designations. Each producer and NPIP Official State Agency can choose to elevate such standards beyond the national NPIP standard. Such state-level definitions and standards are subsequently incorporated into respective state's animal health related administrative rules. Flock or state-level NPIP status and standards are often used or referenced to govern interstate movement of poultry in lieu of certificates of veterinary inspection. In the current, ever evolving, and continuously improving US Pork Industry, perhaps it is time to proceed with such efforts to create a more formal and universally recognized system for defining such standards for demonstrating freedom of specified swine diseases from a given premises, network of related premises under a common management structure or ownership, compartment, region, or state? Establishing such a system may be useful in providing referenceable guidance to industry led efforts to implement continuity of business programs (Secure Pork Supply Plan), establishing recognized regions or compartments within the US, and facilitating larger-scale efforts to improve the health of US swine over the course of time. Domestic consumption of pork per capita has remained relatively unchanged in the US over the past 40 years. Pork exports have increased 10-fold in the past 20 years. A growing list of endemic diseases continue to be of consequence to the pigs within our care and costly to the US Pork Industry. Establishing a more formal and universally recognized approach towards defining such standards and definitions for swine health may be useful in providing guidance to larger-scale efforts to improve and protect the health of US swine and in enhancing the US Pork Industry's longer-term competitiveness and market access in the global marketplace.

Eliminate a disease

We are approaching 20 years since the last time a disease of high consequence was eliminated from the US domestic pig population. I would estimate that something close to 50% or more of today's practicing veterinarians have not been directly involved in a formal state-wide or national disease elimination effort. Such systems, infrastructure, and likely a fair-bit of institutional knowledge for doing so have long since been diminished. Preparedness is a term commonly used in reference to an entity's ability to effectively respond to and perform in a time of need. My experience suggests that preparedness is best defined by the practices, systems, and activities that one executes on a routine basis. I do not believe one can simply "flip the switch" and suddenly all such systems, practices, and capabilities will be in place and a go. Whether it be preparing for an animal health emergency or running a race of significant distance, my sense is that it is the daily routines that have the largest impact on the outcome. Perhaps it is time we go for a jog and stretch our legs?

Conclusion

The US and greater North American Pork Industry and our swine veterinary profession have so very many things to be proud of, as is most evident by the ever-improving levels of productivity and the overall growth in the industry observed. Our systems of pork production today are an exemplary model of sustainable agriculture that competes well in delivering high quality food products in a global marketplace. The opportunities, challenges, and risks have not likely ever been greater. Our profession is well positioned to build upon current and historical successes and play an ever-increasing leadership role in helping the producers and industry we serve meet the challenges of tomorrow. As was the case with those that came before our time, each of the individual and collective efforts, actions, and choices we make today, are and will continue to have a profound impact on US Pork Industry and swine veterinary profession inherited by those to follow.

References

1, National Pork Board. Pork Quick Facts. pork.org/facts. Accessed November 2017.

2. USDA. National Animal Statistics Service. nass.usda.gov. Accessed November 2017.

3. Harris D.L. Multi-site pig production. 2000. Iowa State University Press. Ames, IA.

4. Greenwood M. Iowa State Swine Day and National Hog Farmer. June 2017.

5. Lowe J et al. Role of transportation in spread of Porcine Epidemic Diarrhea Virus, United States. Emerging Infectious Disease. 2014. 20(5):872-874.

6. Pasick J et al. Investigation into the role of potentially contaminated feed as a source of the first detected outbreaks of Porcine Epidemic Diarrhea in Canada. Transboundary and Emerging Diseases. 2014.61(5):397-410.

7. Dee et al. An evaluation of contaminated complete feed as a vehicle for porcine epidemic diarrhea virus infection of naïve pigs following consumption via natural feeding behavior: proof of concept. BMC Veterinary Research. 2014. 10:176.

8. Kochhar H. Porcine epidemic in diarrhea in Canada: As emerging disease case study. Canadian Veterinary Journal. 2014. 55(11):1048-1049.

9. Anderson L et al. Pseudorabies (Aujesky's Disease) and its eradication. A review of the US Experience. USDA APHIS Technical Bulletin. 2008. No 1923.

10. USDA APHIS. Hog cholera and its eradication: a review of the US experience. 1981. USDA National Agricultural Library.

11. Lowe J et al. Factors that influence mechanical transmission of porcine reproductive and respiratory syndrome virus at the time of unloading animals into slaughter plant lairage. Journal of Swine Health and Production. 2017. 25(1):19-23.

12. National Poultry Improvement Plan. http://www.poultryimprovement.org. Accessed November 2017.

Effect of antimicrobial administration on the prevalence and abundance of selected antibiotic resistance genes in perinatal piglets

J. Lowe[1,2]; B. Blair[1]; M. Zeineldin[1]; A. Megahed[1,3]

[1]Integrated Food Animal Management Systems, Department of Veterinary Clinical Medicine, College of Veterinary Medicine, University of Illinois at Urbana-Champaign; [2]Production Animal Consultation, LLC, Oakley, Kansas; [3]Department of Animal Medicine, Internal Medicine, Faculty of Veterinary Medicine, Benha University, Moshtohor-Toukh, Kalyobiya, Egypt

Introduction

The emergence of antibiotic resistance genes (ARG) is an urgent public health issue creating pressure to reduce the use of antimicrobials. While the antibiotic resistance profiles of pathogens and opportunistic bacteria cultured from swine have been characterized, the ARG from the whole gut microbiota have received far less attention. In this study, high-throughput next generation sequencing platforms were used to quantify the effect of antimicrobial administration on the prevalence and abundance of selected ARG (Tet O, Tet W, Tet C, Sul 1, Sul 11, bla ctx and erm B) in perinatal piglets.

Materials and methods

The study procedure was reviewed and approved by the Ethical Committee for Institutional Animal Use and Care of the University of Illinois at Urbana-Champaign. A randomized complete block design was used. Forty-eight litters were blocked to one of six treatments (N = 8) by farrowing day and dam parity group. Within litter all pigs received the same treatment. Pigs were weighed, and treatments were administered at 24 hours of age, after litters have been balanced for size. Treatments were as follows: Control (saline 1cc), Tulathromycin (2.5 mg/kg IM, TUL)), Ceftiofur Crystalline free acid (5.0 mg /kg IM, CCFA), Ceftiofur hydrochloride (5 mg/kg IM, CHC), Oxytetracycline (22 mg/kg IM, OTC) and Procaine Penicillin G (33,000 units/kg IM, PPG). Two pigs per litter were individually identified and weights and deep fecal swabs were collected at days 0 (prior to treatment), 5, 10, 15 and 20 after treatment. Once collected, all samples were placed on dry ice and held at -80°C until analysis. Weights of all piglets were recorded on study days 0 and at day 21. All dead piglets were recorded throughout the study.

Individual fecal swabs were thawed for DNA extraction. Genomic DNA was extracted from fecal swabs using the MoBio Power Fecal®DNA Isolation Kit according to the manufacturer instruction. The quality and concentration of DNA were checked by agarose gel electrophoresis and ultraviolet-absorbance (ND1000, Nanodrop, Thermo Fisher Scientific Inc.). DNA was stored at -20°C pending further analysis.

Genomic DNA was subject to Fluidigm Access Array Amplification (Fluidigm Corporation, South San Francisco, CA, USA). Prior to amplification all DNA samples were measured on a Qubit (Life technologies, Grand Island, NY, USA) using the High Sensitivity DNA Kit (Agilent Technologies, Santa Clara, CA, USA) (Zeineldin et al., 2017b) (Zeineldin et al., 2017a)especially in recently weaned, intensely raised and newly transported feedlot cattle. While the importance of upper airway structure and function in the susceptibility of the lower respiratory tract to colonization with potential pathogens is well established, the role of the mucosal microbiota in respiratory health is less well defined. The objective of this study was to characterize the nasopharyngeal microbiota of feedlot cattle at entry into a commercial feedlot, during initial management processing, and to compare the dynamics of change in these microbial communities between clinically healthy calves and those that develop BRD within the first month after entry. Deep nasopharyngeal swabs were collected from randomly selected healthy calves (n\u00a0=\u00a066. Samples were diluted to 2 ng/ul concentrations. A mastermix for amplification was prepared using the Roche High Fidelity Fast Start Kit and 20x Access Array loading reagent according to Fluidigm protocols. Mastermix was aliquoted to 48 wells of a PCR plate. In a separate plate, primer pairs were prepared and aliquoted. A total of 7 primer sets targeting 7 different ARGs, were used in this study (Table 1). Following the sample and primers loading stage, the IFC plate was loaded on the Fluidigm Biomark HD PCR machine and samples were amplified using the Access Array cycling program without imaging. Amplified PCR products were run on a Fragment Analyzer (Advanced Analytics, Ames, IA) and amplicon regions and expected sizes were confirmed. Samples were then pooled in equal amounts according to product concentration. The pooled products were then size selected on a 2% agarose E-gel (Life Technologies) and extracted from the isolated gel slice with Qiagen gel extraction kit (Qiagen). Cleaned size selected products were run on an Agilent Bioanalyzer to confirm appropriate profile and determination of average size. The final pooled Fluidigm libraries were transferred to the DNA Services lab at the W. M. Keck Center for Comparative and Functional Genomics at the University of Illinois at Urbana-Champaign for Illumina sequencing. The Illumina MiSeq platform (Illumina,

Table 1: Primers targeting selected antibiotic resistance genes used in this study.

Target gene	Primer	Sequence 5'-3'	Amplicon size (bp)	Reference
Tet O	tetO_F	ACGGARAGTTTATTGTATACC	1920	(Ng et al., 2001)
	tetO_R	TGGCGTATCTATAATGTTGAC		
Tet W	tetW_F	GAGAGCCTGCTATATGCCAGC	1920	(Ng et al., 2001)
	tetW_R	GGGCGTATCCACAATGTTAAC		
Tet C	tetC_F	GCGGGATATCGTCCATTCCG	1191	(Ng et al., 2001)
	tetC_R	GCGTAGAGGATCCACAGGACG		
Sul I	sul_I_F	CGCACCGGAAACATCGCTGCAC	840	(Pei et al., 2006)
	sul_I_R	TGAAGTTCCGCCGCAAGGCTCG		
Sul II	sul_II_F	TCCGGTGGAGGCCGGTATCTGG	816	(Pei et al., 2006)
	sul_II_R	CGGGAATGCCATCTGCCTTGAG		
erm B	ermB_f	GGTTGCTCTTGCACACTCAAG	738	(Martineau et al., 2000)
	ermB_R	CAGTTGACGATATTCTCGATTG		
bla CTX	RTCTXM_F	CTATGGCACCACCAACGATA	876	(Birkett et al., 2007)
	RTCTXM_R	ACGGCTTTCTGCCTTAGGTT		

San Diego, CA, USA) was used to sequence the selected ARG gene according to the Illumina instructions.

The sequence run generated Illumina base call (bcl) files were converted into demultiplexed, compressed fastq files using bcl-2fastq 1.8.4 (Illumina, CA). A secondary pipeline was used to decompress the fastq files, generate quality score plots (FastX Tool Kit), and to report the number of reads per sample per library. The bcl files were also processed in bcl2fastq 1.8.4 without demultiplexing, and then sorted by initial PCR-specific primer using a custom in-house pipeline. After preprocessing, raw sequence data was filtered to remove low quality reads.

To avoid human errors in ARG-like sequence classification, we developed a customized version of the Antibiotic Resistance Gene Database (ARG-ANNOT) incorporated all sequences of the ARG that used in this study. The customized ARG-ANNOT database was used to align the different the ARG raw sequences reads obtained from the Illumina sequencing into different types according to the used primers. Additionally, the number of ARG sequences depth and coverage in each ARG were also counted automatically. The number of hits to each AMR gene was complied with samples in columns and AMR genes in rows. To avoid the bias due to different prokaryotic DNA portions in the samples, normalization by the ARG reference sequence length by the 16S-rRNA gene sequence length was conducted in this study. ARG 'abundance' was expressed as 'copy of ARG per copy of 16S-rRNA gene', which is the same as that of qPCR results reported in most previous literature. The relative abundance of ARG was calculated using the following equation according to (Li et al., 2015):

$$\text{Abundance} = \sum_{1}^{n} \frac{N_{\text{ARG-like sequence}} \times L_{\text{reads}} / L_{\text{ARG reference sequence}}}{N_{\text{16S sequence}} \times L_{\text{reads}} / L_{\text{16S sequence}}}$$

where ^{N}ARG-like sequence is the number of the of sequence read normalized hits to one specific AMR gene; ^{L}ARG reference sequence is the sequence length of the corresponding AMR reference gene; ^{N}16S sequence is the number of the 16S sequence identified from the sequencing data; ^{L}16S sequence is the average length of the 16S sequence in the reference database for the 16S sequence identification; n is the number of the mapped ARG reference sequence belonging to the ARG type; ^{L} reads is the sequence length of the Illumina reads (300 nt) that was used in the present study.

Abundance data was subjected to ANOVA analysis with AMR gene frequency at the dependent variable and sampling day and antibiotic treatment as fixed effects. For significant models

Results

The average weight gain and mortalities percent did not differ significantly between the groups at 21 days of age ($P > 0.05$). Across all the samples, the predominant ARG were erm B (33.9%), tet W (11.6 %) and sul II (9.01 %). The observed changes in the selected ARG demonstrated antimicrobial-specific variations in both duration and extent. Antibiotic treatment did not influence the frequency of Sul I, Sul II, or tet C genes over the observation period. Across all the samples, the relative abundance of Tet W and erm B were increased over time with days 15 and 20 being greater than days 0 and 5. Day 10 was greater than day 0 but was similar to day 5 for both genes. For time periods 5, 10, 15 and 20 days, TUL treatment increased the frequency of tet O, tet W and erm B compared to controls. All other antibiotics were similar to controls for those three ARG. In conclusion, our results showed that the different antimicrobial treatment of new-born

Figure 1: Changes in relative abundance of AMR genes following treatment with five different antibiotics at one day of age, expressed as a percentage of AMR gene abundance in untreated control animals. **A)** TETW, **B)** TETO, **C)** ERMB

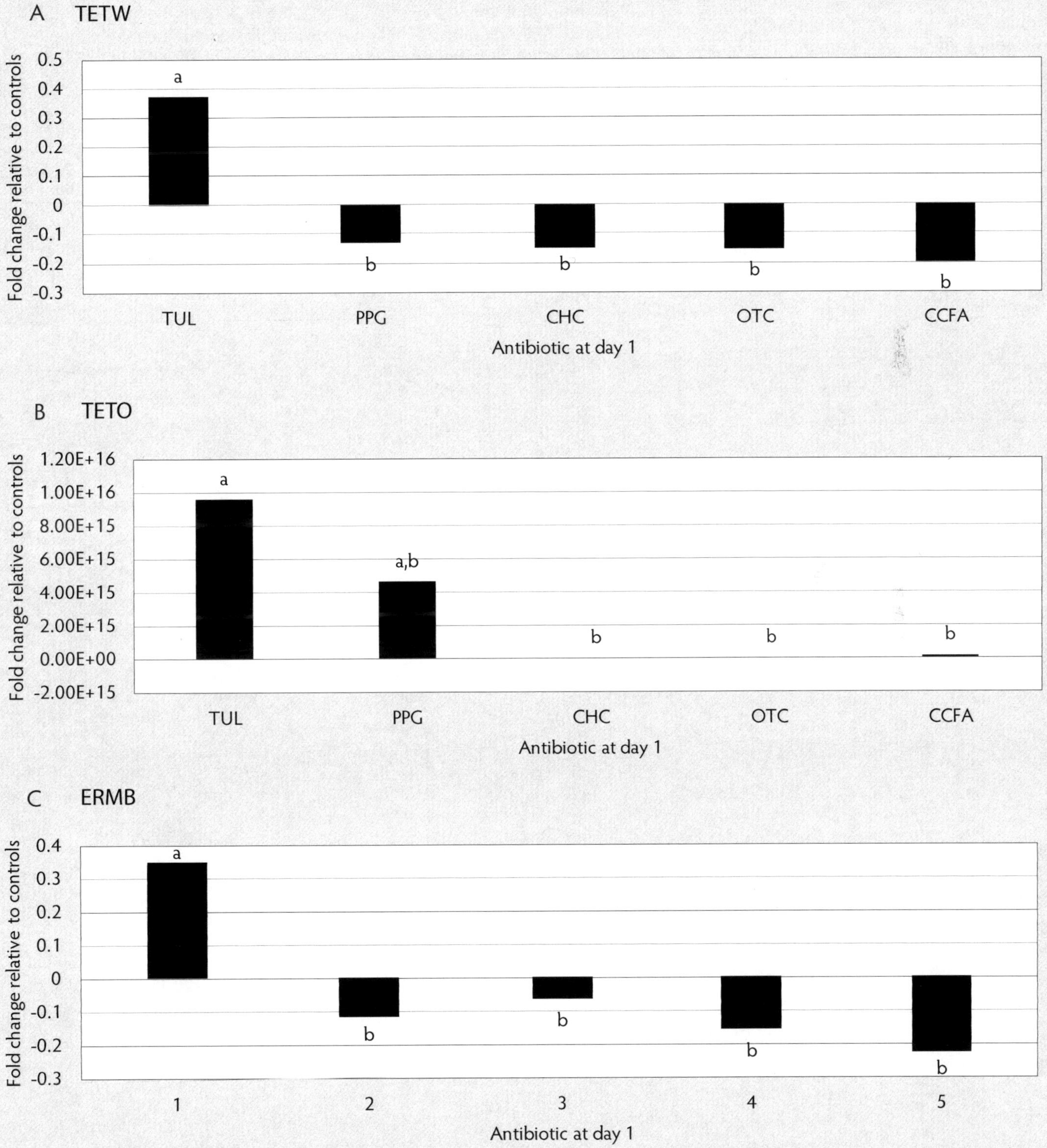

[a,b] Values with different letters are significantly different, $P < 0.05$.

piglets failed to reduce mortality and increase the average daily weight gain. Further, single dose of antimicrobial treatment especially PPG and TUL of newborn piglets could promote selection of ARG in swine gut. Therefore, we suggest rejection of routine antimicrobial administration at birth.

References

1. Birkett, C. I., Ludlam, H. A., Woodford, N., Brown, D. F. J., Brown, N. M., Roberts, M. T. M., et al. (2007). Real-time TaqMan PCR for rapid detection and typing of genes encoding CTX-M extended-spectrum -lactamases. *J. Med. Microbiol.* 56, 52–55. doi:10.1099/jmm.0.46909-0.

2. Li, B., Yang, Y., Ma, L., Ju, F., Guo, F., Tiedje, J. M., et al. (2015). Metagenomic and network analysis reveal wide distribution and co-occurrence of environmental antibiotic resistance genes. *ISME J.* 9, 2490–2502. doi:10.1038/ismej.2015.59.

3. Martineau, F., Picard, F. J., Lansac, N., Ménard, C., Roy, P. H., Ouellette, M., et al. (2000). Correlation between the resistance genotype determined by multiplex PCR assays and the antibiotic susceptibility patterns of Staphylococcus aureus and Staphylococcus epidermidis. *Antimicrob. Agents Chemother.* 44, 231–8. Available at: http://www.ncbi.nlm.nih.gov/pubmed/10639342 [Accessed November 15, 2017].

4. Ng, L.-K., Martin, I., Alfa, M., and Mulvey, M. (2001). Multiplex PCR for the detection of tetracycline resistant genes. *Mol. Cell. Probes* 15, 209–215. doi:10.1006/mcpr.2001.0363.

5. Pei, R., Kim, S.-C., Carlson, K. H., and Pruden, A. (2006). Effect of River Landscape on the sediment concentrations of antibiotics and corresponding antibiotic resistance genes (ARG). *Water Res.* 40, 2427–2435. doi:10.1016/j.watres.2006.04.017.

6. Zeineldin, M., Lowe, J., de Godoy, M., Maradiaga, N., Ramirez, C., Ghanem, M., et al. (2017a). Disparity in the nasopharyngeal microbiota between healthy cattle on feed, at entry processing and with respiratory disease. *Vet. Microbiol.* 208. doi:10.1016/j.vetmic.2017.07.006.

7. Zeineldin, M. M., Lowe, J. F., Grimmer, E. D., de Godoy, M. R. C., Ghanem, M. M., Abd El-Raof, Y. M., et al. (2017b). Relationship between nasopharyngeal and bronchoalveolar microbial communities in clinically healthy feedlot cattle. *BMC Microbiol.* 17, 138. doi:10.1186/s12866-017-1042-2.

Microbial killing capacity of aqueous and gaseous ozone on different surfaces contaminated with manure

Ameer A. Megahed[1,2]; **Brian Aldridge**[1]; **James F. Lowe**[1]

[1]Department of Veterinary Clinical Medicine, College of Veterinary Medicine, University of Illinois at Urbana-Champaign, Illinois; [2]Department of Animal Medicine, Internal Medicine, Faculty of Veterinary Medicine, Benha University, Moshtohor-Toukh, Kalyobiya, Egypt

Background

A high reactivity and low production of harmful residues make ozone (O3) an attractive alternative disinfectant for improving farm hygiene and biosecurity, and for aiding in the prevention of infectious diseases. To explore the practical potential of using ozone we designed a study to (1) characterize the microbial killing capacity of aqueous and gaseous O3 on manure-based pathogens (MBP) associated with various materials (plastic, metal, nylon, rubber, and wood); (2) determine the effect of microbial load on the killing capacity of aqueous O3.

Materials and methods

In a crossover design, 14 strips (7.5 × 2.5 cm) of each material were randomly assigned between 3 groups, treatment (n = 6), positive control (n = 6), and negative control (n = 2). The material strips were soaked in diluted dairy cattle manure harboring an inoculum level of 10^6 - 10^7 microbes, for 60 minutes. The treatment strips were exposed to aqueous O3 of 2, 4, and 9 ppm, and gaseous ozone of 1 and 9 ppm for 2, 4, and 8 minutes exposure. Following ozone exposure, quantitative bacterial cultures were performed on each strip using 3M™ Petrifilm™ rapid aerobic count plate (RAC) and plate reader.

Results and discussion

On smooth surfaces (plastic and metal) aqueous O3, at 4 ppm or greater, reduced MBP to a safe level (> 5 log10 reduction) within 2 minutes. On the same surfaces, gaseous ozone, at 9 ppm for 4 minutes, inactivated 3.3-log10 of MBP. Aqueous O3 of 9 ppm was also effective in reducing MBP to a safe level on nylon and rubber surfaces. On more complex surfaces (e.g. wood) both aqueous and gaseous O3, at up to 9 ppm, did not reduce MBP to a safe level. Overall, the bacterial load was a strong predictor for reduction in cell counts ($P < 0.0001$, R2 = 0.72). We conclude that aqueous O3 of 4 and 9 ppm for 2 minutes are efficient in reducing MBP to a safe level on smooth, and moderately rough surfaces, respectively. However, O3 alone is not an adequate means of controlling bacterial populations on complex surfaces.

Evaluation of the survival of viral pathogens in contaminated feed ingredients using transboundary shipment models

Scott Dee[1]; Fernando V. Bauermann[2]; Megan C. Niederwerder[6]; Aaron Singrey[2]; Travis Clement[2]; Marcelo de Lima [2,3]; Gilbert Patterson[4]; Steve Dritz[5]; Mike Tokach[7]; Jason Woodworth[7]; Cassandra Jones[7]; Jon DeJong[1]; Gordon Spronk[1]; Jane Christopher-Hennings[2]; Bob Rowland[5]; Eric Nelson[2]; Diego Diel[2]
[1]Pipestone Applied Research, Pipestone Veterinary Services, Pipestone, Minnesota; [2]Animal Disease Research and Diagnostic Laboratory, South Dakota State University, Brookings, South Dakota; [3]Universidade Federal de Pelotas, RS, Brazil; [4]Center for Animal Health in Appalachia, Lincoln Memorial University, Harrogate, Tennessee; [5]Department of Diagnostic Medicine/Pathobiology, College of Veterinary Medicine, Kansas State University, Manhattan, Kansas; [6]Kansas State University Veterinary Diagnostic Laboratory, Kansas State University, Manhattan, Kansas; [7]Department of Animal Sciences and Industry, College of Agriculture, Kansas State University, Manhattan, Kansas

Abstract

This study evaluated survival of important viral pathogens of swine or their surrogates in contaminated feed ingredients during simulated transboundary transportation. Based on global significance, 11 viruses were selected, including Foot and Mouth Disease Virus (FMDV), Classical Swine Fever Virus (CSFV), African Swine Fever Virus (ASFV), Influenza A Virus of Swine (IAV-S), Pseudorabies virus (PRV), Nipah Virus (NiV), Porcine Reproductive and Respiratory Syndrome Virus (PRRSV), Swine Vesicular Disease Virus (SVDV), Vesicular Stomatitis Virus (VSV), Porcine Circovirus type 2 (PCV2) and Vesicular Exanthema of Swine Virus (VESV). To model the survival of FMDV, CSFV, PRV, NiV, SVDV and VESV, surrogate viruses with similar physical properties and stability were used, and those consisted of Senecavirus A (SVA) for FMDV, Bovine Viral Diarrhea Virus (BVDV) for CSFV, Bovine Herpesvirus Type 1 (BHV-1) for PRV, Canine Distemper Virus (CDV) for NiV, Porcine Sapelovirus (PSV) for SVDV and Feline Calicivirus (FCV) for VESV. Remaining assessments involved the actual pathogen. Controls included complete feed (positive and negative controls) and stock virus positive controls (virus only, no feed matrix). Virus survival was evaluated using either a Trans-Pacific or Trans-Atlantic transboundary model, involving representative feed ingredients, transport times and environmental conditions, with samples tested by PCR, VI and/or swine bioassay. Select viruses (SVA, FCV, BHV-1, PRRSV, PSV and PCV) maintained infectivity during transport, while others (BVDV, VSV, CDV and IAV-S) did not. Survival was maximized in conventional soybean meal, lysine hydrochloride, and vitamin D. The ASFV survival phase is currently underway and results will be presented at the conference. These results demonstrate survival of certain viruses in specific feed ingredients ("high-risk combinations") under conditions simulating transport between countries. This work supports previously published data on the survival of Porcine Epidemic Diarrhea Virus in feed and provides further evidence indicating that contaminated feed ingredients may serve as risk factors for foreign animal and endemic diseases.

Spatial autocorrelation and what it means for swine surveillance

M. Rotolo; C. Wang; M Haddad; Y. Sun; M. Hoogland; R. Main; J. Zimmerman
Iowa State University, Ames, Iowa

Description of the problem

Spatial autocorrelation is based on Tobler's first law of geography: *everything is related to everything else, but near things are more related then distant things*. This simple, intuitive concept has huge implications for the way we conduct disease surveillance. In a recent study, spatial patterns associated with the spread of PRRSV were explored using oral fluid sampling data.

Oral fluids were collected from every occupied pen (108 pens; ~25 pigs per pen) in 3 commercial wean-to-finish barns on one finishing site for 8 weeks for a total of 972 OF samples. Oral fluid samples were completely randomized, tested for PRRSV by RT-rtPCR, and then analyzed for spatial autocorrelation using threshold distance as the spatial weight matrix. Moran's *I*, a quantitative measure of spatial autocorrelation calculated using GeoDa 1.10, showed positive global spatial autocorrelation in the distribution of PRRSV RT-rtPCR results within barns. LISA (Local Indicators of Spatial Association) analysis identified PRRSV clusters within barns and showed that the spatiotemporal pattern of clusters differed among barns.

Discussion

Contagious diseases move from pig-to-pig and pen-to-pen, i.e., spatially. With this in mind, the fact that PRRSV exhibited spatial autocorrelation is not surprising. However, this outcome is important because of its implications for the surveillance methods we have relied on since the mid-1980s.

Since the PRV eradication program, surveillance has been based on the assumption that disease follows a "hypergeometric distribution." From this it can be shown that 30 samples from a population are enough to achieve a 95% probability of detection if 10% of the population is infected. However, hypergeometric distribution assumes that the target (disease) is randomly distributed within the population. This assumption is violated when the target demonstrates a spatial pattern, i.e., shows positive spatial autocorrelation. Spatial autocorrelation probably appeared as the swine industry modernized. That is, as the industry shifted from small, outdoor herds to larger, confined populations.

Our surveillance methods have not evolved with the industry. In particular, the presence of spatial autocorrelation signals the need to reevaluate and explore new surveillance methodologies that account for positive spatial autocorrelation.

Pathogenesis of a porcine parainfluenza virus-1 isolate (USA/MN25890NS/2016) in conventional and CDCD piglets

M. Welch, BS; J. Park, PhD; K. Harmon, PhD; J. Zhang, MD, PhD; P. Piñeyro, DVM, MVSc, DVSc, PhD; L. Gimenez-Lirola, MS, PhD; P. Gauger, DVM, PhD
Department of Veterinary Diagnostic and Preventive Animal Medicine, College of Veterinary Medicine, Iowa State University, Ames, Iowa

Introduction

Researchers from the University of Hong Kong first detected a novel paramyxovirus, provisionally designated porcine parainfluenza virus type-1 (PPIV-1), from slaughter swine in 2013.[1] Detection of this new paramyxovirus spurred several veterinary diagnostic labs in the United States (US) to develop PPIV-1 specific real-time reverse transcriptase polymerase chain reaction (RT-qPCR). However, it had been unclear whether PPIV-1 contributes to the porcine respiratory disease complex (PRDC), since the virus had been detected in both affected and non-affected herds.[2,3] Additionally, the pathogenesis of PPIV-1 could not be determined because obtaining a PPIV-1 isolate in cell culture had previously been unsuccessful. The objective of this study was to evaluate the pathogenesis of a PPIV-1 isolate (USA/MN25890NS/2016) in weaned and caesarian derived, colostrum deprived (CDCD) piglets. The clinical signs, replication, shedding, transmission, gross and histological lesions, and antibody response were analyzed in this study.

Materials and methods

Conventional pig challenge study

Fifty, three-week-old piglets obtained from a commercial swine facility were randomly allocated into three groups. This included 30 pigs in the challenged group (Ch), 15 pigs in the negative control group (Neg), and five pigs in the contact group (Cont). Piglets were pre-screened using a whole virus ELISA for PPIV-1 and confirmed as PPIV-1 seronegative. Piglets also were confirmed negative for influenza A virus, porcine reproductive and respiratory syndrome virus, *Mycoplasma hyopneumoniae,* and porcine circovirus type 2 by RT-qPCR prior to the start of the study. Thirty piglets were inoculated intranasally (1 mL per pig; 0.5 ml per nostril) and intratracheally (2 mL per pig) with $1 \times 10^{4.5}$ tissue culture infectious dose 50 per milliliter ($TCID_{50}$/mL) of PPIV-1 isolate USA/MN25890NS/2016. Rectal temperatures, nasal swabs, and oral fluid samples were collected at day post-inoculation (DPI) 0, 5, 7, 9, 14, 16, and 21. All swab samples were placed in 2 ml minimum essential media (MEM) as previously described.[5] The Cont pigs were placed in a pen two feet away from the challenged pigs at 2 DPI, or 0 day post-contact (DPC), sharing the same air space and evaluated for indirect transmission. Sera was collected from all animals on DPI 3, 7, 9, 14, and 21.

Pigs were observed daily for evidence of clinical signs including coughing, lethargy, and respiratory distress. The clinical scores were based on severity with zero being normal and three being severe with dyspnea and exaggerated abdominal breathing. The lungs were examined at necropsy for percent purple-red consolidation using a previously described method.[6,7] The percent consolidation was determined for intermediate, apical, cardiac, and diaphragmatic lung lobes bilaterally. The total percentage of consolidation for the entire lung was calculated based on weighted proportions of each lobe to the total lung volume.

Ten Ch and 5 Neg piglets were necropsied at DPI 2, 5, and 21, with Cont piglets necropsied at 21 DPI. Fresh tissue samples were collected at necropsy including each nasal turbinate, one section of lung, proximal trachea, and distal trachea. Swabs were collected from both the proximal and distal trachea and placed into two mL of cell culture medium for RT-qPCR. Bronchoalveolar lavage fluid (BALF) was obtained using 50 mL of cell culture medium. A section of each nasal turbinate, proximal trachea, distal trachea, right apical, right cardiac, right diaphragmatic (two sections), and intermediate lung lobes were fixed in 10% formalin and evaluated for evidence of histological lesions.

CDCD pig challenge study

Results from the conventional pig pathogenesis study were replicated in CDCD piglets using the same virus, challenge dose ($1 \times 10^{4.8}$ $TCID_{50}$/mL) and inoculation method. Fourteen piglets PPIV-1 negative for virus and antibody were randomly allocated into two groups. This included ten Ch piglets and four Neg piglets. There was no Cont group in this study. Nasal swabs were collected at DPI 0-7, 11, 13, 15, 16, 17, 19, 21, 24, and 27. Rectal temperatures and clinical observations were recorded on DPI 0-7 using the same method as the conventional pig study. Sera were collected on DPI 0, 3, 5, 7, 9, 13, 16, 20, 23, and 27 DPI. The same samples were collected at necropsy for both studies. However, the CDCD piglet study was longer in duration, and piglets were necropsied on DPI 2, 5, and 27 to better evaluate seroconversion.

Statistical analysis

All statistics were conducted with the statistics program R, supported by the R Foundation for Statistical Computing. A linear mixed model was used to obtain lsmeans and contrasts from the qRT-PCR data. The histology data were analyzed using the Mann-Whitney U test and a Bonferroni correction for multiple comparisons. A P-value ≤ 0.05 was considered significant.

Results

Clinical signs

No significant clinical signs of respiratory disease (sneezing, coughing, or dyspnea) were observed from 0-7 DPI in the conventional piglets. Mild coughing was observed sporadically after 9 DPI. However, no moderate or severe respiratory clinical signs (dyspnea or severe barking-type cough) were observed throughout the study. Weight did not vary significantly between groups except at 21 DPI. The control group was significantly heavier than the contact group ($P = 0.009$) and marginally heavier than the challenge group ($P = 0.0688$) at 21 DPI. The CDCD piglets demonstrated no clinical signs for the duration of the study, and no coughing was observed.

Viral shedding and distribution in conventional piglets

Nasal swabs, oral fluids, and necropsy samples were tested by PPIV-1 RT-qPCR and results are summarized in Figures 1-4. PPIV-1 was detected in nasal swabs from 29/30 pigs in the Ch group and 5/5 pigs in the Cont group one day post-exposure and post-contact, respectively. All pigs were infected and shedding two days post-exposure/contact. Viral shedding was consistently observed 1–9 DPI in the Ch group and began to decrease by 14 DPI in both nasal swabs and oral fluids. Maximum viral shedding detected in nasal swabs occurred 2-7 DPI, with average viral load ranging from $10^{6.4}$ to $10^{7.1}$ $\log_{10}$ genomic copies/mL (GC/mL). The Cont pigs showed a similar pattern of infection, but the initial magnitude of shedding was lower. Elevated Cts (greater than 30) were detected sporadically until day 21 in nasal swabs and necropsy samples.

A linear mixed model was used to compare RT-qPCR Ct values across the different sample types collected at necropsy from the Ch pigs. Samples included bronchoalveolar lavage fluid, turbinates, and tracheal swabs. Multiple pairwise comparisons of lsmeans blocked by necropsy day showed significantly higher viral loads at 2 DPI in the trachea compared to all other samples collected the same day (P-values ranging 0.001 to 0.01). Viral titers were narrowly distributed with only a 95 and 23 fold difference between the strongest and weakest titers at 2 and 5 DPI respectively. Virus was intermittently detected at low levels in nasal swabs, BALF, turbinates, and oral fluids at 21 DPI.

Viral shedding and distribution in CDCD piglets

Similar to the conventional pig study, nasal swabs, oral fluids, and necropsy samples were tested by PPIV-1 RT-qPCR. The nasal swab results are summarized in Figure 5. All challenged piglets (10/10) were shedding virus at 1 DPI. Genomic copies detected in the Ch group started to decrease at approximately 9 DPI and became undetectable after 15 DPI. Maximum viral shedding was similar to the conventional pig study, ranging from $10^{6.4}$ to $10^{7.7}$ $\log_{10}$ genomic copies per ml of nasal swab homogenate from 2-7 DPI. A statistical comparison of CDCD piglet lsmeans was conducted across necropsy samples similar to the conventional pig study. The sample size was much smaller, however similar

Figure 1: Post-challenge PPIV-1 RT-qPCR $\log_{10}$ genomic copies/mL of nasal swab homogenate and percent positive challenged pigs (Ch)

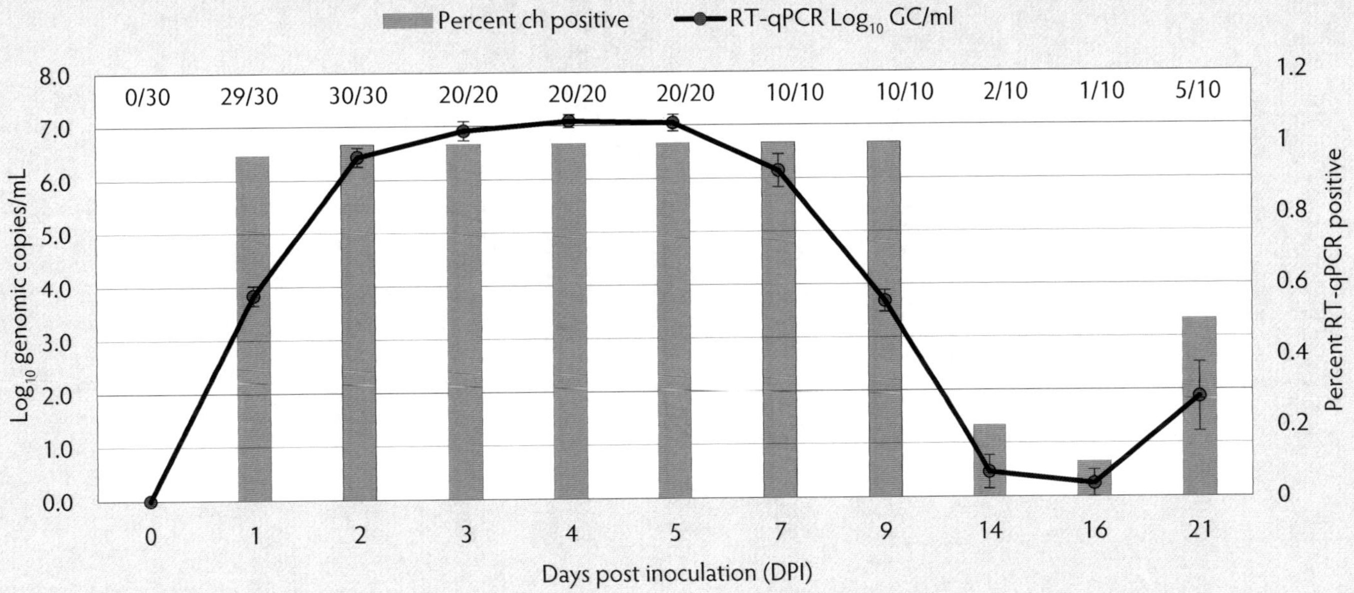

Figure 2: Post-contact PPIV-1 RT-qPCR log_{10} genomic copies/mL of nasal swab homogenate and percent positive contact pigs (Cont)

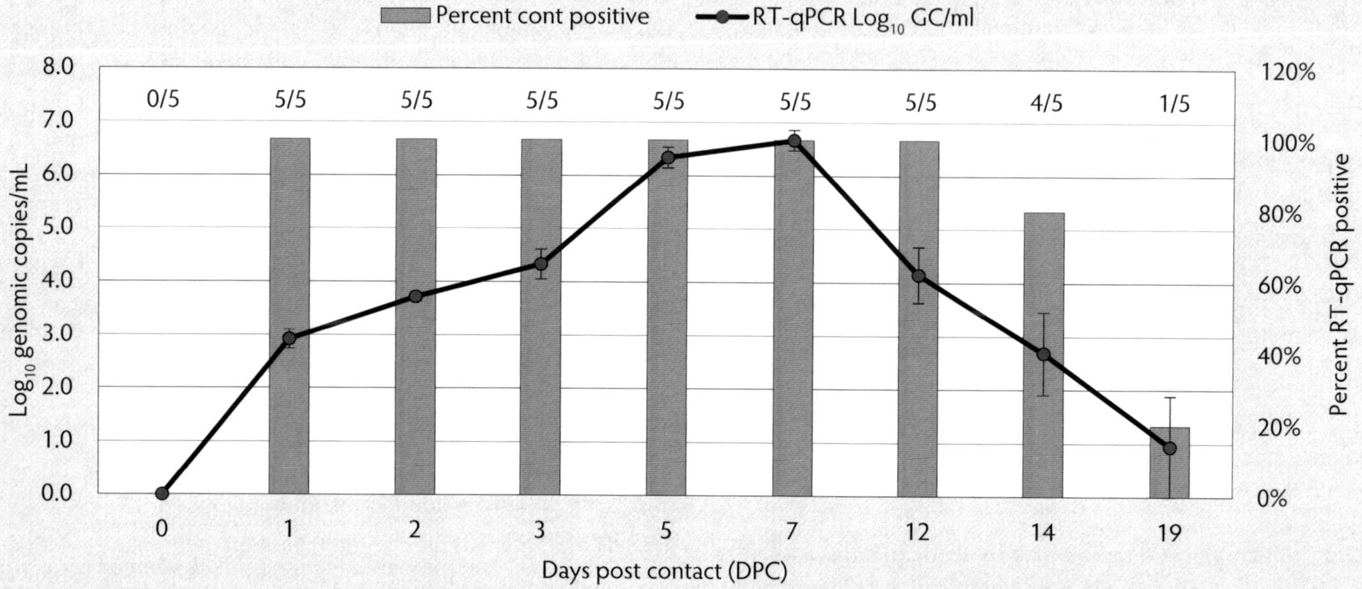

Figure 3: Post-challenge PPIV-1 RT-qPCR log_{10} genomic copies/mL of nasal swab, proximal trachea swab, distal trachea swab, bronchoalveolar lavage fluid or turbinate homogenate collected at 2, 5 and 21 days post inoculation from challenged pigs (Ch)

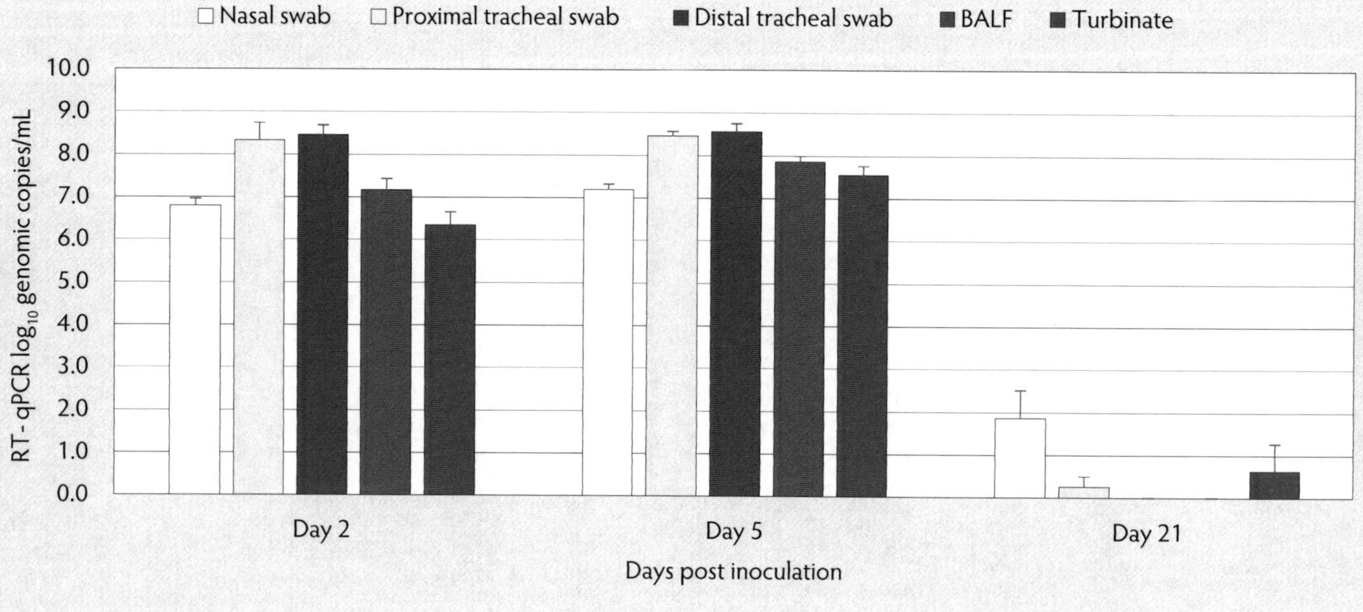

differences were found across sample types. Proximal and distal trachea demonstrated significantly higher viral loads compared to other samples at 2 and 5 DPI. Collectively, nasal swab and BALF samples demonstrated lower PPIV-1 viral loads compared to the other sample types.

Macroscopic and microscopic lesions

Macroscopic lung lesions were minimally demonstrated in the conventional pigs challenged with PPIV-1. Mean Ch and Neg macroscopic lung lesions were 1% at 2 DPI. At 5 DPI, mean percent consolidation was 4% for Ch pigs. No lung lesions were observed in the Neg group. Only the Ch pigs demonstrated lung lesions at 21 DPI that averaged 4% lung consolidation. Median histopathology lung lesion scores demonstrated in the Ch pigs were not significantly different compared to the Neg group measured by the Wilcoxon rank sum test.

A non-validated immunohistochemistry (IHC) procedure was conducted on formalin fixed Ch and Neg lung, trachea and nasal turbinate at 5 DPI. PPIV-1 signals were detected in abundance in the epithelium of pulmonary bronchi and bronchioles, tracheal epithelium and the turbinate epithelium. PPIV-1 antigen was not detected by IHC in the Neg group.

Conventional piglet serology results

All pigs were confirmed seronegative by ELISA prior to the start of the study. Conventional pig serum samples on 21 DPI were tested by PPIV-1 specific serum virus neutralization (SVN) test and indirect fluorescent antibody (IFA) assay. The SVN started with a tenfold dilution of serum with subsequent two-fold serial dilutions. Reciprocal titers above 20 were considered significant by SVN. Seroconversion occurred at 21 DPI, and SVN reciprocal titers in the Ch and Cont groups ranged from 20-80. All Neg pigs had serum titers at or below 20. Serum samples were tested for IFA antibody at dilutions of 1:100 and 1:300. All animals in the Ch and Cont groups were positive by IFA at 21 DPI, with titers ranging from 100 to ≥ 300. The Neg pigs tested by IFA all had Ab titers < 100. PPIV-1 antibody assay validation is in progress for both CDCD and conventional piglets using serum, oral fluids and nasal swabs.

Conclusions and discussion

Experimental PPIV-1 infection in both conventional and CDCD piglets demonstrated no mortality and minimal morbidity in spite of viral replication under the conditions of this study. All naïve animals became infected upon PPIV-1 challenge or exposure and shed virus in nasal swabs until 14 DPI and 16 DPI in the conventional and CDCD piglets, respectively. Clinical signs were non-existent to mild, and there were minimal lung lesions associated with PPIV-1 in the conventional pigs under the conditions of this study. Significant viral loads were found in trachea samples compared to the lung and nasal turbinate. However, viral RNA was detected at high levels throughout the upper and lower respiratory tracts suggesting viral replication occurs throughout the respiratory tracts, but may replicate to higher levels in the upper respiratory tract. Although a PPIV-1 cell culture isolate did not cause significant respiratory clinical signs or lung lesions, infection was established and detected by IHC. Experimental conditions may not replicate field infections that often include co-factors that may manifest in more severe clinical signs and disease. Therefore, additional studies are necessary to evaluate the pathogenesis of different strains of PPIV-1 and its potential role as a secondary pathogen in the porcine respiratory disease complex. Future work will investigate the pathogenesis of potential co-infections of PPIV-1 with other viruses such as influenza A virus and bacterial pathogens including *Haemophilus parasuis* or *Streptococcus suis.* Development and validation of additional diagnostic tests, including immunohistochemistry, serum virus neutralization, virus isolation, and immunofluorescent assays are still in progress.

References

1. Lau SK, Woo PC, Wu Y, Wong BH, Lau CC, Fan RY, Cai JP, Tsoi HW, Yuen KY. 2013. Identification and characterization of a novel paramyxovirus, porcine parainfluenza virus 1, from deceased pigs. *J Gen Virol* 94:2184–2190.

2. Palinski RM, Chen Z, Henningson JN, Lang Y, Rowland RR, Fang Y, Prickett J, Gauger PC, Hause BM. 2016. Widespread detection and characterization of porcine parainfluenza virus 1 in pigs in the USA. *J Gen Virol* 97:281–286.

3. Gauger PC, Lin S, Bade S, Park J, Harmon K, Zhang J, Piñeyro PE. 2016. Porcine Parainfluenza Virus Type 1 (PPIV-1) Disease, Diagnostics, and Clinical Significance, *ISU James D. McKean Swine Disease Conference Proceedings,* Ames, Iowa.

4. Reed LJ, Muench H. 1938. A simple method of estimating fifty percent endpoints. *Am J Infect Control* 1938:493–497.

5. Vincent AL, Ma W, Lager KM, Richt JA, Janke BH, Sandbulte MR, Gauger PC, Loving CL, Webby RJ, Garcia-Sastre A. 2012. Live attenuated influenza vaccine provides superior protection from heterologous infection in pigs with maternal antibodies without inducing vaccine-associated enhanced respiratory disease. *J Virol* 86:10597–10605.

6. Gauger PC, Vincent AL, Loving CL, Henningson JN, Lager KM, Janke BH, Kehrli ME Jr, Roth JA. 2012. Kinetics of lung lesion development and pro-inflammatory cytokine response in pigs with vaccine-associated enhanced respiratory disease induced by challenge with pandemic (2009) A/H1N1 influenza virus. *Vet Pathol* 49:900–912.

7. Halbur PG, Paul PS, Frey ML, et al. 1995. Comparison of the pathogenicity of two U.S. porcine reproductive and respiratory syndrome virus isolates with that of the Lelystad virus. *Vet Pathol* 32:648–660.

Figure 4: PPIV-1 RT-qPCR log_{10} genomic copies/mL of oral fluid collected from challenged and contact pigs post-challenge or post-contact

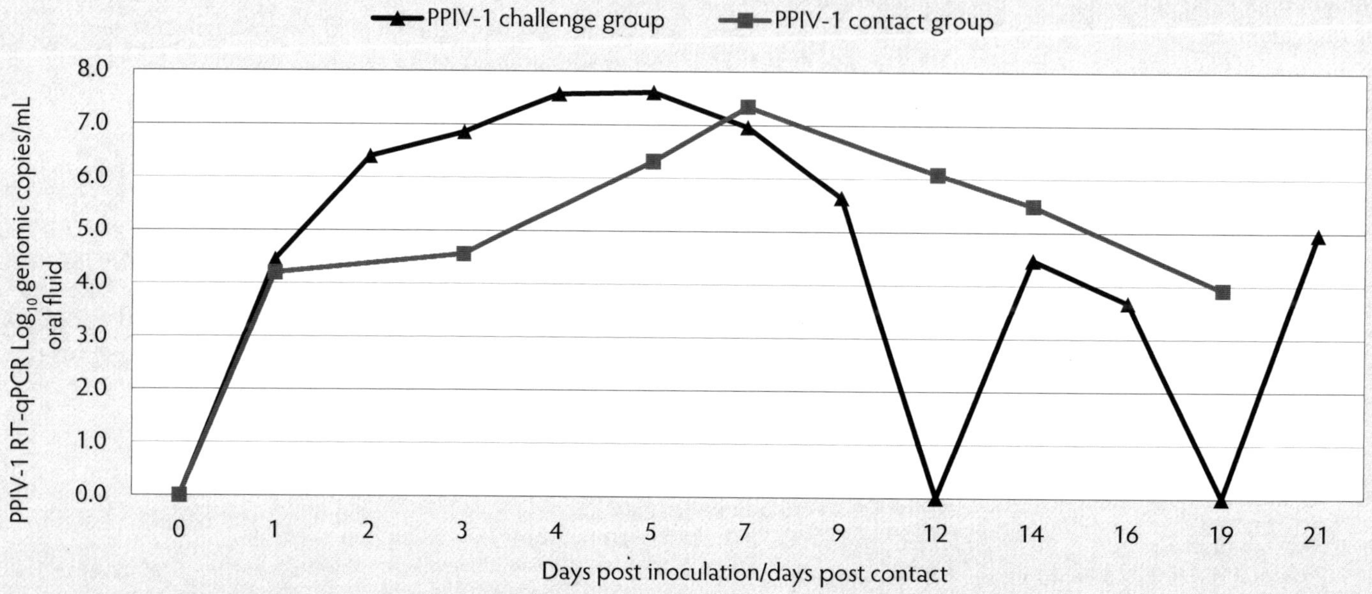

Figure 5: PPIV-1 RT-qPCR log_{10} genomic copies/mL of nasal swab homogenate collected from CDCD challenged pigs

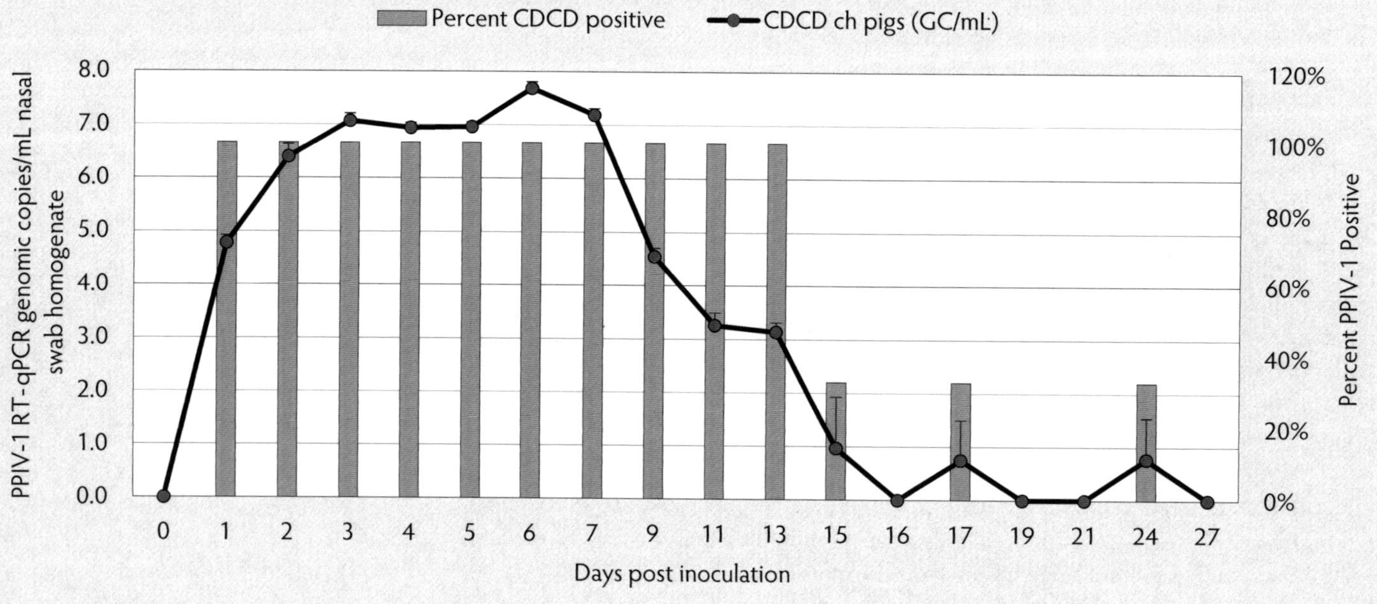

Breeding herd factors associated with influenza in piglets at weaning

F. Chamba[1]; A. Schelkopf[2]; M. Allerson[3]; B. Morrison[1]; M. Culhane[1]; M. Torremorell[1]
[1]Department of Veterinary Population Medicine, College of Veterinary Medicine, University of Minnesota, St. Paul, Minnesota; [2]Pipestone Veterinary Services, Pipestone, Minnesota; [3]Holden Farms Inc, Northfield, Minnesota

Introduction

Piglets in breeding herds can maintain, diversify and transmit influenza at weaning[1]. Because breeding herds play a central role in the dissemination of influenza, there is a need to further understand which herd factors are associated with influenza in piglets at weaning. In this study, we evaluated the association of breeding herd characteristics and management practices with influenza detection in piglets at weaning.

Materials and methods

A total of 83 sow farms located in the Midwestern US were voluntarily enrolled in the study and agreed to share influenza testing and breeding herd information. Farms were sampled monthly by collecting oral fluids, nasal or oropharyngeal swabs from piglets prior to weaning. Samples were tested for influenza A virus (IAV) matrix gene by RT-PCR. A sample was considered positive if the cycle threshold (ct) value was 38 or lower. A survey was administered to each farm and information on breeding herd characteristics (herd size, number of employees, female replacement rate, type of farrowing flow, use of air filtration, number of farms within a radius of 1 and 3 miles, and parity segregation) and breeding herd management practices (practice of early weaning, herd closure, influenza sow vaccination protocol and vaccine type, presence of personnel with influenza-like illness, use of personnel temperature screening methods at entry, and influenza vaccination policies for personnel) were obtained. Gilt and piglet management practices included gilt influenza status at entry, gilt influenza vaccination, gilt development unit (GDU) location and flow, gilt isolation before entry, cross-fostering of piglets older than 10 days of age, and holding-back piglets at weaning. Data was analyzed using mixed logistic regression models adjusted by farm, season and correlation between samplings.

Results

We obtained diagnostic and breeding herd data between July 2011 and March 2017. There were 12,814 samples from 2989 submissions of which, 72%, 20% and 8% were oral fluids, oropharyngeal and nasal swabs respectively. Overall, 23% of samples and 30% of submissions tested influenza positive. Sixty-seven of 83 (81%) breeding herds completed the survey. All vaccinated farms, 72% (58/81), reported the use of influenza inactivated commercial vaccines. Among all the herd factors that we evaluated (24), only influenza sow vaccination protocol and gilt influenza status at entry were associated with influenza positive piglets at weaning (P-values = 0.0140 and 0.0001 respectively). Both, prefarrow and whole herd vaccination protocols were statistically similar (P-values = 0.4303) and both were statistically better than no influenza sow vaccination (P-values = 0.0039).

Conclusions and discussion

Lack of influenza sow vaccination and influenza positive gilts at entry resulted in higher influenza detection in piglets at weaning. Non-significant effect of herd closure, air filtration and farm density suggests that IAV persistence within breeding herds is common, complex and needs to be further studied. Our results are relevant to guide influenza control strategies in breeding herds. Decreasing IAV circulation in pigs may reduce IAV genetic diversity, production loses and the emergence of new strains with pandemic potential.

Acknowledgments

This study was funded by the Minnesota Rapid Agricultural Response Fund (RARF) and producers participating in the influenza surveillance program.

References

1. Diaz, Andres, Douglas Marthaler, Marie Culhane, Srinand Sreevatsan, Moh Alkhamis, and Montserrat Torremorell. 2017. 'Complete Genome Sequencing of Influenza A Viruses within Swine Farrow-to-Wean Farms Reveals the Emergence, Persistence, and Subsidence of Diverse Viral Genotypes', J Virol: JVI. 00745-17.

Is influenza vaccination a key driver of influenza genetic diversity in piglets?

J. Nirmala[1]; A. Perez[1]; M. Culhane[1]; M. Allerson[1]; S. Sreevatsan[2]; M. Kinney[3]; A. Mohr[4]; M. Torremorell[1]
[1]College of Veterinary Medicine, University of Minnesota, St. Paul, Minnesota; [2]Department of Pathobiology and Diagnostic Investigation, Michigan State University, East Lansing, Michigan,[3]Department of Statistics, University of Minnesota, St. Paul; [4]Liberal Arts Technologies and Innovation Services; University of Minnesota, St. Paul, Minnesota

Introduction

Influenza A virus (IAV) is endemic in the US pig population. IAVs may persist and evolve within populations, especially in piglets, giving rise to new strains that hamper effective mitigation strategies. Piglets play a critical role in maintaining IAVs in breeding herds and disseminating IAVs between farms. Piglets can also sustain infections and promote IAV variability despite the presence of maternal immunity, thereby negatively impacting agricultural sustainability and overall competitiveness of swine production in the US.

Vaccination is one of the strategies most commonly used to control influenza in breeding herds. However, questions remain regarding the effect of vaccination on driving virus genetic change that could render vaccines ineffective. In this study, we evaluated the genetic diversity of influenza in vaccinated and non-vaccinated herds and assessed whether vaccines may drive genetic and antigenic change.

Materials and methods

Samples (n = 9,090) collected from vaccinated and non-vaccinated breeding herds (n = 52) over a period of 6 months across seasons were evaluated by qRT-PCR for the presence of IAV. Some (n = 391) nasal swab samples with cycle threshold (CT) values ranging from 16 to 35.7 were whole-genome sequenced by Hi Seq 2500 rapid mode 250 bp paired end reads. A combination of both de novo and reference genome based assembly/alignment was performed. Resulting sequences were annotated using FLAN annotation tool (Bao et al., 2007) and then subsequently classified into subtypes utilizing the tools in the Influenza Research Database (IRD) or by constructing phylogenetic trees using maximum-likelihood with RAXML v7.6.8. Complete and functional sequences were analyzed using an Analysis of Molecular Variance Analysis (AMOVA) implemented in the R (version 2.4.1) "POPPR" package (Kamvar, et al., 2014) to assess the null hypothesis that genetic variation was different in IAV-vaccinated and –unvaccinated farms.

Results and conclusion

Classification of the viral samples based on antigenic segments 4 (hemagglutinin) and 6 (neuraminidase) revealed the co-circulation in farms of different viral subtypes. Co-circulation (either continuous or sporadic) of multiple subtypes (H1N1, H1N2, H3N2 and H3N1) was detected in 96%percent of the farms (n = 22 of 23) exhibited, whereas only one farm exhibited circulation of a single subtype. Despite observing a significant and dynamic variability among herds and IAVs, AMOVA analysis of all the antigenic and non-antigenic segments at both the nucleotide and amino acid level revealed no differences in genetic and antigenic diversity between vaccinated and non-vaccinated herds. Thus, we failed to identify an evident association between vaccination and diversity and variability of IAV strains. We speculate that influenza diversity in piglets may be influenced by a number of epidemiological factors and cannot be explained by the use of vaccination alone.

References

1. Bao Y et al. FLAN: a web server for influenza virus genome annotation. Nucleic Acids Res. 2007; 35: W280-284.

2. Kamvar ZN et al. *Poppr*: an R package for genetic analysis of populations with clonal, partially clonal, and/or sexual reproduction. PeerJ. 2014; 2:e281 https://doi.org/10.7717/peerj.281.

Novel approaches for influenza surveillance in swine breeding herds

Jorge Garrido Mantilla; Julio Alvarez; Marie Culhane; Montserrat Torremorell
College of Veterinary Medicine, University of Minnesota, St. Paul, Minnesota

Introduction

Surveillance for influenza A virus (IAV) is central to the control of influenza in pigs and the prevention of zoonotic infections. However, routinely and readily detecting influenza infections can be challenging particularly in endemically infected herds where antibody levels are high, clinical signs are mild and virus levels fluctuate. Furthermore, control of IAV infections usually requires the successful recovery of virus isolates from collected samples with subsequent characterization, vaccine seedstock production, and diagnostic assay development from the isolates. In this study, we developed and compared novel sampling strategies to detect and isolate IAV in swine breeding herds.

Materials and methods

A longitudinal prospective study was designed to compare eight distinct sampling procedures to detect and isolate IAV in pigs: nasal swabs (NS), nasal wipes (NW), oropharyngeal swabs (OS), oral fluids (OF), surfaces wipes (SW), sow udder skin wipes (UW), environmental particle deposition (EPD) and air. Sampling was conducted in piglets or their environment prior to weaning in 6 breeding herds. Three piglets from 10 litters of approximately 18-21 days of age distributed in 3 farrowing rooms were selected in each farm. Each pig was sampled individually by first collecting a nasal swab, then a nasal wipe and lastly an oropharyngeal swab. Oral fluids and wipes of surfaces accessible to piglets from the litters where individual pigs had been sampled were also collected. Air samples and samples from surfaces positioned to capture the settlement of airborne particles were collected simultaneously from the same air space where piglets were sampled. Lastly, a wipe from the udder skin from lactating sows was also collected from the same litters where the pigs were sampled. All samples were tested (OS, NS, NW in pools of 3) by IAV matrix gene rRT-PCR (Spackman et al., 2002) and results considered positive if cycle threshold value was < 35. A subset of rRT-PCR positive samples were cultured for virus isolation using MDCK cells (Detmer et al, 2011). Agreement between the test results for each sample type was measured using the kappa statistic, and the optimum sample type for IAV detection was identified using McNemar test ($P < 0.05$).

Results

IAV was detected in 4 out of 6 breeding herds. Out of the 40 samples collected in IAV positive herds, 78% (31/40) of OS, 55% (22/40) of NS and 53% (21/40) of NW were rRT-PCR positive ($P = 0.012$). Seventy eight percent (31/40) of UW were positive compared with 60% (24/40) of SW and 86% (6/7) of OF. IAV was detected in 100% (40/40) of air samples compared with 88% (35/40) of EPD. Out of the RT-PCR positive samples IAV was isolated from 78.4% (40/51) of OS, 62.9% (34/54) of NS, 64.9% (37/57) of NW, 47.6% (10/21) of SW, 75.0% (18/24) of UW, 16.6% (1/6) of OF, 29.1% (7/24) of EPD and 34.6% (9/26) of air samples. Sample type agreement measured with the kappa coefficient ranged from moderate to substantial. Use of UW and OS and to a lesser extent OF and SW significantly increased the likelihood of detecting IAV by rRT-PCR compared to NS (baseline category) ($P < 0.01$).

Discussion

In this study, the oropharyngeal swab was the optimum sample type for both detecting and isolating IAV from pigs. Sampling the udder skin of lactating sows was also a sensitive method of identifying positive litters and yielded viable IAV. Collection of OF was limited given the reluctance of piglets to chew on the ropes and did not yield a viable isolate. Sampling the environment also appeared to be a good approach to detect IAV since IAV was readily detected from surfaces and air, although environmental samples yielded fewer isolates than oropharyngeal swabs and udder skin wipes.

This study provides new information on sampling approaches to detect influenza in breeding herds and highlights the role of the environment and the lactating sows as important sources of IAV in farms.

Acknowledgments

Pipestone Veterinary Services, PIC, SSUI, Holden farms, Schwartz farms. Swine Disease Eradication Center – University of Minnesota.

References

1. Edwards et al. Utility of snout wipe samples for influenza A virus surveillance in exhibition swine populations. Influenza and other respiratory viruses 2014; 574:579.

2. Neira et al. Characterization of Viral Load, Viability and Persistence of Influenza A Virus in Air and on Surfaces of Swine Production Facilities. *Plos one* 2016.

3. Romagosa et al. Sensitivity of oral fluids for detecting influenza A virus in populations of vaccinated and non-vaccinated pigs. *Influenza and other respiratory viruses* 2012; 110:118.

Case report: Identification of *Mycoplasma hyopneumoniae* infection in a breeding herd through tracheo-bronchial swab monitoring

F. Vangroenweghe[1]; E. Willems[2]; O. Thas[3]; D. Maes[4]
[1]Elanco Animal Health Benelux, Antwerp, Belgium; [2]TopigsNorsvin International, Vught, The Netherlands; [3]Department of Mathematical Modelling, Statistics and Bio-informatics, Faculty of Bio-engineering Sciences, Ghent University, Ghent, Belgium; [4]Department of Swine Herd Health and Reproduction, Faculty of Veterinary Medicine, Ghent University, Merelbeke, Belgium

Background and objectives

Mycoplasma hyopneumoniae (M hyo), the primary pathogen of enzootic pneumonia, occurs worldwide and causes major economic losses to the pig industry. The pathogen adheres to and damages the ciliated epithelium of the respiratory tract. Affected pigs show chronic coughing, are more susceptible to other respiratory infections and have a reduced performance.[1] Piglets become infected with M hyo during the suckling period and have been shown positive from weaning onwards.[2-7] Moreover, once infected with M hyo, animals can excrete the pathogen over a long period of time, with total clearance lasting less than 254 days post-infection).[8] This implies that infected gilts could carry M hyo across their first pregnancy well into their first lactation cycle, infecting their offspring with M hyo in early life.

Therefore, dedicated programs to monitor for freedom of M hyo has been developed within breeding companies that currently deliver high health breeding animals to their end customers, mainly multiplier herds producing pigs that enter the fattening unit. Serology is the preferential approach in order to screen for *Mycoplasma hyopneumoniae* (M hyo), using an ELISA test.[9-14] In case of positive serology, further decisions on farm health status and the related consequences should be based on additional confirmation tests. When a specific pathogen free (SPF) pig breeding herd becomes infected with M hyo, the interval to emergence of clinical symptoms is a matter of concern, because transfer of subclinically infected breeding animals to their customers constitutes a potential risk of spreading the infection.[15] Clinical diagnosis of enzootic pneumonia can be verified by serological analysis.[15] However, in SPF programmes, the herd prevalence of M hyo infections is often low and the positive herd predictive value of a serological result decrease progressively with the decreasing herd prevalence.[16] Moreover, recent research has shown that currently used ELISA test only start showing a seroconversion from 21 days post-infection onwards.[17] The percentage of animals seroconverting in the early stages of M hyo detection using one of the commercially available M hyo ELISAs remains relatively low (16-22% at 21 days and 35-45% at 28 days post-infection),[17] which requires an elevated number of samples collected within a monitoring program in order to reliably detect the presence or circulation of M hyo within the monitored herd.

In the Danish SPF program, the final verification of herd infection with M hyo is consequently performed by demonstration of the agent.[9] A recent comparative study on diagnostic sampling approach for M hyo*o* detection showed that laryngeal swabs were a reliable option to obtain an early detection of M hyo, followed by brocho-alveolar lavage fluids and nasal swabs.[17] Other innovative sampling techniques, such as tracheo-bronchial swab (TBS) sampling[6,7,18] have been introduced in combination with PCR detection of M hyo to reliably detect the pathogen from the lung of infected animals. The objective of the current case report is to show the added value of TBS sampling in the confirmation of a dubious M hyo-seropositive situation.

Materials and methods

Case description

A cluster of two high health breeding farms in Eastern-Europe (2000-sow herd, farm A; 3600-gilt rearing facility, farm B) had been negative for M hyo for 10 years using a standard serological monitoring schedule (2×/year; 25 samples per farm). Besides freedom for M hyo, the farms were also negative for *Pasteurella multocida* DNT+, *Sarcoptes scabiei* var. *suis*, *Hematopinus suis*, *Brachyspira hyodysenteriae*, PRRSV and *Actinobacillus pleuropneumoniae*. The cluster was positive for M hyo using the conventional ELISA serology (Idexx HerdChek Mhyo ELISA, indirect ELISA; Idexx Laboratories). Looking for confirmation with a second ELISA test (IDScreen *Mycoplasma hyopneumoniae*, indirect ELISA; IDVet or BioChek M hyo Ab kit; BioChek), however, samples showed initially more negative serology, but in a later phase confirmed the positive serology. Moreover, throughout the entire monitoring period, no coughing, necropsy lesions or lesions at slaughter could be detected in order to obtain a definite confirmation on the M hyo health status. The evolution of S/P ratios of the first screening ELISA test and the % of positive samples per sampling date and location are given in Figure 1A.

Diagnostic approach with TBS

Since health status of breeding farms should be reliable in order to guarantee a continuous outflow of high health animals, other techniques such as TBS were used to confirm the health status

Figure 1A: Evolution of the % of positive animals (based on S/P-ratio > 0.40) in regular monitoring ELISA test (IDEXX HerdChek Mhyo ELISA; IDEXX Laboratories) in both herd A (sow herd, dotted line, n = 30 per timepoint) and herd B (gilt rearing herd, straight line, n = 20 per timepoint).

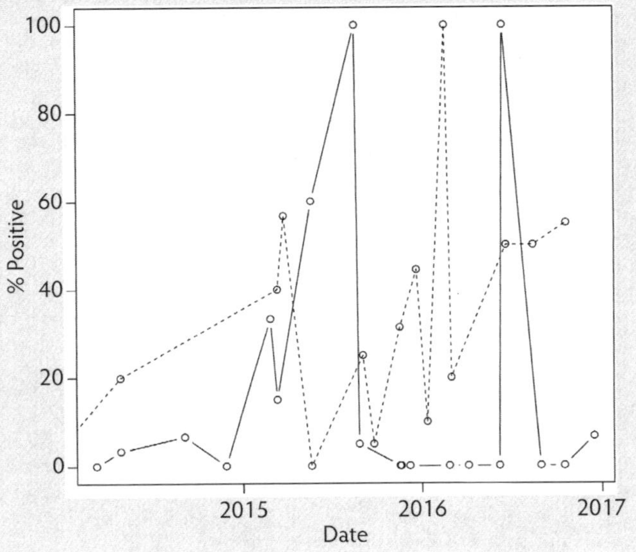

for M hyo.[6,7] Tracheo-bronchial sampling was performed as previously described.[6,7] Briefly, TBS samples were obtained following restraint of the piglets with a nose snare, and subsequent use of a mouth opener. The aspiration tube used (CH12 × 50 cm; Medinorm) was inserted through the mouth and glottis down to the trachea-bronchial bifurcation where mucus was collected through gentle swab movement. The tip of the swab was collected in a sterile 10 mL polystyrene tube (MLS), mixed with 1 mL sterile saline and kept at 3-5°C until analysis within 48 h of sampling.

Analysis of tracheo-bronchial swabs

The material collected by TBS was processed in an M hyo p183 real-time-PCR.[19] Briefly, nucleic acid was extracted from TBS using an RNA/DNA isolation kit (MagMAX Pathogen RNA/DNA Kit, Life Technologies) and an automated nucleic acid isolation processor (MagMAX Express 96 processor, Life Technologies) based on magnetic bead technology. One microlitre of TBS was centrifuged for 5 min at 16,000 *g*, the pellet suspended in 400 µL lysis buffer, and 400 µL of the suspension was used as the sample. If no pellet was observable, 300 µL of the TBS was used as the sample. Bead mix and lysis/binding solution were added and the mix transferred onto a 96-well plate in the processor. Nucleic acid isolation was performed according to the manufacturer's instructions. The PCR results were reported as negative or positive for the presence of M hyo.[19] The detection limit for M hyo was from 10 ng/µL to 2.5 fg/µL. The detection limit for the PCR was validated for TBS spiked with dilutions of M hyo strain J of at least 5 fg/µL.

Sampling for determination of freedom of disease

Therefore, a statistical calculation for the minimal number of samples to be taken to show 'freedom of disease' in both locations was performed (`http://epitools.ausvet.com.au`) with the assumption that a maximum of 2% of the animals were positive to M hyo. Based on this calculation, a total of 186 samples were collected at different ages: piglets at 4 and 8 weeks of age, gilts at 22 and 26 weeks of age in the gilt rearing location, and gilts and sows in the breeding location.

Statistical analysis

To determine potential association between animals positive for M hyo serology in farm A and B, logistic regression was used for a range of different lag times. The strongest association was selected by means of the Akaike's Information Criterion (AIC), and the resulting parameter estimates and P-values were reported. To account for model selection (selecting model with smallest AIC) a bootstrap procedure was implemented in which the outcome variable (positive serology in B) was resampled to mimic the null hypothesis. In each bootstrap run (1000 runs) the model selection procedure was applied and the parameter estimate of the optimal model was retained. The set of 1000 parameter estimates served as the null distribution.

Results

The results of the AIC demonstrated a significant positive association between serologically M hyo-positive animals (% positive) in farm A and farm B 30 days earlier (OR = 20.51, 95% CI [1.79;246.06]; $P = 0.0147$). Collected TBS samples were negative until 8 weeks of age, but older gilts (22 and 26 weeks of age) before and after introduction in the breeding herd were shown M hyo-positive (Figure 1B).

Discussion

The current case report demonstrated that diagnosis of an M hyo infection at low prevalence is difficult using the standard monitoring tools, such as clinical observation, serology[10] and slaughterhouse checks for typical lung lesions.[1] Therefore, collection of TBS samples to confirm the M hyo health status of a breeding herd has been shown to create additional information in order to take crucial decisions concerning health management within a breeding company. Analysis of the M hyo infection kinetics based on serology showed an initial seroconversion (2014) at the sow herd, followed at least 6 months later (2015) by an M hyo seroconversion in the gilt rearing facilities. TBS samples demonstrated no early infection at 4-8 weeks of age, however, older rearing gilts were found at least partially (10-25%) positive. This positive status was further detected in the gilts already present in the sow herd itself with 100% positivity from 28 weeks onwards. The confirmation of the positive M hyo status resulted in a subsequent discussion on M hyo elimination plans and schedules which were feasible and suitable for the specific farm. Several approaches, with or without depopulation, have been used with

Figure 1B: Tracheo-bronchial swab (TBS) sampling results of different age categories at the gilt rearing herd (4-8-22-26 weeks of age; straight line) and older gilts (27-31-35 weeks of age; dotted line) after introduction into the sow herd. The figure shows the percentage of M hyo-positive TBS samples using a nPCR test for M hyo.

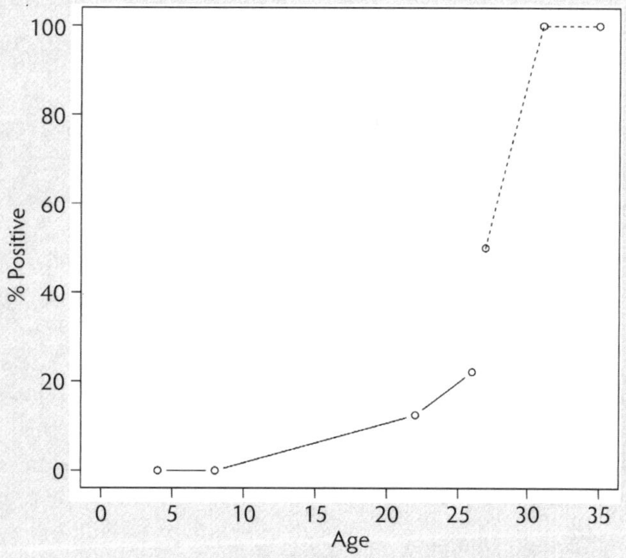

success in the recent past.[20,21] Eradication of M hyo from a herd includes several interventions such as antimicrobial treatment, vaccination, all-in all-out per compartment and overall internal biosecurity measures to reduce infection pressure and finally omit the pathogen from the herd.

Conclusions

In conclusion, PCR testing of TBS samples confirmed the presence of M hyo in a herd that was serologically negative or doubtful for M hyo and without typical lesions and clinical signs of M hyo infection, and allowed the farmer to implement proper health management practices to control or eliminate M hyo from the breeding herd.

References

1. Maes D., Segales J., Meyns T., Sibila M., Pieters M., Haesebrouck F. (2008) Review: Control of *Mycoplasma hyopneumoniae* infections in pigs. Veterinary Microbiology 126, 297–309

2. Fano E., Pijoan C., Dee S., Deen J. (2007) Effect of *Mycoplasma hyopneumoniae* colonization at weaning on disease severity in growing pigs. Canadian Journal of Veterinary Research 71, 195–200

3. Sibila M., Nofrarias M., Lopez-Soria S., Segales J., Riera P., Liopart D., Calsamiglia M. (2007) Exploratory field study on *Mycoplasma hyopneumoniae* infection in suckling pigs. Veterinary Microbiology 121, 352–356

4. Villarreal I., Vranckx K., Duchateau L., Pasmans F., Haesebrouck F., Jensen J. C., Nanjiani I. A., Maes D. (2010) Early *Mycoplasma hyopneumoniae* infections in European suckling pigs in herds with respiratory problems: detection rate and risk factors. Veterinarni Medicina 55, 318–324

5. Fablet C., Marois C., Dorenlor V., Eono F., Eveno E., Jolly J.P., Le Devendec L., Kobisch M., Madec F., Rose N. (2012) Bacterial pathogens associated with lung lesions in slaughter pigs from 125 herds. Research in Veterinary Science 93, 627-630.

6. Vangroenweghe F.A.C.J., Labarque G.L., Piepers S., Strutzberg-Minder K., Maes D. (2015a) *Mycoplasma hyopneumoniae* infections in peri-weaned and post-weaned pigs in Belgium and The Netherlands: prevalence and associations with climatic conditions. The Veterinary Journal 205, 93-97

7. Vangroenweghe F., Karriker L., Main R., Christianson E., Marsteller T., Hammen K., Bates J., Thomas P., Ellingson J., Harmon K., Abate S., Crawford K. (2015b) Assesment of litter prevalence of *Mycoplasma hyopneumoniae* in preweaned piglets utilizing an antemorten tracheobronchial mucus collection technique and real-time polymerase chain reaction assay. Journal of Veterinary Diagnostic Investigations 27, 606-610

8. Pieters M., Pijoan C., Fano E., Dee S. (2009) An assessment of the duration of *Mycoplasma hyopneumoniae* infection in an experimentally infected population of pigs. Veterinary Microbiology 134, 261–266

9. Morris C.R., Gardner I.A., Hietala S.K., Carpenter T.E., Anderson R.J., Parker K.M. (1995) Seroepidemiologic study of natural transmission of *Mycoplasma hyopneumoniae* in a swine herd. Preventive Veterinary Medicine 21, 323–337

10. Sørensen V., Ahrens P., Barfod K., Feenstra A.A., Feld N.C., Friis N.F., Bille-Hansen V., Jensen N.E., Pedersen M.W. (1997) *Mycoplasma hyopneumoniae* infection in pigs: duration of the disease and evaluation of four diagnostic assays. Veterinary Microbiology 54, 23–34

11. Andreasen M., Nielsen J. P., Bækbo P., Willeberg P., Botner A. (2000) A longitudinal study of serological patterns of respiratory infections in nine infected Danish swine herds. Preventive Veterinary Medicine 45, 221–235

12. Martelli P., Terreni M., Guazetti S., Cavirani S. (2006) Antibody response to *Mycoplasma hyopneumoniae* infection in vaccinated pigs with or without maternal antibodies induced by sow vaccination. Journal of Veterinary Medicine B 53, 229-233

13. Meyns T., Van Steelant J., Rolly E., Dewulf J., Haesebrouck F., Maes D. (2011) A cross-sectional study of risk factors associated with pulmonary lesions in pigs at slaughter. Veterinary Journal 187, 388–392

14. Reynolds S.C., St Aubin L.B., Sabbadini L.G., Kula J., Vogelaar J., Runnels P., Peters A.R. (2009) Reduced lung lesions in pigs challenged 25 weeks after the administration of a single dose of *Mycoplasma hyopneumoniae* vaccine at approximately 1 week of age. The Veterinary Journal 181, 312-320

15. Thomsen B.L., Jorsal S.E., Andersen S., Willeberg P. (1992) The Cox regression model applied to risk factor analysis of infections in the breeding and multiplying herds in the Danish SPF system. Preventive Veterinary Medicine 12, 287-297

16. Sørensen V., Barfod K., Feld N.C. (1992) Evaluation of a monoclonal blocking ELISA and IHA for antibodies to *Mycoplasma hyopneumoniae* in SPF-herds. Veterinary Record 130, 488-490

17. Pieters M., Daniels J., Rovira A. (2017) Comparison of sample types and diagnostic methods for in vivo detection of *Mycoplasma hyopneumoniae* during early stages of infection. Veterinary Microbiology 203, 103–109

18. Fablet C., Marois C., Kobish M., Madec F., Rose N. (2010) Estimation of the sensitivity of four sampling methods for *Mycoplasma hyopneumoniae* detection in live pigs using a Bayesian approach. Veterinary Microbiology 143, 238-245

19. Strait E.L., Madsen M.L., Minion F.C., Christopher-Hennings J., Dammen M., Jones K.R., Thacker E.L. (2008) Real-time PCR assays to address genetic diversity among strains of *Mycoplasma hyopneumoniae*. Journal of Clinical Microbiology 46, 2491–2498

20. Rautiainen E., Oravainen J., Virolainen J.V., Tuovinen V. (2001) Regional eradication of *Mycoplasma hyopneumoniae* from pig herds and documentation of freedom of the disease. Acta Veterinaria Scandinavia 42, 355-364

21. Heinonen M., Laurila T., Vidgren G., Levonen K. (2011) Eradication of *Mycoplasma hyopneumoniae* from a swine finishing herd without total depopulation. The Veterinary Journal 188, 110-114

Genotypic and phenotypic differences between LA-MRSA ST5 and clinical MRSA ST5 from humans

Samantha J. Hau[1]; Anna Allué-Guardia[2,3]; Brigida Rusconi[2,3]; Jisun Sun[4]; Peter R. Davies[4]; Timothy S. Frana[1]; Mark Eppinger[2,3]; Tracy L. Nicholson[5]

[1]Department of Veterinary Diagnostic and Production Animal Medicine, College of Veterinary Medicine, Iowa State University, Ames, Iowa; [2]South Texas Center for Emerging Infectious Diseases, University of Texas at San Antonio, San Antonio, Texas; [3]Department of Biology, University of Texas at San Antonio, San Antonio, Texas; [4]Department of Veterinary Population Medicine, College of Veterinary Medicine, University of Minnesota, St Paul, Minnesota; [5]National Animal Disease Center, Agricultural Research Service, United States Department of Agriculture, Ames, Iowa

Introduction

Staphylococcus aureus is a commensal bacterium that is found as a component of the microbiota in 25-30% of healthy humans in developed countries.[1] Additionally, it can cause a range of disease in humans and other species, including skin and soft tissue infections, invasive disease, and toxin mediated disease. Control of *S. aureus* is complicated by the development of antimicrobial resistance, especially methicillin resistance. Methicillin resistant *S. aureus* (MRSA) isolates have become a significant burden in both the hospital (HA-MRSA) and community settings (CA-MRSA), and are recognized by both the World Health Organization and Centers for Disease Control as a significant drug-resistant threat.

More recently, the association of MRSA with swine and other livestock species (LA-MRSA) has raised public health concerns due to the potential that LA-MRSA is contributing to the CA-MRSA burden and livestock may represent the largest source of MRSA outside of hospital settings.[2,3] Studies investigating LA-MRSA indicate that many sequence types (STs) are animal adapted, including ST398 and ST9, the most prevalent lineages of LA-MRSA found in swine from Europe and Asia, respectively.[4,5] However, this is not the case for MRSA ST5 isolates, which have also been reported in swine.[6,7] The MRSA ST5 lineage is a globally disseminated cause of both community and hospital acquired infections, leading to additional public health concerns surrounding swine associated LA-MRSA.[8]

The public health concerns associated with LA-MRSA are two-fold: direct and indirect risks. Direct risks encompass the potential of LA-MRSA to act as a source of infection in humans. Indirect risks of LA-MRSA involve the potential for LA-MRSA isolates to act as a source of mobile genetic elements (MGEs) harboring virulence factors or antimicrobial resistance genes that can be transferred to other bacteria. To investigate the potential direct and indirect risks associated with LA-MRSA ST5, we compared a collection of human clinical MRSA ST5 and swine associated LA-MRSA ST5 isolates using genetic analyses and phenotypic screening for antimicrobial resistance and adherence to human keratinocytes and swine skin.

Materials and methods

Draft genome sequences were generated for 82 LA-MRSA ST5 isolates and 71 clinical MRSA ST5 isolates from humans with no known swine contact using methods previously described.[9] Single nucleotide polymorphisms (SNPs) were identified by comparison to the MRSA ST5 reference strain Mu50, as previously described.[10] Identified SNPs were utilized in a core genome phylogenetic analysis to detect genome relatedness and included three MRSA ST5 reference strains, two HA-MRSA ST5 isolates (Mu3 and Mu50) and one poultry MRSA ST5 isolate (ED98). Additionally, MGEs encoding antimicrobial resistance genes and virulence factors were detected using online databases including ResFinder and VirulenceFinder and through the use of PCR and BLAST searches of the draft genome sequences.[11,12] The information was illustrated using the online tree building tool Evolview.[13]

Genetic data was paired with phenotypic screening to determine the impact of identified genes. Antimicrobial resistance was evaluated using the microbroth dilution method for antibiotics deemed relevant in the clinical and agricultural setting. Additionally, adherence capacity was evaluated *in vitro* for both swine skin and human epidermal keratinocytes.

Results

SNPiIdentification and phylogenetic analysis

SNP identification resulted in 759 total SNPs. These SNPs were utilized in a phylogenetic analysis that revealed the majority of human clinical MRSA ST5 isolates and swine associated LA-MRSA ST5 isolates clustered into two distinct clades (Figure 1). All but three clinical MRSA ST5 isolates clustered within the clinical MRSA clade, with only one clinical MRSA ST5 isolate identified within the swine associated clade (UCSF14436). The additional two clinical MRSA ST5 isolates formed a separate clade (UCSF14054 and UCI06). The isolates within the swine associated clade were divided into four subclades that contained clonal isolates. Each subclade represented a group of isolates from an individual farm or production system. These isolates possessed a unique fingerprint enabling trace back of isolates to the farm from which they were obtained.

Figure 1: The phylogenetic comparison of LA-MRSA ST5 and clinical MRSA ST5 isolates is depicted as a circular phylogenetic tree and annotated with details from the mobile genetic element analysis. The SCC*mec* type has been indicated in the inner-most circle. Virulence factor analysis includes the β-hemolysin converting bacteriophage and is represented by the black star in the second internal ring. Antimicrobial resistance elements are represented by the two outermost rings, with tetracycline resistance genes and macrolide resistance genes being indicated by the circles.

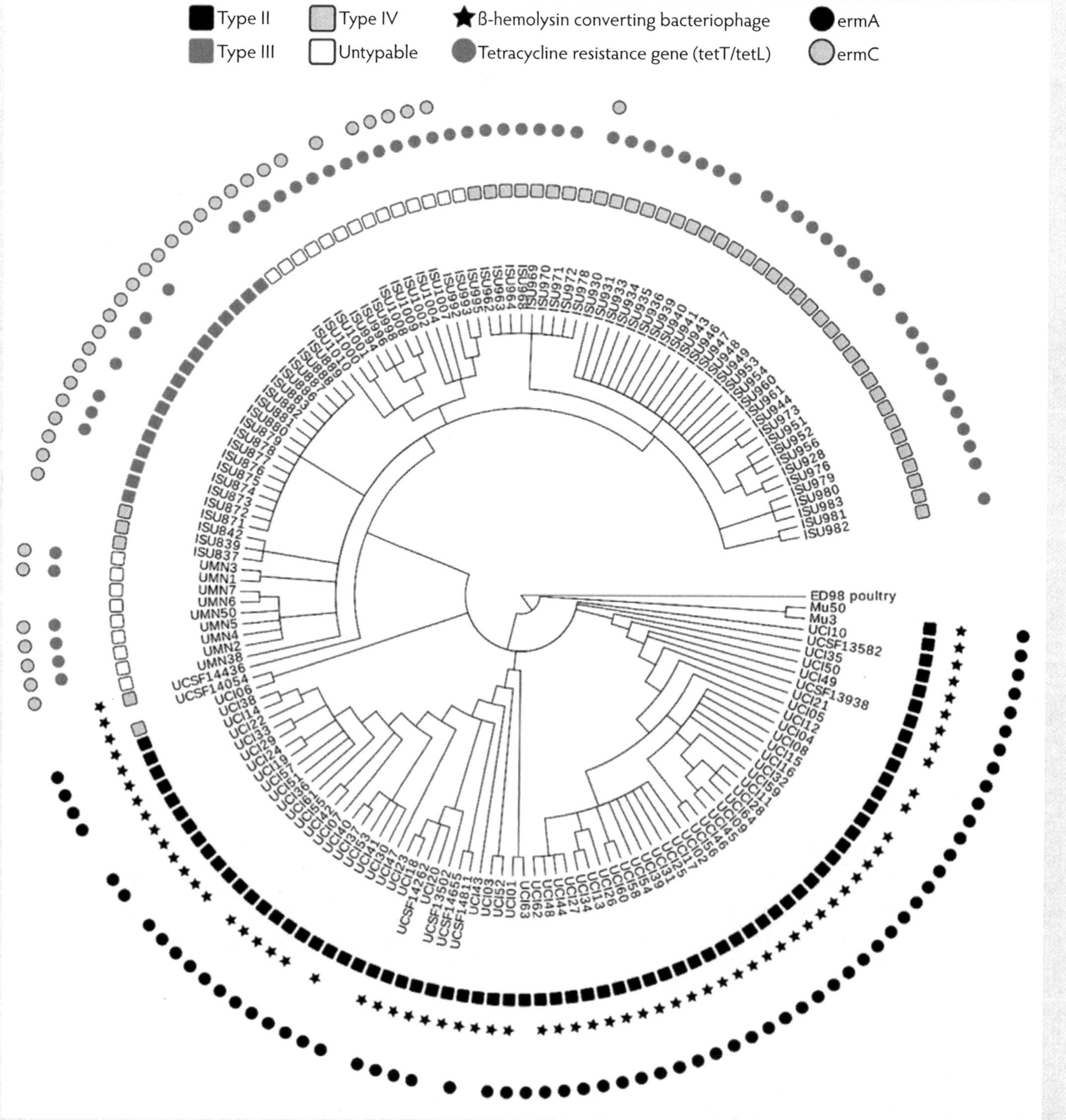

Mobile genetic element analysis

Mobile genetic element analysis of clinical and swine associated MRSA isolates indicated the MGE complement of these isolates differed between clinical and swine associated isolates (Figure 1). The antimicrobial resistance genes identified were consistent with selection pressures found in the environment the isolates were obtained from. For example, LA-MRSA ST5 isolates were more likely to harbor tetracycline resistance genes and clinical MRSA ST5 isolates were more likely to harbor genes conferring resistance to fluoroquinolones. The isolates also harbored different genes conferring resistance to macrolide and licosamide antibiotics and the SCC*mec* elements harbored by each subset were different (Figure 1). Additionally, virulence factors identified in the clinical MRSA ST5 subset were absent from the swine associated LA-MRSA ST5 isolates. These included the immune evasion genes harbored within the β-hemolysin converting bacteriophage (Figure 1) and the genes involved in survival on the skin harbored within the arginine catabolic mobile element.

Phenotypic analysis

All isolates were screened for phenotypic antimicrobial resistance. No phenotypic resistance to vancomycin and linezolid, primary drugs used for MRSA treatment, was noted in either subset of isolates. The phenotypic resistance of the isolate subsets was consistent with the antimicrobial resistance genes identified and with antimicrobial use in their respective sectors. A subset of isolates was also screened for *in vitro* adherence capacity. There was no difference found in the ability for clinical MRSA ST5 and LA-MRSA ST5 isolates to adhere to human keratinocytes; however, LA-MRSA ST5 isolates adhered better to swine epithelial tissues (Figure 2).

Discussion

In this study, we attempted to address the potential direct and indirect risks associated with LA-MRSA ST5 isolates through phylogenetic analysis, detection of antimicrobial resistance and virulence genes found on MGEs, and comparison of genetic findings with phenotypic analysis of antimicrobial resistance and *in vitro* adherence capacity. The distinct clades identified during phylogenetic analysis indicated minimal epidemiological association between the two subsets of isolates. We infer the direct risks of clinical infections due to swine associated LA-MRSA ST5 isolates are minimal in regions lacking high density swine production. Additionally, although LA-MRSA ST5 isolates adhered equivalently to human keratinocytes *in vitro*, virulence factors that contribute to survival on skin and evasion of the host immune response were absent in LA-MRSA ST5 isolates. The absence of these virulence genes is expected to result in reduced capacity of swine associated LA-MRSA ST5 isolates to cause human infections as compared to clinical MRSA ST5 isolates.

The indirect risks of genetic exchange between swine associated LA-MRSA ST5 isolates and other bacteria was assessed by comparing the complement of MGEs in MRSA ST5 isolates from

Figure 2: Adherence to human epidermal keratinocyte (HEK) is represented as colony forming unit (CFU) per HEK for swine associated LA-MRSA ST5 isolates and clinical MRSA ST5 isolates from humans with no swine contact (left). Adherence to swine skin biopsies is represented as CFU per mm² of swine skin. (*) denotes statistical significance of $P < 0.05$

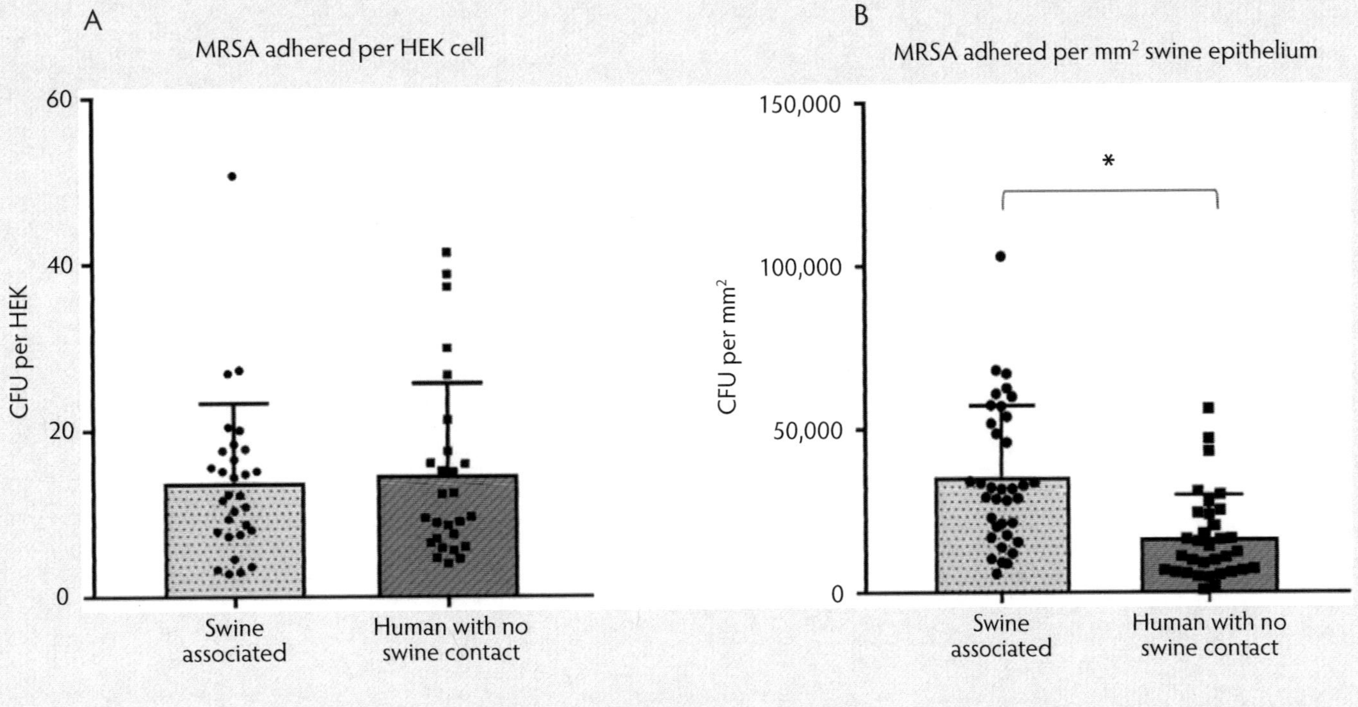

different sources. The lack of shared MGEs indicates genetic exchange between the studied isolate groups is unlikely; however, because antimicrobial resistance genes identified in LA-MRSA ST5 isolates were found on MGEs, there is a potential for transfer to other bacteria. Specific selection pressures were visible for phenotypic and genotypic antimicrobial resistance in both clinical and swine associated MRSA ST5 isolates, which may limit the transfer and dissemination of some MGEs harboring antimicrobial resistance genes from the agricultural setting. Overall, our data indicate genetic exchange between swine associated and clinical MRSA ST5 isolates is unlikely and LA-MRSA ST5 isolates are not contributing to the burden of clinical MRSA infections in regions without high density swine production.

References

1. Gorwitz, R. J. et al. Changes in the prevalence of nasal colonization with Staphylococcus aureus in the United States, 2001-2004. *The Journal of infectious diseases* 197, 1226-1234, doi:10.1086/533494 (2008).

2. Voss, A., Loeffen, F., Bakker, J., Klaassen, C. & Wulf, M. Methicillin-resistant *Staphylococcus aureus* in pig farming. *Emerging infectious diseases* 11, 1965-1966, doi:10.3201/eid1112.050428 (2005).

3. Schijffelen, M. J., Boel, C. H., van Strijp, J. A. & Fluit, A. C. Whole genome analysis of a livestock-associated methicillin-resistant *Staphylococcus aureus* ST398 isolate from a case of human endocarditis. *BMC genomics* 11, 376, doi:10.1186/1471-2164-11-376 (2010).

4. Andreoletti, O. et al. Scientific Opinion of the Panel on Biological Hazards on a request from the European Commission on Assessment of the Public Health significance of meticillin resistant *Staphylococcus aureus* (MRSA) in animals and foods. *The EFSA Journal* 993, 1-73 (2009).

5. Wagenaar, J. A. et al. Unexpected sequence types in livestock associated methicillin-resistant *Staphylococcus aureus* (MRSA): MRSA ST9 and a single locus variant of ST9 in pig farming in China. *Veterinary microbiology* 139, 405-409, doi:10.1016/j.vetmic.2009.06.014 (2009).

6. Frana, T. S. et al. Isolation and characterization of methicillin-resistant *Staphylococcus aureus* from pork farms and visiting veterinary students. *PloS one* 8, e53738, doi:10.1371/journal.pone.0053738 (2013).

7. Molla, B. et al. Epidemiology and genotypic characteristics of methicillin-resistant *Staphylococcus aureus* strains of porcine origin. *Journal of clinical microbiology* 50, 3687-3693, doi:10.1128/JCM.01971-12 (2012).

8. Monecke, S. et al. A field guide to pandemic, epidemic and sporadic clones of methicillin-resistant *Staphylococcus aureus*. *PloS one* 6, e17936, doi:10.1371/journal.pone.0017936 (2011).

9. Hau, S. J., Bayles, D. O., Alt, D. P. & Nicholson, T. L. Draft Genome Sequences of 14 *Staphylococcus aureus* Sequence Type 5 Isolates from California, USA. *Genome Announc.* 5, e00098-00017, doi:10.1128/genomeA.00098-17 (2017).

10. Eppinger, M. et al. Genomic epidemiology of the Haitian cholera outbreak: a single introduction followed by rapid, extensive, and continued spread characterized the onset of the epidemic. *mBio* 5, e01721, doi:10.1128/mBio.01721-14 (2014).

11. Zankari, E. *et al.* Identification of acquired antimicrobial resistance genes. *The Journal of antimicrobial chemotherapy* 67, 2640-2644, doi:10.1093/jac/dks261 (2012).

12. Joensen, K. G. *et al.* Real-time whole-genome sequencing for routine typing, surveillance, and outbreak detection of verotoxigenic *Escherichia coli*. *Journal of clinical microbiology* 52, 1501-1510, doi:10.1128/JCM.03617-13 (2014).

13. He, Z. *et al.* Evolview v2: an online visualization and management tool for customized and annotated phylogenetic trees. *Nucleic Acids Res* 44, W236-241, doi:10.1093/nar/gkw370 (2016).

Processing fluids for PRRSV monitoring and surveillance systems

Will A. López[1], DVM; Jose Angulo[3], DVM; Clayton Johnson[2], DVM; Jeffrey J. Zimmerman[1], DVM, PhD, ACVPM; Daniel C. L. Linhares[1], DVM, MBA, PhD

[1]Veterinary Diagnostic and Production Animal Medicine, Iowa State University, Ames, Iowa; [2]Carthage Veterinary Service, Ltd.; [3]Zoetis, Parsippany, New Jersey

Introduction

Ongoing challenges to detect and manage disease outbreaks in the US threaten the economical sustainability of the swine industry in this country. Specifically, porcine reproductive and respiratory syndrome virus (PRRSV) is a serious and expensive problem, costing to the USA swine industry more than $ USD 1 billion per year.[1] A critical step to control and eliminate porcine reproductive and respiratory syndrome virus (PRRSV) in production systems is to interrupt the transmission cycle in breeding herds, with the goal of producing PRRSV-free piglets at weaning.[2] Thus, effective PRRSV monitoring is essential to measure progress towards PRRS control and elimination efforts. The objective of this study is to describe the use of "processing fluids" to strengthen monitoring and surveillance systems in breeding herds. PRRSV replication in testicular epithelial cells and macrophages has been reported in the past,[3] which supports the theory that 'processing fluids' may be a suitable sample for PRRSV monitoring.

Materials and methods

Study design

"Processing fluids" are defined as an aggregate (population) sample of serosanguinous fluids recovered at the time of piglet tail docking and castration.[4] One set of 12 processing fluids samples were collected from 4 commercial breed-to-wean herds, and 30 matching individual piglet blood serum samples were collected on the same day from the same piglet population for detection of PRRSV RNA by rRT-PCR (Group 1). Additionally, a second set of processing fluids samples (n = 20) were collected from 5 breed-to-wean herds, and 30 matching tail blood swabs were collected on the same day from the same piglet population for detection of PRRSV RNA by rRT-PCR and porcine circovirus type 2 (PCV2) DNA by PCR (Group 2). Five processing fluids samples from Set 1 were tested for anti-PRRSV antibodies by IDEXX ELISA and submitted for PRRSV open reading frame (ORF)-5 sequencing.

Diagnostic testing

All the samples were tested at the Iowa State University Veterinary Diagnostic Laboratory (ISU-VDL) using assays routinely used for swine serum samples. Sera and blood swabs were tested in pools of 1:5, and processing fluids were tested individually. All processing fluids (n = 32) pooled serum samples (n = 72) and pooled tail blood swabs (n = 120) from Sets 1 and 2, were tested for the presence of PRRSV RNA using the Applied Biosystems TaqMan® kit for North American and European PRRS virus RNA detection. Processing fluids from Set 2 (n = 20) and its correspondent matching tail blood swabs (n = 120), were tested for the presence of PCV2 DNA using the Applied Biosystems VetMAX™ Porcine PCV2 Quant Kit. Additionally, 5 matching sets of samples from Set 1 (sets 5, 7, 8, 9, 10) were selected to be tested for PRRSV antibody detection using the IDEXX PRRS X3 Ab ELISA test and 5 matching sets of samples (2, 3, 4, 5, 7) were submitted for PRRS ORF-5 sequencing (Sanger method).

Results

The frequency of PRRSV RNA detection by rRT-PCR in processing fluids surpassed that of the matching 30 individual blood samples (83.3% vs. 66.6%). Similarly, higher frequency of PRRSV RNA detection was observed in processing fluids when compared with the matching 30 individual tail blood swabs (55% vs. 10%). Additionally, PCV2 DNA was detected at a greater frequency in processing fluids than in the matching 30 individual tail blood swabs (60% vs. 15%).

The overall PRRSV RNA frequency of detection by rRT-PCR in processing fluids samples from both sample sets together was of 66% (21 positives out of 32 samples), while the frequency of detection of the corresponding individual samples (blood sera & tail blood swabs) pooled in groups of five (reference method) had a lower PRRSV RNA frequency of detection of 31% (10 positives out of 32 sets of samples) using the same PCR assays. The frequency of detection of PRRSV RNA by rRT-PCR in the blood sera and tail blood swabs in terms of total pools tested was of 7% (13 positive pools out of 192 total pools tested).

Processing fluids samples submitted for serology tested positive for anti-PRRSV antibodies in 100% of the cases (n = 5), yielding S/P ratios ranging from of 0.42 up to 2.50. Regarding the attempts to obtain PRRS ORF-5 sequences (n = 5), it was possible to achieve it in all the samples submitted for that purpose. All 5 cases had a nucleotide sequence similarity greater than 99% when comparing processing fluids and blood samples taken from the same population. Wild type PRRSV was identified in all the cases for ORF-5 sequencing procedure.

Discussion

The volume retrieved per processing fluid sample was enough to allow for multiple diagnostic tests.[4] The tests performed in processing fluids demonstrated that this novel sampling method possess a higher sensitivity for the detection PRRSV RNA and PCV2 DNA when compared to the individual piglet sampling protocol used for monitoring breeding herds (PRRS detected in 21 of 30 samples and PCV2 detected in 12 of 20 samples). The detection of -PRRSV antibodies (detected in 5 of 5 samples) and the ability to obtain PRRSV ORF-5 sequences (5 of 5 samples) was also demonstrated. This supports the hypothesis that antibodies and PRRSV RNA are relatively stable and in high quantities in processing fluids collected under field conditions. Results of this study showed the ability to detect PRRSV RNA by rRT-PCR, and PCV2 DNA by qPCR in processing fluids with a higher herd sensitivity (greater probability to detect PRRSV-positive pigs) than matching 30 individual blood sera, or 30 tail blood swabs when these sample types were pooled in groups of five. Therefore, these findings indicated that processing fluids were more effective for monitoring PRRSV and PCV2 in suckling pigs that current methods. Thus, this sampling methodology represents a large improvement for screening infectious diseases in breeding herds and it may be a powerful tool for strengthening monitoring and surveillance systems in swine operations. Further studies are needed to better understand the impact of processing fluids sample size (number of litters per processing fluid sample) on the probability of detecting antigens and/or antibodies under low prevalence scenarios.

References

1. Holtkamp DJ, Kliebenstein JB, Neumann EJ, et al. Assessment of the economic impact of porcine reproductive and respiratory syndrome virus on United States pork producers. *J Swine Heal Prod*. 2013;21(2):72-84. %3CGo.

2. Corzo CA, Mondaca E, Wayne S, et al. Control and elimination of porcine reproductive and respiratory syndrome virus. *Virus Res*. 2010;154(1-2):185-192. doi:10.1016/j.virusres.2010.08.016.

3. Sur JH, Doster AR, Christian JS, et al. Porcine reproductive and respiratory syndrome virus replicates in testicular germ cells, alters spermatogenesis, and induces germ cell death by apoptosis. *J Virol*. 1997;71(12):9170-9179. https://www.ncbi.nlm.nih.gov/pubmed/9371575.

4. López WA, Angulo J, Zimmerman JJ, Linhares D. PRRS monitoring in breeding herds using processing fluids. *J Swine Heal Prod*. 2017;In press.

PRRS outbreak investigation pilot program: Lessons learned

Derald Holtkamp, DVM, MS; Kimberlee Baker; Rita Neat; Christine Mowrer; Daniel Linhares, DVM, PhD
Iowa State University, Ames, Iowa

Introduction

It has been three decades since porcine reproductive and respiratory syndrome (PRRS) was first recognized in 1987 (Keffaber, 1989). Despite years of research and experience, the virus that causes PRRS still causes productivity losses worth $664 million annually (Holtkamp et al., 2013). While progress has been made, 20 to 40% of US breeding herds continue to experience outbreaks each year. Experimental research studies can tell us how a viurs can be transmitted from one herd to another (Perez et al, 2015), but cannot tell us the most frequent routes by which the virus is transmitted. Only systematic observation can answer that question. The objective of this project is to establish a program to assess the most frequent routes of viral transmission and help producers and veterinarians prioritize where to focus their bio-exclusion efforts to reduce the frequency of outbreaks in breeding herds over time.

Materials and methods

Twenty-nine farrow-to-wean breeding herds are enrolled in the PRRS Outbreak Investigation Program, which was initiated in 2013. Eighteen of the herds are also enrolled in one of 3 PRRSV regional control projects. A database of information on each herd is updated each summer. The database includes information on the current PRRSV status, PRRSV vaccination programs, PRRSV related diagnostics, weekly production data and biosecurity related information collected with a risk assessment survey. When a PRRS outbreak occurs, an Investigation Coordinator is notified by a regional PRRS control project coordinator, herd veterinarian or directly by the producer. To prepare for an investigation, the coordinator gathers information from the database, the herd veterinarian and the farm manager. The information is entered into a standard outbreak investigation form. The Investigation Coordinator and Investigation Facilitator, a veterinarian not affiliated with the farm, conduct the investigation with the herd veterinarian and producer. The Outbreak Investigation Form is used as a guide for an open-ended discussion of the case and the risk events that occurred in the 4 weeks preceding the first clinical signs. The investigators qualitatively assign each risk event a low, medium, or high risk level for the likelihood of PRRS introduction into the herd. The assignment is based on 1) frequency and timing of the risk event; 2) likelihood that one or more carrying agents associated with the event were contaminated or infected with PRRSV; 3) likelihood of failure to mitigate the contamination or infection; and 4) likelihood that infectious PRRSV was transmitted from the carrying agent to swine in the breeding herd. The results of the investigation are compiled in a summary report, which includes a concise executive summary, and detailed findings from each section of the survey.

Results and discussion

Entry of employees and removal of culls were the risk events rated high most frequently. Of 17 investigations conducted since 2014, entry of employees and removal of culls was each rated high 8 times. Employee entry was the most frequent risk event on all 17 farms investigated. For several farms investigated, employees were at risk of becoming contaminated with PRRSV while performing other jobs, such as managing a finishing site, performing maintenance on other swine farms, driving feed trucks, transporting pigs to and from other farms, managing a feed mill and loading market hogs. For removal of culls, the risk of livestock trailers, cutting boards, other equipment and drivers being contaminated was high in several cases. On several farms investigated, hauling of culls was contracted to a third party that also transported pigs for other producers whose identities were unknown. In some cases, the contractors used trailers and drivers to haul market hogs, weaned pigs and replacement gilts as well. In almost every case, the producer relied on the contractor to establish protocols for cleaning and decontaminating the trailers with no oversight or auditing. Under these circumstances, the producers had no knowledge of where the trailer and driver had been, what types of swine it had hauled or how well it had been cleaned and decontaminated prior to arriving at the farm. In 5 of the 17 cases, an operational connection to a farm with a nearly identical sequence, timing of the event or location of first clinical signs strongly implicated a single risk event ("smoking gun"). In 3 of the 5 cases with a smoking gun, a repair event was identified as the cause of the outbreak.

Acknowledgements

The PRRS Outbreak Investigation Program is funded by the Iowa Pork Producers Association with pork checkoff funds.

References

1. Keffaber K.K. 1989, Reproductive failure of unknown etiology. Am. Assoc. Swine. Pract. Newsl. 1: l-9.

2. Holtkamp D.J., etal. 2013. Assessment of the economic impact of porcine reproductive and respiratory syndrome virus on United States pork producers. J Swine Health Prod. 21(2):72-84.

3. Perez A, etal. 2015. Lessons learned and knowledge gaps about the epidemiology and control of porcine reproductive and respiratory syndrome virus in North America. J. Am. Vet. Med. 246 (12):1304-1317.

Evaluating natural planned exposure protocols on rotavirus shedding patterns in gilts and the impact on their suckling pigs

Amanda V Anderson[1]; Jamie Madigan[2], BS; Derald Holtkamp[1], DVM, MS; Francisco Dominguez[3], PhD; Jeremy S Pittman[4], DVM, MS, DABVP; Douglas Marthaler[3], PhD

[1]Iowa State University, Ames, Iowa; [2]North Carolina State University, Raleigh, North Carolina; [3]Kansas State University, Manhattan, Kansas; [4]Smithfield Hog Production Division- North Region, Waverly, Virginia

Introduction

Rotavirus is a prevalent pathogen in United States swine populations, and occurrence of rotavirus A (RVA), rotavirus B (RVB), and rotavirus C (RVC) has been reported as 70.1, 34.2, and 62.2%, respectively.[1] Herds may experience a pre-weaning mortality increase of 3-20% and a decreased weaning weight of 0.5-1.0 pounds.[2] The fastidious nature of RVB and RVC and the lack of cross protection between and within RVA, RVB, and RVC makes it difficult to prevent clinical disease. Natural planned exposure is used to protect pigs against rotavirus infections, yet literature is lacking to support its efficacy and safety. The objective of this study was to determine the pattern of rotavirus shedding in gilts after natural planned exposure and to assess the impact on shedding of rotavirus and weaning weight of their piglets.

Materials and methods

Initially, 70 bred gilts were randomly divided into 3 treatment groups and a control group. Based on farrowing dates, 11-12 gilts per treatment remained enrolled in the study through lactation, for a total of 46 litters. The feedback material was created using the master seed method and on-farm protocols that were described in Dr. Jeremy Pittman's presentation at the 2016 James D McKean Swine Disease Conference.[3] Natural planned exposure to RVA and RVC was administered in a feed and water slurry and given orally according to different protocols for each treatment group. Treatment group 1 was given natural planned exposure at five, four, and three weeks prefarrow; treatment group 2 at five and three weeks prefarrow; and treatment group 3 at five weeks prefarrow only. The control group did not receive natural planned exposure at any time. Fecal samples were collected biweekly from 5 weeks prefarrow to 2 weeks prefarrow, after which weekly sampling occurred until weaning. At farrowing, five piglets per gilt were tagged and enrolled in the trial. Fecal swabs were collected from the tagged pigs four times from birth to weaning. Sow fecal samples were pooled by three (within treatment group), and piglet fecal swabs were pooled by litter. All of the samples were tested by RT-PCR for RVA, RVB, and RVC. At weaning, piglet weights were recorded and adjusted to a 21-day approximation.

Sow results

Prior to the initial administration of natural planned exposure, 2 gilts were positive for RVA while all of the gilts were negative for RVC. The first feedback administration induced RVA shedding in 26 of the 35 gilts in treatment groups (Table 1). However, only 7 of the 35 treated gilts were shedding RVC. At 4 weeks prefarrow, RVA shedding decreased and RVC shedding increased in the treatment groups. Peak RVA shedding occurred at 3.5 weeks prefarrow for treatment group 1, with all of the gilts in this group shedding RVA, yet only a single gilt was shedding RVC. At farrowing, RVA shedding was observed in all of the treatment groups but not in the control group. Shedding of RVA continued until weaning, and shedding of RVC only occurred at week 2. At weaning, neither RVA or RVC shedding occurred. Lastly, 17 of the 46 sows were actively shedding rotavirus at one or more sampling point while in the farrowing room.

Piglet results

Fecal swabs were collected from each of the tagged piglets within 24 hours of farrowing, and all of the sampled piglets were negative for rotavirus at this time (Table 2). At week 1, 3 and 15 piglet pools were positive for RVA and RVC, respectively. Prevalence of RVC in the piglet pools increased while only a few pools tested positive for RVA at each sampling point. All the piglet pools were negative for RVB throughout farrowing. Upon entry into the nursery, all piglet pools (except 1 piglet pool) became positive for RVA and RVC, and 28 of 46 pools were positive for RVB. At the first sampling point in the nursery, the RVA CT values were approximately 12-14, while the RVB and RVC CT values were in the 20s (data not shown). Subsequent sampling in the nursey yielded 100% positive piglet pools for RVA, RVB, and RVC, with CT values in the 20s. In the farrowing room, the piglet pools from control gilts contained more positive RVC piglet pools than each of the treatment groups, specifically treatment group 1. Generally, piglet pools testing positive for RVA or RVC in early farrowing tested positive for the respective rotavirus at later sampling points, as well. Additionally, many piglet pools that were negative at 1 week of age remained negative until the animals were moved to the nursery. With the exception of 3 piglet pools, all of the litters that came from sows which were shedding rotavirus in farrowing room

Table 1: Gilts positive for rotavirus by PCR.

	Sampling period																							
	5 WPF		4.5 WPF		4 WPF		3.5 WPF		3 WPF		2.5 WPF		2 WPF		1 WPF		Farrow		1 Week		2 Weeks		3 Weeks	
RV Type	A	C	A	C	A	C	A	C	A	C	A	C	A	C	A	C	A	C	A	C	A	C	A	C
Group 1	0	0	10	1	10	4	12	1	1	1	1	0	1	0	0	0	1	0	1	0	2	3	0	0
Group 2	0	0	9	3	6	6	5	0	2	0	5	0	4	0	1	1	3	0	1	0	0	3	0	0
Group 3	0	0	7	3	5	4	5	0	3	2	0	0	0	0	0	0	3	0	1	0	1	0	0	0
Total (Treatments)	0	0	26	7	21	14	22	1	6	3	6	0	5	0	1	1	7	0	3	0	3	6	0	0
Control	2	0	1	0	0	1	0	0	1	0	0	0	0	0	0	0	0	0	0	4	0	1	0	0
Total (All)	2	0	27	7	21	15	22	1	7	3	6	0	5	0	1	1	7	0	3	4	3	7	0	0

were positive for rotavirus at one or more sampling points prior to their entry into the nursery.

The average 21-day adjusted piglet weaning weights for treatment group 1, treatment group 2, treatment group 3, and the control group were 14.7, 13.5, 13.7, and 13.2 lbs, respectively. There was a significant difference ($P < .05$) between treatment group 1 and the control, which correlates to an average of 1.4 additional pounds gained per piglet at weaning. Additionally, a linear increase in pre-weaning mortality rate was observed, with the lowest pre-weaning mortality occurring in treatment group 1 and the highest pre-weaning mortality occurring the control group.

Discussion and conclusions

The treatments failed to induce viral shedding of RVA and RVC in all of the gilts. The CT values of the natural planned exposure slurry varied slightly for RVA and RVC and may suggest that a higher concentration of virus is needed in the natural planned exposure mixture to achieve shedding in all of the animals. Further research on proper dosage levels is needed. Additionally, shedding of both RVA and RVC continued past farrowing, suggesting a need for further research on the proper timing of natural planned exposure in order to limit viral exposure of piglets in the farrowing room. There was a significant difference in weaning weights of piglets born to gilts that received three natural planned exposure administrations compared to the controls, which is consistent with previous reports on the impact of rotavirus in suckling pigs.[2] Coupled with a higher mortality rate in control piglets, these results suggest that three administrations of natural planned exposure improves production parameters in herds affected by rotavirus. This relatively inexpensive method for rotavirus control results in significant economic benefits for swine producers. Individual piglet PCR data, viral sequencing, and serology results are expected in the future and will help in drawing additional conclusions.

Table 2: Litters positive for rotavirus by PCR.

	Sampling period																
	Farrow*		1 Week*		2 Weeks*		3 Weeks*		4 Weeks			5 Weeks			6 Weeks		
RV type	A	C	A	C	A	C	A	C	A	B	C	A	B	C	A	B	C
Group 1	0	0	1	2	1	6	1	5	12	8	12	12	12	12	12	12	12
Group 2	0	0	0	5	0	7	0	7	12	9	12	12	12	12	12	12	12
Group 3	0	0	1	1	1	6	2	6	11	5	11	11	11	11	11	11	11
Total (treatments)	0	0	2	8	2	19	3	18	35	22	35	35	35	35	35	35	35
Control	0	0	1	7	0	9	0	7	10	6	10	11	11	11	11	11	11
Total (all)	0	0	3	15	2	28	3	25	45	28	45	46	46	46	46	46	46

* All pools were negative for RVB at these sampling point.

Acknowledgements

We would like to thank Boehringer Ingelheim for their sponsorship of this project, Smithfield Hog Production for providing the animals and facilities for the trial, the farm employees for their assistance, and Kansas State University for completing the diagnostic testing.

References

1. Vlasova AN, Amimo JO, and Saif LJ. 2017. Porcine rotaviruses: Epidemiology, immune responses and control strategies. *Viruses* 9(3): 48.

2. Groth D. 2014. Clinical management of rotavirus. American Association of Swine Veterinarians. Dallas, Texas. pp. 561-562.

3. Pittman J. 2016. Field experiences with interventions for rotavirus control. James D McKean Swine Disease Conference. Ames, Iowa. pp. 30-36.

Evaluation of the persistence of Senecavirus A during elimination program in a sow farm

Deb Murray[1]; Levi Johnson[1]; Matthew Sturos[2]; Fabio Vannucci[2]
[1]New Fashion Pork, Jackson, Minnesota; [2]University of Minnesota, St. Paul, Minnesota

Introduction

Senecavirus A (SVA) has been identified and isolated from pigs showing vesicular lesions, acute lameness and sudden neonatal death. Since July 2015, over 230 cases of SVA have been reported by the Swine Health Monitoring Project. The major economic impacts include but are not limited to: (ii) temporary closure of pork processing plants until the diagnosis of foreign animal diseases has been ruled out, (i) reallocation of resources for diagnostic investigation of FMD and other vesicular diseases and (iii) sporadic mortality of neonatal piglets. The objective of the present study is to evaluate the persistence of SVA during an elimination program established after an outbreak of neonatal mortality and vesicular disease in a sow farm.

Methods

The elimination protocol for SVA elimination was established after an outbreak in a 2000-head sow unit, filtered, PRRSv negative, and with high biosecurity standards. The farm was also negative for SVA by IFA serology and PCR in oral fluids. The outbreak was first characterized by acute mortality in the farrowing room followed by vesicular disease in sows in the farrowing and gestation and heat-checking boars. After a foreign animal disease investigation, the diagnosis of SVA was confirmed. An exposure protocol was performed by allowing the nose-to-nose contact of the clinically-affected heat-checking boar with sows. Additionally, vesicular fluids and saliva were collected from affected animals, diluted in PBS and inoculated intranasally (1 ml containing 10^5 SVA) in 86 animals that were introduced from a negative farm before closure for SVA elimination. Feces, oral swabs and serum were collected from piglets (n = 30) born from affected at weaning age and tested by PCR and serology. SVA persistence was monitored weekly through PCR testing in air sample and environmental swabs (farrowing and gestation crates). Feces, oral swabs and serum were also collected in 54 of the 86 SVA-exposed animals. Furthermore, after the outbreak a new disinfection protocol started being applied using hydrogen peroxide.

Results and discussion

Seven days after forced exposure, five animals showed clinical signs characterized by blisters and off-feed. All 54 animals tested by PCR were positive on feces and oral swabs. Viremia was observed in 87% (47/54) of the animals. There were no positive animals by serology (IFA).

Twenty-four days post-exposure (dpe), 52% (28/54) were PCR positive on feces and 15% (8/54) PCR positive on oral swabs. Viremia was not observed at 24 dpe, but SVA antibodies were detected by IFA in 96% (52/54). Fecal and oral shedding was not observed after 24 post-exposure. These finding confirmed the success of the intranasal exposure using vesicular fluids and saliva. The dynamic of the infection was similar from those reported in animals experimentally-infected with SVA.

Air samples collected in gestation and farrowing rooms were positive by PCR only up to two and four weeks after the first clinical signs of vesicular disease, respectively. In contrast, environmental swabs collected from farrowing and gestation crates were consistently positive by PCR up to 10 weeks post clinical signs. Farrowing room became negative sooner probably due to the ability to apply a more effective washing when they are empty, compared to the gestation room.

Weaning piglets were negative by oral and fecal PCR since the first sampling (3 weeks post-clinical signs). SVA antibodies were detected by IFA in 80% (24/30) of the piglets at the weaning age (3 weeks of age) and were not present at the end of the nursery (7 weeks of age). This decay is suggestive of maternal-derived antibodies.

Conclusions and implications

After a 3 month closure the farm re-entered naive gilts/sows which have remained negative to date. Elimination programs in sow farms have been largely based on the ability of weaning negative piglets in order to determine the status of the farm. However, the present study has shown a different dynamic of infection and persistence in a farm experiencing a SVA outbreak.

Shedding and persistence of Senecavirus A in boars: Natural exposure and experimental infection with an historical and a contemporary strain

Matthew Sturos[1]; Deborah Murray[2]; Darwin Reicks[3]; Jean Paul Cano[4]; Stephanie Rossow[1]; Fabio Vannucci[1]
[1]Veterinary Diagnostic Laboratory, University of Minnesota, St. Paul, Minnesota; [2]New Fashion Pork, Jackson, Minnesota; [3]Reicks Veterinary Research & Consulting, St. Peter, Minnesota; [4]Pig Improvement Company, Hendersonville, Tennessee

Background

Senecavirus A (SVA), also known as Seneca Valley virus or SVV, is in the family *Picornaviridae* and has been demonstrated to cause vesicular disease in pigs.[1] SVA is also associated with transient neonatal losses on some infected sow farms but experimental reproduction of this syndrome has not been reported.[1] The routes of transmission and sources of virus introduction into naïve farms often remain speculative. Boar semen is a possible source for virus introduction and it has been previously reported to be PCR-positive from at least one naturally-infected boar but has not been subject to thorough investigation.[2] This abstract will describe SVA viral RNA shedding in semen and tissue distribution at necropsy from two naturally-infected boars, as well as provide data from experimentally-inoculated boars.

Naturally-infected boars

Two teaser boars developed vesicular disease on a Minnesota sow farm that experienced an outbreak of vesicular disease in January 2017. Samples from the farm were negative for foot and mouth disease virus but positive for SVA. Semen was collected from one or both of these boars at approximately weekly intervals, starting nine days after the development of vesicular disease on the farm until the time of euthanasia and necropsy, which occurred 92 days after the outbreak. Semen and necropsy samples were tested for SVA by real-time RT-PCR (PCR) at the University of Minnesota Veterinary Diagnostic Laboratory (UMN VDL). Semen was PCR-positive for both boars at all sampling points during the investigation period, with threshold cycle (Ct) values of Ct 20 up to Ct 32. The lowest Ct values were obtained between day 42 and day 55 after the development of lesions. The virus could only be isolated from one semen sample collected nine days after the outbreak. Necropsy included a comprehensive examination of the reproductive tract, including testes, epididymis, spermatic cord, accessory sex glands, and penis. Testes of both boars were PCR-positive (Ct 18 and Ct 23), as well as the tonsils (Ct 26 and Ct 29). Histologic lesions were restricted to the testes and included multifocal interstitial lymphocytic orchitis with degeneration and atrophy of seminiferous tubules. SVA RNA was identified within the areas of inflammation in the testes by in-situ hybridization (ISH, see Figure 1). SVA RNA was also identified within interfollicular and subepithelial areas of the tonsil using ISH.

Figure 1: Boar testis. Senecavirus A in-situ hybridization (SVA ISH) with hematoxylin counterstain. Lymphocytes are infiltrating the interstitium adjacent to degenerate and atrophic seminiferous tubules. Bright red indicates positive ISH signal for SVA RNA.

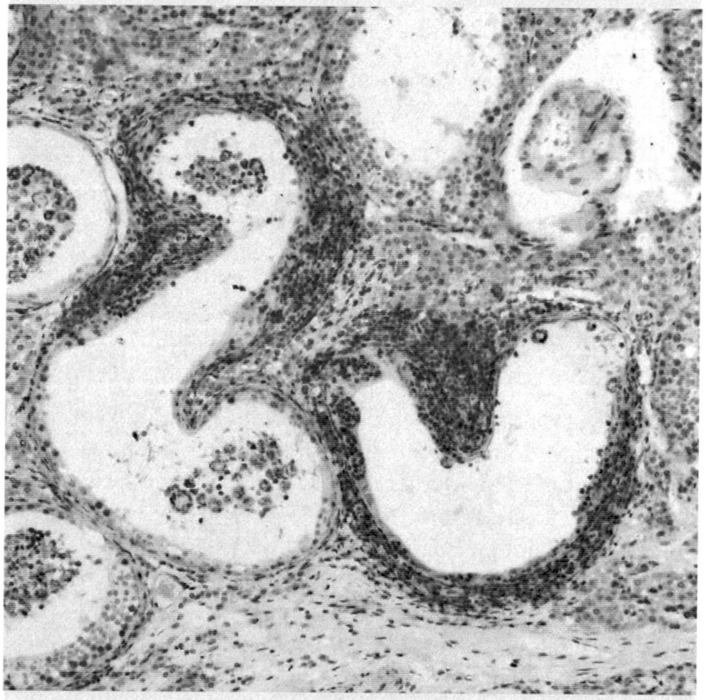

Experimentally-infected boars

Materials and methods

Twelve mature boars were inoculated intra-nasally with SVA. One group of six boars was inoculated with an isolate obtained in 1999 from an aborted porcine fetus (historical group) and one group of six boars was inoculated with an isolate obtained in 2017 from boar semen (contemporary group). Both groups were inoculated with 6.0×10^8 TCID50 per boar. An additional two boars were sham inoculated (control group). Semen, serum, oral swabs, fecal swabs, and lesion swabs, if applicable, were collected on days 1, 2, 3, 5, 7, 14, 22, 28, 35, and 42 post-inoculation (PI).

Boars were euthanized and necropsied on day 49 PI. All clinical samples and tissues collected at necropsy were tested by PCR at the UMN VDL. Serum was also tested for SVA-specific IgG antibodies by IFA at the UMN VDL.

Results

Results for PCR testing of the clinical samples are summarized in Table 1. On serology, there were 2/6 contemporary group boars seropositive on day 5 PI, 3/6 boars in each group were seropositive on day 7, and all boars were seropositive from day 14 until the end of the study. Two boars in the contemporary group developed vesicular/ulcerated dermal lesions on the snout or feet during the first week (snout lesions appeared on day 2 and day 3 PI, the foot lesion appeared on day 5 PI), which resolved after approximately one week. None of the boars in the historical group developed lesions. At necropsy, all boars from both groups were PCR-positive on tonsil (Ct 21-28) and pooled retropharyngeal and mandibular lymph nodes (Ct 22-32). Testis was PCR-positive for 4/6 contemporary group boars (Ct 33-34) and for 2/6 historical group boars (Ct 32-34). There were no histologic lesions in the testes, lymph nodes, or tonsils. Control group boars were PCR-negative in clinical and necropsy samples and did not seroconvert.

Conclusions and discussion

Mature boars are susceptible to infection SVA strains isolated from both 1999 and 2017 by intranasal inoculation as well as by natural exposure. SVA RNA is shed in semen from both naturally-infected and experimentally-inoculated boars and virus was successfully isolated the semen of one naturally-infected boar. Shedding of viral RNA in semen for over three months and the demonstration of SVA RNA in the testes and tonsils of the naturally-infected boars have not been previously reported and suggests that persistent infection may occur in boars. Viral RNA shedding in semen appears to be a repeatable phenomenon with the 2017 SVA isolate, with all boars in the contemporary group PCR-positive in semen for at least one collection each. The duration of shedding in semen was much shorter for the experimentally-infected boars than for the naturally-infected boars but the cause for this difference is unknown. The pattern of shedding in semen in the experimentally-infected boars closely follows the pattern of viremia in these boars, while the strong PCR-positive results on testes, chronic orchitis, and intra-testicular ISH-positivity in the naturally-infected boars indicate that replication of SVA may occur within the testes in some cases. Shedding occurred in some experimentally-infected boars which did not exhibit clinical signs or develop vesicular lesions which shows that shedding is possible in subclinical infections. It is not known if semen from infected boars is capable of causing infection if used for inseminating susceptible females but the isolation of virus from semen of one of the naturally-infected boars indicates that viable virus can be present in semen. These investigations provide valuable information about SVA infection in boars but also highlight the need for additional research.

References

1. Segalés J, Barcellos D, Alfieri A, Burrough E, Marthaler D. Senecavirus A: An Emerging Pathogen Causing Vesicular Disease and Mortality in Pigs? *Vet Pathol.* 2016. doi:10.1177/0300985816653990.

2. Canning P, Canon A, Bates JL, et al. Neonatal Mortality, Vesicular Lesions and Lameness Associated with Senecavirus A in a U.S. Sow Farm. *Transbound Emerg Dis.* 2016;63:373-378. doi:10.1111/tbed.12516.

Table 1: Proportion of experimentally-infected boars PCR positive for SVA by sample type and group over time

Sample type	Group	Days post-inoculation									
		1	2	3	5	7	14	22	28	35	42
Serum	Historical	6/6	6/6	6/6	4/6	3/6	0/6	0/6	0/6	0/6	0/6
	Contemporary	6/6	6/6	6/6	6/6	5/6	1/6	0/6	0/6	0/6	0/6
Semen	Historical	0/6	1/6	0/6	0/6	0/6	0/6	0/6	0/6	0/6	0/6
	Contemporary	1/6	4/6	1/6	6/6	2/6	1/6	1/6	0/6	0/6	0/6
Oral swab	Historical	5/6	6/6	6/6	6/6	6/6	3/6	1/6	0/6	0/6	0/6
	Contemporary	6/6	6/6	6/6	6/6	6/6	5/6	4/6	3/6	1/6	2/6
Fecal swab	Historical	1/6	2/6	5/6	5/6	6/6	6/6	3/6	5/6	1/6	0/6
	Contemporary	2/6	2/6	6/6	6/6	6/6	6/6	6/6	5/6	5/6	3/6

Developing a useful model to evaluate the judicious use of a ceftiofur crystalline free acid to control umbilical infections and hernias in a commercial production system

Cassandra Fitzgerald[1], BS; Pete Thomas[2], DVM; Richard Swalla[3], DVM; Deborah Amodie[3], MS
[1]Iowa State University, Ames, Iowa; [2]Iowa Select Farms, Iowa Falls, Iowa;
[3]Zoetis, Parsippany, New Jersey

Introduction

Umbilical infection and herniation is financially costly at market and increases the risk of morbidity and mortality during the growing period. Reported umbilical herniation prevalence occurs between 0.4 and 6.7% in commercial swine herds.[1] Previous work demonstrated that the use of a ceftiofur crystalline free acid (CCFA, Excede®, Zoetis) within 16 hours of birth reduced the prevalence of both umbilical infections and hernias. The Iowa Select Farms production system currently practices legal extra-label use of CCFA on pigs less than 24 hours old to control umbilical infection and herniation in the population, and is committed to responsible antibiotic use. The objective of this study was to develop a model the production system can utilize that: (1) is feasible and easily carried out at periodic intervals, (2) demonstrates incidence of umbilical infections prior to injection of CCFA in at least 10% of the pigs sampled from each gilt flow (3) confirms CCFA is effective against bacteria found in the navel, and (4) supports the clinical benefits of CCFA in pigs by using umbilical scoring and joint swelling at around 21 days of age as measurable outcomes.

Materials and methods

This two-phase study was conducted in three conventional 4000-4500 head sow herds selected because they were supplied by three different gilt flows. All three herds weaned pigs around 21 days of age. For phase one, an aseptic method was used to swab the inside of the umbilicus of ten dead or euthanized pigs less than 24 hours old that had not received CCFA, and samples were delivered to the Iowa State University Veterinary Diagnostic Laboratory for bacterial culture and sensitivity. For phase two, 40 litters from each farm that were farrowing within three days of each other were selected and enrolled in the study (120 total litters). Litters were randomly assigned a gender, and within each litter, two pigs of the same gender were selected and assigned to the treatment group (0.1 cc of CCFA, within 24 hours of birth) or control group (no CCFA, within 24 hours of birth). Around 21 days of age each umbilicus was scored using a previously validated scoring system, and the presence of swollen joints evaluated. Results were assessed using a Chi-Square analysis by treatment.

Results

In phase one, Farm A and Farm C had 30% and Farm B had 20% of swabs with positive bacterial culture. The percentage of isolates sensitive to antibiotics potentially used on farm are as follows: 82% to ceftiofur, 100% to ampicillin, 91% to enrofloxacin, 45% to florfenicol, 0% to oxytetracycline, and 45% to penicillin. In phase two the control group had a numerically higher percentage of infection (56%) than the treated group (45%) at 21 days of age. Joint swelling was significantly higher in the control group (9%) than the treated group (1%) ($P = 0.0092$).

Discussion

Phase one provided documentation of clinical disease to justify use of CCFA in these three farms as a control and treatment of disease and demonstrated the sensitivity of bacteria cultured to CCFA. Ampicillin and Enrofloxacin had higher sensitivities but ampicillin is not labeled for pigs and should be used as a multiple day treatment. Enrofloxacin cannot legally be used for control of a disease in swine. Phase two demonstrated the clinical benefits of using CCFA at day 1 as measured by the decrease in umbilical infections and joint swelling. In addition, it is important to consider possible actions to control umbilical infections and hernias before looking at blanket antibiotic treatment such as: ensuring good sanitation in the farrowing rooms, providing proper piglet handling and day 1 care, and adequate timely colostrum consumption. However, these interventions may not be enough in all cases which then it is important to develop a strategy to control umbilical infections and hernias that is still judicious. The novel model developed in this study may be used to support the judicious use of antimicrobials for control of umbilical infections and hernias in any farm or production system and may be adapted for other clinical diseases.

References

1. Straw B, Bates R, May G. Anatomical abnormalities in a group of finishing pigs: prevalence and pig performance. *J Swine Health Prod.* 2009;17(1):28–31.

The effect of increased feed intake during late gestation on litter characteristics of commercial sows

K. O'Brien[1], BS; A.Ornelas[2]; F. Zarate[2], MS; R. Moreno[2], PhD;
R. Robbins[2], DVM, PhD; J. Flohr[2], PhD; R. Schmitt[2] PhD
[1]University of Illinois College of Veterinary Medicine, Urbana-Champaign, Urbana, Illinois;
[2]Seaboard Foods, Guymon, Oklahoma

Introduction

Increasing feed to sows during late gestation, termed bumping, is commonly practiced in the US swine industry. Previous studies concluded bumping can positively improve sow reproductive performance by significantly increasing the total born and born alive per litter.[1] Additionally, studies have concluded that feed bumping in late gestation can increase piglet birthweight.[2] More recent research,[3] found bumping multiparous sows in late gestation can lead to increased stillborns, as well as, possible farrowing problems, crushing piglets, and lowered prolificacy in subsequent parities.[4] Furthermore, restricting feed intake prior to farrowing may significantly reduce the risk for postpartum dysgalactia syndrome.[4] Therefore, eliminating the bumping practice may positively impact litter performance and reduce feed cost. The objective of this study was to determine if feed bumping sows, during late gestation, effects litter characteristics.

Materials and methods

A total of 2,657 sows (PIC 1040; parity 1 to 8), from five consecutive breed groups, were enrolled. All sows were profiled by parity and initial sow caliper score[5] during week 13 of gestation. During week 14, sows were randomly assigned to one of three treatments; (1) No feed bump, (2) 1 week: 1 lb feed increase during week 15 (pre-farrow), and (3) 2 week: 1 lb feed increase during week 14 (SOP). The gestation diet was corn and soybean meal based, with 40% wheat middlings and was formulated to a standardized ileal digestible (SID) lysine level of 0.62%. According to normal farm procedures, sows were moved into farrowing at day 113 of gestation and all sows received 2 lbs of lactation feed (Corn and soybean meal based with 1.10% SID lysine) twice daily until farrowing. Born alive, stillborns, and mummies were recorded for each litter to calculate total born per litter. If a sow was removed from the trial, the reason was recorded. To assess condition change during lactation, two wean groups, 370 sows, enrolled in the trial were calipered within 48hrs of weaning. Sows were weaned and were bred back according to normal farm procedures, and then tracked through first ultrasound to determine conception rate. The results were analyzed using mixed effect models with sow as the experimental unit and breed group as a random effect in R.[6] Non-gaussian distributed response criteria were evaluated using discrete Poisson or binomial probability distributions.

Results

As shown in Table 1, feed bumping had no effect on the total born per litter between treatment groups ($P < 0.56$). Additionally, born alive ($P < 0.72$), stillborns ($P < 0.24$), and mummies ($P < 0.38$) as a % of total born did not vary significantly across treatment groups (Table 1). Also seen in Table 1, pre-weaning mortality was significantly different among feed bump treatments ($P < 0.001$). Group 2 had the lowest pre-wean mortality with a mean of 16.61% (Table 1). Group 1 had the highest pre-wean mortality, 19.03 % (Table 1). Post-weaning caliper scores did not demonstrate a significant difference ($P < 0.44$) between the three groups. Treatment group did not have a significant effect on post-weaning key performance indicators (Table 1).

Discussion and application

Limitations of this field study were the inability to capture birth weight within 24 hours of parturition and wean weight because this was beyond the scope of on-farm data collection capabilities. The results show a significant response of sow feed bump treatment on pre-weaning mortality. Biologically, it is challenging to discern why the lowest pre-weaning mortality would occur when a 1 week feed bump was administered relative to a 2-week feed bump or no feed bump. Results show that late gestation feed bumps of 1 lb occurring 1 or 2 weeks pre-farrow had a null effect on litter or subsequent sow reproductive performance. Removal of the bump could result in a reduction of 17-34 lbs of gestation feed annually per sow. This would have a significant financial impact for the farm, as well as, reduce labor requirements.

References

1. Cromwell, G., D. Hall, A. Clawson, G. Combs, D. Knabe, C. Maxwell, P. Noland, D. Orr, and T. Prince. 1989. Effects of Additional Feed during Late Gestation on Reproductive Performance of Sows: A Cooperative Study1. J. Anim. Sci. 67:3-14. doi:10.2527/jas1989.6713

2. Goncalves,M., K. Gourley, S. Dritz, M. Tokach, N. Bello, J. DeRouchey, J. Woodworth, R. Goodband.2015. Maternal dietary energy rather than lysine intake during late gestation positively influences piglet birth weight. Animal Production Science. 55(11):1505. doi:10.1071/ANv55n12Ab120

3. Goncalves,M., K. Gourley, S. Dritz, M. Tokach, N. Bello, J. DeRouchey, J. Woodworth, R. Goodband.2016. Effects of amino acids and energy intake during late gestation of high-performing gilts and sows on litter and reproductive performance under commercial conditions. J. Animal. Sci. 94: 1993-2003. doi: 10.2527/jas2015-0087

4. Kraeling,R., S. Webel. 2015. Current strategies for reproductive management of gilts and sows in North America. Journal of Animal Science and Biotechnology. 6(1): 1-14. doi: 10.1186/2049-1891-6-3

5. Miller, D. 2012. Sow Body Condition Caliper Unveiled. National Hog Farmer. 57(6):22-23.

6. Bates, D., M. Maechler, B. Bolker, S. Walker. 2015. Fitting Linear Mixed-Effect Models Using lme4. Journal of Statistical Software. 67:1. p1-48.

Table 1: Sow and litter characteristics for each treatment group

Feed bump treatment:	No bump	Pre-farrow bump	SOP	SEM	Probability, *P* <
Sow information					
Sows, n	887	886	884	---	---
Parity	3.70	3.73	3.74	0.07	0.90
Caliper score	10.45	10.50	10.49	0.14	0.89
Sow removals					
NIP, % of total sows	0.53	1.28	1.07	0.43	0.23
Early weans, % of total sows	4.86	4.31	4.43	0.88	0.84
Nurse sows, % of total sows	4.17	3.72	3.96	0.67	0.89
Culls, % of total sows	0.68	1.47	1.36	0.40	0.24
Mortality, % of total sows	3.83	3.72	3.28	0.64	0.80
Litter characteristics					
Total torn, n	14.60	14.78	14.66	0.19	0.56
Born alive, % of total born	91.40	91.19	91.54	0.30	0.72
Stillborn, % of total born	6.49	6.54	6.08	0.23	0.24
Mummies, % of total born	2.09	2.26	2.36	0.19	0.38
Weaning information					
Average lactation length, days	17.0	17.1	17.2	0.20	0.49
Pre-weaning mortality[1]	19.03[a]	16.61[c]	18.02[b]	0.68	**0.001**
Caliper score[2]	9.65	9.71	10.03	0.28	0.44
Post-weaning information					
Wean to estrus interval, days	5.31	5.61	5.33	0.23	0.36
Breed back, %	87.7	87.8	88.8	1.3	0.73
Conception rate, %	93.1	93.6	93.8	0.9	0.82

[1] Means with different superscripts, $P < 0.05$
[2] Only two wean groups were calipered (370 sows)

Evaluation of antibiotics at weaning on swine respiratory disease in the early wean-finish phase

Joel Steckelberg[1]; Ethan Spronk[2], DVM; Noel Garbes[3], DVM; Deborah Amodie[3], MS
[1]College of Veterinary Medicine, Iowa State University, Ames, Iowa; [2]Eichelberger Farms, Inc., Wayland, Iowa;
[3]Zoetis, Parsippany, New Jersey

Introduction

In the early wean-finish phase, bacterial pathogens like *Streptococcus suis* and *Haemophilus parasuis* can cause significant morbidity and mortality.[1] Industry practice is to control respiratory disease in young, growing pigs with an antibiotic treatment at weaning. The objectives of this study were to determine the effectiveness of using an antibiotic at weaning to control respiratory disease in the early wean-finish phase and to compare two antibiotics, ceftiofur and enrofloxacin, on early wean-finish health.

Materials and methods

A 3,100 head sow farm weaning porcine reproductive and respiratory syndrome virus (PRRSV) negative pigs, based on weaned pig serum tested monthly by PCR, with a high incidence of *S suis* related pneumonia and septicemia post-weaning was chosen for the project. 325 litters were randomly assigned into three treatment groups: ceftiofur (Ce), enrofloxacin (En), and negative control (NC). At weaning, the Ce group received Excede® (Zoetis) and the En group received Enroflox® 100 (Norbrook Laboratories) according to label instructions. The NC group did not receive antibiotic. On average, pigs were weaned at 24.5 days and weighed 13.4 pounds. At the wean-finish barn, each pen was randomly assigned to a single treatment group. Mortalities, treatments, and pulls into hospital pens were recorded by pen, the experimental unit, by a blinded contract grower. Individual Pig Care (IPC) scores were obtained on days 2, 7, 11, 21 and 35 post-placement to measure the percentage of acute (A), subacute (B), and chronic (C) pigs for each treatment group. 3,844 pigs were included in this study (1,275 pigs in the Ce group, 1,289 pigs in the En group, and 1,280 pigs in the NC group). There was an average of 54 pigs per pen. Oral fluids (OF) were collected at placement and at the conclusion of the study. Two spikes (S1 and S2) in mortality were investigated to determine the health challenges present during the study. For S1, complete tissue sets from each treatment group were collected and sent to the Iowa State University Veterinary Diagnostic Laboratory. For S2, OF were collected. The study ended five weeks after the last group of pigs started the study. Percentages were transformed by the arcsine (square-root) transformation to stabilize variance and normalize data. Transformed mortality, treatments, and pulls were analyzed by a linear mixed model approach. IPC data was analyzed by a linear mixed model approach for repeated measures. Treatment was tested at the 5% level of significance (two-sided test). Treatment Least Square Means (LSMeans) were calculated for each group and compared by the two-sided Student's t-test at the 5% level of significance. LSMeans were back-transformed.

Results

The average parities of the litters in the Ce, En, and NC groups were 3.3, 3.1, and 3.0, respectively. The OF collected at placement were negative for PRRSV and Influenza A virus of swine (IAV-S). Tissue results from S1 showed all treatment groups had *S. suis* septicemia and were negative for PRRSV and IAV-S. OF from S2 were positive for IAV-S (H3N2). OF collected at the end of the study were negative for IAV-S and 8 out of the 12 samples were positive for PRRSV after Fostera® (Zoetis) PRRS vaccination one week prior (genomic sequence was 99.7% similar to Fostera®). Mortality percentage and percent pulled into hospital pens were not statistically different between the three treatment groups (Table 1). The Ce group had a statistically higher percent treated when compared to the En group (Table 1). There was a statistically lower percentage of pigs in the Ce and En groups that were acutely (A) sick (Table 1).

Discussion

In this flow, as seen in Table 1, antibiotics at weaning had no significant effect on early-wean finish mortality and had mixed effect on morbidity when previously healthy pigs were challenged with *S suis*. Pigs receiving antibiotics at weaning had a significant reduction in acutely sick pigs showing some health improvement during the early wean-finish phase.

References

1. Zimmerman J et al. *Diseases of Swine*, 10th edition. Wiley-Blackwell, 2012, pg 22.

Table 1: Mortality, treatment, pulls, and IPC results

Treatment group	% Mortality	% Treated	% Pulled	% A	% B	% C
NC	0.91[a]	21.45[a,b]	3.68[a]	2.54[a]	0.11[a]	0.0058[a]
Ce	1.06[a]	27.69[a]	4.23[a]	1.45[b]	0.07[a]	0.0005[a]
En	1.52[a]	17.78[b]	2.90[a]	1.17[b]	0.06[a]	0.0038[a]

[GMean with the same letter grouping within a column are not significantly different][1]

Comparison of oral fluids, environmental swabs, and pooled feces for detection of dysentery-associated *Brachyspira* spp. in pigs

W. Webb[1], BS; H. Warneke[2], MS; J. Seate[3], DVM; R. Jones[4], DVM; A. Scheidt[3], DVM, MS; P. Gauger[2,5], DVM, MS, PhD; K. Harmon[2,5], PhD; E. Burrough[2,5], DVM, PhD, Diplomate ACVP

[1]College of Veterinary Medicine, Iowa State University, Ames, Iowa; [2]Veterinary Diagnostic Laboratory, Iowa State University, Ames, Iowa; [3]Boehringer Ingelheim Vetmedica, Inc.; [4]Livestock Veterinary Services, Kingston, North Carolina; [5]Veterinary Diagnostic and Production Animal Medicine, Iowa State University, Ames, Iowa

Introduction

Brachyspira spp. are anaerobic spirochetes that colonize the large intestine of many species including humans and swine. There are several recognized species of *Brachyspira* that colonize pigs, including three species capable of causing swine dysentery (*B hyodysenteriae, B hampsonii, and B suanatina*) and the agent of porcine intestinal spirochetosis (*B pilosicoli*).[1-3] *Brachyspira* are classified as strongly and weakly beta-hemolytic, with strongly hemolytic species such as *B hyodysenteriae* considered more virulent.[4] All known agents of swine dysentery (SD) are strongly beta-hemolytic. These bacteria typically colonize the crypts of the colon and cecum, affecting goblet cells and causing an excessive amount of mucus production. This is accomplished by chemotaxis and motility of flagella to penetrate the mucus.[1,2] The classical sign of SD, mucohemorrhagic diarrhea, is the result of cytotoxic factors produced by the bacteria that cause necrosis of superficial colonic epithelium and mucus excretion. While multiple factors are likely involved, the hemolysin encoded by the *tlyA* gene has been shown to impact virulence.[5]

Brachyspira spp. are especially difficult to culture and isolate because they are slow growing anaerobes. Particular care is taken while collecting, transporting, and culturing samples to maintain viability of the bacteria. Specifically, samples should remain cool and moist prior to culture.[1] Samples are typically cultured on special blood agar that contains antibiotics (spectinomycin, colistin, and vancomycin) to inhibit and prevent overgrowth of other fecal bacteria.[6] Methods are being developed in efforts to increase the speed and confidence of SD diagnosis including real-time polymerase chain reaction (PCR) assays and serological tests such as enzyme-linked immunosorbent assays (ELISA). Currently, PCR is the more definitive of the two, as serological tests often have lower specificity and ELISA tests may be more useful for identifying affected and/or exposed herds.[1] The objective of the present study was to compare routine laboratory testing for *Brachyspira* spp. at the Iowa State University Veterinary Diagnostic Laboratory (ISU VDL) on three composite sample types (oral fluids, environmental swabs, and pooled feces) and to determine the associated diagnostic sensitivity and specificity of direct culture and PCR on these samples for detecting pens of pigs infected with agents of swine dysentery.

Materials and methods

Samples

Samples were collected from seven finishing farms in North Carolina from June 12, 2017 through July 19, 2017. The farms were either experiencing clinical signs of SD, historically positive for agents of SD but currently asymptomatic, or historically negative. At each farm, an oral fluid, environmental bootie, and pooled fecal sample were collected from 10 separate pens. Samples were shipped overnight to the ISU VDL and processed the next day. Oral fluids were collected routinely by hanging a cotton rope in each pen for several minutes and then extracting the fluid into a snap-cap tube. Environmental booties were collected routinely by wearing cloth booties over the footwear of the collector while walking around the pen. Once removed, booties were placed into whirl-top bags. During processing at the laboratory, 10 ml of phosphate buffered saline (PBS) was added to the whirl-top bags, the cloth was manually squeezed, and the resulting fluid was collected. Pooled feces were collected using a plastic spoon and scooping material from 5 abnormal or loose stools from the same pen into a single zip-top plastic bag.

Direct culture

At sample processing, swabs were made from each composite sample with Starplex Modified Liquid Stuart's single rayon sterile swab. In the laboratory, swab samples were plated onto two different trypticase-soy agar 5% bovine blood plates containing colistin, vancomycin, and spectinomycin (CVS) or colistin, vancomycin, spectinomycin, spiramycin, and rifampin (CV2SR).[6] Plates were incubated anaerobically at 41 ± 1 degrees Celsius for 48 hours. Plates were checked for evidence of strong hemolysis / suspect *Brachyspira* colonies every 2 days following initial incubation and reincubated for up to 6 days. Suspects were restreaked to new plates of CVS or CV2SR for isolation. Matrix-assisted laser desorption/ionization time-of-flight mass spectrometry (MALDI-TOF MS) was used for species identification of *Brachyspira* suspects. Isolated colonies were extracted as previously described prior to MALDI-TOF MS[7] and a score of 2.00 was used as a cutoff value for reliable species identification.

PCR

At sample processing, swabs made from feces and environmental booties were rinsed in a snap-cap tube containing approximately 0.9 mL PBS. After initial processing, approximately 1.0 ml of oral fluid was used for PCR testing. This aliquot was forwarded to the Molecular Diagnostics Section of the ISU VDL for routine nucleic acid extraction and a duplex real-time PCR (*B hyodysenteriae* or *B hampsonii*). For each *Brachyspira* spp., the PCR targeted two gene sequences, *nox* and *tlyA,* and a Ct value below 35 cycles for both *nox* and *tlyA* was considered positive for either species.

Enrichment

An enrichment medium was created using trypticase soy broth (TSB, Becton, Dickinson and Company, Sparks, MD) containing the same antibiotics as CV2SR for Herds A-C. Samples from Herds B-C were also enriched in brain heart infusion (BHI, Oxoid Ltd, Hants, UK) broth containing the same antibiotics. BHI broth containing the same antibiotics and 10% fetal calf serum (GE Healthcare Life Sciences, Hyclone Laboratories, Logan, UT) was used for Herds D-G. Eight ml of prepared broth was dispensed to individual sterile snap-cap tubes.

At sample processing, a portion of each sample was added to the TSB or BHI broth. Approximately 0.5 ml of oral fluid, environmental fluid from the whirl-top bag, and 1.0 gram of feces were added to separate 8.0 ml tubes of prepared broth. Samples enriched in TSB and BHI without fetal calf serum (Herds A-C) were incubated anaerobically at 41 ± 1 degrees Celsius for 48 hours. Enriched samples from Herds D-G were incubated anaerobically at 37 ± 1 degrees Celsius for 48 hours and shaken at 115 rotations per minute. After 48 hours, swabs were taken from the enrichment broth off all samples and were plated for *Brachyspira* culture as described above. An aliquot from each enrichment sample was forwarded to the Molecular Diagnostics Section of the ISU VDL for routine nucleic acid extraction and duplex PCR as described above.

Statistical analyses

A Fisher's Exact Test was used to compare the frequency of detection by each assay for each sample type and a one-way ANOVA was used to compare mean PCR Ct values. A Wilcoxon signed rank test was used to compare differences in PCR Ct values for direct and enriched sample pairs. A commercial statistical software package was used for all analyses (JMP Pro 11, SAS, Cary, NC) and $P \leq 0.05$ was considered statistically significant. Kappa coefficients are reported and interpreted as follows: values > 0.8 suggest excellent agreement, 0.6-0.79 substantial agreement, 0.4-0.59 are fair agreement, and < 0.4 indicate poor or no agreement.

Results

Direct culture vs. PCR

Two out of seven finishing farms, Herds D and G, were negative for all samples submitted on both traditional culture and PCR assays for *B hyodysenteriae*. None of the samples submitted were positive for *B hampsonii* or *B suanatina* on culture or PCR. The clinical status of the herds used in this study is presented in Table 1. A composite table of all sample types with direct culture and PCR results are summarized in Table 2. The probability of obtaining a positive result on both tests is significantly more likely than a positive on culture and negative on PCR ($P < 0.01$). Degree of agreement analysis showed substantial agreement between direct culture and PCR across all sample types (Kappa coefficient = 0.76). Furthermore, pooled feces results show excellent agreement (Kappa coefficient = 0.83). Environmental sample results were lower, but with an excellent agreement between culture and PCR (Kappa coefficient = 0.80). However, oral fluids showed only fair agreement between culture and PCR (Kappa coefficient = 0.50). Frequency of detection by selective culture or PCR by sample type was significantly different ($P < 0.01$ and $P = 0.04$, respectively). Oral fluids are more likely than other sample types to be negative on both culture and PCR, whereas pooled fecal samples are most likely to be culture positive.

PCR

Mean *nox* PCR Ct values, summarized in Figure 1, were lowest in pooled fecal samples (mean: 35.60 ± 4.96) when compared to oral fluids (mean: 37.53 ± 3.01) and environmental swabs (mean: 36.20 ± 3.97). This difference was statistically significant between pooled feces and oral fluids ($P = 0.02$). No statistical significance was detected between mean Ct values of oral fluids and environmental swabs ($P = 0.13$) or environmental and pooled feces ($P = 0.66$). Similarly, *tlyA* Ct scores summarized in Figure 2 show oral fluids had the highest mean score (mean: 37.01 ± 3.35). Pooled feces again were significantly lower (mean: 35.16 ± 5.35) than oral fluids ($P = 0.04$). Environmental swabs average Ct value (mean: 35.63 ± 4.35) was nearly the same as pooled feces ($P = 0.81$).

Diagnostic specificity and sensitivity

A pen was considered positive when any sample type from that pen was positive by either assay in order to determine the diagnostic specificity and sensitivity of pooled feces, oral fluid, and environmental culture and PCR for determining pen status as reported in Table 3. The diagnostic tests used in this study are routine assays

Table 1: Summary of clinical status by herd.

Herd	Clinical status
A	Active disease
B	Active disease
C	Active disease
D	Subclinical
E	Active disease
F	Active disease
G	Negative

Table 2: Summary of results comparing culture vs. PCR across sample types.

Sample type		Feces		Environmental		Oral fluids		Total	
B hyodysenteriae culture		Pos.	Neg.	Pos.	Neg.	Pos.	Neg.	Pos.	Neg.
B hyodysenteriae PCR	Pos.	29	0	19	5	7	8	55	13
	Neg.	6	35	1	45	2	53	9	133

at the ISU VDL and were considered 100% specific in detecting positive. When compared to fecal culture, fecal PCR was the next most sensitive (92.11% and 76.32%, respectively). Fecal PCR had the second highest predictive value of a negative test result of the three composite sample types tested (78.05%), when compared to fecal culture (91.43%). Oral fluid culture was the least sensitive diagnostic assay in this comparison (23.68%). Environmental culture and PCR were similarly sensitive (52.63% and 63.16%, respectively). The predictive value of a negative test result was lower for oral fluids and environmental swabs than fecal culture and PCR.

Enrichment culture and PCR

All enrichment samples from Herds A-C were culture negative. Samples from Herds D and G were negative for *B hyodysenteriae* with and without enrichment. For Herds E and F, enrichment resulted in 7/20 environmental samples, 13/20 fecal samples, and 2/20 oral fluids positive by culture. Culture after enrichment for Herd E showed excellent agreement for the three collection methods when compared to culture of direct samples (Kappa coefficient = 0.92). However, Herd F showed poor or no agreement between culture with and without enrichment (Kappa coefficient = 0.37). One pooled fecal sample from Herd F tested negative by selective culture and PCR without enrichment, but tested positive on culture after enrichment.

The results for PCR targeting *nox* and *tlyA* gene sequences showed an increase in the mean Ct values for feces with enrichment for Herds D-G (mean: 36.78 ± 4.79 and 36.35 ± 5.42, respectively) when compared to the mean Ct value for feces without enrichment (mean: 35.60 ± 4.96 and 35.16 ± 5.35, respectively). This pattern was also seen for both gene sequences with oral fluid and environmental samples. Overall, there was a significant mean difference for the *nox* (0.64, $P < 0.01$) and *tlyA* (0.47, $P < 0.01$) gene sequences among sample types.

Figure 1: PCR assay of *B hyodysenteriae* nox gene sequence Ct scores by sample type. Green diamonds reflect the mean and 95% confidence interval.

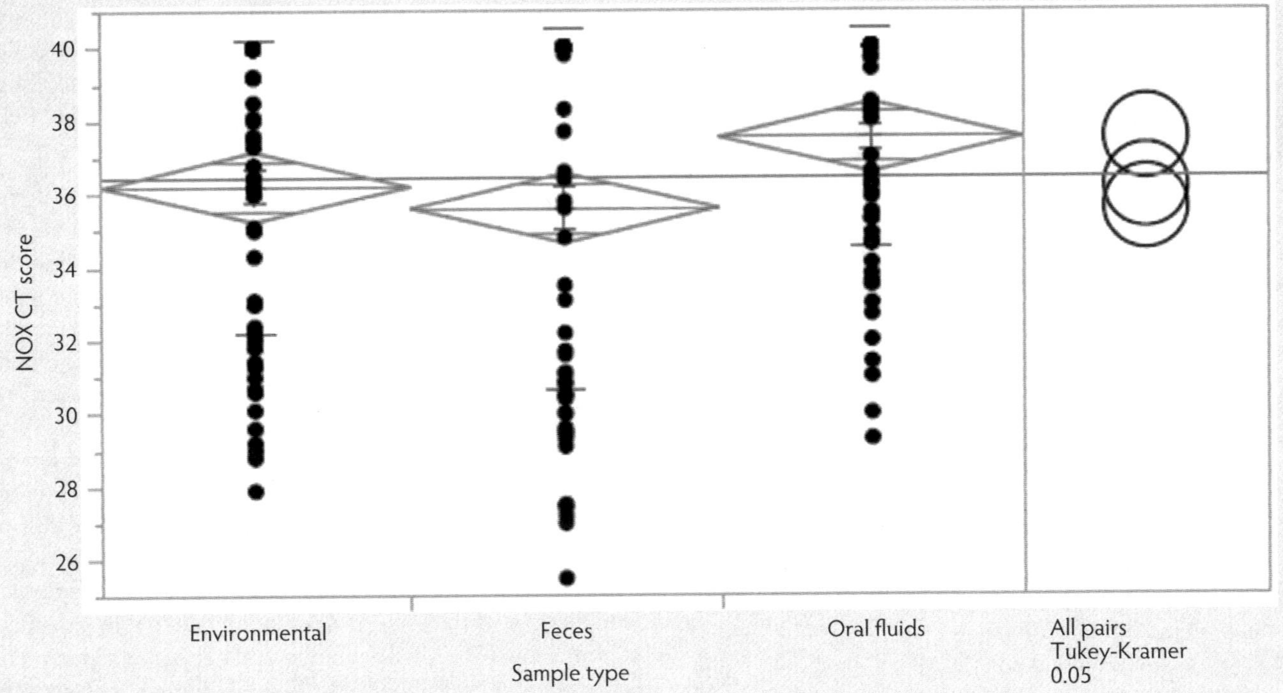

Figure 2: PCR assay of *B hyodysenteriae tlyA* gene sequence Ct scores by sample type. Green diamonds reflect the mean and 95% confidence interval.

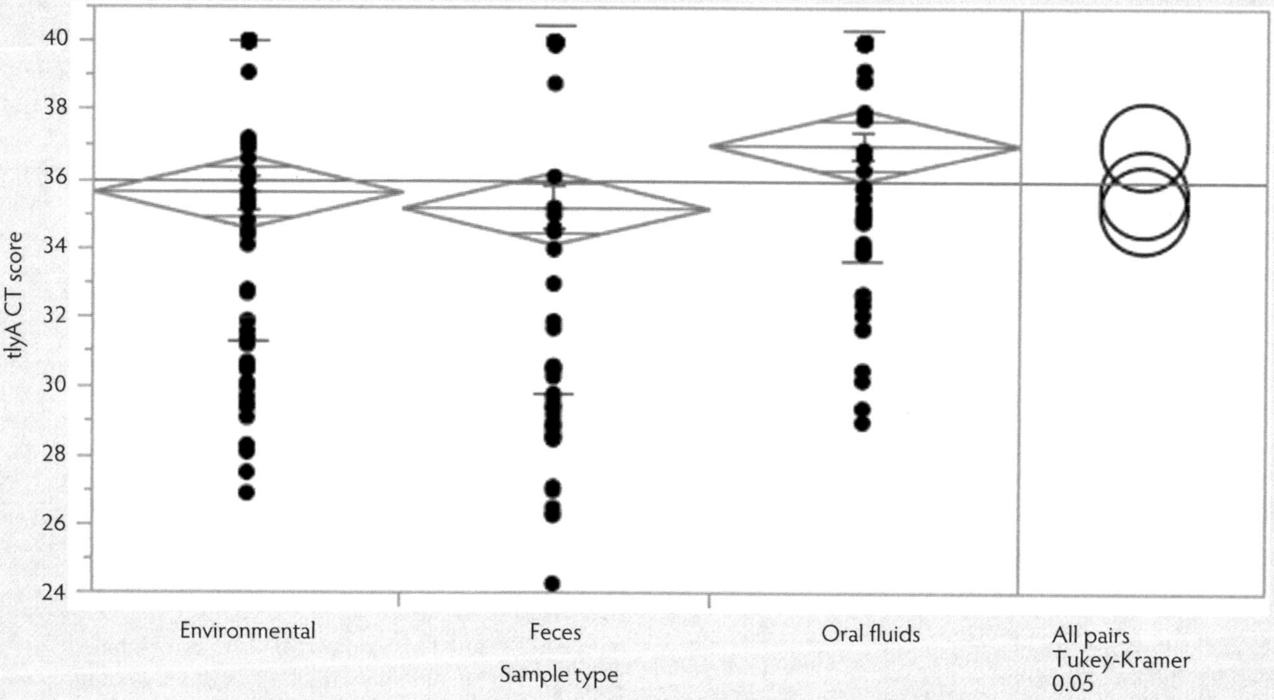

Discussion and conclusion

This study confirms traditional selective culture of feces is the most sensitive diagnostic tool for identifying agents of SD on a pen level which is consistent with previous studies[8] showing culture to be more sensitive than PCR for agents of SD. Additionally, this study also reveals that pooling 5 abnormal stool samples does not appreciably reduce the sensitivity of detection as pooled feces yielded the highest diagnostic sensitivity by culture. Results from this study also suggest fecal PCR is the next best option, which is significant given that while feces are known to harbor viable *Brachyspira* organisms for up to a week at room temperature,[3] loss of viability during transport to laboratory is always a concern with *Brachyspira* spp. and PCR assays are robust to this issue. A similar pattern is seen with environmental swabs which are predominantly feces and debris from the pen slats. While oral fluids have been shown to be an effective sample type for culture in experimental studies,[9] the difference between positive results with oral fluids and the other two composite sample types in the present study, particularly fecal culture, suggests the composition of oral fluids is not ideal to sustain agents of SD during transport of diagnostic samples to the laboratory.

The mean PCR Ct values and analysis of means for proportion further indicate oral fluids are more likely to be negative of the three sample types by both culture and PCR. However, these data did reveal two cases where oral fluid PCR was positive while fecal culture was negative, one case in Herd B and one in Herd C, suggesting the organism may have been non-viable in the feces. This would not be unexpected given the elevated temperatures likely experienced during transport to the lab in July. The low predictive value of negative tests for oral fluids and environmental samples indicates a relatively high likelihood of falsely negative results if only one sample type is submitted for PCR without fecal culture to confirm. Accordingly, parallel testing involving PCR to detect non-viable spirochetes and culture to detect low levels of viable *Brachyspira* spp. is the most sensitive approach for detection of agents of SD.

The enrichment medium was adjusted during the study after all samples from the first three herds were culture negative after enrichment. After adjusting the medium to contain calf serum and incubation at a lower temperature, forty-eight hours of enrichment before culturing did not significantly improve the frequency of detection by culture for any of the sample types. However, one pooled fecal sample from Herd F was positive on culture only after enrichment, suggesting a low number of organisms present in the original sample and maintenance of the organism in the enrichment medium. Plated samples contained an overgrowth of contaminant bacteria, and proved more difficult to isolate *Brachyspira* spp. Enrichment for forty-eight hours did not decrease the PCR Ct values in feces or across sample types, which further suggests the bacteria did not grow in the enrichment medium. However, given enrichment Ct values increased for all sample types, the change might suggest dilution of DNA sent for

Table 3: Diagnostic specificity and sensitivity for determining pen status by diagnostic test and sample type.

Sample type-Diagnostic test	Specificity	Sensitivity	PV+	PV-
Fecal – culture	100%	92.11%	100%	91.43%
Oral fluid – culture	100%	23.68%	100%	52.46%
Environmental – culture	100%	52.63%	100%	64%
Fecal – PCR	100%	76.32%	100%	78.05%
Oral fluid – PCR	100%	39.47%	100%	58.18%
Environmental - PCR	100%	63.16%	100%	69.57%

PV+ = predictive value of a positive test
PV- = predictive value of a negative test

PCR. Further study is warranted to determine if an alternate enrichment medium would improve the diagnostic sensitivity of the PCR assay used in this report.

In summary, the results of this study demonstrate that PCR on pooled feces is an additional diagnostic tool to identify positive cases of SD with a strong positive predictive value; however, fecal culture remains important to confirm negative PCR results. The clinical significance of this study is the potential for monitoring herds without evidence of SD by PCR using fecal or environmental swabs; however, the low negative predictive value of oral fluids in this study suggests this sample type is not acceptable for surveillance unless done in concert with selective culture of feces. Therefore, this study confirms fecal culture is an essential diagnostic tool for determining the presence or absence of strongly beta-hemolytic isolates of *B hyodysenteriae* and for surveillance for other agents of SD.

Acknowledgements

This research was supported in part by the Boehringer Ingelheim Vetmedica Professorship in Food Animal Infectious Diseases.

References

1. Burrough ER: 2017, Swine Dysentery: Etiopathogenesis and Diagnosis of a Reemerging Disease. *VeterinaryPathology* 54:22-31.

2. Hampson DJ, Nagaraja TG, Kennan RM, Rood JL: 2010 Gram-negative anaerobes. In: Gyles CL, Prescott JF, Songer JG, Thoen CO, editors. *Pathogenesis of Bacterial Infections of Animals*. 4th edition. Ames, IA: Wiley-Blackwell. p. 521 – 522.

3. Hampson DJ: 2012 Brachyspiral colitis. In: Zimmerman JJ, Karriker LA, Ramirez A, Schwartz KJ, Stevenson GW, editors. *Diseases of Swine*. 10th ed. Ames, IA: Wiley-Blackwell. p. 680 – 696.

4. Burrough ER, Strait EL, Kinyon JM, et al.: 2012, Comparative virulence of clinical *Brachyspira spp.* isolates in inoculated pigs. *Journal of Veterinary Diagnostic Investigation* 24:1025-1034.

5. Ter Huurne A.A., Muir S., van Houten M., van der Zeijst B.A., Gaastra W., Kusters J.G: 1994, Characterization of three putative *Serpulina hyodysenteriae* hemolysins. *Microbial Pathogenesis* 16:269–282.

6. Stanton T: 2011, Isolation and Maintenance of *Brachyspira* species. *Current Protocols in Microbiology* 12D.1.1-12D.1.14.

7. Warneke H, Kinyon J, Bower L, et al.: 2014, Matrix-assisted laser desorption ionization time-of-flight mass spectrometry for rapid identification of Brachyspira species isolated from swine, including the newly described "*Brachyspira hampsonii*." *Journal of Veterinary Diagnostic Investigation* 26(5) 635-639.

8. Wilberts BL, Warneke HL, Bower LP, Kinyon JM, Burrough ER, 2015, Comparison of culture, polymerase chain reaction, and fluorescent in situ hybridization for detection of *Brachyspira hyodysenteriae* and "*Brachyspira hampsonii*" in pig feces. *Journal of Veterinary Diagnostic Investigation* 27(1) 41-46.

9. Wilberts BL Arruda PH, Kinyon JM, Frana TS, Wang C, Magstadt DR, Madson DM, Patience JF, Burrough ER, 2014, Investigation of the impact of increased dietary insoluble fiber through the feeding of distillers dried grains with solubles (DDGS) on the incidence and severity of *Brachyspira*-associated colitis in pigs. PLoS One 9(12) e114741.

Comparison of biosecurity aspects between swine breeding herds with low and high incidence of porcine reproductive and respiratory syndrome virus outbreaks

Kimberlee L. Baker[1], BS; Gustavo Silva[1], DVM, MS; Robert Morrison[2], DVM, MBA, PhD; Carles Vilalta[2]; Derald Holtkamp[1], DVM, MS; Daniel Linhares[1], DVM, MBA, PhD

[1]Iowa State University College of Veterinary Medicine, Ames, Iowa;

[2]University of Minnesota College of Veterinary Medicine, St. Paul, Minnesota

Introduction

Porcine reproductive and respiratory syndrome (PRRS) compromises the health of millions of pigs and costs the industry $664 million annually. To avoid the economic and animal welfare implications of a PRRS outbreak, swine producers adopt biosecurity measures targeted at decreasing the frequency of PRRS outbreaks. There is a critical need to better understand the effects of biosecurity aspects on frequency of PRRS outbreaks in breeding herds. The objective of this study is to describe key differences in the biosecurity aspects of breeding herds with low PRRS incidence, compared to those with high PRRS incidence.

Materials and methods

Swine production systems in the US were invited to enroll a breeding herd set that met the following criteria: (1) weekly PRRSV status since 2013 was available; (2) willingness to complete a biosecurity survey; and (3) breed-to-wean premises. A herd set was defined as 3 breeding herds within the 25th percentile of PRRS outbreaks (low PRRS incidence) and 3 breeding herds above the 75th percentile (high PRRS incidence) within a production system. Herd selection was completed by ranking each production system's herds based on the number of PRRS outbreaks since 2013 and then randomly selecting 3 farms from both the 25th and 75th percentiles using the RAND function in Microsoft Excel. The biosecurity aspects of each breeding herd were assessed using a 346 question biosecurity survey that contained multiple choice and short answer questions about herd demographics, swine density, PRRS outbreak history, frequency of risk events, and biosecurity practices related to swine transport, people movement, carcass disposal, supply deliveries, and other risk events. A conditional logistic regression model with Lasso penalty was applied to the data using the *clogitL1* R package to determine which biosecurity aspects were significantly different between the low and high PRRS incidence farms.

Results and discussion

Fourteen herd sets (84 herds) were enrolled in the study representing 13 states. Mean herd size was 3453 breeding females (range: 543-7200) for the low PRRS incidence group and 4099 breeding females (range: 1000-10852) for the high PRRS incidence group. Four general areas of biosecurity separated the low and high PRRS incidence farms: (1) monthly event frequency, (2) downtime requirements, (3) swine density, and (4) operational connections to other swine sites. Rendering was the most significant difference between the groups: 64.3% of high PRRS incidence herds used rendering compared to 31% of low PRRS incidence herds. Mean monthly rendering frequency was 12.7 for high PRRS incidence herds and 5.7 for low PRRS incidence herds. High PRRS incidence farms had a higher monthly frequency of visits from visitors (range: 1-24) than low PRRS incidence farms (range: 0.5-8). Low PRRS incidence farms had significantly longer downtime requirements for visitors and manure removal personnel than high PRRS incidence farms. High PRRS incidence farms were located in areas with significantly higher densities of wean-to-finish swine. Interestingly, a higher number of boars and finishing pigs within a 3-mile radius were significantly associated with a low PRRS incidence. This may be accounted for, in part, by the higher level of biosecurity practiced at boar studs. Operational connections to other swine sites were also important as several operational connection related variables were associated with high PRRS incidence. These observations will enable the swine industry to more effectively allocate resources to specific aspects of biosecurity which may help reduce the animal welfare and economic impacts of PRRS in the future.

Putting to test a diagnostic protocol for confirmation of *Mycoplasma hyopneumoniae* elimination in sow farms

B. Chapman[1], BS; R. Evelsizer[2], DVM; M. Pieters[3], DVM
[1]Virginia Maryland College of Veterinary Medicine; [2]Elanco Animal Health;
[3]College of Veterinary Medicine, University of Minnesota

Introduction

Mycoplasma hyopneumoniae (Mhp) is one of the major bacteria affecting the respiratory tract of pigs. Clinical signs associated with *Mhp* infection are non-productive cough and decreased production performance.[1] Although *Mhp* infection is detrimental for pigs at the finishing stage, elimination of the bacterium from animals in the reproductive herd is crucial to decrease pathogen transmission and to avoid disease development in downstream pigs.[2] Once a herd has undergone a *Mhp* elimination protocol it is vital to ensure the program has been successful. In order to determine the elimination of the pathogen, a diagnostic protocol and epidemiological model were proposed to evaluate the elimination programs.[3] However, to the best of our knowledge, this diagnostic protocol had not been tested in the field. Therefore, this investigation was designed to evaluate the feasibility of the diagnostic protocol to detect *Mhp* in farms with expected low prevalence.

Materials and methods

Four sow farms (1-4) in four different US states were enrolled in this study. Farms were selected based on their elimination protocol, which consisted of medication/vaccination with a varying herd closure length. The herd closure length for Farms 1 and 2 was 240 days, Farm 3 did not have a traditional herd closure, but conducted off sight breeding, and Farm 4 had a 180-day herd closure. The number of samples to be collected were based off a 99% sensitivity and a 95% detection probability.[3] The expected prevalence of *Mhp* of 5% or less, total herd population and an affordable budget for each farm were also taken into account. Based on the model, all farms were visited one to three times, and a total of 180 laryngeal swabs were collected at each farm. Samples were systematically collected from females in the last groups exposed to *Mhp*. Samples were submitted for testing in pools of five and processed through real time PCR (VetMax *Mhp*). Farm size and sample frequency are shown in Table 1. In addition, the protocol for positive pool cases is shown.

Results and discussion

Mhp was detected by PCR (low bacterial load – high Ct value) in one pool in two farms, and in two pools in one farm in this study, even though clinical signs were not observed in animals at any of the farms. Each farm followed a different protocol for retest or treatment of the gilts in the positive pools. Due to an increasing trend in *Mhp* elimination programs, there is a large demand for diagnostic protocols to determine the success in eliminating the pathogen. Producers and practitioners require diagnostic protocols that are adapted to field needs, while being highly accurate. The protocol used in this study yielded detection of *Mhp* in several farms and was feasible within a budget. It is our recommendation that this protocol is used at the completion of elimination programs to aid the decision making process of successful disease elimination.

References

1. Sibila, Marina, Maria Pieters, Thomas Molitor, Dominiek Maes, Freddy Haesebrouck, and Joaquim Segalacs. "Current Perspectives on the Diagnosis and Epidemiology of *Mycoplasma Hyopneumoniae* Infection." *The Veterinary Journal* 181.3 (2009): 221-31.

2. Maes, D., J. Segales, T. Meyns, M. Sibila, M. Pieters, and F. Haesebrouck. "Control of *Mycoplasma hyopneumoniae* Infections in Pigs." *Veterinary Microbiology* 126.4 (2008): 297-309.

3. Fitzgerald, C., A. Sponheim, E. Fano, D. Polson, and M. Pieters. "Stochastic Modeling to Determine Number of Laryngeal Swab Sample Pools and Sample Collections Needed for Low *Mycoplasma hyopneumoniae* Group Prevalences." (2016): n. pag. *American Association of Swine Veterinarian.*

Table 1: Herd and sample collection and processing to confirm negative status for *M hyopneumoniae*

Farm #	Farm size	Number of pools positive or suspect	Number positive in re-test *If needed	Follow up
1	750	1/36	1/12	Females from the positive pool were retested and were spot treated with medication
2	5000	1/36	N/A	Females from the suspect pool were culled
3	1200	2/36	N/A	Females in suspect pools were spot treated with medication
4	2500	0/36	N/A	N/A

Evaluation of washing-disinfectant-TADD conditions required to produce a negative PCR result when monitoring the presence of PRRSV on metal surfaces

Rita Neat; Jianqiang Zhang; Hai Hoang; Derald Holtkamp
Iowa State University, Ames, Iowa

Introduction

Because contaminated livestock trailers are a significant risk for transmitting viruses between herds, various methods of washing, disinfecting, and TADD conditions have been evaluated for their effectiveness in inactivating porcine reproductive and respiratory syndrome virus (PRRSV) on contaminated surfaces. In practice, environmental samples are sometimes tested by PCR for monitoring compliance with trailer sanitation and decontamination protocols. However, information on when to expect negative PCR results after adequate trailer sanitation is lacking. The objective of this study was to evaluate whether there are conditions associated with washing-disinfectant-TADD procedures that will consistently produce a negative PCR result for the purpose of monitoring compliance with trailer sanitation and decontamination protocols for PRRSV on metal surfaces.

Materials and methods

144 diamond plate aluminum coupons were allocated to 48 different treatment groups with 3 replicates each. Coupons were contaminated with a matrix of 2 mL of phosphate buffered saline (PBS) or PRRSV-negative feces and then spiked with 2 mL of PRRSV VR-2385 at the titer of 1 × 106 TCID50/mL, followed by treatment with disinfectant. Negative controls used unspiked PBS or feces followed by sham disinfectant (PBS). Pre-treatment swabs were collected prior to disinfection and placed in 2.5 mL PBS. Disinfectants evaluated included an accelerated hydrogen peroxide (AHP) disinfectant licensed in the US (Intervention®, Virox Technologies Inc., Oakville, Ontario, Canada) at 1:16 or 1:64 dilutions, an AHP disinfectant licensed in Canada (Prevail®, Virox Technologies Inc., Oakville, Ontario, Canada) at 1:40 dilution, or a quaternary ammonium and glutaraldehyde combination disinfectant (Synergize®, Neogen Corporation, Lansing, MI, USA) at 1:256 dilution. Disinfectant was applied for 5 or 60 minutes of contact time at either 20°C or -10°C. 10 mL of liquid was then pipetted out of the coupon, modeling liquid draining out of a trailer at a trailer wash. All coupons were heated until the surface temperature of the coupon reached 71°C and then held for 10 minutes to simulate TADD under field conditions. Post-treatment swabs were collected and placed in 3.5 mL Dey-Engley Broth. Samples were submitted to Iowa State University Veterinary Diagnostic Laboratory for testing for PRRSV by a quantitative real-time RT-PCR. A multivariate linear model was used to analyze post-treatment genomic copies for significant explanatory variables. Pairwise comparisons of pre-treatment genomic copies were performed.

Results and discussion

Post-treatment swabs for all treatment groups, except negative control groups, were positive by PRRSV RT-PCR. Disinfectant and matrix (feces or PBS) had significant effects on the reduction of genomic copies from pre-treatment to post-treatment swabs. The least squares mean for reduction of genomic copies for matrix ranged from 2.0×10^8 (feces) to 3.0×10^8 (PBS). The least squares mean for reduction of genomic copies for disinfectants ranged from 1.1×10^8 (no disinfectant) to 3.7×10^8(AHP 1:64). While there was a significant reduction of genomic copies from pre-treatment to post-treatment, under the conditions evaluated in this study, consistently negative PCR results after treatments were not found. The results suggest that conditions where PCR results would be expected to be negative are rare. Therefore, for the purpose of monitoring compliance with trailer sanitation and decontamination protocols for PRRSV, alternatives to PCR should be explored.

Litter porcine reproductive and respiratory syndrome (PRRS) status at processing and relationship to PRRS status at weaning using different pools of sampling tissues

Mikalah Smith[1], BS; John Waddell[2], DVM, MBA; Greg Hartsook[2], BS; Aaron Lower[3], DVM; Laura Greiner[3], PhD; Luis Ochoa[4], PhD

[1]Iowa State University CVM; [2]Boehringer Ingelheim Vetmedica, Duluth, Georgia; [3]Carthage Veterinary Services, Carthage, Illinois; [4]Carthage Innovative Swine Solutions, Carthage, Illinois

Introduction

Time to negative production (TTNP) of pigs at weaning is a common metric in porcine reproductive and respiratory syndrome virus (PRRSV) elimination projects. This has driven sampling at processing to understand if positive weaned pigs are the result of vertical transmission from the sow, or lateral transmission within the litter. PRRSV spreads trans-placentally or through colostrum, however the ability of maternal PRRSV antibodies to protect piglets through passive transfer is dose dependent.[1] Even when piglets were protected from viremia, tissues of these young pigs contained replicating PRRSV and continued to shed the virus, contributing to lateral transmission within the litter.[1] This information is important in light of breeding herd stability, which is defined as the absence of vertical and horizontal transmission of virus thereby producing weaned piglets not infected with PRRSV.[2] The swine industry continues to innovate diagnostic methodologies such as the use of processing fluids, to understand PRRS viremia status of suckling piglets. There is recent interest and research reporting the use of processing fluids (PF) (liquid accumulated at the bottom of the pail during castration and tail docking) as a successful diagnostic sampling technique.[3] The sampling of pigs at processing using processing fluids as an alternative diagnostic method provides an opportunity to improve PRRSV monitoring in the farrowing house in an effort to combat a disease process that despite over 25 years of experience and research remains a costly disease to the swine industry.[4] Therefore, the objectives of this study were to determine whether the PRRSV status of a litter at litter processing is correlated with PRRSV status at weaning, and to determine the correlation between testicular fluid samples and pooled tail swab samples within the litters at the time of processing.

Materials and methods

A PRRSV negative (Category IV) 5,000 head breed-to-wean herd became infected with a virulent PRRSV with a 1-7-4 RFLP pattern. Testing was performed in two rounds of 50 litters in the first round and 51 litters in the second, with 101 litters in total enrolled in the study tested between 16 and 21 weeks post-infection. At approximately 3 days of age, pigs (n = 1,009) from 101 litters were ear tagged with an individual ID number and tails were docked. A Copan Liquid Amies Elution swab (Copan Diagnostics Incorporated, Corona, CA) was used to collect an individual blood sample from the tail stump. Serology testing from blood swabs for PRRSV was used in other studies evaluating novel diagnostic methodologies, such as oral fluids, and were used as the standard representing the suspected true status of the sample.[5,6] Males were castrated and testicles from each litter were placed into a sealable plastic bag to collect a testicular fluid sample. This study evaluated testicular fluid alone to eliminate potential environmental contamination of the tail samples. At weaning (approximately 18 days of age), blood was collected from each of the piglets via jugular venipuncture. Tail swabs, testicular fluids and serum were pooled by litter and tested for PRRSV by PCR. The data was then analyzed via Kappa Agreement statistical technique.

Results and discussion

The data is presented as PCR results either positive or negative for PRRSV. The results are compared between tail swab samples at processing and serum at weaning (Table 1) as well as tail swab samples at processing compared to testicular fluid at processing (Table 2). The percentage of litters with the same PRRSV status at processing (tail swabs) and weaning (serum) was high, with contingency tables showing substantial agreement (93%; Kappa 0.806) (Table 3). Similarly, substantial agreement was observed in 90 % of the litters (Kappa 0.688) having consistent results for PRRSV status at processing (i.e., positive or negative) regardless of sampling pool used (testicular fluids or swabs) (Table 4). However, in 6% of litters, pooled tail swab samples tested positive while pooled testicular fluid samples tested negative for PRRSV (Table 5). Individual tail swabs from barrows in these litters all tested negative for PRRSV. In 4% of litters, pooled tail swab samples were negative and testicular fluid was positive (Table 6). Individual tail swabs showed three of these litters had at least one positive pig with high Ct values (> 34.0) that yielded a negative test for PRRS when pooled. Only one of these litters was suspected as a false positive of testicular fluid (no positive individual tail swabs). Pooling of tail swabs by litter resulted in a false negative result in 3% of litters (i.e. there was at least 1 positive individual tail swab in each of these litters). While pooling increases the risk of a false negative result, these results suggest that risk may be higher using testicular fluids as only barrows are being tested. The 1% of litters considered a false positive (i.e. testicle fluid positive litters with no individual positive tail swabs) can be attributed to possible environmental contamination or limitations of the sampling method. The majority of litters maintained their status through weaning and litters born positive via vertical transmission maintain that status. Under the conditions of this study between 16-21 weeks post-outbreak, this data suggests samples obtained at processing are indicative of a litters' PRRSV status at weaning and pools of both testicular fluids and tail swabs can be utilized to understand farm PPRSV status at processing. Potential applications of these findings include surveillance diagnostics in a presumed negative/stable herd and directing management decisions such as establishing internal biosecurity protocols.

References

1. Osorio FA et al. Passive transfer of virus-specific antibodies convers protection against reproductive failure induced by a virulent strain of porcine reproductive and respiratory syndrome virus and establishes sterilizing immunity. *Virology.* 2002; 302: 9-20.

2. Cano JP et al. PRRSV vertical transmission dynamics in an endemically infected sow-herd. *American Association of Swine Veterinarians.* 2008; 105-108.

3. Vilalta C et al. Use of processing fluids for PRRSV diagnostics. Dr. Bob Morrison's Swine Health Monitoring Project. 9/29/2017.

4. Holtkamp D.J et al. Assessment of the economic impact of porcine reproductive and respiratory syndrome virus on United States pork producers. *J Swine Health Prod.* 2013;21(2):72-84.

5. Reicks D. L. et al. Sampling of adult boars during early infection with porcine reproductive and respiratory syndrome virus for testing by polymerase chain reaction using a new blood collection technique. *J Swine Health Prod. 2006. 258-264.*

6. Kittawornrat A. et al. Porcine reproductive and respiratory syndrome virus (PRRSV) in serum and oral fluid samples from individual boars: will oral fluid replace serum for PRRSV surveillance? *Virus Research.* 2010. 154: 170-176.

Table 1: PCR results comparing PRRSV status at processing (tail swabs) and at weaning (serum) pooled by litter.

		Tail swabs	
		+	-
Serum	+	20	6
	-	1	73

Table 2: PCR results of PRRSV at processing comparing tail swabs and testicular fluid sampling pooled by litter.

		Tail swabs	
		+	-
Testicle fluid	+	15	4
	-	6	75

Table 3: Summary and kappa analysis of PCR results comparing PRRSV at processing (tail swabs) and at weaning (serum) pooled by litter.

Result	Processing vs. weaning
Agreement (maintained status)	93% (Kappa 0.806)
Positive- negative	1%
Negative- positive	6%

Table 4: Summary and kappa analysis of PCR results comparing PRRSV status at processing comparing tail swabs and testicular fluid sampling pooled by litter.

Result	Tail swab vs. testicle fluid
Agreement (maintained status)	90% (Kappa 0.688)
Positive- negative	6%*
Negative- positive	4%**

* Individual PCR results for barrows in these litters were all negative for PRRSV.

** PCR results for individuals in 3% of litters in this category contained at least 1 positive swab with high CT values (> 34.0) that yielded a negative test for PRRSV when pooled.

Table 5: PCR results for PRRSV of pooled tail swabs compared to pooled testicle fluid.

Tail swabs pooled	Testicle fluids pooled	Pooled serum at weaning	Percent
Positive	Negative	Negative	1% (n = 1)
Positive	Negative	Positive	5% (n = 5)

Table 6: PCR results for PRRSV of pooled tail swabs compared to pooled testicle fluid.

Tail swabs pooled	Testicle fluids pooled	Pooled serum at weaning	Percent
Negative	Positive	Positive	2% (n = 2)
Negative	Positive	Negative	2% (n = 2)

Vaccination shedding of a porcine reproductive and respiratory syndrome (PRRS) modified live virus in nursery pigs

E. Koep[1], BS; B. Leuwerke[2], DVM, MS; S. Holst[2], DVM
[1]Iowa State University, Ames, Iowa; [2]Swine Vet Center, St. Peter, Minnesota

Introduction

Production losses from porcine reproductive and respiratory syndrome (PRRS) costs the US swine industry $664 million annually, making PRRS the most significant swine disease in the US.[1] Despite sound biosecurity practices, PRRSV has been shown to travel up to 9.1 km through the air, making protection from PRRS aerosol transmission a challenge.[2] Vaccination is employed to populations in high-risk areas to mitigate the effects of wild-type virus infection.[3-5] However, PRRS MLV vaccine virus in the breeding herd can be transmitted to naïve piglets pre-weaning or transmitted to non-vaccinated groups in finishing. Due to virus shedding and transmission, some producers have employed partial population vaccination to achieve economic savings. To decrease the potential of infecting negative animals with vaccine virus, more knowledge is needed on the shedding pattern of PRRS MLV vaccines. This study aimed to characterize the shedding dynamics of Prime Pac™ PRRS+ by evaluating total viral shedding, time to seroconversion, and viral clearance after vaccination.

Materials and methods

A 7,200 head (3 rooms of 2,400 each) wean-to-finish site receiving 21-day old weaned pigs from an AASV category 4 PRRSV negative sow farm[6] was chosen for this study. Prior to pigs arriving, Swiffer™ wipes were used on a cross-section of water cups and feed pans and submitted for PRRS PCR to ensure no PRRSV contamination of the housing facilities. Upon arrival (day 0), 72 pigs were randomly selected, ear tagged, and comingled within the room. Prior to vaccination, serum was obtained from each of the 72 pigs and submitted for PRRS PCR and ELISA. Thirty-six tagged pigs were then randomly allocated to remain unvaccinated (NVx group), while the rest of the population was vaccinated with Prime Pac™ PRRS+ vaccine according to label directions (Vx group). Serum was then obtained from each tagged pig and pen oral fluids were collected on 7, 14, 21, 35, 49, 63, 76, and 91 days post vaccination (DPV). Air samples were collected weekly until 35 DPV using a liquid cyclonic collector and minimum essential media. Serum was tested individually by PRRS quantitative PCR (qPCR) and ELISA. Oral fluids were tested by PRRS PCR and ELISA. Air samples were tested by PRRS qPCR. Genetic sequencing was performed on 7, 14, 21, and 63 DPV to ensure absences of a wild-type PRRSV.

Results

All samples were PCR and ELISA negative on day 0. Figure 1 displays the qPCR and ELISA results for the study duration. At 7 DPV, 86.1% of Vx pigs and 8.3% of NVx pigs were PCR positive. Figure 2 exhibits the prevalence of the virus between the two groups. Assuming a prevalence threshold of 80% PCR positives, the Vx group exceeded 80% by 7 DPV, while the NVx group reached 80% at 63 DPV. The incidence rate of the NVx group peaked at 25.7% on 63 DPV. Figure 3 displays the average ORF 6 RNA copies/mL from the positive qPCR results and shows the number of qPCR positive pigs used for calculating the average virus quantity. The ORF 6 RNA copies/mL averaged 5.74×10^6 higher in the NVx group from 14 to 76 DPV. At 91 DPV, the Vx group had a higher average ORF 6 RNA copies/mL, but the NVx group had 3.5 times more qPCR positive pigs. Vx pigs became ELISA positive at 14 DPV and reached > 90% by 35 DPV. Similar ELISA positive rates in the NVx group were 7.9 weeks behind the Vx group. All oral fluids were PCR positive from 7 to 63 DPV, and ELISA positive starting at 14 DPV. One air sample was positive on 21 DPV with 2.21×10^5 ORF 6 RNA copies/mL. All of the sequenced samples were > 99.3% similar to Prime Pac™ PRRS+ vaccine, indicating that there was likely no wild-type PRRSV introduced.

Discussion

NVx pigs became PCR and ELISA positive at a lower incidence rate compared to the Vx group. This study demonstrates the value of whole site vaccination with an MLV PRRSV vaccine compared to an attempt to reduce costs by partial population vaccination. The Vx group rapidly became 97% PCR positive by 21 DPV; while the NVx group had an average incidence rate of 13% throughout the study period. The NVx group reached a similar prevalence of ELISA positive animals almost 8 weeks after the Vx group. Throughout the study, the NVx group consistently had 1 log more ORF 6 copies/mL than the Vx group. This signifies a higher viral load and potential shedding level over time by partial population PRRSV vaccination compared to vaccinating the entire population with PRRSV.

Figure 1: qPCR and ELISA results over time.

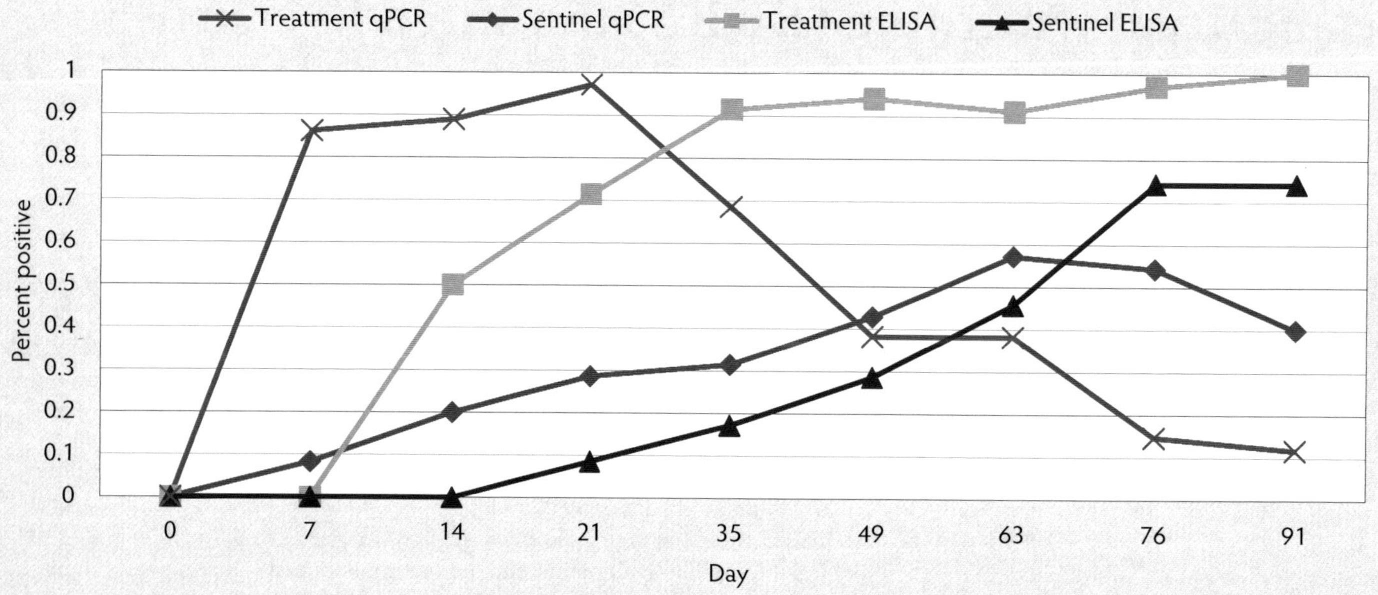

References

1. Holtkamp DJ, Kliebenstein JB, Neumann EJ, et al. Assessment of the economic impact of porcine reproductive and respiratory syndrome virus on United States pork producers. *J Swine Health Prod.* 2013;21(2):72-84.

2. Otake et al. Long-distance airborne transport of infectious PRRSV and *Mycoplasma hyopneumoniae* from a swine population infected with multiple viral variants. Veterinary Microbiology. 2010:145:198-208.

3. Lager, KM, Mengeling WL, Wesley, RD. Evidence for local spread of porcine reproductive and respiratory syndrome virus. *J Swine Health Prod.* 2002;10(4):167-170.

4. Yeske et al. Evaluation of airborne shedding of modified-live porcine reproductive and respiratory syndrome virus from growing pigs. AASV. 2012.

5. Dee et al. Evidence of long distance airborne transport of porcine reproductive and respiratory syndrome virus and *Mycoplasma hyopneumoniae. Veterinary Research.* 2009:40(4):39.

6. Holtkamp DJ, Polson DD, Torremorell M, et al. Terminology for classifying swine herds by porcine reproductive and respiratory syndrome virus status. *J Swine Health Prod.* 2011;19(1):44–56.

Figure 2: PRRSV qPCR prevalence between the study groups throughout the trial.

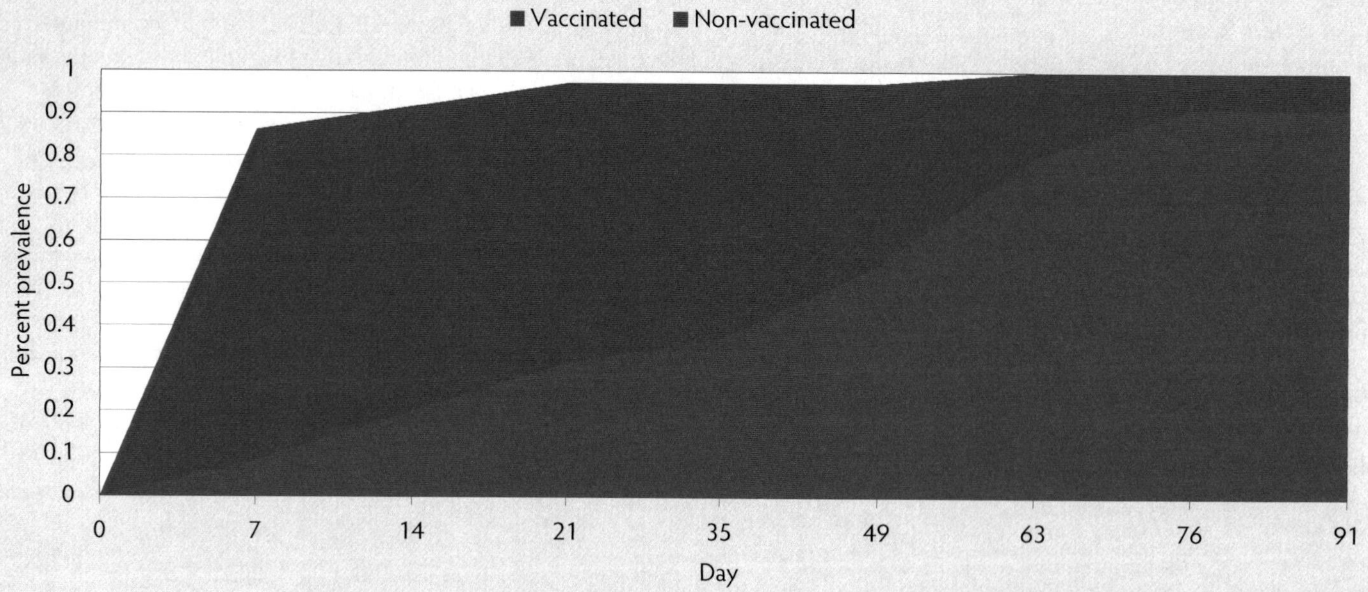

Figure 3: The average quantity of virus as shown by the line graph and left axis; the number of qPCR positive pigs used to calculate the virus quantity is shown by the bar graph and right axis.

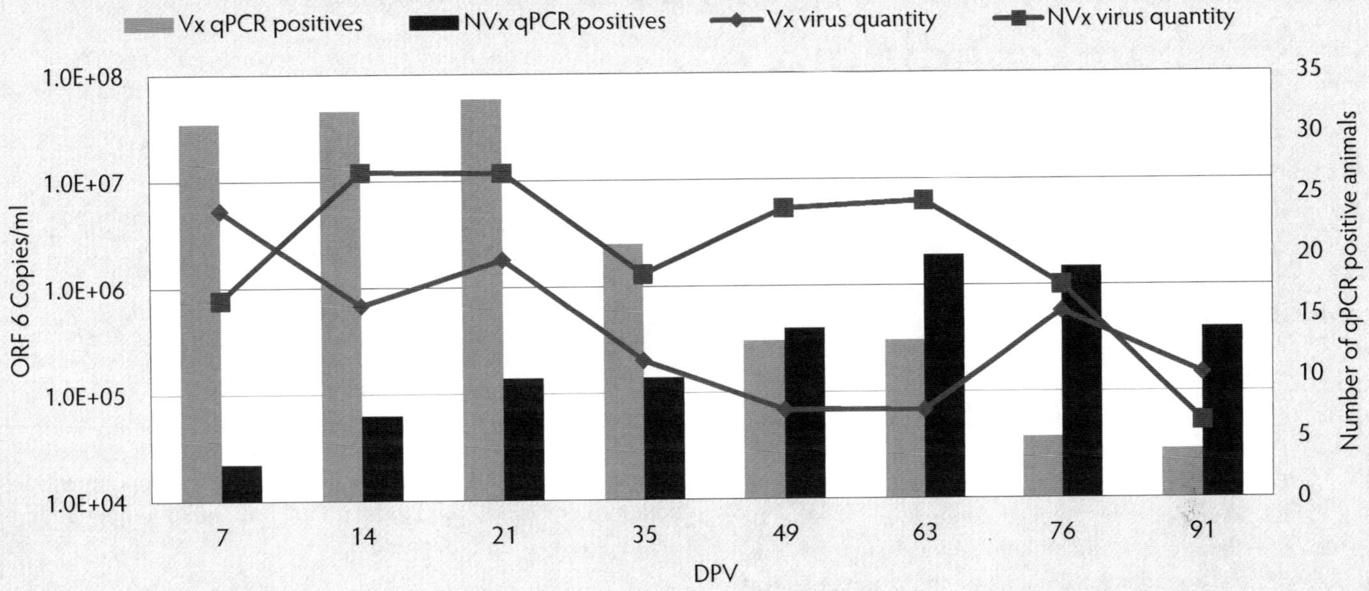

Evaluation of Senecavirus A transmission in laboratory and wild type mice strains under experimental conditions

E. R. Houston[1]; G. Temeeyasen[2]; L. G. Gimenez-Lirola[1]; R.L. Magtoto[1]; Rolf Rauh[3]; P. E. Piñeyro[1]
[1]Department of Veterinary Diagnostic and Production Animal Medicine, College of Veterinary Medicine, Iowa State University, Ames, Iowa; [2]Faculty of Veterinary Science, Department of Veterinary Microbiology, Chulalongkorn University, Bangkok, Thailand; [3]Tetracore, Inc., Rockville, Maryland

Introduction and objectives

Senecavirus A (SVA) is a non-enveloped, single-stranded RNA virus of the genus *Senecavirus* in the family *Picornaviridae*.[1] The virus possesses four structural proteins (VP1-4) composing its icosahedral capsid, and seven non-structural proteins.[2] Seneca Valley Virus (SVV) was discovered as a contaminant in cultured cells in 2002 and demonstrated oncolytic properties. SVA has been documented retrospectively in serum from United States' swine back to 1988.[3] SVV was renamed SVA, and contemporary strains have recently been identified as the causative agent of outbreaks of porcine idiopathic vesicular disease (IVD) and epidemic transient neonatal losses (ETNL) in swine production systems. Increased incidence was observed in the United States beginning in 2014.[4-6] The disease has also been reported in Canada,[7] Brazil,[8] China,[9] Colombia,[10] and Thailand.[11] Outbreaks of SVA IVD are characterized by abrupt onset of vesicles that progress to ulcerative dermatitis on the nostril and coronary bands in a high percentage of sows or finisher pigs. Animal present with transient fever, off feed, and most importantly lameness. However, SVA can be difficult to diagnose because both clinical signs and lesions are transient. Outbreaks in 2015 and 2016 showed recurrent seasonal patterns, with increased cases load in mid-summer tapering through the fall. Lesions associated with SVA are clinically indistinguishable from those observed in foot-and-mouth disease (FMD), vesicular stomatitis (VS), swine vesicular disease (SVD), and vesicular exanthema of swine (VES). Clinically unaffected animals at sites experiencing outbreaks showed SVA-specific antibody response, suggesting that the virus may circulating subclinically.[12]

Swine are the only natural hosts of SVA and no information is available regarding the mechanism of SVA transmission. However, mice have been used as model to explore the oncolytic properties of SVA. In addition, field studies have demonstrated the presence of SVA neutralizing antibodies in wild type mice.[3] More recent studies have detected SVA nucleic acid in mouse samples collected from farms during an outbreak of SVA IVD.[13]

Due to the increased incidence of cases of SVA IVD, we believe it is important to understand the mechanism of transmission, prevention, and spread of the virus. Therefore, the objective of this study was to evaluate the role of mice as a potential vector for SVA under experimental conditions.

Materials and methods

In all experiments, mice were monitored every 6 hours for the first 72 hours post-infection and daily thereafter. Body weight, clinical signs, and feces were collected daily from individual mice to evaluate virus shedding. Serum samples were collected to determine viremia and seroconversion, and a full set of tissues were collected at euthanasia to evaluate the presence of lesions and detection of viral RNA by PCR.

Determining a susceptible mouse strain

Six mice were obtained for each mouse strain: BALB/C, C57BL/6, SJL/J, Swiss, and SCID. All mice arrived at eight weeks of age with the exclusion of SCID mice at four week of age, and they were allowed a one week acclimatization period before the experimental trial. Mice were housed by strain and sex in groups of three, and they were fed on standard mouse feed and water *ad libitum*. Wood shaving bedding and standard enrichment was provided. Each mouse was inoculated subcutaneously in the rear left foot pad (RLFP) with 100 µL of SVA at 10^9 $TCID_{50}$/mL in RPMI1640 media. One animal from each strain was designated the sentinel animal, which was mock infected with 100 µL of media and housed with infected animals for the duration of the study (Table 1). Direct contact transmission was evaluated with this sentinel mouse. Clinical signs of systemic infection including dyspnea, cutaneous rush, and lameness were recorded daily. At 14 dpi, mice were anesthetized for blood collection by cardiac puncture followed by euthanasia using CO_2. Then, a necropsy was performed and a full set of tissues collected (heart, lungs, spleen, liver, intestine, kidney, and brain). All tissues were stored at -80°C for PCR evaluation and fixed in 10% neutral buffered formalin for histopathological examination.

In a second experiment, we evaluated the effect and potential transmission of SVA in wild type mice of the species *Peromyscus*. Three cages with two infected and one sentinel mice each (2:1 infected: sentinel ratio) per treatment, including SC and IP inoculation and negative control groups were inoculated with 100 µL of SVA at 10^9 $TCID_{50}$/mL or RPMI1640 media. Clinical signs, body weight, and feces were collected daily for 28 days. At 28 dpi, mice were anesthetized prior to blood collection by cardiac puncture and euthanized using CO_2. A necropsy was performed and a full set of tissues collected (heart, lungs, spleen, liver, intestine, kidney, and brain) for SVA PCR detection and histopathological examination.

Table 1: Six mice were obtained for each mouse strain: BALB/C, C57BL/6, SJL/J, Swiss, and SCID. All mice arrived at eight weeks of age with the exclusion of SCID mice at four week of age, and were allowed one week of acclimatization. One animal per strain was mock infected and housed with infected animals to explore transmission

Mouse strain	Age	Infected	Mock infected
SCID	4 weeks	5	1
C57BL	8 weeks	5	1
BALB/C	8 weeks	5	1
SJL	8 weeks	5	1
SWISS	8 weeks	5	1

Evaluating infectious route

In order to determine the most efficient inoculation route 10 BALB/c mice per group were inoculated subcutaneous, intranasal, and intraperitoneal with 100 µL of 10^9 $TCID_{50}$/mL SVA. Mice were monitored as described above. Feces and weight measurement were collected daily. Blood was collected from the submandibular vein at 0, 3, 5, 7, 10, 14, 17, 21, and 28 dpi. At 28 dpi all mice were euthanized by CO2 inhalation, and an unrestricted volume of blood was collected thereafter by cardiac puncture. A necropsy was performed and tissues were collected in 10% neutral buffered formalin for histopathology and stored at -80 for PCR. The serological response to SVA was evaluated by VP1-ELISA and IFA. Serum and feces were tested by RT-PCR. This section of the experiment was run in duplicate.

Evaluating the antibody response

Ninety-six well microtiter plates were coated with 100µL of SVA recombinant VP1 (rVP1) at 0.18 µg/mL/well in a carbonate buffer. Plates were incubated 16 h at 4°C, then washed five times with phosphate buffered saline (PBS) containing 0.1% Tween 20 (PBST) and blocked with a 1% (w/v) bovine serum albumin (BSA) solution (Jackson ImmunoResearch Inc.) for 2 h at 25°C. Mouse serum samples (tested in triplicate) were diluted at 1:50 in serum diluent and then 100 µl of the mixture was transferred to an ELISA plate followed by 1 h incubation at 37°C. Plates were then washed 5 times with 350 µl of PBST, and 100 µl of 1:5000 goat anti-mouse IgG (Fc) horseradish peroxidase (HRP)-conjugated (Pierce Protein Biology) was added to each well. The plates were incubated for 1 h at 37°C. The washing cycle was repeated, then 100 µl of tetramethylbenzidine-hydrogen peroxide substrate (TMB, Surmodics) was added to each well and the plates incubated for 5 min at room temperature to visualize the reaction. Thereafter, 100 µl of stop solution (Surmodics) was added to each well and the plate was read (450 nm) using an ELISA plate reader (BioTek ELx800, Biotek ®) operated with a commercial software (GEN5TM, Biotek ®).

For the whole virus (WV) ELISA, ninety-six well high binding microtiter plates were coated with 100 µL of viral particles diluted 1:400 in PBS, incubated for 16 h at 4°C, and processed as described above for the rVP1 ELISA. Samples were diluted 1:50 and 100 µL of each diluted sample per well was loaded and incubated for 1 h at 37°C followed by five washes with PBST. Then, 100 µL of a 1:10,000 dilution of HRP conjugated goat anti-mouse IgG (Fc) was added to each well and incubated for 30 min at 37°C. The reaction was visualized, terminated, and measured as described above for the rVP1 ELISA.

For IFA evaluation, NCI-H1299 cells were propagated in 96-well plates and infected at 40-50% confluency with $1x10^3$ $TCID_{50}$/mL SVA. After 24 h, cells were fixed with 200 µL cold 80% acetone for 10 min at 4°C. Cells were washed with 200 µL PBS, then incubated with 50 µL 1:50 diluted mouse serum for 1 h at 37°C. Wells were washed three times with 200 µL PBS and incubated with 50 µL of 1:100 FITC labeled anti-mouse IgG (KPL Scientific), diluted 1:100, for 30 min at 37°C. Plates were washed three times with PBS and examined under a fluorescence microscope (SpectraMax i3x MiniMax 300).

SVA detection by PCR

For PCR on fecal specimens, two fecal pellets were added to 0.2 ml PBS and agitated manually for 3 s prior to the viral RNA extraction. Samples (PBS diluted feces, tissue homogenate, or serum) were manually extracted using the QIAamp Viral RNA Mini Kit (Qiagen) following manufacturer's recommendations. In order to control the integrity of the PCR reaction and the extraction, 6 µL of 800 copies per µL of the Internal Control (IC) RNA was added to the lysis buffer (Tetracore, Inc.). The viral RNA was eluted in 90 µL of elution buffer, and 7 µL of the extracted samples were tested with 18 µL of the real-time RT-PCR EZ-SVA reagents (Tetracore, Inc.). This real time RT-PCR was designed to target a conserved sequence within the 3D region of SVA genome. The reactions were performed on an ABI 7500 Fast instrument (Life Technologies) in standard mode using the following conditions: 1 cycle of 48°C for 15 min, 1 cycle of 95°C for 2 min and 45 cycles of 95°C for 5 s and 60°C for 40 s. For analysis of PCR data, the baseline was set to automatic. Threshold settings were set to the maximum of the positive control for SVA and IC. For internal test validation, the IC for the sample required a Ct value of less than 40.

Statistical analysis

Data were represented as mean and standard deviation of the mean. ELISA and IFA testing between mouse strains were assessed using an unpaired t-test. ELISA, IFA, and viremia results for inoculation rout experiments were evaluated using unpaired t-test with Holm-Sidak corrections for multiple comparisons. Statistical significance was set to alpha = 0.05. All analyses were performed using commercially available software GraphPad Prism® 7 (GraphPad Software, Inc.).

Results

SVA infection in experimentally infected mice is not mice strain specific

Only BALB/c mice developed mild clinical signs including a rough coat, poor body condition, and a slight and transient decrease in body weight that last for the first 3 dpi. However, no other significant clinical changes were observed on the strains used in this study. All infected mouse strains and sentinel mice showed variable levels of viral shedding in feces. Virus shedding in feces was detected in inoculated mice from 1 dpi to 10-14 dpi. Sentinel animals shed virus intermittently from 4-13 dpi (Figure 2). Viral RNA was detected in all tissues (brain, heart, lung, liver, spleen, kidney, intestines) from inoculated mice at 14 dpi, regardless mouse strain. Viral RNA was only detected in the kidney of two sentinel mice (Table 2). No histologic lesions were observed after 14 dpi. Except for SCID and sentinel mice, all infected mice developed SVA IgG antibody response detected by SVA-rVP1 and WV ELISAs, and by IFA (Figure 1).

Infection of *Peromyscus* yielded the same pattern observed in laboratory mice. No evidence of clinical signs were observed regardless the infectious route. Viral shedding was observed from 1 to 12 dpi in infected mice and intermittently in sentinel mice. Viremia was detected from 3-10 dpi in inoculated mice, but not in sentinel mice. SVA RNA was detected in all tissue evaluated in inoculated, but not in sentinel mice. No significant lesions were observed histologically. All inoculated mice showed seroconversion detected by ELISA and IFA.

SVA infectious route affects viral replication, shedding and seroconversion

No differences in clinical signs or body weight were observed regardless the infectious route. Mice inoculated SC and IP, but not IN, demonstrated viremia with a peak at 3 dpi. Viremia was short in mice infected SC (1-5 dpi) with a peak at 3 dpi, while the viremia in IP inoculated animals lasted until 10 dpi (Figure 3). Seroconversion was detected by IFA in SC and IP inoculated mice at 5 and 7 dpi respectively, and remained positive until the end of the study. (Figure 4).

Conclusion

These results show that both laboratory and wild type mice are susceptible to SVA infection under experimental conditions. Infection was achieved successful using both SC and IP technique. The virus was shed through feces and it was detected in multiple tissues. The lack of detection of SVA in tissues from sentinel mice could be a result of the low infectious dose and viral load occurring during natural infection. Seroconversion and viremia are detected in the absence of clinical signs and histological lesions. Viremia was transient and infectious route dependent. More importantly, subclinical infection and transmission to sentinel mice supports previous field observations suggesting that mice could be a reservoir or vector for SVA in swine herds.

Based on information generated in this study we can conclude that mice could act as reservoir and vector for SVA in swine farms. However, further work is warranted to study the relative risk associated with mouse infection on swine farms. Acknowledgement of the role of mice in the transmission of SVA could push changes in on-farm protocols to decrease incidence and spread of disease in confined swine population. In addition to implementation of control programs, exposure to mouse droppings should be avoided in order to prevent transmission between herds. In this study, all animals were in direct contact and transmission was demonstrated to sentinel mice in addition to viral elimination in feces. However, since SVA was detected in kidney, urine should also be considered another elimination route and perhaps another method of transmission amongst animals. Therefore, these results reinforce the need for aggressive pest control measures and eradication programs. In this scenario, with mice

Table 2: PCR was run on tissues collected at necropsy. This chart shows the number of positive samples compared to the number of samples tested (in parenthesis) for each tissue type

Mouse strain	Brain	Heart	Lung	Liver	Spleen	Kidney	Intestine
SCID	1 (5)	5 (5)	5 (5)	5 (5)	5 (5)	5 (5)	5 (5)
C57BL	4 (5)	4 (5)	4 (5)	4 (5)	4 (5)	5 (5)	5 (5)
BALB/C	4 (5)	5 (5)	5 (5)	5 (5)	5 (5)	5 (5)	5 (5)
SJL	3 (5)	5 (5)	5 (5)	5 (5)	5 (5)	5 (5)	5 (5)
SWISS	4 (5)	5 (5)	5 (5)	5 (5)	5 (5)	5 (5)	5 (5)
Sentinel	0 (5)	0 (5)	0 (5)	0 (5)	0 (5)	2 (5)	0 (5)

Figure 1: Serological response was evaluated at 14 days after subcutaneous injection with 10^6 $TCID_{50}$/mL SVA. Antibodies were measured using indirect whole virus and SVA-rVP1 IgG ELISA, as well as IFA. These figures show mean and standard deviation for five mice per strain. Negative control values are the average of one sentinel animal from each strain. No significant difference in optical density was observed between the C57BL, BALB/c, SJL, SWISS mice and sentinel animals. The SCID mice did not seroconvert.

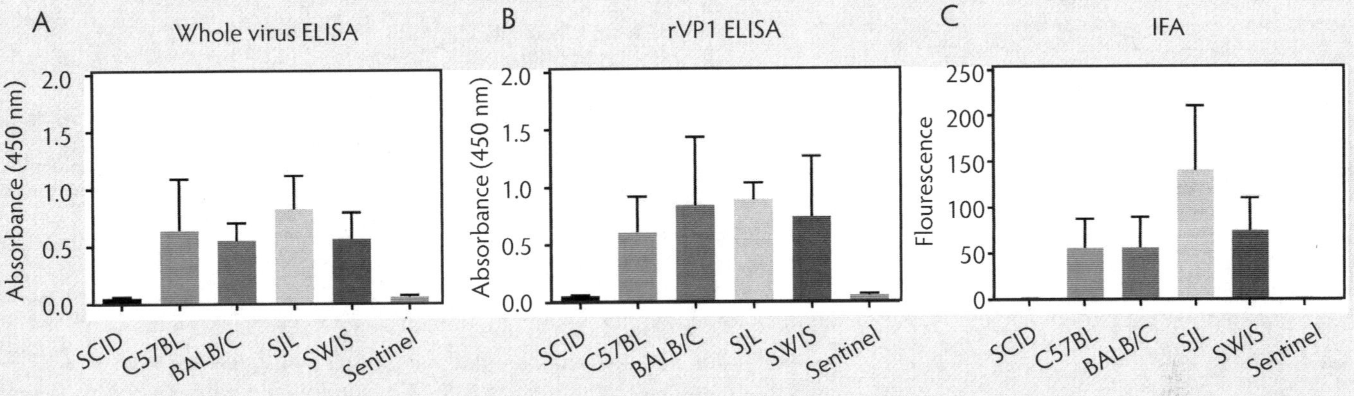

acting as SVA carrier, PCR on mouse samples from trapped animals may play a role in future surveillance for SVA at swine sites. The risk of outbreak could be evaluated onsite based on mouse trapping and environmental swabbing approaches. The establishment of surveillance programs based on mice samples may be predictive of viral introductions into commercial herds and it could help to predict breaks of SVA associated vesicular disease.

References

1. Adams MJ, Lefkowitz EJ, King AM, Bamford DH, Breitbart M, Davison AJ, Ghabrial SA, Gorbalenya AE, Knowles NJ, Krell P, Lavigne R, Prangishvili D, Sanfacon H, Siddell SG, Simmonds P, Carstens EB. 2015. Ratification vote on taxonomic proposals to the International Committee on Taxonomy of Viruses (2015). Arch Virol 160:1837-1850.

2. Hales LM, Knowles NJ, Reddy PS, Xu L, Hay C, Hallenbeck PL. 2008. Complete genome sequence analysis of Seneca Valley virus-001, a novel oncolytic picornavirus. J Gen Virol 89:1265–1275.

3. Knowles NJ, Hales LM, Jones BH, Landgraf JG, House JA, Skele KL, Burroughs KD, Hallenbeck PL. Epidemiology of Seneca Valley virus: identification and characterization of isolates from pigs in the United States, abstr G2, p. In (ed).

Figure 2: Viral shedding was evaluated by PCR of feces, collected daily. Infected mice shed viral particles beginning 1 dpi and continuing to 10-14 dpi. Viral particles were detected intermittently in the feces of sentinel animals.

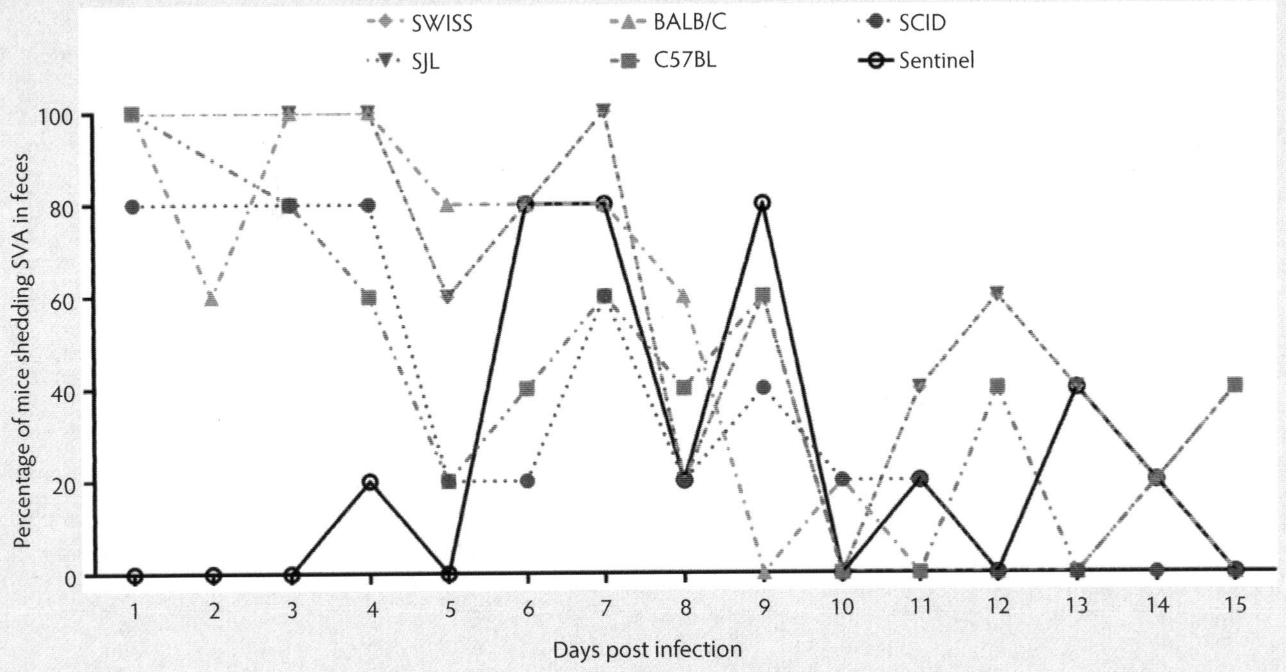

Figure 3: Viremia was detected in serum of animals inoculated SC or IP, peaking at 3 dpi. Viremia persisted for 7-10 dpi. Control mice and those inoculated IN remained negative throughout the study.

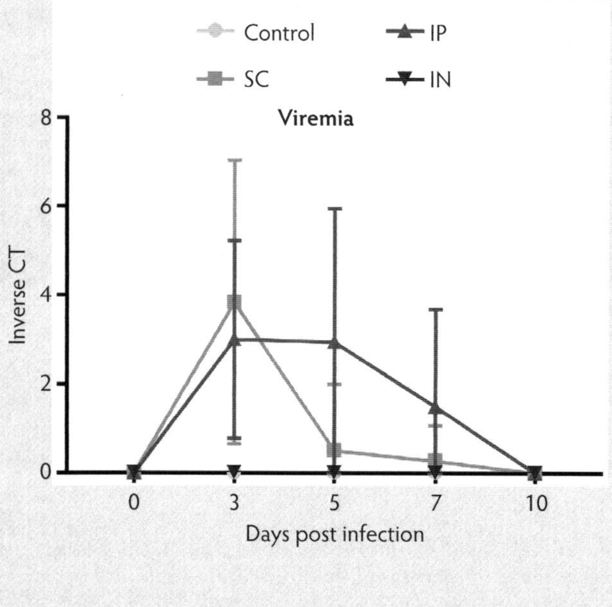

Figure 4: SVA Inoculation by SC or IP route resulted in seroconversion detected by IFA from 7 (SC) and 10 (IP) dpi to the end of the study.

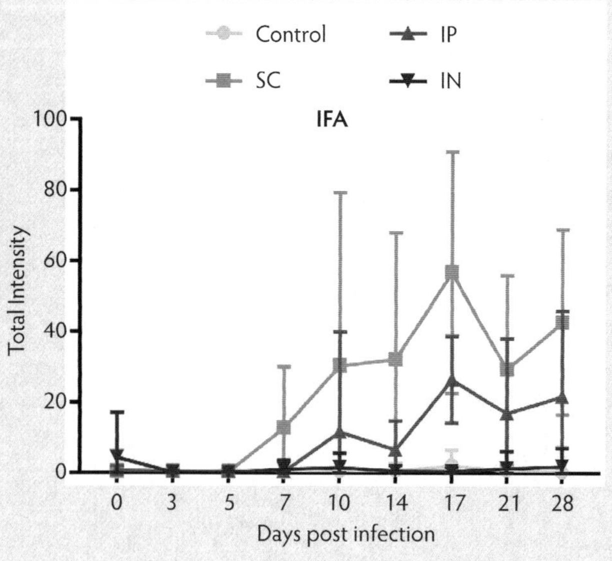

4. Baker KL, Mowrer C, Canon A, Linhares DCL, Rademacher C, Karriker LA, Holtkamp DJ. 2016. Systematic Epidemiological Investigations of Cases of Senecavirus A in US Swine Breeding Herds. Transboundary and Emerging Diseases 64(1) 11-18.

5. Zhang J, Piñeyro P, Chen Q, Zheng Y, Li G, Rademacher C, Derscheid R, Guo B, Yoon KJ, Madson D, Gauger P, Schwartz K, Harmon K, Linhares D, Main R. 2015. Full-Length Genome Sequences of Senecavirus A from Recent Idiopathic Vesicular Disease Outbreaks in U.S. Swine. Genome Announc 3(6).

6. Guo B, Piñeyro P, Rademacher C, Zheng Y, Li G, Yuan J, Hoang H, Gauger P, Madson D, Schwartz K, Canning P, Arruda B, Cooper V, Baum D, Linhares D, Main R, Yoon K-J. 2016. Novel Senecavirus A in Swine with Vesicular Disease, United States, July 2015. Emerg Infect Dis 22(7):1325-1327.

7. Pasma T, Davidson S, Shaw SL. 2008. Idiopathic vesicular disease in swine in Manitoba. The Canadian Veterinary Journal 49(1):84-85.

8. Vannucci FA, Linhares DCL, Barcellos DESN, Lam HC, Collins J, Marthaler D. 2015. Identification and Complete Genome of Seneca Valley Virus in Vesicular Fluid and Sera of Pigs Affected with Idiopathic Vesicular Disease, Brazil. Transboundary and Emerging Diseases 62(6):589-593.

9. Wu Q, Zhao X, Bai Y, Sun B, Xie Q, Ma J. 2016. The First Identification and Complete Genome of Senecavirus A Affecting Pig with Idiopathic Vesicular Disease in China. Transboundary and Emerging Diseases 64(5): 1633-1640.

10. Sun D, Vannucci F, Knutson TP, Corzo C, Marthaler DG. 2017. Emergence and whole-genome sequence of Senecavirus A in Colombia. Transboundary and Emerging Diseases 64(5): 1346-1349.

11. Saeng-chuto K, Rodtian P, Temeeyasen G, Wegner M, Nilubol D. 2017. The first detection of Senecavirus A in pigs in Thailand, 2016. Transboundary and Emerging Diseases.

12. Gimenez-Lirola LG, Rademacher C, Linhares D, Harmon K, Rotolo M, Sun Y, Baum DH, Zimmerman J, Pineyro P. 2016. Serological and Molecular Detection of Senecavirus A Associated with an Outbreak of Swine Idiopathic Vesicular Disease and Neonatal Mortality. J Clin Microbiol 54(8):2082-2089.

13. Joshi LR, Mohr KA, Clement T, Hain KS, Myers B, Yaros J, Nelson EA, Christopher-Hennings J, Gava D, Schaefer R, Caron L, Dee S, Diel DG. 2016. Detection of the Emerging Senecavirus A in Pigs, Mice and Houseflies. Journal of Clinical Microbiology 54(6): 1536-1545.

Utilizing piglet-processing fluids to detect PRRSV by PCR in a low-prevalence population

Jacob Baker; Deb Murray; Amanda Sponheim; Fred Sylvia; Dale Polson

Introduction

Recent work indicated that fluids gathered during piglet processing may serve as an effective diagnostic tool for PRRSV on a sow farm. However, no research has been published investigating how varying degrees of dilution might affect the detection of a single PRRSV positive piglet. Therefore, the objective of this study was to determine the dilution effect of tail and testicle processing fluid in extreme low-prevalence populations.

Materials and methods

Part one: Tails and testicles were collected from a known PRRSV positive farm. Swabs (FLOQSwab™ with 3 mL UTM™ medium) of the incision sites were taken at the time of tail and testicle collection to determine which samples were positive and to estimate cycle time by PRRSV PCR (Applied Biosystems). Processing fluid was collected from a known PRRSV negative farm in 10 litter increments, averaging 11 pigs per litter. Processing fluid from all litters was confirmed negative by PRRSV PCR. Processing fluid from 1 positive tail and pair of testicles (Ct 22.06) was added to PRRSV negative processing fluid collected from 10 litters. This mixture was vortexed and 4 mL removed for testing via PCR. Negative processing fluid from an additional 10 litters was then added, the mixture vortexed, and 4 mL removed for testing. This process was repeated to achieve one positive male piglet in 30, 40, and 50 litters worth of negative piglets. Immediately following dilutions, 360 PCRs (72/dilution) distributed over six plates, with each dilution represented on each plate, were run.

Part two: PRRSV PCR positive tail and testicles (Ct 22.14) were added to a Ziploc® bag containing known PRRSV PCR negative tails and testicles from ten litters. The tissues were refrigerated for 24 hours, then one mL of processing fluid was removed for testing via PRRSV PCR. This process was repeated, each time adding tails and testicles from 10 more negative litters, ultimately resulting in one positive male piglet in 50 litters worth of piglets. One PCR was run for each dilution. This process was repeated to investigate the difference in the detectability of a PRRSV positive female piglet (Ct 34.28) and a PRRSV positive male piglet (Ct 33.56) and to determine the limitations of the test when attempting to detect a piglet infected with a low amount of virus.

Results and implications

All PCRs were positive for every dilution in part one of the study. These results indicated that piglet-processing fluids may be an effective method of PRRSV detection, even in an extreme low-prevalence scenario, when the PRRSV positive piglet has a Ct value $\leq$ 22. In part two, dilutions of 10, 20, 30, and 40 litters tested positive by PRRSV PCR, when inoculated with one PRRS positive male piglet (Ct 33.56). However, the 50 litter dilution was negative for PRRSV by PCR. In part two, all dilutions inoculated with a tail from one positive female piglet (Ct 34.28) tested PRRSV PCR negative. Results from part two indicated that detection of one PRRSV positive piglet may be highly dependent on the positive piglet's Ct value. In addition, a correlation was observed between the sex of the piglet and detection. This observation is consistent with concurrent research that has indicated the detection of one PRRSV positive piglet may also be highly dependent on the sex of the piglet (C. Vilalta, University of Minnesota personal communication). Due to the high rate of detection observed with this sampling method, it is currently being used to monitor genetic multipliers and PRRSV positive farms as they approach stability.

Geographic distribution and genetic diversity of porcine circovirus type 3 from clinical samples in the US swine farms

Zhen Yang[1]; Todd P. Knutson[1], PhD; Fangzhou Chen[1], MS; Douglas Marthaler[2], PhD; Albert Rovira[1], DVM, MS, PhD

[1]College of Veterinary Medicine, University of Minnesota; [2]College of Veterinary Medicine, Kansas State University

Introduction

PCV3 was first reported in pigs in the United States in 2016 using next-generation sequencing (NGS) and later described in China, South Korea, Poland, Brazil and Italy in 2017.[1-8]
The role of PCV3 as a causative agent of disease is still unclear; however, PCV3 has been associated with cases of wasting, porcine dermatitis and nephropathy syndrome, myocarditis, systemic vasculitis, and abortion.[1,2,4] Previous studies suggested that PCV3 may be endemic in swine herds.[1,2] In this study, we investigated the geographic distribution, prevalence, and genetic diversity of PCV3 in cases submitted to the Minnesota Veterinary Diagnostic Laboratory, United States.

Materials and methods

Between February 2016 and June 2017, 729 swine samples, with variable clinical signs and specimen types were tested for PCV3, using an in-house PCV3 qPCR. Samples with a threshold (Ct) value below 37 were defined as PCV3 positive. Any case, site or farm with at least one PCV3 positive sample was considered as positive. Descriptive statistics, including clinical signs and the rates of PCV3 positive samples, cases and systems were analyzed by Excel, and the PCV3 geographic distribution were visualized by R software. Sixteen tissue homogenates were selected for Next Generation Sequencing using paired-end, 300bp reads on the Illumina MiSeq, and another fifteen samples were selected for Sanger sequencing. Alignments of the open reading frame (ORF) 1, ORF2, and whole PCV3 genome were created using global sequences.

Results

The 729 samples came from 121 cases in 108 sites of 30 systems in 13 states. The positive PCV3 rates among samples, cases, sites, systems and states levels were 26% (191/729), 52% (63/121), 51% (55/108), 70% (21/30) and 77% (10/13) respectively. By eliminating samples from one system that severely biased our sampling population, the rate of positive PCV3 among samples, cases, sites and systems became 12% (65/551), 38% (27/71), 39% (27/69) and 69% (20/29), which may be closer to the true prevalence in our diagnostic samples. Some PCV3 positive cases were co-infected with PRRSV (26% (49/191)) and PCV2 (18% (35/191)). Of 87 positive samples with veterinarian descriptions, their clinical signs were described as respiratory disease (50%), gastrointestinal (20%), lameness (14%), and abortion (11%).

One PCV3 complete genome and eight incomplete genomes were obtained from NGS whereas six PCV3 complete genomes were obtained from Sanger sequencing. The whole genome nucleotide identities between our seven PCV3 genomes and five previous U.S PCV3 strains were 98.6-99.8%. Our PCV3 genomes had higher nucleotide identities if the samples were from same farms or systems. PCV3 strains in the US and other countries shared 97.5%-99.8% of nucleotide identity and were classified into two clusters in the phylogenetic tree.

Conclusion and discussion

Our data indicates that PCV3 is widespread and exists in multiple US swine systems in different states. Co-infections with other viruses were commonly found and the role of PCV3 in these cases remains unclear. Additionally, PCV3 was also detected in clinically healthy pigs. Tissue samples with a Ct value over 30 were less likely to generate a whole genome sequence by either of our current sequencing methods. Genetic analysis of PCV3 suggested current PCV3 in U.S were highly similar. A broad non-biased surveillance study should be done to reflect the prevalence and genetic evolution of PCV3 in the US.

References

1. Palinski, R. et al. A Novel Porcine Circovirus Distantly Related to Known Circoviruses Is Associated with Porcine Dermatitis and Nephropathy Syndrome and Reproductive Failure. J. Virol. 91, e01879-16 (2017).

2. Phan, T. G. et al. Detection of a novel circovirus PCV3 in pigs with cardiac and multi-systemic inflammation. Virol. J. 13, 184 (2016).

3. Shen, H. et al. Genome characterization of a porcine circovirus type 3 in South China. Transbound. Emerg. Dis. 1–3 (2017). doi:10.1111/tbed.12639

4. Ku, X. et al. Identification and genetic characterization of porcine circovirus type 3 in China. Transbound. Emerg. Dis. 703–708 (2017). doi:10.1111/tbed.12638

5. Kwon, T., Yoo, S. J., Park, C.-K. & Lyoo, Y. S. Prevalence of novel porcine circovirus 3 in Korean pig populations. Vet. Microbiol. 207, 178–180 (2017).

6. Stadejek, T., Woźniak, A., Miłek, D. & Biernacka, K. First detection of porcine circovirus type 3 on commercial pig farms in Poland. Transbound. Emerg. Dis. 3, (2017).

7. Tochetto, C. et al. Full-Genome Sequence of Porcine Circovirus type 3 recovered from serum of sows with stillbirths in Brazil. Transbound. Emerg. Dis. (2017). doi:10.1111/tbed.12735

8. Faccini, S. et al. Detection and genetic characterization of Porcine circovirus type 3 in Italy. Transbound. Emerg. Dis. 1661–1664 (2017). doi:10.1111/tbed.12714.

Use of a hurricane fogger for *Mycoplasma hyopneumoniae* inoculation in nursery age gilts

M. Nickel[1], BS; E. Toohill[2], DVM; J. Lehman[2], DVM, MS
[1]Iowa State University, Ames, Iowa; [2]The Maschhoffs, Carlyle, Illinois

Introduction

Mycoplasma hyopneumoniae (*Mhp*) is a widespread respiratory pathogen in the swine industry. It can result in significant economic losses due to reduced growth rate, feed efficiency and increased late finishing morbidity, mortality, and increased medication costs. Replacement gilts infected with *Mhp* early in life will shed less *Mhp* to piglets in farrowing.[1] *Mhp* exposure of gilts is typically done via intra-tracheal inoculation, which is labor intensive. The objective of this study was to evaluate a *Mhp* inoculation method with similar efficacy to the intra-tracheal method using a mechanical fogger.

Materials and methods

Animals verified to be *Mhp* positive by PCR via tracheo-bronchial swab with a 14 FR, red rubber, urinary catheter were humanely euthanized and their lung tissue was collected and frozen at -80°C for a minimum of 24 hours. The tissue was then thawed and placed in Friis Medium (Teknova, Hollister, CA) (60:40 ratio), blended, strained with pantyhose, and stored in 50 mL Falcon Tubes at -80°C.[2] Just prior to use, the homogenate was thawed in a cold water bath, mixed with 1-2 L of Friis and added to Dyna-Fog® Hurricane™ Fogger holding tanks. Five foggers were used in each room. Four were placed on pen feeders, while the fifth was manually moved throughout the room. Two groups of 1,300 gilts originating from a *Mhp*-negative sow farm were utilized for this study. For each group, a subset of animals were individually tagged at weaning for serial testing. Group 1 contained a subset of 22 tagged gilts. Group 2 contained 25 tagged gilts. The groups were two weeks apart in age and were allocated at weaning (21 days of age) into a nursery facility. Each group was housed in two separate nursery rooms within the same building (4 rooms total). All tagged gilts were individually tested for *Mhp* by PCR prior to fogging via tracheo-bronchial swab with a red rubber catheter. The timing of aerosol exposure was different for each group. Group 1 was exposed at 4 weeks of age using a 16 mL Friis-homogenate dose per gilt. Group 2 was exposed at 9 weeks of age with an 8 mL Friis-homogenate dose per gilt. During exposure, room ventilation was shut down to enhance homogenate inhalation.[2] All foggers were run continuously until empty (15-30 minutes). Two technicians, each wearing P95 Maintenance-Free Dual Cartridge Respirators remained in the room to manage foggers and monitor pigs for heat stress. After the foggers were empty, the ventilation remained off for an additional 45-60 minutes, until room temperature reached 88-92°F. Tagged pigs were serially tested for *Mhp* by PCR individually via tracheo-bronchial swab with a red rubber catheter until a positive *Mhp* by PCR result was detected.

Results and discussion

Intra-tracheal inoculation on the trial farm produced *Mhp* by PCR positive gilts 4 weeks post-exposure (internal data). In this study, 38% of Group 1 gilts tested *Mhp* by PCR positive by 5 weeks post-exposure, whereas 95% of Group 2 gilts tested *Mhp* by PCR positive by 4 weeks post-exposure (Table 1). Additional work on gilt age, homogenate dose, and heat stress during fogging is warranted to determine the best protocol for successful exposure. In conclusion, this study demonstrated that hurricane fogger *Mhp* inoculation is not only possible, but was more successful when completed at an older age, required less labor, and was more easily accomplished when compared to intra-tracheal inoculation.

References

1. Pieters, M., Pijoan, C., Fano, E., Dee, S., 2008. An assessment of the duration of *Mycoplasma hyopneumoniae* infection in an experimentally infected population of pigs. *Journal of Veterinary Microbiology* 134, 261-266.

2. Toohill, E. "Achieving Day 0 in Large Swine Operations." Allen D. Leman Swine Conference, 18 September 2017, Saint Paul, Minnesota.

Table 1: Number (no.) & percent (pct.) Mhp positive by weeks post-fogging

Group	Age at exposure (weeks)	Dose per gilt (mL)	Parameter	Weeks post-fogging					
				0	4	5	7	8	11
1	4	16	No. Positive	0/22	ns*	8/21	14/20	17/21	21/21
			Pct. Positive	0%		38%[a]	70%	81%	100%
2	9	8	No. Positive	0/25	20/21	ns*	ns*	22/22	ns*
			Pct. Positive	0%	95%[b]			100%	

* ns = no sample collected.
[a,b] Percent positives at 4 or 5 weeks were significantly different by Fisher's Exact ($P = 0.0002$).

A comparison of the use of tails versus testicles in the production of processing fluids for detection of porcine reproductive and respiratory syndrome virus

M. Hood[1], BS; S. Hough[2], DVM; J. Angulo[3], DVM; G. Nemechek[3], DVM; W. Lopez[4], DVM; D. Linhares[4], DVM, PhD
[1]North Carolina State University, Raleigh, North Carolina; [2]Prestage Farms, Clinton, North Carolina;
[3]Zoetis, Parsippany, New Jersey; [4]Iowa State University, Ames, Iowa

Introduction

The current industry standard porcine reproductive and respiratory syndrome virus (PRRSV) monitoring in suckling piglets involves testing via pooled serum samples, which may lack sensitivity to detect PRRSV circulation at low prevalence (near to zero) due to the limited number of samples taken. To improve PRRSV detection in breeding herds, the swine industry needs a method of detecting virus that tests more pigs, more frequently while remaining practical, accurate, and affordable. Processing fluids could be the answer to this issue. Processing fluids are an aggregate sample of fluids composed of the serosanguinous drainage from piglet testicles and tails at the time of castration and tail docking, usually within the first week of age.[1] Because not all farms castrate pigs, the objective of this study was to compare the use of tails only versus testicles during the collection of processing fluids for use in PRRSV detection.

Materials and methods

Seven breed-to-wean farms in North Carolina were enrolled in the study. Farms were selected based on their PRRSV status with an emphasis on PRRS positive unstable farms[2] based on routine serum diagnostics. To collect the processing fluids, two buckets were prepared for processing – one to receive tails cut via clippers, the other to receive testicles. Each bucket contained a gallon plastic bag liner with a hairnet affixed to catch the tail/testicles, both attached to the bucket with a rubber band (Figure 1). The fluids from the testicles were collected in the bag and poured into a 50 ml tube. Processing fluids from tails were collected in a similar manner. The hairnet once containing the tails was placed into a bag along with 10 ml of nuclease-free water. Once the hairnet was saturated with water, the fluid was squeezed into the bag and poured into a 12 ml tube. Both tubes were frozen on site awaiting shipment. Samples were tested by qPCR for detection of PRRSV RNA followed by an analysis on cycle threshold (Ct) values in positive results. A total of 32 paired samples (N = 64) were collected over the three-week period from the seven farms.

Results and discussion

Testicular fluids had a greater frequency of PRRS qPCR positive results (53.1%) compared to tails (31.3%) ($P = 0.0431$) as shown in Table 1. The odds of a testicular fluid sample testing positive for PRRS was 2.5 times higher than that of a tail sample. Ct values from testicular fluids were lower and statistically significant when compared to Ct values from tail fluids ($P = 0.044$) (Figure 2). As an interesting observation, 85% of the positive samples were positive for Type 1 (EU) PRRSV. Because some pigs were vaccinated with an MLV PRRS vaccine at the time of processing, Type 2 (US) PRRSV positive samples were ORF-5 sequenced. One farm testing positive for US PRRS was found to contain vaccine virus. The method of collecting processing fluid samples from tails was less effective for PRRSV detection than collection from testicles. This may be due to the necessity of adding nuclease free water and subsequent virus dilution. However, positive samples were identified from the tail samples despite having decreased sensitivity compared to testicle samples.

References

1. López WA, Angulo J, Zimmerman JJ, Linhares D. PRRS monitoring in breeding herds using processing fluids. *J Swine Heal Prod.* 2017; In press.

2. Holtkamp et al. Terminology for classifying swine herds by PRRSV status. *Journal of Swine Health and Production.* 19:1, 2011.

Figure 1: Processing fluids sample collection.

Table 1: Two by two table from positive and negative PRRS qPCR results

	PRRS qPCR results (%)		
Group	**Positive**	**Negative**	**Total**
Testicles	17 (53.1%)	15 (46.9%)	32
Tails	10 (31.3%)	22 (68.8%)	32
Total	27	37	64

Figure 2: Box-and-whisker plot for PRRS qPCR Ct values.

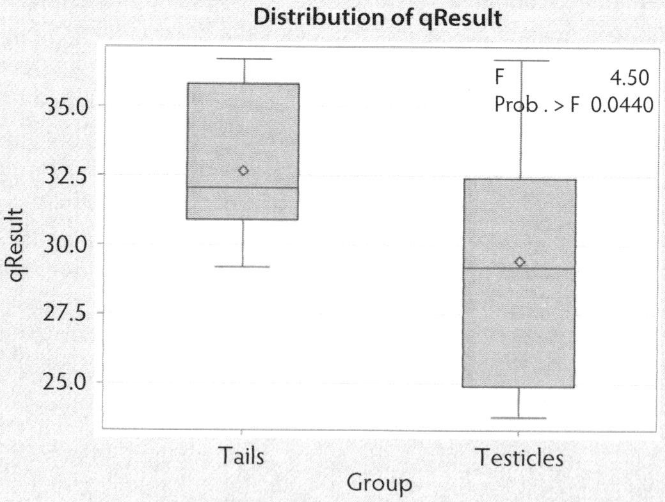

Evaluating the effects of an isotonic protein solution in nursery pigs

Stephanie Betbeze[1], BS; Ava Firth[2], BS, DVM, MVS, MANZCVS, DACVECC, DECVECC; Chris Sievers[3], DVM, MS; Henry Johnson[3], DVM
[1]Lincoln Memorial University-College of Veterinary Medicine, Harrogate, Tennessee;
[2]Tonisity International, Dublin, Ireland; [3]Swine Vet Center, St. Peter, Minnesota

Introduction

Within the first four weeks of life, a pig undergoes rapid gastrointestinal development that is critical for digestion, absorption, growth, and immunity.[1] Therefore, the stress of weaning can have lifelong effects on porcine growth and performance.[2] Enhanced feeding management practices around the time of weaning can be beneficial to ease nutritional stressors and negative impacts during this period.[2] The objective of this study was to investigate the effect of supplementation with an isotonic protein solution in the peri-weaning period, using growth parameters, feed efficiency, mortality, and antibiotic usage in the 6-week post-weaning period as the major outcomes.

Materials and methods

Pre-wean phase

Within a 5,000-head sow farm with a stable health status, 178 sows and gilts were randomly allocated and blocked by parity to control or treatment groups. Within 24 hours of farrowing, piglets within each litter were tagged and individually weighed. On days 2 and 3 pre-weaning, piglets in treatment litters were offered 500 mL/day of a 3% isotonic protein solution in a metal creep feeder. The day before weaning at ~20 days of age, treatment litters were offered 2 lbs of a gruel mixture made with 3% isotonic protein solution and dry, bagged creep feed. Every treatment litter was monitored for consumption by checking each metal creep feeder and recording any leftovers. Control litters did not receive any supplementation.

Post-wean phase

Upon arrival to nursery, weaned pigs were placed according to their treatment group to ~50 pigs/pen blocking by weight, gender, and location; totaling 949 control pigs and 936 treatment pigs within 38 pens. At the time of placement, pigs with critical health status that were recumbent, unable to walk, or chronically lame were separated to designated fallback pens and excluded from data collection and statistical analysis. Each treatment pen was given 2 plastic gruel pans and offered 0.25 lbs/pig of gruel mixture twice daily for the next 3 days. Gruel was prepared using 3% isotonic protein solution on the day of weaning, 1.5% solution the second day, and 0.75% solution the third day. The gruel mixture was made using the ratio of 1 gallon of solution to 6 lbs dry bin feed to create an 'oatmeal' consistency. Gruel pans within each treatment pen were monitored for consumption. Control pens were not offered a gruel pan nor gruel. All pens were provided dry bin feed and water according to standard practices and delivered using an automated FeedLogic system. Feed consumption per pen was recorded by the Feedlogic system. Individual pig weights were taken on the day of weaning and again at 6-weeks post-weaning. Pen weights were taken at 7, 12, 21, 34 and 42 days after weaning. After placement, pens were evaluated for scours and given a pen scour score twice weekly. Routine monitoring and evaluation for any evidence of disease or mortality was performed on a daily basis. Antimicrobial treatments (medications) were administered and recorded by barn staff according to standard operating protocols following veterinary direction and were not controlled by study protocol. The reasons for administration of antibiotics to individual pigs were judged and recorded by trained barn staff using a standardized list of causes for morbidity. A chi-square test was used to analyze the association of medications with frequency of antimicrobial treatments. Individual weights at 42 days post-wean were analyzed using a GLM using treatment, gender and wean weight as covariates. Pen weights, average daily gain and feed conversion ratio were analyzed using the pen as the experimental unit. All statistical analysis was performed using R software.†

Results

Consumption during pre-wean phase

88% of the litters consumed all of the 3% isotonic protein solution on days 2 and 3 pre-weaning. 30% of the litters consumed all of their gruel on the day before weaning; other litters consumed partial amounts.

Impact on antibiotic usage

The treatment group had decreased morbidity as evidenced by a 33% reduction ($P < 0.0001$) in individual antimicrobial

† R Core Team (2015). R: A language and environment for statistical computing. R Foundation for Statistical Computing, Vienna, Austria. ISBN 3-900051-07-0, URL `http://www.R-project.org`

treatments needed (Table 1). Significant differences were seen among treatment and control groups in number of pigs treated for being 'gaunt' ($P < 0.0001$), in addition to a tendency ($P = 0.075$) for the treatment group to have fewer pigs treated for respiratory signs (Table 1).

Impact on weight gain

The treatment group at 6-weeks post-weaning had significantly higher individual weights ($P < 0.001$) and higher average daily gain ($P < 0.001$) compared to the control group (Table 2). Data analysis of pen weights, pen average daily gain and pen feed conversion is pending at the time of publication.

Discussion

Supplemented pigs within treatment group exhibited a significant reduction in morbidity and improved growth. The effects of this protocol may have been due to a variety of factors, including improved hydration or the provision of microenteral nutrients, or the substantial increased feed intake immediately post-wean. This study showed that using an isotonic protein solution in a combination of liquid and gruel in the peri-weaning phase presents a non-antibiotic option to help pigs transition through weaning, which may allow for both a reduction in overall antibiotic usage as well as decreases in antibiotic costs. Further production parameters of gain, feed intake, and feed efficiency are pending analysis and may show further benefit.

References

1. Pluske, J. R. 2016. Invited review: Aspects of gastrointestinal tract growth and maturation in the pre- and postweaning period of pigs. J. Anim. Sci. 94(Suppl3):399-411. doi:10.2527/jas.2015-9767

2. Kick AR, Tompkins MB, Flowers WL, Whisnant CS, Almond GW; Effects of stress associated with weaning on the adaptive immune system in pigs; J Anim Sci. 2012 Feb;90(2):649-56. Epub 2011 Sep 16.

Table 1: Effect of treatment group on medications administered

	Control	Treatment	*P* value
Total number of pigs medicated (all causes)	239 (25.2%)	163 (17.4%)	< 0.0001
Total number of individual medications given	348 (36.7%)	228 (24.4%)	< 0.0001
Number of pigs medicated for reason 'gaunt'	164 (17.26%)	110 (11.75%)	< 0.0001
Number of pigs medicated for reason 'respiratory'	24 (2.53%)	13 (1.39%)	0.075

Table 2: Effect of treatment group on individual pig weights

	Control	Treatment	*P* value
Average wean weight (lb)	12.1	11.7	0.1537
Average 6-week post-wean weight (lb)	42.6	44.0	< 0.001
Average daily gain (lb/d)	0.769	0.799	< 0.001

Effect of increasing GleptoForte dosage in newborn pigs on sow and litter performance

H. Williams[1], MS; J. DeRouchey[1], MS, PhD; J. Woodworth[1], MS, PhD; M. Tokach[1], MS, PhD; S. S. Dritz[2], MS, PhD, DVM; R. Goodband[1], MS, PhD; A. Holtcamp[3], DVM
[1]Department of Animal Science, Kansas State University, Manhattan, Kansas; [2]Department of Diagnostic Medicine/Pathobiology, College of Veterinary Medicine, Kansas State University, Manhattan, Kansas; [3]Ceva Animal Health, LLC, Lenexa, Kansas

Introduction

Newborn piglets are more susceptible to iron deficiency due to inadequate iron stores at birth and rapid growth rate before weaning. This can cause decreased number of circulating red blood cells, lethargy, and mortality.[1] The importance of iron supplementation in newborn piglets within 3 days of birth has been well established.[2,3] The injection of 200 mg of iron is commonplace in the swine industry at time of piglet processing. This iron injection results in improved growth rate and iron status of piglets, but there is some concern over the level provided with one injection as opposed to giving a booster before weaning.[3] Gelptoforte (Ceva Animal Health, LLC., Lenexa, KS) is an injectable iron that contains gleptoferron and is utilized to prevent anemia in newborn piglets. Little data is available that confirms the appropriate level of iron injection needed with modern genotypes. Therefore, the purpose of this study was to evaluate the effects of increasing dosage of Gleptoforte in newborn pigs on sow and litter performance.

Materials and methods

A total of 336 newborn pigs (DNA 241 × 600, initially 3.83 ± 0.114 lb BW) were used in a 21-d farrowing study evaluating the effects of increasing Gleptoforte (Ceva Animal Health, LLC., Lenexa, KS) dosage on suckling pig performance and blood parameters. A total of 28 litters were utilized with number of pigs per sow equalized on each day of farrowing. On the day of processing (approximately d 3 after birth), all piglets were weighed and six barrows and six gilts per litter were allotted to treatment in a completely randomized design for a total of 56 piglets per treatment. Treatments consisted of a negative control receiving no iron injection and increasing levels of iron from Gleptoforte to achieve either 50, 100, 150, 200, or 200 mg plus a 100 mg booster at d 11 of farrowing. Each 1 ml of Gleptoforte contained 200 mg of iron, thus injection dosage was 0, 0.25, 0.50, 0.75, 1.0, or 1.0 plus the 0.50 ml booster for each treatment, respectively. Piglets were weighed at processing, d 11, and weaning to calculate ADG during farrowing. One barrow per treatment per litter was utilized for blood collection via jugular venipuncture on d 3 (day of processing), d 11, and weaning (d 21). Blood criteria measured included: Hemoglobin (Hgb), Hematocrit (Hct), Serum Fe, and Total Iron Binding Capacity (TIBC). The lactation feed contained 110 mg/kg added iron from ferrous sulfate.

Results and discussion

From d 3 to 11, ADG of piglets increased (quadratic; $P = 0.002$) with increasing dosage of Gelptoforte. (Table 1) Day 11 ending BW of piglets increased (quadratic; $P = 0.018$) with increasing dosage of Gleptoforte. From d 11 to 21, ADG of piglets increased (quadratic; $P = 0.001$) with increasing dosage of Gelptoforte. Overall, ADG and d 21 ending BW improved (quadratic; $P = 0.001$) with increasing dosage of Gleptoforte. However, there was no evidence of difference in performance in all three periods between the 200 mg and 200 mg + 100 mg injection of Gleptoforte which suggests the booster injection of 100 mg iron at d 11 did not yield additional benefits compared to the pigs that received only the 200 mg at d 3. There was no evidence of difference observed for any hematological criteria measured on d 3 prior to the iron injection. (Table 2)For Hgb, a treatment × day interaction (Table 2) ($P = 0.001$) was observed with Hgb increasing (quadratic; $P = 0.001$) on d 11 and 21 with the 0 mg treatment having the lowest and the 200 mg treatment having the highest Hgb values, respectively. On d 21, a significant difference ($P = 0.011$) was observed between the 200 mg and 200 mg + 100 mg treatments with the 200 mg + 100 mg treatment having a greater Hgb value. For Hct, a treatment × day interaction ($P = 0.001$) was observed with Hct increasing (quadratic; $P = 0.001$) on d 11 and 21 with the 0 mg treatment having the lowest and the 200 mg treatment having the highest Hct values, respectively. On d 21, a significant difference ($P = 0.046$) was observed between the 200 mg and 200 mg + 100 mg treatments with the 200 mg + 100 mg treatment having a greater Hct value. For Serum Fe, a treatment × day interaction ($P = 0.001$) was observed in which serum Fe increased (linear; $P = 0.001$) on d 11 and increased (quadratic; $P = 0.001$) on 21 with the 0 mg treatment having the lowest and the 200 mg treatment having the highest Serum Fe values, respectively. On d 21, a significant difference ($P = 0.019$) was observed between the 200 mg and 200 mg + 100 mg treatments with the 200 mg + 100 mg treatment having a greater Serum Fe value. For TIBC, a treatment × day interaction ($P = 0.001$) was observed. TIBC increased (quadratic; $P = 0.001$) on d 11 and increased (linear; $P = 0.001$) on 21 with the 0 mg treatment having the lowest and the 200 mg treatment having the highest Serum Fe values, respectively. There was no evidence for difference of TIBC between the 200 mg and 200 mg + 100 mg treatments on any of the days tested. In summary, a

Table 1: Effects of Gleptoforte dosage on suckling pig performance[1]

	Dosage, mg/ml[2]							Probability, $P <$		
	0	50	100	150	200	200 + 100[3]	SEM	Linear[4]	Quadratic[5]	200 vs. 200 + 100[6]
BW, lb										
d 3[7]	3.8	3.8	3.8	3.9	3.8	3.9	0.114	0.793	0.943	0.556
d 11[8]	7.2	7.9	7.8	7.9	7.8	7.7	0.214	0.012	0.018	0.702
d 21	10.4	12.5	12.9	12.7	12.7	12.6	0.322	0.001	0.001	0.800
ADG, lb										
d 0 to 11	0.31	0.37	0.37	0.36	0.36	0.35	0.013	0.002	0.002	0.409
d 11 to 21	0.34	0.49	0.54	0.51	0.52	0.52	0.018	0.001	0.001	0.881
d 0 to 21	0.32	0.42	0.45	0.43	0.43	0.43	0.014	0.001	0.001	0.611

1 A total of 336 suckling pigs (DNA 241 × 600) were used in a 21 d suckling experiment with 12 pigs per sow and 2 pigs within each sow individually treated for a total of 56 pigs per treatment.
2 Gleptoforte (Ceva Animal health, LLC., Lenexa, KS) dosage administered 3 d after farrowing.
3 Pigs were administered 200 mg at 3 d after farrowing and 100 mg 11 d after farrowing.
4 Linear comparison of 0 mg to 200 mg dosage.
5 Quadratic comparison of 0 mg to 200 mg dosage.
6 Pairwise comparison between mean of 200 mg and 200 + 100 mg treatments.
7 Represents 3 d after farrowing.
8 Represents 11 d after farrowing.

lack of iron injection resulted in the poorest growth and blood parameters of iron status of suckling piglets as expected. However, the administration of 100 mg of Gleptoforte resulted in the greatest growth performance. Finally, the administration of 200 mg + 100 mg of Gleptoforte resulted in improved hematological criteria but did not influence suckling piglet growth performance compared to 200 mg alone.

References

1. Mateos GG et al. Trace Minerals. *Redefining Mineral Nutrition*. 2012; 41:43.

2. Jolliff, J.S. and Mahan, D.C. Effect of injected and dietary iron in young pigs on blood hematology and postnatal pig growth performance. J. Animal Science. 2011; 4068-4080.

3. Starzyński, RR et al. Iron supplementation in suckling piglets: how to correct iron deficiency anemia without affecting plasma hepcidin levels. *PloS one*. 2013;e64022.

Table 2: Effects of Gleptoforte dosage on suckling pig hematological criteria[1]

	Dosage, mg/ml[2]							Probability, $P <$		
	0	50	100	150	200	200 + 100[3]	SEM	Linear[4]	Quadratic[5]	200 vs. 200 + 100[6]
Hgb (g/dl)[7]										
d 38	8.4	8.3	8.3	8.3	8.2	8.4	0.250	0.719	0.850	0.613
d 119	5.7	8.3	9.9	10.1	10.7	10.5	0.235	0.001	0.001	0.703
d 21	4.6	6.8	9.3	11.3	12.0	12.8	0.217	0.001	0.001	0.011
Hct (%)[7]										
d 3	28.0	27.1	27.6	27.4	27.4	28.0	0.806	0.809	0.749	0.699
d 11	20.0	29.2	34.3	35.8	36.5	36.2	0.660	0.001	0.001	0.722
d 21	16.0	23.4	30.9	37.3	38.8	40.9	0.715	0.001	0.001	0.046
Serum Fe (μg/dl)[7]										
d 3	26	24	30	29	25	24	3.82	0.816	0.463	0.838
d 11	19	29	101	149	162	157	8.73	0.001	0.558	0.675
d 21	22	15	25	53	86	113	7.85	0.001	0.001	0.019
TIBC (μg/dl)[7]										
d 3	252	248	216	236	242	223	13.78	0.454	0.166	0.351
d 11	698	536	442	417	406	421	22.77	0.001	0.001	0.669
d 21	726	667	519	479	415	398	27.43	0.001	0.3446	0.670

1 A total of 336 suckling pigs (DNA 241 × 600) were used in a 21 d suckling experiment with 12 pigs per sow and 2 pigs within each sow individually treated for a total of 56 pigs per treatment. All barrows were bled at each of the timepoints to measure hematological criteria.
2 Gleptoforte (Ceva Animal Health, LLC., Lenexa, KS) dosage administered 3 d after farrowing.
3 Pigs were administered 200 mg at beginning of trial and 100 mg 11 d after farrowing.
4 Linear comparison of 0 mg to 200 mg dosage.
5 Quadratic comparison of 0 mg to 200 mg dosage.
6 Pairwise comparison between mean of 200 mg and 200 + 100 mg treatments.
7 Trt × day interaction ($P < 0.001$).
8 Represents 3 d after farrowing.
9 Represents 11 d after farrowing.

Effect of ButiPEARL™ Z level and feed form on nursery pig growth performance for the first 6 weeks after weaning

A. L. P. de Souza; B. Kremer; T. Marsteller
Kemin Industries, Des Moines, Iowa

Introduction

ButiPEARL™ Z is a novel, encapsulated butyric acid and zinc (BPZ) product developed, manufactured and sold by Kemin Industries. Previous trials show compelling evidence of BPZ's effect on mitigating heat stress in swine by recovering lost average daily gain (ADG) associated with increased intestinal permeability.[1] Certain short-chain fatty acids and essential minerals have critical roles in maintaining the intestinal barrier. Specifically, butyric acid increases the expression of tight junction proteins and the proliferation of epithelial cells[4,5] in addition to butyric acid being an epithelial energy source. Zinc, the other component of BPZ, plays an important role in the upregulation of tight junction proteins in intestinal cells and increases the microbial diversity of the intestine.[2,3] Together, butyric acid and zinc have the ability to yield a gut health product with synergistic modes of action. The dose of BPZ has been established in a previous trial at 1 lb/ton for best response to stress pressure.[6] The response of BPZ in pelleted feeds has been consistent and release of product confirmed at 4 h post ingestion. However, when applied to meal form diets, no response has been detected mainly due to release of product being above 8 h post ingestion.[7] Thus, this study was designed to evaluate the effect of a new formula of BPZ on pelleted or mash feed diets on the growth performance of pigs after weaning while submitting animals to moderate heat stress, 110% stocking density compared to the standard used in the testing facility and an extra vaccination during nursery.

Materials and Methods

A total of 396 mixed gender pigs (5.99 kg ± 0.2) were weaned at 18-20 days and allotted 11 pigs per pen (2.73 sq ft/pig), based on weight, in a completely randomized design. Pens were balanced with barrows and gilts and had one self-feeder with 4 holes and a bowl drinker to provide ad libitum access to feed and water. The rooms were air ventilated with a metal tri-bar flooring and concrete aisles. Temperatures were set at start at 90°F, with reduction in temperature of 0.2°F in the first week, 0.3°F in weeks 2 and 3, 0.4°F in week 4 and 0.5°F in week 5 until the end period. When transitioning to diet phase 3, the animals were vaccinated with Boehringer Ingelheim Enterisol® Ileitis (Ileitis, *Lawsonia intracellularis*). Pens were not cleaned from the previous group to add environmental and pathogen stress levels. Treatments were feed form (mash or pellet) and ButiPEARL Z (0 and 0.75 kg/t), presented through all 42 days of trial in a feeding program with 3 phases (0-7 days, 7-21 days and 21-42 days). Zinc oxide was used as standard in the feed at 3,000 ppm in the first phase and at 2,000 ppm in the second phase, last phase diet used only standard zinc requirements. Diets were formulated to lower levels of lactose. Pig weights, count and feed disappearance was collected weekly throughout the trial periods. The data was analyzed using JMP software (Cary, NC), using a complete factorial approach. Significance was observed at $P < 0.05$ and tendency when P-value was between 0.05 and 0.10.

Results and discussion

During the first week of trial, three pigs presented exudative epidermitis and had to be removed from their pens. The statistical analysis did not show significance due to the small change in stocking density. No interactions among feed form and ButiPEARL Z were observed. Feed form, pelleted compared to mashed feeds, started showing statistical differences for end avg wt, ADG and average daily feed intake (ADFI) (lbs) from period 0-14 days ($P < 0.01$). From 0-14 days, end avg wt 17.1 vs 16.5, ADG 0.276 vs 0.237, ADFI 0.421 vs 0.397; from 0-21 days end avg wt 22.8 vs 20.8, ADG 0.46 vs 0.368, ADFI 0.631 vs 0.516; from 0-28 days end avg wt 29.6 vs 26.5, ADG 0.587 vs 0.482, ADFI 0.811 vs 0.643; from 0-35 days end avg wt 39.2 vs 35.3, ADG 0.746 vs 0.641, ADFI 1.059 vs 0.841; from 0-42 days end avg wt 506 vs 46.0, ADG 0.894 vs 0.791, ADFI 1.314 vs 1.070. Feed to gain (F:G) ratio was significantly different than mashed feeds for periods 0-35 days 1.421 vs 1.313 ($P = 0.042$) and from 0-42 days 1.471 vs 1.353 ($P < 0.001$). ButiPEARL Z showed tendency to impact ADG ($P = 0.055$) and F:G ($P = 0.056$) at 0-14 days, 0.271 vs 0.242 lb/d and 1.50 vs 1.675, respectively. In the longer time periods, ButiPEARL Z feeding resulted in improvements of ADG and F:G. From 0-21 days, impacting end avg wt (22.2 vs 21.4, $P = 0.002$), ADG (0.438 vs 0.391, $P < 0.001$), and F:G (1.339 vs 1.443, $P = 0.004$). During period 0-28 days, results showed end avg wt 28.6 vs 27.5 lbs ($P = 002$), ADG 0.555 vs 0.514 lb/d ($P = 001$) and ADFI 0.744 vs 0.71 lb/d ($P = 0.051$) when ButiPEARL Z was fed. For period 0-35 days, end avg wt 38 vs 36.5 lbs ($P = 0.004$), ADG 0.716 vs 0.67 lb/d ($P = 0.005$) and ADFI 0.975 vs 0.925 lb/d ($P = 0.030$) showing significant values upon feeding ButiPEARL Z. For the final, complete period of 0-42 days, ButiPEARL Z was significantly efficient to change end avg wt 49 vs 47.6 lbs ($P = 0.050$) and showed tendency to significance for ADG 0.858 vs 0.826 lb/d ($P = 0.066$) and ADFI

Table 1: Feed Form animal performance by period, cumulative

Feed form	**Pellet**	**Mash**	***P*-value**
End avg wt, kg	5.98	5.96	0.627
	0-7 days		
End avg wt, kg	6.31	6.38	0.416
ADG, kg	0.047	0.060	0.083
ADFI, kg	0.104	0.102	0.853
G:F	1.124	0.545	0.559
Mortality, %	0.00	0.00	-
	0-14 days		
End avg wt, kg	7.74	7.46	0.006
ADG, kg	0.125	0.107	0.009
ADFI, kg	0.191	0.169	0.003
G:F	0.636	0.624	0.728
Mortality, %	0.00	1.06	0.167
	0-21 days		
End avg wt, kg	10.35	9.45	<0.001
ADG, kg	0.209	0.167	<0.001
ADFI, kg	0.286	0.234	<0.001
G:F	0.728	0.711	0.337
Mortality, %	0.51	1.06	0.544
	0-28 days		
End avg wt, kg	13.41	12.03	<0.001
ADG, kg	0.266	0.219	<0.001
ADFI, kg	0.368	0.292	<0.001
G:F	0.723	0.747	0.042
Mortality, %	0.51	1.06	0.544
	0-35 days		
End avg wt, kg	17.78	16.03	<0.001
ADG, kg	0.338	0.291	<0.001
ADFI, kg	0.480	0.381	<0.001
G:F	0.704	0.762	<0.001
Mortality, %	0.51	1.06	0.544
	0-42 days		
End avg wt, kg	22.94	20.87	<0.001
ADG, kg	0.405	0.359	<0.001
ADFI, kg	0.596	0.486	<0.001
G:F	0.680	0.739	<0.001
Mortality, %	0.51	1.06	0.544

Table 2: ButiPEARL™ Z animal performance by period, cumulative

ButiPEARL Z, kg/t	0	0.75	*P*-value
End avg wt, kg	6.01	5.93	0.174
	0-7 days		
End avg wt, kg	6.38	6.31	0.280
ADG, kg	0.053	0.054	0.852
ADFI, kg	0.103	0.102	0.767
G:F	2.304	0.437	0.254
Mortality, %	0.00	0.00	-
	0-14 days		
End avg wt, kg	7.56	7.65	0.351
ADG, kg	0.110	0.123	0.055
ADFI, kg	0.178	0.182	0.503
G:F	0.597	0.667	0.056
Mortality, %	0.56	0.51	0.947
	0-21 days		
End avg wt, kg	9.71	10.09	0.002
ADG, kg	0.177	0.199	<0.001
ADFI, kg	0.254	0.266	0.129
G:F	0.693	0.747	0.004
Mortality, %	1.06	0.51	0.544
	0-28 days		
End avg wt, kg	12.49	12.95	0.002
ADG, kg	0.233	0.252	0.001
ADFI, kg	0.322	0.337	0.051
G:F	0.723	0.747	0.435
Mortality, %	1.06	0.51	0.544
	0-35 days		
End avg wt, kg	16.56	17.24	0.004
ADG, kg	0.304	0.325	0.005
ADFI, kg	0.420	0.442	0.030
G:F	0.726	0.737	0.194
Mortality, %	1.06	0.51	0.544
	0-42 days		
End avg wt, kg	21.61	22.20	0.050
ADG, kg	0.375	0.389	0.066
ADFI, kg	0.529	0.552	0.060
G:F	0.710	0.707	0.701
Mortality, %	1.06	0.51	0.544

Table 3: Cumulative performance results for weight (avg wt), average daily gain (ADG), average daily feed intake (ADFI), feed to gain ratio (F:G), and mortality.

Feed Form	**Pellet**		**Mash**			
ButiPEARL Z FR, kg/t	**0**	**0.75**	**0**	**0.75**	**RMSE***	***P*-value**
End avg wt, kg	6.1	5.9	6.0	5.9	0.408	0.419
			0-7 days			
End avg wt, kg	6.4	6.2	6.4	6.4	0.486	0.379
ADG, kg	0.048	0.046	0.058	0.062	0.048	0.704
ADFI, kg	0.107	0.100	0.099	0.104	0.030	0.216
G:F	-1.068	0.368	0.554	0.537	4.791	0.269
Mortality, %	0.00	0.00	0.00	0.00	-	-
			0-14 days			
End avg wt, kg	7.7	7.8	7.4	7.5	0.616	0.727
ADG, kg	0.119	0.132	0.101	0.113	0.043	0.930
ADFI, kg	0.190	0.192	0.167	0.172	0.036	0.878
G:F	0.599	0.679	0.595	0.655	0.266	0.814
Mortality, %	0.00	0.00	1.11	1.01	2.254	0.947
			0-21 days			
End avg wt, kg	10.2	10.5	9.3	9.6	0.722	0.819
ADG, kg	0.198	0.220	0.157	0.178	0.031	0.941
ADFI, kg	0.279	0.294	0.230	0.238	0.048	0.629
G:F	0.709	0.747	0.678	0.746	0.100	0.358
Mortality, %	1.01	0.00	1.11	1.01	2.716	0.618
			0-28 days			
End avg wt, kg	13.2	13.6	11.7	12.3	0.913	0.323
ADG, kg	0.259	0.274	0.207	0.230	0.034	0.383
ADFI, kg	0.360	0.376	0.284	0.299	0.050	0.960
G:F	0.719	0.727	0.727	0.769	0.064	0.177
Mortality, %	1.01	0.00	1.11	1.01	2.716	0.618
			0-35 days			
End avg wt, kg	17.4	18.1	15.7	16.4	1.458	0.910
ADG, kg	0.328	0.348	0.280	0.301	0.045	0.902
ADFI, kg	0.468	0.493	0.371	0.392	0.065	0.841
G:F	0.701	0.707	0.754	0.770	0.045	0.623
Mortality, %	1.01	0.00	1.11	1.01	2.716	0.618
			0-42 days			
End avg wt, kg	22.6	23.3	20.6	21.1	1.903	0.842
ADG, kg	0.398	0.413	0.352	0.366	0.051	0.912
ADFI, kg	0.584	0.608	0.475	0.496	0.078	0.922
G:F	0.681	0.679	0.741	0.737	0.043	0.870
Mortality, %	1.01	0.00	1.11	1.01	2.716	0.618

* RMSE – Root Mean Square Error

Table 4: Laboratory results for proximate analysis and ButiPEARL™ Z content in feeds tested.

	Phase of treatment		
Nutrient	**1**	**2**	**3**
Crude protein, %	22.89 ± 0.82	22.71 ± 0.49	20.51 ± 0.58
Crude fat, %	4.41 ± 0.10	4.44 ± 0.22	4.33 ± 0.10
Calcium, %	0.88 ± 0.11	0.76 ± 0.03	0.63 ± 0.03
Phosphorus, %	0.69 ± 0.02	0.66 ± 0.02	0.53 ± 0.01
Sodium, %	0.33 ± 0.02	0.16 ± 0.01	0.19 ± 0.01
ButiPEARL Z, ppm			
Trt pellet 0 kg/t BPZ	0.38	<LOQ	< LOQ
Trt pellet 0.75 kg/t BPZ	1.86	1.78	1.90
Trt mash 0 kg/t BPZ	<LOQ	<LOQ	< LOQ
Trt mash 0.75 kg/t BPZ	1.55	1.54	1.44

1.218 vs 1.167 lb/d ($P = 0.06$). The significant values on the use of ButiPEARL Z for pigs fed pelleted and mashed feeds were observed under the same conditions of compounded stress.

References

1. Kemin Internal Document, 16-00070.

2. Katouli, M., L. Melin, M. Jensen-Waern, P. Walgreen, and R. Mollby. 1999. The effect of zinc oxide supplementation on the stability of the intestinal ora with special reference to composition of coliforms in weaned pigs. J. Appl. Microbiol. 87:564-573.

3. Zhang, B., Y. Shao, D. Liu, P. Yin, Y. Guo, and J. Yuan. 2012. Zinc prevents *Salmonella enterica* serovar Typhimurium-induced loss of intestinal mucosal barrier function in broiler chickens. Avian Pathol. 41:361-367.

4. Peng, L., Z. R. Li, R. S. Green, I. R. Holzman, and J. Lin. 2009. Butyrate Enhances the Intestinal Barrier by Facilitating Tight Junction Assembly via Activation of AMP-Activated Protein Kinase in Caco-2 Cell Monolayers. J. Nutr. 139:1619-1625.

5. Kotunia, A, J. Wolinski, D. Laubitz, M. Jurkowska, V. Rome, P. Guilloteau, and R. Zabielski. 2004. Effect of sodium butyrate on the small intestine development in neonatal piglets fed by articial sow. J. Physiol. Pharmacol. 55:59-68.

6. Kemin Internal Document, 17-00818.

7. Kemin Internal Document, 17-00821.

Is it possible that an antibiotic may actually reduce antibiotic resistance? Flavomycin review

D. Nolan, DVM; C. Sparks, PhD
Huvepharma, Inc., Peachtree City, Georgia

Introduction

Bambermycin (flavophospholipol), also known as moenomycin, is a phosphoglycolipid antimicrobial produced by a number of *Streptomyces* strains. Bambermycin was the first member of the phosphoglycolipid family discovered during routine soil screening in Bamberg, Germany by A.G. Hoechst in 1960. In the United States bambermycin is sold as a feed additive by Huvepharma, Inc. under the brand names: Flavomycin® for increased rate of weight gain and improved feed efficiency in swine, broilers and turkeys, and GAINPRO® for increased rate of weight gain or increased rate of weight gain and improved feed efficiency in multiple production classes of cattle. Neither Bambermycin nor any of the phosphoglycolipids have a therapeutic use in humans and therefore its growth promotion (production) label claims were not impacted by the voluntary label modifications as a result of the FDA's GFI #209 (2012) and GFI #213 (2013).[3,4] Bambermycin is not absorbed from the gastrointestinal tract and has been shown to actively suppress certain microorganisms considered as detrimental in the gut (*Staphylococcus spp.* and *Enterococcus faecalis*) while promoting the propagation of those considered to be beneficial for the overall gut health (*Lactobacillus spp.* and *Bifidobacterium*).[6] Flavophospholipol interferes with bacterial wall synthesis but its activity is distinctly different than the β-lactam antibiotics. The moenomycins are the only known class of antibiotics to inhibit peptidoglycan glycosytransferase.[5] Because of the different site and mechanism of activity, bambermycin does not lead to β-lactam resistance, nor is it impacted by pre-existing β-lactam resistance. Although only labeled for growth and efficiency in swine, Bambermycin has been repeatedly shown to reduce the presence of, or the spread of, antimicrobial resistance in both in vitro and in vivo studies.[1,2,6,7,8] This leads to a potentially beneficial mechanism that may have the practical impact of restoring the efficacy of antimicrobial therapy to a site or herd that previously had antimicrobial resistance. Furthermore, it has the possibility to reduce the resistance in the bacterial population as a whole by reducing the transfer of antimicrobial resistance to other bacterial populations.

Discussion

In the 1970s and 1980s in vitro studies demonstrated that plasmid containing strains of *E coli*, *Salmonella spp.* and several *Enterobacteriaceae spp.* appeared to lose plasmids and convert from drug resistant phenotypes to a susceptible or wild phenotype through the reduction of the number and ratio of cells with the resistance plasmids. Several study discussions have used the phrase "plasmid-curing effect" to describe this observation while others have argued that the term is misleading because flavophospholipol selectively inhibits the growth of bacteria harboring resistant plasmids.[6] Although there continues to be discussion as to the exact mechanisms that bambermycins may reduce resistance, there is much agreement that the result of reduced resistance is a beneficial trait, especially if realized in the field under production conditions. Perhaps the most encouraging study that demonstrated a significant reduction in resistance was conducted in 2000 with swine under field conditions by A. van den Bogaard, et. al., in the Netherlands.[8] This finishing study found significant reductions in the prevalence of multi-drug resistant *E coli* after 12 weeks of feeding, post challenge with a multi-drug resistant strain of *E coli*, in the bambermycin group (9 mg/kg) compared to the negative control (no antimicrobial GP) and the positive control (avoparcin 15 mg/kg). There was however an increase in the prevalence of resistant *Enterococcus faecalis* in the bambermycin group. The author concluded that the major decrease in resistance of intestinal *E coli* outweighed the small increase in the resistance of the *E faecalis* detected. In 2009 Door et. al. published in the *Journal of Swine Health and Production* a study that demonstrated a reduction in proportions of DT104 and DT193 drug resistant *Salmonella* post challenge when nursery pigs were fed a ration containing 2.2 mg/kg bambermycin.[2] In 2005 Ragland et. al shared a paper in the 2005 AASV proceeding that demonstrated a trend in reduction of resistant *E coli* in nursery pigs fed diets containing 4.4 mg/kg bambermycin.[7]

Conclusion

Bambermycin has the potential to be a tool that may contribute to the reduction of antimicrobial resistance in swine. As the public concern over antibiotic resistance continues to grow exponentially, the swine industry must proactively develop tools and standards for measurement and improvement, or reduction, of antimicrobial resistance. Further investigation of bambermycin and other feed ingredients, along with production practices, is warranted for the reduction of antibiotic resistance bacterial populations in the environments where we raise our pigs and process pork for human consumption.

References

1. Butaye, P. et al. Antimicrobial growth promoters used in animal feed: effects of less well known antibiotics on gram-positive bacteria. *Clinical Microbiology Review* 16:175-188.

2. Door, P. et al. Flavophospholipol: Effect on multi-drug resistant *Salmonella enterica* serovar Typhimurium in swine. *Journal of Swine Health and Production* 2009 308-315.

3. FDA GFI #209. Available at: https://www.fda.gov/downloads/AnimalVeterinary/GuidanceCompliance Enforcement/GuidanceforIndustry/UCM216936.pdf.

4. FDA GFI#213. Available at: https://www.fda.gov/downloads/AnimalVeterinary/GuidanceCompliance Enforcement/GuidanceforIndustry/UCM299624.pdf.

5. Ostash, B. et al. Moenomycin family antibiotics; chemical synthesis, biosynthesis, biological activity. *Natural Product Report*, 2010 November 21:27(1) 1594-1617

6. Pfaller, M. Flavophospholipol use in animals: Positive implications for antimicrobial resistance based on its microbiologic properties. *Diagnostic Microbiology and Infectious Disease* 56 (2006);115-121.

7. Ragland, D et al. Flavomycin supplementation of nursery pig diets. Proceedings: *American Association of Swine Veterinarians 2005* 183-187.

8. van den Boogaard, A, et al. Effects of Flavophospholipol on Resistance in Fecal *Escherichia coli* and Enterococci of Fattening Pigs. *Antimicrobial Agents and Chemotherapy*, Jan 2002 110-118.

Mucosal IgA production in response to a killed, Carbigen-adjuvanted vaccine administered intranasally

G. Weaver[1]; D. Rudrappa[1]; J. Kula[1]; D. Guggenbiller[1]; K. Kinsley[1]; J. Newberry[1]; R. Nimmo[1]; M. Weaver[2]
[1]Phibro Animal Health and Nutrition, Teaneck, New Jersey; [2]WW Research, Webster City, Iowa

Introduction

For the past several years, alternative routes of vaccine administration have been considered as a possible means to improve the overall immune response. Protective immunity against mucosal pathogens will require novel vaccine strategies to induce mucosal immune responses tailored to anatomic location and to the threat of the invading pathogen.[1] The combination of alternative immunization routes and the use of appropriate antigen delivery systems appears to be a rational approach for eliciting protective immunity at mucosal surfaces.[2] Human medicine has also recognized that most infections are caused by pathogens which have a mucosal portal of entry, hence the need to develop mucosal vaccines against the myriad of respiratory and enteric infections, especially during early childhood.[3] Other reasons for investigating the mucosal route of administration include potential for cross-protection,[1] a better point of contact immune response due to heightened levels of mucosal antibody production,[5] and the ability to circumvent interfering maternal immunity. The results of a field study comparing Emulsigen™-adjuvanted systemically administered vaccine with the same antigens adjuvanted with Carbigen™ adjuvant and administered intranasally were presented at AASV in 2016.[4] No statistical differences were found between the production parameters measured and IgG and IgA levels in serum. The goal of this study was to evaluate the level of mucosal IgA produced by a killed, Carbigen™-adjuvanted IAV-S vaccine compared to placebo control.

Materials and methods

Carbigen™ is a terminally sterilized, carbomer-based adjuvant suspension containing a proprietary emulsified component. It has the potential to elicit higher levels of humoral antibody and cellular immunity, more rapid onset of immunity and enhanced protection with a single vaccine dose as compared with conventional aluminum-based adjuvants. Carbigen may be combined with bacterial and viral antigens and administered parenterally or intranasally.[6]

The commercial cross pigs used in this study were weaning age (approximately 19-21 d. on arrival) and were sourced from a PRRS and IAV-S negative breeding herd. They were housed in the same room (single airspace) and divided by pen into five animals per treatment group (five animals per pen). The pens consisted of wire mesh flooring and vertical iron rod gating with the animals having *ad libitum* access to feed and water. Water supply was provided by water lines with nipples and feed was delivered by trough-style feeders. The feed was a commercially-formulated, age-appropriate ration that was pelleted and bagged prior to arrival at the study site. The room was fully power ventilated containing both ceiling and end-wall inlets. Supplemental heating was provided with a propane heater. The animals had five days of acclimation prior to initiation of the study.

Vaccine

The vaccine used in the study was a fully-licensed IAV-S vaccine (FluSure XP) adjuvanted with a 10% Carbigen™ formulation that has been approved for use in pigs. The control vaccine was phosphate buffered saline (PBS). The experimental vaccines were formulated as described in Table 1.

Experimental design

Animals were randomly assigned to one of two treatment groups of five animals per treatment group. T01 animals were injected on Days 0 and 21 intramuscularly with 2ml of PBS as formulated according to Table 1. T02 animals were vaccinated with 1ml Vaccine+Carbigen™ IN on Day 0 and 1ml Vaccine + Carbigen™ IM on Day 21 also as formulated according to Table 1. On Day 0, 1ml of vaccine was instilled into one nostril of each animal. Blood samples from both treatment groups were collected on Days 0, 21, and 42. Nasal swab samples were collected on Day 0 (baseline), at approximately 10 hours, and at 24 hours (Day 1),

Table 1: Formulation

Treatment		Description
T01	PBS only	Formulation phosphate buffered saline Dose: 2mL (including adjuvant)
T02	IAV-S vaccine + Carbigen™	Formulation FluSureXP Zoetis, Inc. adjuvanted with Carbigen™ Dose: 1ml IN/1ml IM (including adjuvant)

and on Days 3, 7, and 21 of the study. One nasal swab was used to collect mucous from both nares of each animal swabbed. The study was conducted according to the design as shown in Table 2.

Laboratory testing

Nasal swabs were frozen at -20°C and delivered to Phibro Animal Health Diagnostic Laboratory (PVO), Omaha, NE. The presence of mucosal IgA in the pooled treatment samples (five animals) was evaluated at PVO using an antigen-specific ELISA. Results are shown in Figure 1. Blood samples were processed for serum and frozen to -20°C. An aliquot of each serum sample was submitted to Iowa State University VDL for an HI (Zoetis antigens) test specific for H1N1 and H3N2. Results are shown in Table 3.

Statistical analysis

The serum HI titers were compared using the Student's t-test. No statistical differences were observed. The antigen-specific IgA ELISA titers were not analyzed but were numerically compared with all values adjusted to Day 0 value (pre-vaccination baseline) acting as a blank.

Results and discussion

Measurements of antibodies in secretions are notoriously difficult to assay and to standardize due to variables including sample processing, viscosity of specimen, and interference from enzymes and desquamated cells.[2] In this study, the antigen-specific mucosal IgA levels peaked on Day 3 in the Carbigen™-vaccinated animals as compared to the Control group. These titers waned dramatically at the subsequent time points as shown in Figure 1. HI titers to the IAV-S vaccine strains used in the study are summarized in Table 3.

Even though nasal secretion results for individual animals in this model may not be entirely valid, the peak value of the pooled sample noted on Day 3 post-vaccination trended favorably for the Carbigen™ treatment group. In a previous study by another research group, a similar result was reported in swine using a different antigen (PCV2) and another delivery system.[5] These findings suggest that mucosal surface vaccination may hold significant advantages over conventional parenteral routes commonly used in commercial pork production. Further research to investigate novel mucosal surface antigen-delivery systems is warranted.

References

1.Igor M. Belyakov, Jeffrey D. Ahlers. What Role Does the Route of Immunization Play in the Generation of Protective Immunity against Mucosal Pathogens? The Journal of Immunology. 2009;183(11):6883-6892.

2.Bradley J. Sedgmen, Els N.T. Meeusen, Shari A. Lofthouse. Alternative routes of mucosal immunization in large animals. Immunology and Cell Biology. 2004;82:10-16.

3.C.Czerkinsky, J.Holmgren. Topical Immunization Strategies. Mucosal Immunology. 2010;3:545-555.

4.R.Jones, B.Heins, G.Weaver, J. Kula, B. Lin, W. Johnson. Field evaluation of the intranasal route as an alternative for administering Streptococcus suis and Haemophilus parasuis bacterins. Proc. AASV. New Orleans, LA. 2016;179-183.

5.Yu-feng Liu, Lu Chen, Pan-pan Ma, Hang-tao Chang, Xia Yang, Chuan-quing Wang. Intranasal immunization with recombinant adenovirus expressing the major epitopes of capsid protein of porcine circovirus 2 induces mucosal and systemic response. 22nd IPVS. Seoul, Korea. 2012;928.

6.Carbigen™ Technical Bulletin. MVP Technologies. Omaha, NE. Copyright 2012.

Table 2: Study design

Treatment	No.of pigs	Dose/Route of inj.	Study days for procedure		
			Injection	Collect blood	Swab nose
T01 Negative control	5	2ml IM	D0 D21	D0 D21	D0 10 hrs. 24 hrs. D3 D7 D21
T02 10% Carbigen™	5	1ml IN 1ml IM	D0 D21	D0 D21 D42	D0 10 hrs. 24 hrs. D3 D7 D21

Table 3: HI titers (Zoetis antigens)

	H1N1		H1N2		H3N2	
Treatment	**T01**	**T02**	**T01**	**T02**	**T01**	**T02**
Day 0	5 neg	5 neg	5 neg	5 neg	1 pos	5 neg
Day 21	5 neg	5 neg	5 neg	5 neg	5 neg	5 neg
Day 42	5 neg	5 neg	5 neg	5 neg	5 neg	5 neg

Figure 1: Development of specific IgA to IAV-S following IN vaccination

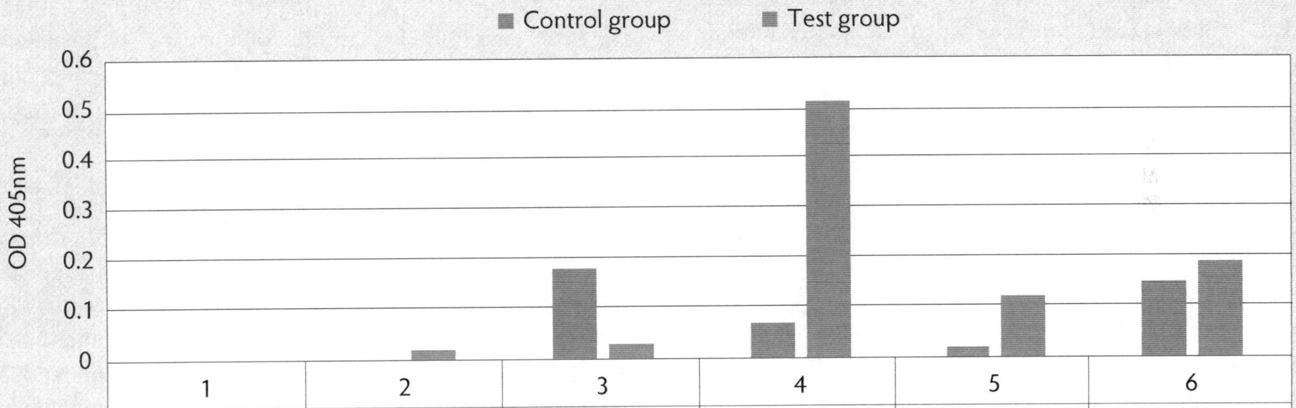

	1	2	3	4	5	6
Control group	0	0	0.18	0.07	0.02	0.15
Test group	0	0.02	0.03	0.51	0.12	0.19

Time post vacination (hrs)

Evaluation of a killed virus vaccine (MJPRRS®) as part of a PRRS virus mitigation program

K Kinsley, DVM; D Guggenbiller, DVM, MBA; J Newberry, DVM, MBA, MS; R Nimmo, PhD; G Weaver, DVM
Phibro Animal Health Corporation, Teaneck, New Jersey

Introduction

Much emphasis has been placed on producing PRRSV PCR negative pigs at the time of weaning in current industry definitions of PRRSV stability.[9] As our ability to identify the presence of smaller numbers of positive pigs in a population improves through new sampling techniques, the bar is raised as to the definition of PRRSV Negative, and therefore PRRSV Stable, production.[8] Information exists that demonstrates PRRSV load as being an important factor associated with growing pig performance and overall health, not to mention the reproductive ramifications associated with PCR identification of PRRSV in a breeding herd.[11] As our definition of PRRSV Negative/Stable becomes more stringent, our ability to maintain reproductive health, respond to viral changes and introductions more rapidly and wean populations of pigs that are PRRSV PCR positive, but still perform, will need to improve. Previous literature evaluations suggest that killed virus PRRSV vaccines are capable of further reducing the viral load as compared to non-vaccinated or modified-live vaccinated animals and improving downstream performance.[4,7,12,14] Coupled with MJPRRS® grouping and vaccine production technology this observation is further validated with heterologous virus vaccination and challenge studies. As the definition of PRRSV Negative/Stable changes, making it more difficult to achieve truly negative status, viral burden will likely begin to carry added economic weight. Programs that take into consideration virus diversity and viral burden reduction rather than singular live virus substitution will benefit PRRSV control regions as well as individual farms and their downstream production.

As techniques such as composite sampling (family oral fluids and processing fluids) continue to evolve our ability to identify smaller and smaller quantities of PRRSV will improve. Additionally, it will be possible to monitor viral mutations and identify emerging and clinically aggressive strains nearly in real-time.[9] The goal of this monitoring is, and should be, to provide producers and veterinarians the best ability to rapidly respond to these newly identified viruses. MJPRRS grouping and vaccine production technology provide a means of further classifying these viral changes thus allowing a means of response via vaccine intervention.

Previous AASV presentations have discussed the basis and process for MJPRRS grouping technology, but there have been few studies demonstrating response of vaccination in a heterologous challenge.[5] Viral homology has been loosely defined in the literature as nucleotide sequences that range from 97–99.3% similar.[2,10] Much of this work has been done with modified-live vaccine as the base virus and serial pig-to-pig passages as the means of monitoring viral variation.[1,2,13] While there is no consensus on the quantity of genetic difference required to suggest that two viruses are not related, rate of change in nucleotide sequence of greater than 1% over a defined period of time (100 days - 1 year) may be suggestive of two virus isolates that are not closely related.[3]

The MJPRRS grouping definition of homology is quite simple as viruses are neatly grouped into 1 of 25 immuno-groups (D-1–D-8, and D-4b; S-1–S-8; and E-1–E-8).[5] While there are similarities as the virus can transition from one MJPRRS group to another, and at times multiple viruses can be identified in the same population, complete naivety to a virus group is frequently reported in the form of increased clinical signs within the population. In two studies, heterologous viruses were used as a challenge virus as compared to the experimentally produced MJPRRS vaccine serials. Viruses are related by MJPRRS grouping definition (D-7 challenge and a D-7 antigen in the vaccine), however the challenge virus and vaccine viruses differ by source and nucleotide sequence of >1.0%.

Materials and methods

Study 1 – growing pigs

One hundred sixty-eight (168) 4-week old commercial pigs were acquired and randomly assigned to one of three treatment groups: Treatment 1 (T1) – PRRSV naïve, Treatment 2 (T2) – Modified-live vaccine (MLV), and Treatment 3 (T3) – MLV-primed followed by MJPRRS (D4567 S5) vaccination. All animals were PRRSV, *APP*, *Mycoplasma hyopneumoniae*, and PEDV naïve upon arrival. Following randomization and serum sampling (day 0), T1 animals were moved to a separate site to maintain PRRSV naïve status until challenge (day 28). Animals in T2 and T3 were vaccinated with a single intramuscular (IM) 2 ml dose of Ingelvac® PRRS-ATP vaccine. On day 14, the off-site T1 animals were serum sampled and processed prior to handling those in T2 and T3 followed by all T3 receiving a 1 ml IM dose of MJPRRS. After all animals had been processed on day 28, T1 animals were returned to the PRRSV positive challenge site and assigned to their host pen (preassigned to house two pigs from each treatment group in one of 28 pens). All animals were individually challenged with an intranasal (IN) dose of a D-7 / 1-7-4 RFLP challenge virus. Animals were clinically assessed for the first 14 days post-challenge, and on day 42 all treatments were serum sampled and processed. Animals were again processed on day 56 and preselected animals from each treatment were euthanized for gross lung lesion scoring and the study was terminated. All samples were split and stored at -20°C until study termination and processed at the same time through the University of Minnesota Veterinary Diagnostic Laboratory.

Study 2 – pregnant gilts

Thirty-three (33) 4 to 5-week pregnant animals were acquired and randomly assigned to one of three treatment groups. Treatment 1 (T1) – PRRSV naïve, Treatment 2 (T2) – Modified-live vaccine (MLV), and Treatment 3 (T3) – MLV-primed followed by MJPRRS (D478 S12) vaccinations. On day -4, all gilts were received and T2 and T3 animals were moved to a secondary site for future PRRSV priming. On day 0, all gilts in all treatments were confirmed pregnant and serum sampled. T2 and T3 animals were additionally vaccinated IM with a 2 ml dose of Ingelvac® PRRS-ATP vaccine. On day 28, beginning with T1 all animals in all treatments were confirmed pregnant and serum sampled. Animals in T2 received an IM, 2 ml dose of sterile saline and T3 animals received their first MJPRRS vaccination. On day 49, all animals were again confirmed pregnant and serum sampled beginning with T1. T2 animals received a second IM, 2 ml dose of sterile saline and T3 animals received their second MJPRRS vaccination. Ten days later, day 59, approximately three weeks prefarrowing, 9 confirmed pregnant and serum sampled animals from each of the three treatments were randomly assigned to a farrowing crate where all were challenged via IN and IM routes with a D-7/1-7-4 RFLP challenge.

As litters began to farrow, farrowed gilts were again serum sampled and up to 5 randomly selected piglets per litter were tagged and serum sampled within 24 hours of birth. Approximately 24 days after the first animals began to farrow, all litters 20-days of age and older were weaned. At this time gilts were again serum sampled and surviving tagged piglets were serum sampled. Three days later all remaining litters were weaned with gilts and surviving tagged piglets being processed in the same manner. All samples were split and stored at -20° C until study termination and processed at the same time through the University of Minnesota Veterinary Diagnostic Laboratory.

Results and discussion

Study 1 – growing pigs

At the time of composition, statistical analyses on data from study 1 had not yet been completed. However, numerically T2 and T3 animals demonstrate advantages over T1 in many parameters – individual animal treatment, relative viral loads, and gross lesions at necropsy.

Virus load is numerically reduced in either T2 or T3 animals versus T1 in the period post-challenge (Figure 1). A key component of regional PRRSV monitoring and control projects is the reduction of viral load per animal resulting in less viral shedding into the environment. Vaccination of growing pigs with MLV and/or MJPRRS prior to PRRSV exposure can reduce the overall environmental PRRSV contamination.

Again describing the post-challenge period, the percentage of individual animals requiring antibiotic intervention and percentage of animals upon necropsy with grossly observable lung lesions by treatment numerically supports vaccination (Figure 2). A lower percentage of individual animals per treatment in T2 or T3 required antibiotic intervention compared to T1 animals. Upon necropsy, a lower percentage of T2 or T3 animals had grossly visible lung lesions. While antibiotic treatments are unlikely to be eliminated following a PRRSV infection with any vaccination protocol, reduction of antibiotic use is always beneficial. Lesion-free lungs are not sold; pounds of pork are. Reduction in the percentage of animals demonstrating lung lesions upon necropsy is

Figure 1: Average CT by treatment post challenge (Study 1)

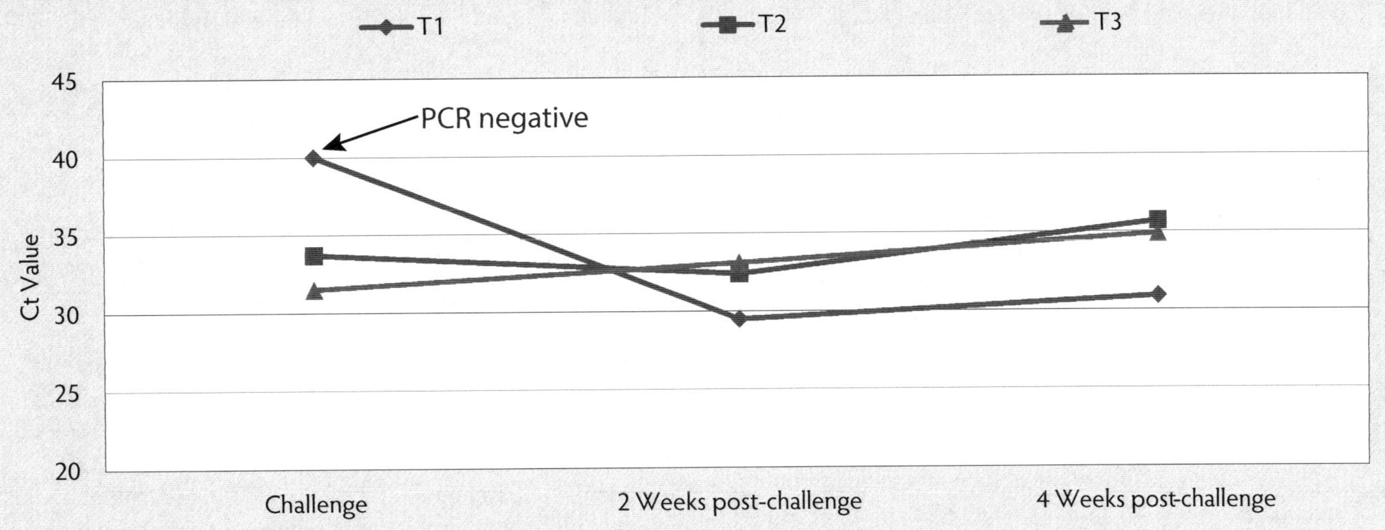

Figure 2: Percent of each treatment's pigs (Study 1)

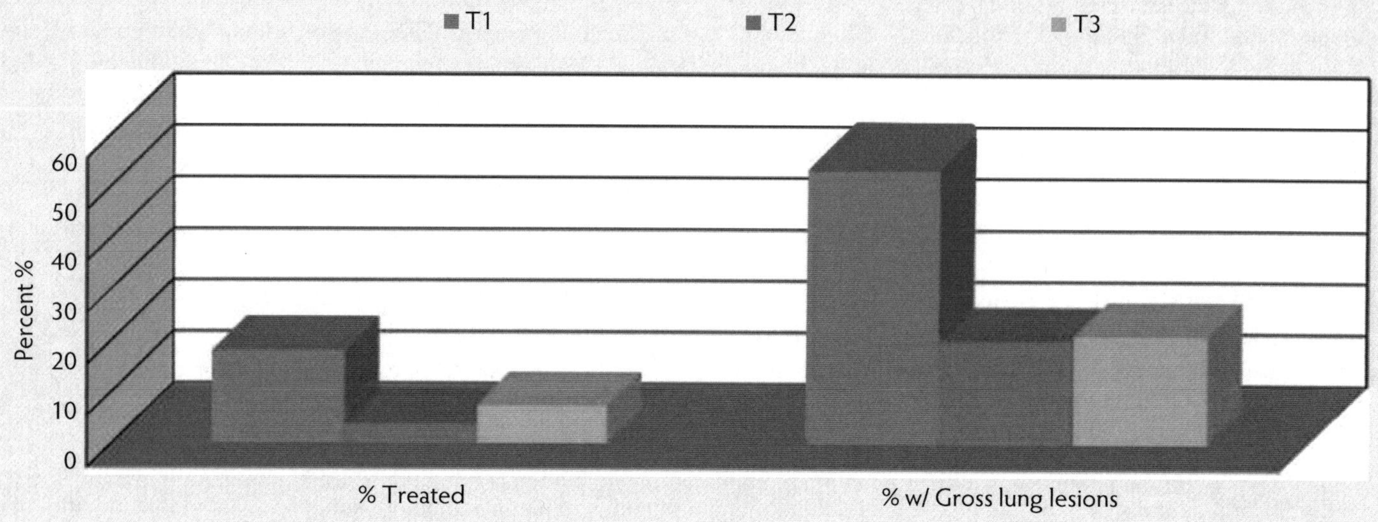

of negligible value if not also correlated with increased pounds of pork at market, an increased growth rate and efficiency, or reduced treatments, group morbidity, and/or group mortality.

Animal growth and feed efficiency, or lack thereof, is often a direct measure of population health. Total gain during the 4-week period post-challenge and average daily gain (ADG) is shown between treatments (Table 1). Total gain and ADG post-challenge numerically favor T3 over either T1 or T2. Whether this improvement of animals additionally receiving MJPRRS vaccination is a function of improved immune recognition of the viral challenge is uncertain. As was the case with the challenge used in both studies, through the MJPRRS grouping system 1-7-4 viruses have been most commonly identified as a D-7 virus. Those 1-7-4 viruses that have been identified as something other than a D-7 virus group have been a D-2, D-4, D-5, or D-6. Ingelvac® PRRS-ATP, an MJPRRS group D-4 virus, at times could be anticipated to line up immunologically with a presentation of the 1-7-4.

Study 2 – pregnant gilts

A careful observer will have identified that 27 of 33 gilts participated in the challenge phase of the study. A description of the gilts lost prior to challenge is given (Table 2). All treatments challenged and farrowed nine animals. While not an ideal virus priming and MJPRRS vaccination protocol, pre-challenge losses were anticipated due to delivery of a live PRRSV through vaccination. Following vaccination (day 0) with a MLV vaccine, gilts were monitored via PCR for continued virus circulation. Samples were collected and evaluated prior to challenge on day 0 (study initiation), day 28 (1st MJPRRS vaccination), and day 49 (2nd MJPRRS vaccination) (Table 3). All T1 animals were ELISA and PRRSV PCR negative at all time points mentioned prior to challenge. Each T2 and T3 animals mounted an ELISA response following MLV vaccination. At day 28, animals in both T2 and T3 were PRRSV PCR positive. Twenty-one days after the first MJPRRS vaccination (D478 S12 combination) all T3 animals tested PRRSV PCR negative. On day 49, the same testing point, T2 animals remained PRRSV PCR positive.

Approximately three weeks post-challenge gilts began farrowing and gilts were sampled within 24 hours of giving birth. At this same time, up to 5 piglets (if available) per litter were tagged and serum sampled. A single litter PRRSV PCR pool was run for evaluation of piglet samples (Table 4). Statistically fewer PRRSV PCR positive litters are born to MJPRRS (T3) vaccinated gilts compared to naïve challenged (T1) gilts ($P < 0.005$). There is a trend towards significance for fewer litters to be born PCR positive to Ingelvac® ATP vaccinated animals compared to naïve challenged (T1) gilts. Statistical comparisons cannot be applied with a non-continuous variable Ct, but there is a substantial reduction in the relative virus quantity between MJPRRS vaccinated (T3) and naïve (T1) litters. Not only are there fewer litters born PCR positive, those that are born positive to T3 animals are carrying far lower quantities of virus at birth. This reduction in numbers of positive litters and magnitude of average viral load should contribute to lower environmental contamination due to PRRSV.

With all gilts randomly assigned to a farrowing crate the expectation may be that the PRRSV status of all litters would merge towards the middle by the time of weaning. The number of PRRSV PCR litters at the time of weaning as well as the average Ct of positive pools were compared (Table 5). Once again, the number of litters weaning PRRSV PCR positive from MJPRRS (T3) vaccinated animals are significantly less ($P < 0.005$) than then number of litters weaned PCR positive from naïve (T1) challenged animals. Ingelvac® PRRS-ATP (T2) vaccinated litters appear as PRRSV PCR positive at the time of weaning with the

Table 1: Performance by treatment post-challenge (Study 1)

Treatment	Gain post-challenge (pounds)	ADG post-challenge (pounds)
T1	53.7	1.92
T2	53.7	1.92
T3	54.5	1.95

Table 2: Gilt status to the time of farrowing (Study 2)

Treatment	Day 0	Aborted	Day of farrowing
T1	11	0	9[A]
T2	11	2	9
T3	11	1	9[B]

[A] Two gilts randomly selected for removal from study prior to challenge.
[B] One gilt humanely euthanized for lameness prior to challenge

Table 3: Gilt PRRSV ELISA and serum PCR status pre-challenge (Study 2)

Study day	0		28		49	
Treatment	Average ELISA S/P	PCR status (Pos/Neg)	Average ELISA S/P	PCR status (Pos/Neg)	Average ELISA S/P	PCR status (Pos/Neg)
T1	0	Negative	0	Negative	0	Negative
T2	0	Negative	2.16	Positive	2.17	Positive
T3	0	Negative	2.13	Positive	2.16	Negative

Table 4: Farrowing litter pool PRRSV PCR results (Study 2)

Treatment	Positive litters (Average Ct)	Negative litters	Chi-squared testing pair-wise		
T1	9 (21.1)	0		χ^2-Value	*P* value
T2	6 (27.9)	3	T1 v T2	3.56	< 0.06
T3	3 (31.3)	6	T1 v T3	8.86	< 0.005

Table 5: Weaning litter pool PRRSV PCR results (Study 2)

Treatment	Positive litters (Average Ct)	Negative litters	Chi-squared testing pair-wise		
T1	7 (19.1)	0		χ^2-Value	*P* value
T2	7 (23.6)	2	T1 v T2	2.21	< 0.14
T3	5 (26.1)	3	T1 v T3	3.56	< 0.005

same regularity as naïve (T1) challenged litters ($P < 0.14$). The assumption being that all litters would move toward a common middle ground with respect to the number of PRRSV PCR positive litters and the magnitude of viremia per litter was not confirmed. Not only can fewer PRRSV PCR positive litters be identified, but a substantial numerical difference is still apparent when comparing relative viral load as determined by PRRSV PCR Ct.

As demonstrated by other manufacturers of inactivated PRRSV vaccines, there is a benefit to utilize a combination of MLV priming and a killed PRRSV vaccine.[4,7,12,14] The magnitude of each vaccine effect may be due to the technology of vaccine preparation, delivery, and mechanism by which high value isolates can be identified. An MJPRRS vaccination program as described is capable of reducing the number of PRRSV PCR positive litters at the time of farrowing. By being able to identify and select viruses of high immunological value, despite being heterologous by conventional definition, may help to better prepare regions looking to further reduce the environmental PRRSV loads and thereby reduce the risk to neighboring herds. In addition to reducing the environmental contamination due to PRRSV, MJPRRS grouping technology can help to identify and classify emerging, clinically aggressive viruses as well as provide a means of addressing this new introduction to the region.

As improvements continue to be made in our ability to detect lower and lower levels of PRRSV prevalence on farm, the financial incentives to drive viral levels below the thresholds of detection or out of the population completely will become much greater. There is value in identifying means of stepping closer to such goals to avoid being penalized monetarily. MJPRRS Grouping and Vaccine Technology offer a potential solution to this economic challenge. When coupled with a MLV primer, PRRSV prevalence can be positively manipulated through the use of a MJPRRS vaccination program resulting in fewer litters born PRRSV PCR positive and lower levels of viremia in those that are born PCR positive. Routine monitoring, sequencing, and MJPRRS grouping in conjunction with family or processing fluid testing may serve as an early identifier of new viral introductions or harmful internal mutations.

References

1. Amonsin A et al. Comparative analysis of complete nucleotide sequence of porcine reproductive and respiratory syndrome virus (PRRSV) isolates in Thailand (US and EU genotypes). Virology Journal. 2009, 6:143; available at: https://doi.org/10.1186/1743-422X-6-143.

2. Cha SH et al. Instability of the restriction fragment length polymorphism pattern of open reading frame 5 of porcine reproductive and respiratory syndrome virus during sequential pig-to-pig passages. J of Clin Microbiology. 2004, Vol 42 No 10; 4462-7.

3. Christopher-Hennings J et al. Porcine reproductive and respiratory syndrome (PRRS) diagnostics: interpretation and limitations. JSHAP. 2002, Vol 10 No 5; 213-8.

4. Joisel F et al. Advances in PRRS vaccination strategy: vaccination schedules alternating MLVs and KVs increase efficiency. AASV, 2016; 123-9.

5. Kinsley K et al. Reviewing a novel approach to classifying porcine reproductive and respiratory syndrome virus MJPRRS® Grouping Technology. AASV, 2016; 227-31.

6. Kolb J et al. A summary of four large scale system-based PRRS control programs. AASV. 2015; 167-70.

7. Lawrence P et al. Use of an autogenous PRRSV vaccine in conjunction with MLV elicits broadly neutralizing antibodies. AASV. 2017; 141-3.

8. Lopez W et al. Monitor herds for PRRS using processing-fluids samples. National Hog Farmer. 2017; available at: http://www.nationalhogfarmer.com.

9. Morrison RB et al. PRRS epidemiology: Data from the Swine Health Monitoring Project. AASV. 2016; 367-70.

10. Murtaugh MT et al. Use and interpretation of sequencing in PRRSV control programs. Leman. 2012; 49-55.

11. Niederwerder MC et al. Microbiome associations in pigs with the best and worst clinical outcomes following co-infection with porcine reproductive and respiratory syndrome virus (PRRSV) and porcine circovirus type 2 (PCV2). Vet Micro. 2016, Vol 188; 1-11.

12. Nodelijk G et al. A quantitative assessment of the effectiveness of PRRSV vaccination in pigs under experimental conditions. Vaccine. 2001, Vol 19; 3636-44.

13. Opriessnig T et al. Comparison of molecular and biological characteristics of a modified live porcine reproductive and respiratory syndrome virus (PRRSV) vaccine (Ingelvac PRRS MLV), the parent strain of the vaccine (ATCC VR2332), ATCC VR2385, and two recent field isolates of PRRSV. J of Virol. 2002, Vol 76 No 23; 11837-44.

14. Plana-Duran J et al. Efficacy of an inactivated vaccine for prevention of reproductive failure induced by porcine reproductive and respiratory syndrome virus. Vet Micro. 1997, Vol 55; 361 – 70.

15. Xiao Z et al. The level of virus-specific t-cell and macrophage recruitment in porcine reproductive and respiratory syndrome virus infection in pigs is independent of virus load. J of Virol. 2004, Vol 78 No 11; 5923-33.

Field studies to evaluate the efficacy of BOAR BETTER®, a new pheromone spray for sows

Michael Kuhn,[1,2] DVM, MBA; John J. McGlone,[2,3] PhD; Garrett Thompson,[2] MS
[1]Westminster, Colorado, [2]Animal Biotech, Lubbock, Texas, [3]Texas Tech University, Lubbock, Texas

Introduction

Pheromones are chemicals that impact the behavior and or physiology of animals through stimulation of multiple sensory organs including the volmeronasal organ, olfactory bulb and other olfactory receptors. They are secreted by special glands that provide a method of communication among individuals. Pheromones are not species specific and can have completely different impacts on other species (semiochemicals). They can directly impact the physiology of the receiving animal causing both hormonal and behavioral changes.

Pheromones have been used in sows in commercially available formulations to help stimulate estrus behaviors. These older products containing a single pheromone compound have provided mixed results and have experienced limited acceptance in commercial swine operations. Research conducted by Animal Biotech[1] used solid-phase micro-extraction (SPME) and gas chromatography-mass spectrometry (GC-MS) to compare compounds found in boar saliva to those in sow saliva. It was determined that boar saliva contained not two but three compounds not found in sow saliva. This triple combination of compounds was patented as Boar Saliva Analog (BSA™) and contains not only Androstenone and Androstenol but also Quinoline. A study conducted on a commercial farm in Texas[2] showed that while each of these compounds have the capability to stimulate estrus behaviors, the combination of all three stimulates a stronger behavior than any of the three alone or in pairs. A final study[3] demonstrated that BSA stimulated a physiological response when administered to sows. In that study, evaluation of serum collected from sows beginning at the time of BSA administration demonstrated significant increases in Luteinizing Hormone showing that not only does BSA stimulate a behavioral response but also a physiological one.

The BSA combination was further refined into its final, commercial formulation called BOARBETTER®. BOARBETTER is recommended for use in sows beginning 4 days post weaning when, in conjunction with boar exposure approximately 4 mL of BOARBETTER is sprayed on the sows' snout to improve estrus detection, breeding rates, conception and farrowing rates.

Field studies conducted in conjunction with Texas Tech University[4] demonstrated 6.8% increase in breeding rate and 8.9% increase in farrowing rate among sows administered BOARBETTER vs control sows. Additional field trials were developed to further evaluate BOARBETTER.

Materials and methods

Field trials were conducted with several commercial breeding herds to document reproductive performance when using BOARBETTER under a variety of field conditions. A consistent study design was developed to provide general guidance to the study investigators and to allow a meta-analysis of results overall. Basic study design involved application of treatment (BOARBETTER or Control) to weaning groups by day or by week. Individual animals could not be alternated due to cross contamination of neighboring sows. Investigators and farm staff were not blinded due to the blue coloring of BOARBETTER. Most farms alternated 3 weeks of each treatment for a total of 6 weeks of data collection. Farms ranged in size from approximately 600 to 5,000 sows. Farms utilized either traditional artificial insemination (AI) or post-cervical artificial insemination (PCAI). All farms utilized BOARBETTER in conjunction with traditional boar exposure for estrus detection and insemination.

Specifically, BOARBETTER was administered beginning 4 days post weaning to sows' noses using a garden-type sprayer calibrated to deliver approximately 4 mL of BOARBETTER. Farms utilizing PCAI administered a second dose of BOARBETTER after all sows in the group had been mated. Farms utilizing traditional AI administered a second dose of BOARBETTER just prior to insemination if AI did not occur at the same time as heat detection. All animals not mated by 7 days post-weaning were considered breeding failures and were recorded as such for data analysis.

Three groups of sows were removed from data analysis:

1. Sows mated prior to 4 days post-weaning. These animals were not exposed to BOARBETTER and were removed from both the BOARBETTER groups and the Control groups.
2. Nurse sows. Due to challenges in data capture and management in various record keeping systems, sows listed as nurse sows were excluded from both treatment groups.
3. Early-wean sows. Sows with 16 or fewer days of lactation were removed from both treatment groups because these animals may not cycle normally.

Data analysis involved measurement of the following parameters (also see Table 1):

Breeding rate – Number of sows mated as a percentage of weaned sows eligible to be mated

Conception rate – Number of sows confirmed pregnant as a percentage of sows mated

Table 1: Parameters measured

Key	Item	Example value	Measure	Formula	Example value
A	Weaned sow eligible to be mated	100			
B	Sows mated	95	Breeding rate	100*(B/A)	95%
C	Sows pregnant at ultrasound (d 35-45)	90	Conception rate	100*(C/B)	94.7%
		90	Conception success rate	100*(C/A)	90%
D	Sows farrowed	85	Farrowing rate	100*(D/C)	94.4%
		85	Farrowing success rate	100*(D/A)	85%

Conception success rate – Number of sows confirmed pregnant as a percentage of weaned sows eligible to be mated

Farrowing rate - Number of sows farrowed as a percentage of sows confirmed pregnant

Farrowing success rate – Number of sows farrowed as a percentage of weaned sows eligible to be mated

In addition, piglets born alive, stillborn and mummy numbers were quantified.

Farms were enrolled in field studies beginning in April of 2017 and continuing through August of 2017. Farm staff were trained by Animal Biotech personnel on the proper use of BOARBETTER. Product administration and data collection were completed by farm staff.

Results and discussion

Due to the start date and duration of the studies, complete results were not available at the time of publication.

Partial results from 2 farms were available prior to publication and are contained in Table 2. Complete results will be shared during the meeting in San Diego.

Previous results have shown improvements in breeding and farrowing rates of 6.8 and 8.9% respectively. These results can lead to significant reproductive improvement and economic returns. Early results from two field studies confirm earlier results demonstrating improvements in breeding, conception and farrowing results.

In general, BOARBETTER was reported as easy to use and useful in helping identify animals in estrus, particularly with inexperienced breeding technicians.

References

1. Study #AB2015-1
2. Study #AB2015-2
3. Study #AB2017-7
4. Study #AB2017-12

Table 2: Least squares means reproductive results from 2 field studies

Study number	ABT2017BBFS-1	ABT2017BBFS-2	Average
Number of sows at this farm	600	500	
Control breeding rate	96.7%	87.0%	91.8%
BOARBETTER breeding rate	95.6%	100.0%	97.8%
Breeding rate difference	**-1.1%**	**13.0%**	**6.0%**
Control conception rate	87.6%	92.5%	90.1%
BOARBETTER conception rate	87.4%	97.6%	92.5%
Conception rate difference	**-0.3%**	**5.1%**	**2.4%**
Control conception success rate	84.8%	80.4%	82.6%
BOARBETTER conception success rate	83.5%	97.6%	90.5%
Conception success rate difference	**-1.3%**	**17.1%**	**7.9%**
Control farrowing rate	78.7%	85.0%	81.8%
BOARBETTER farrowing rate	83.9%	87.8%	85.9%
Farrowing rate difference	**5.3%**	**2.8%**	**4.0%**
Control farrowing success rate	76.1%	73.9%	75.0%
BOARBETTER farrowing success rate	80.2%	87.8%	84.0%
Farrowing success rate difference	**4.1%**	**13.9%**	**9.0%**
Control born alive	11.68	11.26	11.47
BOARBETTER born alive	12.00	12.26	12.13
Born alive difference	**0.32**	**1.00**	**0.66**
Control stillborn	0.51	0.46	0.48
BOARBETTER stillborn	0.43	0.48	0.46
Stillborn difference	**-0.07**	**0.02**	**-0.03**
Control mummified	0.12	0.08	0.10
BOARBETTER mummified	0.06	0.04	0.05
Mummified difference	**-0.06**	**-0.04**	**-0.05**

Impact of sire-specific mortality on commercial swine profitability

D. W. Newcom[1]; C. Witte[2]; M. Brubaker[2]; S. E. Lawrence[2]
[1]National Swine Registry, West Lafayette, Indiana; [2]PureTek Genetics, Albion, Indiana

Introduction

Mortality, both pre- and post-weaning, is one of the most important economic drivers in any commercial swine production system. Genetic improvement programs have been extremely adept at making genetic improvement for economically important post-weaning traits such as growth, feed efficiency, and carcass composition, and for maternally-oriented traits such as litter size, litter weight, and rebreed interval. However, most have no measure of an individual's offspring's ability to survive to market. USDA records indicate the number of pigs marketed as a percentage of the total born has steadily decreased over the last 25 years.[1] As regulations are introduced to reduce antibiotic usage in food animals and more systems venture into antibiotic-free programs, robustness and survival will be key to the profitability of commercial swine production.

The purpose of this analysis is to describe the variation that exists in survival across progeny of known sires in a commercial setting, and assess the economic impact of these differences.

Materials and methods

Data from a Commercial Test Herd (CTH) was utilized in this study. Commercial pedigreed Yorkshire × Landrace females were bred to pedigreed purebred Duroc boars via single-sire inseminations. Piglets were identified at birth and tracked through the system to market. Cross-fostering was performed to balance litters, but was not tracked for individual pigs. Removals were recorded and noted in on-farm herd management software. Pigs were reared in commercial facilities where weight, 10th rib off-midline backfat and loin muscle depth was collected and lean percentage was calculated. Groups were off-tested on a single day to minimize extra labor.

Test groups with incomplete removal records or missing performance test data, and sires with fewer than 50 pigs born were removed prior to analysis. A total of 39,939 piglet birth records, sired by 220 Duroc sires, from 2,568 dams and 3,603 litters were utilized in the study. Pigs were performance tested in 52 test groups under commercial finishing conditions, with 32,874 pigs marketed. Age at removal (AR) was calculated as days from birth to removal and averaged for each sire. Performance test measures for days to market (DAYS), backfat (BF), loin muscle depth (LD), and lean percentage (LP) were adjusted to 127 kg. Pigs born, pigs marketed, and percent marketed were calculated for each sire.

Current genetic merit for each sire, in the form of Commercial Terminal Sire Index (CTSI), was utilized to determine economic impact of genetic differences for performance traits. The CTSI is a proprietary bioeconomic index derived from a combined purebred/crossbred evaluation that includes post-weaning performance measures from both purebred Duroc animals and Duroc-sired commercial progeny. Value per CTSI point is assumed at \$0.10 per pig produced, based on calculations from the Swine Testing and Genetic Evaluation System STAGES™ of the National Swine Registry (NSR). Value for mortality/removals was estimated under two scenarios: 1.) Fixed - assuming a cost of \$42.00 for an average pig removal (24 d), cost for each pig removed was fixed, regardless of age at removal, and 2.) Variable - assuming the cost of a pig at birth, weaning, and nursery of \$12.00, \$40.00, and \$65.00, respectively, cost by sire were adjusted based on the average age at removal of their commercial offspring. The total value per sire was calculated as the sum of the genetic value and mortality value. Two additional scenarios were investigated to determine the impact of removing the lower 16% of the sires, based on ranking the sires by either Percent Marketed or CTSI. Economic values are expressed as a deviation from the average of both CTSI and Percent Marketed across sire groups, for cohorts of 100 pigs born (approximately one week of semen production).

Results and discussion

Descriptive statistics for survival to market are shown in Table 1. As the data shows, sires vary greatly in their progeny's ability to survive to market; almost 40% between the best and worst sires. The average age at removal appears to be shortly after weaning (24 d), which would indicate many of the pigs removed are lost prior to weaning. However, there is an almost 60 d difference between the best and worst sire for age at removal, indicating some sires' progeny are being removed much later in the growing period, when additional costs of feed, vaccine, and housing are accumulated. The least cost associated with a pig is at birth, so it would be more economically advantageous for pigs to be removed as early as possible.

These data would indicate a slight trend for sires of higher genetic merit to have commercial progeny with lower survival rate to market (Figure 1). Since the genetic estimates are current, and not from the time the sires were initially selected, this could be causing this phenomena. Investigating the initial CTSI value at selection versus percent marketed would shed light on this conclusion, but was not available in this study. Although, it stands to reason that high-performing animals would expend most of their energy to performance and may not have the needed reserves to fend off infection, disease, or adverse environmental factors.

Table 1: Descriptive statistics for sire-group averages (220 Duroc sires) for survival to market

	Average	St. Deviation	Minimum	Maximum
Pigs born	181.5	75.2	52	528
Pigs marketed	149.4	63.2	33	436
Percent marketed, %	82.4	7.3	54.3	93.4
Removal age, d	24.1	10.3	8.9	66.5

Figure 1: Genetic merit versus percent marketed for commercial progeny of 220 Duroc sires

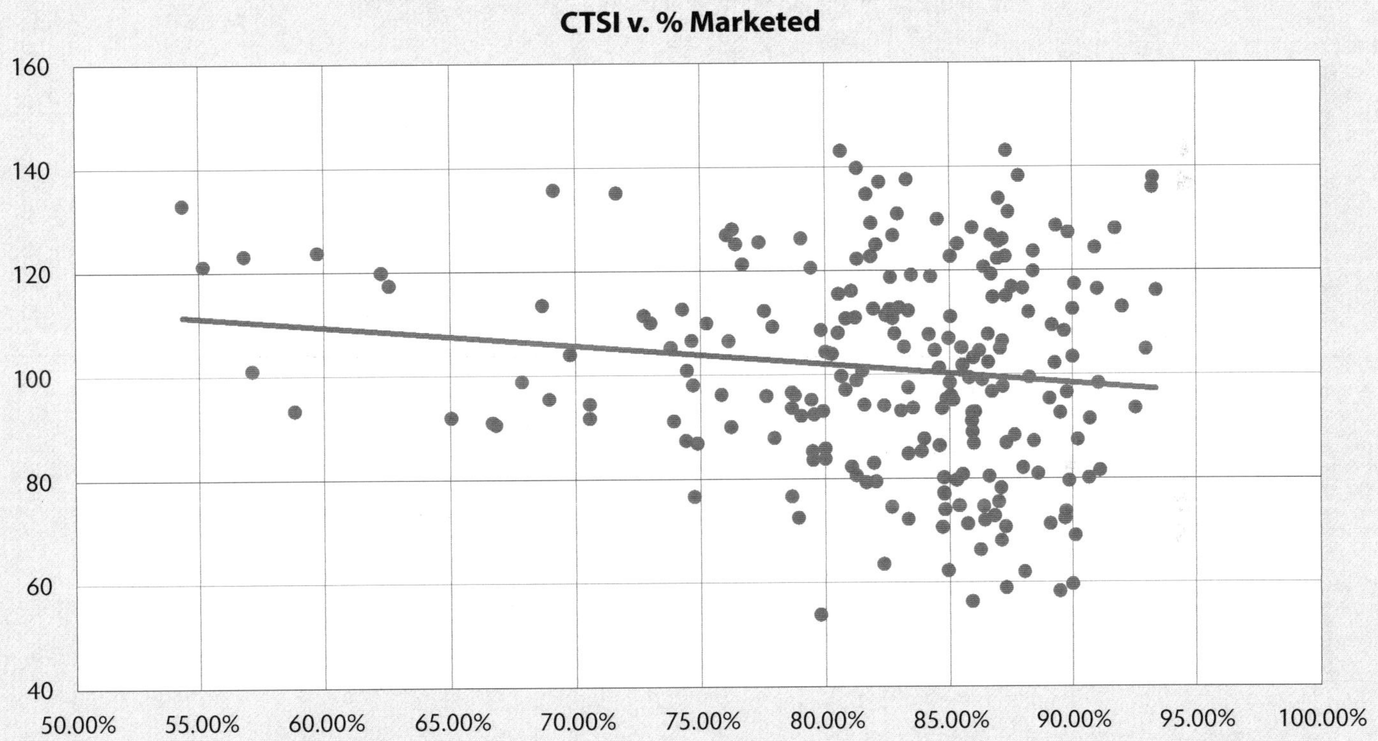

Figures 2 and 3 illustrate the relationship of the total value per sire per 100 pigs born with genetic merit and percent marketed, respectively, based on cost of removals adjusted for age at removal described above. Similar results were found for the fixed cost of removal analysis, only varying for individual sires at both tails of the age at removal distribution. Investigating these graphs together leads the authors to believe there is a closer relationship between percent marketed and total value than there is between genetic merit and total value. This conclusion is further supported by Table 2. When removing the bottom 16% of the sires based on percent marketed, there is a four-fold improvement in total value compared to removing the bottom 16% of sires based on genetic merit (~$0.80/pig versus ~$0.20/pig).

These data would suggest there is a slight relationship between genetic merit and percent marketed, with animals of higher genetic merit tending to produce commercial progeny with a lower rate of survival to market. Results also indicate more improvement in profitability can be made by increasing the percent of pigs going to market than by selecting for genetic merit alone. While breeding stock companies and seedstock suppliers have made rapid progress in performance traits in terminal breeds/lines, this study illustrates the need to include some measure of survival/mortality/percent marketed in the breeding goal. This type of analysis is not trivial, and solid removal records are critical to the success of including survival in any genetic program.

Figure 2: Genetic merit versus total value per sire per 100 pigs born

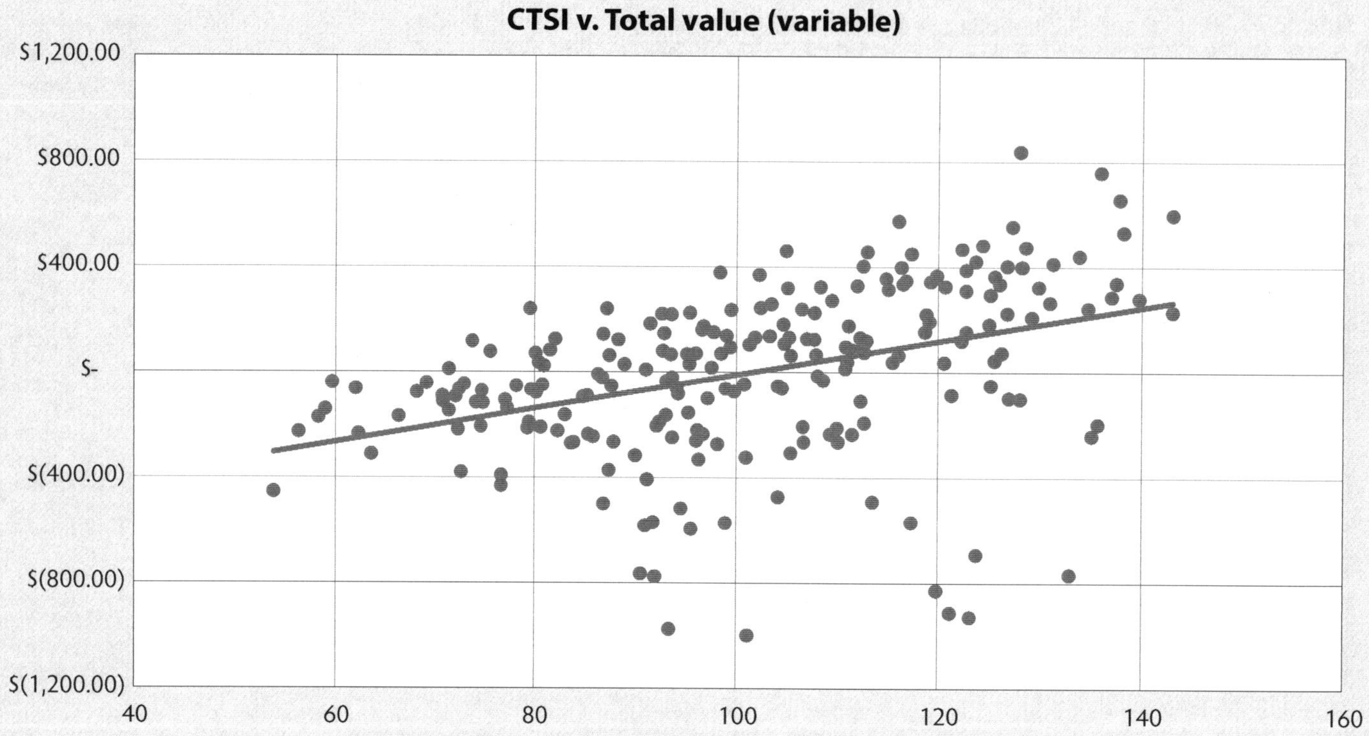

Figure 3: Percent marketed versus total value per sire per 100 pigs born

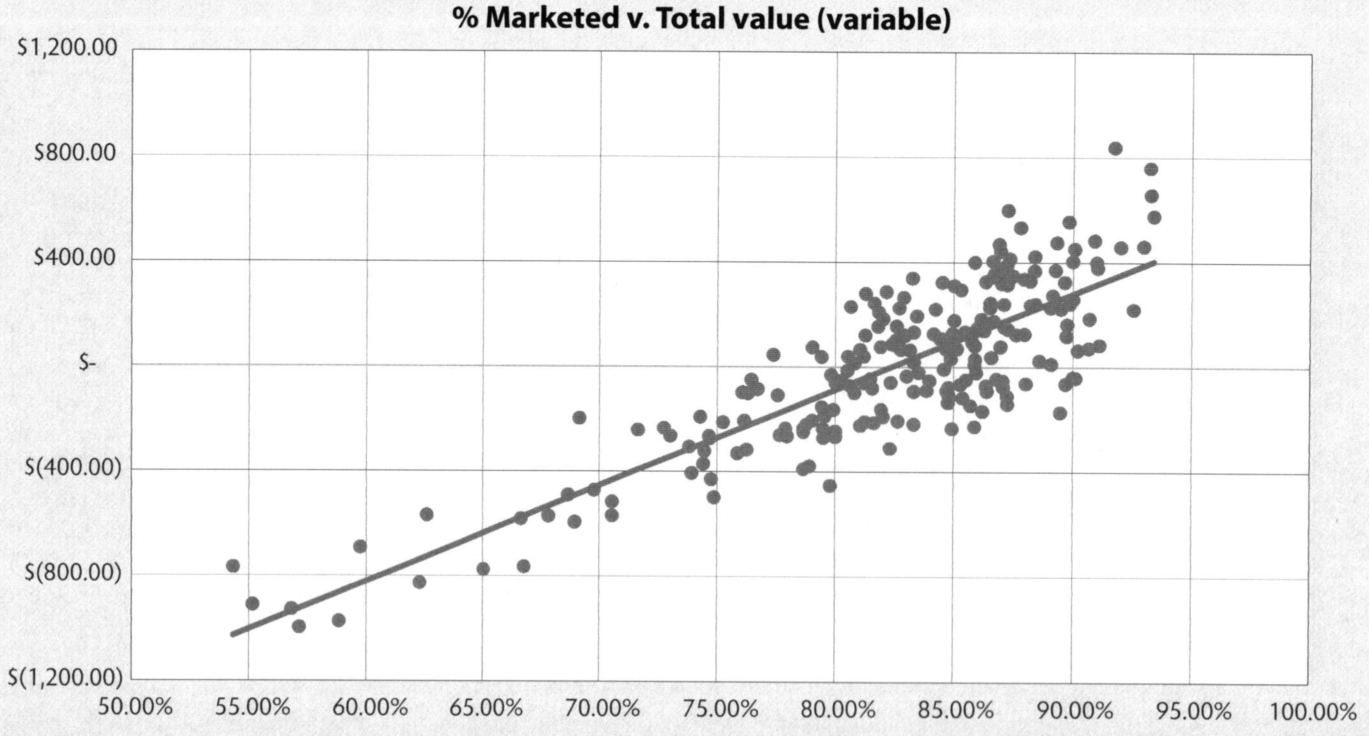

Table 2: Value of genetic merit and removals from commercial progeny groups of 220 Duroc sires

Scenario[1]		CTSI[2]	Percent marketed[3]	Value/100 pigs born – fixed[4]	Value/100 pigs born – variable[5]
All sires	Mean	101.2	82.4%	($1.94)	($1.76)
	St. dev.	20.0	7.3%	$334.31	$318.23
	Minimum	53.9	54.3%	($1,059.48)	($997.46)
	Maximum	143.0	93.4%	$799.23	$838.37
Mortality	Mean	100.6	84.6%	$89.53	$82.23
	St. dev.	20.5	4.1%	$239.91	$236.61
	Minimum	53.9	75.2%	($485.49)	($453.61)
	Maximum	143.0	93.4%	$799.23	$838.37
Genetic	Mean	106.3	81.9%	$20.69	$21.35
	St. dev.	16.4	7.6%	$350.16	$332.39
	Minimum	78.2	54.3%	($1,059.48)	($997.46)
	Maximum	143.0	93.4%	$799.23	$838.37

1 Scenarios: All sires included; Mortality – bottom 16% of sires removed, ranked by Percent Marketed; Genetic - bottom 16% of sires removed, ranked by genetic merit (CTSI)

2 CTSI = Commercial Terminal Sire Index, proprietary index of genetic merit based on a combined purebred/crossbred genetic evaluation, value assumed as $0.10/index point/pig produced

3 Progeny marketed per Duroc sire, as a percentage of the number born

4 Combined value of genetic merit and percent marketed for each sire group as a deviation from the group mean, based on a cohort of 100 commercial pigs born, and a fixed cost of $42.00 per pig removed

5 Combined value of genetic merit and percent marketed for each sire group as a deviation from the group mean, based on a cohort of 100 commercial pigs born, and a variable cost of $12.00, $40.00, and $65.00 for a pig at birth, weaning, and nursery, respectively, with cost per sire adjusted based on the average age at removal of their commercial offspring

References

1. Lemmon, M. US swine industry structure and disease control: A "wicked" problem. 2017. American Association of Swine Veterinarians Annual Meeting. Denver, CO.

Effect of Tonisity Px™ administration to sows and their piglets

Ava M. Firth[1], BS, DVM, MVS, MANZCVS, DACVECC, DECVECC; Guillermo López Cano[2], DVM MSc; Ahmad R. Seradj[3], PhD; Alberto Morillo Alujas[2], BS DVM MS PhD MBA
[1]Tonisity International, Dublin, Ireland; [2]Tests and Trials S.A., Monzón, Huesca, Spain;
[3]Department of Animal Science, ETSEA, University of Lleida, Spain

Introduction

Tonisity Px™ (Px) is a highly palatable isotonic protein solution that provides microenteral nutrition to the intestinal cells. Px is normally given as a 3% solution to neonatal and weaning-age pigs, but anecdotal reports have suggested that Px given as a top-dressing on sow feed could be helpful in increasing sow feed intake and associated productivity parameters in both sows and piglets. The objective of this study was to test this hypothesis and explore possible mechanisms of action.

Materials and methods

Setting

A study using Px was conducted in a sow farm located in Spain, during July and August 2017. The farm had experienced a PRRS outbreak in the previous month and had chronic issues with neonatal rotavirus diarrhea. The average daily temperature in the area during this period was 30°C (86°F), with maximum temperatures reaching 35°C (95°F). The farrowing rooms were not air-conditioned. The historic average of live-born pigs was 14.7 pigs/litter, but in the month previous to this study the average live-born was only 12 pigs/litter due to the PRRS outbreak. The average number of live-born pigs was 11.7 pigs/litter in this study.

Treatments

103 sows (Tai Zumu × Landrace × Large White) and their litters (1208 piglets) were enrolled in the study and randomized to one of three groups, which were Control (C), sows receiving 120 mL of Px (P120) or sows receiving 500 mL of Px (P500). Sows receiving Px were given either 120 mL or 500 mL of 3% Px solution once daily for 7 days pre-farrowing and once daily for 7 days pre-weaning. Sows were fed according to the farm's standard feed curve three times daily, and were given the Px solution as top-dressing on their second feeding on treatment days.

All litters in the 3 treatment groups were given 500 mL of Px/litter/day in an open pan, once daily from day 2-8 of age. All sows and piglets had access to an automatic drinker. Creep feed was offered to all litters beginning at 10 days of age.

The usual farm treatment for neonatal diarrhea consisted of a combination injection of amoxicillin and gentamicin. Any treatments were administered by barn staff according to standard operating protocols and were not controlled by the study protocol.

Data collection

At entry to the farrowing house, individual sow weight and back fat was measured. Each sow's feed intake was measured daily, and samples of feed were taken daily to allow calculation of dry matter intake throughout lactation. Sow weight and back fat was measured again on the day of weaning.

Within 24 h of birth, piglets were individually ear-tagged and weighed. Blood samples for serum immunoglobulin assay were taken from 20 piglets from primiparous sows in each treatment group at 24 h of age. Piglets were again weighed at 6 days before weaning, and at weaning (~ 21 days of age). Average daily gain (ADG) at weaning was calculated based on actual days of age. Piglet mortalities (with probable cause of death and corpse weight) were recorded daily. Fecal scores per litter and individual antibiotic injections given to either sows or piglets were recorded daily.

Statistical analysis

Statistical analysis of sow-related variables including body weight and back fat was performed using a generalized linear model with the sow as the experimental unit, using parity class (primiparous, multiparous) and treatment group as co-factors. Statistical analysis for piglet body weight, ADG, creep feed intake, fecal score and survival rate was performed using a mixed model with the sow (litter) as the experimental unit and including sow parity class as a co-factor. The number of piglets receiving one or more antibiotic injections (Amoxicillin and Gentamicin) per treatment (C, P120 and P500) and parity class were tabulated in contingency tables and the probabilities were calculated using Fisher's exact test. Odds ratios were also calculated for antibiotic usage in piglets using a chi-square test. All statistical analysis was performed using SAS software (version 9.4).

Results

Antibiotic usage

Results of the study showed statistically significant ($P < .01$) differences in the number and percentage of pigs in each group that were given antibiotic injections. Both treatment groups had significantly fewer treated pigs compared to the control group (Table 1 and Figure 1). The difference in antibiotic requirements between the control group (38.8% of those piglets) and the P120 group (24.4% of those piglets) represents a 37.1% reduction in

Table 1: Percentage of piglets receiving antibiotic injections

	C	P120	P500
Total number of pigs in group	415	406	387
Number and % of pigs receiving antibiotic injections	161 (38.8%)[a]	99 (24.4%)[ab]	56 (14.5%)[b]

[a,b] Different superscripts within rows denote significant differences between treatment groups at $P < .01$

antibiotic usage. The difference in antibiotic requirements between the control group (38.8% of those piglets) and the P500 group (14.5% of those piglets) represents a 62.6% reduction in antibiotic usage.

Odds ratios were also calculated for the likelihood of antibiotic injections in each treatment group. Piglets in the control group were 3.75 times more likely to receive an antibiotic injection compared to piglets in the P500 group. Similarly, piglets in the control group were 1.97 times more likely to receive antibiotic injections than piglets in the P120 group. Finally, piglets in the P120 group were 1.91 times more likely to receive antibiotic injections than piglets in the P500 group. All of these odds ratio differences were statistically significant at $P < .01$. See Table 2.

Piglet weight gain

There was no significant difference in piglet birth weight between the groups. However, at weaning, piglets in the P500 group weighed 400 g (0.88 lb) more than piglets in the C group ($P < .05$). Piglets in the P120 group weighed 200 g (0.44 lb) more than piglets in the C group ($P = .31$) (Table 3).

Other piglet parameters

No significant differences were detected in serum immunoglobulin levels, survival rate or average fecal score between treatment groups.

Sow parameters

The average lactation period was not significantly different between sows in the treatment groups, ranging from 20.0 to 21.1 days. There were no significant differences between treatment groups in sow weight loss during lactation, dry matter intake or back fat loss.

Conclusions and discussion

Supplementing sow feed with 3% Px solution in the pre-farrowing and pre-weaning stages reduced antibiotic usage in piglets by a range of 37- 62%. Similarly, weaning weights were increased by 200 - 400 g (0.44 - 0.88 lb) which is a marked increase in productivity. Both effects followed a dose-response curve, where weaning weight and antibiotic usage showed a larger effect in sows receiving 500 mL Px compared to sows receiving 120 mL Px, and in turn, compared to control sows.

Figure 1: Percentage of piglets treated with antibiotics

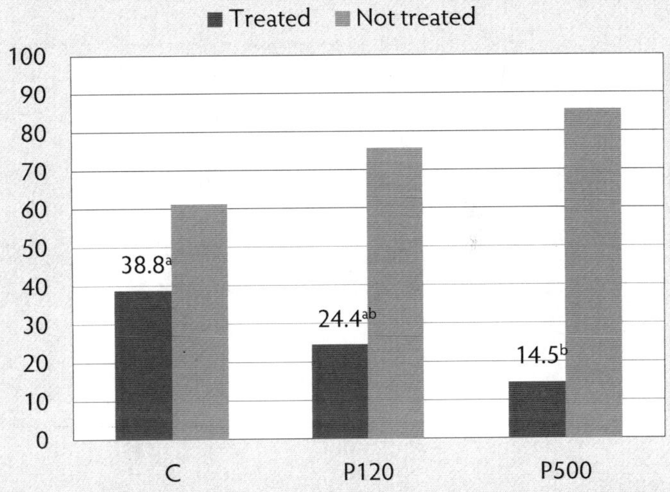

[a,b] Different superscripts denote significant differences between treatment groups at $P < 0.01$.

This study was conducted under challenging environmental and health conditions which are encountered in many production operations. The piglet productivity gains and reduction in antibiotic usage were both statistically and practically significant. The economic and veterinary benefits of this reduction in antibiotic usage are highly relevant to todays' challenges in swine production.

Further studies to determine the optimum dose and timing of Px supplementation for sows and to further explore the possible mechanisms of action are being conducted.

Table 2: Odds of receiving antibiotic treatment

Treatments	Odds ratio	95% CI	*P* value
C vs P500	3.75	2.653 - 5.291	< .01
C vs P120	1.97	1.455 - 2.654	< .01
P120 vs P500	1.91	1.326 - 2.739	< .01

Table 3: Pig weights and average daily gain†

	C	P120	P500
Birth weight (kg)	1.4 ± 0.04	1.4 ± 0.04	1.4 ± 0.04
Weaning weight (kg)	4.4 ± 0.15[a] (95% CI 4.1 – 4.7)	4.6 ± 0.15 kg[ab] (95% CI 4.3- 4.9)	4.8 ± 0.15[b] (95% CI 4.5 – 5.1)
ADG (g/day)	148.4 ± 7[a]	162.7 ± 7[ab]	172.5 ± 7[b]

† Least squares means
[ab] Different superscripts within rows denote significant differences between treatment groups ($P < .05$)

Production performance evaluation of Porcilis™ Ileitis in a commercial finisher using a *Lawsonia intracellularis* seeder pig challenge model

Brett O'Brien[1], VMD, MS; Scott Dee[2], DVM, PhD, Dipl ACVM; Joel Nerem[2], DVM; Roy Edler[2], MS; Dan Hanson[2], BS; Brad Thacker[1], DVM, PhD, Dipl AVBP; Robyn Fleck[1], DVM; David Renter[3], DVM, PhD

[1]Merck Animal Health, Madison, New Jersey; [2]Pipestone Applied Research, Pipestone, Minnesota; [3]Kansas State University, Manhattan, Kansas

Introduction

Ileitis (*Lawsonia intracellularis*) is one of the most common enteric diseases of finishing pigs in the United States with estimated costs of more than $100 million annually.[1] The costs incurred by ileitis are primarily a result of increased mortality, decreased average daily gain (ADG), reduced feed utilization, increased market weight variation, and increased medication costs. In order to help mitigate the impact of this disease, veterinarians and producers utilize commercially available vaccines like Porcilis™ Ileitis and Enterisol® Ileitis in their swine operations. Porcilis Ileitis entered the US marketplace in late 2015 as the first inactivated *Lawsonia intracellularis* vaccine administered parenterally in swine. Initial licensing data and subsequent field applications have demonstrated the efficacy of Porcilis Ileitis in protecting pigs against ileitis.[2] The objective of this study was to compare the production performance between pigs vaccinated with Porcilis Ileitis or Enterisol Ileitis, and non-vaccinated controls that were challenged with *Lawsonia intracellularis* using a seeder pig challenge model.

Materials and methods

Pigs (n = 2103) from two PRRSV negative commercial sow farms were enrolled after weaning into a commercial wean-to-finish barn at three weeks of age. The source sow farms were located on the same site and shared similar genetics. Both sow farms weaned IAV-S positive pigs but were no longer in the acute phase of their outbreaks. The pigs were tagged, weighed, and sexed at enrollment. 240 pigs in the middle weight tertile for the group were randomly selected as seeder pigs. The remaining enrolled pigs were blocked by sow farm, gender, and weight and randomly assigned to one of three treatment groups (T01:Enterisol Ileitis, T02:Controls, and T03:Porcilis Ileitis) resulting in 624 animals per treatment group.

At enrollment, pigs in the same treatment groups were housed by pen at a stocking density of 40 pigs per pen. Pens were randomly assigned to treatment group such that there was equal representation of treatment pens in both rooms of the barn. The Porcilis Ileitis treatment group was vaccinated IM with 2 mL of Porcilis Ileitis two days after enrollment. At 7 weeks of age, animals were rehoused such that the Enterisol pigs were housed in one room while the Porcilis, control, and seeder pigs were housed in the other room. Enterisol Ileitis was then administered per label via the water to the Enterisol treatment group. Pigs were vaccinated per normal farm protocol for PCV2, *Mycoplasma hyopneumoniae*, PRRSV, and *Erysipelas*. At ten weeks of age, seeder pigs were challenged with 50 mL of a pre-quantified gut mucosal homogenate containing *Lawsonia intracellularis* (provided by Dr. Nathan Winkelman) by oral gavage.[3] Five seeder pigs were then moved to each pen of 40 pigs to start the process of exposing the principal pigs. Two weeks post-challenge (study day 0), all pigs were commingled such that each of 78 pens contained 8 pigs per treatment group and 3 seeder pigs for a total of 27 pigs per pen. A review of the study timeline is shown in Table 1.

Throughout the entire study, all feed was non-medicated and every feed delivery was tested to confirm the absence of antibiotics. Caretakers were blinded to treatment. All animals were weighed on study days -63 (allocation), 0 (commingling), 43, and 88 (two days prior to first cut). Treatment records were maintained throughout the entire study in order to determine treatment rates. Mortalities and hospital pen removals were recorded throughout the study. At the time of final weight, pigs were categorized as either a 1) Full value pig (≥ 230 lbs. at final market in good health with no physical abnormalities); 2) Light cull (< 230 lbs. at final market); or 3) Defect cull (physical abnormality such as inguinal hernia, scrotal hernia, and tail bite or lameness).

Nasal swabs were obtained on a subset of pigs at allocation for IAV-S PCR and strain subtyping. Oral fluids were obtained (6 ropes/room) monthly and tested at the ISU VDL by PCR for *Lawsonia*, IAV-S, *M hyopneumoniae*, and PCV2. Tissues were submitted from selected mortalities for a complete diagnostic workup in order to characterize other health challenges circulating within the study's population. In addition, an approximately 5 inch section of distal ileum from pigs of all treatment groups that died post-challenge was preserved in formalin and submitted for determination of *Lawsonia* infection by histological examination and *Lawsonia* IHC. Paired serum and fecal samples were collected from 30 randomly selected pigs per treatment group. Serum was collected on study days -63, -14, and 43 and tested by *Lawsonia* ELISA. Feces was collected on study days -35, -14, 0 and 22 and tested by *Lawsonia* qPCR.

Table 1: Overview of study design

Study Day	Age of pigs (wks.)	Trial activity	Housing
-63	3	Allocation, Porcilis vaccination	In pen by treatment, all treatments in each room
-35	7	Rehouse pigs, Enterisol vaccination	In pen by treatment, Enterisol in same room
-14	10	Challenge and introduce seeder pigs	In pen by treatment, Enterisol in same room
0	12	Commingle and start weight	All treatments in each pen in both rooms
43	18	Mid-finishing weight	All treatments in each pen in both rooms
88	24.5	Final weight	All treatments in each pen in both rooms

Descriptive statistics were used for the data collected prior to commingling (d0). Weights and ADG were analyzed using generalized linear mixed models using JMP (SAS, Cary, NC). Mortality, full value pigs, culls and treatment rates were compared using a generalized linear mixed model (SAS). A repeated measures analysis was performed on fecal shedding data. Statistical significance was assigned at $P < 0.05$ for all analyses.

Results

Nasal swabs and oral fluids confirmed that influenza A (H1N2) was circulating in piglets at the onset of the trial and extended through 10 weeks of age, at the time of *Lawsonia* challenge, but were negative for IAV-S by 15 weeks of age. Oral fluid results confirmed PCV2 was endemic throughout the study but no clinical PCVAD was observed. All oral fluid results were *M hyopneumoniae* negative throughout the study.

The *Lawsonia* seeder pigs challenge model was effective in eliciting clinical disease in this study. Clinical signs of ileitis were present in the study population starting at two weeks post-challenge as noted by bloody, watery diarrhea, wasting, and skin pallor. *Lawsonia* PCR results on oral fluids were 100 percent positive at all sampling points following *Lawsonia* challenge. Likewise, ileum samples collected from mortalities post-challenge yielded a confirmatory diagnosis of *Lawsonia* infection based on histopathology and IHC in both the control and seeder pigs present (44% of controls were positive and 62% of seeder pigs). There was no evidence of *Lawsonia* infection in tissues submitted from any vaccinated pigs.

The serological data shown in Figure 1 demonstrates that the Porcilis vaccinated pigs were *Lawsonia* ELISA positive following vaccination with an amnestic response observed post-challenge. The Enterisol and Control groups were *Lawsonia* ELISA negative at the time of challenge but seroconverted after challenge. The fecal shedding results are shown in Figure 2. Enterisol pigs were shedding *Lawsonia* (suspected vaccine strain) at the time of challenge. Control pigs shed significantly higher amounts of *Lawsonia* in feces five weeks post-challenge than either of the vaccinated groups. There was no statistical difference in the levels

Figure 1: *Lawsonia* ELISA (means).

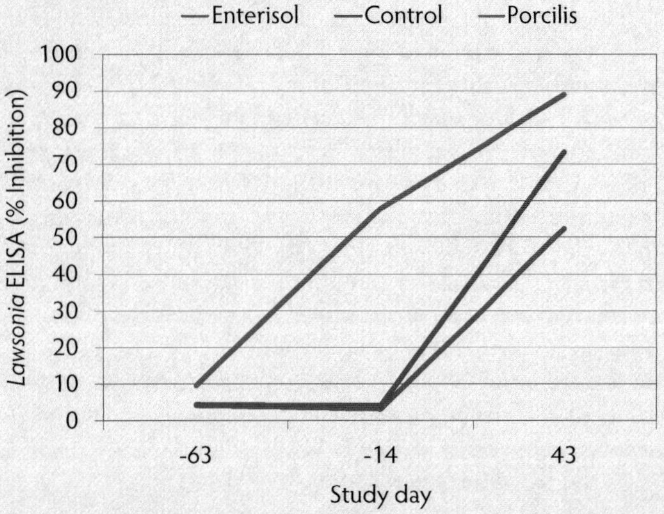

Figure 2: *Lawsonia* fecal qPCR data (median values)

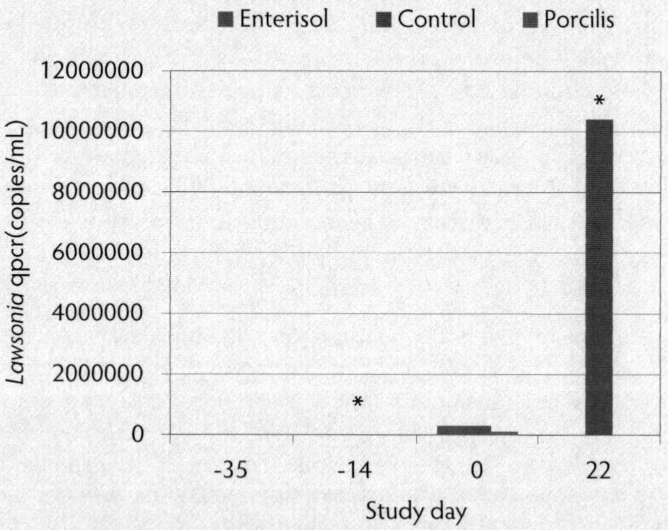

* Significant difference at $P \leq 0.05$

Table 2: Production performance data from study day 0 to 88 (model-adjusted means)

Parameter	Porcilis Ileitis	Enterisol Ileitis	Control	*P*-value
Weaning weight (lbs.)	13.0	13.0	13.0	0.27
Commingling weight (lbs.)	77.6[ab]	76.3[b]	77.9[a]	0.04
Final weight (lbs.)	253.7[a]	254.2[a]	247.2[b]	< 0.0001
ADG comm. to final (lbs/d)	1.98[a]	2.00[a]	1.90[b]	< 0.0001
Mortality (%)	1.6	1.6	2.6	0.42
Hospital pen removals (%)	0.7[b]	1.0[b]	3.2[a]	0.001
Full value pigs (%)	97.9	98.4	96.9	0.23
Light cull (%)	1.3	1.0	2.0	0.27
Defect Cull (%)	0.8	0.7	1.0	0.43
Treated pigs (%)	1.8	1.3	2.9	0.10

[a,b] Means in rows with different superscripts are significantly different ($P \leq 0.05$).

of *Lawsonia* shed in feces between the Porcilis or Enterisol vaccinated pigs post-challenge.

The production performance data for the treatment groups is summarized in Table 2. Vaccinated pigs had significantly higher final weights than controls but there was no significant difference between the two vaccinated groups (T01 = 254.2 lbs, T02 = 247.2, and T03 = 253.7, $P < 0.0001$). Likewise, both vaccinated groups outperformed controls for ADG from study day 0 (commingling) to the end of the study with Porcilis pigs averaging 1.98 lbs./day, Enterisol pigs averaging 2.00 lbs./day, and Control pigs averaging 1.90 lbs./day ($P < 0.0001$). There was no significant difference between ADG results of the two vaccinated treatment groups. There were fewer pigs pulled to the hospital pen in vaccinated pigs versus control pigs (T01 = 1.0%, T02 = 3.2%, T03 = 0.7%, $P = 0.001$). There were no significant differences between any of the three treatment groups in % mortality, % full value pigs, % culls, and treatment rates although the controls were numerically worse for all of these parameters.

Discussion

Porcilis Ileitis was effective in minimizing the clinical impact of *Lawsonia intracellularis* when compared to controls as evidenced by improved ADG, fewer hospital pen removals and reduced fecal shedding. Pigs vaccinated with Porcilis Ileitis were 6.5 pounds heavier at the time of marketing than unvaccinated pigs in this study. This study demonstrates that Porcilis Ileitis which is a given parenterally was just as effective in controlling the impact of *Lawsonia* on performance in a pig population as Enterisol which is given orally. Also, with Porcilis Ileitis there are no restrictions on administration of feed and water medications around the time of vaccination as is required with Enterisol administration. The ability to treat pigs when needed can be especially important in early health challenged flows. As this study demonstrates, producers can have confidence that Porcilis Ileitis is an effective tool to help reduce the clinical and production impacts of *Lawsonia intracellularis* in their pigs.

References

1. McOrist S. Defining the full costs of endemic porcine proliferative enteropathy. Vet J 2005:170(1); 8-9.

2. Thacker B., et al. Field Experiences with Porcilis™ Ileitis: The newest tool for controlling *Lawsonia intracellularis*. AASV.2017;191-194.

3. Winkelman N., et al. Validation of an ileitis mucosal homogenate seeder pig model to study the natural transmission of *Lawsonia intracellularis* in grow-finish pigs. AASV. 2017;170-174.

Efficacy of swine influenza virus replicon particle vaccination including cellular and humoral responses after challenge

Pravina Kitikoon, DVM, PhD; Channing Sebo, DVM; Mark Mogler, PhD; Chandra Morgan, BS; Heather Dempsey, BS; Brad Thacker, DVM, PhD; Erin Strait, DVM, PhD
Merck Animal Health, De Soto, Kansas

Introduction

Influenza A virus in swine (IAV-S) is an important respiratory disease affecting all ages of pigs worldwide. Infection is typically associated with low mortality but high morbidity resulting in reduced growth rate, decreased feed efficiency and increased medication costs, which have negative economic impacts to the swine industry.[1] Vaccination is frequently used as an intervention strategy to control disease induced by IAV-S. Available commercial and autogenous vaccines consist of killed viruses (KV) containing different virus subtypes and strains. The KV vaccines primarily induce serum antibodies against the major envelope glycoproteins, hemagglutinin (HA) and neuraminidase (NA) which provide protection against disease.

The protective capacity of vaccine-induced antibodies is affected by the rate of viral antigenic drift that causes mismatches of the vaccine strains to field strains. Accordingly, timely updates of IAV-S vaccine strains can improve vaccine efficacy. Currently in the US, the ongoing IAV-S surveillance program provides genetic information of currently circulating strains and a virus repository to facilitate strain updates for commercial vaccines. However due to the nature of the conventional KV vaccine platform, the time from when a vaccine strain is selected, propagated, inactivated and formulated to the availability of the final updated KV vaccine can be lengthy. In contrast, the alphavirus replicon technology can quickly produce a non-replicating RNA particle (RP) vaccine that is safe. The RP vaccine platform does not require virus isolation, propagation and inactivation resulting in a more rapid strain update process that will enhance vaccine efficacy by better vaccine strain matching to the field situation.[2]

The objective of the RP vaccine efficacy study reported here was to compare the immunogenicity and efficacy of the H1-RP vaccine to a prototype adjuvanted H1N2-KV vaccine when tested against a homologous H1N2 IAV-S challenge. In addition, interferon gamma cytokine was measured to evaluate the cell mediated immunity (CMI) response to both vaccine platforms. We also report here recent field experiences with regard to antibody levels following mass vaccination of sow herds and our current efforts to develop an overarching information management system to capture and analyze IAV-S sequencing data and corresponding demographic data in order to effectively design and utilize RNA particle IAV-S vaccines in the field.

Materials and methods

RP vaccine efficacy study

Thirty, three-week-old pigs, serologically naïve to IAV-S H1N2 strain A/swine/IL/A01475495/2014 (H1N2) and from a herd free of *Mycoplasma hyopneumoniae* and porcine reproductive and respiratory syndrome virus were assigned to 3 treatment groups. At 4 weeks of age (WOA), pigs were intramuscularly (IM) administered one of the following test articles: Hemagglutinin 1 (H1)-RNA particle (RP) without adjuvant, killed virus (KV) vaccine blended with 50% X-Solve adjuvant or saline. The KV vaccine group served as a positive control, industry standard, for immunological and efficacy comparisons and the saline group served as the naïve-challenge, non-vaccinated control group. A booster dose of the same preparations was administered IM at 7 WOA, 3 weeks after the first vaccination. The H1 gene in the H1-RP and KV vaccine were generated based on the challenge strain A/swine/IL/A01475495/2014 (H1N2). At 10 WOA, pigs were challenged intra-tracheally (IT) with the H1N2 virus. All pigs were necropsied at 5 days post challenge (DPC).

Vaccine efficacy was assessed by evaluating nasal shedding, virus levels in lungs by virus titration in cell culture and percent lung lesions at 5 DPC.[3] Nasal swabs were collected at 1, 3 and 5 DPC. Serum samples were collected prior to each vaccination, prior to challenge and at necropsy. Serological response to vaccination was evaluated by the hemagglutination-inhibition (HI) test.[3] Titers are reported as the reciprocal of the highest serum dilution that inhibited hemagglutination. For CMI response evaluation, whole blood was collected from all pigs prior to challenge for peripheral blood mononuclear cell (PBMC) isolation. Isolated PBMCs were stimulated with H1N2 virus and evaluated for interferon gamma secretion as an indication of cellular immune response to vaccination.[4] All results were analyzed using analysis of variance (ANOVA) with $P \leq 0.05$ considered significant.

Sow herd mass vaccination field experience

A bivalent prescription RP vaccine was prepared using sequences derived from two different H1N2 variants (S and N) that were endemic in the production system. The vaccine was administered twice, 4 weeks apart to all sows in the system. In order to evaluate post-vaccination serological responses, serum was obtained from three-week-old pigs just prior to weaning in two sow herds and at three time points to determine sow serostatus prior to mass vaccination, after the first vaccination (just prior to the second vaccination) and four weeks after the second vaccination. Serum was tested by HI using reference strains that were highly homologous to the two field strains that were used to design the RP vaccine. The highest dilution measured was 1:640. Separate tests were performed for each variant. Data was analyzed by Kruskal-Wallis non-parametric ANOVA and Dunn's All-Pairwise Comparison Test using Statistix™ 10 (Analytic Software, Tallahassee, FL).

Information management system development

Creating prescription RNA particle IAV-S vaccines requires a sequence of the HA protein. Samples such as nasal swabs, tissue and oral fluids are collected from clinically affected animals, submitted to state diagnostic laboratories for testing and HA sequences are generated. These sequences are provided to Merck Animal Health (MAH) electronically and an analysis is generated in order to build a custom prescription RP vaccine. Sequences are analyzed in a variety of ways including homology comparisons at the amino acid level, dendogram trees and putative amino acid matching along with geospatial analysis to provide regional information to the overall analysis. In the past, each analysis was a labor-intensive process that was not scalable or widely available. Recognizing the need for better data management for our customers, a custom, solution- driven program was built. The overall goal of the new software was to facilitate the transfer of sequencing data and analyze the data in a fast, accurate and easily accessible manner. This analysis tool is not limited to IAV-S. The program is adaptable to any pathogen including emerging diseases and current uses include rotavirus, porcine reproductive and respiratory syndrome (PRRS), porcine parainfluenza virus 1 (PPIV1) and circoviruses.

Results and discussion

RP vaccine efficacy study

The level of H1N2-induced pneumonia, virus titer in lungs at 5 DPC and nasal shedding at 1, 3 and 5 DPC are summarized in Table 1. The severity of pneumonia, virus load in lungs and nasal shedding of virus were not different between the H1-RP and H1N2-adjuvanted KV vaccinated pigs. Both vaccinated groups had significantly reduced lung lesions and virus in lungs at necropsy compared to the naïve-challenge control group. Nasal shedding in both the H1-RP and H1N2-adjuvanted KV vaccinated pigs at all time points was significantly lower than the naïve-challenge control group.

Serologically, all pigs were seronegative based on HI to the H1N2 challenge virus at the beginning of the study. Pigs vaccinated with the H1N2-adjuvanted KV seroconverted by 3 weeks after the first vaccination. At 3 weeks post second vaccination, both H1-RP and adjuvanted H1N2-KV vaccinated pigs had high HI titers (≥ 320). The naïve-challenge control pigs were seronegative at all time points prior to challenge.

For evaluating CMI responses just prior to challenge, PBMCs were collected from all pigs and stimulated with the H1N2 challenge virus to evaluate for interferon gamma secretion in the cell culture supernatant. The presence of interferon gamma is an indication of cell mediated immunity that is induced from a recall response to H1N2 virus exposure. Pigs vaccinated with H1-RP had 2.7- and 16-fold higher levels of interferon gamma than pigs vaccinated with adjuvanted-H1N2 KV and naïve-challenge control pigs, respectively.

Sow herd mass vaccination field experience

The HI antibody titers before and following each vaccination are shown in Table 2. Antibody titers to both strains increased following the first vaccination. Titers after the second vaccination were mostly numerically higher but statistically not different to the titers induced by the first vaccination. Farm differences were detected as Farm B had significantly ($P < 0.05$) or numerically higher titers at all 6 time point/reference strain combinations.

Information management system development

The new data management system has generated a large, integrated database of pathogen and strain information for several diseases. Our sequence information gathering began in 2015 and there are now over 2,900 sequences in the database. Approximately 48% of the sequences are IAV-S field isolates. With IAV-S, H1's are more prevalent than H3's and over half of the H1's are H1N2-deltas.

Table 1: Average of percent lung lesions, virus in lungs, nasal shedding and immune responses in pigs vaccinated with H1-RP or adjuvanted-H1N2 KV and naïve-challenge control pigs

Group (n = 10)	% Lung lesion	Virus in lungs (log_{10} $TCID_{50}$/mL ± STD)	Nasal shedding (log_{10} $TCID_{50}$/mL ± STD)			HI titer to H1N2 (Reciprocal geomean titer)			IFN-g prior to challenge (pg/mL ± SEM)
			1 DPC	3 DPC	5 DPC	0 DPV1	0 DPV2	0 DPC	
H1-RP	0.4[a]	0.9 ± 0.0[a]	0.5 ± 0.0[a]	1.0 ± 1.1[a]	0.5 ± 0.0[a]	< 10	10	368[b]	877.4 ± 277.4
KV	0.2[a]	0.9 ± 0.0[a]	0.5 ± 0.0[a]	1.0 ± 1.1[a]	0.5 ± 0.0[a]	< 10	39	557[b]	318.6 ± 183.0
Saline	12.3[b]	2.7 ± 1.2[b]	3.2 ± 1.3[b]	4.6 ± 0.4[b]	3.0 ± 0.6[b]	< 10	< 10	< 10[a]	54.8 ± 33.0

Means with different letters within columns are statistically different ($P \leq 0.05$).
DPC = days post challenge, DPV1 = days post 1st vaccination, DPV2 = days post 2nd vaccination

Approximately 30% of the sequences in the database are rotaviruses with rotavirus C sequences dominating this group followed by rotavirus A and then rotavirus B. PRRS, PPIV and circovirus sequences are also captured in the database. The maximum amino acid homology variation within AIV-S and rotavirus types is presented in Table 3.

Conclusions

RP vaccine efficacy study

The intent of this study was to compare the antibody and cell mediated immune responses to H1-RP versus a prototype adjuvanted-H1N2 KV vaccine and assess vaccine efficacy against a homologous H1N2 virus infection in weaned pigs. The study outcome demonstrated that H1-RP vaccine was as efficacious as the adjuvanted-H1N2 KV vaccine as it significantly reduced the level of pneumonia, virus in lungs and nasal shedding compared to the naïve-challenge control group. It was also shown that H1-RP vaccine can induce equivalent serum HI antibody compared to the adjuvanted, H1N2 KV vaccine.

Cell mediated immunity is known to play a major role in viral clearance and recovery from IAV infection.[5] This study evaluated CMI response to vaccination by assessing the level of interferon gamma prior to challenge. It was demonstrated that H1-RP vaccinated pigs had 16-fold higher levels compared to the naïve-challenge control pigs and almost 3-fold higher levels compared to the adjuvanted-H1N2 KV vaccinated pigs. This finding further demonstrated the RNA particles have a strong tropism towards dendritic cells resulting in a more balanced immune response which is another advantage over the KV vaccine platform where CMI responses can be limited.[2,5]

Sow herd mass vaccination field experience

Vaccination with the prescription RP vaccines increased titer levels for each of the two reference strains in both herds. The magnitude of response appears to be positively correlated with the titer levels prior to vaccination. The reason for this difference in titer levels prior to vaccination is unknown but could be related to the timing of exposure to the field strains tested. In most situations, sow herds have pre-existing immunity to IAV-S and vaccination results in boostering of this pre-existing immunity. In this case, the pre-existing titers were relatively low in both herds. The boostering of these low titers demonstrates the value of using a prescription RP vaccine that is specifically designed to address

Table 2: IAV-S antibody levels in three-week-old pigs from dams before and after mass sow herd vaccination

		Reciprocal geomean titers by farm		
H1N2 reference	**Vaccination status**	**Farm A**	**Farm B**	**Both farms**
	Pre	16.3$^{a;c}$	40.0$^{b;c}$	18.0^{c}
S	Post 1st	70$^{a;d}$	189.1$^{b;d}$	89.8^{d}
	Post 2nd	121.3$^{a;d}$	237.0$^{b;d}$	122.2^{d}
	Pre	10.3$^{a;c}$	15.6$^{a;c}$	12.7^{c}
N	Post 1st	37.9$^{a;d}$	120.1$^{b;d}$	68.9^{d}
	Post 2nd	74.6$^{a;d}$	105.6$^{a;d}$	88.5^{d}

^{a,b} Values with different superscripts within the same row (between farm comparison) were significantly different ($P \le 0.05$).
^{c,d} Values with different superscripts within the same column (between vaccination status comparison) were significantly different ($P \le 0.05$).

Table 3: IAV-S and rotavirus maximum amino acid homology variation in sequences from MAH data base

Pathogen	Type	No. of sequences	Maximum homology variation
	H3N2-Clade 4 and human-like	370	23.81%
	H1N1-gamma	285	5.13%
IAV-S	H1N1-beta	20	9.19%
	H1N1-pandemic	130	19.82%
	H1N2-delta 1 and 2	575	32.14%
	Type A	300	33.84%
Rotavirus	Type B	120	49.60%
	Type C	440	38.55%

the IAV-S strains that are circulating in the sow herd in order to increase the level of herd immunity.

Information management system development

With the abundant diversity of IAV-S present throughout the United States swine population and lack of cross protection among strains, the need for custom prescription RP vaccines is evident. Building and managing custom strain matched vaccines is labor intensive and requires a high level of detail. The creation of the new data management program assists customers in not only building vaccines, but tracking pathogen spread within their system and region. In addition, the program can provide a valuable tool for detecting emerging strains and pathogens.

References

1. Vincent AL et al. Swine Influenza Viruses: A North American Perspective. *Adv V Res.* 2008: 127-154.

2. Vander Veen RL et al. Alphavirus replicon vaccines. *Ani Health Res Rev.* 2012: 1-9.

3. Kitikoon P et al. Pathogenicity and Transmission in Pigs of Novel A (H3N2) Influenza Virus Isolated from Humans and Characterization of Swine H3N2 Viruses Isolated in 2010-2011. *J Virol.* 2012: 6804-6814.

4. Olson ZF et al. Factors affecting induction of peripheral IFN-γ recall response to influenza A virus vaccination in pigs. *Vet Immunol Immunopath.* 2017: 57-65.

5. Tamura S et al. Mechanisms of Broad Cross-Protection Provided by Influenza Virus Infection and Their Application to Vaccines. *Jpn J Infect Dis.* 2005: 195-207

6. Sebo C et al. RNA particle vaccines: Bringing solutions to emerging and rapidly mutating pathogens. AASV. 2016; 215.

Efficacy and field safety of Prime Pac PRRS™ RS, a new tool against porcine reproductive and respiratory syndrome virus

E. Strait, DVM, PhD; F. Roerink, MS; S. Knetter, PhD; C. Morgan, BS;
B. O'Brien, VMD, MS; B. Thacker, DVM, PhD
Merck Animal Health, De Soto, Kansas

Introduction

PRRS remains one of the most costly swine diseases in the United States with estimated costs of more than a billion dollars annually.[1] In light of the significant impact on production that producers face with PRRSV, it is important for them to have tools, such as modified-live vaccines, that can help mitigate the impact of the disease in their systems. A new tool that can be used in the fight against PRRSV is Prime Pac PRRS RS which is a modified-live virus vaccine labeled for "the vaccination of healthy swine 3 weeks of age or older against Porcine Reproductive and Respiratory Syndrome (PRRS) virus, respiratory form." Licensing study data summarized here demonstrate the protection Prime Pac PRRS RS provides pigs in the face of a heterologous PRRSV challenge while also being a safe product under field conditions.

Materials and method

Efficacy study

The efficacy study was performed in two locations using the same protocol. At each site, 64 PRRSV negative pigs were enrolled at 3 weeks of age and were randomly assigned to one of two treatment groups with even distribution of treatments within each litter. During the pre-challenge period, treatment groups were housed in separate rooms. One group received 1 mL of Prime Pac PRRS RS intramuscularly at 3 weeks of age while the positive controls received a placebo injection at the same time.

All pigs were challenged intranasally with a heterologous PRRSV strain (NADC 20, kindly supplied by Dr. Lager from the NADC, USDA-ARS) four and a half weeks after vaccination. Pigs were commingled immediately prior to challenge such that each room contained an equal number of pigs from the two treatment groups.

All pigs were necropsied 14 days post-challenge. Gross lung lesion scores, the primary study variable, were assigned by a pathologist that was blinded to treatment. Macroscopic lung lesions were given a score to estimate the percentage of the lung affected by pneumonia.[2] Three tissue sections were collected from each set of lungs for histopathology and immunohistochemistry (IHC) evaluation. Microscopic lesion severity and IHC staining intensity were scored on a 0 to 4 scale (Lesion severity: 0 = no lesions, 1 = mild, 2 = moderate multifocal, 3 = moderate diffuse and 4 = severe; IHC: 0 = no PRRSV antigen positive cells, 1 = 1-10 cells, 2 = 11-30 cells, 3 = 31-100 cells, and 4 ≥100 cells per section). The maximum score per animal was used in the analysis.

Individual body weights were recorded one day prior to vaccination, immediately prior to challenge and at necropsy. Blood samples and nasal swabs were collected at enrollment, immediately prior to NADC 20 challenge (0 DPC) and at 3, 7, 10, and 14 days post-challenge and tested by qPCR to determine the level of PRRSV viremia and shedding profiles. Blood samples were also tested by IDEXX PRRSV X3 Ab test to determine serostatus before and after challenge.

Statistical analysis was performed by one way ANOVA on lung lesion scores and average daily weight gains; Kruskal-Wallis non-parametric ANOVA on viremia data; two-sided Cochran-Armitage Trend Test on histopathology and IHC scores; and Fisher's Exact Test on binary variables. Statistical significance was assigned at $P \leq 0.05$.

Field safety study

A field safety evaluation was conducted in three locations (Texas, Iowa, Minnesota) using a total of 677 healthy pigs that were on average 3 weeks of age. Pigs were vaccinated with 1 mL of Prime Pac PRRS RS intramuscularly in the neck. All vaccinated pigs were monitored closely for systemic or local reactions for at least one hour post-vaccination followed by daily observations for the subsequent 14 days. Injection sites were palpated one day after vaccination for evaluating localized reactions. In addition, pigs were observed daily for overall heath and all mortalities were necropsied to determine cause of death.

Results

Efficacy study

Prime Pac PRRS RS vaccinated pigs demonstrated a significant reduction in lung lesions versus controls after the NADC 20 PRRSV challenge (31.8% vs. 9.3%, combined average), demonstrating a 71% reduction overall (Figure 1, $P < 0.0001$). More than two-thirds of the controls had lung lesion scores greater than 20% in contrast to markedly fewer vaccinates with scores of that magnitude (Site 1 and 2 – 12.5% and 6.3%). Likewise, on histological examination of lung tissue, the maximum microscopic lesion scores and IHC scores in vaccinates were significantly lower than controls (Table 1). In both sites, the percentage of control pigs with moderate-diffuse to severe histopathology scores 3 and 4

Figure 1: Average lung lesion scores in Control and Prime Pac PRRS RS vaccinated pigs

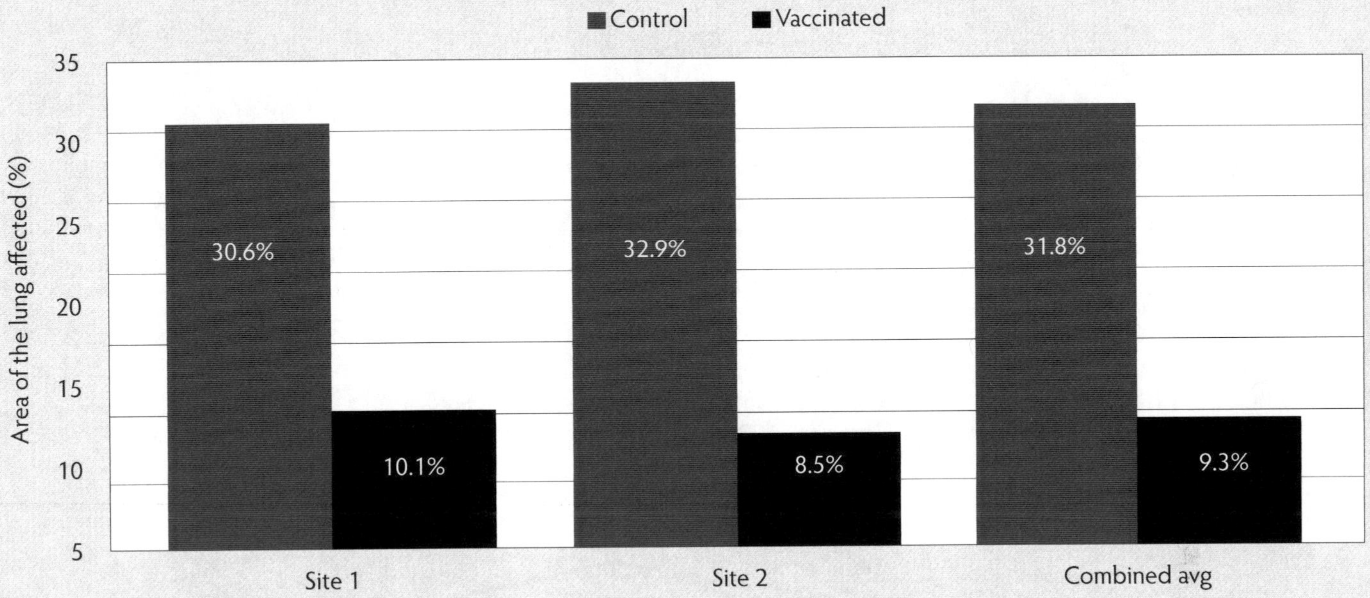

Table 1: Histopathology and immunohistochemistry scores from both sites

Variable	Treatment group	Percentage of pigs with scores 0-4				
		0	1	2	3	4
Lung IHC scores	Control	0.0%	12.5%	15.6%	42.2%	29.7%
	Vaccinates	45.3%	28.1%	17.2%	7.8%	1.6%
Lung histopath scores	Control	0.0%	1.6%	9.4%	20.3%	68.8%
	Vaccinates	0.0%	15.6%	59.4%	20.3%	4.7%

was greater than the vaccinated pigs (89% vs. 25%). For the IHC results, 88% of controls and 27% of vaccinates had >10 staining cells per section scores 2 to 4.

All pigs were confirmed PRRSV negative at enrollment via PRRSV PCR and ELISA. At the time of challenge, almost all of the vaccinated pigs were still viremic with presumably vaccine virus while control animals remained negative. Vaccinated pigs seroconverted by the time of challenge while controls remained seronegative until challenge and then seroconverted thereafter. Viremia levels are shown in Figure 2. The groups were 100 percent viremic three days post-challenge with peak viremia occurring at seven days post-challenge. Vaccinates had significantly less virus in serum compared to controls at all sampling points after challenge ($P < 0.01$). There was an overall 77% reduction in peak viremia levels in vaccinated pigs in comparison to the controls. The amount of virus in nasal swabs was much less than in serum at all sampling points. However, at the end of the study there were significantly fewer pigs shedding PRRSV in vaccinated pigs than controls (42% vs. 77%, $P < 0.01$).

As shown in Figure 3, Prime Pac PRRS RS vaccinated pigs outperformed the controls on average daily gain (ADG) during the post-challenge period. Prime Pac PRRS RS vaccinated pigs grew on average 95 grams more per day than the unvaccinated controls after challenge with NADC 20, which equates to almost a three pound growth advantage by the end of the 14-day period ($P < 0.0001$).

Field safety study

There were no injection site or anaphylactic reactions observed following vaccination and no adverse events were attributable to vaccination. Throughout the study, four mortalities occurred, but all were attributed to causes unrelated to vaccination with Prime Pac PRRS RS.

Figure 2: Median levels of PRRSV in serum after challenge with NADC 20

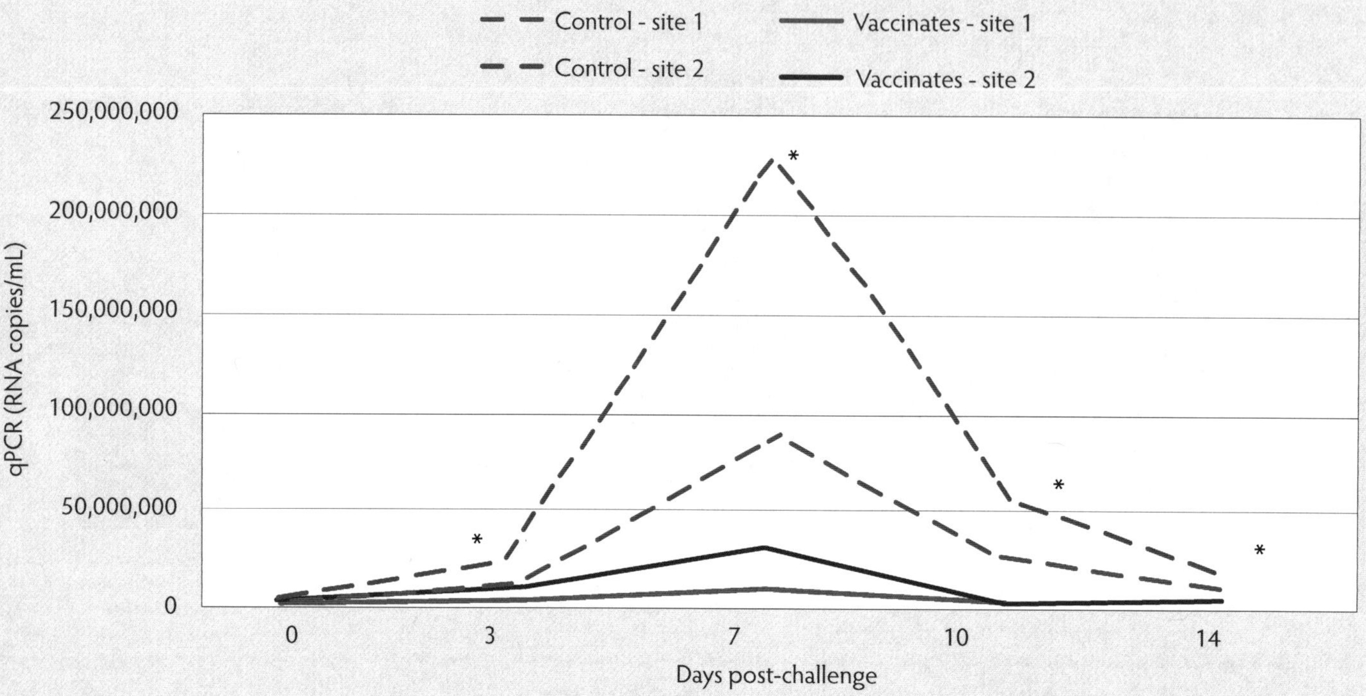

Figure 3: Post-challenge average daily gain (ADG)

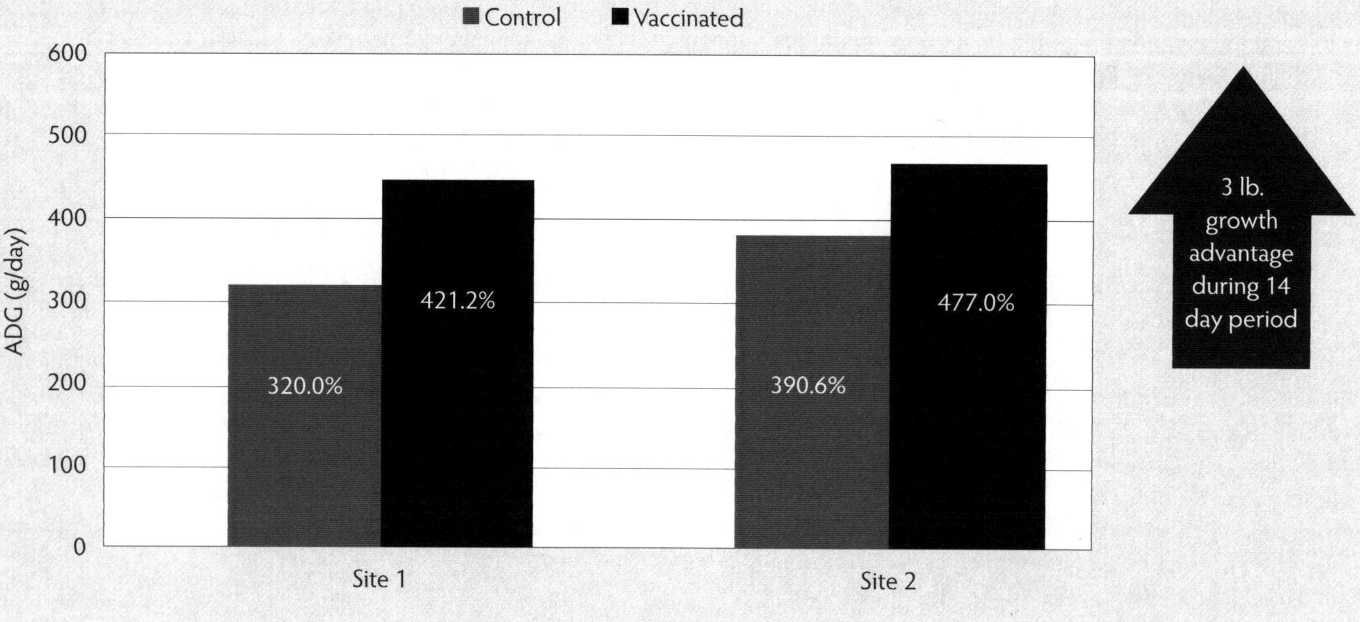

Conclusions

Prime Pac PRRS RS can be an effective management tool shown to reduce the respiratory impact in the face of a heterologous PRRSV challenge. Vaccinated pigs in the efficacy studies showed improved growth performance post-challenge compared to controls. In addition, vaccinated pigs had reduced viral loads during peak viremia. Reduced viral load has been reported to reduce the risk of wild type virus shedding.[3] Post-vaccination clinical monitoring indicated that Prime Pac PRRS RS is safe to use in 3-week-old pigs. Prime Pac PRRS RS is a safe product that can be incorporated into vaccination protocols as a tool to help mitigate the impact of PRRSV introductions in swine herds.

References

1. Holtkamp, D., et al. Assessment of the economic impact of porcine reproductive and respiratory syndrome virus on the United States pork producers. *J Swine Health and Prod.* 2013. 72-84.

2. Halbur, P.G., et al. Comparison of the Pathogenicity of Two US Porcine Reproductive and Respiratory Syndrome Virus Isolates with that of the Lelystad Virus. *Vet Path.* 1995. 32:648-660.

3. Zimmerman, J. PRRS virus transmission. AASV. 2007. 479-484.

4. Data on file, Merck Animal Health.

Swine models: An integrated approach to providing management and nutritional solutions

N. S. Ferguson[1]; M. Swantek[2]; D. Woods[1]
[1]Trouw Nutrition, Guelph, Ontario, Canada; [2]Trouw Nutrition USA, Highland, Illinois

Introduction

Since the 1970's a number of swine growth models have been proposed to help improve our understanding of, and decision-making in growing pigs.[1-10] However, the application of these models has had limited commercial success due in part to low precision and accuracy in predicting animal performances under a wide range of commercial conditions, and to poor execution into daily business activities. Assimilation of an accurate biological swine model that predicts the growth and feed intake of individual pigs within a population, with a dynamic feed formulator and optimization capabilities, affords the opportunity to provide meaningful management and nutritional advice in a constantly changing production environment.

Integrated model

Swine production operates in a volatile market and for this reason, an integrated management system (NutriOpt® Swine Model) was developed and has been applied in commercial practice. The key components of the model include a stochastic, mechanistic animal growth model, a formulator and an optimization procedure. The theory and biology of the animal growth component are based on scientific evidence published over the last 30 year. The model incorporates the interactions between individual animals and their health status, feed, and both the physical and social environment. With these interactions, it is possible to predict individual voluntary feed intake and subsequent body tissue growth of individual pigs.

In practice, the NutriOpt® Swine Model enables users to optimize for a multitude of criteria such as improved economic or biological performance. Having the ability to examine in detail not only the biological responses of the animal but also the economic responses to management and nutrition changes, is a powerful aspect of simulation modelling that sheds the past paradigm that improving performance would necessitate a benefit in profitability. The use of an integrated model considers the cause and effect relationship of formulating swine diets for specific outcomes and how we can make meaningful changes of the correct magnitude at the correct time to yield optimal results. A key to accurate prediction of feed intake and growth in pigs, is understanding the social interaction between individual pigs and how their ability to cope with social stressors, including health challenges, impacts how much feed they desire to eat and their ability to grow to their potential.

Social stressors

There are a number of medium to long term social stressors that prevent pigs from growing to their potential, including health and well-being, pig space (stocking density) and feeder space.[11-14] For example, the amount of feeder space available to a group of pigs can become a constraining factor when not every pig can satisfy its desired intake. Similarly, high stocking density and poor health can both prevent an individual pig from exhibiting their potential growth. It is known that socially dominant individuals perform better than their subordinates when exposed to these stressors,[15,16] furthermore, it would appear that larger pigs, for the same age, tend to be more dominant and less affected by social stress than smaller pigs. This "ability to cope" with stress is unique to an individual pig and is correlated to its genetic potential. This means individuals with a lower ability to cope will have, for example, proportionally less time at the feeder than socially dominant and larger pigs when feeder space is limiting. They will also respond in a similar way to insufficient pig space or health challenge within the barn.

Health

The health status of an individual pig is an important determinant of whether the animal is able to express its genetic growth potential as well as defining its reduced growth capacity. In addition, the health condition of the animal will influence the nutrient requirements due to the different amino acid ratios required for immune activation, as well as increased maintenance costs for fever and recovery of body tissue damaged by the health challenge. The NutriOpt® model uses a relative health and well-being scoring system based on mortality, incidence and severity of coughing and fecal consistency, environment and air quality and animal welfare. Completing a health score sheet will provide an estimate of the health status ranging from severely challenged through to optimal with the majority of herds falling between. Although simplistic by design, the health score facilitates the average health status of the batch of pigs, from which individual pigs will have a variance about this mean. The accuracy of predicting feed intakes and growth rates of a batch of pigs is significantly enhanced when including the contribution of health status. While the model not designed for specific health challenges (e.g PRRSV), it will allow for more detailed analysis of the impact health has on animal performance, requirements and economics under commercial conditions.

® Registered Trademark of Trouw International B.V.

Commercial application

For successfully commercial application of an integrated model, it should be able to be used strategically to drive significant nutritional, production and/or manufacturing changes with large economic consequences for the provider and the customer. However, it should also have the capability to be used incrementally on-farm to improve performance or profitability or help identify nutritional and management problems and solutions. Some examples of commercial benefits to applying the NutriOpt® model include a) farm specific performance and profit improvements; b) identifying the financial and performance consequences of production changes (e.g changing genetics); c) adopting nutritional strategies to maximize profits or minimize costs with changes in commodity and pig prices; d) identify optimum shipping weights that capitalize on ideal market weights for specific packing plants; e) nutrient requirements for specific health status, genetics and genders, including immune-castrated animals; and f) measure changes in carcass characteristics and fatty acid profiles relative to ingredient changes in diet formulations.

In conclusion, the NutriOpt® swine model is a new generation stochastic optimization swine model that provides a powerful decision-making tool for implementing optimum management and nutritional solutions and where appropriately used can bring significant cost savings and profit enhancement to both small producers and large integrated customers.

References

1. Whittemore and Fawcett. Theoretical aspects of a flexible model to simulate protein and lipid growth in pigs. *Animal Production*. 1976;22:87.

2. Black et al. Simulation of energy and amino acid utilization in the pig. *Res. Dev. Agric.* 1986;3:121.

3. Pomar et al. Computer simulation of swine production systems. 1: Modeling the growth of young pigs. *J. Anim. Sci.* 1991;69:1468.

4. Ferguson et al. Preferred components for the construction of a new simulation model of growth, feed intake and nutrient requirements of growing pigs. *South African Journal of Animal Science*. 1994;24:10.

5. Moughan et al. Modelling growth in the pig. *Wageningen Press, Wageningen, Netherlands*. 1995

6. Knap. Simulation of growth in pigs: evaluation of a model to relate thermoregulation to body protein and lipid content and deposition. Anim. Sci. 1999;68:655.

7. Birkett and de Lange. A computational framework for a nutrient flow representation of energy utilization by growing monogastric animals. Br. J. Nutr. 2001;86:661.

8. Green and Whittemore. Architecture of a harmonized model of the growing pig for the determination of dietary net energy and protein requirements and of excretions into the environment. *Anim. Sci.* 2003;77:113.

9. Wellock et al. Modelling the effects of thermal environment and dietary composition on pig performance: model logic and concepts. *Anim. Sci.* 2003a;77:256.

10. Van Milgen et al. InraPorc: a model and decision support tool for the nutrition of growing pigs. *Anim. Feed Sci. Tech.* 2008;143:387.

11. Kournegay and Notter. Effects of floor space and number of pigs per pen on performance. *Pig News Info*. 1984;5:23.

12. Nielsen. Effect of group size on feeding behavior, social behavior, and performance of growing pigs. *Appl. Anim. Behav. Sci.* 1995;47:235.

13. Gonyou and Stricklin. Effects of floor area allowance and group size on the productivity of growing/finishing pigs. *J. Anim. Sci.* 1998;76:1326.

14. Ferguson et al. The effect of stocking density on the responses of growing pigs to dietary lysine. *Anim. Sci.* 2001;73:459.

15. Randolph et al. Effects of group size and space allowance on performance and behavior of swine. *J. Anim. Sci.* 1981;53:922.

16. Botermans. Feeding environment for growing-finishing pigs. Effects of competition for feeding frequency on performance, behavior, injuries, plasma cortisol and exocrine pancreatic secretion. *Acta Universitatis Agriculturae Sueciae – Agraria*. 1999;192:1.

Trends of mycotoxin contamination in US corn

E. G. Hendel[1], VMD, PhD; S. Masching[2], DVM; T. Jenkins[2], PhD; P. N. Gott[1], PhD; G. R. Murugesan[1], DVM, PhD
[1]BIOMIN America, Inc., Kansas City, Kansas; [2]BIOMIN Holding GmbH, Getzersdorf, Austria

Introduction

Mycotoxins are harmful secondary fungal metabolites and are of key concern to food and feed safety globally. In addition to compromised performance, mycotoxins negatively impact animal health. Although classic symptoms such as decreased feed intake and vomiting are utilized in the field as an indicator for exposure, even at low levels mycotoxins act as predisposing factors for diseases by causing inflammation,[1] immune suppression,[2] and modulating the gastrointestinal environment.[1] BIOMIN's annual corn survey indicated increasing trends of prevalence and co-occurrence of toxins produced by *Fusarium* ssp. fungi in US corn in prior years.[3] The data herein presents the most current information regarding mycotoxin contamination of samples from the 2017 corn harvest in comparison to the previous five years, as well as comparisons with international data.

Materials and methods

Corn samples for use in multiple livestock industries (cattle, swine, and poultry) were collected from mid-August to October 2017 from the United States and submitted for analysis utilizing the LC-MS/MS method. Six major mycotoxin groups were analyzed: aflatoxins, type A trichothecenes, type B trichothecenes (i.e. DON; vomitoxin), fumonisins (FUM), zearalenone (ZEN) and ochratoxin-A. The data of this sample pool of 2017 US corn were compared to previous years to examine trends of mycotoxin contamination (Figure 1, 2, and 3). International data from Europe, Brazil, and China submitted for analysis over this time period utilizing multiple methodologies is provided as a general comparison (Table 1).

Results and discussion

A high proportion of samples of the 2017 corn harvest from the US contained at least one detected mycotoxin type (96%, 101 samples), similar to the situation in 2016 (96% positive, 549 samples). However, this prevalence is slightly higher when compared to the sample pool of 2016 in October (2017: 96%, 2016: 83%, 77 samples). The prevalence of co-occurrence in samples thus far is lower than 2016 (≥ 2 mycotoxins, 2017: 43%, Oct. 2016: 58%, Overall 2016: 77%). The most prevalent mycotoxin, DON, has similar prevalence and average contamination levels compared to last year's corn crop (2017: 86%, 1505 ppb 2016: 87%, 1687 ppb). Although prevalence of FUM and ZEN are less than then overall sample pool from 2016, it is similar to FUM and higher to ZEN when compared to the 2016 sample pool in October (2017: FUM, 36% ZEN: 31%, October 2016: FUM: 34%, ZEN, 14% Overall 2016: FUM: 77%, ZEN, 56%). The percent of positive samples above threshold levels for the three fusarium toxins are reduced from ZEN and B-Trich from 2016, but still high for FUM, and still elevated compared to 2015. (2017: 26%, 24%, 57%, 2016: 44%, 47%, 58%, 2015: 28, 17, 30%, % positive samples over 200 ppb ZEN, 1000 ppb DON, and 1000 ppb FUM respectively).

The preliminary results from the 2017 corn harvest suggest a continued risk due to mycotoxins produced by Fusarium fungal species. The prevalence of ZEN and FUM in US corn seems to be less compared to corn from international markets and emphasizes the importance of evaluating imported corn for mycotoxin risk prior to use. Because of continued risk of multi-mycotoxin contamination in samples thus far, multiple mitigation strategies are needed beyond adsorption, including biotransformation and supporting the immune system and liver function.

References

1. Pinton et al. Toxicity of Deoxynivalinol and its acetylated derivatives on the intestine: differential effects on morphology, barrier function, tight junction proteins, and mitogen-activated protein kinases. *Toxilogical Sciences*. 2012; 130(1): 180-190

2. Pierron et al. Impact of mycotoxin on immune response and consequences for pig health. *Animal Nutrition*. 2016; 2: 63-68.

3. E.G. Hendel, P.N. Gott, T.Jenkins and G.R. Murugesan. Mycotoxin survey in 2016 US Corn, *Biomin World Mycotoxin Survey 2016*. 2016.

Figure 1: Percent positive samples for Afla, ZEN, DON and FUM are presented by year (due to small numbers of positive samples, T2 and OTA are not represented graphically).

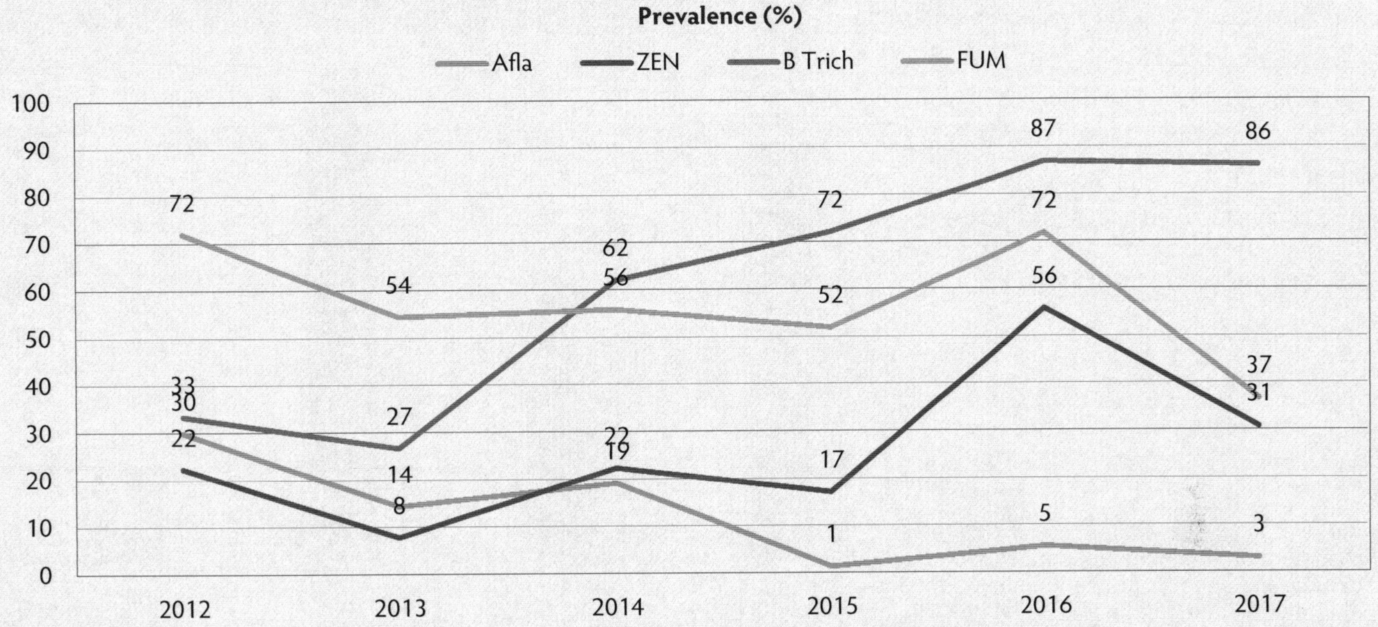

Figure 2: Percent of samples below the limit of detection, with one, or more than one mycotoxin for each year

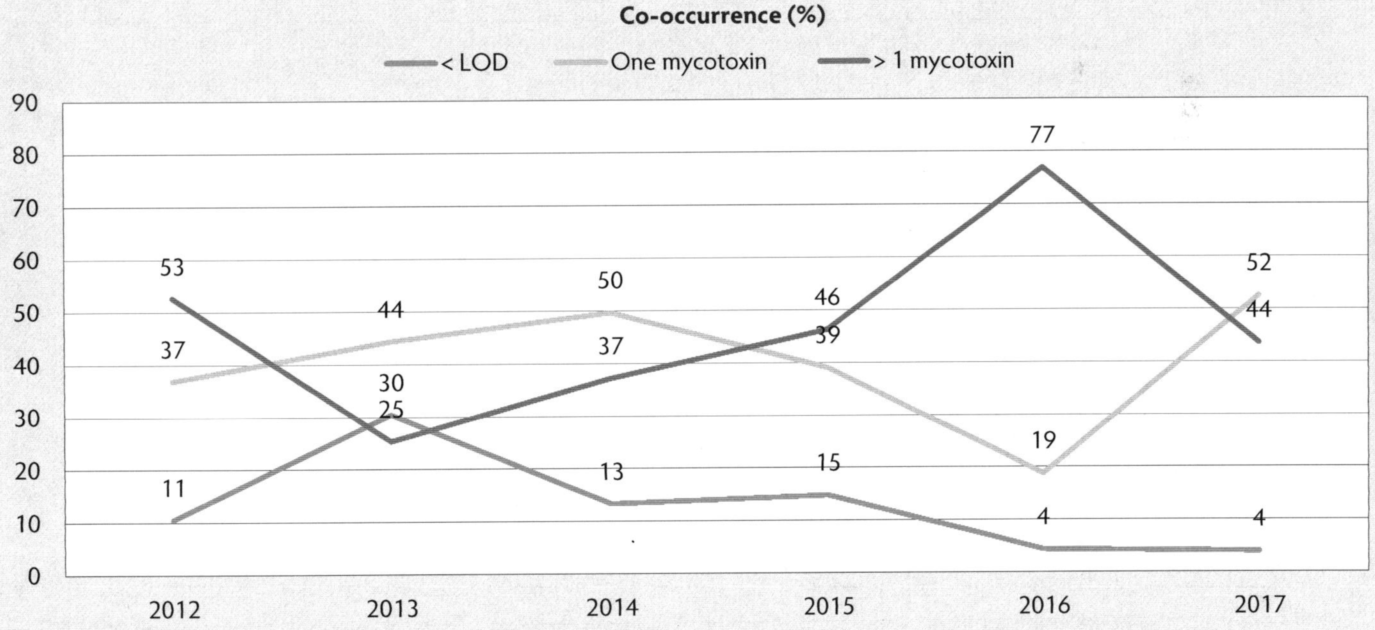

Figure 3: Number of US corn samples for each year from 2012 through the 2017 harvest.

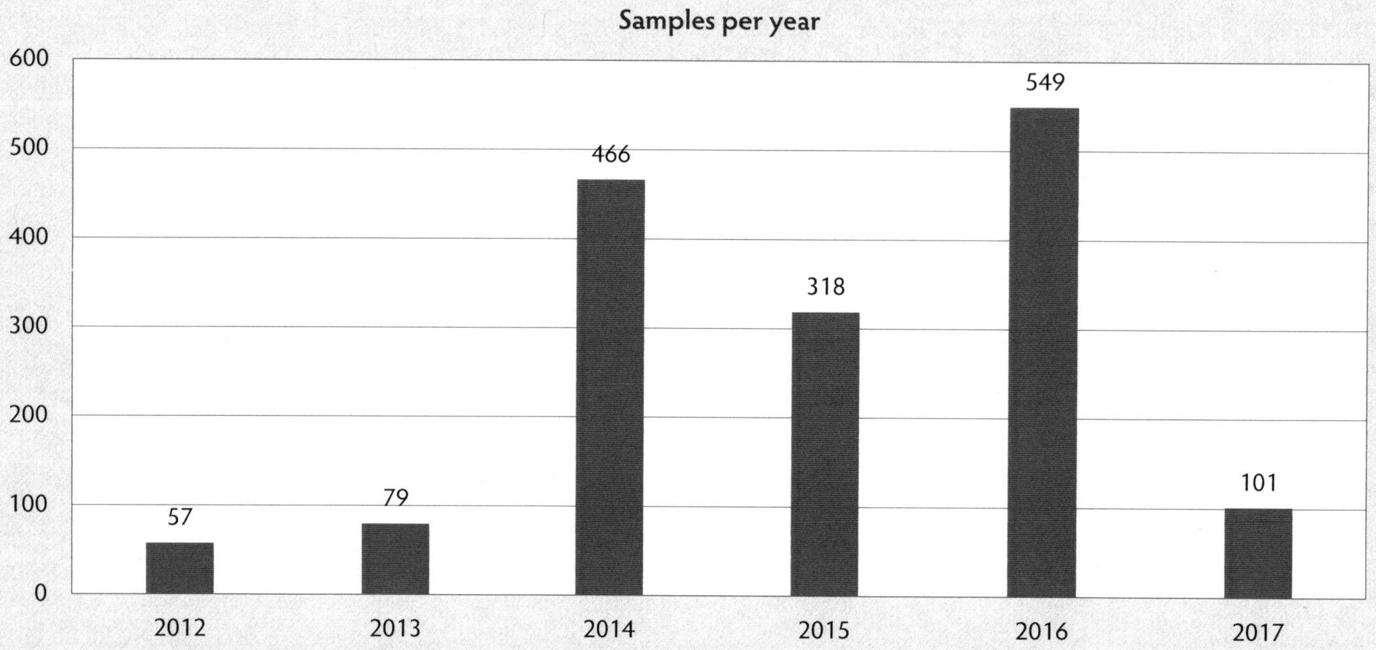

Table 1: Occurrence and contamination levels of 2017 corn

Region	Parameter	Afla	Zen	DON	FUM	T2	OTA
USA N = 101	Prevalence (%)	3	31	86	37	0	0
	Median (ppb)	5	104	509	1200	-	-
	St. dev (ppb)	9	738	5853	10685	-	-
	90th percentile (ppb)	15.48	401	1978	5320	-	-
Europe N = 195	Prevalence (%)	12	42	74	66	11	7
	Median (ppb)	149	101	782	1473	294	12
	St. Dev (ppb)	149	101	782	1473	294	12
	90th percentile (ppb)	204	224	1668	2751	377	17
Brazil N =1653	Prevalence (%)	25	52	82	92	13	5
	Median (ppb)	5	48	500	2060	25	1
	St. dev (ppb)	26	353	820	12212	7	0
	90th percentile (ppb)	36	333	1450	8370	35	1
China N = 98	Prevalence (%)	75	64	99	95	0	0
	Median (ppb)	41	25	340	3120	-	-
	St. Dev (ppb)	135	109	361	3404	-	-
	90th percentile (ppb)	344	160	846	7359	-	-

Polymeric formulations are efficient and versatile adjuvants for swine vaccines

S. B. Xu[1]; J. Ben Arous[2]; H. Imbault[2]
[1]SEPPIC Inc; [2]SEPPIC, Paris, France

Introduction

Diverse synthetic polycationic and polyanionic polymers have been shown to have potent vaccine adjuvants properties. In particular, cross-linked polyacrylates (carbomers) have been shown to be efficient for swine vaccination and have been included in commercial vaccines.[1] Montanide™ Gel adjuvants are a range of ready-to-use polyacrylate based formulations designed to improve safety and efficacy of aqueous vaccines. They have been shown to be safe and efficient adjuvants for swine vaccines.[2-4] Here we present results to show that these adjuvants can be used to improve the efficacy of very diverse swine vaccines, based on inactivated, recombinant or attenuated antigens.

Materials and methods

Trial 1: Swine influenza inactivated vaccine

In a first study, inactivated vaccines against Swine influenza virus (SIV) were formulated with the polymer adjuvant Montanide™ Gel 01 (Gel) or without adjuvant. A non vaccinated group was included as a negative control. At day 0 and day 21, 10 seronegative pigs (15 kg) were vaccinated in each group intramuscularly in the neck with 2ml of the corresponding inactivated SIV vaccine. Safety properties of the vaccines (systemic and local reactions) were assessed at vaccination, during the trial and at slaughter. Vaccine efficacy was scored both by serological analysis and application of a challenge procedure. Antigen specific antibody titration against SIV were performed at day 0 (before vaccination), day 21 (before revaccination), day 42 (before virulent challenge) and day 56 (at slaughter) by haemagglutination inhibition (HI). At day 42, 5 animals in each group were infected by swine influenza virus (strain A/swine/California/07/09) by intranasal administration of a dose of 8.3 lg EID50 in 2ml. Clinical signs and hyperthermia after challenge were measured for 14 days, and nasal viral loads were scored. Bacterial over-infections of the lungs were scored at slaughter at day 56.

Trial 2: PRRS live attenuated vaccine

In a second study, freeze-dried live PRRS vaccines (North American genotype) were formulated extemporaneously in diluent with no adjuvant, or in diluent containing polymer adjuvant Montanide™ Gel 01 (Gel) at 10%. Each vaccine contained either 50% or 100% of the commercial antigenic dose (4.3 log TCD50/ml). At day 0, 10 PRRS negative pigs (15 kg) were vaccinated in each group intramuscularly in the neck with 2ml of vaccine. A non vaccinated group was used as negative control. Blood samplings were performed at D0, 7, 14, 21 and 30 after vaccination and antibody titers were assessed by antigen specific ELISA. A challenge procedure was performed on D30 using a highly virulent PRRSV strain (Irkutsky 2007 isolate of North American genotype: titer 5.0 log TCD50/ml). Clinical signs (behavior and rectal T°) were recorded daily for 15 days after challenge. Pigs sera were analysed by PCR for the presence of PRRS virus before challenge and 15 days post challenge. Post-mortem examination of injection site and lungs was performed after natural death or euthanasia of animals 15 days after challenge.

Results

Trial 1: Swine influenza inactivated vaccine

All tested vaccines were safe. Body temperature increase after injection was <1°C in all groups. No other reactions were observed during the trial. Specific antibody titers against SIV were significantly superior in group vaccinated with Gel adjuvanted formulations compared to the non adjuvanted vaccine. Protective HI titers (> 6 log2) were reached before D21 in the group vaccinated with Gel adjuvanted vaccine (Table 1). Protection against SIV infection was also improved in the group vaccinated with Gel adjuvanted formulation, as hyperthermia after infection, lung lesions at slaughter and nasal viral shedding after virulent challenge were reduced compared both to non vaccinated and non adjuvanted groups.

Trial 2: PRRS live vaccine

All vaccines were safe, and animals did not show any side effect after vaccination. Antibody titers were not significantly different between Gel adjuvanted vaccine and non adjuvanted vaccine groups. However, protection to PRRS challenge, as measured by the number of viremic pigs after virulent challenge and lung lesion score after slaughter, was significantly superior for adjuvanted formulations containing 100% of antigen load compared to the non adjuvanted vaccine (Table 2). Moreover, despite lower antibody titers, the protection to challenge given by the adjuvanted vaccine containing only 50% of the antigen load was equivalent to the protection given by the non-adjuvanted vaccine.

Conclusions

These results show that relevant polymer adjuvants such as Montanide™ Gel are efficient adjuvants compatible with both inactivated and attenuated viral vaccines for swine. Therefore, such adjuvants could allow the formulation of multivalent combined

Table 1: Specific antibody titers after vaccination against SIV (HI titers, log2)

	D0	D21	D42	D56 Non infected	D56 Infected
CONTROL	< 2	< 2	< 2	< 2	9.4 ± 0.3
ANTIGEN	< 2	4.7 ± 0.3	5.6 ± 0.2	5.6 ± 0.2	9.4 ± 0.5
GEL	< 2	6.7 ± 0.3	7.7 ± 0.2	7.4 ± 0.3	8.8 ± 0.3

Table 2: Protection to PRRS virulent challenge.

Vaccine group	Lung lesions score	Number of viremic pigs 15 days after challenge	Duration of hyperthermia (> 40°C), days
Gel adjuvanted vaccine (100% antigen)	3 ± 2*	1/10*	5 ± 2
Non adjuvanted vaccine (100% antigen)	12 ± 18	3/10*	7 ± 3
Gel adjuvanted vaccine (50% antigen)	5 ± 8	3/10*	7 ± 3
Challenge control Non vaccinated	18 ± 22	10/10*	11 ± 2
Naïve control Non vaccinated non challenged	0.4 ± 0.5	0/10	0

* Significantly different from the not vaccinated challenged group

inactivated/live vaccines. Such combined vaccines could lead to the reduction of the number of injections given to pigs in the field.

References

1. Y Burakova et al. Adjuvants for animal vaccines. Viral Immunology 2017, June 15.

2. Q Xue et al. First comparison of adjuvant for trivalent inactivated Haemophilus parasuis serovars 4, 5 and 12 vaccines against Glässer's disease. Vet Immunol Immunopathol. 2015 Dec 15;168(3-4):153-8.

3. X Li et al. Montanide™ Gel01 ST Adjuvant Enhances PRRS Modified Live Vaccine Efficacy by Regulating Porcine Humoral and Cellular Immune Responses. World Journal of Vaccines, 2013, 3, 1-9

4. J Ben Arous et al. Adjuvant formulations designed to improve swine vaccine stability: Application to PCV2 vaccines. Procedia in Vaccinology 7 (2013) 34 – 39.

Increasing intestinal mucin production using algae extracts

M. Gallissot[1], MSc; L. Miel[1], MSc; C. Rousseaux[2], PhD; M.A. Rodriguez[1], MSc;
C. Holloway[1], MSc; P. Nyvall-Collen[1], PhD
[1]Olmix SA, Brehan, France; [2]Intestinal Biotech Development, Lille, France

Introduction

In the small intestine the mucus layer and transmembrane proteins act as the first line of defense against pathogens and toxins present in the lumen. The mucus layer is a 3D, viscoelastic network structured by mucin glycoproteins. It protects the epithelial cells from dehydration, mechanical damage, and autodigestion, as well as preventing pathogen absorption and system dissemination. The mucus layer also hosts a wide range of innate immune molecules to help further eliminate pathogens. The various transmembrane proteins control the paracellular space between epithelial cells, by mitigating the types of products that can pass between the cells. They prevent the passage of undesirable substances and control ion flux and cell polarity. Both the mucus layer and the transmembrane proteins play important roles in controlling animal and gut health. Previous work[1] has suggested that an extract from the algae *Ulva* sp. can stimulate the excretion of mucin proteins from colonic goblet cells, but the mode of action was not identified. Thus, one objective of the present study was to determine the mechanism for how mucus production was being stimulated. A second objective was to determine if the extract would improve tight junction protein expression given their complementary role in protecting the animal from pathogen absorption. The hypothesis was that the *Ulva* sp extract from Olmix (MSP®$_{MUCIN}$) was stimulating the mRNA expression of both goblet cells and tight junction proteins.

Materials and methods

Two colonic epithelial cell lines were used: the enterocyte-like Caco-2 cells (ATTC® HTB-37™, Molsheim, France) and the mucus-secreting HT-29 MTX cells (INSERM U560, Lille, France). Both cell lines were grown separately in Dulbecco's Modified Eagle's Medium (DMEM 10% FCS, PAA Laboratories, Les Mureaux, France) and antibiotics (streptomycin 100 µg mL and penicillin 100 units/mL, Sigma Aldrich, St Quentin Fallavier, France). Differentiation was completed 15 days post-confluency for Caco-2 cells and 21 days post-confluency for HT-29 MTX cells. The MSP®$_{MUCIN}$ was tested in triplicate by addition in the cell cultures after complete differentiation, for 1h or 4h, at 3 different concentrations (0.01, 0.1 and 1mg/ml) in standard conditions. In inflammatory conditions, a pre-stimulation of the cells with LPS (from *E. coli* O111:B4) was performed before the addition of MSP®$_{MUCIN}$. After 1 or 4h of contact with the algae extract, the cells were extensively washed and harvested for mRNA extraction. The gene expression of several mucins (MUC1, MUC2, MUC4, MUC5AB and MUC5AC) and tight junction proteins (ZO-1, ZO-2, claudin-2) was evaluated by mRNA quantification by qRT-PCR. Total mRNA was isolated from cells using Rneasy kit (Macherey Nagel, Hoerdt, France) according to the manufacturer's instructions. RNA quantification was performed using spectrophotometry. After treatment at 37°C for 30 min with 20-50 units of RNase-free DNase I (Roche Diagnostics Corporation, Indianapolis, IN, USA), oligo-dT primers (Roche Diagnostics Corporation, Indianapolis, USA) and the High Capacity cDNA Reverse Transcription Kit (Applied Biosystems) was used to synthesize single-stranded cDNA. The mRNA was quantified using SYBR green Master Mix (Applera, Courtaboeuf, France) with mice specific oligonucleotides in a GeneAmp Abiprism 7000 (Applera, Courtaboeuf, France). In each assay, calibrated and no-template controls were included. SYBR green dye intensity was analyzed using the Abiprism 7000 SDS software (Applera, Courtaboeuf, France). All results were normalized to the unaffected housekeeping gene GAPDH. All the results were expressed as a fold of the mean of activation of triplication compared to the mean of triplicate in medium alone (control).

Results and discussion

Results showed that both in standard and inflammatory conditions, the MSP®$_{MUCIN}$ induced early expression of MUC1 (52 to 115 times that of control) and MUC5AC (104 to 687 times that of control) genes, at 0.1 and 1mg/ml (Figure 1). After 4 hours, this effect was maintained for MUC1, and enhanced for MUC5AC. MUC2 gene expression was also increased, to a lesser extent, after 4 hours of stimulation with 0.1 and 1 mg/ml of MSP®$_{MUCIN}$ (2 to 7 times that of control) (Figure 2). The gene expression of tight junction proteins was moderately stimulated by the MSP®$_{MUCIN}$. Zonula Occludin 1 (ZO1) and Zonula Occludin (ZO2) at 1 and 4 hours, respectively, showed the best response, especially under standard conditions (Figure 3). The expression of claudin2 was not affected by the algae extract.

Therefore, the tested algae extract showed some capacity to induce the expression of both membrane-bound mucins (MUC1) involved in cell-cell and cell-matrix interactions, as well as cell-signaling; and gel-forming mucins (MUC5AC and MUC2), which are particularly involved in host-protection and control of pathogen dissemination, in both standard and inflammatory conditions. The tested extract was also able to stimulate the expression of scaffolding proteins of the tight junctions (ZO1 and ZO2),

Figure 1: *In vitro* gene expression of mucin proteins exposed to different concentrations of MSPMUCIN during 1 hour. **A)** by Caco-2 cells, in physiological conditions. **B)** by HT-29 MTX cells, in physiological conditions. **C)** by Caco-2 cells, in inflammatory conditions. **D)** by HT-29 MTX cells, in inflammatory conditions.

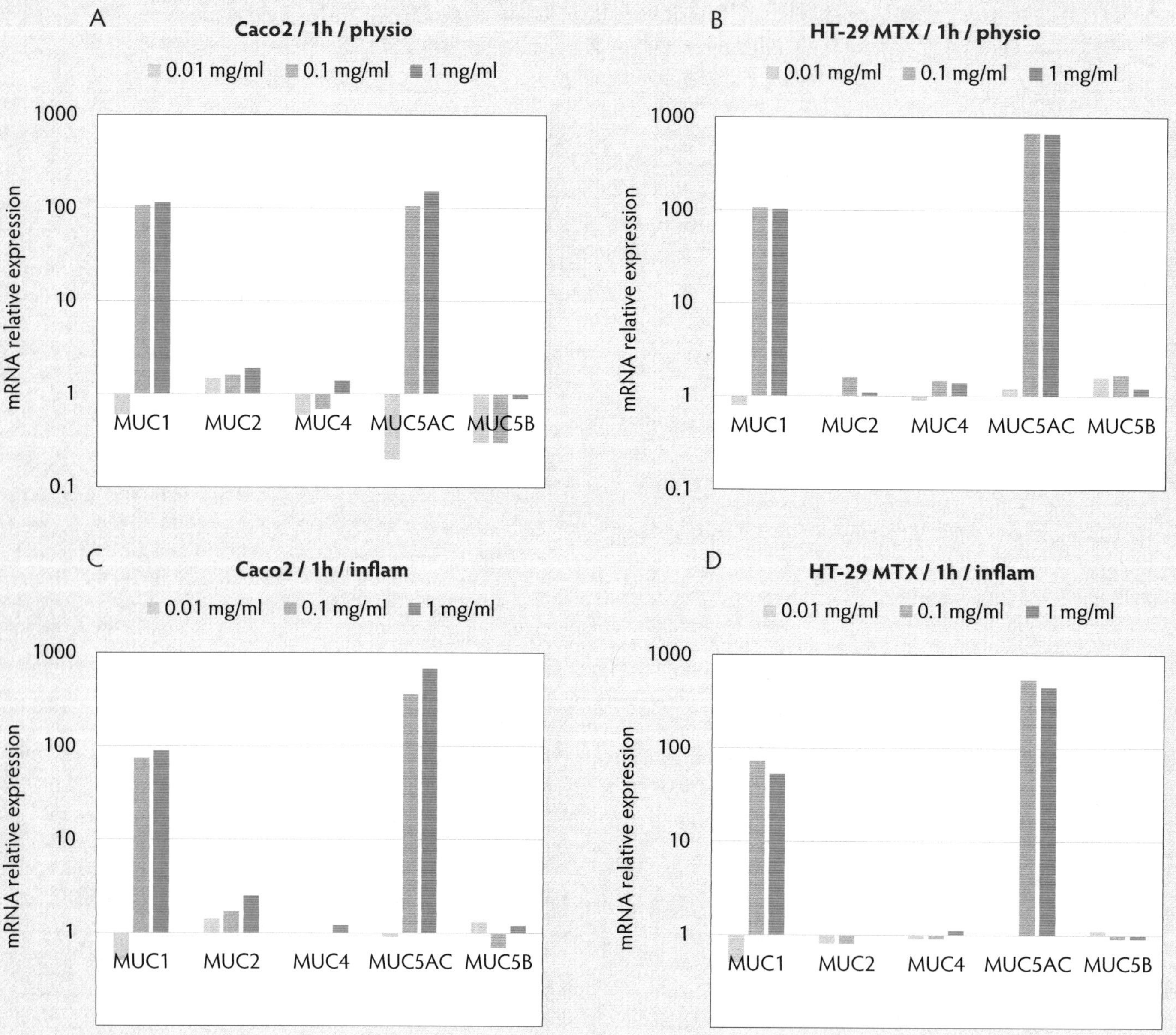

Figure 2: *In vitro* gene expression of mucin proteins exposed to different concentrations of MSP_{MUCIN} during 4 hours. **A)** by Caco-2 cells, in physiological conditions. **B)** by HT-29 MTX cells, in physiological conditions. **C)** by Caco-2 cells, in inflammatory conditions. **D)** by HT-29 MTX cells, in inflammatory conditions.

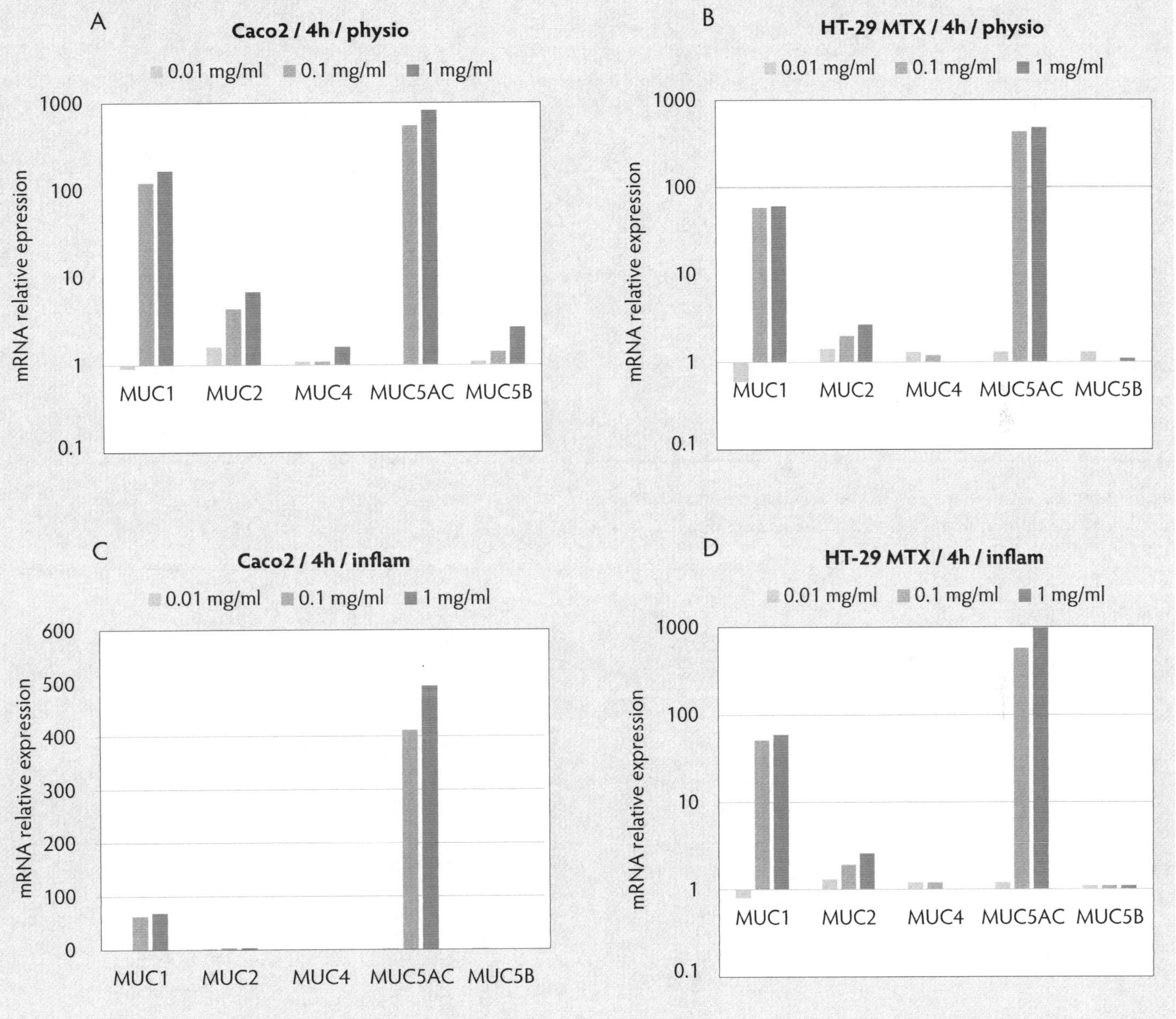

Figure 3: *In vitro* gene expression of tight junction proteins by Caco-2 cells exposed to different concentrations of MSP_{MUCIN} during 1 hour **(A)** or 4 hours **(B)** in physiological conditions.

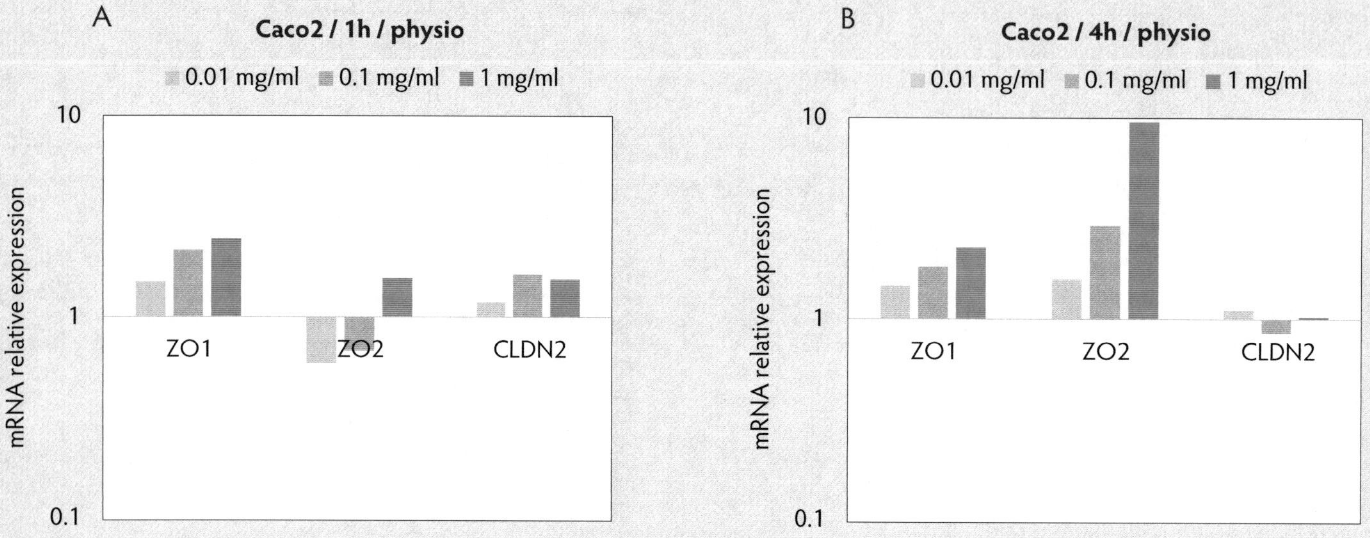

which control cell-cell contacts and the polymerization and contractility of the tight junctions. Consequently, the prophylactic use of the $MSP^{®}_{MUCIN}$ should be considered as a method to reinforce the intestinal barrier of piglets and prevent gastrointestinal related health troubles.

References

1. Barcelo A *et al.* Mucin secretion is modulated by luminal factors in the isolated vascularly perfused rat colon. Gut. 2000;46:218-224.

Seneca Valley virus VI positive flies *(Musca domestica)*: What questions does this pose for the swine industry?

Mark Engesser[1], DVM; Phil Gauger[2], DVM, MS, PhD; Jianqiang Zhang[2], MD, MS, PhD; Grant Allison[3], DVM; Gene Spellman[4], BS

[1]ME Swine Service, Rose Creek, Minnesota; [2]Vet Diagnostic and Production Animal Medicine, Veterinary Diagnostic Laboratory, Iowa State University, Ames, Iowa; [3]Walcott Vet Clinic, Walcott, Iowa; [4]Central Life Sciences, Schaumburg, Illinois

Introduction

Vesicular disease cases associated with Seneca Valley virus (SVV) (Senecavirus A) have recently increased in the swine industry.[1] Vesicular disease is a broad term used to cover several diseases associated with similar clinical signs, such as open or closed blisters located on the snout and/or at the junction where the skin and the hoof wall meet (Coronary band), sudden lameness with redness and swelling at or around the coronary band, and pigs that are suddenly off feed, lethargic and/or have a fever up to 105 degrees Fahrenheit.[2] Foot and mouth disease (FMD) is the major disease of concern. FMD is on the foreign animal disease list and can have a major impact due to being very contagious and having the potential to close foreign trade. Diseases associated with vesicular disease are guilty of being FMD until proven otherwise through diagnostic testing. Idiopathic vesicular disease caused by the Seneca Valley virus is one such disease. Seneca Valley virus belongs to the same virus family as FMD, *(picornaviridae)* and will have similar clinical symptoms to FMD. Major disruptions in pig flow for producers and packers are the result of pigs shipped exhibiting signs of vesicular disease, until a non-FMD diagnosis is confirmed. A recent general veterinarian opinion and observation was that SVV has been seen more often in warmer summer months, versus other viral pathogens. What is the significance of this general observation? Is there something about warmer months associated with an increase in SVV infections? In this discussion, flies were mentioned and a question was posed: can flies carry SVV? Seeking this answer became the basis for this preliminary study testing whether flies can harbor live viable Seneca Valley virus.

Figure 1: Captivator Fly Trap

Materials and methods

Permission was obtained to sample a 4,800 head finisher in Southern Minnesota positive for both SVV and porcine reproductive and respiratory syndrome virus (PRRSV) 24 days post diagnosis (viral swab, oral fluid). Flies were collected in Captivator® jug liquid traps (Starbar®, Central Life Sciences, Figure 1) to reduce any cross-contamination from the environment that may occur during sample collection. Traps were hung six to a room with a negative control that prevented flies from entering the device but allowed room air to circulate in the traps. Flies randomly enter these traps by their own volition. Attractant was solubilized with sterile water. Traps were placed at approximately 8:30 a.m. and removed approximately 12:30 p.m. Flies were netted out of the traps into labeled zip-lock bags and delivered same-day to the Iowa State University Veterinary Diagnostic Lab (ISU VDL). A minimum of 100 flies per trap/bag was targeted. Liquid from the trap was also collected for subsequent PCR testing. Flies were processed as a 10% homogenate solution using phosphate buffered saline. Homogenates were clarified by centrifugation and supernatants were analyzed by real-time reverse transcription polymerase chain reaction (rRT-PCR) per routine protocols at the ISU VDL for PRRSV and SVV. The remaining supernatants were stored at -80°C. Seneca Valley virus has a propensity to grow in cell culture and virus isolation was conducted in H1299 cells (ATCC CRL-5803) per routine procedures at the ISU VDL.

Results and discussion

One out of six traps were PCR positive for PRSSV with cycle threshold of 31.6 (Ct 31.6). Two out of six traps were PCR positive for SVV (Ct 30.5 and Ct 30.4). All fly attractant liquid samples were rRT-PCR negative. Virus isolation attempts were conducted on the two PCR positive SVV fly samples. The Ct 30.5 sample was VI positive and confirmed by PCR. The initial passage of cell culture material (P0) had a Ct of 20, and the subsequent passage (P1) had a Ct of 17. These results indicate that the fly sample contained viable SVV (Table 1). A piglet bioassay opportunity for SVV is the next goal. Flies may bypass all current known biosecurity measures and have multiple routes of entry into swine farms. Flies are ubiquitous, can fly up to four miles in search of harborage,[3] live and breed in close proximity to swine, and can be transported by farm vehicles (Figure 2). Insect vectors such as flies have been investigated as a possible source of porcine reproductive and respiratory virus PRRSV.[4] There also is a proven model for transmission of a corona virus by a flying vector.[5] Are flies a potential risk factor in the transmission of SVV? Should current industry biosecurity protocols now include more intensive fly control? More questions arise and further research is needed. This preliminary study showed through positive virus isolation flies can harbor live viable SVV 24 days after diagnosis. At present, there is no standard protocol for fly control in the industry. New technology can help develop strategies to lessen this risk from flies, both in a convenient and economical, e.g. feed-thru, method.

References

1. National Assembly of State Animal Health Officials (NASAHO) Senecavirus A Working Group Report. October 12, 2017.

2. Securepork.org Wall Chart Swine Vesicular Disease SVD http://www.securepork.org/Resources/WallChart_SVD_130709.pdf.

3. Nazni WI, Luke H et al. Determination of the flight range and dispersal of the house fly, Musca domestica (L.) using mark release recapture technique. *Trop. Biomed* 2005 Jun;22(1):53-61.

4. Otake S, Dee SA et al. Transmission of porcine reproductive and respiratory syndrome virus by houseflies, (Musca domestica Linnaeus). *Vet Rec* 2003. 152: 73-76

5. Diseases of Swine. 9th Edition 2006 pg 494-495.

Table 1: Diagnostics

Virus	Date	Sample	Result	Comment
PRRS	8/04/17	Oral fluid	positive	
PRRS	8/28/17	Flies	positive (Ct 31.6)	Trap (1 of 6 positive)
PRRS	8/28/17	Liquid	negative	Liquid attractant in traps
SVV	8/04/17	Oral swab	positive	
SVV	8/28/17	Flies	positive (Ct 30.5 & 30.4)	Trap (2 of 6 positive)
SVV	8/28/17	Liquid	negative	Liquid attractant in traps
SVV	8/28/17	Flies (Ct 30.5)	positive (Ct 20.0)	Virus isolation passage "0"
SVV	8/28/17	Flies	positive (Ct 17.0)	Virus isolation passage "1"

Figure 2: Fly on vehicle dashboard

Use estimates of in-feed antimicrobials in swine production in the United States, 2012

E. Bush, DVM, MS; K. Bjork, DVM, MSPH, PhD; C. Kopral, PhD
USDA:APHIS:VS, Center for Epidemiology and Animal Health, Fort Collins, Colorado

Introduction

Antimicrobial use (AMU) in feed is one of several management practices used to prevent and control disease in swine. Objective, scientifically valid, and nationally representative data on AMU are useful for risk and cost/benefit analyses and for providing context for antimicrobial resistance (AMR) surveillance programs, such as the National Antimicrobial Resistance Monitoring System. AMU data from serial cross-sectional studies provide estimates for trends in AMU over time, as changes in demographics, husbandry practices, and the regulatory environment occur. Collecting AMU data can also enable an objective assessment of the effectiveness of stewardship and judicious-use campaigns.

The USDA's National Animal Health Monitoring System (NAHMS) collected and published data on AMU in the US swine industry during five National studies conducted in 1990, 1995, 2000, 2006, and 2012. AMU data from the NAHMS Swine 2006 study were used to provide a national estimate of the total kilograms of antimicrobials used (by class of antimicrobial) by the swine industry in 2006 and by reason for use.[1] The following is a similar analysis using AMU data from the NAHMS Swine 2012 study.

Methods

Data collection via NAHMS Swine 2012 study

The Swine 2012 study was conducted in the Nation's top 13 swine-producing States which represented 90.8 percent of the US swine inventory on operations with 100 or more swine on June 1, 2012, and 88.9 percent of US operations with 100 or more swine.[2] A stratified random sample of 4,600 operations was selected from the USDA's National Agricultural Statistics Service (NASS) list frame. A total of 2,119 producers completed the NASS general swine farm questionnaire during phase I of the study. Respondents who completed the first questionnaire were asked to participate in phase II of the study by completing a second questionnaire to be administered by State and Federal veterinarians from September 5 through November 17, 2012. A total of 474 swine sites completed the second questionnaire. Antimicrobial-use data from these 474 sites were extrapolated to represent a population of about 48 million weaned market pigs on about 16,585 production sites with weaned market hogs.

Calculation of total kg of antimicrobials used

The equation for calculating total kilograms of antimicrobials used for each antimicrobial/reason combination was the same as that applied and reported for the Swine 2006 AMU data:[1]

$$\text{Use} = N \times F \times T \times D$$

Eighteen antimicrobial or antimicrobial combination products included in the NAHMS 2012 Swine study represented all feed-grade antimicrobials approved and marketed for use in US swine production. Respondents were asked to report usage data for each antimicrobial or antimicrobial combination products used. Feed additives were listed alphabetically in the study questionnaire to minimize reporting bias. Usage data reported by fewer than three respondents were excluded from the analysis to ensure data confidentiality.

N is the number of animals in a group for the given phase. The NAHMS Swine 2012 study captured the number of animals—at the site level—that entered any production phase during the 6 months from December 1, 2011, through May 31, 2012; the average age of pigs entering and leaving each phase; and the number of pigs that died during the 6-month period. Pig-entry numbers from the 6-month period were converted to the number of pigs in individual groups, because feed consumption, timing, and duration of AMU was expressed in days during a defined portion of the production cycle. Therefore, N = (the number of pigs entering during 6-month period) – ([0.5 × mortalities])/ (number of groups).

F is the estimated feed consumed per day. For the nursery phase, a constant feed-consumption value of 1.5 lb/pig/day was used.[3] For the grower/finisher and wean-to-finish phases, feed consumption was calculated using a linear model based on a starting feed intake of 2.5 lb/pig/day at an entry age of 65 days and culminating at 7.0 lb/pig/day at the end of the finishing period at 180 days of age (ADFI = 0.0368 + 0.039 × [days of age]). Feed intake was calculated for each specific antimicrobial/reason combination. Pig age was established as the midpoint between the age of animals when first receiving antimicrobials in their feed and the age of animals when antimicrobials were removed from their feed. The pig age when a given antimicrobial was removed from the feed was calculated by adding the duration of use to the starting age.

T is the duration in days that pigs received antimicrobials in feed for the given phase. Duration of use for each antimicrobial, reason, and production phase combination was taken directly from the reported days in feed. The median in-feed time for a particular antimicrobial/reason combination was used to impute any missing duration values.

D is the dose of antimicrobials in feed. The dose for a particular antimicrobial/reason combination was also taken directly from the study's second questionnaire. The median dose for an antimicrobial/reason combination was used to impute any missing values. Total doses used for combination products were allocated to the different antimicrobials, according to their proportions in the product labels.

The total kilograms used for each antimicrobial was calculated for each production phase and production site and was based on the reported use characteristics of the individual site. It was assumed that an antimicrobial used on a site was administered to 100 percent of the pigs in the production phase. Based on these values, national estimates were generated by production phase for nursery, grower/finisher, and wean-to-finish phases.

The national estimates for the number of days in the three production phases were used to calculate the length of a production cycle. Age inputs, as a mean of all site sizes, were 19.3 days of age for entering the nursery, 65.1 days of age for entering grow/finisher phase, and 179.7 days for leaving the grower/finisher unit and going to market. Subtracting the nursery-phase entry age from the market age gave a production-cycle estimate of 160.4 days. This number was divided into 365 days to calculate the number of production cycles completed per year, assuming no gap between pig groups. The rounded result is 2.25 production cycles per year. Extrapolation to yearly usage was accomplished by multiplying the sum of the final drug-use estimates for nursery and grower/finisher phases by 2.25. The calculated number of production cycles per year for wean-to-finish sites was 2.4.

Results

Model inputs calculated by antimicrobial/reason combinations for nursery, grower/finisher, and wean-to-finish production phases have been published in USDA:APHIS:VS Technical Brief reporting on this analysis. Similarly, national AMU estimates for nursery, grower/finisher, and wean-to-finish phases are published in the same Technical Brief. In the nursery phase, 5, 10, and 7 antimicrobials were used in feed for growth promotion, disease prevention, and disease treatment, respectively. In the grower/finisher phase, 7, 11, and 9 antimicrobials were used for growth promotion, disease prevention, and disease treatment, respectively. In the wean-to-finish phase, one, eight, and four antimicrobials were used for growth promotion, disease prevention, and disease treatment, respectively.

The overall 2012 annual estimates for in-feed AMU in the U.S. swine industry are reported in Table 1. Antimicrobials are reported by category, as defined in Appendix A of the Food and Drug Administration Center for Veterinary Medicine (FDA/CVM) Guidance for Industry #152 (GFI #152).

FDA classifies antimicrobials as "important," "highly important," or "critically important" for human health. None of the antimicrobials designated as "important" in GFI #152 Appendix A is labeled for use in swine feed, so this designation is not included in the table. The total kilogram estimate for antimicrobials not listed in GFI #152 Appendix A comprises 33.2 percent of the total in-feed AMU estimate. Antimicrobials classified as highly important comprised 59.7 percent of the 2012 total use estimate, and those classified as critically important were estimated at 7.2 percent of the total 2012 annual use estimate. Of the overall 2012 annual AMU estimates, 32.7 percent were used for growth promotion, whereas 55.4 and 11.9 percent were used for disease prevention and disease treatment, respectively.

Discussion

The NAHMS Swine 2012 study captured AMU data for a 6-month period before the second study questionnaire was administered. For estimating the total kilograms of each antimicrobial used in a year, it was assumed that AMU patterns applied to the entire year. The second questionnaire asked producers to report AMU by primary reason it was included in feed. It did not ask for all the reasons that an antimicrobial was used in feed for that production phase.

The NAHMS Swine 2012 study was a cross-sectional study in which participation was voluntary and, therefore, prone to selection bias. In the context of this study, selection bias could result from producers' nonrandom refusals to participate. The capture of data via a questionnaire, such as the reporting of AMU, is subject to recall bias and misclassification errors.

The estimates of total kilograms of antimicrobials used in swine feed in the United States reported herein are an update of those published for 2006.[1] Overall, the change from 2006 to 2012 in total annual kilograms used (1.3 million to 0.7 million kg) represented a nearly 50-percent reduction in AMU in swine feed.

In contrast, when looking at AMU as the percent of sites including any antimicrobial in feed for nursery or grower/finisher units (USDA, 2016), there was no significant change between 2006[4] and 2012. There are several reasons behind the apparent discrepancies in trends for these two measures of AMU. Some design changes in 2012 occurred, particularly the way in which large sites were sub-selected for participating companies. Another design change was to the data collection instrument. Whereas in 2006 AMU data was captured for nursery-age pigs which included sites with a nursery unit or wean-to-finish sites, in 2012 AMU data was captured separately for nursery, grower/finisher and wean-to-finish production units.

Changes in how producers use antimicrobials also help explain the differences. For one, there was a broad shift in the reason for administering antimicrobials in feed and water of grower/finisher pigs. Specifically, there was a significant decrease in AMU in feed for disease treatment. This was accompanied by a significant increase in AMU in water for disease treatment. AMU in feed for disease prevention significantly increased and became the most common reason for including antimicrobials in feed. Secondly, and perhaps not unrelated, there was a change in specific usage patterns regarding dose and duration of AMU in feed, with both

Table 1: National estimates of total kilograms of in-feed antimicrobials used for all production cycles in 2012, by antimicrobial and by reason

FDA/CVM Guidance 152 Appendix A	Drug	Growth promotion	Prevention	Therapy	Reason* yearly basis
Antimicrobials not listed	Bacitracin	73,370	2,856	424	76,650
	Bacitracin zinc	83,563			83,563
	Bambermycin	439	439		878
	Carbadox				
	as Carbadox alone	1,101	6,098	4,508	11,706
	as Carbadox/oxytetracycline		4,774		4,774
	Sulfamethazine				
	as CTC/sulfamethazine/pen (ASP)		2,202	871	3,073
	as Tylosin/sulfamethazine	106	2,924		3,030
	Tiamulin	208	42,402	5,604	48,214
Antimicrobials listed as highly important	Chlortetracycline				
	as Chlortetracycline alone	7,542	265,559	52,906	326,008
	as CTC/sulfamethazine/pen (ASP)		2,202	871	3,073
	Lincomycin		17,403	2,291	19,694
	Neomycin				
	as Neomycin/oxytetracycline		262		262
	Oxytetracycline				
	as Oxytetracycline alone		5,679	2,317	7,996
	as Carbadox/oxytetracycline		4,774		4,774
	as Neomycin/oxytetracycline		203		203
	Penicillin				
	as CTC/Sulfamethazine/pen (ASP)		1,101	436	1,537
	Virginiamycin	48,178	5,284	343	53,805
Antimicrobials listed as critically important	Tilmicosin		2,082	715	2,796
	Tylosin				
	as Tylosin alone	14,003	18,156	12,053	44,212
	as Tylosin/sulfamethazine	106	2,924		3,030

* Growth promotion, disease prevention or disease therapy

being lower on average in 2012 than 2006 for several antimicrobials. Thirdly, there was a shift in which antimicrobials were most commonly used, which happened to be away from lower potency drugs, such as chlortetracycline, and towards higher potency drugs which are used at lower grams per ton of feed.

Antimicrobials classified by FDA/CVM as critically important to human health made up a smaller proportion of antimicrobials used in 2012 (7.2 percent of total estimate) than reported in 2006 (18.2 percent of total estimate). We estimated that two-thirds of antimicrobials fed to swine were considered medically important by FDA/CVM. Of the total kilograms of antimicrobials sold or distributed for use in food-producing animals in 2012—excluding ionophores—over 88 percent were medically important.[5]

References

1. Apley MD, Bush EJ, Morrison RB, Singer RS, Snelson H. 2012. Use estimates of in-feed antimicrobials in swine production in the United States. Foodborne Path and Dis 9:272–279.

2. USDA–APHIS–VS 2016. Swine 2012 Part II: Reference of Swine Health and Health Management Practices in the United States, 2012. National Animal Health Monitoring System (NAHMS) Fort Collins, CO. #N676.0216

3. DeRouchey JM, Dritz SS, Goodband RD, Nelssen JL, Tokach MD. 2007. KSU swine nutrition guide: general nutrition principles for swine. Kansas State University Agricultural Experiment Station and Cooperative Extension Service Publication MF2298.

4. USDA–APHIS–VS. 2007. Swine 2006 Part II: Reference of Swine Health and Health Management Practices in the United States, 2006. National Animal Health Monitoring System (NAHMS) Fort Collins, CO. #N479.1007

5. Food and Drug Administration Center for Veterinary Medicine. 2014. Summary report on antimicrobials sold or distributed for food-producing animals in 2012.

Development of multiplex PCRs for a full swine respiratory panel

E. van Esch, DVM
BioChek BV, Reeuwijk, Netherlands

Introduction

In swine, respiratory diseases are one of the most important diseases. There are many pathogens involved, both bacterial and viral, and seldom it is only one pathogen which is causing clinical or subclinical problems. For that reason the porcine respiratory disease complex was introduced as a concept and major pathogens here are PRRSV, PCV2, SIV, App, Hps and Mycoplasma's. Since none of these are showing very typical signs under field conditions, one should always test for all.

Since the PCR technique has evolved more over time, the access to PCRs for veterinarians and their customers is massively increased. PCRs have become a standard diagnostic tool for them to gather accurate and meaningful information in order to make the correct and timely decision on how to approach PRDC.

This paper is describing the setup of different multiplex PCRs for the porcine respiratory disease complex and their use will be discussed in this paper as well.

Materials and methods

From out of literature a search was to done to identify the most unique targets for PRRSV, PCV2, SIV, App, Hps and Mycoplasma. In GenBank (`https://www.ncbi.nlm.nih.gov/genbank/`) blasts were done to design the specific primers and probes and unique in house post-blast software was used to optimize these primer and probe sets. To these particular primer/probe sets plasmids were designed in order to check the first performance of the PCR set up. These plasmids are also used as positive controls in the PCR's

For validation of the PCRs different materials were used. Included for sensitivity were different reference strains, vaccine strains, cultures, tissues, blood/serum, oral fluids and FTA cards. To check specificity of the assays a panel different non relevant bacteria, viruses and related pathogens were used. Samples were prepared in such ways that these were suitable to be used in extraction methods.

Different extraction methods were evaluated, varying from crude methods such as heat boil and extraction buffers, to more sophisticated methods as spin column based methods and bead based methods. For evaluation of the PCRs a total NA extraction method was used (RBCBioscience, MagCore® Automated Nucleic Acid Extractor and MagCore® Viral Nucleic Acid Extraction kit 201)

Different PCR cyclers (BioRad CFX, Applied Biosystems® 7500, Agilent Stratagene Mx3005P) were used as well in order to validate the PCRs on the different machine and to set thresh holds for the different PCR cyclers. All PCRs are running on the same PCR cycler protocol in respect of timing and temperature, this is allowing to set up multi assay plates and limit the number or PCR runs. Also in all PCRs an internal control is present in order to validate negative results. For all PCRs standards were made in order to quantify results.

Results

In Table 1 the major performance parameters are shown.

As can be seen in Table 1 all PCRs showed a high specificity and high sensitivity. Also most PCR are highly sensitive as can be seen from the Lower Level of Detection (LLOD), expressed as copies per reaction (reaction volume was 25 µL). This means that all PCRs are able to detect their specific targets below infectious dose of all pathogens. During evaluation a total NA extraction method was used, this will influence the sensitivity the most. When using a more specific extraction method sensitivity will be increased.

Discussion and conclusion

PCRs have found their way into veterinary diagnostics and nowadays PCRs have become a common tool for veterinarians and farmers to gather information on the swine herds. Depending on the level of information, different routes can be followed to obtain the required information and one has to choose what is needed. In Figure 1 the diagnostic flow for the porcine respiratory disease complex (PRDC) is shown.

In the flow chart the diagnostic decision tree starts with fact that in PRDC most of the time there are no very typical clinical signs found in the farm which is identifying the pathogen or pathogens involved. This means in order to make a good diagnosis, a whole diagnostic panel for the most common pathogens should be used. The major differentiation between viral or bacterial pathogens will determine the biggest outcome. Bacterial infections can be countered by antibiotics, while viral infections will be treated in a different way. The second step will be to choose for a more general approach or a more specific approach. This will be determined by what is known already on the farm and/or what is needed to know. In most cases a general screening will be the first choice, determining if viral or bacterial pathogens are the most important ones. Since all PCRs are able to quantify the pathogen load, based on this load more information is coming in so one can evaluate the relevance of the pathogens better. This approach is a

Table 1: Overview of performance parameters PCRs

PCR and target	Specificity (%)	Sensitivity (%)	Lower level of detection (copies/reaction)
Mycoplasma multiplex			
Mycoplasma hyopneumoniae	>99	>98	100
Mycoplasma hyosynoviae	>99	>99	50
Mycoplasma hyorhinis	>99	>83	100
Actinobacillus pleuropneumoniae multiplex			
apxIV	>99	>99	50
apxI	>99	>99	50
apxIII	>99	>99	50
Haemophilus parasuis multiplex			
pan Hps	>99	>99	50
vtaA10 virulence marker	>99	>99	100
PRRSV multiplex			
serotype 1	>99	>93[1]	25
serotype 2	>99	>99	50
PCV2			
pan PCV2	>99	>99	50
SIV			
Influenza A	>99	>99	100
Screening viral[2] multiplex			
PRRSV total	tbd	tbd	tbd
PCV2	tbd	tbd	tbd
SIV	tbd	tbd	tbd
Screening bacterial multiplex			
Mycoplasma hyopneumoniae	>99	>98	100
Actinobacillus pleuropneumoniae apxIV	>99	>99	50
Haemophilus parasuis pan Hps	>99	>99	50

[1] Depending on extraction method, PRRSV serotype showed at least 93% sensitivity. This can be increased by specific RNA extraction methods, during evaluation total NA extraction was used.
[2] At time of preparing this paper the screening viral PCR was under evaluation and no data available at that time

cost effective way to gather the most critical information and the time needed here is limited. From that point onwards, one could need more in depth information on the pathogens. For example for App it will be interesting to know if one is dealing with low or high pathogenic strains, or for PRRSV if serotype 1 or serotype 2 strains are involved. From the first screening it will be possible to evaluate only that what is important in more specific target PCRs in order to make better judgments.

Based on what information is already available on the farm one could also choose to approach the problem directly with more specific PCR's. The biggest danger here is to overlook other possible causes, so one must be very sure that choosing the more specific route is well supported by other available recent information. In this way the needed information can be obtained with limiting the number of tests.

The availability of a full respiratory PCR panel for PRDC and with a differentiation between a more general approach or a more specific approach will help to gather information in a timely and cost effective way so veterinarians and farmers can make the right decisions. With these developed PCRs one has now the availability of tools to gather the needed information. Testing the full panel of pathogens can now be done in an effective way so there is no limitation to gather this information. Only testing for all possible pathogens will lead to the correct conclusion. Evaluation

Figure 1: Diagnostic flow chart porcine respiratory disease complex

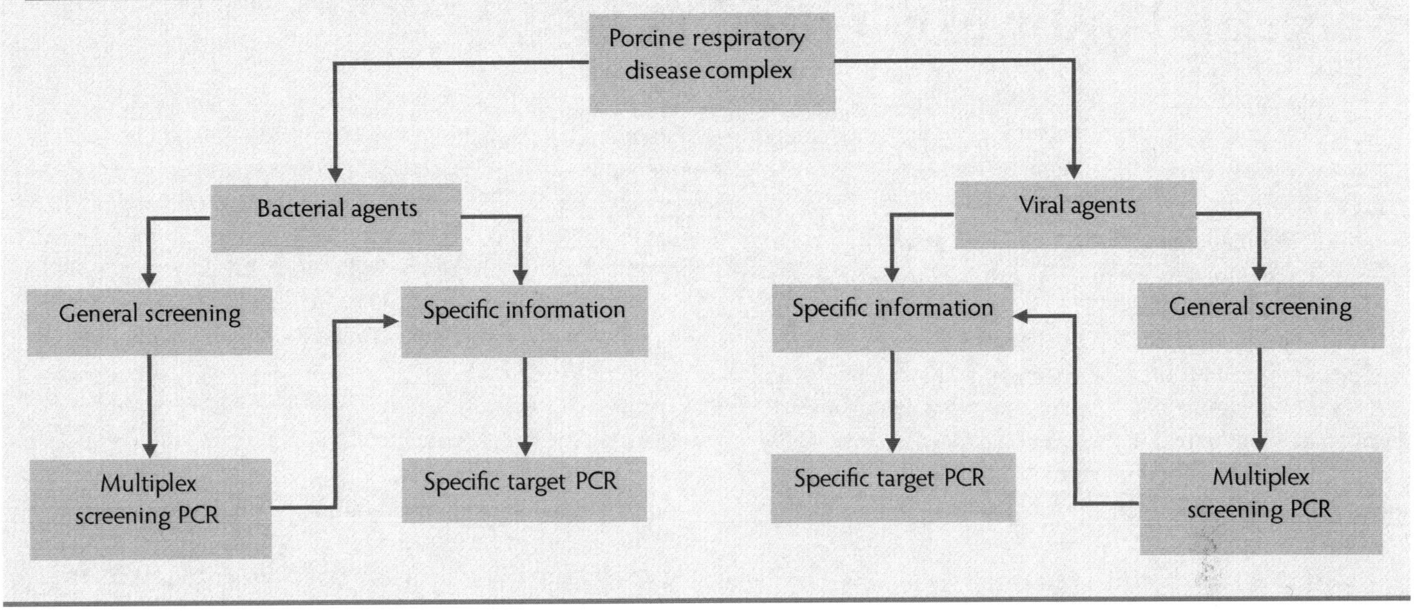

of results based on specific markers of interest and pathogen load will make the diagnosis more easier and solid in order to plan the best approach on the farm in order to control PRDC.

Note

All these PCRs will be available for the North American market as reagent packs from BioChek. In the rest of the world these PCRs will be available as diagnostic kits.

How heat placement in farrowing affects nutrient intake and pre-wean mortality (PWM)

B. Strobel, PE, MS, MBA
Osborne Industries, Inc. Osborne, Kansas

Introduction

The period from birth to weaning is critical for both sow and piglet, and each require separate, independent environmental zones. A sow's thermal-neutral zone (TNZ) is 17-18°C with upper critical temperature (UCT) of 21°C. Feed conversion has been proven to be improved in rooms with depressed temperatures.[1] Room temperatures above the UCT will decrease feed intake of all lactating sows, but first parity sows are the most sensitive. Daily feed intake decreases by 0.44 kg per 0.6°C when temperatures rise above 19°C,[2] which has negative effects on both the sow herself and her litter of piglets.

With pre-wean mortality (PWM) at 12-18% and sow mortality at 7.32%,[3] providing a micro-environment for the piglet litter is critical for early development and livability. This environment for piglets should consist of a draft-free area and temperature near 35°C. Supplemental heat in the piglet zone must be kept away from the sow so piglets do not lay close to her for warmth, rather to suckle at one or more of the teats.[4] Effective farrowing should allow piglets to eat for growth, rather than warmth, and for sows to maximize feed intake for body condition and milk production.

Nutritional requirements of sows and piglets

The appetite of a lactating sow is lower in early lactation when compared to late lactation. Feed intake increases up to the third week postpartum. Factors impacting nutrient intake are parity, barn temperature, back fat at farrowing, and feeding system. Data from several studies have shown that back fat levels of less than 14 mm at weaning may delay return to estrus. Depending on the parity of the sow, this weight loss could consist of 40-60% fat.[5] Litter weaning weights are lower and weaning-to-estrus intervals are longer and more variable on farms using a gradual increase feeding program in early lactation as compared to farms that employ a rapid increase in early lactation feed intake.[6] This highlights the importance of maximizing feed intake during each week of lactation in order to optimize piglet growth rate, limit sow weight loss and optimize subsequent reproductive performance. Lactation feed intake generally increases from the first to the sixth parity, with the major increase from first to second parity (15-20%). For each 1.0 kg daily deficit in feed requirement, the sow will lose 0.7 kg of body weight.[7] It is important to limit non-productive days (NPD) to keep sows and gilts in optimal condition for re-breeding. Proper placement of piglet supplemental heat will ensure the sow's appetite is not suppressed, and will help ensure optimal performance of both her and her piglets.

The nutritional goal for piglets is to eat for growth, not for warmth. Piglets nurse from a mother's teat about every 90 minutes when newborn to three to four hour intervals within a week later.[8] Creep areas with inadequate heating allows piglets to become chilled and seek the sow for warmth. Laid-ons can account for 55% of PWM.[4]

Farrowing environments

Modern farrowing rooms typically have capacity for 20-80 sows at one time, managed all-in/all-out. Farrowing crates and turn-around pens are two commercially-acceptable ways to house a sow and her litter. In each housing style, sows and piglets each have their own defined zones which they occupy during the turn. See Figures 1 and 2. Widths of piglet and sow zones are typically around 0.6 m. The key components of a piglet's environment are thermal, social, dietary, and management,[9] with thermal environment being the interaction of air temperature, moisture, and airflow. Gradual environmental changes tend to be less stressful than abrupt changes, especially for piglets, so temperatures in each zone should be lowered gradually throughout the turn via a temperature-dependent variable output controller.[10]

Throughout the year, both piglet and sow zones must be controlled independently. Excess heat should be kept off sows. In summer, ventilation should be used to remove excess heat from

Figure 1: Typical farrowing crate layout

1. Sow space
2. Unheated creep area
3. Heated area
4. Sow feeder
5. Water
6. Pen dividers

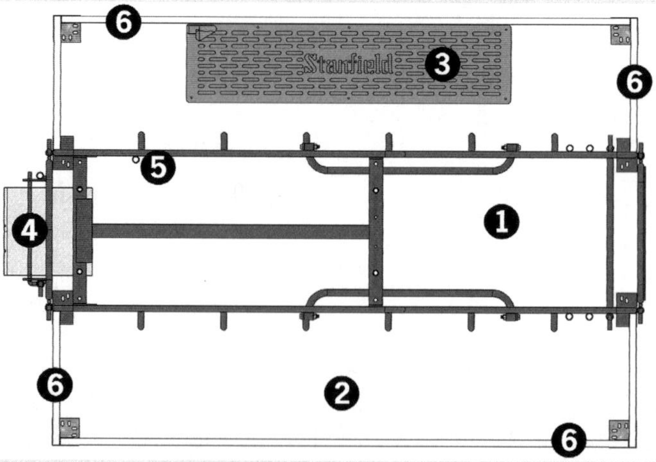

the building. Drip-cooling is necessary in the sow zone, however, piglets should be kept dry to prevent scours and other health issues. In winter, ventilation should be used to remove moisture without drafting the litter. A goal with heat placement in farrowing is to provide 0.03 m^2 of uniform heated area per piglet,[10] so a 0.4 m^2 area can heat 12 piglets. Typically, heat lamps provide heat in a round pattern that reaches the sow while heat mats supply rectangular floor heat across a larger footprint in the piglet zone only. A typical heated pad lies midway along the creep divider and projects out 0.3 m and is 1.2 m long. At maximum power output, a 0.4 m^2 heated area is typically about 100W or 8.5W/piglet. Patterns are established for eating at the sow, sleeping in the heated zone, and exercising in other areas of the crate or pen. Once the patterns are established, piglets continue with these habits as they become better able to regulate their body temperature.

Room environment throughout the turn

At birth, piglets should be dried and introduced to zones in the farrowing crate: heated area and teat order in the nursing area.[11] The heated, draft-free area, or thermal-neutral zone (TNZ), keeps piglets from being chilled. It is important to engrain in piglets to not lay near the much larger sow who can unknowingly lay on them, especially challenged pigs desiring to be warm. Piglets spend approximately 60% of the time nursing, and lying next to each other in a non-heated zone on the creep floor between supplemental heat and the sow[8], however, with well-defined zones and early training, the remaining time can be spent in the heated zone. 35°C is the desired starting temperature of the piglet micro-environment, which is slightly less than the 39°C body temperature of the piglet. Refer to Figure 3 for the interaction of micro-environment temperature, day of age, and piglet weight.

At Day 3, most producers allow ad-lib feeding for sows as they become more active. The primary responsibility is keeping sows comfortable to maximize feed intake. This translates into desired body condition and ample milk for the litter. Heat from the piglet zone must be kept off sows and ideal room temperatures at somewhere less than 21°C.[7] Sows can eat six to eight meals throughout the day if stimulated by someone in the room or noise at the feeder. Intake of 10-14 kg of feed per day is the target for optimum sow nutrition. The interaction of nutrition with her genetics, being used to eating 2.7 kg in gestation in one to two meals per day, to eating up to six times in lactation is a four-fold increase in nutrient intake.

Around Day 17, creep feeding a high-protein diet on a mat or in a small feeder in the creep area can begin. This supplements and transitions them from liquid meals to bulk solids. Room and creep temperatures are near equal until the end of the turn.[7]

Conclusion

The farrowing room houses two stages of production: the sow and piglet. Providing the optimum temperature for the sow improves feed intake, litter weight, and decreases NPD. Additionally, strategic heat placement within the piglet zone reduces PWM and allows producers to wean nearly as many piglets as were born.

Figure 2: Typical farrowing pen layout

1. Sow space
2. Heated area
3. Anti-lay bar
4. Sow feeder
5. Water
6. Pen dividers

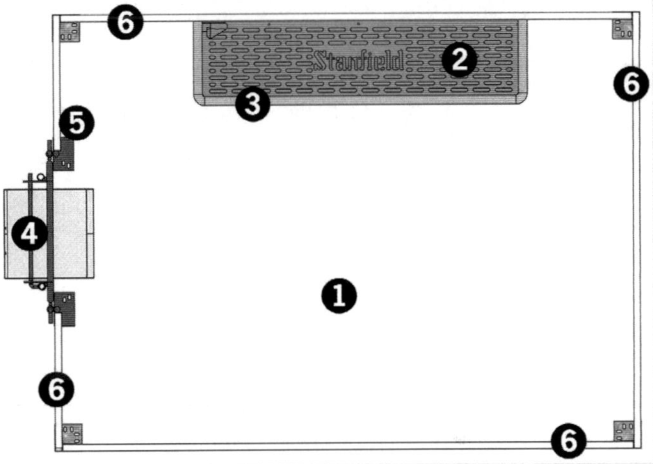

Figure 3: Changes in heating requirements as piglets grow

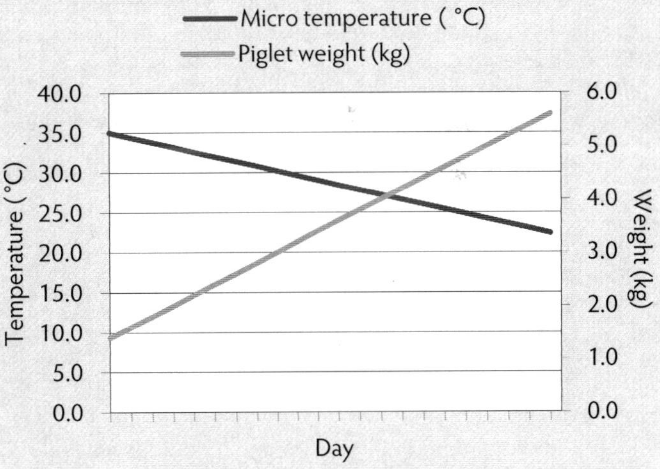

References

1. Nienaber, J.A, Shanklin, M.D., Hahn, G.L., and R.C. Manak. Performance of Neonatal and newly-weaned pigs as affected by temperature and diet. Transactions of the ASABE. Vol 28 (5) 1626-1633. 2010.

2. Johnston, L. Feeding management for lactating sows. University of Minnesota. 2011.

3. Rix, M, Ketchem, R. Focus on farrowing to save more pigs. National Hog Farmer. October 11, 2011.

4. Yin, G., Liu, H., Li, X., Quan, D., and J. Bao. Effect of farrowing environment on behavior and physiology of primiparous sows with 35-day lactation. 2016.

5. Genest, M., and S.D'Allaire. 1995. Feeding strategies during the lactation period for first parity sows. Can. J. An. Sci. 75: 461-467.

6. Aherne, F. Feeding the lactating sow. 2010.

7. Malmkvist, J., Pedersen, L.J., Kammersgaard, T.S., and E. Jurgensen. Influence of thermal environment on sows around farrowing and during the lactation period. 2012.

8. Hrupka, B.J., Leibbrandt, V.D., Crenshaw, T.D., and N.J. Beneve. The effect of farrowing crate heat lamp location on sow and pig patterns of lying and pig survival. 2011.

9. Lammers, P.J., Stender, D.R., and M.S. Honeyman. Environmental needs of the pig. 2007.

10. Stinn, J., and H. Xin. Heat lamp vs heat mat as localized heat source in swine farrowing crate. 2014.

11. Johnson, A.K., and J.J. Marchant-Forde. Welfare of pigs in the farrowing environment. 2009.

Improving the speed of health-based decision making

E. J. Lowe[1], DVM; J. T. Holck[2], DVM, MS, MBA; D. D. Polson[1], DVM, MS, PhD; B. Cowan[3], MS; B. Martínez-López[4], DVM, MPVM, PhD
[1]Boehringer Ingelheim Animal Health, Duluth, Georgia; [2]Feed His People, LLC, Gilbert, Iowa; [3]GlobalVetLINK, Ames, Iowa; [4]University of California-Davis, Center for Animal Disease Modeling and Surveillance, Davis, California

Introduction

Diagnostic testing has become an increasingly important part of not only diagnosis of clinical episodes, but also for continuous monitoring and surveillance in pork production systems. Diagnostic test results have a direct impact on farms and flows as they are used routinely to assess process implementation, red or green light animal movements, justify product use, and inform expectations of future group production performance. The challenge for a veterinarian upon receipt of diagnostic case test results is to review the results, understand what is occurring in the sampled and tested population/site and communicate that understanding quickly to production personnel to drive appropriate on-farm actions. GlobalVetLINK has partnered with the University of California-Davis to automatically transmit aggregated diagnostic information from LabLINK HIMS™ into the Disease BioPortal™. This partnership will better enable veterinarians to improve the time from the receipt of test results to on-farm action.

The current process

There are several steps between receipt of diagnostic test results and communication to production personnel, including: aggregation, analysis and forming expert understanding or interpretation of the results. In this paper we will use a Porcine Reproductive and Respiratory Syndrome Virus (PRRSV) genetic sequence result as an example to illustrate the process (Figure 1).

1. Receive

This step is fairly simple as US diagnostic laboratories have developed routine and efficient ways to communicate diagnostic test results to submitting veterinarians via web portals and email distribution. However, with industry consolidation, production systems and veterinary practices have become more geographically diverse. Currently two challenges exist. First, samples may be submitted to different veterinary diagnostic laboratories within a practice or production system, thus results come from multiple sources rather than a single source. Second, this geographic diversity has increased the reliance on multiple veterinarians or production personnel to submit samples for the same farm, flow or production system, thus results come from different submitters as well as different lab sources. Each submitter may have a different way of entering a case into a diagnostic lab. This often means that the same information such as contact information or case history is not included or included in the same way, or that all of the correct people are not affiliated with the case in order to receive the results directly from the diagnostic lab.

These issues are not insurmountable. Web based submission tools and processes have not only streamlined the submission process but have driven increased uniformity of submission information in those veterinary practices and production systems that have implemented them. It has also largely addressed the affiliation concerns with cases to all of the appropriate entities. LabLINK HIMS™ is a tool available since 2014 from GlobalVetLINK that can support the standardized and uniform receipt of diagnostic case results.[1] LabLINK HIMS can be utilized to view diagnostic case results from multiple labs in one location and to correct or include submission information on individual cases within a web-based diagnostic database. Finding all results in one location can save time in this receipt step while standardizing submission information can save time in future steps in the process.

2. Aggregate

Upon receipt, results often need to be aggregated over time and across sites or flows for analysis and interpretation. Using the PRRSV sequence example, the veterinarian reviews the comparison to the commercially available vaccines routinely provided by the diagnostic laboratory. The veterinarian then typically asks "Is this the same virus that has been on this farm or is it new to this farm?" and "If it is new, where did it come from?". To answer these questions, the sequence and related metadata must be aggregated with other sequences for analysis. Historically this process has been as simple as asking the lab to add this sequence to the veterinarian's, or production system's, existing "dendrogram", meaning the post-alignment sequence tree. This has been generally effective for answering the first question, "Is it new to this farm?" if there are historical sequences from the farm previously compiled and on file. If not, historical cases must be searched and a list of sequences compiled for the farm. This process can be time consuming and incomplete, as the search process will only be as good as the search criteria used against the information included in prior submissions and the cases available to the veterinarian to search.

The "Is this new?" question has been made easier with web based submissions and LabLINK HIMS as they not only help in the Receive step, but also in this Aggregate step. If the attributes included on the submission form are uniform and consistent, searching for relevant historical cases in the diagnostic case list at diagnostic labs or in the already aggregated dataset in LabLINK HIMS is fast and likely more complete.

Currently the second question, "Where did it come from?", requires more time to aggregate the new sequence with other sequences that may be relevant to explore the answer to this question. Many veterinarians "shop the sequence around" to other colleagues who have sequence datasets in order to aggregate their sequence with other sequences outside of their own database for comparison. To clarify, veterinarian A may take this new sequence and send it to veterinarian B to add to their dataset and see if they have something similar. Veterinarian B then goes through the process to analyze, interpret and communicate back to veterinarian A. The time for veterinarian B to return results to veterinarian A can be significant. This process may involve many different veterinarians to aggregate the same sequence in an area or region in an attempt to find a site or sites with one or more closely related sequences – the potential "smoking gun". Veterinarian A now has to aggregate the information from veterinarian B, C, D, and any more, interpret the information and communicate the findings and related proposed response actions to production personnel.

The speed of answering this second question has been improved through the creation of data sharing networks and use of the Disease BioPortal. Several veterinary practices and productions systems use Disease BioPortal for sequence analysis, interpretation and communication. These dataset administrators currently upload their information via .csv file into the tool.[2] Administrators can also elect for others to access their sequence information for real-time aggregation of sequence information. One example of this is the Illinois Swine Health Network (ISHN) where ten dataset administrators have granted access to ISHN to create these area based sequence comparisons for the participants upon request.[3] Because the Disease BioPortal allows for permissioned aggregation of multiple datasets of like sequences (species, disease agent, gene) in the cloud, a single entity can now be permissioned to aggregate sequences (and permissioned meta-data) for veterinarians A, B, C, D and more; thereby improving the speed of the process.

3. Analyze

Analytical tools are used in this step to assess the aggregated results. This is where the math is done. In our PRRSV sequence example this step is the sequence alignment process. The sequence alignment file is used to generate the percent identity matrix file and phylogenetic tree (two of the more commonly used sequence output methods), which then are used in the next Interpretation step. The alignment step is currently performed by veterinary diagnostic laboratories, as well as some veterinary practices and production companies using available PC-based sequence alignment software or web-based sequence alignment tools.

The speed of the sequence alignment step is dependent on the number of sequences being aligned and the computational power of the PC or server(s) being used. As the number of sequences increases, the time required for alignment will also increase. However, the results of these more inclusive (larger) alignments will enhance interpretation. Disease BioPortal currently has an automatic alignment feature for sequence alignments of up to 2,000 sequences. For larger datasets, or for veterinarians that prefer something other than the default alignment method, Disease BioPortal has the ability to export the .fasta file with the sequence information for alignment outside of the platform. Resulting files from the sequence alignment can then be uploaded back into Disease BioPortal for use in interpretation.

4. Interpret

The Analysis step results in visual tools to aid in this Interpretation step. For sequence analysis results, .pdf based phylogenetic trees and Excel based percent identity matrices (a.k.a., "homology tables") are typically provided by the veterinary diagnostic laboratory. As sequencing has become more common and the size of sequence datasets have become substantially larger, the related phylogenetic trees and percent identity matrices have become larger and more cumbersome to effectively interpret. Common challenges include: inconsistency in the naming of sequences over time, the amount of time it takes to sort down the identity matrix for what is really important, the inability to search the .pdf file for the sequence of interest and the lack of association with desired metadata (e.g., site name, date and flow).

Disease BioPortal is used to better enable interpretation of genetic sequence results.[2,4] The platform allows the user to standardize data inputs and the names of sequences. It has a search feature within the phylogenetic tree and can zoom to the sequence of interest as well as a phylogenetic chart component that ranks the most closely related sequences to the sequence of interest. These allow the user to zoom quickly to what is important for interpreting the new sequence, even in a very large dataset. Disease BioPortal also allows the user to associate other metadata related to the sequence as a portion of the label in the phylogenetic tree, or as an additional column in the identity matrix. In addition, the mapping features within Disease BioPortal combined with the phylogenetic information allow for spatio-temporal interpretation of sequences if site locations are included in the dataset.

5. Communicate

The final and most important step is communicating the interpretation of the results to production personnel to then act on. This may be accomplished verbally, over the phone or in person at the next farm visit, or by e-mail to a larger group of production personnel. This step may be time consuming if there is a long lag time between the steps that the producers are involved in, the receipt of the sequence and the communication of the interpretation.

The visual tools available in Disease BioPortal can be used to enable more effective communication to production personnel and support the veterinarian's interpretation of the sequence result. Static visuals from Disease BioPortal can be sent via e-mail as screenshots, or can be assembled easily into a custom built report and sent via .pdf or as an interactive html report. Alternatively, producers can access live BioPortal Dashboards to zoom in, or filter the data to answer their own questions via password protected web link.

Summary

In reviewing the process between the receipt of case results to the communication of understanding to production personnel we've described how LabLINK HIMS and Disease BioPortal can independently improve the speed to health-based decision making. LabLINK HIMS assists with receipt of results in a single location and assists with standardization to make aggregation faster and easier. Disease BioPortal allows for easier aggregation of results within a dataset and with other datasets with permission as well as for better visualization and more context (geography, production system, flow, etc.) to also be included in the interpretation. It also can improve communication of the interpretation to production personnel. The partnership between GlobalVetLINK and the University of California-Davis now allows for the automatic transmission of aggregated diagnostic information from LabLINK HIMS directly into Disease BioPortal. This partnership aids veterinarians and production systems by further improving the speed of health based decision making.

References

1. Holck et al. Herd health management solutions: the future is now. AASV. 2016; 209-21.

2. Gauger et al. Disease BioPortal: Current applications for Influenza A virus in swine. AASV. 2017; 27-29.

3. J Lowe et al. Leveraging data across a large geographic area to increase context and understanding of PRRSV transmission. ESPHM. 2017. HHM-OP-01.

4. E Lowe et al. Porcine reproductive and respiratory syndrome: Is it the resident virus or is it a new introduction? ESPHM. 2017. VVD-027.

Figure 1: Process for health-based decision making: Genetic sequence result as an example

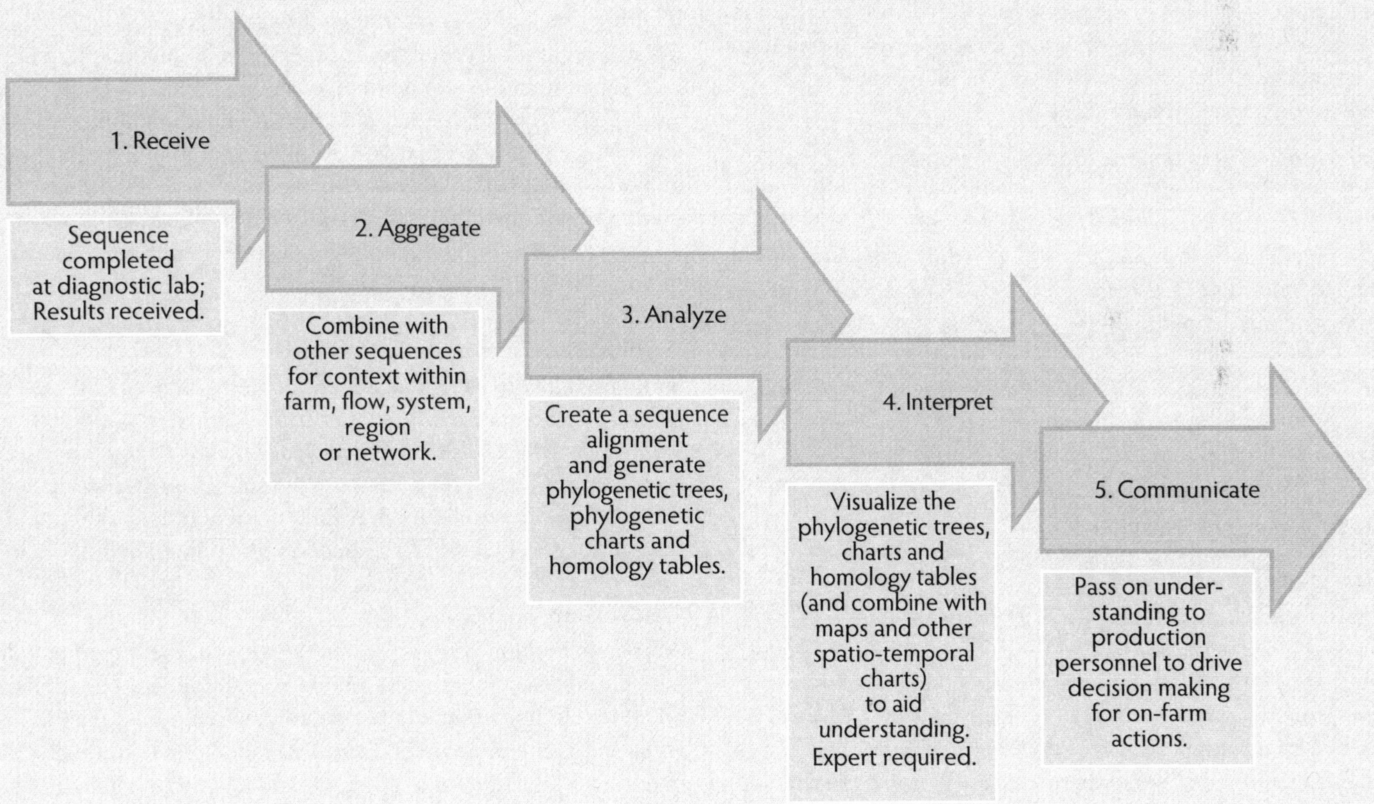

The impact of zinc amino acid complex supplementation on the porcine response to subclinical *Lawsonia intracellularis* infection

F. Leite[1], DVM, MS, PhD Candidate; E. Vasquez[1], DVM; F. Vannucci[1], DVM, MSc, PhD; A. Rendahl[1], PhD; J. Torrison[1], DVM, PhD, DACVPM; A. Mueller[2], DVM; N. Winkelman[2], DVM; C. Gebhart[1], MS, PhD; Z. Rambo[3], PhD; R. Isaacson[1], MS, PhD

[1]College of Veterinary Medicine, University of Minnesota, St. Paul, Minnesota;
[2]Swine Services Unlimited Inc, Rice, Minnesota; [3]Zinpro Corporation, Eden Prairie, Minnesota

Introduction

Zinc is an essential trace mineral for animals and a critical component of many structural proteins, enzymatic processes and transcription factors.[1] Additionally, zinc also has been shown to impact both the innate and adaptive immune systems.[2] At the cellular (enterocyte) level in the pig, it has been demonstrated that a zinc-amino acid complex (Availa Zn) can improve trans-epithelial resistance and help maintain epithelial cell morphology in the presence of a stressor such as heat stress.[3]

L intracellularis is an obligate intracellular pathogen that replicates inside of ileal enterocytes and causes a disease called proliferative enteropathy (PE).[4] The overall effect of infection is to cause diarrhea and reduction in animal growth. This disease has two clinical forms: porcine intestinal adenomastosis (PIA) and proliferative hemorrhagic enteropathy (PHE). PIA is the most common form of this disease which generally has a subclinical manifestation leading to decreased animal production performance.[4] The goal of this study was to investigate the impact of Availa Zn and Availa Zn LQ on the host response to subclinical PE.

Materials and methods

Study design and animal diets

In this study, a total of 4 treatment groups were used in a randomized block design. A total of 72 4-week-old pigs were randomized by weight and gender and assigned to one of four treatment groups. There were 18 pigs per treatment divided among 6 pens with three animals per pen (Table 1). The pens were equally distributed in to two barns. All pigs were fed their respective diets ad libitum 21 days prior to challenge with *L intracellularis*. The treatment groups consisted of a negative and positive control group which received 125 ppm of zinc sulfate, a group which received 75 ppm of zinc sulfate and 50 ppm of zinc amino acid complex (Availa Zn) and a fourth group which received 125ppm of zinc sulfate and zinc amino acid complex (Availa Zn LQ) provided in water (average daily water Zn disappearance of 0.1mg per pig). All groups except the negative control group were challenged with a mucosal homogenate from a finishing pig with acute PHE. The pig that was used for the challenge homogenate was screened for other pathogens and none were found. The challenge dose of *L intracellularis* was calculated to be 8×10^7 organisms per animal and was administered to pigs by oral gavage. Groups of 6 randomly chosen pigs per treatment were euthanized at 14, 21 and 28 days post challenge.

Animal production measurements

To investigate the impact of zinc supplementation with Availa Zn on the production performance of animals when infected with *L intracellularis*, measures of average daily gain (ADG), average daily feed intake (ADFI), feed to gain ratio (F:G) and water disappearance were obtained. Animals were weighed weekly and recording of feed and water provided was recorded daily.

Clinical observations

Animals were clinically monitored throughout the experimental period and evaluated for demeanor score, abdominal score and fecal score. All scores used for evaluation varied from zero to three increasing in severity of signs.

Table 1: Experimental design and treatments.

Treatment	Pens/treatment	Pigs/pen	Zn level (ppm) & source	*Lawsonia* challenge
Positive control	6	3	125 SO_4	Yes
Availa ZN in feed	6	3	75 SO_4 + 50 Availa ZN in feed	Yes
Availa ZN LQ in water	6	3	125 SO_4 + Availa ZN LQ in water	Yes
Negative control	6	3	125 SO_4	No

Gross and microscopic pathology evaluation

At 14, 21 and 28 days post infection (dpi) groups of animals (n = 6) were euthanized and postmortem examination of the ileum, jejunum, cecum and colon of all animals was carried out to identify the presence of any gross pathologic changes including those characteristic of PE. Each lesion was scored as either mild (score of 1), moderate (score of 2) or severe (score of 3). For histopathology and evaluation of microscopic lesions characteristic of PE, a section of the ileum proximal to the ileocecal valve was collected from all animals. Histological evaluation was performed by hematoxylin and eosin (HE) stain and immunohistochemistry (IHC) performed using murine anti-*L intracellularis* specific polyclonal antibody, to evaluate the extent of proliferative lesions and quantify the amount of *L intracellularis* present, respectively. The presence of *L intracellularis* specific antigen was evaluated blindly with a five-grade IHC scoring scale (grade 0 equal to no lesions; and grade 4, equal to all enteric crypts containing antigen). The extent of lesions was also evaluated blindly and on a 3 point grading scale representing the distribution of crypt dysplasia as: 1 for focal lesions, 2 for multifocal lesions, and 3 for diffuse lesion distribution.

Antibody response

To investigate the antibody response to *L intracellularis*, serum samples were collected at 14, 21 and 28 days post infection. Serum samples were serially diluted from 1:30 to 1: 1920 in phosphate-buffered saline (PBS) containing 2.5% fetal bovine serum, 1.0% rabbit serum and 0.08% tween 80 and screened for antibodies against *L intracellularis* following a previously described protocol.[5]

Statistical testing

Statistical tests were performed in R version 3.3.3 (2017-03-06) using a mixed model and differences in lsmeans evaluated for statistical significance. The pen was considered the experimental unit. P values below 0.05 were considered significant. Trends were described as statistical results with P values below 0.1. To compare positive and negative samples from animals for seroconversion, gross lesions, IHC and HE lesions, the N-1 chi-squared test was used.

Results

Clinical measurements of disease

There was only one animal with a fecal score above zero following challenge with *L intracellularis* that had a score of one indicating that it had a loose stool. This animal was in the positive control group. There were no cases of diarrhea or clinical signs characteristic of PPE throughout the experimental period, thus subclinical disease was induced.

Production measurements

No statistical differences were found in any measure of animal production. Average daily gain varied from an overall average across the treatments of 1.24 lbs per day at day 14 of the trial to 2.03 lbs per day at day 49. There was no statistical difference between treatments when comparing different treatments within a timepoint, however there was a numerical reduction (ranging from 0.1lbs to 0.27lbs after challenge) in average daily gain in the positive control group supplemented with zinc sulfate. There was a trend for the ADG in pigs receiving Availa Zn in feed to have a slightly lower average daily weight gain compared to pigs in the negative control group at 28 dpi, with a difference of 0.51 lbs ($P < 0.1$). There was also a trend for the group treated with Availa Zn in feed to have a higher feed to gain ratio compared to the negative control group at 28 dpi. These ratios were of 2.37 and 2.01, respectively ($P < 0.1$). These trends were not statistically significant.

Gross lesions

Gross lesions were observed at 21 and 28 dpi and none were observed at 14 dpi. On both 21 and 28 dpi pigs in both Availa Zn treatment groups had fewer animals with gross lesions compared to pigs in the positive control group. At 21 dpi 33% of animals which received either Availa treatment had gross lesions (2 of 6 animals) while 66% of animals in the positive control group had gross lesions (4 of 6 animals). At 28 dpi, the Availa Zn group had the least number of animals with gross lesions (33% (2 of 6)) while 50% of pigs in the Availa Zn LQ treatment had positive gross lesions (3 of 6). All pigs in the positive control group had gross lesions at this time point (6 of 6 animals). The reduction in the number of animals with gross lesions in the Availa Zn group compared to the positive control group at this timepoint was statistically significant ($P < 0.05$). When analyzing the extent of lesions, both Availa Zn treatments led to a significant reduction in lesion severity at 28 dpi ($P < 0.05$). The average lesion score was of 1.33 in the positive control group compared to 0.33 in the Availa Zn feed group and 0.5 in the Availa Zn LQ group (Figure 1).

Immunohistochemistry and microscopic lesions

Trends observed in microscopic lesion severity (HE score) were similar to those for IHC score. This is expected since lesion severity is associated with the amount of *L intracellularis* present in tissue. When analyzing the number of animals that had microscopic PE lesions, Availa Zn given in feed led to a significant reduction in the number of animals with lesions ($P < 0.05$). At 28 dpi, only 17% (1 of 6) of animals in the Availa Zn group had lesions compared to 83% (5 of 6) of pigs in the positive control and Availa Zn LQ groups (Figure 2).

Serum antibody response

Evaluating the number of animals with a positive serologic response, at 14 dpi both Availa zinc treatments led to a significant increase in the number of animals with anti-*L intracellularis* antibodies compared to the positive control group ($P < 0.05$) (Figure 3). This difference was of 94% seroconversion in the Availa Zn group (17 of 18 animals) and 89% seroconversion (16 of 18 animals) in the Availa Zn LQ group compared to 50%

Figure 1: Gross lesion score in ileum at different days post infection. Error bars represent standard error of mean. Different letters indicate statistical significance ($P < 0.05$).

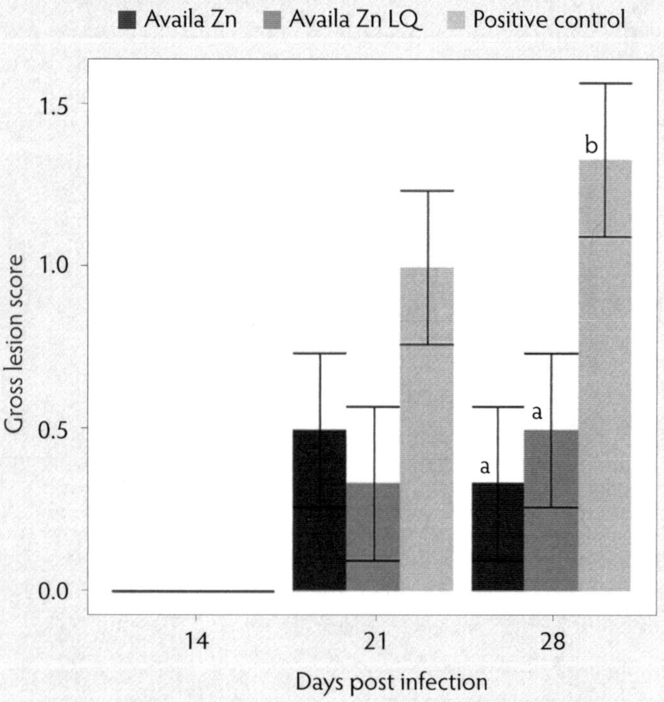

Figure 2: Percent of animals with microscopic PPE lesions evaluated by H&E staining. Different letters indicate statistical significance ($P < 0.05$).

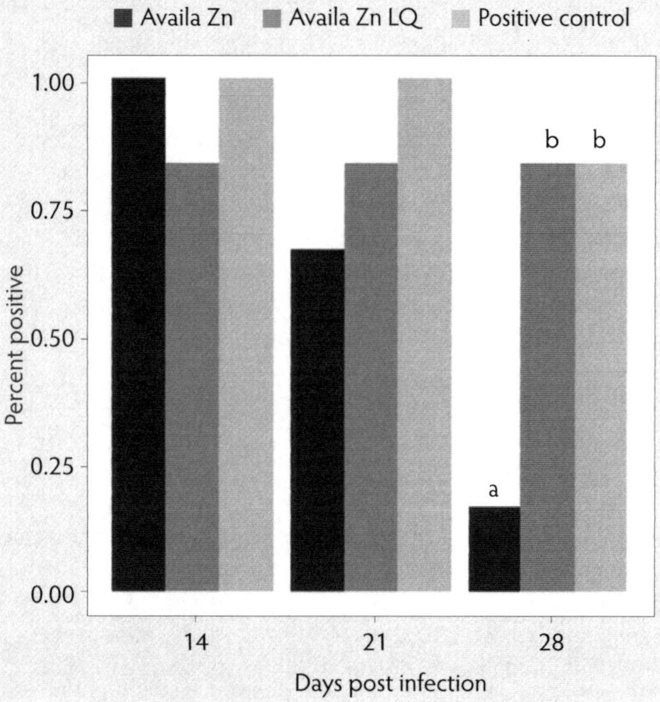

Figure 3: Percent of animals with antibodies in serum against *L. intracellularis*. Different letters indicate statistical significance ($P < 0.05$).

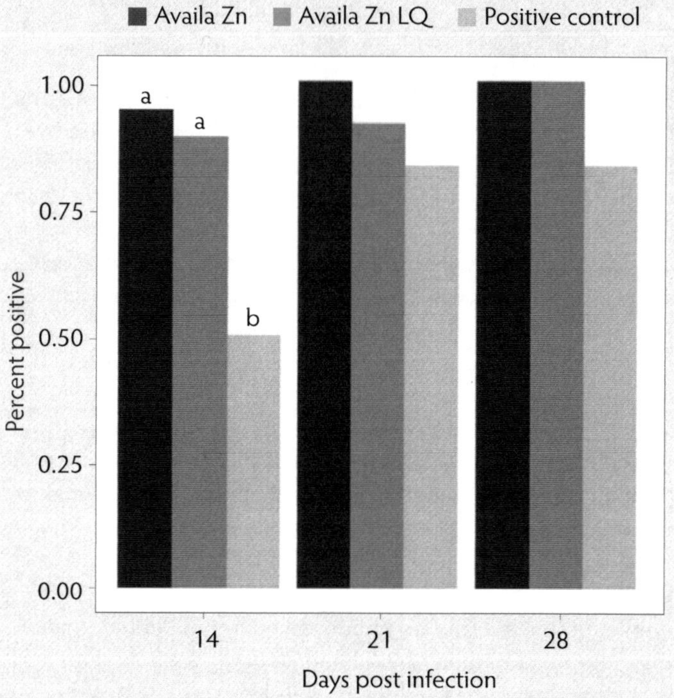

seroconversion in the positive control group (9 of 18 animals). Significant differences in antibody titers were not observed between the treatment groups, demonstrating that treatment with Availa Zn reduced the time to seroconversion but did not affect the magnitude of the response.

Discussion

This study was performed to investigate the effect of Availa Zn and Avalia Zn LQ on the host response to *L intracellularis* infection. Significant differences were observed using traditional methods of disease measurement: gross lesions and microscopic PE lesions. Availa Zn given in feed or Availa Zn LQ given in water both led to an overall reduction in the number of animals with gross lesions and a significant decrease in the severity of lesions at 28 dpi ($P < 0.05$). In terms of microscopic lesions, Availa Zn given in feed led to a large and significant decrease (from 83% to 17%) in the number of animals with lesions at 28 dpi ($P < 0.05$). This treatment also led to a trend of having less *L intracellularis* present in ileum tissue at 28 dpi, demonstrating its positive effect on resolving infection.

There were no statistically significant differences between treatments in ADG, ADFI, F:G ratio and water disappearance. Two factors that likely contributed to this were the lack of a clinical impact on overt disease since the challenge dose of *L intracellularis* only induced mild disease and the low number of animals used in this trial. Two trends were identified indicating that Availa Zn in feed may lead to decreased weight gain and increased feed to gain ratio. These differences should be confirmed with a greater number of animals. It is known that *L intracellularis* does reduce growth performance of pigs and had the disease been more robust it is possible that improved growth would have been observed with the Availa Zn treated pigs.

Zinc is known to have an impact on immune function.[1,2] Interestingly, both Availa Zn and Availa Zn LQ treatments led to an earlier onset ($P < 0.05$) of seroconversion compared to supplementation with zinc sulfate alone. Zinc is a trace mineral that is not efficiently stored in the body, thus daily intake is required to achieve adequate steady-state concentrations to support all its functions.[2] It is likely that Availa Zn, which is a zinc amino acid complex, led to superior supplementation than zinc in the sulfate form alone since total zinc levels were the same in both treatments (125 ppm).

This study demonstrated the impact of Availa Zn and Availa Zn LQ on the host response to *L intracellularis* infection. A decrease in gross and microscopic lesions was observed, which was more pronounced with the group that received Availa Zn given in feed. Both Availa Zn treatments also led to an earlier seroconversion compared to the group that received challenge without Avila Zn supplementation. These results indicate that Availa Zn and Availa Zn LQ aid the pig immune response to infection and lead to decreased number of PE lesions and their severity.

References

1. Tuerk MJ et al. Zinc deficiency. *Curr. Opin. Gastroenterol.* 2009, 136–143.

2. Bonaventura P et al. Zinc and its role in immunity and inflammation. Autoimmun. Rev. 2015, 277–285.

3. Pearce SC et al. Dietary organic zinc attenuates heat stress–induced changes in pig intestinal integrity and metabolism. J. Anim. Sci. 2015, 4702–4713.

4. Vannucci FA et al. Recent Advances in Understanding the Pathogenesis of *Lawsonia intracellularis* Infections. Vet. Pathol. 2014, 465–477.

5. Guedes RMC et al. Validation of an immunoperoxidase monolayer assay as a serologic test for porcine proliferative enteropathy. J. Vet. Diagn. Invest. 2002, 528–30.

Effect of avilamycin (Kavault®) on incidence and severity of diarrhea in presence of pathogenic *Escherichia coli* in nursery pigs

D. C. Beckler[1], DVM, MS; R. Evelsizer[2], DVM; C. L. Puls[2], PhD; J. M. Hammer[2], DVM; N. B. Betts[2],MS; S. N. Carr[2], PhD
[1]Gut Bugs, Inc., Fergus Falls, Minnesota; [2]Elanco Animal Health, Greenfield, Indiana

Introduction

Enterotoxigenic *Escherichia coli* (ETEC) is the major causative agent of post-weaning colibacillosis. Without appropriate treatment/prophylaxis nursery pigs can have significant reduction in growth performance and increased morbidity and mortality.[1] In recent years, a new treatment (avilamycin; Kavault, Elanco Animal Health, Greenfield, IN) was approved in the United States for the reduction in the incidence and overall severity of diarrhea in the presence of pathogenic *E coli* in weaned pigs.[2] Avilamycin is in the orthosomycin class of antibiotics, used only in veterinary medicine, and not currently deemed medically important to human medicine. The mode of action is relatively unclear but it has been hypothesized that avilamycin inhibits the expression of fimbrial proteins by enterotoxigenic *E coli*, thus reducing the bacteria's ability to adhere to enterocytes.[3,4] Without intimate fimbrial attachment, membrane bound and secreted toxin are not able to act on enterocytes and cause hypersecretion of fluids and electrolytes. Clinical registration studies for avilamycin showed effectiveness in pigs experiencing significant infection, as 91% of control animals showed signs of diarrhea.[2] However, there has been little research performed in the US to validate results observed in the clinical registration studies. Therefore, the objective of this study was to evaluate the effects of feeding 80 ppm (73 g/ton) avilamycin for up to 42 days on the incidence and severity of diarrhea and subsequent growth performance in nursery pigs when compared to an untreated control.

Materials and methods

This study was carried out using a randomized complete block design (blocking factor = body weight) with 2 dietary treatments: 1) Control (no medication) and 2) 80 ppm avilamycin. A total of 240 crossbred barrows were used on the study and were housed in 40 pens with 6 pigs per pen. Pen was the experimental unit for all measures.

The pigs in the facility prior to this study were challenged with a F18+ ETEC strain and exhibited disease symptoms (e.g., diarrhea) commonly associated with post-weaning colibacillosis. The test facility was not cleaned or disinfected prior to pig arrival. Diagnostic samples taken during the current study confirmed infection and presence of ETEC. Trained personnel checked pigs daily. Clinical observations are detailed below. General categories included Diarrhea, Depression, Gauntness, and Perianal Inflammation. Clinical scores were collected daily for each pig and summarized by pen.

Clinical diarrhea was measured on a 4-point scale: 0 = none; 1 = semi-solid feces/cow pie consistency; 2 = semi watery feces/oatmeal consistency; and 3 = profuse watery feces.

Depression was measured on a 3-point scale: 0 = bright, alert, and responsive; 1 = slow to rise, delayed response to stimulus, stands with head down, listless; and 2 = severe depression, reluctant to rise or recumbent.

Gauntness was measured on a 3-point scale: 0 = normal abdominal fill, flank is full and rounded; 1 = decreased gut fill, flank is flat; and 2 = severely gaunt, flank is hollow.

Perianal Inflammation was measured on a 4-point scale: 0 = normal; 1 = mild pink colored anus; 2 = moderate anus pink to red in color, some swelling or slight elevation of anus, occasional tail twitching; and 3 = severe anus flame red in color, anus and perineal area swollen, much tail twitching.

Pigs had *ad libitum* access to feed and water during the study. A 3-phase corn-soy based dietary program formulated to meet or exceed NRC (2012)[5] recommendations was used with Phase 1 being fed from day 0 to day 7, Phase 2 being fed from day 7 to 21, and Phase 3 being fed from day 21 to 42. Pharmacological levels of Zn (2804 and 3048 ppm, respectively) and Cu (156 and 200 ppm, respectively) were fed during the first two dietary phases. All feed additions to the feeder and the amount of feed remaining in the feeders at the time of pig weighing were recorded. Body weights were measured on day 0 (start), 21, and 42 (end) of study.

The indicator response variables of elevated score (Diarrhea >0, Diarrhea > 1, Depression > 0, Gauntness > 0, or Perianal > 0) by time period (day 0-21 or day 22-42) were analyzed using PROC GLIMMIX in SAS using a binomial distribution. The count of days with elevated scores (Diarrhea > 0, Diarrhea > 1, Depression > 0, Gauntness > 0, or Perianal > 0) by time period (day 0-21 or day 22-42) was analyzed using PROC GLIMMIX in SAS using a Poisson distribution. The model included the fixed effect of dietary treatment and random effect of block. Differences were deemed statistically significant if $P \leq 0.05$.

Growth performance measures were analyzed using PROC

MIXED in SAS using a normal distribution. The model included the fixed effect of dietary treatment and random effect of block. Differences were deemed statistically significant if $P \leq 0.05$.

Results and discussion

Pigs exhibited relatively high levels of diarrhea incidence (55.0 vs. 83.3% for avilamycin and control, respectively) in the first 21 days of study, suggesting that the lack of cleaning and disinfecting the facility did lead to a substantial infection rate in the animals. In the first 21 days of study, feeding avilamycin significantly reduced ($P < 0.05$) the incidence (34.0%) and duration, and numerically reduced the severity of diarrhea, compared to the controls (Table 1). Similarly, from day 21 to 42 of study, feeding avilamycin reduced diarrhea incidence, duration, and severity of diarrhea compared to controls. There were very few observations of depression, gauntness, and perianal inflammation in the study, and there was little to no difference between dietary treatments for these parameters (Table 1).

There was no effect ($P > 0.05$) of dietary treatment on growth performance measures in the first 21 days of study (Table 2). However, from day 21 to 42, pigs fed avilamycin grew 5.3% faster ($P = 0.17$) and had improved ($P < 0.05$) F:G (-2.5%) and G:F (2.9%) compared to controls. Pigs fed avilamycin were 1.4 lb heavier than controls at the end of study, although differences were not significant ($P > 0.05$). For the overall 42 day study period, pigs fed avilamycin had higher ADG (4.0%), ADFI (2.8%), and G:F (2.7%), and lower F:G (1.4%), although differences were not statistically significant ($P > 0.05$) (Table 2).

The results of the current study confirm expectations and substantiates previous research. Previous reports showed that feeding avilamycin resulted in a 31.2% reduction in the incidence of diarrhea.[2,4] In the present study, total diarrhea incidence levels were lower than previous studies, but feeding avilamycin still resulted in substantially lower diarrhea incidence levels (34% lower). Previous research, however, showed that feeding avilamycin also resulted in lower depression and gauntness scores compared to the controls.[2,4] In the present study, there was little to no depression or gauntness observed, likely due to a milder challenge than was reported in previous studies, and this is likely the reason for no difference between dietary treatments for these parameters. Nonetheless, the results of this study suggest that avilamycin, when fed for 42 days, is effective at reducing the incidence and overall severity of diarrhea, and validate the results reported in previous research evaluating avilamycin.

References

1. Amezcua, R., R. M. Friendship, C. E. Dewey, C. Gyles, and J. M. Fairbrother. 2002. Presentation of postweaning *Escherichia coli* diarrhea in southern Ontario, prevalence of hemolytic *E. coli* serogroups involved, and their antimicrobial resistance patterns. Can. J. Vet. Res. 66:73-78.

2. FDA. 2015. Freedom of Information Summary. NADA 141-438 Kavault. Available at: https://www.fda.gov/downloads/AnimalVeterinary/Products/ApprovedAnimalDrugProducts/FOIADrugSummaries/UCM449239.pdf. Accessed 08 November 2017.

3. Rostagno M and G. Pelger. 2014. Exposure to the antibiotic avilamycin inhibits *E coli* fimbriae and attachment. Pg 246 in the Proceedings of the 23rd IPVS Congress.

4. Marsteller, T., B. Carter, K. Eggers, E. Christianson, M. Hammer, R. Evelsizer, M. Rostagno, and J. Hammelman. 2016. Effectiveness of avilamycin (Kavault™) for reducing the incidence and severity of colibacillosis in post-weaned pigs. Pg 154-157 in the Proceedings of the 47th AASV Annual Meeting.

5. NRC. 2012. Nutrient requirements of swine. 11th ed. Natl. Acad. Press, Washington, DC.

Table 1: Least-squares means for the effects of avilamycin[1] inclusion level on clinical disease scores of nursery pigs exhibiting diarrhea associated with *E coli.*

Item	Avilamycin[1] inclusion level, ppm				
	0	SEM	80	SEM	*P*-value
No. of pens	20	-	20	-	-
Diarrhea[2]					
% incidence > 0, day 0-21	83.33[a]	0.0340	55.00[b]	0.0454	< 0.001
% of days > 0, day 0-21	11.45[a]	0.0071	5.46[b]	0.0048	< 0.001
% incidence > 1, day 0-21	7.52	0.0289	4.46	0.0211	0.32
% incidence > 0, day 21-42	40.00[a]	0.0447	14.17[b]	0.0318	< 0.001
% of days > 0, day 21-42	3.20[a]	0.0040	1.01[b]	0.0021	< 0.001
% of days > 1, day 21-42	1.68[a]	0.0033	0.90[b]	0.0021	0.02
Depression[3]					
% incidence > 0, day 0-21	0.33	0.0072	0.33	0.0072	1.00
% of days > 0, day 0-21	0.00031	0.0	0.00030	0.0	0.95
Gauntness[4]					
% incidence > 0, day 0-21	0.034	0.0009	0.003	0.0	0.06
% of days > 0, d 0-21	0.00057[a]	0.0	0.00012[b]	0.0	0.003
Perianal inflammation[5]					
% incidence > 0, day 0-21	2.57	0.0187	1.50	0.0125	0.48
% of days > 0, day 0-21	0.125	0.0010	0.046	0.0005	0.16

[a,b] Means within a row with different superscripts are statistically different ($P \leq 0.05$).

[1] Avilamycin: Trade name = Kavault® (Elanco Animal Health, Greenfield, IN).

[2] Diarrhea measured on 4-point scale: 0 = none; 1 = semi-solid feces/cow pie consistency; 2 = semi watery feces/oatmeal consistency; and 3 = profuse watery feces.

[3] Depression measured on 3-point scale: 0 = bright, alert, and responsive; 1 = slow to rise, delayed response to stimulus, stands with head down, listless; and 2 = severe depression, reluctant to rise or recumbent.

[4] Gauntness measured on 3-point scale: 0 = normal abdominal fill, flank is full and rounded; 1 = decreased gut fill, flank is flat; and 2 = severely gaunt, flank is hollow.

[5] Perianal inflammation measured on 4-point scale: 0 = normal; 1 = mild pink colored anus; 2 = moderate anus pink to red in color, some swelling or slight elevation of anus, occasional tail twitching; and 3 = severe anus flame red in color, anus and perineal area swollen, much tail twitching.

Table 2: Least-squares means for the effects of avilamycin[1] inclusion level on growth performance of nursery pigs exhibiting diarrhea associated with *E coli*

Item	Avilamycin[1] inclusion level, ppm 0	80	Pooled SEM	*P*-value
No. of pens	20	20	-	-
BW, lb				
Day 0 (start)	11.7	11.7	0.39	0.93
Day 21	25.5	25.6	0.71	0.93
Day 42 (end)	53.4	54.8	1.22	0.34
ADG, lb/d				
Day 0 to 21	0.66	0.66	0.023	0.98
Day 21 to 42	1.32	1.39	0.032	0.17
Day 0 to 42	0.99	1.03	0.025	0.26
ADFI, lb/d				
Day 0 to 21	0.83	0.83	0.026	0.96
Day 21 to 42	2.10	2.15	0.054	0.48
Day 0 to 42	1.45	1.49	0.039	0.51
G:F, lb:lb				
Day 0 to 21	0.799	0.801	0.0094	0.88
Day 21 to 42	0.630[b]	0.648[a]	0.0071	0.03
Day 0 to 42	0.679	0.691	0.0061	0.10
F:G, lb:lb				
Day 0 to 21	1.26	1.25	0.016	0.92
Day 21 to 42	1.59[a]	1.55[b]	0.018	0.04
Day 0 to 42	1.47	1.45	0.013	0.10

[a,b] Means within a row with different superscripts are statistically different ($P \leq 0.05$).
[1] Avilamycin: Trade name = Kavault® (Elanco Animal Health, Greenfield, IN).

Use of tilmicosin (Pulmotil AC®) water medication for *Mycoplasma hyopneumoniae* elimination

P. Yeske[1], DVM, MS; R. Evelsizer[2], DVM; M. Pieters[3], DVM, PhD; C. Sievers[1], DVM, MS
[1]Swine Vet Center, St. Peter, Minnesota; [2]Elanco Animal Health, Greenfield, Illinois;
[3]University of Minnesota, St. Paul, Minnesota

Introduction

There are many examples of herd elimination *Mycoplasma hyopneumoniae* that can be performed.[1] Previous *in vitro* studies indicate Tilmicosin (Pulmotil AC®) is effective against *Mycoplasma hyopneumoniae*.[2] The decision to embark on an elimination program was made on a swine herd with an acute outbreak of PRRSv, which had been positive stable for *Mycoplasma hyopneumoniae.* It was decided to add a herd closure plan in place for the PRRS elimination; the project was extended to include *Mycoplasma hyopneumoniae.* The other incentive was that this would allow for the entire production system to be *Mycoplasma hyopneumoniae* negative.

Materials and methods

The selected herd was an 800-sow farrow-to-wean farm. It was closed to new animal introduction for 240 days. Replacement gilts for the herd were already from a negative source. The sow herd was vaccinated with M+Pac Merck® at the beginning of the closure and every 90 days until the closure was completed to booster herd immunity prior to elimination. Just prior to the medication program, the entire gestation facility was washed down, as farrowing rooms are washed with each turn, no extra washing was performed.

Tilmicosin (Pulmotil AC®) at 200 mg per liter (200 ppm) was used as water medication for 7 days to the entire sow herd. Tilmicosin (Pulmotil AC®) was mixed fresh daily and administered with a proportioner water medicator. Tilmicosin (Pulmotil® 90 premix) was mixed in the feed for 21 days at 181 gm/ton in the Lactation diet and 363 gm/ton in the gestation diet. All piglets were treated with a dose of Tulathromycin (Draxxin®) at the beginning of the water medication period, and again 14 days after the initial injection. After the initial day of the medication program, all pigs were treated at birth, and once again 14 days later. This protocol was performed during 4 weeks. A statistical sample of sows were tested using laryngeal swabs for detection of *Mycoplasma hyopneumoniae* genetic material by PCR (Strait et al., 2008) once the elimination was completed.

Results and discussion

One hundred eighty samples were collected and pooled in 5s. All pools were negative except for 1 pool that resulted suspect in PCR (Ct value > 37). An additional 60 samples were collected and 1 pool was positive. These sows were resampled 1 week later and were all negative. The herd was declared negative and open to new animal introductions based on PCR testing and sentinel gilts, and finishing pigs will be followed downstream 6 months following the elimination. At this point it is too early to report. Based on these results, Pulmotil AC® at 200ppm in water medication administered to the sow herd and piglet injection with Draxxin® are another tool that can be used for Mycoplasma elimination in farrow-to-wean herds.

References

1. Holst S, Yeske P, Pieters M; Elimination of *Mycoplasma hyopneumoniae* from breed-to-wean farms: A review of current protocols with emphasis on herd closure and medication; AASV/ SHAP / v23n6p321.

2. Wu CC, Shryock TR, Lin YL, et al. Testing antimicrobial susceptibility against *Mycoplasma hyopneumoniae in vitro; Swine Health and Production* 1997;5(6):227-330.

3. Strait EL, Madsen ML, Minion FC, Christopher-Hennings J, Dammen M, Jones KR, Thacker EL. Real-time PCR assays to address genetic diversity among strains of Mycoplasma hyopneumoniae. Journal of clinical microbiology. 2008 Aug 1;46(8):2491-8.

Controlling bacterial pathogens in the growing pig: New technology

Keith Wilson, DVM; Eric Bumgardner, MS; Paulraj Lawrence, PhD
Biological Development and Research Department, Newport Laboratories Inc., Worthington, Minnesota

Introduction

Bacterial pathogens have been an endemic problem in the US swine industry since the introduction of intensive agriculture practice. The adoption of new husbandry practices in the latter half of the 20th century in which thousands of pigs could be raised within a single vertically integrated system has dramatically increased swine production and aided in the overall health of the United States swine herd.[1] However, these changes have also exacerbated the impact that disease outbreaks might have on swine production systems when they are not well controlled.[2] Until very recently producers have largely succeeded in controlling bacterial pathogens with antibiotics. However, the emergence of antibiotic resistant bacteria, new federal regulations, and changes in consumer behavior are driving producers to look toward alternative methods of control.[3-5] The advent of the earliest vaccines pre-dates the discovery of antibiotics by more than a century. Over the course of time vaccines and the technology used to produce them has continually improved. Indeed, vaccines have been used to eradicate some of the most grievous illnesses from circulation in both humans and animals. Specific to animal agriculture, vaccines have proven instrumental in the world-wide eradication of rinderpest virus and in the elimination of pseudorabies virus from the United States commercial swine herd.[6,7] Therefore vaccines are an attractive alternative to which producers might turn when evaluating control strategies that can reduce their reliance on antibiotics.

The field of veterinary diagnostics has continued to evolve alongside the field of vaccinology. Over the course of the past three decades several revolutionary technologies have had a strong impact on the field. Most recently, next-generation sequencing has emerged as a new technology that has had an increasingly large impact. As the cost of sequencing has decreased dramatically the availability of this type of testing has increased. At present, the cost of performing whole genome sequencing (WGS) on a bacterial isolate is often the same or less than performing standard diagnostic testing. In addition, many researchers have begun to use metagenomics for pathogen discovery as it is an attractive method to identify previously unknown pathogens.[8] This method can be used to more thoroughly interrogate a sample so that multiple potential causes of disease might be identified.

Standard diagnostic testing often will reveal one or more of the usual suspects that produce a particular disease etiology. However, we have observed that multiple pathogens are usually present that remain unidentified simply because they were not suspected. Once isolates have been identified to the species level further work is often required to determine whether each strain is unique or whether isolates are similar to one another. Methods such as serotyping, multi-locus sequence typing, and DNA fingerprinting are all employed with this objective in mind.

In the work presented here we describe a new paradigm of pathogen detection and characterization in which metagenomics is used to identify potential disease causing pathogens within a herd. Once the veterinarian or producer has a better idea what types of pathogens might be affecting their pigs, isolates are obtained and characterized by WGS. Data from WGS is then used to identify unique strains of each pathogen that might be included in a custom made vaccine to help control disease within a herd or production system. WGS capsule typing methods, virulome characterization, and antibiotic resistance profiling are all employed depending on the pathogen identified. We have found that this strategy often provides producers and veterinarians with new knowledge regarding the types of pathogens that might be affecting their herd and can be successfully employed in controlling the pathogens that are identified.

Materials and methods

Metagenome study

In the metagenome study discussed in this report samples were obtained from each of 23 different farms overseen by a large veterinary service. At each site, fecal samples were taken from clinically normal pigs and from pigs manifesting symptoms of enteric disease. In total, 220 samples were submitted for analysis. Nucleic acid was extracted from each sample. Following extraction, libraries were prepared for sequencing on a MiSeq instrument using a Nextera XT library preparation kit (Illumina) and 2 × 300 bp chemistry. Once sequenced, the metagenome of each isolate was analyzed using Kraken. A sample was considered positive for a pathogen if greater than 100 reads were detected that mapped to an organism. Virus genomes were assembled using commercial software and reference sequences where available. Genomes of novel circoviruses were assembled *de novo*. Circular contigs were sorted from the assembly and circovirus genomes were identified based on the homology of amino acid sequences from open reading frames to known circovirus genomes.

Bacterial genome sequencing and analysis

Sequencing of isolates was performed by extracting genomic DNA using commercially available kits, preparing libraries using Nextera XT, and sequencing on a MiSeq instrument using 2 × 150 bp chemistry. Capsule typing was accomplished using previously described methods for *S suis* and *H parasuis*.[9,10] Where *de novo* assembled contigs were used in analysis they were assembled using Spades.[11] For *A pleuropneumoniae*, reference loci from each of the 16 currently described serotypes were downloaded from public databases. Reads from each isolate were then mapped against the capsular database to identify the serotype with the greatest locus coverage. The presence or absence of virulence factors was established by mapping reads to custom databases of previously published virulence factors for each species. Antibiotic resistance genes were identified using the comprehensive antibiotic resistance database[12] or by BLAST search of contigs against an antibiotic resistance gene database obtained from ResFinder.[13]

Results and discussion

Metagenome study

In our metagenome study we were able to successfully identify the primary causes of enteric disease within the producer's production system. Results varied substantially at the farm level in terms of the overall contribution of each pathogen to disease at that site. Overall, the most commonly identified enteric pathogens were *C perfringens*, *P difficile*, *E coli*, Porcine kobuvirus, *Campylobacter sp*., Sapporo virus, Porcine astrovirus 3, and Rotavirus type C. Several novel circoviruses were also identified and complete genome sequences for these viruses were readily obtained from the metagenome data. It was also apparent that this system has been substantially affected by a novel Porcine kobuvirus strain. Analysis of complete genome sequences from these samples revealed that the viruses affecting the herds within this system belong to two clusters. One cluster is typical of the viruses that have been isolated in North America whereas the other represents a new Porcine kobuvirus cluster that has diverged significantly from the sequences that have been reported previously.

Characterization of bacterial isolates

Characterization of the virulomes of *C perfringens*, *P difficile*, and *E coli* isolates using whole genome sequencing led to the discovery that all *C perfringens* isolates taken from within this system were very similar except that differences in the presence or absence of beta 2 toxin and hyaluronadase were identified.

P difficile isolates varied in the types of adhesins encoded within the genome. In addition, we identified one strain that encoded a significantly different flagellar operon and a glycopeptide resistance gene. Screening of *E coli* isolates revealed that many of them could be typed as O32:H10 isolates however each isolate varied substantially in the number and type of virulence factors encoded.

In a separate series of studies we have employed the use of WGS as a method for determining the capsule type of *S suis*, *H parasuis*, and *A pleuropneumoniae* strains. We have been able to correlate WGS data to the results of traditional serotyping and have demonstrated that when loci are similar to reference strains we are able to obtain results that match traditional testing. The capsular loci of the *H parasuis* and *A pleuropneumoniae* that we have typed thus far are very similar to the sequences available for reference strains. In the case of *S suis*, many isolates have novel capsular loci that appear to be the result of genome rearrangements and horizontal gene transfer events. Traditional testing has been unable to detect these changes and can report them as "non-typeable" strains or can incorrectly report them as a known serotype when only a portion of the locus is present. In addition to capsule typing we are able to query the genome of these isolates for the presence of known virulence factors and antibiotic resistance genes. We have observed significant variation between strains from all three of these species and in some cases have noted that presence or absence may correspond to the body cavity that is colonized by a particular strain.

In vivo studies

The objective of these studies has been to develop custom made vaccines that can be used to solve disease problems unique to a farm or production system. The *S suis* vaccines developed using this method have been successful in reducing replacement gilts mortality from 10% to 2%. We have also had the opportunity to test our *S suis* vaccines in controlled studies and have repeatedly demonstrated complete protection against lethal challenges. The results obtained thus far have been extraordinarily promising and have been successful in providing new information and new solutions to the veterinarians overseeing the production systems that have utilized this approach.

References

1. United States Department of Agriculture Emerging Issues in Pigs Report.2017. Available at `https://www.aphis.usda.gov/animal_health/emergingissues/downloads/1pigs.pdf`. Accessed 01 November 2017.

2. Graham JP et al. The animal-human interface and infectious disease in industrial food animal production: rethinking biosecurity and biocontainment. *Public Health Rep*. 2008;123:282-299.

3. Food and Drug Administration. Veterinary Feed Directive Fact Sheet.2017. Available at `www.fda.gov/animalveterinary/developmentapprovalprocess/ucm449019.htm`. Accessed 20 October 2017.

4. Aarestrup FM et al. Antimicrobial resistance in swine production. *Animal Health Res Rev*. 2008;9:135-148.

5. Pew Charitable Trust Report. Antibiotic-free food animal production: a profitable path from the farm to the table. Available at `http://www.pewtrusts.org/~/media/legacy/uploadedfiles/phg/content_level_pages/issue_briefs/hh-ifibantibioticfreeproductionpdf.pdf`. Accessed 01 November 2017.

6. Mariner JC et al. Rinderpest eradication: Appropriate technology and social innovations. *Science* 2012;337:1309-1312.

7. US Department of Agriculture, Animal and Plant Health Inspection Service, Veterinary Services, Centers for Epidemiology and Animal Health, National Surveillance Unit: National pseudorabies surveillance plan final draft. 2008. Available at `https://www.aphis.usda.gov/vs/nahss/swine/prv/prv_surveillance_plan_final_draft_04_16_08.pdf`. Accessed 01 November 2017.

8. Lecuit M et al. The diagnosis of infectious diseases by whole genome next generation sequencing: a new era is opening. *Front Cell Infect Microbiol*. 2014;4:25.

9. Lawrence PK et al. Map-based serotyping of Streptococcus suis field isolates for vaccine formulation. *Proc. AAVLD Ann Meet*. 2015.164.

10. Bumgardner EA et al. Serotyping and selection of *H. parasuis* field isolates for vaccine formulation using whole genome sequencing. *Proc. AAVLD Ann Meet.* 2017.

11. Bankovich et al. Spades: a new genome assembly algorithm and its applications to single-cell sequencing. J Comput Biol. 2012;19:455-477.

12. McArthur et al. The comprehensive antibiotic resistance database. *Antimicrob Agents Chemother.* 2013;57:3348-3357.

13. Zankari et al. Identification of acquired antimicrobial resistance genes. *J Antimicrob. Chemother.* 2012;67:2640-2644.

Streptococcus suis: Research overview and field experiences

Paulraj Lawrence, PhD; Eric Bumgardner, MS
Biological Development and Research Department, Newport Laboratories Inc., Worthington, Minnesota

Introduction

Streptococcus suis is a significant swine pathogen and an emerging zoonotic agent in humans.[1] In swine, disease caused by this pathogen can result in bronchopneumonia, arthritis, septicemia, and fatal meningitis, particularly in young pigs during the nursery phase. Historically, producers have been able to control *S suis* through the use of antibiotics.[2] However, recent changes in the regulation of antibiotic use and consumer behavior is driving producers toward other methods of control.[3] Under these circumstances vaccination is a key prevention strategy to control bacterial pathogens including *S suis* in the swine herd. However, vaccine cross protection amongst serotypes is very low under field or experimental conditions.[4] Over the years, many researchers have tried to identify the primary virulence factors that are involved in *S suis* pathogenesis so that more effective vaccines could be developed. Efforts have focused on proteins such as suilysin, extracellular factor, muraminidase release protein, murein associated protein, and surface antigen one, as well as the *S suis* capsule.[5-7] Vaccination with capsular conjugate vaccines has been demonstrated to lead to complete protection *in vivo*.[7] Our internal studies corroborate these observations.[8] If the capsule is the most important factor in selecting strains for inclusion in *S suis* vaccines then it is imperative to identify the capsular type unequivocally so that the vaccine strains can match the strains causing disease in a farm system. Currently, there are 35 described serotypes of *S suis*[9] although it has been proposed that serotypes 20, 22, 26, 32, 33, and 34 are better described as other *Streptococcus* species.[10,11] Furthermore, many field strains remain non-typeable using traditional serotyping or PCR methods. Additionally, there have been several recent reports describing strains with novel capsular loci.[12-14] In the current study, we describe a new method of capsular typing based on whole genome sequencing (WGS) and report the discovery of several *S suis* strains with novel capsular loci. In fact, these strains appear to be quite common in modern swine production facilities.

Materials and methods

Four-hundred sixteen *S suis* field isolates were obtained from twenty large veterinary clinics and commercial production systems from the years 2014-2017. Genomic DNA was extracted from all isolates using commercially available kits and libraries were prepared using a Nextera XT kit (Illumina). Whole genome sequencing was performed on the isolates on a MiSeq instrument using 2 × 150 bp chemistry. Serotypes were determined using a previously described method.[8] Resolution of serotypes 1 and 14 as well as of 2 and 1/2 was accomplished using a method recently described by Athey *et al.*[15]

Results and discussion

The results of our analysis suggest that serotypes 2, 1/2, and 1 are the most prevalent typeable *S suis* strains in circulation within the United States swine herd although serotypes 7 and 3 were also commonly identified (Figure 1). However, the significance of any one serotype appears to vary substantially between farms. Isolates were identified as having novel capsular loci if less than 70% of the locus from previously typed reference strains could be identified in the test strain. The novel loci illustrated in Figure 2, were obtained by *de novo* assembly of reads from a representative serotype 9-like novel strain into contigs and comparing contigs encompassing the capsular loci against reference strains using the BlastX feature in Easy Fig 2.0.

Strains with novel capsular loci made up a full 20% of the isolates tested. Within this heterogenous group the majority of isolates appeared to express capsular loci with some similarity to serotypes 3,5,9, 26, and 27 but strains that had greater similarity to other rarely seen serotypes were also identified. It is possible that these strains are representative of those strains that had previously been described as "non-typeable" using more traditional methods. Strains with novel loci bearing consistent levels of similarity to one or more reference strains were identified in isolates obtained from multiple locations and at multiple time points indicating that these strains have circulated extensively and have been present in the swine herd for some time. Overall, the WGS method allows for a complete characterization and unambiguous typing of the capsular locus, ensuring that even when a novel strain is identified on a farm, a matching vaccine strain can be selected for inclusion in an autogenous vaccine. Vaccines designed using this strategy offer better homologus and possibly heterologous protection in the field.

References

1. Goyette-Desjardins G et al. *Streptococcus suis*, an important pig pathogen and emerging zoonotic agent–an update on the worldwide distribution based on serotyping and sequence typing. *Emerg. Microbes Infect*. 2014;3:e45.

2. Gottschalk et al. Susceptibility of *Streptococcus suis* to penicillin. *J Vet Diag Invest*. 1991;3:170-172.

3. Food and Drug Administration. Veterinary Feed Directive Fact Sheet.2017. Available at www.fda.gov/animalveterinary/developmentapprovalprocess/ucm449019.htm. Accessed 20 October 2017.

4. Simonson R et al. *Streptococcus suis* immunology: Stimulation of immunity using a commercial bacterin (Strep BAC™). *Proc AASP Ann Meet*. 1989: 83-89.

Figure 1: Prevalence of *S. suis* capsular types in the U.S. swine herds based on whole genome sequencing

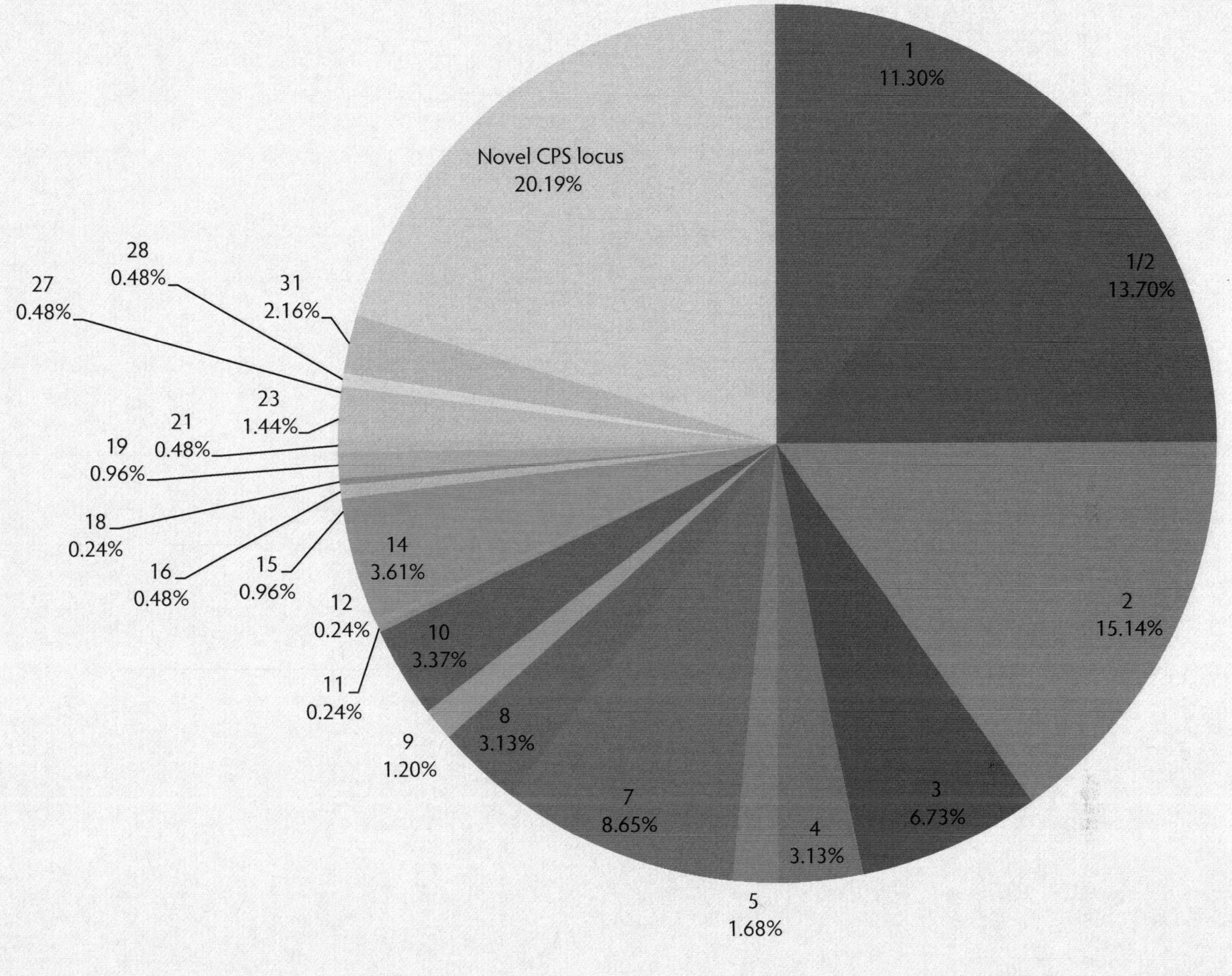

5. Baums CG et al. *Streptococcus suis* Bacterin and Subunit Vaccine Immunogenicities and Protective Efficacies against Serotypes 2 and 9. *Clin Vaccine Immunol.* 2009;16:200-208.

6. Li et al. Identification of a surface protein of *Streptococcus suis* and evaluation of its immunogenic and protective capacity in pigs. Infect Immun. 2006;74:305-312.

7. Goyette-Desjardins G et al. Protection against *Streptococcus suis* serotype 2 infection using a capsular polysaccharide glycoconjugate vaccine. Infect Immun. 2016;84:2059-2075.

8. Lawrence PK et al. Map-based serotyping of *Streptococcus suis* field isolates for vaccine formulation. P*roc. AAVLD Ann Meet*. 2015.164.

9. Higgins R et al. Description of six new capsular types (29-34) of *Streptococcus suis*. *J Vet Diagn Invest*. 1995; 7:405-406.

10. Hill JE et al. Biochemical analysis, cpn60 and 16S rDNA sequence data indicate that *Streptococcus suis* serotypes 32 and 34, isolated from pigs, are *Streptococcus orisratti*. *Vet Microbio*l. 2005; 107:63-69.

11. Tien LHT et al. Reappraisal of the taxonomy of *Streptococcus suis* serotypes 20, 22, 26, and 33 based on DNA-DNA homology and sodA and recN phylogenies. *Vet Microbiol*. 2013;162:842-849.

12. Zheng H et al. Eight novel capsular polysaccharide synthesis gene loci identified in nontypeable *Streptococcus suis* isolates. *Appl. Environ. Microbiol*. 2015; 81:4111-4119.

13. Pan Z et al. Novel variant serotype of *streptococcus suis* isolated from piglets with meningitis. *Appl. Environ. Microbiol*., 2015;81:976-985.

14. Qiu Z et al. Novel capsular polysaccharide loci and new diagnostic tools for hight-throughput capsular gene typing in *Streptococcus suis*. *Appl. Environ. Microbiol*. 2016; 21:7102-7112.

15. Athey et al. Determining *Streptococcus suis* serotype from short-read whole-genome sequencing data. *BMC Microbiol*. 2016;16:162.

Figure 2: Novel capsular types 1 and 2 isolated from the Midwest swine herds.

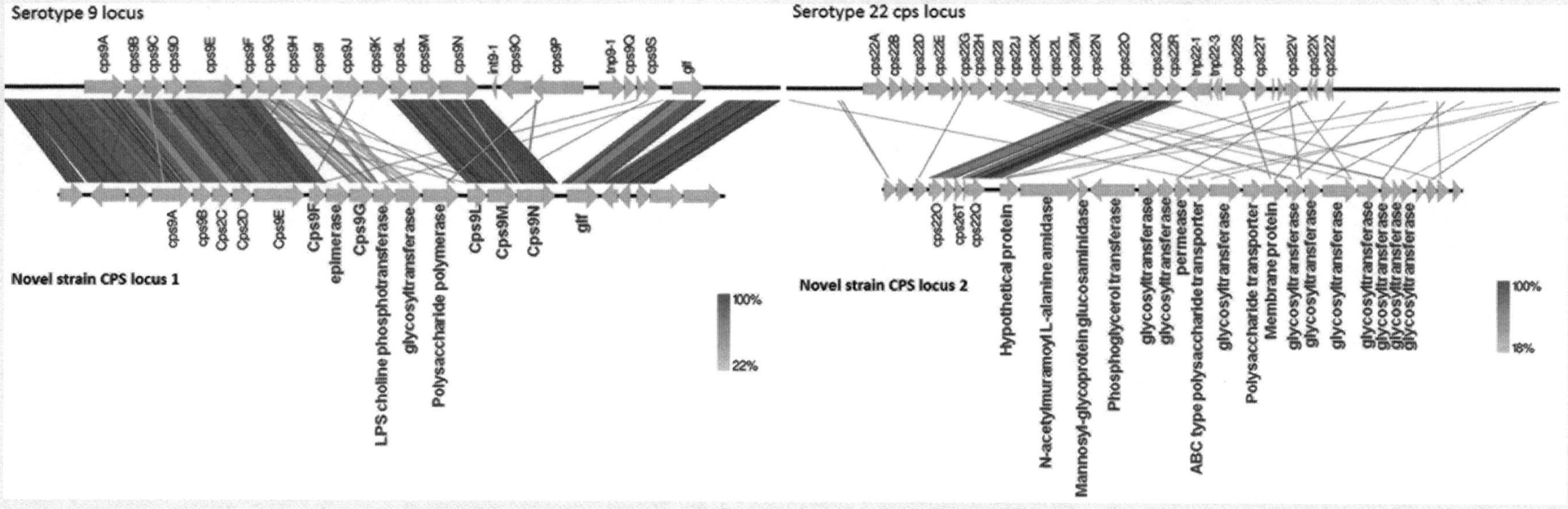

Molecules produced by probiotics prevent enteric colibacillosis in pigs

R. Nordeste[1]; A. Tessema[2]; S. Sharma[2]; Z. Kovač[3]; C. Wang[2]; R. Morales[2]; M.W. Griffiths[2]
[1]MicroSintesis Inc., Charlottetown, Prince Edward Island, Canada; [2]Canadian Research Institute for Food Safety, University of Guelph, Guelph, Ontario, Canada; [3]Department of Veterinary Clinical and Animal Sciences, Faculty of Health and Medical Sciences, University of Copenhagen, Copenhagen, Denmark

Introduction

Enteric colibacillosis is a common disease of nursing and young pigs caused by enterotoxigenic strains of *Escherichia coli*. Fimbria or pili expressed by these strains of *E coli* allow them to adhere to and colonize the absorptive epithelial cells of the jejunum and ileum. The common antigenic types of pili associated with pathogenicity are K88, K99, 987P, and F4.[1] Pathogenic strains produce enterotoxins that cause fluid and electrolytes to be secreted into the intestinal lumen, which results in diarrhea, dehydration, and acidosis.[2] Less common strains are able to produce shiga toxin Stx2e, which may result in edema.[3] Infection in neonates is commonly caused by K88 and 987P strains, whereas post-weaning colibacillosis is nearly always due to the K88 strain.

Amid ever-increasing concerns about the impact of antimicrobial resistant pathogens on human and animal health, research has focused on alternative strategies to combat infectious diseases. Of these, probiotics have been extensively studied.[4] Because of potential problems associated with the delivery of whole cells, attention has shifted to the use of metabolites of probiotic.[5] Griffiths and colleagues were among the first research groups to investigate the efficacy of bioactive molecules isolated following the growth of probiotics against many enteric pathogens, including *E coli* O157:H7,[6-10] *Salmonella,*[9-12] *Campylobacter,*[13] and *Clostridium difficile*.[14] These bioactives, termed proteobiotics, down-regulate genes involved in adhesion to and invasion of epithelial cells through interference of cell-cell communication pathways.[6,15] The efficacy of these molecules has been demonstrated in mouse models but their application as mediators of infection in production animals needs to be assessed. Thus, the objective of this study was to determine if bioactive molecules obtained from *Lactobacillus acidophilus* could be used to alleviate infection by *E coli* K88 in piglets.

Materials and methods

Study design

The study was conducted as a completely randomized block design. Each treatment had 6 pens with 6 piglets per pen. All piglets were placed in the same room and pens were randomly blocked within the room as 4 pens per block according to Standard Operating Procedures (SOP) of the facility. There were 6 blocks in this study. The following randomization procedures were adopted.

Allocation of pigs to pens

Piglets (approximately 18 days +/-2 days of age upon arrival at the test facility) were obtained from commercial sow farm(s) located in Ontario, Canada and were a commercial cross breed (Newsham Genetics). There were 36 pigs (18 gilts and 18 barrows) per treatment, resulting in 144 piglets in total for the study. Each pen had 6 piglets (3 gilts and 3 barrows). Only piglets with no physical abnormality or conditions were used in the trial and intact male piglets and ridglings were excluded.

On day -7, piglets were ear-tagged and randomly assigned to pens according to their body weight. There were 4 pens in each block. Each pen had 6 piglets (3 gilts and 3 barrows). Piglets were distributed across the range of body weights (small/medium or medium/heavy). If piglets were not able to reach feed and/or water or if they had a diarrhea or demeanor score of 3 for 3 consecutive days they were removed from the study. Each pen within each block was numbered and randomly assigned to one treatment.

The size of each pen was approximately 1.2 × 1.7 m with plastic slatted flooring. Feed and water was provided *ad libitum*. The stocking density was approximately 0.33 m^2 per piglet at the time of placement. Temperature and humidity of the facility were in accordance with piglets' age and size throughout the study. Minimum/maximum temperatures and humidity were recorded daily. The facility including water lines was cleaned and disinfected before the arrival of piglets.

Study protocol

There was a 7 d acclimation period (day -7 to day 0) before the piglets were challenged with *E* coli K88 on day 0 of the study. At this time, all piglets were orally challenged with 3 ml of an inoculum containing approximately 10^9 CFU/ml of *E coli* K88. The inoculum was prepared at Nutreco Canada Agresearch according to their SOP (SOP# RN #301). The piglets were observed twice daily from the time of placement until the last day of the study. The observation periods were at least 4 hours apart. Any abnormalities observed in the piglets were recorded. Any piglets that died or were euthanized during the study were weighed and necropsied to determine the possible cause(s) of death.

All piglets were individually scored for fecal consistency and appearance/demeanor from day 0 to 7 of the study using the scoring system documented in Table 1. The person responsible for fecal and demeanor scoring was blinded to treatments.

Table 1: Criteria used to assess fecal and appearance/demeanor scores

Fecal score	Description
0	No diarrhea, normal stools
1	Pasty feces
2	Liquid feces with some evidence of solids
3	Severe watery diarrhea, with or without blood
Appearance/ demeanor score	**Description**
0	Normal pig, good body condition, normal activity and attention
1	Pig shows mild signs of gauntness, some hollowing at the flanks, skin is duller, pig may be less active and less attentive
2	Pig shows significant signs of gauntness, skin is pale and hair is starting to become more apparent, pig shows mild signs of lethargy and less attentive to surroundings, more reluctant to move
3	Pig shows major signs of gauntness, gray pallor to skin, hair is pronounced, back may be humped up, pig shows definite signs of lethargy, and dull demeanor

Results

Impact of bioactive of fecal scores

The fecal scores observed for pigs in Treatment 1 (control) and Treatment 2 (130 mg/kg dose) were significantly different ($P = 0.004$) (Table 2). When a fecal score of ≥ 2 is used as an indication that a pig is sick, the odds of a pig being sick in the control group (Treatment 1) are 2.862 (1.359, 6.030) times higher than the odds of a pig being sick if administered the bioactive at a 130 mg/kg dose (Treatment 2).

There was also a significant difference ($P = 0.029$) in fecal scores among pigs in Treatment 1 (control) and those in Treatment 4 (520 mg/kg dose). The odds of a pig being sick in the control group are 2.142 (1.066, 4.304) times higher than the odds of an animal becoming sick when fed the bioactive at a 520 mg/kg dose (Treatment 4).

A pig's overall probability of being sick over all 8 days of the trial (day 0 to day 7 inclusive) for the control group (Treatment 1) was 8.585% (5.546%, 13.05%); whereas the probability of a pig becoming sick in the Treatment 2 group (130 mg/kg dose) was 3.176% (1.828%, 5.461%) and in Treatment 4 (520 mg/kg dose) it was 4.199% (2.512%, 6.939%).

Pigs in Treatment 3 (260 mg/kg dose) were 1.381 times less likely to become ill compared to animals in the control group and their probability of becoming sick was 6.367%. This indicates that animals in this group are protected against illness but not significantly so.

Demeanor and appearance score

Similar results were observed when the demeanor and appearance score was used as an index of infection (Table 3). Again a score of ≥ 2 was used as an indication that a pig is sick. In general, there was a significant effect of treatment ($P = 0.0124$) on susceptibility of pigs to infection, but a pen effect ($P = 0.0415$) was also observed when results were analysed using the Wald test. There was also the suggestion of a gender effect ($P = 0.0809$), indicating that

Table 2: Effects of treatments on average fecal scores

Treatment	Description	D0	D1	D2	D3	D4	D5	D6	D7
1	Challenged + 0 mg/kg	0.06	0.39	1.16[a]	1.01	1.28	0.83[a]	0.52	0.33
2	Challenged + 130 mg/kg	0.06	0.25	0.67[ab]	0.77	0.67	0.17[b]	0.13	0.07
3	Challenged + 260 mg/kg	0.06	0.08	0.81[ab]	1	1.23	0.63[ab]	0.37	0.33
4	Challenged + 520 mg/kg	0.03	0.25	0.56[b]	0.57	0.82	0.23[ab]	0.17	0.10
SEM		0.04	0.08	0.14	0.18	0.22	0.16	0.11	0.09
P-value		0.952	0.082	0.045	0.277	0.167	0.026	0.070	0.075

[a,b] Means within a column with no common letters are significantly different ($P < 0.05$).

Table 3: Effects of treatments on average demeanor scores

Treatment	Description	D0	D1	D2	D3	D4	D5	D6	D7
1	Challenged + 0 mg/kg	0.06	0.06	0.17	1.07	1.40^{a}	1.23^{a}	0.74	0.37
2	Challenged + 130 mg/kg	0	0.03	0.06	0.70	0.77^{b}	0.70^{b}	0.43	0.23
3	Challenged + 260 mg/kg	0.03	0.06	0.19	0.97	1.17^{ab}	1.33^{a}	0.93	0.57
4	Challenged + 520 mg/kg	0	0	0.03	0.70	0.83^{b}	0.80^{b}	0.30	0.10
SEM		0.03	0.04	0.07	0.10	0.16	0.16	0.22	0.15
P-value		0.569	0.732	0.346	0.047	0.045	0.037	0.210	0.181

^{a,b} Means within a column with no common letters are significantly different ($P < 0.05$).

demeanor may be affected by the sex of the animal. The odds of being sick for males was estimated to be 0.583 (0.318, 1.069) that of the odds for females being ill. When compared to animals in Treatment 1 (control), those administered a 130 mg/kg dose of the bioactive (Treatment 2) were significantly protected from illness ($P = 0.0207$), with the odds ratio for the Treatment 1 group over Treatment 2 being 14.366 (1.414, 145.956). Indicating that animals receiving the bioactive were about 14-times less likely to exhibit severe symptoms of illness. When compared to Treatment 1 (control), animals in group Treatment 4 (520 mg/kg dose) were significantly less likely ($P = 0.0831$) to show signs of infection. The odds ratio (control over 520 mg/kg dose) was 5.542 (0.842, 36.459). Both fecal and demeanor/appearance scores peaked at 4 days following challenge, indicating that they are good indices of infection.

Impact of bioactive on severity and duration of clinical symptoms of diarrhea

This analysis was done by counting the number of days a pig had a fecal score of 3 before returning to a score of 1 or 0. Although there were insufficient animals in each treatment group satisfying these criteria to allow a valid statistical analysis to be carried out, trends were observed (Table 4). Far fewer pigs in Treatment 2 (n = 1) and Treatment 4 (n = 2) exhibited severe watery diarrhea than found in the control group Treatment 1 (n = 9). The length of time that they showed this symptom was also shorter for animals in Treatment 2 (0.03 days) and Treatment 4 (0.1 day) than for those receiving no treatment (T1: 0.6 days). Piglets in the Treatment 2 and Treatment 4 group were about 3 and 2-times less likely to develop severe diarrhea, respectively, than their counterparts in the control group.

Discussion

Enteric colibacillosis is a common disease of nursing and weaned pigs caused by enterotoxigenic strains of *E coli* (ETEC), which are able to colonize the small intestine. This is achieved by the fimbriae or pili of the bacteria attaching to receptors on epithelial cells of the jejunum and ileum. The common antigenic types of fimbriae associated with pathogenicity are K88, K99, 987P, and F41.[16] Pathogenic strains produce enterotoxins that cause fluid and electrolytes to be secreted into the intestinal lumen, which results in diarrhea, dehydration, and acidosis.[17] Infection in neonates is commonly caused by K88 and 987P strains, whereas post-weaning colibacillosis is nearly always due to the K88 strain.[18,19]

With the increasing concerns surrounding the use of antibiotics in animal production, novel alternatives for maintaining animal health are needed. However, completely novel strategies to control infections that involve the use of molecules to modulate bacterial virulence are gaining credence.[20] We have coined the term proteobiotics for molecules produced by probiotics that down regulate virulence genes in both Gram positive and Gram negative bacteria.[6,11,13,14] In *E coli* it has been demonstrated that proteobiotics suppress production of quorum sensing molecules (AI2) leading to suppression of genes involved in the Type 3 Secretion System of the bacterium and the consequent alleviation

Table 4: Effect of bioactive on severity and length of infection.

	Treatment 1 (0 mg/kg)	Treatment 2 (130 mg/kg)	Treatment 3 (260 mg/kg)	Treatment 4 (520 mg/kg)
Pigs that scored a 3 (severe)	9	1	6	2
Average days spent sick	0.5721	0.0294	0.2353	0.100
Odds ratio of untreated pigs becoming sick compared with treatment group	-	2.862	1.381	2.142

of infection in a mouse model.[6,7] The aim of the current study was to demonstrate the efficacy of this treatment in a more relevant model: that is ETEC infection of pigs.

The proteobiotic was fed to pigs at three different dose levels calculated based on the results obtained in the mouse study.[7] The low and high dose treatments produced a significant alleviation in the symptoms of the disease; determined by measuring fecal and demeanor scores. It is interesting to note that at the 4th day following challenge almost half the pigs (44%) that did not receive the proteobiotic were shedding *E coli* K88 in their feces; whereas only about 16% of animals receiving the low dose treatment were shedding the pathogen (data not shown). The respective percentages 7 days following infection were 6 and 0%. This suggests that not only does the treatment alleviate symptoms but it also may help prevent dissemination of the infection.

For those pigs receiving the proteobiotic that did show signs of illness, the duration of the symptoms was shortened compared with the control animals, and their risk of becoming ill was between 2- to 3-times less. The fact that not all the pigs in the groups not administered the proteobiotic suffered diarrhea may be explained by their susceptibility to infection. The attachment of *E coli* K88 to the wall of piglet intestines depends on the availability of K88 receptors, which can be found at the intestinal cell brush border and the mucus layer. Due to the existence of several different piglet phenotypes that determine the presence and type of these receptors, it can be considered that piglets generally do not uniformly respond to an *E coli* K88 challenge. There is also a piglet phenotype that is resistant to binding of *E coli* K88, as piglets with this phenotype do not possess the required receptors.[17]

References

1. Francis, D. H. (1999). "Colibacillosis in pigs and its diagnosis." *Journal of Swine Health and Production* 7(5): 241-244.

2. Jin, L. and X. Zhao (2000). "Intestinal receptors for adhesive fimbriae of enterotoxigenic *Escherichia coli* (ETEC) K88 in swine – a review." *Applied Microbiology and Biotechnology* 54(3): 311-318.

3. Paul, N. (2015). "Review: VIrulence nature of *Escherichia coli* In neonatal swine." *Online Journal of Animal and Feed Research* 5(6): 169-174.

4. Alvarez-Olmos, M. I. and R. A. Oberhelman (2001). "Probiotic agents and infectious diseases: A modern perspective on a traditional therapy." *Clinical Infectious Diseases* 32(11): 1567-1576.

5. Shenderov, B. A. (2013). "Metabiotics: novel idea or natural development of probiotic conception." *Microb Ecol Health Dis* 24: 20399.

6. Medellin-Peña, M. J., et al. (2007). "Probiotics Affect Virulence-Related Gene Expression in Escherichia coli O157:H7." *Applied and Environmental Microbiology* 73(13): 4259-4267.

7. Medellin-Pena, M. J. and M. W. Griffiths (2009). "Effect of Molecules Secreted by Lactobacillus acidophilus Strain La-5 on Escherichia coli O157:H7 Colonization." *Applied and Environmental Microbiology* 75(4): 1165-1172.

8. Zeinhom, M., et al. (2012). "Yogurt containing bioactive molecules produced by *Lactobacillus acidophilus* La-5 exerts a protective effect against enterohemorrhagic *Escherichia coli* in mice." *Journal of Food Protection* 75(10): 1796-1805.

9. Tellez, A. M., et al. (2011). "A peptidic fraction from milk fermented with *Lactobacillus helveticus* protects mice against *Salmonella* infection." *International Dairy Journal* 21: 607-614.

10. Bondue, S., et al. (2016). "Cell-free spent media obtained from *Bifidobacterium bifidum* and *Bifidobacterium crudilactis* grown in media supplemented with 3'-sialyllactose modulate virulence gene expression in *Escherichia coli* O157:H7 and *Salmonella* Typhimurium." *Frontiers in Microbiology* 7.

11. Bayoumi, M. A. and M. W. Griffiths (2010). "Probiotics Down-Regulate Genes in Salmonella enterica Serovar Typhimurium Pathogenicity Islands 1 and 2." *Journal of Food Protection* 73(3): 452-460.

12. Bayoumi, M. A. and M. W. Griffiths (2012). "In vitro inhibition of expression of virulence genes responsible for colonization and systemic spread of enteric pathogens using Bifidobacterium bifidum secreted molecules." *International Journal of Food Microbiology* 156(3): 255-263.

13. Mundi, A., et al. (2013). "Cell-free preparations of *Lactobacillus acidophilus* strain La-5 and *Bifidobacterium longum* strain NCC2705 affect virulence gene expression in *Campylobacter jejuni*." *Journal of Food Protection* 76(1740-1746): 1740.

14. Yun, B., et al. (2014). "*Lactobacillus acidophilus* modulates the virulence of *Clostridium difficile*." *Journal of Dairy Science* 97: 4745-4758.

15. Kim, Y., et al. (2012). "*Bifidobacterium* spp. influences the production of autoinducer-2 and biofilm formation by *Escherichia coli* O157:H7." *Anaerobe* 18: 539-545.

16. Wilson, R. A. and D. H. Francis (1986). "Fimbriae and enterotoxins associated with *Escherichia coli* serogroups isolated from pigs with colibacillosis." *American Journal of Veterinary Research* 47(2): 213-217.

17. Debreuil, J. D., et al. (2016). "Animal enterotoxigenic *E. coli*." *EcoSal Plus* 7(1).

18. Wittig, W., et al. (1995). "Prevalence of the fimbrial antigens F18 and K88 and of enterotoxins and verotoxins among *Escherichia coli* isolated from weaned pigs." *Zentralblatt für Mikrobiologie* 283(1): 95-104.

19. Zhang, W., et al. (2007). "Prevalence of virulence genes in *Escherichia coli* strains recently isolated from young pigs with diarrhea in in North Central US." *Veterinary Microbiology* 123(1-3): 145-152.

20. Mühlen, S. and P. Dersch (2016). "Anti-virulence strategies to target bacterial infections." *Current Topics in Microbiology and Immunology* 398: 147-183.

Genetic selection for health and welfare using meat inspection data

P. K. Mathur[1], MVSc, PhD; J. M. Eggert[2], MS, MBA, PHD, Dipl. ACAG
[1]Topigs Norsvin Research Center, Beuningen, The Netherlands; [2]Topigs Norsvin USA, Burnsville, Minnesota

Introduction

Animal health and welfare are monitored during meat inspection in many slaughter plants around the world. Carcasses are examined by meat inspectors and remarks are made with respect to different diseases, injuries and other abnormalities. The primary purpose is to check carcasses for safety of meat as a food for human consumption. In many countries, it is regulated by local government agencies such as Canadian Food Inspection Agency (CFIA) in Canada, Food Safety and Inspection Service (FSIS) in USA and European Food Safety Agency (EFSA) for European countries. Carcasses are checked by trained inspectors and remarks are made for different diseases and abnormalities.

Many of the slaughter plants are very large and process thousands of carcasses every day. This generates large volume of data on various aspects of animal health and welfare. Some slaughter plants use these data to provide feedback to the farms that supply the pigs. Apart from this, the data could be very useful for taking preventive measures for diseases at the farms and enhancing animal welfare.[1,2] However, there are limited examples of using the data for this purpose.

In this investigation, data were obtained from slaughter plants in Germany. These pigs had radio-frequency identification (RFID) tags and the plants had devices to read them during the fast-moving slaughter pig chains. The data could also be linked to individual performance and pedigree records. The main objective of this study was to investigate the possibility of genetic selection to reduce the risk of most prevalent diseases and indicators of suboptimal animal welfare. As part of this main objective, effects of various non-genetic factors were also examined to help in farm management.

Materials and methods

Data

The data for this study consisted of records on carcass condemnations or slaughter remarks on 140,375 pigs from slaughter plants in Germany from July 2011 to July 2016. The data consisted of 69 slaughter remarks. The most common remarks were pneumonia (15.4%), joint disorders (9.8%), pleuritis (4.7%), pericarditis (2.3%) and liver lesions (2.2%). These traits were analyzed in detail. A total of 27.1 (%) pigs had either one or more such remark. Therefore, an additional trait was considered as presence or absence of any of these remarks (yes = 1 or no = 0) and denoted as "condemnation" in this manuscript. These pigs originated from 13 farms in Germany and the Netherlands. Pedigree records and records on finishing and carcass quality trait were available on 142,324 pigs from these farms and were used for genetic analysis.

Statistical analysis

The data were analyzed in two steps. In the first step, effects of non-genetic factors were studied using a generalized linear model in PROC GLM of SAS. The model for each trait included fixed effects of the farm of origin, sex of the pig and date of slaughter.

In the second step, data were analyzed with a sire model using ASReml 4.0.[3] In addition to above fixed effects, this model also included random effects of the sire and litter of the dam. It was assumed that, although the traits are observed and recorded on a binary scale (0 and 1), they follow an underlying normal distribution with a threshold.[4] Hence, the data were analyzed on liability scale using log link function.[3]

A genome wide association study was conducted using the GCTA method.[5] For this study, genotypes of the sires were available. These genotypes were obtained using llumina Porcine SNP60 Beadchip. These were then imputed to 660 K using the pigs genotyped with Axiom porcine 660 K array from Affymetrix (Affymetrix Inc., Santa Clara, CA, USA) as reference. After quality control 446,117 SNPs were used. All genotyped animals that had a frequency of missing genotypes below the threshold of 0.05 were excluded.

A selection index was developed for identification of boars for use by producers. This index was calculated as sum of product of the breeding values with their respective economic weights. It was quite difficult to get the most current estimates of gain and losses to producers and slaughter plants for individual slaughter remarks. The information is either not available or it is not shared. Hence, the economic weights were approximated with available information. The economic values for the slaughter remarks were based on Bondt, et al.[6] and adapted to the incidences of the slaughter remarks in current study and current prices.

Results

Slaughter remarks and growth

Relationship between slaughter remarks and growth is shown in Figure 1. The average daily gain from birth to slaughter was divided into classes of 10g/day. The incidences in each class are presented as least-squares means accounting for differences in farm, sex and slaughter date. In general, there is tendency for the lower condemnation remarks with increase in growth rate, except

Figure 1: Relationship between growth and incidences (%) of different remarks.

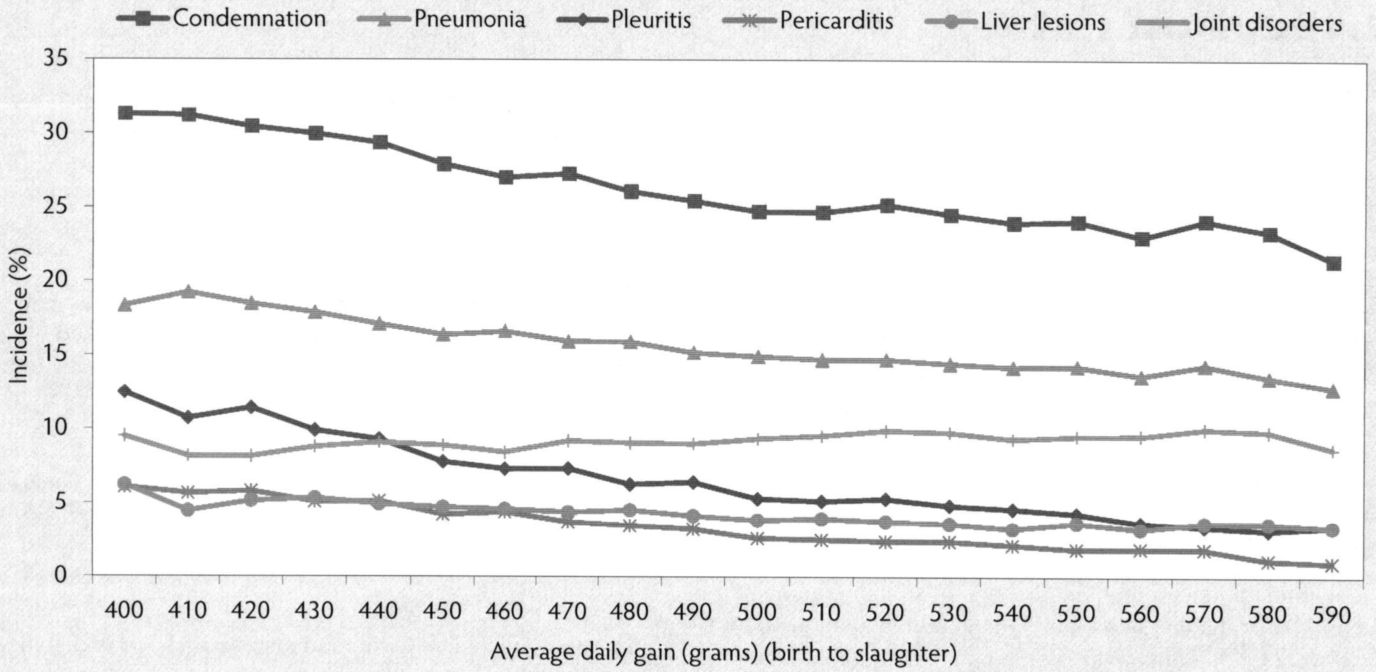

for joint disorders that show a slightly increasing trend. These results show that faster growth rate is not necessarily associated with unfavorable effects on welfare. In fact fast growing pigs have fewer incidences of the diseases and therefore lead to less pain and discomfort to the animals.

Estimates of variance components and heritability

As this analysis was conducted on a liability scale using log link function, the residual variance is considered to be a fix value of 3.29. The genetic variance was highest for liver lesions and joint disorders followed by pneumonia, pleuritis and pericarditis. However, the common litter effect was lowest for joint disorders compared to other traits, explaining less than 1% of the genetic variance. This variance accounted for 3.7, 5.7, 8.1 and 5.8 per cent of the total phenotypic variance in pneumonia, pleuritis, pericarditis and liver lesions, respectively. This suggests higher predisposition to these traits by common maternal environment. The heritability estimates for joint disorders, pneumonia, pleuritis, pericarditis, and liver lesions were 0.10, 0.09, 0.14, 0.24 and 0.17, respectively. These estimates are not very high but not very low as well.

Associated genomic regions

Results of the genome wide association study (GWAS) to further verify the existence of genetic variation for these traits are given in Figure 2. A majority of the most significant regions appear to be on the tail ends of the chromosomes that could be an artifact of the method used for the GWAS. Therefore, some examples of the SNP effect were examined in detail which suggests that the SNPs had significant effects in reality as well.

Phenotypic and genetic correlations among slaughter remarks

In general, there were very low phenotypic correlations among the slaughter remarks. The highest phenotypic correlation was between pleuritis and pericarditis ($r_p = 0.30 \pm 0.003$; $P < 0.01$). This was followed by correlation between pleuritis and pneumonia ($r_p = 0.09 \pm 0.003$; $P < 0.01$). This is expected, as inflammation due to infection of heart and lungs could also lead to inflammation of the pleura surrounds the heart and lungs.

The genetic correlations between the slaughter remarks were of similar magnitude or even higher. However, they were associated with larger standard errors. The correlations among slaughter remarks were close to zero and nonsignificant for most of the traits. It was highest between pleuritis and pericarditis ($r_g = 0.43 \pm 0.025$; $P < 0.01$) and even larger than the phenotypic correlation, suggesting similar genetic predisposition for both the traits. However, the genetic correlation between pneumonia and pleuritis ($r_g = 0.05 \pm 0.027$; NS) was lower than phenotypic correlation and not significantly different from zero. This suggests the phenotypic association was mainly due to the transmission of infection between the lungs and pleura.

Genetic correlations with production traits

Estimates of genetic correlations with different production traits are given in Table 1. These include traits of finishing and carcass quality. A majority of these correlations were quite low and non-significant. They were in favorable direction for pneumonia, pleuritis, pericarditis and unfavorable for liver lesions. The genetic

Figure 2: Manhattan plots showing genome-wide association results for pneumonia, pleuritis, pericarditis and joint disorders. SNPs are plotted on the x-axis according to their position on each chromosome against statistical association with these traits on the y-axis. The horizontal line indicates genome-wide significance ($P < 1.0^{-6}$)

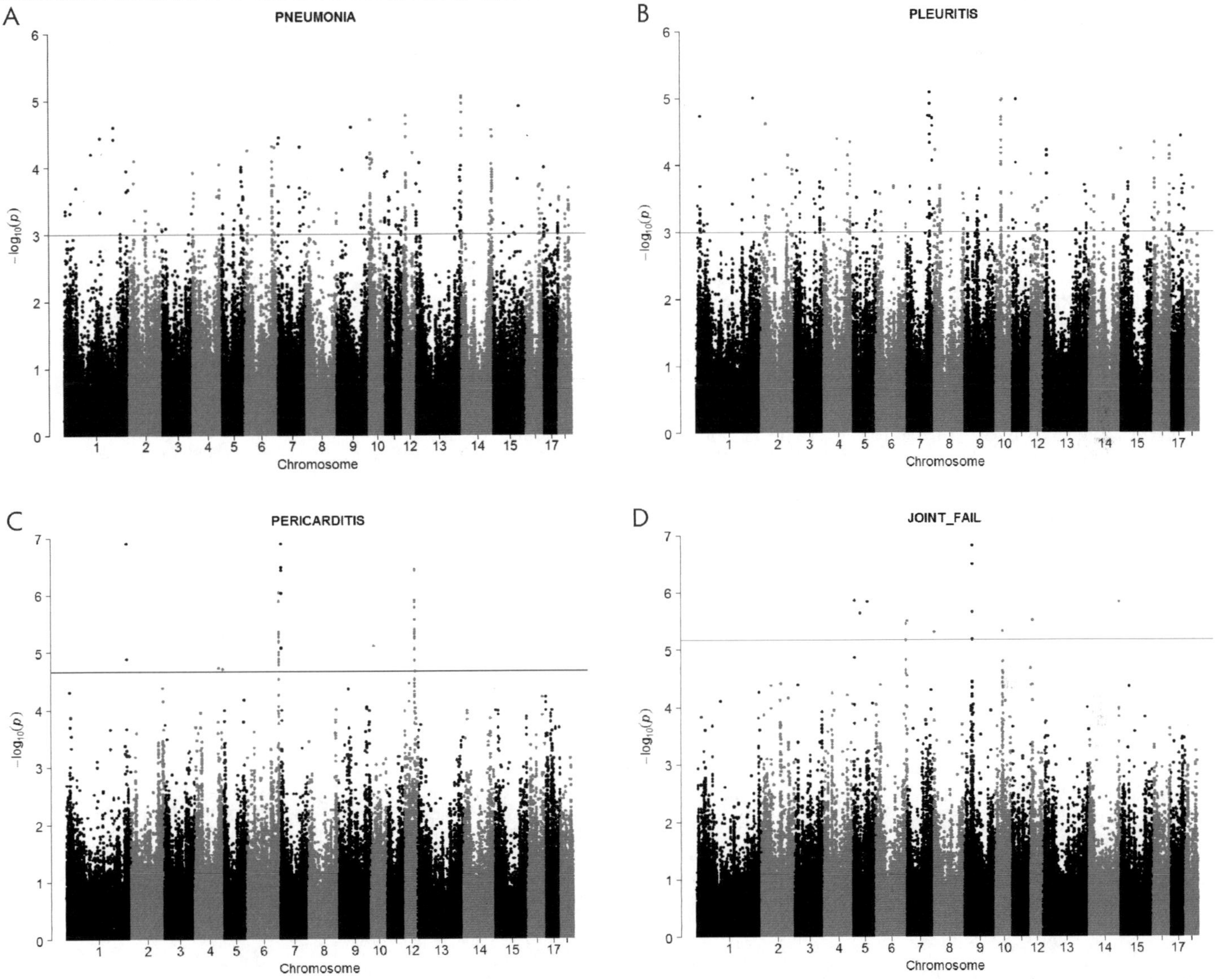

correlations of joint disorders with average daily gain from birth to slaughter and with carcass backfat were slightly larger, suggesting that joint disorders are associated with adverse effects on these traits. The genetic correlation of backfat with joint disorders and condemnation were low but unfavorable to selection for reducing backfat. These correlations actually suggest that leaner pigs could have higher joint disorders and overall higher condemnations. The correlation between growth and condemnation was in favorable direction, suggesting pigs with faster growth tend to have lower risks associated with these indicators of suboptimal health and welfare.

Discussion

In this study, it was possible to obtain data on individual pigs from the slaughter plants. This is not very common particularly in North America and many other regions as the slaughter pigs do not have individual identifiers and there are no devices to read them in the slaughter plants. In those plants, the observations are recorded on a group, e.g. based on slap tattoo of a batch of pigs. However, with increasing awareness in food safety and animal welfare many slaughter plants are putting them in place. This will allow easier linkage to farm records and more analysis on various aspects of disease prevention and genetics.

Table 1: Correlations with finishing and carcass quality traits

	Pneumonia	Pleuritis	Pericarditis	Liver lesions	Joint disorders	Condemnation
Average daily gain	-0.07 * (0.027)	-0.10** (0.027)	-0.05 (0.028)	0.02 (0.028)	0.09** (0.027)	-0.07* (0.027)
Backfat	-0.03 (0.028)	0.02 (0.028)	0.02 (0.028)	-0.02 (0.028)	-0.11** (0.027)	-0.06* (0.027)

NS = $P > 0.05$, * = $P < 0.05$, ** = $P < 0.01$.

There are several concerns regarding accuracy and usefulness of data from slaughter plants as the data are recorded in fast moving chains by different inspectors in different plants. The concerns include a lack of ability of the meat inspector to detect affected carcasses in the fast-moving slaughter chain, lack of standardization among the inspectors and conditions in the slaughter plant, and that only apparently healthy pigs are sent to slaughter.[7]

The genetic variation is partly normal to expect. Some families have better joints, lungs, hearts, livers etc. than others that make them less liable to the pain and suffering due to poor animal welfare at the farm. Genetic selection for the better families can reduce the incidence of infections and, with that, reduce the chance that other animals are infected. However, genetic selection for more robust animals should not reduce the focus on management improvements. The genetic correlations between carcass condemnations and finishing traits were small but mostly negative, suggesting the possibility of enhancing pig health and welfare simultaneously with genetic improvement in finishing traits.

To our knowledge, this is the first example of using meat inspection data for genetic selection to enhance animal welfare in pigs. A system has been put in place to obtain these data by direct links with the organizations responsible. This is allowing routine genetic evaluations and selection of pigs using the welfare index to reduce the risk of unfavorable slaughter remarks and enhance animal welfare in pig farms. In this way, this is also an example of a cooperation along the pork value chain, from processing back to production and genetics as a consorted effort of the value chain partners is required to enhance animal welfare.

References

1. Cleveland-Nielsen, A.; Christensen, G.; Ersbøll, A. Prevalences of welfare-related lesions at post-mortem meat-inspection in danish sows. *Preventive veterinary medicine* 2004, 64, 123-131.

2. Harley, S.; More, S.; Boyle, L.; O'Connell, N.; Hanlon, A. Good animal welfare makes economic sense: Potential of pig abattoir meat inspection as a welfare surveillance tool. *Irish veterinary journal* 2012, 65, 11.

3. Gilmour, A.; Gogel, B.; Cullis, B.; Welham, S.; Thompson, R. Asreml user guide release 4.1 structural specification. *Hemel Hempstead: VSN International Ltd* 2015.

4. Falconer, D.; Mackay, T. Introduction to quantitative genetics. John willey and sons. *Inc., New York* 1989, 313-320.

5. Yang, J.; Lee, S.H.; Goddard, M.E.; Visscher, P.M. Gcta: A tool for genome-wide complex trait analysis. *The American Journal of Human Genetics* 2011, 88, 76-82.

6. Bondt, N.; van den Elzen, G.; Hoste, R.; van Wagenberg, C.; Vermeij, I.; van der Fels, J. *Terugdringen van slachtafwijkingen in de varkensvleesketen*; 9052429227; LEI: 2004.

7. Correia-Gomes, C.; Smith, R.P.; Eze, J.I.; Henry, M.K.; Gunn, G.J.; Williamson, S.; Tongue, S.C. Pig abattoir inspection data: Can it be used for surveillance purposes? *PloS one* 2016, 11, e0161990.

Evaluation of two serological assays for *Lawsonia intracellularis* antibodies

Erika Vasquez; Lacey Marshall Lund; Fabio Vannucci; Devi Patnayak; Connie Gebhart
Veterinary Diagnostic Laboratory, College of Veterinary Medicine, University of Minnesota, St. Paul, Minnesota

Introduction

Lawsonia intracellularis, a gram negative, obligately intracellular bacterium, is the causative agent of porcine proliferative enteropathy. This disease negatively impacts the swine industry due to the decrease of production parameters in affected animals. Serological diagnosis on a herd level, detecting specific antibodies against *L intracellularis*, is considered the most sensitive and cost-effective approach to diagnose porcine proliferative enteropathy and monitor exposure to *L intracellularis*. Detecting the presence of the pathogen and determining its prevalence in the herd improves disease control efforts. Among the serological assays available, the Immuno-Peroxidase Monolayer Assay (IPMA) has been shown to have high sensitivity and specificity, 90.7% and 100%, respectively, and can be used to generate endpoint titers.[1] Limitations of the IPMA include the need to cultivate *L intracellularis* in monolayers for the assay and the fact that the results must be interpreted manually. A blocking serological assay, the Enzyme-Linked Immunosorbent Assay (ELISA; Svanovir® *L intracellularis*/Ileitis-Ab), does not generate end-point titers but does give a percentage inhibition (PI) value and is commercially available. A recent version of this blocking ELISA reported to have 96.5% sensitivity and 98.7% specificity for detection of *L intracellularis.*[2] The objective of this study was to evaluate both the commercial blocking ELISA and the IPMA using serum samples from pigs with known status, including not only positive and negative status, but also aged and vaccinated animals.

Materials and methods

A total of 254 serum samples were selected to represent three groups including 1) a negative control group, with strict negative serum samples, including aged animals defined as older than three months of age; 2) a positive control group, with samples from naturally- and experimentally-infected animals; and 3) a vaccine trial group, with serum samples from modified live, orally-vaccinated pigs (Enterisol Ileitis®). The numbers of animals in each group are shown in Table 1. The aged pig sera were obtained from known negative pigs older than three months of age. Negative status of the negative aged group was confirmed by weekly *L intracellularis* fecal PCR.

Table 1: Diagnostic sensitivity (SE) and specificity (SP) based on the known status of each group and agreement between IPMA and ELISA

Group		n	IPMA		ELISA		Agreement		
			SE %	SP %	SE %	SP %	r	Kappa	% agreement
Negative	All	134	--	97	--	87	--	--	87
	Without aged	106	--	99	--	98	--	--	97
Positive		63	96	--	74	--	--	--	67
Overall including aged and vaccinated		254	85	98	65	90	0.73	0.63	85
Overall without vaccinated		197	85	97	74	87	0.74	0.62	83
Vaccinated only		57	30	100	0	100	--	--	89
Overall without vaccinated or aged		**169**	**97**	**99**	**74**	**98**	**0.80**	**0.76**	**89**

The positive cutoff for IPMA was 1:30 as described,[1] and endpoint titers were determined for all positive sera. Briefly, serum antibody titers were measured in serial two-fold dilutions from 1:30 to 1:3840. The ELISA was performed according to the manufacturer's instructions where a percent inhibition (PI) of 30 was used as the cutoff for differentiation of seronegative from seropositive results. Serum samples with a PI ≥30% were considered positive and samples with a PI < 30% were considered negative. Note that for the ELISA, the manufacturer recommends that any sample presenting a PI value within the negative range of 20 to 30 should be considered questionable and a second test on a different sample from the same animal should be performed. For this study, any PI results between 20 and 30 were classified as negative.

The diagnostic sensitivity and specificity of the two techniques based on the known status of each group were determined as well as agreement between the two assays with correlation coefficient (r) and Kappa statistics. The positive and negative predictive values were established for each assay. In addition, correlation between the IPMA endpoints and ELISA PI values were further investigated. Approximate equivalence between ELISA PI and IPMA titers were presented on a scatterplot with Loess smoothing added.

Results

Overall the sensitivity and specificity of the serological assays were 65% and 90% for ELISA and 85% and 98% for IPMA, respectively. The correlation coefficient and Kappa for the two techniques were 0.73 and 0.63, respectively (Table 1). However, when results for the aged and the vaccinated groups were excluded, the sensitivity and specificity increased to 74% and 98% for ELISA and 97% and 99% for IPMA, respectively. For the negative control group, the specificity was 87% for ELISA and 97% for IPMA when the sera from the aged and vaccinated groups were included. After excluding these sera, the specificity increased to 98% and 99% for ELISA and IPMA, respectively. For the vaccinated group, the IPMA had only 44% sensitivity and 100% specificity whereas the ELISA did not detect vaccinated pigs at all.

The positive and negative predictive values, assuming a prevalence of 30%, were 74% and 86% for ELISA and 95% and 94% for IPMA, respectively (Figure 1). The correlation coefficient and Kappa were 0.80 and 0.73, respectively, when comparing the two serologic assays using the entire sera set and the sera set without the aged and vaccinated groups (Figure 2). The scatterplot for converting ELISA PI values to approximate endpoint titers is shown in Figure 3.

Discussion

There is a need for a high throughput serological assay that accurately determines the prevalence of proliferative enteropathy and detects changes in the serologic status of the herd. Our objective was to evaluate the sensitivity and specificity of the two commonly used serological assays (IPMA and ELISA) for *L intracellularis* antibodies using serum samples from pigs with known status.

Overall, the sensitivity and sensitivity for each test was acceptable and improved when the aged and vaccinated groups of sera were removed from the analysis. It is likely that background in older pig sera interferes with the specificity of the assays, especially for

Figure 1: **A.** Positive (PPV) and negative (NPV) predictive values for ELISA for a range of prior probabilities of infection assuming 74% sensitivity and 98% specificity. **B.** Positive and negative predictive values for IPMA for a range of prior probabilities of infection assuming 97% sensitivity and 99% specificity.

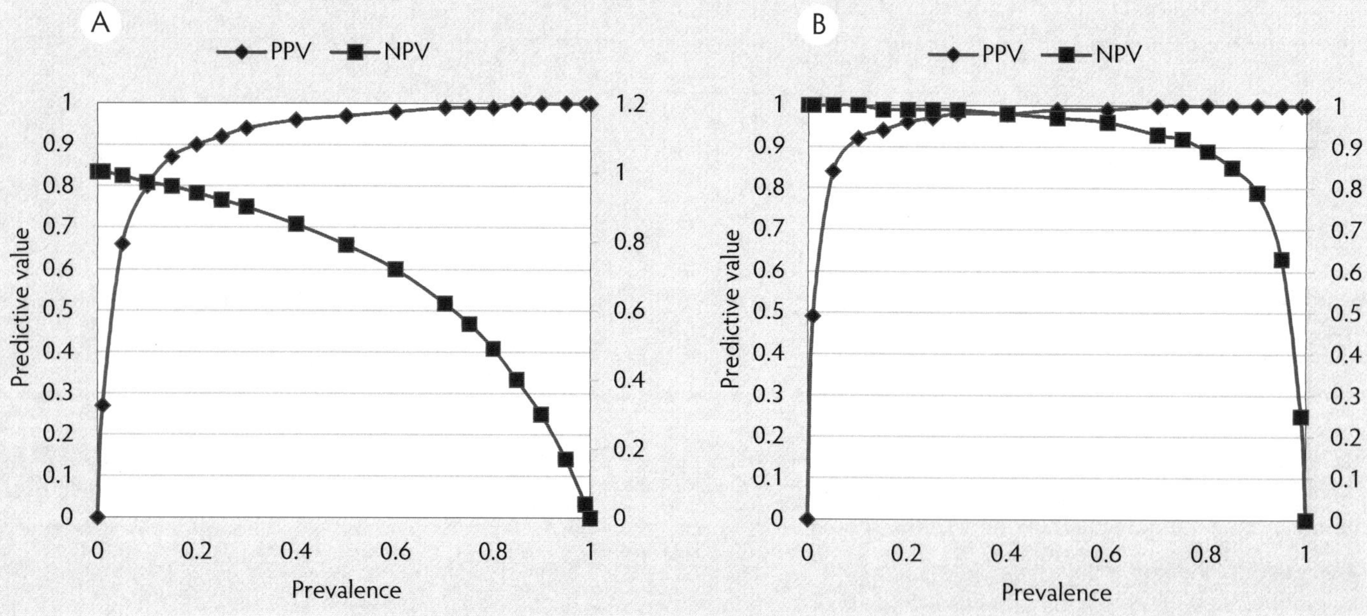

Figure 2: A. Overall agreement between ELISA and IPMA. Correlation coefficient was 0.73 and Kappa value was 0.63. **B.** Agreement between ELISA and IPMA without taking into account sera from aged or vaccinated groups. Correlation coefficient was 0.80 and Kappa value was 0.73.

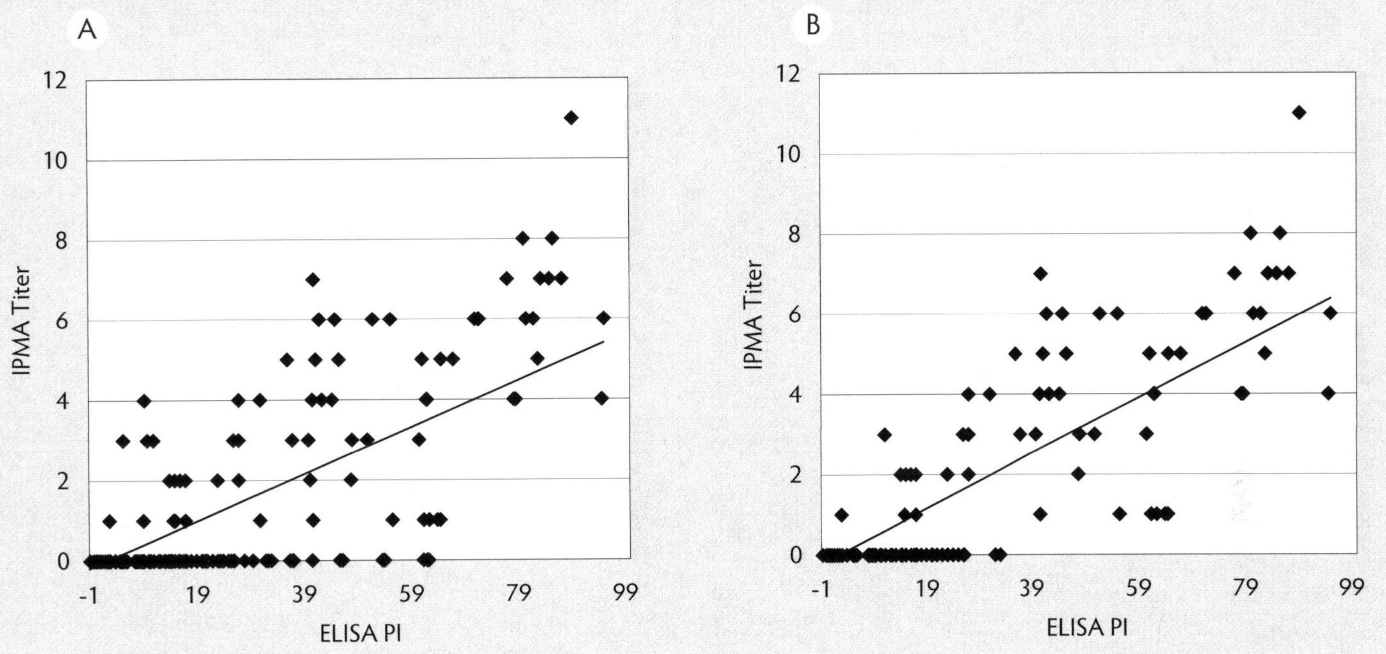

the ELISA. Furthermore, vaccinated pigs were not detected by the ELISA, and so this assay would not be suitable for confirming vaccination status. The IPMA did detect serum titers from some vaccinated pigs, but with a sensitivity of only 30%. Either assay would be an appropriate choice for use in the target age of pigs for diagnosis of proliferative enteropathy, namely growing and finishing pigs. Neither assay would be useful for confirming vaccination status with the modified live oral *L intracellularis* vaccine.

Overall in this study, the two serological assays showed a correlation of 0.80. The correlation coefficient and Kappa improved for both tests after omitting serum sets from aged and vaccinated pigs. For comparative purposes, an approximation of endpoint titer can be made by converting the ELISA PI value to an endpoint titer using a standard curve.

References

1. Guedes RMC et al. Validation of an immunoperoxidase monolayer assay as a serologic test for porcine proliferative enteropathy. J. Vet. Diagn. Invest. 2002, 528–30.

2. Jacobson M et al. Evaluation of a blocking ELISA for the detection of antibodies against *Lawsonia intracellularis* in pig sera. Acta Veterinaria Scandinvica. 2011 .

Figure 3: Approximate equivalence between ELISA percentage inhibition (PI) and IPMA titer. Scatterplot with loess smoothing added

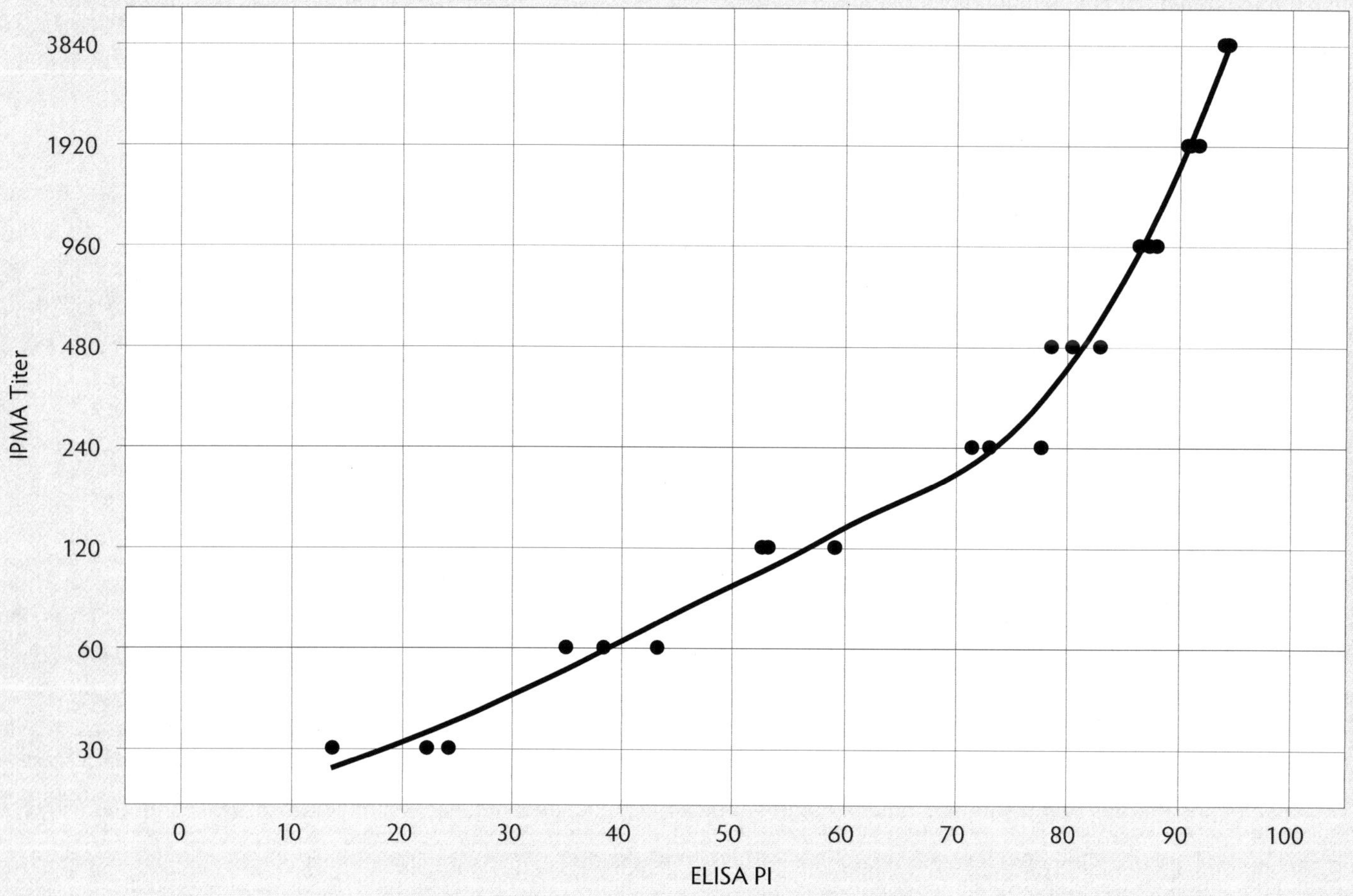

Comparison of DLI® controlled immunity, Enterisol® and Porcilis® Ileitis vaccines in grower pigs challenged with a *Lawsonia intracellularis* mucosal homogenate

Nathan Winkelman[1], DVM; Adam Mueller[1], DVM; Rickie Domangue[2], PhD
[1]Swine Services Unlimited, Inc. Rice, Minnesota
[2]Statistical Consulting Services, Broadway Virginia

Introduction

Ileitis (PPE) caused by *Lawsonia intracellularis* (LI) is the most common grow-finish swine enteric disease in the US Almost all swine production systems have an ileitis control program in place with antibiotics and/or vaccinations. Enterisol® (Boehringer-Ingelheim) oral MLV vaccine is a relatively common method of ileitis control since the early 2000's. DLI® (diluted *Lawsonia intracellularis*) is a method of providing autogenous immunity to pigs via a defined subclinical live LI dose followed approximately two weeks later with a specific antibiotic pulse. Mecadox® (carbodox) at 50 g/ton for 14 days has been shown to eliminate clinical and subclinical LI infections[1] and was used for the antibiotic pulse post-DLI in this study. Porcilis® (Merck Animal Health) is a parenteral ileitis vaccine introduced in 2015 that has also shown efficacy in controlling ileitis.

Study objectives

The objective was to evaluate the comparative efficacy of these three methods of ileitis control vs challenged controls when pigs were orally gavaged with a high dose *Lawsonia intracellularis* mucosal homogenate at 10 weeks of age. They were evaluated for clinical observations, fecal PCR shedding, IPMA (immunoperoxidose monolayer assay) performance (ADG, ADFI, FCR), PPE intestinal gross lesion length and score, and IHC (immunohistochemistry) scores at necropsy.

Materials and methods

One hundred thirty, 3-week old high health weaned pigs were randomized by weight and gender into 26 pens of 5 pigs each as shown in Table 1-Trial Design. Treatments 1-4 had 30 pigs each and Treatment 5, a strict negative control group (unvaccinated, unchallenged), contained 10 pigs that were not included in the final statistical analysis. Treatments groups were separated by room to prevent cross-contamination of DLI and Enterisol bacteria shedding prior to disease challenge. The pen was the statistical experimental unit due to inclusion of FCR performance analysis.

Activity timeline

Porcilis treatment pigs were vaccinated with 2cc IM one day post-weaning on Day 0. Enterisol and DLI treatments were orally vaccinated at 5 weeks old on Day 14. Mecadox at 50 g/ton was fed from Day 28 to Day 42 to the DLI treatment group. All pigs were orally gavaged with a high dose of LI on Day 43.

Clinical observations for fecal consistency score (0-3), fecal color score (0-3), body condition or gauntness score (0-3) and depression score (0-3) were conducted weekly prior to LI challenge and thrice weekly after challenge. Bodyweights and pen feeder weights were taken on Day 0, 14, 28, 42, 56 and 69 for performance data analysis. Individual fecal PCR and serum IPMA samples were collected from 3 pigs per pen on Days 0, 14 and 28 and from all pigs on Days 42, 56 and 69.

Macroscopic PPE Lesion Scores (0-3) and PPE lesion length (cm) were taken from the ileum, jejunum, cecum and colon. An overall PPE Severity Score was than obtained by multiplying length × score and adding each intestinal section together. IHC lesions were scored 0-4 depending on the percentage of heavily infected enterocytes present in an ileal section taken just proximal to the ileal-cecal-colonic junction.

Table 1: Study design

	Treatment	# Pigs	# Pigs/pen	# Pen	Barn #
1	Challenge control	30	5	6	2
2	DLI	30	5	6	1
3	Enterisol	30	5	6	1
4	Porcilis	30	5	6	2
5	Strict negative control	10	5	2	3
	Total	**130**		**26**	

Challenge and vaccine dose

The mucosal homogenate challenge dose to test vaccine efficacy was 2.2×10^8 *Lawsonia intracellularis* per pig (U of MN VDL). The DLI autogenous LI dose given was 3.0×10^5 per pig. The Enterisol dose administered was 2.8×10^6 per pig.

Results and discussion

Animal removals/mortalities

After trial initiation, it was decided to remove eight pigs, two pigs per treatment on both Day 42 (pre-challenge) and Day 49 (post-challenge) to cooperate with a U of MN Ileitis Immunity Study. The results of that study are not reported here. Three pigs with severe ileitis in the Challenged Controls were humanely euthanized on Day 65 and three other pigs were removed due to health reasons unrelated to ileitis during the course of the study.

Performance data

Table 2, Comparison of Performance Measures Day 28-41, 0-41, shows statistical ($P < .05$) differences between treatment groups for ADG and ADFI. Mecadox was fed for 13 days on Days 28-41 to the DLI Immunity group. It is likely the effect of the antibiotic which enhanced ADFI and subsequent ADG during this time period and also the overall Day 0 to 41time period.

Table 3, Comparison of Performance Measures Days 41-56, 56-69, 41-69, and 0-69, shows statistical differences between treatments in all time periods. On Day 41-56 ADFI is already significantly ($P < .01$) lower in the Challenge Control pigs compared to all other treatments. In the Day 56-69 period, all three performance measures were highly significantly ($P < .01$) better in all of the ileitis treatment groups compared to the challenged controls.

Peak negative effects of ileitis on performance, and pathology typically occur 21-28 days after disease challenge. Day 69 was 26 days post-challenge. These negative effects also significantly affected performance for the entire trial 69 day trial period. The performance data is also shown in the tables for the strict negative control pigs but was not statistical analyzed due to lack of equal weight blocks.

Clinical observations

Analysis of clinical observations were not entirely complete for this manuscript except as shown in Table 4, Day 69 Necropsy Parameter Summary by Treatment. In general, clinical observations from Days 41-69 were prevalent in the Challenged Controls and sporadic in the other groups. Table 4 shows the number and percentage of pigs clinically positive with diarrhea, abnormal fecal color, or gauntness on Day 69.

These results emphasize the high subclinical prevalence of this disease in growing pigs. For example, although 100% of the Challenged Controls had low fecal PCR with a mean ct of 25.3 and 100% of pigs had IHC lesions, only 58% were clinically infected (i.e. 42% were subclinically infected). Likewise, although all three vaccine treatments greatly reduced clinical signs of ileitis, none of them prevented subclinical infections. In those three treatments, 51% to 71% of the pigs were still subclinically infected.(Here we define "subclinical" as having either a positive fecal PCR ct ≤ 35 and/or positive IHC scores on Day 69 without clinical signs of abnormal fecal consistency or color , or gauntness scores.)

In another DLI® Controlled Immunity challenge study, both clinical and subclinical ileitis was prevented[2] DLI immunity prevented any subsequent LI shedding in pigs following a high (10^{10}) LI challenge dose. Complete protection was also provided in that no gross or microscopic PPE lesions were present in any pigs afforded DLI protection. It is possible that differences in the autogenous DLI *Lawsonia intracellularis* dose, the age of the pig, the timing of the LI challenge post-vaccination, or the disease challenge itself was a factor to explain the discrepancy between the trials.

Table 2: Comparison of performance measures: Days 28 to 41, 0 to 41[1]

Treatment	Days 28 to 41			Days 0 to 41		
	ADG	ADFI	F:G	ADG	ADFI	F:G
Challenged						
Control	1.56[a]	2.56[a]	1.64	1.04	1.61[a]	1.55
DLI	1.90[b]	3.13[b]	1.65	1.18	1.82[b]	1.56
Enterisol	1.70[c]	2.95[c]	1.75	1.11	1.73[b,c]	1.55
Porcilis	1.75[c]	2.76[d]	1.58	1.11	1.69[a,c]	1.52
SEM	0.05	0.05	0.05	0.03	0.03	0.03
P	<.05	<.05	0.07	0.09	<.05	0.81
Neg. C.	1.75	3.06	1.74	1.21	1.89	1.56

[1] No superscripts in a column indicate global test for differences among treatment means was not significant at indicated *P*. Entries with superscripts indicate global test for differences among means was significant at the 0.05 level. Entries with no common superscript are significant at indicated *P*.

Table 3: Comparison of performance measures: Days 41 to 56, 56 to 69, 41 to 69, 0 to 69[1]

Treatment	Days 41 to 56			Days 56 to 69			Days 41 to 69			Days 0 to 69		
	ADG	ADFI	F:G	ADG	ADFI	F:G	ADG	ADFI	F:G	ADG	ADFI	F:G
Challenged												
Control	1.87	3.50^{a}	1.90	1.00^{a}	3.37^{a}	3.90^{a}	1.49^{a}	3.44^{a}	2.37^{a}	1.21^{a}	2.30^{a}	1.91^{a}
DLI	2.21	4.01^{b}	1.82	1.84^{b}	4.43^{b}	2.46^{b}	2.04^{b}	4.20^{b}	2.06^{b}	1.49^{b}	2.69^{b}	1.80^{b}
Enterisol	2.11	3.78^{a,b}	1.79	1.88^{b}	4.30^{b}	2.30^{b}	2.01^{b}	4.02^{b}	2.00^{b}	1.45^{b}	2.58^{b}	1.78^{b}
Porcilis	2.09	3.82^{a,b}	1.84	1.89^{b}	4.62^{b}	2.48^{b}	2.00^{b}	4.17^{b}	2.09^{b}	1.43^{b}	2.60^{b}	1.81^{b}
SEM	0.09	0.12	0.05	0.14	0.20	0.36	0.09	0.14	0.08	0.04	0.06	0.02
P	0.08	<.01	0.42	<.01	<.01	<.01	< .01	<.05	<.05	<.01	<.01	<.01
Neg. C.	---	4.65	---	----	2.29	---	1.96	3.56	1.83	1.51	2.56	1.70

1 No superscripts in a column indicate global test for differences among treatment means was not significant at indicated *P*. Entries with superscripts indicate global test for differences among means was significant at the 0.05 level. Entries with no common superscript are significant at indicated *P*.

Table 4: Day 69 necropsy parameter summary by treatment

	N	BW	Fecal CT value	IPMA titer	IHC score	PPE severity score	Clinical obs # pig positive (% Pos)
Challenge control	26	97.5	25.30	1:3724	2.21	188	15 (58%)
DLI	24	121.3	29.84	1:1270	1.42	35	3 (8%)
Enterisol	26	117.0	30.63	1:797	1.15	45	5 (19%)
Porcilis	24	117.0	33.11	1:2880	1.04	15	4 (17%)
Neg control	10	119.3	0	0	0	0	0

Fecal PCR shedding

Table 5 shows the % positive fecal PCR (suspects are not included) pigs over the ten week trial. The Challenged Controls and Porcilis pigs remained negative until the Day 42 LI challenge. DLI® and Enterisol pigs were orally "vaccinated" on Day 14 and by Day 28, 66% of the DLI pigs and 11% of the Enterisol pigs were positive shedders. (This fecal shedding is a useful measure of vaccination compliance 10-14 days following DLI administration). Mecadox at 50 g/ton was fed to DLI pigs from Day 28-42 to eliminate fecal shedding by Day 42. Peak LI shedding occurred by Day 56, with fewer % positives and higher LI ct values on Day 69 than on Day 56. All three ileitis control treatments reduced fecal shedding compared to the Challenge Control groups on both Day 56 and Day 69.

IPMA (immunoperoxidose monolayer assay) serology

IPMA serology for *Lawsonia intracellularis* measures IgG antibody titers (U of MN VDL) typically 2-8 weeks following exposure. Titer response is variable depending on the vaccine or DLI immunity control program, LI exposure dose, and also the duration of fecal shedding and severity of the PPE intestinal lesions present in that pig. IPMA titers are typically reported from a low of 1:30 to a high of 1:960. However, in this study titers end points out to >1:7680 were used to determine the degree of immune response following the DLI challenge.

Challenged Control pigs remained seronegative prior to *LI* challenge on Day 42. Titers ranged from 1:30 to 1:960 (geometric mean titer 1:367) on Day 56. IPMA titers ranged from 1:120 to >1:7680 on Day 69 with a mean titer of 1: 3,724 (Table 4).

DLI® controlled immunity pigs remained seronegative prior to challenge. Previous studies have shown an immune response from the subclinical DLI dose in over fifty percent of pigs within four weeks post challenge.[2] At Day 56, titers ranged from negative (16) to 1:240 (1),with a mean of 1:32.Titers ranged from negative (2 pigs) to 1:7680 (1 pig), with a mean titer of 1:1270 on Day 69.

Enterisol vaccinated pigs did not mount immune response prior to LI challenge on Day 42. Following LI challenge, Day 56 titers ranged from negative (18) to 1:240 (2) with a mean titer of 1:39. Titers ranged from negative (4 pigs) to 1:3840 (1 pig), with a mean titer of 1:797 on Day 69.

Porcilis vaccinated pigs seroconverted 83% (5 of 6) by Day 14.

Table 5: % Fecal PCR positive samples over time by treatment (mean ct value)

	0	14	22	28	42	56	69
Challenge control (*)	0	0	0	0	0	100 (21)	100 (25.9)
DLI (0)	0	0	0	66 (33)	0	100 (27)	80 (30)
Enterisol (+)	0	0	0	11 (38)	0	92 (28)	69 (30)
Porcilis	0	0	0	0	0	100 (26)	79 (33)

On Day 42, 56% were positive (10 of 18) with titers ranging from 1:30 (4) to 1:240 (1). All pigs were positive on Day 56 ranging from 1:30 to 1:3840. By Day 69, titers ranged from 1:120 (2) to 1:7680 (4), with a mean of 1:2880.

Seroconversion is a useful tool to determine vaccine compliance in herds using Porcilis vaccine. Following challenge, Porcilis titers were much higher than DLI or Enterisol, but did not reach the levels of the Challenged Controls (Table 4).

IHC results

Table 6 shows the frequency distribution of the IHC ileitis lesion scores by treatment. All challenged controls were positive on IHC analysis. Numerically, the Porcilis treatment had the lowest Mean IHC Score, although no significant differences were shown between T2, T3, and T4.

Since the Chi-square test among the 4 treatments was significant, pairwise comparisons between the Challenged Controls and all the three treatments were highly significant at $P < .0001$. There were no significant pairwise differences between the Enterisol, DLI or Porcilis treatments in the microscopic IHC analysis.

Associations/correlations

The Figure 1 scatterplot with linear regression line shows a direct positive association with the fecal ct value and ADG from Day 56-69 (13 to 26 days after *Lawsonia intracellularis* challenge). The higher the fecal ct value (less bacteria), the better the growth rate. R^2 is 21.1% (R = Pearson Correlation Coefficient = 0.46). The correlation coefficient of .46 indicates 46% of the ADG variation is due to the degree of LI shedding as indicated by the PCR fecal ct value.

Figure 2 depicts a high negative correlation between the fecal ct value and the IHC score, i.e. the lower the fecal ct (more bacteria shed), the more severe the IHC score. The Pearson Correlation Coefficient is -0.78 or 78%. Occasional outliers exist in pigs with an IHC score of 0, yet having low ct values as shown on the lower left of the scatterplot. This indicates a likely false positive IHC. This may occur when the PPE lesion was in a different location other than in the terminal ileum where the sample was taken. PPE lesions can occur in the jejunum, cecum, or colon and are not always present in the ileum.

This strong negative correlation between the ct value of a positive fecal PCR and the quantity of L. intracellularis antigen detected by IHC has been previously reported.3 On the basis of those results at the ISU VDL, PCR ct values ≤ 20.0 had a positive Predictive Value of 100% for the presence of positive IHC lesions. PCR ct values ≥ 30 were associated with a negative predictive value of 95% for those values ($P = .05$). Those ct values in between were less decisive.

Note that *Lawsonia* PCR ct values are not standardized between

Table 6: Frequency distribution of day 69 IHC by treatment[1]

	IHC					
	0	1	2	3	4	Total
Treatment						
Challenge C	0	3	14	6	4	27
	0.00	11.1	51.9	22.2	14.8	
DLI	9	3	7	3	2	24
	37.5	12.5	29.2	12.5	8.3	
Enterisol	12	2	9	2	1	26
	46.2	7.7	34.6	7.7	3.9	
Porcilis	9	6	8	1	0	24
	37.5	25.0	33.3	4.2	0.00	
Negative C.	10	0	0	0	0	10
	100.0	0.0	0.0	0.0	0.0	0.0

[1] Entries are frequencies/percentages of pigs with specified IHC score.

Figure 1: ADG_56_69 vs fecal Ct

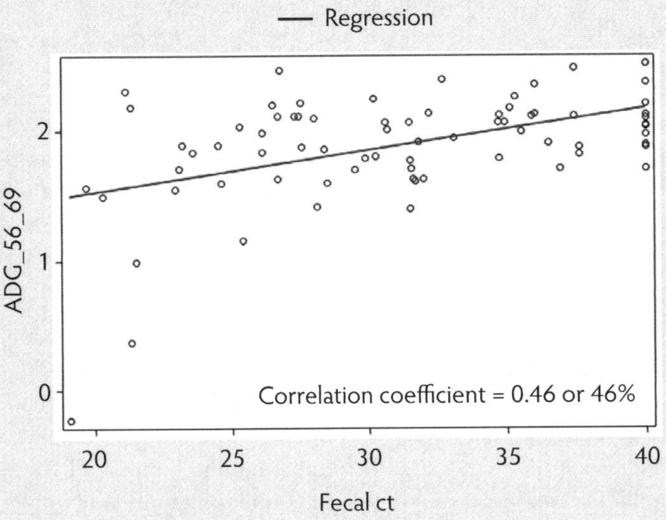

Figure 2: Fecal Ct vs IHC

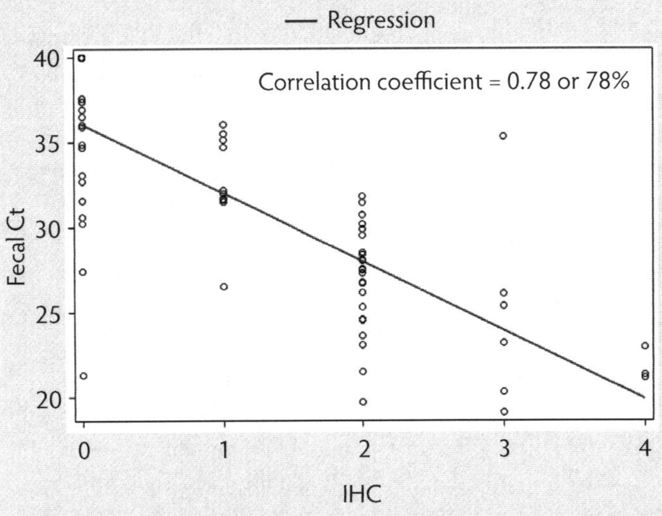

diagnostic labs and are difficult to compare. At the U of MN VDL, LI fecal ct values ≤ 35 are reported as positive, ct > 35 and ≤ ct40 as "suspect," and over ct 40 as "negative." (The sensitivity of the PCR test is 102 bacteria per gram of feces.[4]) Conventional wisdom at this laboratory suggests pigs with a fecal ct values of ≤ 32 would have higher IHC scores of >2 with more severe pathology and poorer performance . Higher Fecal PCR values up to ct40 are less likely to have positive IHC lesions.[5] Intuitively, these correlations indicate diagnostic value for the swine practitioner in determining strong associations with ante-mortem fecal PCR ct values to PPE lesions and to ADG performance.

Conclusions and implications

Compared to the *LI* Challenged Controls, the DLI controlled immunity, Enterisol, and Porcilis all significantly improved all performance parameters, significantly reduced IHC PPE lesion scores, and fecal PCR shedding. Clinical ileitis, diarrhea and gauntness,was effectively controlled by all three treatments. However, none of these control measures effectively prevented or controlled subclinical ileitis in this disease challenge trial. The swine industry still has opportunity for better control and preventative measures for both clinical and subclinical ileitis.

References

1.Kinsley K, Winkelman N, Gebhart C; Elimination of *Lawsonia intracellularis* from Pigs Fed Carbodox (Mecadox[R]). AASV 2006 pp 157-161

2.Winkelman NL, Mueller A, Domangue R, Effectiveness of DLI[R] (Diluted *Lawsonia intracellularis*) "Vaccine" followed by a DenegardR LC, Aivlosin, or Mecadox in Finishing Pigs Challenged with a High Dose LI mucosal homogenate. AASV 2014 pp 273-278

3.Burrough ER, Rotolo ML, Gauger PC, et al, Correlation of *Lawsonia intracellularis* Semi-Quantitive Fecal Polymerace Chain Reaction Assay Results with the Presence of Histologic Lesions of Proliferative Enteropathy and Positive Immunohistochemical Staining; J. Swine Health Prod. 2015; 23(4); 204-207

4.Beckler D, Armbruster G, Rutten-Ramos S, Evaluation of Fecal Shedding by a High-Throughput qPCR Assay in a *Lawsonia intracellularis* Challenge; Proc. AASV (2012) pp 149-153

5.Vannucci F, U of MN VDL (personal communication)

Fat-soluble vitamins: Beyond nutritional needs

Robert L. Stuart, PhD
Stuart Products, Inc. Bedford, Texas

Introduction

Fat-soluble vitamins A, D, E are vital nutrients for all swine. These vitamins are routinely added to swine feeds at or above National Research Council (NRC) recommended levels. Supplementation via drinking water and injection are also utilized.

Published NRC dietary requirements have not been static. In 1941 requirements for vitamin A, vitamin D and carotene were published and the requirement for vitamin E was first published in 1973 and later increased in 1988. In 1988, carotene was removed as a required dietary nutrient. Vitamin D requirements were increased for sows in the 2011 NRC publication. The primary reasons for changes in fat-soluble vitamin requirements are due to a combination of research findings, husbandry practices (confined vs. non-confined), sow productivity, and improved growth.

The primary emphasis for NRC dietary requirements is minimum daily requirements for animal performance. Research has shown that under certain conditions, higher levels of fat-soluble vitamin-intakes may impact other factors besides average daily gain and feed efficiency.

The purpose of this paper is to review information that shows the impact of strategic use of fat-soluble vitamins on other bodily functions besides generally accepted nutritional needs and also the impact of disease and other factors on fat-soluble vitamin status.

Enhanced immunity ensues when higher than NRC fat-soluble levels are fed or injected

There are several excellent reviews on the impact of fat-soluble vitamins on cell-mediated and humoral immunity.[1,2,3] Research as early as 1931, showed that fat-soluble vitamins had a positive effect on the immune system. As swine husbandry practices have changed from non-confinement to confinement and productivity has increased, the need for adequate fat-soluble vitamin supplementation has become even more paramount.

Several experiments in various species have shown that viral and bacterial infections reduce fat-soluble vitamin status, especially vitamin E. Two studies have shown that there was a negative impact of PRRS on vitamin E status in sows and pigs, but not vitamin D status. L. Bruner, 2013, conducted a field evaluation on the effects of PRRS on vitamin E and vitamin D status of PRRS+ and PRRS- sows and pigs (Personal communication). She found that in nine herds, vitamin D status was not reduced by PRRS, but vitamin E status was dramatically reduced by PRRS virus in both sows and nursing pigs. Tousignant found similar results. He found that PRRS infected weaned pigs had no reduction in vitamin D or selenium status, but a 51.3% reduction in vitamin E status at 14 days post-infection (Table 1) (Personal communication).

Jensen et al., 1987, reported that for optimum cell-mediated immunity, pig's serum alpha-tocopherol levels should be maintained at levels above 3 ppm.[4] The conundrum with vitamin E is that newly-weaned pigs do not utilize synthetic vitamin E acetate that is in nursery diets. Kim et al., 2016, evaluated the impact of supplementing newly-weaned pigs with either 50, 100, or 200 I.U. dl-alpha-tocopheryl acetate/ kg diet on inflammatory responses.[5] At the start of the experiment (at weaning) average serum vitamin E was 3.2 ppm. Ten days post-weaning, serum alpha-tocopherol levels were 1.4, 1.7 and 2.2 for pigs fed 50, 100 and 200 I.U., respectively (Figure 1). This study supports findings by Moreira

Table 1: Effect of PRRS challenge on vitamin E and selenium serum status in nursery pigs (Tousignant, 2013, personal communication)

	PRRS positive (n = 9)	PRRS negative (n = 10)	*P*-value[1]
Weaning			
Vitamin E (ppm)	3.48	3.72	0.667
25 OH D (ppb)	88	83	0.75
Selenium (ppm)	141.9	110.9	0.0001
14 DPI[2]			
Vitamin E (ppm)	2.3	3.46	0.098
25 OH D (ppb)	38	40	0.72
Selenium (ppm)	140.8 (12.45)	102.5 (22.86)	0.0004

[1] *P*-values calculated using t-test
[2] DPI = Days post-infection

Figure 1: Effect of dietary dl-alpha tocopheryl acetate levels on plasma vitamin E contents after weaning. All pigs were experimentally infected with an enterotoxigenic strain of *E coli* on d 7. ***$P < 0.001$ (Kim, et. al., 2016)

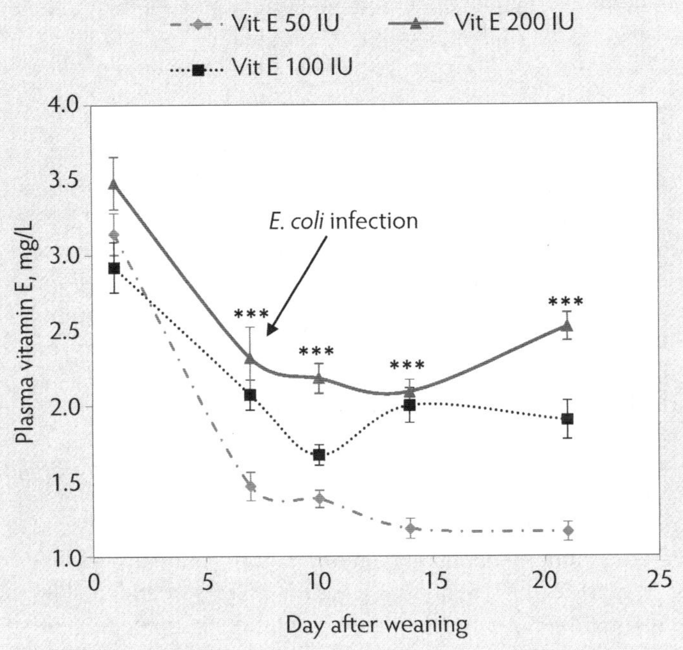

and Mahan, 2002, that nursery pigs are unable to maintain serum vitamin E status after weaning due to lack of ability to utilize synthetic vitamin E acetate at levels up to 200 I.U. per kg (182,000 I.U. per ton).[6]

Hidiroglou et al., 1995,[7] found that pigs injected with d-alpha-tocopherol (VITAL E-500) had dramatically higher plasma alpha-tocopherol concentrations and higher antibody titers to key-hole limpet haemocyanin compared to control pigs. Vitamin E injected pigs also had significantly higher alpha-tocopherol concentrations in spleen, liver, heart, kidney, lung and muscle compared to non-injected pigs.

In a recent experiment at Michigan State University, supplementing the water of newly-weaned pigs with micellized non-esters of *d*-alpha-tocopherol and cholecalciferol reduced oxidative stress. Enzymes superoxide dismutase (SOD) and glutathione peroxide (GPx) were determined in newly-weaned pigs supplemented with vitamin E and D in drinking water (EMCELLE E-D_3 Liquid). SOD was reduced ($P = 0.005$) and GPx activity was reduced due to the water supplementation of vitamins E and D. ($P = 0.0002$).[8]

Another area of interesting research is the impact of vitamin E and selenium deficiencies on enhanced viral pathogenicity. Beck has demonstrated that adequate vitamin E and selenium in rodents reduced viral mutations and subsequent reduction in viral pathogenicity.[9]

Fat-soluble vitamins improve muscle function

No role for vitamin A to improve muscle function has been reported in the literature. Vitamin E has long been shown to be involved in skeletal and cardiac muscle function, and recently smooth muscle function. In the past few years, a growing number of human clinical studies have examined the effects of vitamin D supplementation on muscle function and research on the vitamin D receptor in smooth muscle cells have helped to improve our understanding of the role and actions of vitamin D on smooth muscle function.

In regards to vitamin E deficiencies in nursery pigs, an increase in cardiomyopathy has been associated primarily with vitamin E deficiencies and not selenium.[10] In calves, the fastest growing animals have the lowest vitamin E status.[11] It is interesting that with cardiomyopathy, the fastest growing pigs seem to be affected, therefore supplementing the nursery pig with a biologically available source of vitamin E (EMCELLE TOCOPHEROL) to keep serum vitamin E status above 3-4 ppm is critically important.[12]

Recent university research and field demonstrations has demonstrated that pre-partum injections of vitamins A, D, and E reduces total delivery time approximately one hour (Data on file, Stuart Products, Inc.). Whether the effect is due to enhanced vitamin E and/or vitamin D status has yet to be determined. In the large field study, besides reducing delivery time, stillborns were reduced approximately 30% in the injected sows compared to the non-injected sows. There were 267 sows in the field study. Half were injected with VITAL E-Repro one week pre-farrowing and the other half were not injected. Parities were equally assigned to the two treatments.

Another potential role for elevated levels of vitamin E is sow prolapses. A small survey was conducted in North Carolina to determine if prolapsed sows had different fat-soluble vitamin status compared to cohort sows.The average serum vitamin E status in 10 prolapsed sows was 26% lower when compared to 19 cohort sows (1.30 vs 1.64 ppm), and serum vitamin A and 25 OH D3 were 7 and 6% lower in the prolapsed sows. (Table 2).

Fat-soluble vitamins enhance reproduction

All three fat-soluble vitamins have been shown to impact reproduction in sows.[13] Vitamin A supplementation and/or injection has been shown to improve embryo survival.[14] The form of vitamin A in injectable formulations is critically important. Vitamin A palmitate has been shown to be much more effective on enhancing sow productivity when compared to vitamin A propionate.[15] Jensen and Lauridsen recently presented data on the fat-soluble vitamin needs for highly prolific sows in Denmark.[16] The Danish recommended levels of vitamins A, D and E are dramatically higher than NRC recommendations, especially for the lactating sow (Table 3).[17] As an example, the published vitamin E requirement for lactating sows in Denmark is 165 I.U./kg (150,000 I.U./ton) compared to the NRC requirement of 44 I.U./kg (40,000 I.U./ton).

Table 2: Survey of fat-soluble vitamin status in prolapsed sows and their cohorts (2016, unpublished).

Vitamin	Prolapsed sows (n =10)	Cohort sows (n = 19)
Vitamin E (ppm)	1.30	1.64
Vitamin A (ppm)	0.27	0.29
Vitamin D (ppb)	34.84	36.93

Importance of bioavailability of fat-soluble vitamins

Both naturally-occurring and supplemental fat-soluble vitamins go through various steps prior to absorption that is similar to fat metabolism. Prior to absorption, supplemental fat-soluble vitamins A and E need to be de-esterified by the action of intestinal esterases (pancreatic carboxyl ester hydrolase (CEH).[18] The esters are removed from the vitamin molecules prior to absorption. A second step for fat-soluble vitamins is micellization. Micelles are formed by the action of bile salts to create water-soluble droplets that are absorbed. Supplementing water with micellized, non-esterified vitamin E (d-alpha-tocopherol) assists in maintaining adequate vitamin E status of the post-weaned pig.[12] Yang, et al., 2014, compared the utilization of injectable vs oral sources of vitamins A, D and E (VITAL E-Newborn). Injectable sources were utilized approximately twice that of orally administered vitamins.[19]

van Kempen et al. reported that the bioavailability for oral *d,l-α*-tocopheryl acetate (normally used in complete feeds) was only 12.5% when administered in nursery pigs. Vitamin E dosed as d-*α*-tocopherol at one-eighth the level of the synthetic acetate ester showed both a rapid and a very slow pool, whereas orally dosed synthetic vitamin E acetate showed a single slow pool. The authors stated that the use of synthetic vitamin E-acetate brings into question the efficacy of typical use of vitamin E-acetate in swine diets for alleviating oxidative stress.[20]

Conclusions

Of the three fat-soluble vitamins, vitamin E appears to be the most deficient in all post-weaning phases of swine production, while vitamin D supplementation may be most critical in the nursing pig due to low milk content. Reduced farrowing time due to pre-partum injection of fat-soluble vitamins and reduction in mulberry heart are areas affected by inadequate fat-soluble vitamin intake. Esterification of fat-soluble vitamins A and E is done to improve stability in complete feeds, but there are situations in swine production when it is best not to esterify fat-soluble vitamins, especially vitamin E in nursery diets. Stability of esters will be superior, but at the expense of bioavailability.

In general, the NRC recommended levels for fat-soluble vitamins fat-soluble vitamins are too low. The level for gestating and lactating sows should be dramatically increased. For newly-weaned pigs, the source of vitamin E is more critical than inclusion levels. Pigs fed 200 I.U. synthetic acetate/kg diet were still unable to maintain serum tocopherol levels, while pigs supplemented with micellized d-alpha-tocopherol are able to maintain an optimum serum status. The necessity to supplement with a biologically-available source of vitamin E as found in EMCELLE TOCOPHEROL and EMCELLE ED3 Liquid, or VITAL E-injectables is critical to be assured that vitamin levels are more than adequate.

References

1. Villamor E and W. W. Fawzi. 2005. Effects of vitamin A supplementation on immune responses and correlation with clinical outcomes. Clin Microbiology Reviews. 18:446-464.

2. Lucas, R.M., S. Gorman, S. Geldenhuys, and P. H. Hart. 2014. Vitamin D and Immunity. F1000Prime Reports. 6:118.

3. Moriguchi, S. and M. Kaneyasu. 2003. Role of vitamin E in immune system. J. Clin. Biochem. Nutr. 34:97-109.

Table 3: Comparative fat-soluble vitamin recommended use levels (I.U/kg) for swine feeds in Denmark and US (Jensen and Lauridsen, 2017)

		Sows		Nursery pigs	Pigs	Grow/finish
		Gestating	Lactating	6-9 kg	9-30 kg	30-110 kg
Vitamin A	Denmark	8000	8000	8000	5000	4000
	NRC	4000	2000	2200	1750	1300
Vitamin D_3	Denmark	800	800	800	500	400
	NRC	800	800	220	200	150
Vitamin E	Denmark	40	165	140	140	40
	NRC	44	44	16	11	11

4. Jensen, M., C. Fossum, C., M. Ederroth, R.V.J. Hakkarainen, R.V.J. 1988. The effect of vitamin E in the cell-mediated immune response in pigs. J. Vet. Med. B. 1988;35:549–555.

5. Kim J.C., B.P. Mullan, J. L. Black, R. J. Hewitt, R.J. van Brneveid, and J.R. Ruske. 2016. Acetylsalicylic acid supplementation improves protein utilization efficiency while vitamin E supplementation reduces markers of the inflammatory response in weaned pigs challenged with enterotoxigenic E. coli. E. J. of Animal Sci. And Biotechnology. 7:58.

6. Moreira, I. and D.C. Mahan. 2002. Effect of dietary levels of vitamin E (all-rac-tocopheryl acetate) with or without added fat on weanling pig performance and tissue alpha-tocopherol concentration. J. Anim. Sci. 80:663-669.

7. Hidiroglou, M., T. R. Batra, E. R. Farnworth, and F. Markham. 1995. Effect of vitamin E supplementation on immune status and α-tocopherol in plasma of piglets. Reprod. Nutr. Dev. 35:443–450.

8. Copeland, K. R., D. B. Scales, G. M. Hill, J. E. Link, K. C. Turner, C. J. Rozeboom, and R. L. Stuart. 2017. Effects of vitamins E and D on performance and antioxidant enzymes in nursery pigs. Midwest ASAS Meeting, March 13, 2017.

9. Beck, M. A. 2007. Selenium and vitamin E status: Impact on viral pathogenicity. J. Nutr. 137: 1338 –1340.

10. Pallares, F.J., M. J. Yaeger, B. H. Janke, G. Fernandez and P.G. Halbur. 2002. Vitamin E and selenium concentrations in livers of pigs diagnosed with mulberry heart disease. J Vet Diagn. Invest 14:412–414.

11. Nonnecke, B. J., M.R. Foote, B.L. Miller, D.C. Beitz and R.L. Horst. 2010. Short communication: Fat-soluble vitamin and mineral status of milk replacer-fed dairy calves: effect of growth rate during the preruminant period. J. Dairy Sci. 93:2684-90.

12. Specht, T.A., D.C. Mahan, N.D. Fastinger and R.L. Stuart. 2003. Effects of d-alpha-tocopherol alcohol or acetate in water soluble or emulsified form to the drinking water of weanling pig. J. Anim. Sci. 81 (Suppl. 2): p 88 (Abstract No. 356).

13. Agarwal A., A. Aponte, M. Beena, J Premkumar, A. Shaman and S. Gupta. 2012. The effects of oxidative stress on female reproduction: a review. Reproductive Biology and Endocrinology. 10:1-31.

14. Britt, J. 2014. Vitamin A injection improves embryonic development resulting in uniformity of piglet size at birth for Gilts and Sows. News to Use February, 2014. Stuart Products, Inc.

15. Robertson, J. A. 1997. Investigations of the action of vitamin A and beta-carotene on reproductive performance in pigs. Ph.D. Dissertation Victoria Univ. of Tech., Victoria, Australia.

16. Jensen, S. K. and C. Lauridsen. 2017. Fat Soluble Vitamins - availability and recommendations for highly prolific sows and their progeny. 78th Minnesota Nutr. Conf. Sept. 20-21, 2017. pp 165-175.

17. Seges Pig Research. 2017. 26th edition Danish nutrient standards.

18. Lauridsen, C., M.S. Hedemann, and S. K. Jensen. 2001. Hydrolysis of tocopheryl and retinyl esters by porcine carboxyl ester hydrolase is affected by their carboxylate moiety and bile acids. J. Nutritional Biochemistry. 12: 219-224.

19. Jang, Y.D., M.D. Lindemann, H.J. Monegue, and R.L. Stuart. 2014. The effects of fat-soluble vitamin administration on plasma vitamin status of nursing pigs differ when provided by oral administration or injection. Asian-Australas J Anim Sci. 5:674-82.

20. van Kempen, T. A., M. H. Reijersen, C. de Bruijn, S. De Smet, J. Michiels, M. G. Traber and C. Lauridsen. 2016. Vitamin E plasma kinetics in swine show low bioavailability and short half-life of all-rac -α-tocopheryl acetate. J. Animal Science. 94:4188-4195.

Vaccination and challenge study for an H3N2 swine influenza recombinant DNA vaccine

Tim Miller[1]; Laura Trygstad[1]; Dustin Petrik[1]; Darin Madson[2]; Phillip Gauger[2]
[1]Agrilabs Corporation; [2]Iowa State University

Introduction

An ENABL® LS5 adjuvanted DNA vaccine containing an H3N2 Cluster IV strain of Influenza A Virus of Swine (IAV-Swine) was tested for safety and efficacy in weaning age pigs. For safety, a 200ug dose of the DNA vaccine was inoculated into 22-23 day old weaning pigs and samples were taken at 1, 21, and 60 days post inoculation (DPI) to evaluate the residual vaccine DNA by qRT-PCR gene amplification. The results indicate that the vaccine is safe and does not reside in any tissue beyond threshold of detection for more than 21 days. The data also indicates that the DNA/adjuvant disseminates quickly in that it can be found in the draining lymph node within 24 hours after intramuscular inoculation. To evaluate efficacy, the purified DNA vaccine was inoculated into weaning age pigs (26-27 days of age) with doses ranging from 50 to 175 ug DNA. The dose was administered in 1ml of ENABL® LS5 at day 0 and day 14 with serum collected at 0 (pre-inoculation), 14 and 36 DPI. Twenty-three days after the second vaccination with purified DNA, the pigs were challenged (10ml intratracheal and 2ml intranasal) with $1 \times 10^{5.6}$ $TCID^{50}$/ml of Specific Pathogen Free embryonated chicken egg amplified homologous virus IAV-Swine Cluster IV. Each pig had a matched placebo control (MPC) group, which showed lung pathology typical of IAV infection after challenge with an average lung consolidation percentage of 7.66% (95% CI of 5.33, 9.99). In contrast, at least one pig in every vaccinated group was completely clear of any lung pathology after challenge, and the average lung consolidation percentages for each group vaccinated with 50, 100 and 175µg IAV H3 DNA / dose was 2.68% (95% CI of 0.26, 5.10), 4.47% (95% CI of 1.85, 7.09) and 2.13% (95% CI of -0.39, 4.64), respectively. In addition to the significant reduction in lung consolidation, the shedding of the challenge IAV from bronchoalveolar lavage fluid (BALF) collected five days post-challenge was almost completely eliminated in vaccinated groups compared to the MPC group.

Materials and methods

The vaccine plasmid DNA was derived by fermentation of (*E coli*) Master Seed IAV H3 Lot# MS17Mar16SIV (ML#179705) followed by plasmid purification. This plasmid has a synthetic gene encoding the sequence of the cluster IV, clade F, H3 antigen of IAV-Swine Cluster (H3N2) influenza isolate. The plasmid was mixed with a VaxLiant adjuvant (ENABL® LS5) for swine. Tissue for plasmid DNA retention study were taken by biopsy sampling from the target tissue per the designated time as described in Table 1. At least 25 mg of tissue were collected, a sample ID was assigned, and the sample was placed on ice for processing or storage at -18 ± 5°C. To avoid cross contamination, three targeted tissues were individually isolated and processed. DNA from each sample was extracted individually using the Qiagen DNeasy Blood and Tissue kit, with a protocol specific to the tissue/sample type. Extracted DNA was quantified by UV spectrometry, aliquoted into a minimum of two aliquots, and was stored at 2-7°C until qPCR was performed. The DNA derived from each processed tissue was tested for presence of target plasmid using qPCR. Briefly, the qPCR assay included primers to amplify an endogenous reference gene (ERg) from the swine genome and primers to amplify the target sequence of the vaccine plasmid DNA (pDNA). For each test tissue, the extracted total DNA was added to two separate reactions containing; 1) ERg primers and the associated ERg real-time probe and 2) pDNA primers and the associated pDNA real-time probe. In addition, established controls for the specific tissue-type were run in the same assay and used as the assay suitability control. The samples were analyzed by comparing the cycle threshold (Ct)

Table 1: IAV DNA vaccine plasmid retention study design

Group	Description	# of pigs	Day for tissue collection	Tissues collected and tested via PCR
T01	Control (MPC)	4	1, 21, 60	Injection site (skin and muscle), local lymph node, and spleen 200ug/dose
T02	IAV H3 DNA Vaccine, 1.0mL Dose, LS5	9		

MPC = Matched Placebo Control
LS5 = ENABL® LS5 Adjuvant
All pigs administered 1.0 mL vaccine via intramuscular injection

values from the ERg and pDNA reactions for a single tissue extraction sample. The Ct is the value at which the level of fluorescence becomes greater than background (threshold) and the Ct value is inversely related to the amount of target sequence in the initial sample (extracted tissue DNA).

For efficacy, 60 IAV-seronegative pigs that were 26-27 days of age were injected (15 pigs per group) with either the matched-placebo control (MPC), 50µg IAV H3 DNA vaccine, 100µg IAV H3 DNA vaccine, or 175µg IAV H3 DNA vaccine. Pigs were administered a booster vaccination on day 14 of the study and blood was collected on day 0 prior to the vaccine administration, on day 14 prior to boost vaccination, and on day 36. On study day 37, the pigs were all challenged with virus cluster IV, clade F, H3 antigen of IAV-Swine Cluster (H3N2) by intranasal / intratracheal inoculation. Two liters of stock challenge virus was obtained by amplifying the virus twice in embryonated eggs to a titer of 1x $10^{5.5}$ $TCID_{50}$/ml. The challenge virus was aliquoted and frozen at -80 ± 10°C until use. For challenge, the virus was thawed by running it under cold water and then placed on ice until administration to the pigs. Back titration from the virus used in challenge procedure indicated a titer of 1 × $10^{5.6}$ TCID50/ml. Clinical observations were performed daily during the study, and vaccine efficacy was assessed by the percent of lung lesions present five DPI compared to controls, along with antibody titers and virus shedding in the bronchoalveolar lavage fluid (BALF).

Blood was collected from each pig prior to each vaccination and one day before challenge. Between 5-10 ml of blood was collected via jugular venipuncture using a vacutainer collection system into Serum Separator Tubes (SST). Blood was allowed to clot and then centrifuged to collect serum. Serum was stored at 2-7°C until shipment to ISU VDL for testing by NP ELISA and HAI assay using cluster IV, clade F, H3 antigen of IAV-Swine Cluster (H3N2) as the indicator virus. Geometric means (GMT) were calculated from the HAI titers for each group.

BALF was collected from each animal immediately following euthanasia and lung assessment. The tip of a sterile pipette was placed into the bronchus trachealis supplying the right anterior lung lobus. Dulbecco's Modified Eagle's Medium (DMEM) (25-50 ml) was instilled into the bronchus and aspirated immediately. Samples were placed into sterile transport tubes and stored at -80 ± 10°C until shipped on dry ice to ISU VDL for testing via virus isolation (VI). During necropsy of each animal, samples of nasal turbinate, trachea, and lung lobe were collected using standard histopathological techniques. All samples were stored at room temperature in formalin. No testing has been conducted on these samples.

Results and discussion

The plasmid DNA vaccine was constructed containing an H3-specific hemagglutination gene (HA) from a cluster IV, clade F, H3 antigen of IAV-Swine Cluster (H3N2). This candidate was used to produce the Master Seed Bacteria designated as (*E coli*) Master Seed IAV H3 Lot# MS17Mar16SIV. A representative serial of the IAV H3 DNA vaccine was made, inclusive of VaxLiant adjuvant ENABL® LS5 and the fully assembled vaccine serial was tested in this study. The adjuvant ENABL® LS5 has shown to induce a balanced T and B cell response to DNA vaccines administered by IM inoculation into chickens and pigs. Coupled with the fact that DNA vaccines express antigens only intracellularly, the combined effect of the adjuvant and vaccine technology combine to produce strong cell mediated and humoral immunity. The vaccine is proven safe in thousands of vaccine doses, including doses up to 40× above that considered the minimum immunizing dose in pigs without causing any type of immediate or delayed vaccine related reactions. In addition, tissue retention studies (Table 1) designed to evaluate the persistence or integration of DNA into the genome of target tissue indicate that after 21 days there is no indication the DNA persists or is integrated into the host cellular DNA based on lack of? detection of vaccine DNA derived from various tissues after inoculation and quantitation by qRT-PCR. The same data also shows that in pigs, the DNA vaccine persists at or near the site of vaccination for up to 21 days; long enough for both a priming and boost dose to occur from a single vaccination (Table 2). The results of the vaccination and challenge study indicate that the pigs administered the MPC had no HAI response (all titers ≤10), whereas all vaccinated treatment groups showed positive HAI response. There was also no evidence of plasmid shedding from vaccinates to these comingled MPC controls as none of the MPC pigs seroconverted to IAV during the 36 days prior to challenge. Within the three vaccinated groups, HAI titers were low following the first vaccination with only ~20% of the pigs in each group showing seroconversion by the HAI assay. However, after a booster vaccination, 93.3 to 100% of the pigs in each vaccinated group seroconverted and the geometric mean titers (GMT) were 115.8, 129.3 and 83.8 for the 50, 100 and 175µg IAV H3 DNA vaccinated groups, respectively (Figure 1). Every pig in the matched placebo control (MPC) group had lung pathology typical of IAV infection with an average lung consolidation percentage of 7.66% (95% CI of 5.33, 9.99), a median of 6.39%, a minimum of 0.6% and a maximum of 13.95%. In contrast, at least one pig in every vaccinated group was completely clear of any lung pathology and average percent consolidation for each group vaccinated with 50, 100 and 175µg IAV H3 DNA / dose was 2.68 (95% CI of 0.26, 5.10), 4.47 (95% CI of 1.85, 7.09) and 2.13 (95% CI of -0.39, 4.64), respectively (Figure 2). In addition to the significant reduction in lung consolidation, BALF sample analysis of viral shedding demonstrated that pigs vaccinated with 50, 100 and 175µg IAV H3 DNA / dose had 14%, 46% and 8% viral shedding, respectively, compared to 80% viral shedding in the MPC control group... This data demonstrates that almost all of the MPC group (12 / 15 total pigs) were still shedding virus at the time the study ended. In contrast, only 3 pigs out of 27 total pigs in the 50 and 175µg DNA vaccine groups were still shedding at the end of the study (Table 3). DNA vaccines serve as new tools against IAV genetic drift because IAV HA-based DNA vaccines, unlike

Table 2: IAV DNA vaccine plasmid retention qRT PCR results

Tissue	Day 1 >30,000[1]	Day 1 1000-30,000[1]	Day 21 >30,000	Day 21 1000-30,000	Day 60 >30,000	Day 60 1000-30,000
Lymph node	1/3	0/3	0/3	0/3	0/3	0/3
Skin	3/3	0/3	0/3	1/3	0/3	0/3
Muscle	3/3	0/3	0/3	0/3	0/3	0/3
Spleen	0/3	0/3	0/3	0/3	0/3	0/3

[1] Copies pDNA per µg gDNA

Figure 1: Swine vaccination/challenge study - HAI antibody response to dose titration. Geometric mean HAI titers from pigs responding to dose titration; 2× IM

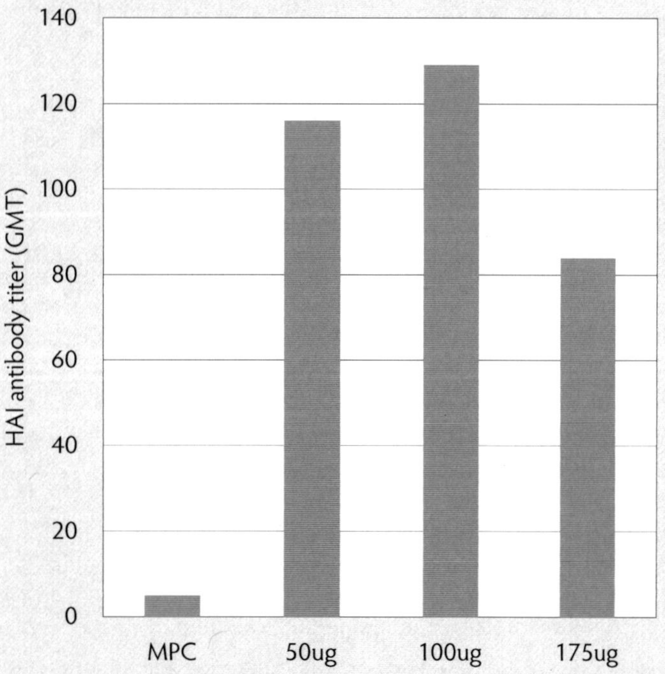

Figure 2: Summary for percent lung consolidation post challenge

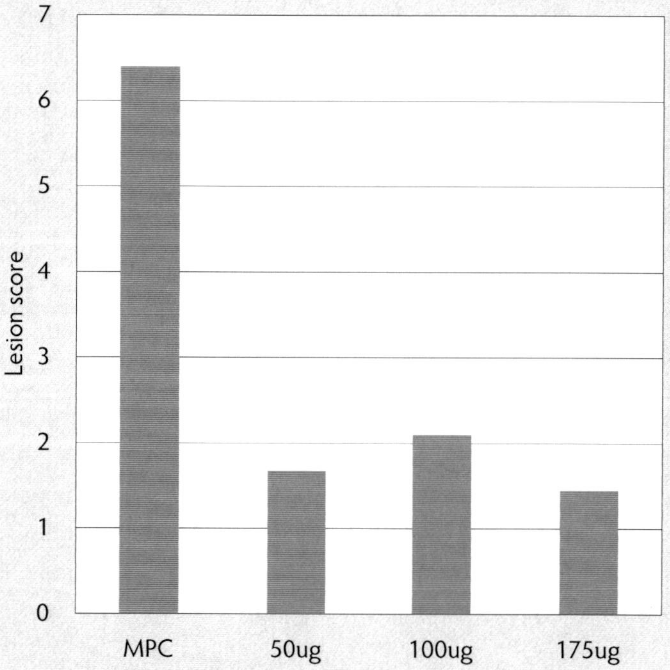

more conventional IAV vaccines, are readily modifiable to adapt to genetic changes in the HA protein allowing vaccines to quickly update antigens to include components of an emerging, problematic field virus. A newly emergent distinct HA gene is rapidly sequenced, synthesized, and substituted into the DNA production platform, thereby specifically addressing disease outbreaks caused by "new or emerging" strains expressing slightly different protein sequences in the HA surface protein. The plasmid backbone DNA was used to carry a synthetic gene encoding the sequence of an H3 antigen identified from a recent IAV in swine H3N2 Cluster IV strain. The H3-HA specific plasmid backbone was mixed with a VaxLiant® adjuvant and inoculated into weaning pigs. The HA specific antibody was evaluated for cross reactivity to several heterologous H3 IAV by HAI. The results demonstrate the use of DNA vaccines with an adjuvant delivery system can be used as a means to quickly provide new, updated and relevant vaccines against emerging disease variants in swine. In this study, the reduced lung consolidation was deemed to be statistically significant and clinically relevant in pigs vaccinated with all three concentrations of the test vaccine compared to pigs vaccinated with the MPC.

Table 3: Virus isolation from BALF post challenge

Vaccine dose administered	Virus isolation $TCID_{50}$/ml $\log_{10}$	Percent BALF positive pigs
MPC (0ug plasmid DNA) control	<1 – 5.5	80
IAV DNA vaccine 50ug	<1 – 1.8	14
IAV DNA vaccine 100ug	<1 – 3.2	46
IAV DNA vaccine 175ug	<1 – 1.5	8

Fast Genetics: The use of technology to increase stability and profitability in the swine production chain

R. Leach[1], PhD; B. Belstra[2], PhD; D. Grossi[3], DVM, PhD; K. Willenburg[2], PhD; M. Duggan[3], MSc
[1]Fast Genetics, Ames, Iowa; [2]Vienna Office, Wisconsin; [3]Fast Genetics, Saskatoon, Canada

Introduction

Fast Genetics is a swine genetics company with global reach. By applying cutting edge technologies to its highly competitive and robust terminal and maternal lines, it has become one of the top genetics companies in North America.

Founded on the principles of high health and competitive animals, Fast Genetics has become an early adopter and pioneer in many technologies, such as the use of: single nucleotide polymorphisms (SNP) associated technologies, feed intake recording equipment (FIRE), and optimum contribution theory mating.

More recently, Fast Genetics is at the vanguard of developing and utilizing sex sorted sperm (SSS). All of the technology that surrounds SSS has recently gone from theory in swine application to being utilized at our multiplier and genetic nucleus (GN) level.

This paper outlines how Fast Genetics has developed and used current technologies to produce the most profitable animal for its customers.

Genotyping

Genotyping animals, especially at the nucleus, is routine for the largest genetics companies since the improved genetic progress is simply too large to ignore. However, as with many technologies, the rate and methodology used can have significant ramifications on downstream benefits. Prime examples of this are: 1) if all animals are genotyped at the nucleus 2) if some form of pre-selection occurs before genotyping and 3) at what age animals are genotyped. All of these factors can affect the rate of genetic gain.

Fast Genetics has maximized genetic gain by genotyping every animal raised on a nucleus farm. In addition, the acquisition of a genotyping lab (Genetic Visions, Middleton, WI) will significantly shorten the time interval between sending an animal's DNA sample and receiving the result. This allows for the selection of animals at off-test with genomic indexes present, and ensures only the most genetically-elite and well-structured animals will move on to the next generation.

Furthermore, a proprietary custom 65,000 SNP chip has been designed using the Fast Genetics lines. The chip design offers more resolution to the genetic evaluation aiding the identification of the most elite animals.

Novel technologies

Developing novel technologies and methodologies are driving forces behind the advancement of the majority of genetics companies. They allow a company to utilize a novel technology first, giving them an advantage over other genetics companies. Two pieces of technology are outlined below. First, we developed a program called the boar stud monitor (BSM) and second, current research investigating the use of embryo transfer (ET).

Boar stud monitor

The BSM is a powerful, proprietary tool that was developed by Fast Genetics. The goal of the program is to accurately predict the declining rate of a boar's index. In order to accomplish this, the BSM utilizes all of the metrics in a boar's index. It also deploys genomic and pedigree information from all animals that are related. The result is that each boar has a predicted date that reflects when the boar is no longer considered elite. This allows for more proficient production, shipment and use of boars. The incorporation of the BSM program has allowed for more efficient use of elite boars and a higher replacement of non-elite animals (Figure 1).

Embryo transfer

The livestock industry has used ET for decades. Briefly, ET is a method of assisted reproduction. It involves removing oocytes from a donor female and placing fertilized embryos into recipient females. Fast genetics has used ET to introduce genetics into its GN farms, since introducing live animals at the GN level results in unwanted health risks. Bringing in genetics via ET allowed Fast Genetics to protect its extremely high health status at the GN level. Having a great deal experience in the use of ET the reproductive physiology and genetics departments have been collaborating to develop a protocol of using ET to capture and re-distribute elite genetics. Early results suggest that ET will be most effective if used within GN farms. However, if elite status animals are produced elsewhere in the production chain, capturing them and introducing donor animal genetics via ET is also a viable option.

Sexed sorted sperm

Sexed semen has been primarily used in the dairy industry as an important component to many breeding programs by allowing for efficient production of replacement heifers from genetically superior animals. Sexed sorted bovine sperm is based on a frozen/thaw model that uses 1/5 of the cells as conventional semen, with similar calving rates. Comparatively, data from sexed sorted trials in pigs has primarily used fresh semen and several hundred millions to billions of cells in an artificial insemination (AI) dose, which is a limiting factor due to the time required to flow cytometrically sort. Research and development in two areas have allowed Fast Genetics to continue to overcome this dose issue:

First: Sorting speed

Through the rapid application of current and proprietary research, the speed at which sperm can be sorted has dramatically increased since 1990, the sort rates have improved by nearly 5000%, which is continuing to increase on a logarithmic scale (Figure 2).

Second: Number of semen cells in a dose

There is an overall trend within the swine industry to reduce the number of cells in a breeding dose. Fast Genetics is at the forefront of this trend and is continuing to reduce the sperm in an AI dose. Through continuing research, Fast Genetics has developed a deep uterine (DU) catheter that deposits cells within close proximity to the site of fertilization. This catheter has been used within the Fast Genetics breeding herds using a single AI dose of less than 500 million cells and produced acceptable fertility levels (Table 1). Furthermore, Fast Genetics is also developing a catheter that has navigational capabilities that can deposit cells in the left and right uterine horns. The navigational catheter will theoretically use a fraction of the cells as compared to a conventional AI dose. Prototypes are still being tested, but initial data looks promising.

The use of all the technologies mentioned in this manuscript (plus others not mentioned) all come together to aid the profit potential of Fast Genetics' customers. Indeed, many of the technologies do this through increasing the rate of genetic gain, as shown in Table 2.

Conclusion

Fast Genetics has an excellent track record of utilizing current and novel technologies to advance the rate of genetic gain. This has enabled the company to grow rapidly and sustainably since 1982.

In recent years, with the use of SNP chips, genomics methodology, FIRE feeders, etc. our customers have benefited greatly from Fast Genetics being an early adopter. Our commitment to the future reflects this, as more research is now being done than ever before.

Figure 1: The number of non-elite boars in 4 studs. When the boar stud monitor was utilized in these studs the number of projected non-elite boars dropped. Thus the percentage of elite boars increased in stud. Stud 2-4 all used the boar stud monitor from December 2015; stud 1 used it from September 2016.

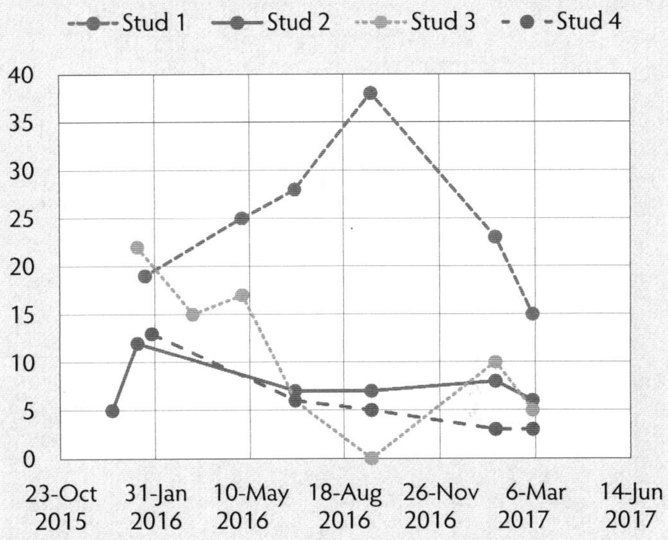

Figure 2: Rapid increase in the speed rate of sorting sexed semen. As new and novel technologies are utilized the rate of sexed semen sorting increases.

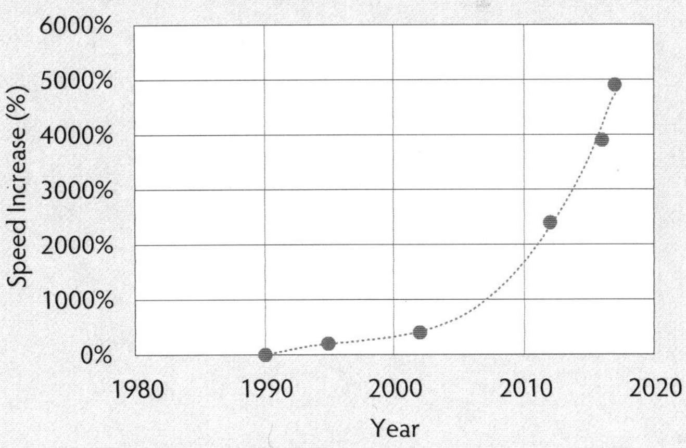

Table 1: Deep uterine versus artificial insemination

Treatment	Farrowing rate	Total born
Control	86.67 ± 0.60	16.51 ± 0.766
DUI	82.4 ± 9.08	15.36 ± 1.28
P value	0.69	0.52

Control: Non synchronized females with two inseminations/estrus (farm protocol; 3B cells).
DUI (deep uterine insemination): Synchronized single AI (600 million cells).

Table 2: An increase in genetic gain is shown. The contribution of three technologies to the genetic gain is shown

Technology	Terminal lines	Maternal lines
Extra FIRE feeders and optimization of data	29%	19%
Genotyping selection candidates	15%	30%
Sexed Semen	Adjustable	16%
Total	44%	65%

Lactobacillus acidophilus fermentation product as an alternative to therapeutic zinc oxide in weaned pig diets on performance and response to *E coli* challenge

X. Guan[2]; B. E. Bass[1]; P. J. van der Aar[2]; C. Piñeiro[3]; J. W. Frank[4]; J. Morales[3]; F. Molist[2]
[1]PMI Nutritional Additives-Land O'Lakes, Arden Hills, Minnesota; [2]Schothorst Feed Research, Lelystad, The Netherlands; [3]PigCHAMP Pro Europa, Segovia, Spain; [4]International Ingredient Corporation, Fenton, Missouri

Introduction

An improvement in health and reduction in the frequency of injectable treatments and mortality was demonstrated in previous research in pigs with chronic K88+ (F4) *E coli*[1] when feeding *Lactobacillus acidophilus* fermentation product (LAFP; SynGenX®, Diamond V, Cedar Rapids, IA, USA) at 1kg/MT in combination with carbadox and zinc oxide. Additionally, the combination of 1 kg/MT LAFP and oxytetracyline, or 2 kg/MT LAFP alone, reduced *E coli* in weaned pigs.[2] Zinc oxide, fed at therapeutic levels, has been used to alleviate diarrhea associated with weaning stress. Concerns about the environmental impact of high inclusion levels of zinc, as well as the potential of zinc to drive antibiotic resistance in bacteria may limit its use in the future. We hypothesize that LAFP can help reduce the impact of *E coli* on pig health and performance, demonstrating similar benefits to therapeutic levels of zinc oxide.

Study 1

Material and methods

In the first of two studies, 288 piglets were weaned at 28 days (d) (7.41 kg body weight [BW]) and randomly assigned to one of four dietary treatments (12 pens per treatment; 3 boars:3 gilts per pen): Negative Control devoid of antibiotics (NC); Positive Control – NC + 3,000 ppm ZnO (PC); NC + 1 kg/MT LAFP in pre-starter and 0.5 kg/MT LAFP in starter (LA1); and NC + 2 kg/MT LAFP in pre-starter and 1 kg/MT LAFP in starter (LA2). Diets were provided for 35 d (Pre-starter from weaning to 12 kg BW; Starter from 12 kg BW to end of study). Pigs and feeders were weighed on d 0, 7, 14, and 35 to determine BW, average daily gain (ADG), average daily feed intake (ADFI), and feed conversion ratio (FCR).

Results

In the first week post-weaning, a trend was observed in ADG and FCR in which PC had higher ADG and lower FCR than NC, with both LA1 and LA2 being intermediate. Significant improvements in ADG ($P = 0.024$) and FCR ($P = 0.013$) were observed using contrast comparisons between NC and the combination of LA1 and LA2. In the starter period (d 14 to d 34) and in the overall nursery period PC had higher ADG and final BW ($P < 0.01$) than NC and LA1, with LA2 being intermediate. Feed-to-gain was higher in NC compared to LA1 and LA2 combined (1.51 *vs.* 1.44 g/g; $P = 0.03$) in starter period and in the overall nursery period (1.44 *vs.* 1.36 g/g; $P = 0.022$). There were no differences in performance between PC and LA2 throughout the study.

Study 2

Materials and methods

In a second study, 48 piglets, 24 boars:24 gilts [Topigs Tempo × (Great Yorkshire × Finnish Landrace)], weaned at 26 d of age (7.1 kg BW) were allocated to 4 treatments: Negative Control devoid of antibiotics (NC2); Positive Control – NC2 + 2,500 ppm ZnO (PC2); NC2 + 1 kg/MT LAFP (SG1); and NC2 + 2 kg/MT LAFP (SG2). Diets were offered to the piglets for 22 d. On d 10, piglets were orally challenged with 5 ml of 8.9 Log of nalidixine resistant *E coli.* On days 11 to 15, 18, 20 and 22, fecal scores were measured and samples were collected to quantify nalidixine resistant *E coli.*

Results

On d 14 ($P = 0.064$), 15 ($P = 0.064$) and 18 ($P = 0.041$), fecal *E coli* excretion was lowest in boars fed SG2, with no differences between gilts and boars fed other diets. On d 20, pigs fed SG2 showed lower fecal excretion of *E coli* than pigs fed with NC2, PC2 and SG1 (3.4 vs. 4.3, 4.3 and 4.4 Log CFU/g feces, respectively; $P = 0.031$). Overall fecal score was lowest for SG2 ($P < 0.05$). Pigs provided PC2 tended to have higher ADG ($P = 0.053$) compared with NC2 and SG1 (466 *vs.* 359 and 385 g/d, respectively), with SG2 intermediate (406 g/d).

Discussion

Inclusion of ZnO at therapeutic levels improved performance compared to negative control fed animals. Feeding LAFP at 2 kg/MT resulted in performance similar to therapeutic ZnO, plus reduced fecal score and shedding of *E coli* following challenge in diets that do not contain antibiotics or other additives. In conclusion, LAFP at 2 kg/MT may be a potential alternative for therapeutic ZnO in piglet diets post-weaning.

References

1. Probst Miller S, Ramirez A, Bass B, Frank J. Quantitative assessment of prevalence and severity to understand impact of *Lactobacillus acidophilus* fermentation product (SynGenX®) in herd with endemic K88 and Salmonella post-wean. AASV. 2016:184.

2. An V, Truong T, Tran T, Chayomchai A. 7th Asian Pig Veterinary Society Congress. 2015:120.

Use of statistical quality control and metrics in PRRS control programs

Jose Angulo, DVM
Zoetis, Parsippany, New Jersey

Introduction

Porcine reproductive and respiratory syndrome (PRRS) is one of the most important swine diseases worldwide. PRRS control may be a difficult task and it requires a systematic approach to achieve mid-long term success. This approach focuses in three main areas: whole herd immunity; bio-management practices and bio-exclusion practices. Current challenges in controlling PRRS are: the relatively high frequency of reproductive outbreaks in farms at different PRRS status, the difficulty to define and predict success outcomes in different PRRS control interventions and a clear need for better metrics that can be applied in the field as a routine manner to evaluate and make decisions around strategies for controlling PRRS. Due to all these challenges PRRS fits a model for applying Statistical Quality Control Improvement philosophy. We define quality as inversely proportional to variability, this means that if variability in a process decreases, the quality of that process increases and by consequence, decisions made based on that process will get improved. Quality improvement is the reduction of variability in processes and quality control improvement involve the set of activities used to ensure that the processes meet requirements and are improved in a continuous basis having Statistical Process Control (SPC) as the major tool.[1] SPC is a framework of ideas and a collection of problem-solving tools useful in achieving process stability through the reduction of variability.[2]

The objective of this review is to propose the use of quality control tools like Statistical Process Control (SPC) to standardize outcomes from interventions to control PRRS in production systems and utilize these metrics as a common language as part of the making-decision process.

Material and methods

A compilation of outcomes from four field PRRS control projects implemented in the united states from 2015 to 2017 in different production systems where Fostera® PRRS was used as the primary immunological tool was summarized. These projects implemented relevant changes in PRRS whole herd immunity strategies, bio-management and bio-exclusion practices. All projects shared a standard structure with an approach of continuous improvement and having as main goals: to develop a PRRS control strategy to improve current PRRS status, improve PRRS stability and mitigate clinical impact in the production system as well as to establish metrics for measuring success in applied PRRS control interventions. During these PRRS continuous improvement projects we developed a PRRS Metrics Measurement System (PRRS MMS) to standardize outcomes and objectively evaluate different interventions within the production system. In addition, a clear and practical process for successfully implement Statistical Process Control (SPC) monitoring was established and tested in different production systems using Shewhart control charts to identify rules violations for detecting nonrandom patterns and before and after capability analysis comparison to measure improvements in reduction of variability and closeness to targets of specific key performance indicators (KPIs) driven by changes in primary interventions. PRRS metrics using PRRS MMS format was used for delivering and visualization of projects' outcomes. Metrics like Time to Baseline Production (TTBP) previously described by Linhares et al.[3] utilizing Exponential Weighted Moving Average (EWMA) control charts, Reproductive Outbreak Window (ROW) rate defined as the number of weeks out of control limits after an outbreak utilizing Shewhart control charts divided by 52 weeks for an annualized outbreak severity and duration measure. A sample of these PRRS continuous improvement projects was selected and abstracted with the intention to show and demonstrate the practical implementation and use of the methodology and metrics proposed in this paper. Minitab statistical software (version 17.1.3) was used for all the Statistical Quality Control tools. Reproductive Outbreak Window (ROW) rate was used as an improvement PRRS metric and it was defined as a mathematical function that assigns a measure to the gap between the output's current state and the goal's desired state.[4] For ROW rate we designed the following formula: ROW = (#weeks out of control * Max KPI) / 52. Where #weeks out of control are the number of weeks identified as Shewhart rule#1 rule violation, Max KPI is the maximum value of the KPI that was reached during the outbreak divided by 52 weeks to annualize the severity and duration of the outbreak event.

Results and discussion

PRRS continuous improvement projects were defined as an ongoing effort to improve current PRRS status seeking for incremental improvements over time. These improvements were identified and agreed within the project's team utilizing a systematic approach. A project charter template[5] was customized for filling specific information for each project in the following sections: Background including improvement opportunities, production system interventions, project scope, communication plan, roles of team members, milestones and deliverables, resources required, measurement of success and lessons learned. The evaluation of changes in PRRS interventions was focused in the following general goals: Improvement of quality of specific Key Performance Indicators (KPIs), minimize the severity and duration of recurrent PRRS outbreaks and improve consistency in PRRSV stability in breeding herds and production performance. We generated a process map for implementation of Statistical Process Control monitoring (Figure 1). Once the steps in the process map were completed an SPC monitoring dashboard was built and kept it updated based on the monitoring frequency previously agreed by the project's team. Figure 2 is showing an example of Shewhart control charts and SPC rules violations identification for Weaned pigs per sow weaned in a 3,200 sow farm and abort rate per 1,000 sows in a 2,100 sow farm before and after implementing these type of projects. Capability analysis was also part of the SPC monitoring program and helped to evaluate KPIs variability against targets before and after project implementation. Figure 3 shows a capability analysis summary report in liveborn before and after implementation of a PRRS control improvement project in 14,000 sow production system, based on this analysis, it was a significant improvement in both, mean (P value < 0.001) and variation (P value 0.002) in liveborn KPI, this improvement meant a reduction of 66% of out of specifications for this production system, this means that before implementing the project, 21.5% liveborn KPI was less than 11.5 compared to the after period which was only 7.23% in a weekly basis. A PRRS metrics measurement system (PRRS MMS) was designed to standardized outcomes from specific PRRS control interventions and helped to generate and compared metrics within the production system. We validated and established the following standard procedure for creating Exponential Weighted Moving Average (EWMA) control charts in Minitab to report Time to Baseline Production (TTBP): the step-process was a follow a) Total weaned pigs is the KPI to build TTBP metric, b) Identify the week of the outbreak and get at least 20-24 weeks prior to the outbreak as a baseline data, c) Generate an EWMA chart with this data only (baseline without the outbreak period). From EWMA chart from baseline, look

Figure 1: Process Map for Statistical Process Control (SPC) monitoring implementation

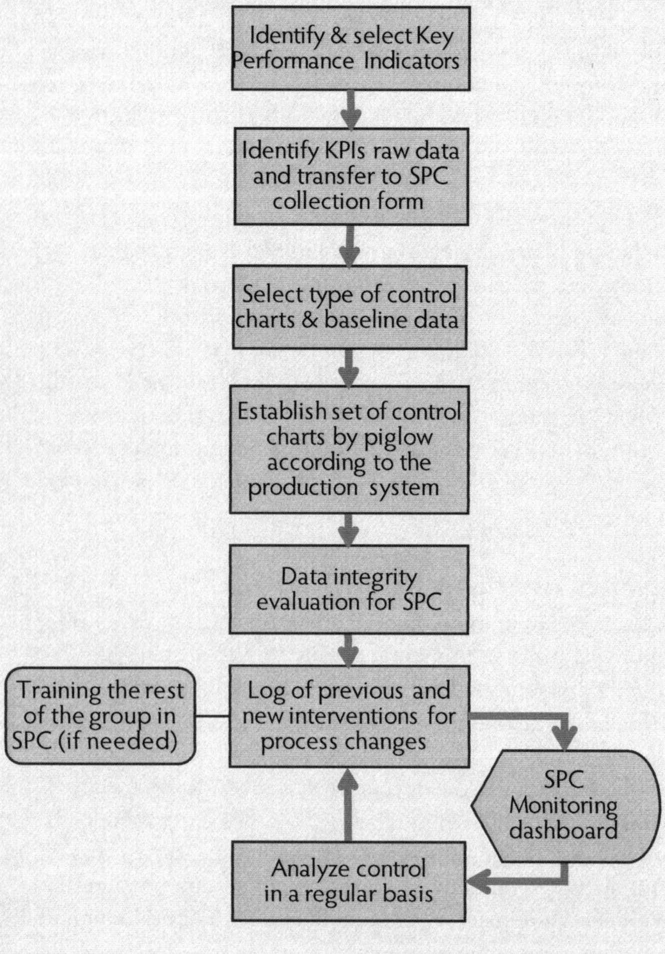

Figure 2: Individual SPC charts for pigs weaned/sow weaned and abort rate/1,000 sows before and after PRRS improvement project implementation in two different sow farms. (MINITAB® 17.1.3)

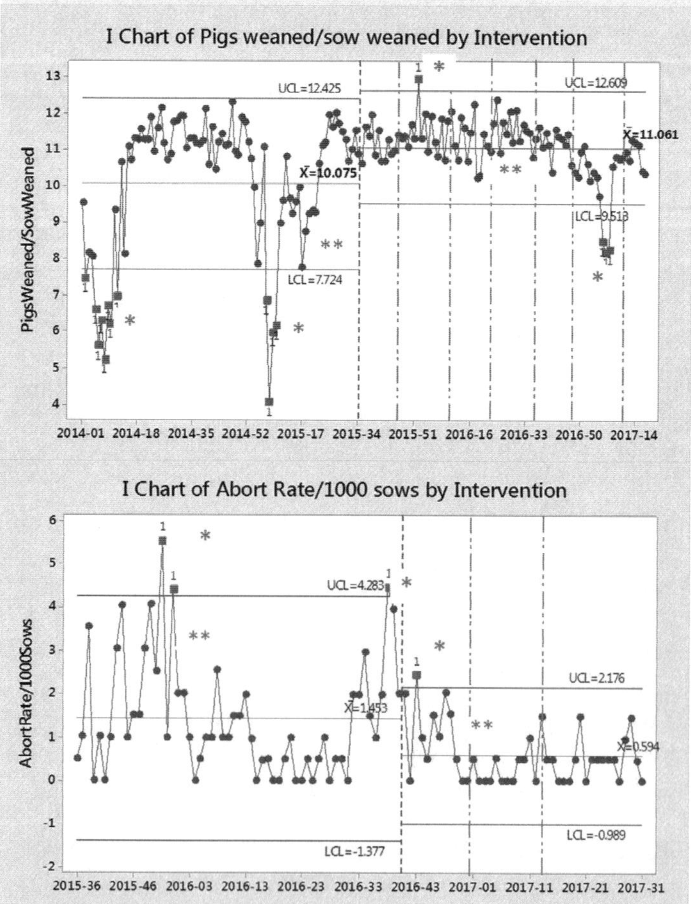

* Rule 1, (point beyond control limits),
** Rule 7 (Mixture or systematic variation)[5]
Mass vaccination with Fostera® PRRS

Figure 3: Before and after capability comparison analysis for Liveborn (MINITAB® 17.1.3)

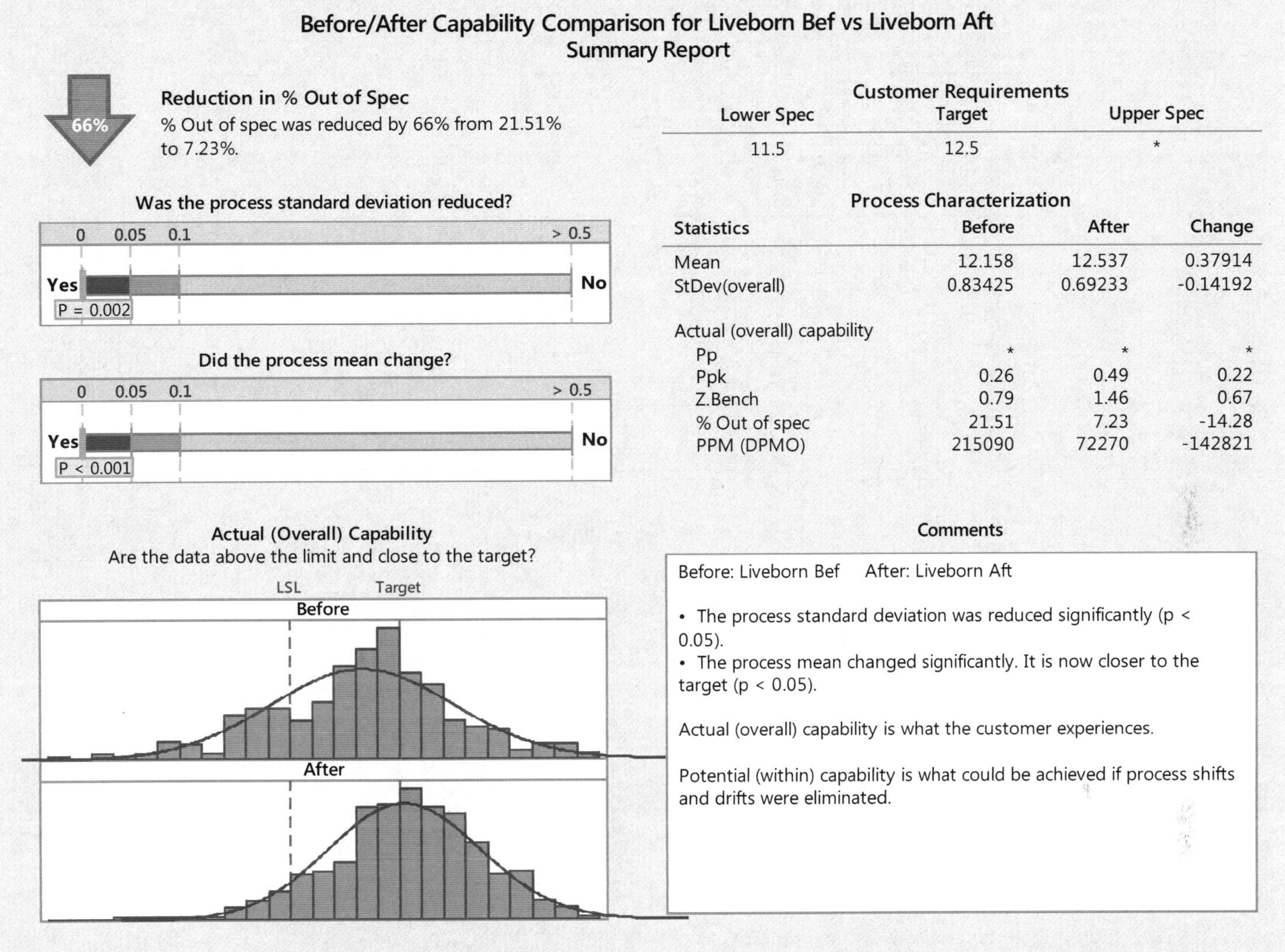

and clean the data from SPC rules violations (signals) to have this EWMA chart in control. d) Obtain the standard deviation and average from in-control baseline data e) add the outbreak data to the baseline and generate the EWMA specifying the standard deviation and average from baseline data only in this way the control limits are based in the baseline data. e) Calculate TTBP counting the weeks out of control limits. Figure 4 shows the EWMA control chart built based on this process. TTBP was 18 weeks in a 2,800 sow farm that faced a new introduction of PRRSV RFLP 1-7-4 with previous whole herd immunity. In addition to the monitoring metrics, we also generated PRRS metrics focused to evaluate severity and duration of new PRRS virus introduction or recirculation of resident virus called reproductive outbreak window rate (ROW), this metric was generated for abort rate per 1,000 sows and Pre weaning mortality KPIs monitored in Shewhart control charts, an example is showed in figure 5 where we calculated the ROW rate in abort rate per 1,000 sows and pre-weaning mortality (PWM) for two breeding herds that had PRRS breaks before and after the implementation of PRRS continuous improvement project, for sow farms A and B, ROW rates were 2.95 and 1.51 for abort rates and 3.39 and 3.13 for PWM respectively during the before periods, these metrics were reduced in both farms during the after period to 0.42 and 0.84 and 1.56 and 1.15. Building this metric helped to quantify the severity and duration of a new PRRS outbreak (Confirmed by diagnostics + SPC rules violation) and its standardization also helped to compare this metric within farms from a production system.

Conclusions

The ultimate goal of this paper was to share the practical application of Statistical quality control tools to generate metrics in PRRS control improvement projects to help swine practitioners to evaluate interventions and generate relevant knowledge within productions systems related to PRRS control interventions and measure significant progress in controlling this important disease. SPC tools are suggested to monitor planned interventions with potential immediate significant economic benefits in production

Figure 4: Time to Baseline Production (TTBP) by EWMA control chart after new WT-PRRSV introduction (MINI AB 17.1.3)

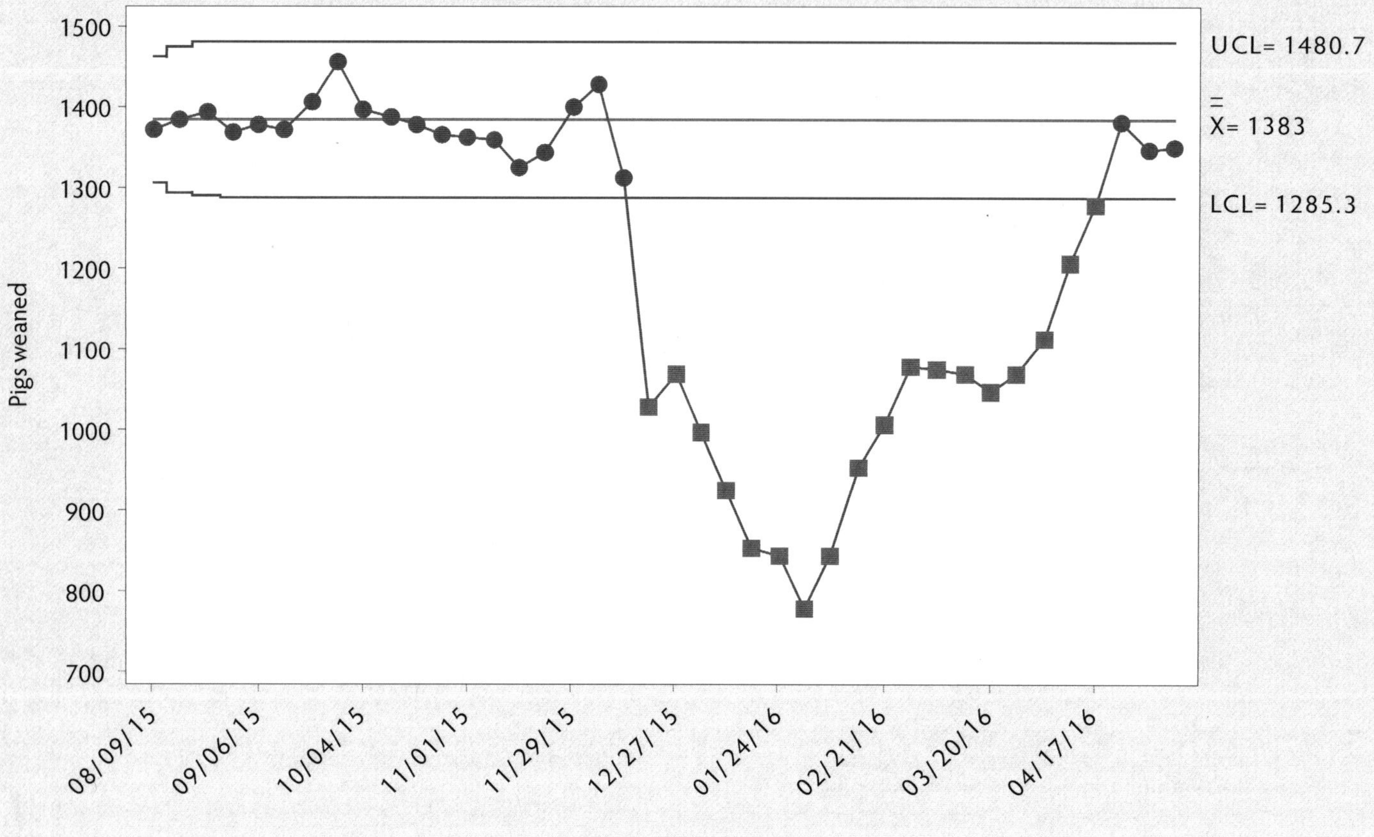

systems. The main goal is to enhance the decision making process towards to more proactively framework. Implementing PRRS control improvement projects may help measure interventions vs baseline data utilizing SPC and metrics in three major areas: current status improvement, severity and duration of new outbreaks and recovery and consistency after outbreaks.

References

1. Montgomery Douglas C. Introduction to Statistical Quality Control. Seventh Edition. Wiley. 2013.

2. De Vries A. Reneau J.K. Application of statistical process control charts to monitor changes in animal production systems. J Anim Sci. 2010. 88: E11-E24.

3. Linhares C.D, Cano J.P, Torremorrel M., Morrison R.B. Comparison of time to PRRSv-stability and produciton losses between two exposure programs to control PRRSv in sow herds. Preventive Veterinary Medicine 116, 111-119. 2014.

4. T. Adams. Metrics for Management. NASA Kennedy Space Center. World conference for Quality Improvement Proceedings. 2015.

5. Lean Six Sigma Black Belt Handbook 2nd ed. Villanova University. 2008.

Figure 5: Reproductive Outbreak Window rates for Abort rate / 1,000 sows and Pre-weaning mortality in two different sow farms before and after implementation of PRRS continuous improvement project. (Dotted line = Mass vaccination with Fostera® PRRS)

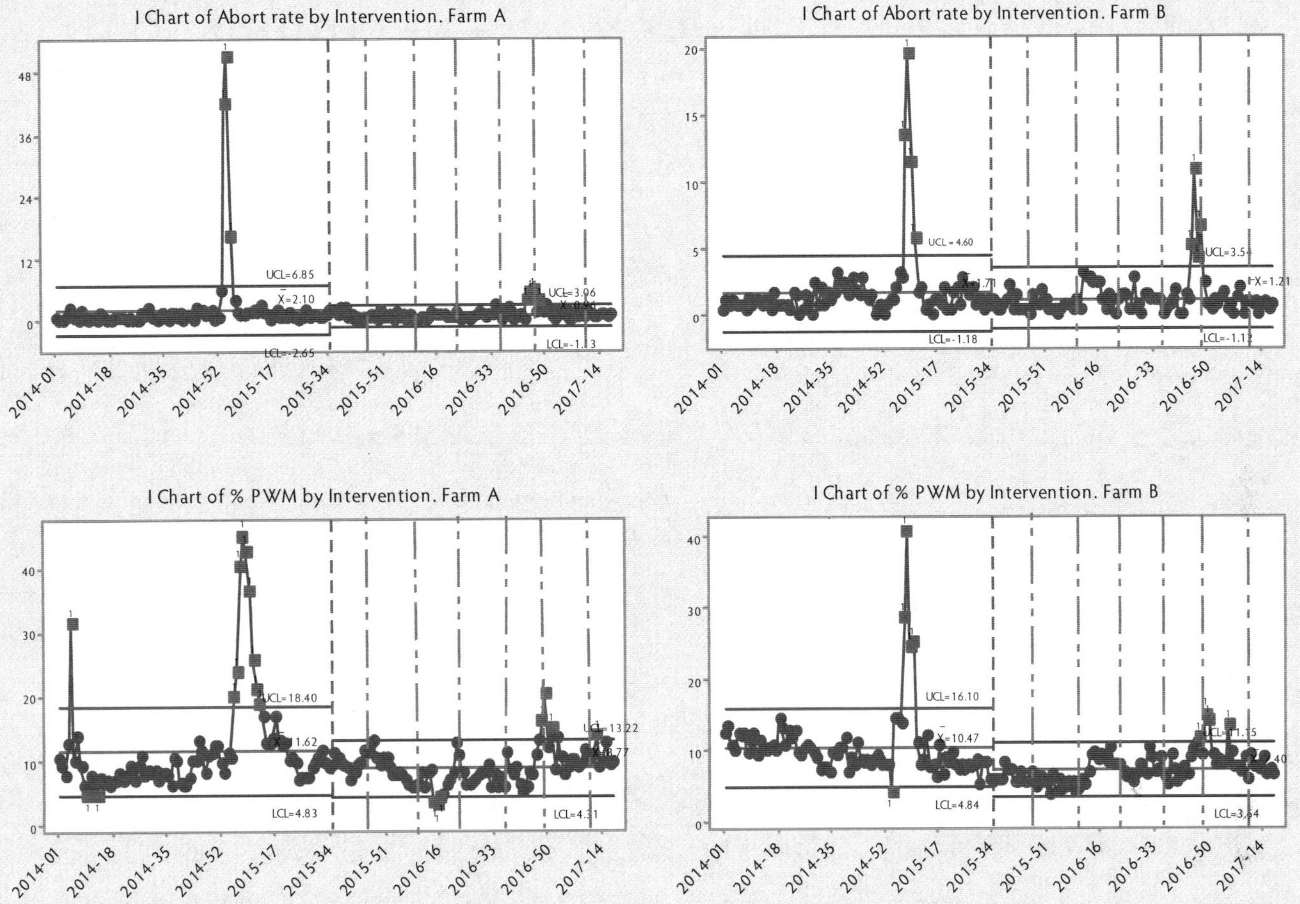

Reproductive Outbreak Window FARM A

Abort Rate / 1,000sows (Before)	
Rule 1 violation #wks	3
KPI Max	51.2
ROW Before	***2.95***
Abort Rate / 1,000sows (After)	
Rule 1 violation #wks	4
KPI Max	5.4
ROW after	***0.42***
PWM %	
Rule 1 violation #wks	9
KPI Max	19.6
ROW Before	***3.39***
Abort Rate / 1,000sows (After)	
Rule 1 violation #wks	4
KPI Max	20.3
ROW after	***1.56***

Reproductive Outbreak Window FARM B

Abort Rate / 1,000sows (Before)	
Rule 1 violation #wks	4
KPI Max	19.65
ROW Before	***1.51***
Abort Rate / 1,000sows (After)	
Rule 1 violation #wks	4
KPI Max	10.88
ROW after	***0.84***
PWM %	
Rule 1 violation #wks	4
KPI Max	40.7
ROW Before	***3.13***
Abort Rate / 1,000sows (After)	
Rule 1 violation #wks	4
KPI Max	15.0
ROW after	***1.15***

Evaluation of a novel PCV + *M hyopneumoniae* vaccine in swine experimentally challenged with PCV2d and *M hyopneumoniae* under commercial conditions

L. Galina, DVM, PhD; B. Cowles, DVM, MS, MBA; D. Amodie, MS; R. Swalla, DVM; M. Mellencamp, PhD
Zoetis, Parsippany, New Jersey

Introduction

Porcine circovirus type 2 (PCV2) is a small, non-enveloped single stranded virus in the *Circoviridae* family.[1] In growing pigs, PCV2 infection is associated a wide range of clinical presentations that are collectively known as PCV2-associated disease (PCVAD).[2] Recent studies have shown a major PCV2 genotype shift in the US herd from PCV2b to PCV2d.[3] Vaccination is essential for herd protection and has been shown to reduce PCV2 infection, viremia and lesions and improve average daily gain compared with non-vaccinated animals. Most commercial vaccines were developed several years ago and are based on PCV2a, which was the predominant virus field strain at that time. The objective of this study was to investigate the ability of a novel, experimental vaccine and current commercial vaccines to protect against an experimental dual challenge of *M hyopneumoniae* and PCV2d.

Materials and methods

Healthy pigs (n = 880), with no known history of vaccination to or infection with *M hyopneumoniae* or PCV2, were selected from a high health status sow herd that was PRRS negative and *M hyopneumoniae* and PCV stable. On study day 0 (3 d. of age), pigs were tagged, weighed, and randomly allocated to 1 of 8 treatment groups in a generalized randomized block design, with blocks determined by litter (n = 110 per treatment). From the 880 pigs, 192 pigs (24 per treatment group) were randomly selected to be necropsied on study day 77.

Animals were housed in a single room at a Midwestern commercial facility. Corn-soy rations with 3% added fat were formulated for maximum lean-gain accretion (six stages) and exceeded NRC requirements. Feed contained ASP 250 for 1 wk. after arrival, followed by Albac. From ~2 wk. prior to *M hyopneumoniae* challenge, pigs received non-medicated diet. Water and feed were provided *ad libitum*.

Pigs were vaccinated against PCV and *M hyopneumoniae* (Table 1). On study day 49, all pigs were challenged intratracheally with 10 mL of a 1:100 dilution of the *M hyopneumoniae* lung homogenate (10^4 CCU). On study day 58, 1 wk. after *M hyopneumoniae* challenge, each pig was challenged intranasally with 2 mL of PCV2d and 1 mL intramuscularly at a dose of $6.0 \pm 0.5 \log_{10}$ $TCID_{50}$ per ml (Figure 1). Blood samples for PCV2 PCR were taken on study days: 0, 15, 57, 64, 71, and 77.

At necropsy (study day 77), lungs were evaluated grossly for lesions attributable to a *M hyopneumoniae*. Percentage of total lung with lesions was calculated using the following formula: $100 \times \{(0.10 \times \text{left cranial}) + (0.10 \times \text{left middle}) + (0.25 \times \text{left caudal}) + (0.10 \times \text{right cranial}) + (0.10 \times \text{right middle}) + (0.25 \times \text{right caudal}) + (0.10 \times \text{accessory})\}$. Liver, heart, lung, tonsils, thymus, spleen, kidney, lymph nodes (superficial inguinal, mediastinal, tracheobronchial, and mesenteric), ileum, and colon were examined histopathologically at ISU Diagnostic laboratory. Lymph nodes and lungs were scored for microscopic lesions from 0-3 (absent, mild, moderate, severe). Other organs were scored 0 (absent) or 1 (present). Duplicate sections of tracheobronchial, mesenteric, inguinal lymph nodes and tonsil were examined for lymphoid depletion and histiocytic replacement. A pig had lymphoid depletion or histiocytic replacement if one or more tissues scored "+". These tissues were tested for PCV antigen by IHC. The results were recorded as 0 (no staining) and 1-3 (different levels of staining). A score 0 were considered as PCV IHC negative and a score of 1 or higher were considered as PCV IHC positive. A pig was considered IHC positive if one or more tissues were IHC positive. A case of PCVAD was defined as having all of the following: (a) clinical signs of wasting/weight loss/ill thrift/failure to thrive, with or without dyspnea or icterus; (b) characteristic histologic lesions; depletion of lymphocytes from lymphoid tissues; and (c) PCV2 infection, as detected by PCR and IHC.

The remaining pigs in each group (n = 86) were weighed periodically until study day 157 (first cut). Final weights on that day were used to determine culls (< 190 lb.), lights (190-230 lb.) and full value pigs (> 230 lb.). Net revenue was defined as: (revenue of primary market pigs + revenue of culls + revenue of lights) – mortality loss. Net revenue of full value pigs was the (final live weight of pigs to primary market/100 × current price/cwt.). Full value pigs received 100% revenue. Light pigs received 90% revenue of full value pigs. Culls received 60% revenue of full value pigs. Mortality loss was calculated as follows: \$35 per death + (# days on feed × est. feed intake × \$0.11 per lb. of feed). Feed cost did not include the cost of feed medications. Cost of vaccines and injectable medications were not included in the financial analysis.

Using SAS Proc Mixed Procedure (SAS 9.3, Cary, NC), weight/ADG were analyzed with a model that considered fixed effects of

Table 1: Treatment groups and vaccination protocols

Treatment	Vaccine	Manufacturer	Number of doses	Vaccination protocol Pig age 3 d	3 week	5/6 week
T01	Saline	NA	2		1 ml	1 ml
T02	Fostera® PCV MH	Zoetis	2		1 ml	1 ml (5 wk)
T03	Fostera® PCV MH	Zoetis	1		2 ml	
T04	Circumvent® PCV-M G2	Merck Animal Health	2	1 ml	1 ml (3 wk later)	
T05	CircoFLEX® + MycoFLEX®	Boehringer Ingelheim Vetmedica	1		2 ml	
T06	Experimental PCV + *M hyopneumoniae*	Zoetis	1		2 ml	
T07	Experimental PCV + *M hyopneumoniae*	Zoetis	2	1 ml	1 ml	
T08	Experimental PCV + *M hyopneumoniae*	Zoetis	2		1 ml	1 ml (6 wk)

treatment, time and interaction of treatment-by-time and random effects of pen, block-within-pen, animal-within-block, animal (pen*block*treatment) and residual error. Initial weight was included as a covariate. Time was the repeated factor. The covariance structure in the repeated measures analysis was investigated using five structural assumptions, namely, compound symmetry (CS), heterogeneous compound symmetry (CHS), first order autoregressive [AR(1)], heterogeneous first order autoregressive [ARH(1)] and unstructured (UN). The assumption giving the minimum value of the Akaike's Information Criterion (AICC) was selected for the final analysis. The animal (pen*block*treatment) was used only as the subject for CS, UN and CSH. For AR(1) and ARH(1) this term was in the random statement and the subject. Treatment was tested at the 5% level of significance (two-sided test). Treatment Least Square Means (LSMeans) was calculated for each group. Comparison of LSMeans was performed by the two-sided Student's t-test at the 5% level of significance.

Microscopic and macroscopic lesions were binominal variables (1 = yes, 0 = no) and analyzed by a Generalized Linear Mixed Model (GLMM) approach using the SAS Proc GLIMMIX procedure with the model that included the fixed effect of treatment and the random effects of pen, block (pen) and the residual error. This analysis utilized a binomial error and logit link. Glimmix results were back transformed. If GLMM did not converge, Cochran-Mantel-Haenszel test was used. The 5% level of significance ($P \leq 0.05$) was used to assess statistical differences.

A log transformation was applied to titer values before analysis. The transformed titers were analyzed using a repeated measures mixed model described above. The 5% level of significance ($P \leq 0.05$) was used to assess all statistical differences. Treatment LSMeans were back-transformed (Geometric means).

Figure 1: Data collection scheme

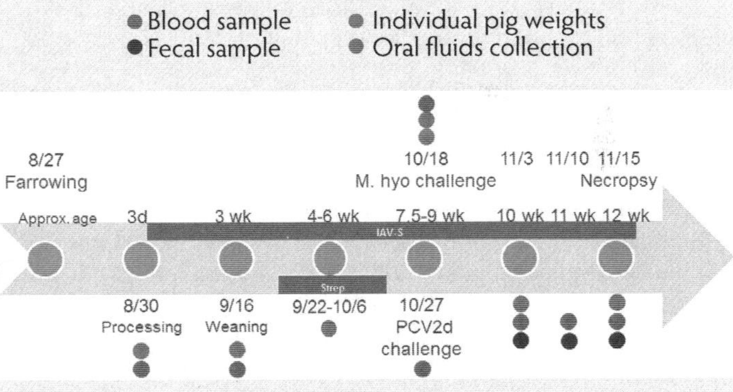

Results and discussion

At weaning, IAV-S and *Streptococcus suis* were isolated in the wean-to-finish commercial barn. Amoxicillin was delivered in water followed by Excede® (Zoetis) as needed for individual pigs. The outbreak was controlled by the time of the experimental challenge.

Results showed that PCV2d viremia was significantly reduced in all vaccine groups compared to the saline control ($P \leq 0.05$) (Figure 2).[4] Post-mortem examination showed T05 CircoFLEX®-MycoFLEX® had significantly higher lung scores than

Figure 2: Effects of PCV2 + *M hyopneumoniae* vaccination on PCV2 viremia, as shown by CT score. Pigs were challenged with PCV2d on study day 58. Saline control (T01) was significantly lower than vaccine groups ($P \leq 0.05$)

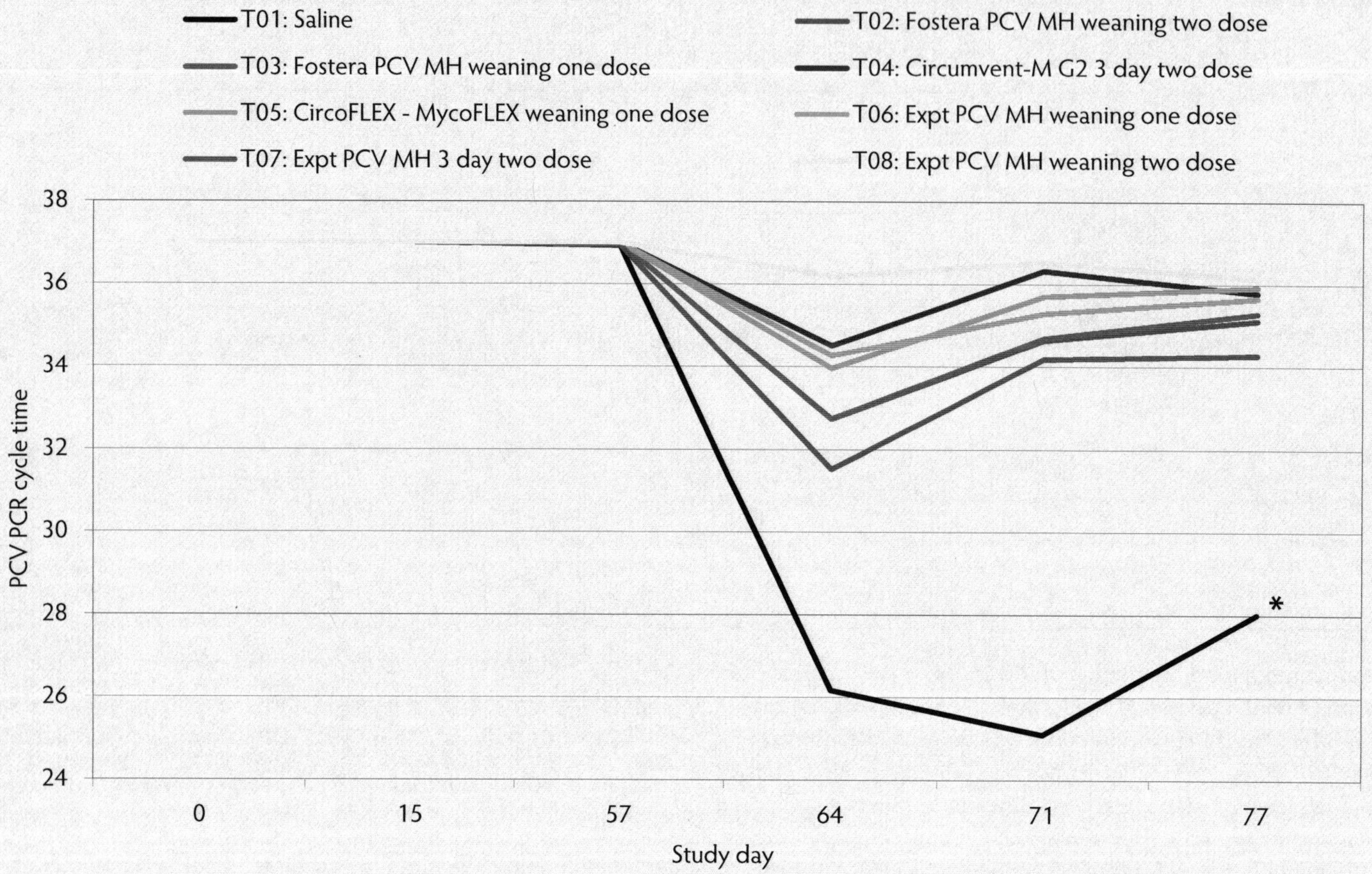

* Saline control (T01) was significantly lower than vaccine treatment groups ($P \leq 0.05$)

the other vaccine groups ($P \leq 0.05$; data not shown). Lymphoid depletion and PCV IHC were significantly lower ($P < 0.05$) in all vaccine groups compared with the saline control (data not shown). Death loss post-challenge in the saline control group was 9.1%, compared with 0.9% to 2.9% in the vaccine groups. Post-mortem evaluation showed that 4/10 saline control animals met the clinical and pathological criteria for PCVAD.

Marketing results showed that the saline group delivered 70.2% full value pigs, while the vaccine groups delivered 79.3% to 91.5% full value pigs. Net revenue modeling showed that vaccines delivered 19.2% to 27.2% greater revenue than saline control (Table 2). Revenue per head placed was $105.09 for the saline control. Vaccinates ranged from $125.25 (Circumvent-M G2) to $133.71 (Fostera PCV MH).

References

1. Tischer I et al. A very small porcine virus with circular single-stranded DNA. *Nature*. 1982;7:64.

2. Opriessnig T et al. Porcine circovirus type 2 associated disease: update on current terminology, clinical manifestations, pathogenesis, diagnosis, and intervention strategies. *J Vet Diagn Invest*. 2007;19:591.

3. Xiao CT et al. Global molecular genetic analysis of porcine circovirus type 2 (PCV2) sequences confirms the presence of four main PCV2 genotypes and reveals a rapid increase of PCV2d. *J Gen Virol*. 2015;96:1830.

4. Zoetis internal study OR 16PRGBIO-01-01.

Table 2: Effects of different PCV2 + *M hyopneumoniae* vaccination protocols on net revenue and revenue per head placed

Treatment	Vaccine	Doses	Net revenue	Revenue per head placed
T01	Saline	2	$9,037.36	$105.09
T02	Fostera® PCV MH	2	$11,499.46	$133.71
T03	Fostera® PCV MH	1	$11,274.15	$131.10
T04	Circumvent® PCV-M G2	2	$10,771.30	$125.25
T05	CircoFLEX® + MycoFLEX®	1	$10,978.38	$127.66
T06	Experimental PCV + *M hyopneumoniae*	1	$10,822.00	$125.84
T07	Experimental PCV + *M hyopneumoniae*	2	$10,849.33	$126.16
T08	Experimental PCV + *M hyopneumoniae*	2	$10,907.48	$126.83

Safety comparative study of two reproductive vaccines administered during lactation

I. Rodriguez-Ballara; G. Ibanez
HIPRA

Introduction

Vaccination with Parvovirus, Erysipelas, and Leptospira is common in most pig farms around the world to prevent infectious reproductive problems.[1,2] The recommended vaccination plan for Parvovirus, Erysipelas and Leptospira is mainly during lactation, thus preventing a future infection during pregnancy. The safety of reproductive vaccines has always been a concern because they are administered at the time of high milk production by the sow, around 2 to 3 weeks after farrowing[3] and, therefore, any safety issues with the administration of reproductive vaccines can affect milk production and piglet perfomance until weaning.[4] The safety difference between commercial reproductive vaccines has already been previously reported.[5] The objective of this study is to compare how the safety difference between two reproductive vaccines can affect the production performance of sows and piglets during lactation.

Key words: lactation, vaccine safety, production performance

Materials and methods

This study was conducted in a herd of 580 sows, working in a four-week batch system, with a total of approximately 110 farrowings per batch. The study was divided in two repetitions, so that the desired number of farrowing sows was obtained. Experimental design: 160 sows were selected in two repetitions (80 sows per repetition) and divided in 2 groups per repetition. Treatments: Group A: vaccinated with a trivalent vaccine (ZOETIS); Group B: vaccinated with ERYSENG® PARVO/LEPTO (HIPRA). Both vaccines were administered by the intramuscular route (2 mL/dose) according to the manufacturer's instructions. The vaccination plan was the same for both groups, one dose 13 days after farrowing. Sows were divided randomly in two groups (A and B) on farrowing day. Piglets from selected sows were identified with numbered ear tags and two different colors that indicated the group they belonged to (744 piglets in Group A and 827 piglets in Group B). Vaccine safety was measured indirectly through the absence/presence of fever (mean rectal temperature [MRT] in °C) at 6, 12 and 24 hours post vaccination (hpv). Sow feed intake was measured two days before vaccination (Day -1, -2), on vaccination day (Day 0) and two days after vaccination (Day +1, +2). Litters were weighed at 7 days of age and at 21 days of age (weaning day). Statistical analysis software (SPSS) was used to check differences between the groups.

Results and discussion

Statistically significant differences were observed in MRT measured at 6 and 12 hpv between both groups. In Group A, 66% and 77% of animals had MRT above the physiological temperature (> 39.5°C) at 6 hpv (39.89 ± 0.21°C vs. 38.72 ± 0.17°C in Group B sows), and at 12 hpv (40.22 ± 0.15°C vs. 39.34 ± 0.21°C in Group B sows), respectively (Mann-Whitney U test, $P < 0.05$) (Figure 1). A statistically significant difference was also observed in sow feed intake between the experimental groups. Thus, Group B consumed, in average, 0.559 kg more than Group A on vaccination day (6.95 ± 0.34 kg/day vs. 6.37 ± 0.36 kg/day in Group A sows); and the day after, Group B consumed 0.392 kg more than Group A (7.29 ± 0.33 kg/day vs. 6.90 ± 0.30 kg/day in Group A sows) (Mann-Whitney U test, $P < 0.05$). At weaning, litters from Group B sows showed an increase of 119 g per piglet in body weight compared with litters from Group A sows (5.63 ± 0.14 g vs. 5.54 ± 0.19 g in Group A litters), although it did not reach statistical significance (t-student test, $P = 0.3$)
(Figure 3).

Conclusion

The vaccine administered to Group A showed safety concerns due to a significant increase in mean rectal temperature, which negatively affected sows feed intake during lactation. Reduced

Figure 1: Mean rectal temperature (°C) on vaccination day, 6h, 12h and 24h after vaccination (± SEM). Different super indexes at each time point show statistically significant differences ($P < 0.05$).

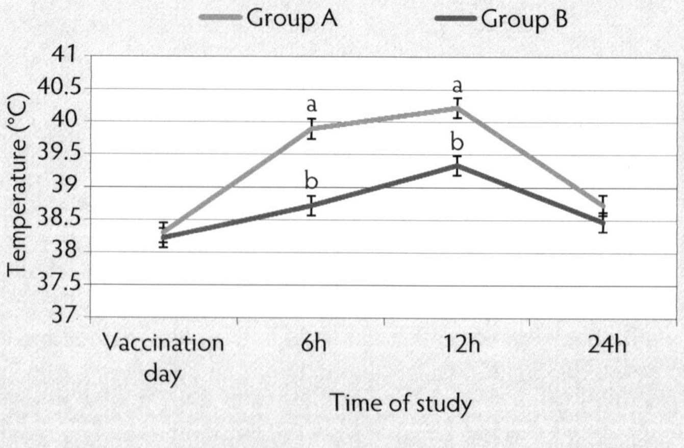

Figure 2: Mean feed consumption per sow one day before vaccination (-1), on vaccination day and one day after vaccination (+1) (± SEM). Different super indexes at each time point show statistically significant differences ($P < 0.05$).

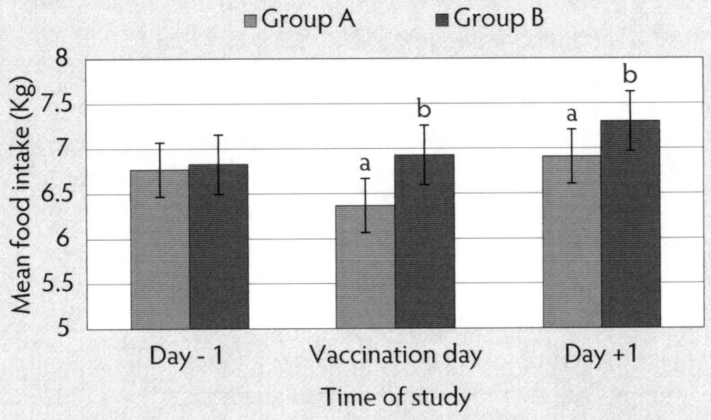

Figure 3: Mean piglet weight at birth and at weaning (± SEM).

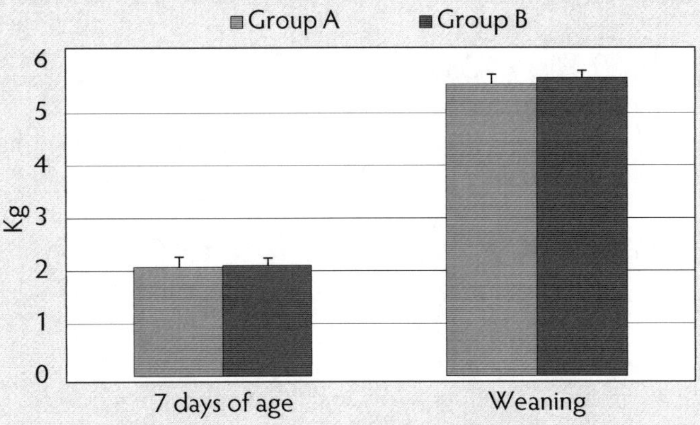

feed intake during lactation causes reduced milk production[6] and this, consequently, could negatively affect piglet performance until weaning. Further studies should determine how the lack of safety of some reproductive vaccines could affect the reproductive performance of the sow during insemination and gestation, since previous studies have already reported the negative impact of food restriction during lactation on the ovarian development of the sow.[7]

References

1. Adler, B. and De la Peña Moctezuma, A. 2010. Vet Microbiol. 140:292-4.
2. Mengeling, WL. et al. 1979. Am J Vet Research. 40(2):204-7.
3. Toner, MS. et al. 1996. Journal of Animal Science. 74:167-72.
4. Papodopoulos, G. 2010. The Veterinary Journal. 184:167-71.
5. Boix, O. et al. 2017. The Asian Pig Veterinary Society Congress. P 258
6. Pluske, JR. et al. 1998. Journal of Animal Science. 76:1165-71.
7. Quesnel, H. et al. 1998. Journal of Animal Science. 76:856-63.

The use of LucraFit® Feed and OutPace® Feed Additive to support the reduction of antibiotics and animal protein in young pig diets

D. McManus[1], DVM; T. Weeden[1], PhD; B. de Rodas[1], PhD; S. Crowder[1], MS; S. Petersen[2], PhD
[1]Purina Animal Nutrition LLC, [2] PMI Nutritional Additives

Introduction

Continuing pressure to reduce the use of antibiotics (AB) in animal feed through legislation and growing public pressure has caused the swine industry to discover and evaluate new feed additives, feeding programs, and management practices to support animal growth and health of pigs raised without the of use antibiotics (Raised without Antibiotics, RWA). Addition of medium chain fatty acids (MCFA), acidifiers, prebiotics, probiotics, and botanicals to swine diets have all been considered as options to support the immune system. Previous research with a commercial blend of MCFA, functional plant components, and acidifiers (OutPace® Feed Additive [OutPace® FA]; PMI Nutritional Additives) has demonstrated effectiveness in optimizing performance of nursery pigs raised without antibiotics.[1] Lucrafit® Feed (LucraFit® F; Purina Animal Nutrition) is designed as a lactose-free feed for nursery pigs, which allows producers to reduce feed costs while maintaining or improving pig performance throughout the nursery period.

A trial was conducted to evaluate two objectives; 1) evaluate the effect of a Lucrafit® F + OutPace® FA feeding program, devoid of highly digestible animal protein sources (lower cost per pound of gain) on performance of nursery pigs, and 2) understand the efficacy of feeding an RWA program incorporating Lucrafit® F and/or OutPace® FA.

Materials and methods

A total of 968 mixed-sex pigs were allotted to one of four dietary treatments (22 pigs per pen with 11 replicates).

Treatment 1 was a nursery feeding program containing highly digestible animal protein source(s) and medication (Phase 1 Denagard®/Phase 2 CTC+Denagard®/Phase 3 Non-Med /Phase 4 CTC+Denagard® per VFD). Treatment 2 was an RWA version of Treatment 1 utilizing OutPace® FA throughout Phase 1-4. Treatment 3 was an animal protein free, medicated nursery program utilizing Lucrafit® F in Phase 1&2 followed by OutPace® FA in Phase 3&4, and Treatment 4 was an animal protein free, RWA nursery program utilizing Lucrafit® F in Phase 1&2 followed by OutPace® FA in Phase 3&4. The dietary treatments are described fully in Table 1.

Table 1: Dietary treatments

Treatment number	Antibiotics	Animal protein sources	Treatment	Feed budget, lbs of feed/pig
1	Phase 1: Denagard® 200 g/ton Phase 2: CTC 400 g/ton /Denagard® 35 g/ton Phase 3: No Meds Phase 4: CTC 400 g/ton - Denagard® 35 g/ton	In Ph 1.	none	Phase 1: 10.0 Phase 2: 10.0 Phase 3: 20.0 Phase 4: 20.0
2	None	In Ph 1.	OutPace®® FA 10 lbs/ton	Phase 1: 10.0 Phase 2: 10.0
			OutPace® FA 5 lbs/ton	Phase 3: 20.0 Phase 4: 20.0
3	Phase 1: Denagard® 200 g/ton Phase 2: CTC 400 g/ton /Denagard® 35 g/ton Phase 3: No Meds Phase 4: CTC 400 g/ton - Denagard® 35 g/ton	No	LucraFit® F	Phase 1: 10.0 Phase 2: 10.0
		No	OutPace® FA 2.5 lbs/ton	Phase 3: 20.0 Phase 4: 20.0
4	None	No	LucraFit® F	Phase 1: 10.0 Phase 2: 10.0
		No	OutPace® FA 5 lbs/ton	Phase 3: 20.0 Phase 4: 20.0

The dietary treatments were designed to be fed in 4 phases during a 56-day period, following a feeding budget of 10 lbs of feed/pig in phase 1 and phase 2, and 20 lbs of feed/pig in phases 3 and 4, respectively. Phase 1 diets were in pellet form (1/8" diameter pellet) and the remaining phase diets were in meal form.

Pigs were housed in a conventional wean-to-finish barn per industry standards for pig space, feeder, and water space. Pigs had free access to feed and water. Production performance parameters including initial and final pig body weight, pen feed consumption, pen weight, and mortality were measured and recorded weekly. Weight gain and Feed:Gain (F:G) ratio was calculated. Data were analyzed using the Fit Model procedure of JMP 13.0 (ANOVA or mixed effect model).

Results and discussion

The trial results for main effect means of dietary treatment are presented in Table 2. At D 56, pigs fed antibiotics in the Control (Trt 1) and LucraFit® F + OutPace® FA (Trt 3) program were heavier than pigs fed a raised without antibiotic program with OutPace® throughout (Trt 2), but similar in weight to pigs fed a raised without antibiotic program with LucraFit® F + OutPace® FA (Trt 4, respectively; $P < 0.05$). For final BW, there is an advantage of feeding LucraFit® F + OutPace® FA versus just OutPace® FA in the nursery period. Overall, pigs fed the Trt 1 or Trt 3 programs with AB had the better ADG to pigs fed Trt 2, with pigs fed Trt 4 having similar ADG as pigs fed either the Trt 1, Trt 3 or Trt 2, (0.933, 0.940, 0.861, 0.895, respectively; $P = 0.0007$). Furthermore, pigs fed AB had greater overall ADFI when compared to pigs fed a RWA program. Pigs fed the Trt 1 had the greater overall ADFI when compared to pigs fed Trt 2, with pigs fed the Trt 3 or Trt 4 programs intermediate (1.419, 1.308, 1.376, 1.354, respectively; $P < 0.0011$). There were no treatment differences for week 1 ADFI, indicating that feeding programs, devoid of highly digestible animal protein sources, did not impact early feed intake of nursery pigs.

Pigs fed the Trt 3 program had better feed efficiency than all other Trts ($P < 0.0001$); pigs fed Trt 3 had a 3.7% improvement over Trt 1. Even in a RWA program, the feeding technologies of LucraFit® F + OutPace® FA allowed for numerically improved feed efficiency compared to Trt 1. There were no differences between treatments for removals.

Due to lower feed costs and improved FCR, Trt 2, Trt 3, and Trt 4 had a lower cost of gain when compared to Trt 1 (Table 3). Pigs fed the Trt 3 program had a greater margin over feed (MOF) to pigs fed Trt 2, but similar to pigs fed Trt 1 or Trt 4 (Table 3, $P = 0.0025$). Furthermore, pigs fed the Trt 4 program had a $1.31/head advantage over the Trt 1 program. In a RWA program comparison, the Trt 4 program had a $2.74/head advantage over the Trt 2 program. Additionally, the LucraFit® F + OutPace® FA when fed with AB protocol had 0.92 lb more gain at the end of the trial, suggesting that days in barn could also be potentially impacted by this feeding program that would result in more savings per head.

Conclusions

There were no treatment differences for week 1, indicating that the feeding programs devoid of highly digestible animal protein

Table 2: Effect of feeding LucraFit® F and OutPace® FA in nursery pig diets on the performance of nursery pigs raised with or without antibiotics during a 56-day period

Variable	Trt 1: AB	Trt 2: ABF + OutPace® FA	Trt 3: AB + LucraFit® F + OutPace® FA	Trt 4: ABF + LucraFit® F + OutPace® FA	*P*-value
Start weight, lbs	12.03	11.86	11.79	11.81	0.87
D 7 Wt, lbs	13.40	13.33	13.33	13.40	0.94
D 56 Wt, lbs	64.58[a]	61.05[b]	65.26[a]	62.98[ab]	0.0014
Day 0-56 Gain, lbs	52.55	49.19	53.47	51.17	
Day 0-7					
ADG, lb/d	0.218	0.206	0.207	0.219	0.92
ADFI, lb/d	0.313	0.318	0.323	0.317	0.78
F:G	1.426	1.818	1.713	1.505	0.41
Day 0-56					
ADG, lb/d	0.933[a]	0.861[b]	0.940[a]	0.895[ab]	0.0007
ADFI, lb/d	1.419[a]	1.308[b]	1.376[ab]	1.354[ab]	0.0011
F:G	1.521[a]	1.518[a]	1.465[b]	1.514[a]	< 0.0001
Removals, %	0.290	0.540	0.397	0.683	0.34

[a,b,c] Means within row with different superscript are different, $P < 0.05$

Table 3: Economics from the experimental period (Summer 2017)

Variable	Trt 1: AB	Trt 2: ABF + OutPace® FA	Trt 3: AB + LucraFit® F + OutPace® FA	Trt 4: ABF + LucraFit® F + OutPace® FA	*P*-value
Total feed cost $/pig	9.534[a]	9.12[b]	8.679[c]	8.502[c]	< 0.0001
Difference		-0.413	-0.855	-1.032	
Cost $ per/lb gain	0.181[a]	0.17[b]	0.177[a]	0.166[b]	< 0.0001
Cost $ of 40 lbs of gain	7.24	6.84	7.08	6.64	
Savings $ /pig/40 lbs gain		-0.40	-0.16	-0.60	
MOF[1], $	$42.13[a]	$38.29[b]	$43.44[a]	$41.03[ab]	0.0025

[a,b,c] Means within row with different superscript are different, $P < 0.05$.

[1] USDA July 7, 2017 report. Weaned Pig value = $34.23 and Feeder Pig (40 lb basis) = $53.06, thus $1.33/lb value. Does not include barn fee, medications, and other associated costs as this cost is presumed to be same per treatment. MOF = Final BW * $1.33 – $34.23.

sources did not impact early feed intake or performance of nursery pigs. Feeding the LucraFit® F followed by OutPace® FA program, devoid of highly digestible animal protein sources, had $1.31/head advantage when compared to the Control program. This was due to 0.68 lb heavier BW at D 56, better overall feed efficiency, and thus less feed cost per head. Throughout the trial, feeding antibiotics had an overall economic advantage, solely driven by a heavier pig on D 56. However, in a raised without antibiotics program, feeding LucraFit® F with OutPace® FA, versus solely OutPace® FA throughout the nursery period, is the best option and equates to a $2.74/hd advantage.

References

1. McManus et al. The effect of feeding MCFA on performance of nursery pigs raised without antibiotics. AASV. 2017, 109-122.
2. LucraFit® and OutPace® are registered trademarks of Land O'Lakes Inc.
3. Denagard® LC is a registered trademark of Eli Lilly and Company.

Management considerations and tactics to maximize sow lactation feed intake

J. C. Pinilla, DVM, MS; S. Casiro, DVM. MS; S. Canavate, DVM; P. Mosqueira, DVM PIC
PIC Global Technical Services, Hendersonville, Tennessee

Introduction

Lactation is a critical phase in the sow's reproductive cycle and has a significant impact on the farm's financial performance. When sows are restricted in feed quantity and/or feed quality, they utilize their body reserves to sustain milk production. When restriction is too severe and/or too prolonged, more vital biological functions can eventually be negatively affected. Lactation weight loss can increase non-productive-days (NPD) by increasing fall outs and/or prolonged wean-to-service intervals (WSI). Other possible outcomes are increased culling and/or mortality rates and a second parity dip in performance.

Considering the impact of weight loss on only reproductive performance, the potential impact should be avoided or, at least, mitigated. Based on published information, when the weaned pig value is $35 and the cost of NPD is $2 per day, the negative impact can exceed $1 per weaned piglet (see Table 1).[1]

In recent years, we have not seen extended WSI as a common consequence of lactation weight loss. However, we have seen improved litter sizes when skip-a-heat is practiced, particularly in gilts weaned from their first lactation. This can be seen as indirect proof that lactation feed intake remains an opportunity in many systems across different geographies.

The goal of this communication is to reinforce the concept of maximumizing lactation feed intake as a means to support farm reproductive efficiency and pig performance from weaning to market. Subsequently, factors that impact lactation feed intake are discussed and general interventions are suggested.

Gilt's body weight at first breeding

Gilt weight and lifetime average daily gain are key indicators of breed eligibility. The optimum body weight at first breeding is in the 300 to 350 lbs range at x days of age which corresponds with an average lifetime daily gain of 1.35 to 1.70 lbs/day. Heavier and/or older gilts will negatively impact cost of production due to higher maintenance nutrient demands..

Heavier gilts tend to have reduced lactation feed intake. This reduction leads to weight loss in lactation, decreased lactation performance resulting in lower litter weaning weights. When body reserves are severely depleted, gilts have a higher likelihood of having a second parity dip or being culled for poor body condition and reproductive failure. Therefore, breeding gilts at the correct weight and stage of development is a crucial first step in optimizing total system economics.

Gestation feed and body condition

The goal of the gestation feeding program is to maintain a body condition consistent with high performance, low removal rate and allow adequate fetal development, whilst remaining focused on the economic viability of the business. Overfeeding in gestation will compromise lactation feed intake and lead to lower colostrum and milk production. Over-conditioned sows will proportionally lose more weight during lactation and have an increased likelihood of low reproductive performance in the subsequent parity. Getting into these body condition swings can ultimately, lead to higher sow removal and mortality rates due to reproductive failure and/or musculoskeletal problems.

Body condition scoring at key milestones needs to be utilized to manage feeding levels and set intervention measures. Routine body condition scoring at weaning as well as at 30, 60 and 90 days of gestation should be the routine check points. As a working target, 85% of the sows should have an ideal body condition by day 30 of gestation and 90% by day 90 of gestation.

Tools such as calipers can provide a more objective estimation of sow body condition. The caliper measures the angularity of a sow's top line from the spinous process to the transverse process, making the condition scoring process more objective. There is increasing evidence that maintaining the herd between 13 and 16 caliper units results in increased litter size at birth and at weaning.

The combination of caliper use, calculation of annualized feed usage and annual sow removal and/or mortality rate would be our suggested ways to provide a complete picture about how a farm is doing on gestation feeding.

Lactation feed strategy

Many farms around the world still utilize a "step-up" lactation feeding program that is defined by a limited feed offer during the first days of lactation. In our opinion, restricted feed in early lactation is a legacy from times when litter sizes were smaller and over feeding could potentially cause udder congestion and hypogalactia. Today, high performing systems show lactation feed usage around 800 lbs per sow per year in a 21 day lactation. To achieve those numbers, full feed from day of farrowing, or even from 2 or 3 days prior to farrowing is necessary.

A significant part of the modern industry has moved away from hand feeding to automated lactation feeding systems. These systems deliver feed to each individual feeder and therefore, feed is always available to the sows. This strategy has proven to be helpful

Table 1: Summary of management factors and potential tactics to improve feed intake in lactation.

Factor	Good management practices
Gilt's body weight at first breeding	• Optimum body weight at service: 300-350 lbs • Target: 90% of gilts bred within 320-340 lbs • ADG from birth to service: 1.35-1.7 lbs/day
Gestation feed usage per sow and body condition	• Asses body condition at weaning • Asses body condition at 30, 60 and 90 days of gestation • 85% of sows in good body condition by day 30 of gestation • 90% of sows in good body condition by day 90 of gestation • Annual use of gestation feed around 1,400 to 1,650 lbs
Feeding during lactation	• Feed ad-libitum from the day of farrowing or even before if benchmarking of body condition is achieved • Achieve a lactation feed intake of >750 lbs per sow/year in 21 day lactations
Water availability/accessibility	• Guarantee a flow rate of >1/2 gal/minute of fresh water • Ensure more than 5 gal/day of water intake
Leg and hoof soundness	• Identify and treat lame gilt/sows before breeding • Select gilts against uneven toes and sound structure
Fever	• Check rectal temperature up to 24hrs after farrowing • Therapeutic intervention if temperature is above 103°F • Keep farrowing spaces clean up to 3 days after farrowing • Have a good sleeving and fever prevention protocol when assisting farrowings
Vaccination during lactation	• Avoid, if possible
Piglets scour	• AI/AO flows to alllowdown time between batches • Clean, disinfect and dry farrowing rooms before loading • Keep farrowing spaces clean up to 3 days after farrowing • Colostrum should be provided 2 times within the first hour • Keep piglets warm: Check proper functioning of mats/lamps • Avoid cold air drafts • Do not step into the crate
Fostering events	• Keep 70-80% of the litters intact • If litter size is up to 14, and there is good udder exposure and nipple size, keep small piglets within the original litter

in increasing average daily feed intake over the last 5-6 years when sows go to farrowing in good body condition and they are in farrowing for less than 5 days prior to the due date. This increased use of technology has not replaced good husbandry: production staff still influences results by checking feed lines, cleaning up feeders, controlling feed clogging and checking sows at least daily by making them stand and implementing interventions or treatments when needed.

Depending on the lactation length, feeding practices and litter size, our general recommendation is to avoid lactating sows to lose more than 7% of their net body weight during lactation.

Environmental temperature

The negative effect of hot weather on feed intake is well documented: feed intake drops by almost 4% for every two degrees above 77 degrees.[2] This highlights the importance to mitigate heat stress during summer, particularly in places with high relative humidity.

An annual service of engines, cleaning of fans belts and blades, maintaining cool cells and a frequent cleaning and review of thermal probes will help to have a consistent and energy-efficient ventilation system while minimizing the risk of malfunctioning. In dry regions, drip cooling for evaporation is a good mechanism to improve sows thermal sensation while taking the necessary precautions to minimize risk to the creep area.

Water availability/accessibility

Increasing water intake is crucial to increase milk yield and weaning weights as the major component of milk is water. Average water intake in lactating sow/gilts is 5 to 8 gal/day. Maximizing water intake should begin with a fresh, clean source from an accessible drinker with a flow rate of above 0.5 gal/minute. Water intake is influenced by feed composition, feed intake and ambient temperature. If full feed is provided but water intake is limited, the actual feed intake will be compromised. In hot weather, water intake can be increased when the water is actually cooler in temperature. Sometimes it is hard to keep water cool, so bleeding the water pipelines could be an option to mitigate that. When extracting water from a well, periodical lab analysis on water quality is strongly recommended.

Leg and hoof soundness

Gilts or sows exhibiting any lameness or hoof problem will spend less time standing to eat and drink than healthier gilts or sows and therefore, will have lower feed intake. New floors can be more abrasive and have sharper edges than older floors which can increase the incidence of lameness in gilts and sows. Also, mixing the animals and fighting after arrival can increase the incidence of lame females. It is vital to closely monitor and identify any gilts or sows with issues (cracks in the horn walls, claws and heel) and treat them according to your herd veterinarian guidelines. In extreme cases, be prepared to early wean them to save the litter. Lameness can be increased by selecting gilts with poor feet and leg structure including such things as uneven toes. Uneven toes, or other structural problems, can affect feed intake by a modification in the biomechanics of standing or walking. Make sure you select replacements against uneven toes prior to the first breeding and control body weight gain in the first gestation.

Fever

An important risk to assisted sows is a fever developing within the first 24 hours after farrowing, particularly when farrowing spaces are not cleaned up of manure. The immune system tries to clear the infection by elevating the body temperature. Feverish animals typically show lethargy and have a reduced appetite and lower milk production. As milk production is compromised, piglets may begin to chew and lacerates the nipples, forcing the farm staff to remove the sow and/or foster the piglets to a healthy sow. Control and prevention of fever is based on three key points: (1) Prevent infection (by removing manure from the farrowing up the day 3 after farrowing); (2) Hygienic and gentle sleeving with a lubricated disposable plastic glove (single use) when needed; and (3) Check rectal temperature 12 and 24 hours after farrowing to establish a therapy prescribed by your herd veterinarian when fever is found in individual sows (rectal temperature above 103 degrees).

Sometimes, infection happens when a piglet is retained in the reproductive tract. If this condition is not identified and the sow is not treated, the sow becomes feverish and again, reduces feed intake and milk production. Ultimately, if left untreated the sow will likely die. Piglet retention should be identified as soon as possible and actions should be taken to release the retained piglet. Oxytocine may be used to promote uterine contraction, and if this fails, manually extraction of the piglet should be performed. Following the sleeving, implement the fever preventive control prescribed by your herd veterinarian for these cases.

Vaccination during lactation

It is a common practice in many countries across the globe to vaccinate sows while in the farrowing house. Even though current vaccines are safe and have reduced side effects, oftentimes they can produce low to moderate hyperthermia, thus reducing feed intake and limiting lactation performance. Our recommendation is to avoid vaccinating sows throughout the lactation whenever possible but this must be done in combination with a comprehensive herd health plan as outlined by your herd veterinarian.

Piglet scours

Milk production is in part regulated by nursing activity. Scouring litters eat less and the interrupted nursing activity may also cause increased issues for the sows such as mastitis. Main causes of scours can include a failure in basic production practices. Having the proper flow, health and sanitation are extremely important to prevent or reduce the incidence of scours Some practical advice to minimize scours includes managing farrowing rooms all in-all out and drying and disinfecting rooms prior to loading them. After sows are moved into farrowing, scrape manure until day 3 after farrowing. Additionally, check that the rubber mat/lamp and provides a warm environment for the newborn piglets and avoid air draft directly to the litters. Ideal temperature in the creep area should be 90-95 degrees and temperatures below 75 degrees in the creep area oftentimes cause diarrhea outbreaks. After farrowing, ensure that each piglet gets colostrum twice within the first hour. Finally avoid stepping into the crate.

Fostering events

To achieve more homogenous weaning weights and even to save more piglets, some farms will cross-foster piglets moving them from one sow to another. If this is performed too often and/or beyond the first 24 hours of life, cross-fostering creates a disruption that stresses not only the sow but also the piglets from both the source and recipient litters. When this happens, gilts and/or sows can be nervous, less prone to lie down on their sides, tend to have more non-productive milk let-downs and can be more aggressive against fostered piglets, particularly the younger parity sows.

We recommend to keep 70-80% of the litters intact by avoiding transferring smaller individuals in litters up to 14 piglets, as long as there are enough functional teats, good udder exposure and nipple size. Otherwise, the remaining can possible be transferred to a so-called runt litter and/or can be transferred to a nurse sow as a fall-behind around day 3 to 5 of life.

Conclusions

A low or restricted feed intake in lactation has the potential to limit expression of the sow farms biological and financial potential. Table 2 summarizes the discussed factors.

Table 2: Farrowing house general recommendations.

	Week of lactation	
	First	**2 and beyond**
Room temperature (°F)	72 to 74 degrees	66 to 68 degrees
Creep area temperature (°F)	90 to 95 degrees	According to piglets behavior
Minimum ventilation (winter)	20 cfm	
Maximum ventilation (summaer)	650 cfm	
Air change (sec.)	40 to 45 sec	

References

1. Thaker, M. Y. and Bilkei, G. 2005. Lactating weight loss influences subsequent reproductive performance of sows. Animal Reproduction Science 88: 309–318.

2. National Research Council. 2012. *Nutrient Requirements of Swine: Eleventh Revised Edition*. Washington, DC: The National Academies Press. https://doi.org/10.17226/13298y.

The chemistry of biosecurity

Jeff Odle

Introduction

Disinfectant efficacy can vary greatly depending on the intrinsic qualities of the microorganism as well as the chemical and physical environment. The pathogens of concern and the amount of microorganisms present, the time of exposure, temperature, pH, hard water and organic challenge, biofilm, and the concentration of the in-use diluted solution all have an impact on the performance of a disinfectant chemistry.[1] Meticulous cleaning prior to disinfection decreases the organic load and thus increases the margin of safety when the antimicrobial is used and shortens the exposure time needed to be effective. Only surfaces that come into direct contact with the antimicrobial for the approved contact time will be disinfected. Each situation where an antimicrobial agent is used needs to be validated to demonstrate effectiveness against the microorganisms under the environmental conditions in which they are used. The objective of this paper is to highlight factors which negatively impact the efficacy of an antimicrobial. This will lead to a more informed decision process when selecting which chemistry to use under challenging conditions.

Innate resistance of microorganisms

Not all pathogens are created equal in that some are much more difficult to kill than others due to protective coatings that act as a barrier to disinfectant uptake by the cell. As shown in Figure 1 below, the pathogen hierarchy.[2]

Spores have a coat and cortex while mycobacteria have a waxy cell wall. Gram-negative bacteria possess an outer membrane which prevents disinfectant entry. Antimicrobials which can kill large enveloped viruses may be ineffective at destroying small non-enveloped viruses. In order to be effective against small non-enveloped

Figure 1: Hierarchy of resistance

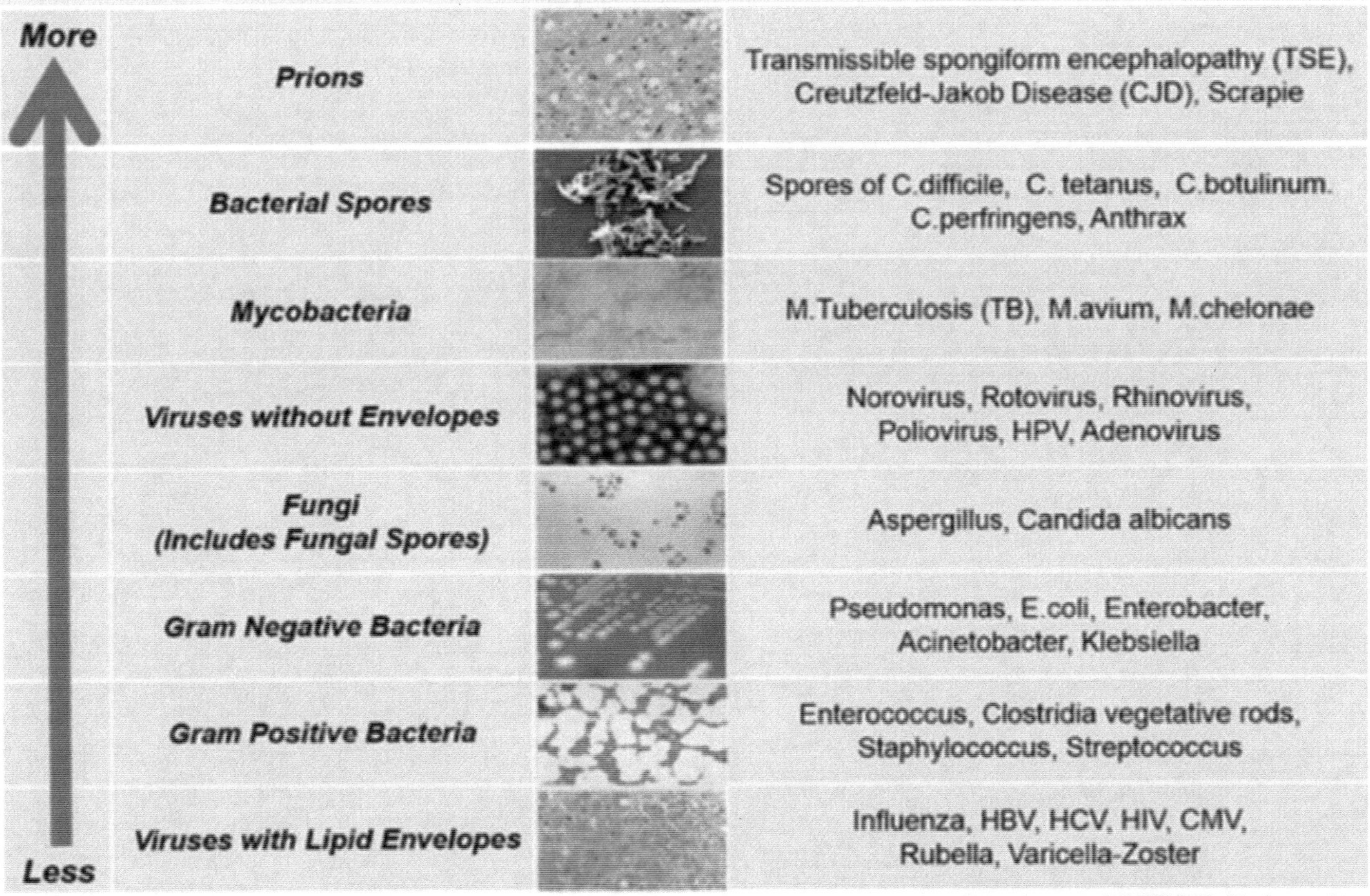

Courtesy of STERIS University

viruses some antimicrobials require longer contact times or higher concentrations. Generally speaking, the higher the concentration of disinfectant the shorter the contact time for effective kill. However, not all disinfectants behave in a similar manner. For example reducing the concentration of a quaternary ammonium by half requires doubling the contact time for effective kill while halving the concentration of a phenol requires a 64 fold increase in contact time.[3]

Just as all pathogens are not the same, neither are all antimicrobials in the speed of kill or effectiveness against the same pathogen. A study[4] performed by the University of Minnesota tested eight commercially available disinfectants, bleach and 3% sodium hydroxide for efficacy against PCV2. Only 4 of the products tested delivered the required 5 $\log_{10}$ reduction in PCV2 but exposure times ranged from 10 minutes to 12 hours as shown in the graph below (Figure 2).

Physical and chemical factors

It is well documented that temperature and pH can have a profound impact on efficacy. An increase in pH improves the antimicrobial activity of glutaraldehyde and quaternary ammonium but, decreases activity of phenols, hypochlorites, and iodine. Glutaraldehyde works well at pH of 9 and temperatures greater than 72°F (22°C) but requires an extended contact time of 2 hours or more at pH 7 or temperatures lower than 39°F (4°C).[3] Union Carbide, the manufacturer of glutaraldehyde, published the effects of temperature on *E coli* at pH 7.5 in Figure 3 and Table1.

The mechanism of oxidizing disinfectants is based on radical-ion reactions. The oxidation of the cell is a continuous reaction, which creates new radicals and ions independent of temperature. The radical ions breakup the protein found inside of the membrane of the microbe. Cold temperatures do not interfere with this reaction and therefore, the efficacy is not affected at low temperatures. Therefore, oxidizing compounds can be used during warm or cold weather conditions without showing a large decrease in reaction rate or efficacy of the disinfectant.[6]

Heavy organic matter can interfere in two ways. First, it acts as a barrier and second it uses up the antimicrobial resulting in a complex which, may have a sub-lethal dose to any remaining pathogens leading to potential resistance. Dried bacteria are significantly more difficult to kill than bacteria in liquid suspension. Once bacterial cells are dried they must be physically wetted or placed at 100% relative humidity to become rehydrated. The concentration of glutaraldehyde needed to be effective against dried non-sporulating bacteria was 50× greater than that needed for the same strain of bacteria in liquid suspension.[1] Chlorine and iodine are most affected by organic contamination. Biofilms are microbial communities that are tightly attached to the surface and are not easily removed. Bacteria within biofilms are 1,000 times more resistant to antimicrobials than bacteria in suspension. Obviously, cleaning plays an important role in reducing the amount of bacteria on the surface making it easier for the disinfectant to be effective.

Water hardness can have an impact on speed of kill because divalent cations such as magnesium and calcium interact with the disinfectant to form insoluble precipitates.

Figure 2:

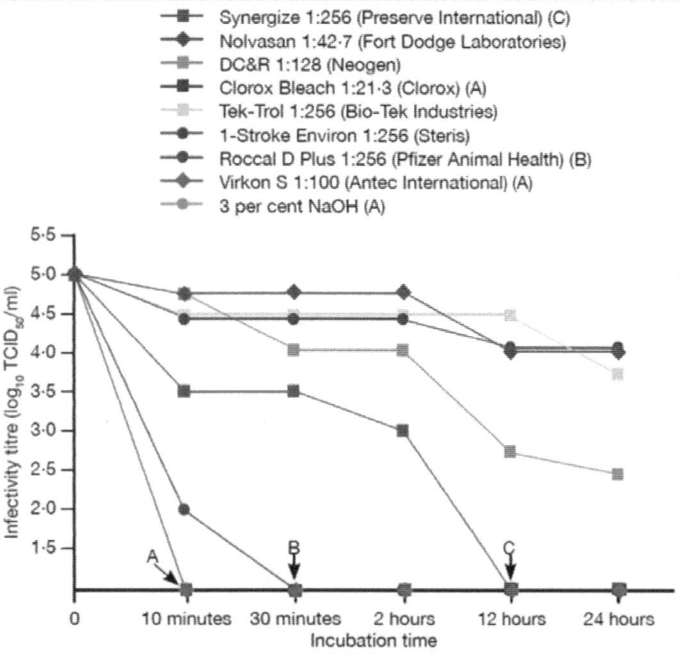

Figure 3: Effect of temperature on efficacy

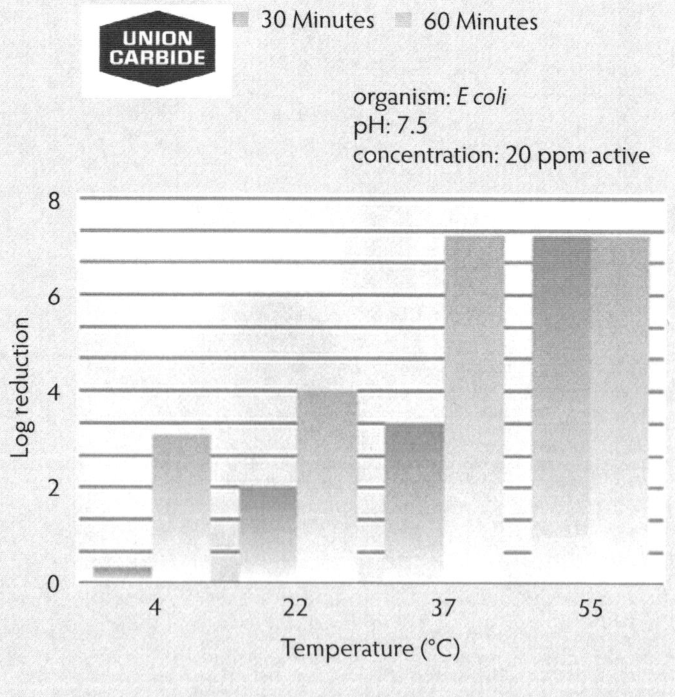

Table 1: Efficacy at cold temperature- Glutaraldehyde

During cold temperatures
• 4°C, 39°F efficacy is 2.8 log reduction, 60 minutes.
• 22°C, 72°F efficacy for a 4 log reduction, 60 minutes.

Conclusion

Understanding the type of organisms and the chemical and physical interactions between the antimicrobial and the environment it is being used in can enhance the effectiveness of a biosecurity program by preventing the spread of disease from farm to farm or within a farm.

References

1. Fraise, A. P. et al: Principles and Practice of Disinfection Preservation & Sterilization 4th edition 2004 by Blackwell Publishing Ltd.

2. Sandra Beauclair. Microbiology essentials for the SPD. KSR Publishing Inc. 2017. https://www.hpnonline.com/1707-microbiology-essentials-for-the-spd/.

3. Block, Seymour, Disinfection, Sterilization, and Preservation, 5th Edition, LLW 2000.

4. Kim, H.B. et al: Efficacy of different disinfectants in vitro against porcine circovirus type 2. The Veterinary Record May 2nd, 2009.

5. Glutaraldehyde The Right Biocide for Many Environments. Union Carbide Corporation. 1999. Form No:066-00009-601QPR. UC-2090.

6. Low Temperature Performance of Disinfectants. Axcentive. The Poultry Site. February 2016. http://www.thepoultrysite.com/articles/3610/low-temperature-performance-of-disinfectants/.

Efficacy of Circo/MycoGard™ to reduce clinical disease related to *Mycoplasma hyopneumoniae* and porcine circovirus type 2

D. Rosener[1], DVM; R. Kaptur[1], DVM; S. Ma[1], PhD; A. Mueller[2], DVM
[1]Pharmgate Animal Health, St. Paul, Minnesota; [2]Swine Services Unlimited, Inc., Rice, Minnesota

Introduction

Mycoplasma hyopneumoniae (Mhp) and porcine circovirus type 2 (PCV2) are two very common pathogens in the swine industry. While these two pathogens individually may not have significant impact on mortality rates, when they combine forces with one another, or with other viruses or secondary bacteria, they can contribute to Porcine Respiratory Disease Complex (PRDC) with devastating mortality, production losses, and major economic consequences. Circo/MycoGard™ is a newly licensed, ready to use, 1-dose, 1 ml, combination vaccine for these pathogens. One unique feature of this new vaccine utilizing baculovirus vector expression is that it features the ORF 2 capsid from PCV2b, while all of the other commercial vaccines contain PCV2a antigen. This paper summarizes two regulatory and one field efficacy trial for Pharmgate Animal Health's new Circo/MycoGard combination vaccine. The former were licensing challenge studies in high health pigs under low stress research conditions. The latter was a natural field challenge in commercial pigs with severe herd health challenges.

Regulatory study A (PCV2)

Study A was a blinded, PCV2 challenge study conducted in Caesarian Derived, Colostrum Deprived (CDCD) piglets at BL-2 isolation facilities. Struve Labs provided the CDCD pigs for the study. Forty-six piglets were randomly divided into the following three groups, blocked by litter: Group 1 (21 pigs) served as the placebo group; Group 2 (21 pigs) served as the vaccinated group; Group 3 (4 pigs) served as the sentinels. The experimental unit was the piglet. The procedure for random assignment to treatment was as follows: Each uniquely numbered piglet was assigned a random number using Microsoft Excel, {function = RAND( )}. A pattern was set for treatment group assignment to provide the desired distribution of piglets and litters for each treatment group. Then, blocking by litter, pigs were sorted by ascending random number for placement into treatment groups. The pigs were housed in individual Horsfall units for the first 10 days of age and fed Struve Labs milk formula three times per day. The CDCD pigs were placed into brooders with 3-4 pigs each at 11 days of age. At 12 days of age (Day -31), Group 1 received saline injections and Group 2 was vaccinated with Circo/MycoGard. They were fed and watered three times a day. Pigs were fed pre-starter dry feed starting at 12 days of age, gradually weaned to a starter feed, and off milk around 30 days of age. All the pigs were kept at this facility until Day -4. At that time, they were moved to a different BL-2 isolation research facility where they were subsequently challenged with PCV2 at 31 days post vaccination (Day 0). All the piglets in Groups 1 and 2 were challenged with both PCV2a and PCV2b isolates. They were challenged with 1 ml of a 50/50 mix of the two different isolates intramuscularly, and were additionally challenged with an intranasal inoculation of 1 ml per nare of the PCV2b isolate. Serum samples were taken at Days 7, 14, 21, and 28 post challenge for qPCR at Iowa State University Veterinary Diagnostic Lab (ISU VDL). Four weeks post-challenge (Day 28), the animals were euthanized, and tissue samples were collected for evaluation.

Results and discussion

The pigs in Groups 1 and 2 had their temperatures monitored from Day -2 to Day 12. There were no significant differences in temperature between Groups 1 and 2 from Day -2 to Day 7 post challenge. The temperatures were, however, significantly higher in the placebo group, when compared to vaccinates, on Day 8 ($P = 0.015$), Day 9 ($P = 0.009$), Day 11 ($P = 0.020$) and Day 12 ($P = 0.009$) post challenge.

One pig from each group died during the post vaccination time period prior to challenge due to causes unrelated to the trial. Two additional pigs from the placebo group died during the post challenge period. These occurred on Days 14 and 27 post challenge and were determined to be associated with PCV2 disease.

Blood samples and lymphoid tissues from all remaining pigs were submitted to ISU VDL for analysis. All tissues were examined and scored by the same pathologist to assure consistency in scoring. There were 13 out of 20 (65%) pigs in Group 1 that exhibited gross pathological signs consistent with the clinical signs of PCV2 infection. Only 2 pigs in Group 2 had mild lymph node hypertrophy. The differences in gross and histopathology, along with the significant temperature difference between the pigs in Groups 1 (placebo) and 2 (vaccinated) demonstrated that Circo/MycoGard helps prevent the clinical signs caused by porcine circovirus type 2 infection.

Circo/MycoGard™ is a trademark of Pharmgate Animal Health, Wilmington, NC.

The weekly serum samples were analyzed via qPCR to detect the number of copies of the PCV2 virus per 1 mL serum. All samples collected on the day of challenge (Day 0) were negative. All samples in Group 1 became positive by Day 7 post challenge, while all but two samples from Group 2 were positive. In addition, 11 animals from Group 2 produced negative qPCR results on Day 28 post challenge while only one Group 1 animal was negative on Day 28. The mean viral load reduction in the vaccinated group as compared to the placebo group was over 97% on Day 7, and over 99% on Days 14, 21 and 28 post challenge. Statistical analysis revealed that viral loads in the vaccinated group were significantly lower than the placebo group on all sample dates after challenge. In addition, the rate of decrease in viral load was significantly greater in the vaccinated group than in the placebo group. These results indicate that Circo/MycoGard aids in the reduction of viremia caused by PCV type 2 infection.

The tonsil (T), mesenteric lymph node (MLN), sub-iliac lymph node (SLN) and tracheobronchial lymph node (TBLN) were collected from all pigs on the day of termination, as well as from the pigs which had died previously. The samples were examined and scored with IHC for PCV2 colonization and for lymphoid depletion at ISU VDL. When comparing the IHC score between the two groups, highly significant differences were found in all the examined tissues; the *P*-values of MLN, SLN, T, and TBLN were 0.0001, 0.0034, 0.0008 and 0.0010, respectively (Table 1). When comparing the lymphoid depletion scores in Groups 1 and 2, significant differences were found in all examined tissues, the *P*-values were 0.0003, 0.0003, 0,005 and 0.0008 for MLN, SLN, T, TBLN, respectively (Table 2). Eighty percent (16 out of 20) of pigs in Group 1 had lymphoid tissue depletion scores, while 85% (17 out of 20) pigs in Group 2 had no lymphoid depletion scores. Three pigs in Group 1 had the maximum score of 12. These results suggest that Circo/MycoGard aids in the prevention of virus colonization of lymphoid tissues and lymphoid depletion caused by porcine circovirus type 2 infection.

Regulatory Study B (Mhp)

Similar to Study A, Study B was a blinded challenge with pigs blocked by litter and randomized to either a placebo group (n = 21 pigs) or a vaccinated group (n = 22). Pigs were vaccinated again at 12 days of age +/- 1. The challenge material was *M hyopneumoniae* lung inoculum (LI37), obtained from the Department of Veterinary Microbiology and Preventive Medicine, College of Veterinary Medicine, Iowa State University. Eighteen (18 ml) of inoculum challenge material was administered intratracheally on day 33 post vaccination. The pigs were terminated on day 37 post challenge and necropsied to score lung lesions. The lesion scores were expressed as the percentage of the lung surface area showing gross lesions typical of enzootic pneumonia. The method for evaluating lung lesions was adapted from Pointon et al.[1] The lung is divided into seven sections: left apical (0.1), right apical (0.1), left cardiac (0.1), right cardiac (0.1), left diaphragmatic (0.25), right diaphragmatic (0.25) and intermediate (0.1). The lung lesions are scored individually for each section

Table 1: Lymphoid tissue PCV2 colonization

Group	IHC score 0	IHC score 1	IHC score 2	IHC score 3	Median/ mean score	*P*-value
Mesenteric lymph node						
Placebo	8 (40%)	6 (30%)	2 (10%)	4 (20%)	1 / 1.1	0.0001
Vaccinates	20 (100%)	0	0	0	0 / 0	
Sub-iliac lymph node						
Placebo	8 (40%)	9 (45%)	1 (5%)	2 (10%)	1 / 0.85	0.0034
Vaccinates	17 (85%)	3 (15%)	0	0	0 / 0.15	
Tonsil						
Placebo	9 (45%)	5 (25%)	3 (15%)	3 (15%)	1 / 1	0.0008
Vaccinates	19 (95%)	1 (5%)	0	0	0 / 0.05	
Tracheobronchial lymph node						
Placebo	8 (40%)	7 (35%)	2 (10%)	3 (15%)	1 / 1	0.0010
Vaccinates	18 (90%)	2 (10%)	0	0	0 / 0.1	

Lymph node IHC score:
0 = Negative
1 = Positive, < 10% of cells with PCV2 staining
2 = Positive, 10-50% of cells with PCV2 staining
3 = Positive, > 50% of cells with PCV2 staining

Table 2: Lymphoid tissue depletion scores

Group	IHC score 0	1	2	3	Median/ mean score	*P*-value
Mesenteric lymph node						
Placebo	8 (40%)	6 (30%)	2 (10%)	4 (20%)	1 / 1.1	0.0003
Vaccinates	19 (95%)	1 (5%)	0	0	0 / 0.05	
Sub-iliac lymph node						
Placebo	6 (30%)	6 (30%)	4 (20%)	4 (20%)	1 / 1.3	0.0003
Vaccinates	17 (85%)	3 (15%)	0	0	0 / 0.15	
Tonsil						
Placebo	10 (50%)	4 (20%)	3 (15%)	3 (15%)	0.5 / 0.95	0.005
Vaccinates	18 (90%)	2 (10%)	0	0	0 / 0.1	
Tracheobronchial lymph node						
Placebo	8 (40%)	5 (25%)	3 (15%)	4 (20%)	1 / 1.15	0.0008
Vaccinates	18 (90%)	2 (10%)	0	0	0 / 0.1	

Lymphoid depletion score:
0 = Negative
1 = Positive – mild
2 = Positive – moderate
3 = Positive – severe

and expressed as a percentage of the diseased lung versus the whole section. When calculating the total lung lesion score, each section is given a coefficient to represent the percentage of each section to the whole lung. The coefficients of each section are shown as the number in parenthesis above.

Results and discussion

The average lung lesion score in the vaccinated pigs was 1.727%, as compared to 6.713% in the nonvaccinated pigs (Figure 1). There were 11 pigs in Group 2 that had no lung lesions, whereas, 18 out of 20 (90%) pigs in Group 1 had lung lesions, with lesion scores ranging from 1.0% to 16.5%. Thus, Circo/MycoGard vaccination significantly reduced lung lesions associated with Mhp as compared to nonvaccinated controls.

Field study (PCV2/Mhp)

While the aforementioned studies were done in Contract Research Organization (CRO) facilities, this study was performed in a commercial farrow to finish unit with numerous health challenges located in a pig dense area of Iowa. While research conducted in BL-2 isolation facilities eliminates a lot of variables and is great for challenge studies, these "pig heaven" facilities with ideal ventilation, low stocking density, etc. are often far removed from the "real world". While the regulatory trials proved efficacy in healthy animals, questions still remained over efficacy under more trying circumstances, e.g., in the presence of maternal antibodies, less than ideal ventilation and stocking densities, and concurrent disease pressures. A field study in a herd with continuous flow nurseries and finishers, and a history of PRDC associated with Mhp, PCV2, and PRRS was conducted to answer these questions. A total of 160 pigs blocked by litter and sex were randomly assigned to either a placebo group or a vaccinated group. The pigs were double ear tagged and either vaccinated with 1 ml Circo/MycoGard (per label) or 1 ml saline at 10 days of age +/- 1. After weaning at 12 days post vaccination, they were housed in mixed pens to keep pathogen exposures and environmental conditions comparable between the two groups. Average Daily Gain (ADG) was monitored on an individual pig basis with weights measured on study days 0, 12, 56, 98, 126, and 159 post vaccination. They were bled on days 0, 56, 98, 126, and 159 post vaccination. Lastly, laryngeal swabs for Mhp PCR testing were taken on days 98, 126, and 159 post vaccination. These pigs were euthanized at 169 days of age (159 days post vaccination) and at that time, in addition to the previously mentioned blood and laryngeal tissue samples being obtained, the tonsil, tracheobronchial and sub-iliac lymph nodes were also harvested.

Results and discussion

Significant growth performance differences are shown in Figure 2. It is important to note that the overall ADG figures include some of the pre-weaning time period and these pigs were also euthanized at a slightly younger age than typical market hogs. However, it cannot be ignored that these growth rates were still less than ideal. This was largely due to the severe health challenges these pigs faced – some planned and some unplanned. In PCV2 vaccine trials, it is often desirable to have a concurrent

Figure 1: Lung lesion scores

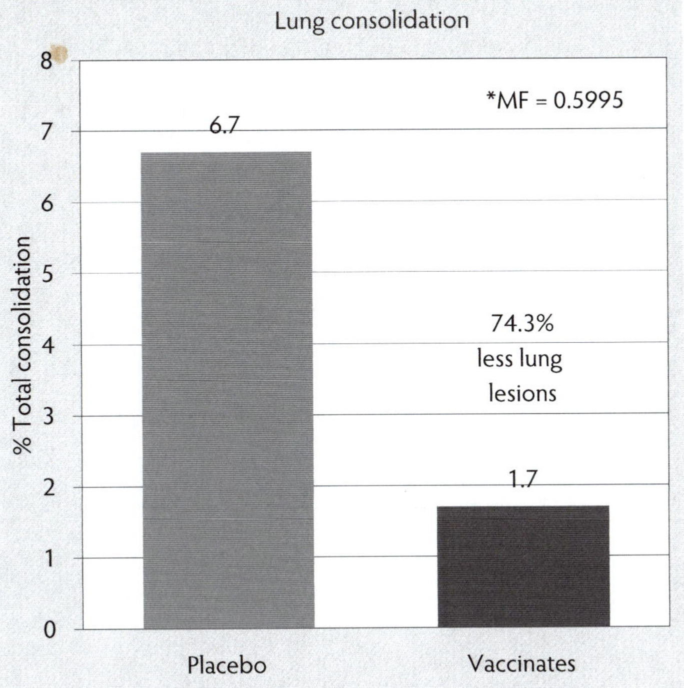

viral challenge like PRRS present to trigger the PCV2 associated diseases (PCVAD).[2] This herd was selected based on its PRRS status and these pigs were not vaccinated for PRRS. Looking at the growth rate chart, most of these pigs seroconverted to PRRS during the post vaccination days 12-56. During the study, the pigs experienced additional health challenges associated with SIV, PEDV and a curtain malfunction. This contributed to more severe challenges across both the vaccinated and none vaccinated groups.

The presence of Mhp was minimal in these pigs, with just one positive lung PCR at the end of the trial and very little found via serology or laryngeal PCR at various sampling points throughout the trial. The day 98 post vaccination serology results for Mhp were negative on 37 of the 39 samples, with one positive and one suspect. Additionally, all 39 laryngeal swabs were negative for Mhp on PCR.

On the other hand, PCV2d was widely identified in the mortalities and in the serum collected at the various sampling points. The majority of the mortalities in the placebo pigs exhibited cachexia, bronchointerstitial pneumonia, enlarged lymph nodes, and lymphoid depletion -- lesions all consistent with PCVAD due to PCV2. The serum samples contained low Ct counts for PCV2, and IHC demonstrated abundant antigen in lymphoid and lung tissues. The PCV2 was sequenced and identified as PCV2d.

Out of 120 pigs sampled on the day of termination, there was only one tonsil and five lymph nodes showing any IHC staining, and that was mild. This was consistent with expectations that PCVAD gradually resolves in most pigs (survivors). Sterilizing immunity (expected within 1-2 months of infection in most pigs) can occur in those pigs which survive a PCVAD insult. Hence, in those pigs that recover, pathologists rarely see much IHC staining in lymphoid tissues at slaughter.[3]

While the vaccinated pigs were also affected by the PRRS, SIV, PEDV, and various secondary bacteria, they handled the challenging health conditions far better than the nonvaccinated pigs expressing the immunosuppressive nature of PCV2.[4] Mortality was 12.5% in the vaccinated pigs vs. 36.3% in the nonvaccinates, a 66% reduction (Fisher's Exact Test $P = 0.0008$). Lung lesion scores were also significantly reduced in the vaccinated pigs, 8% vs. 16.2% (Least Squares Means $P = 0.0077$). Lastly, Circo/MycoGard was found to be safe with no anaphylactic or injection site reactions.

Figure 2: Average daily gain

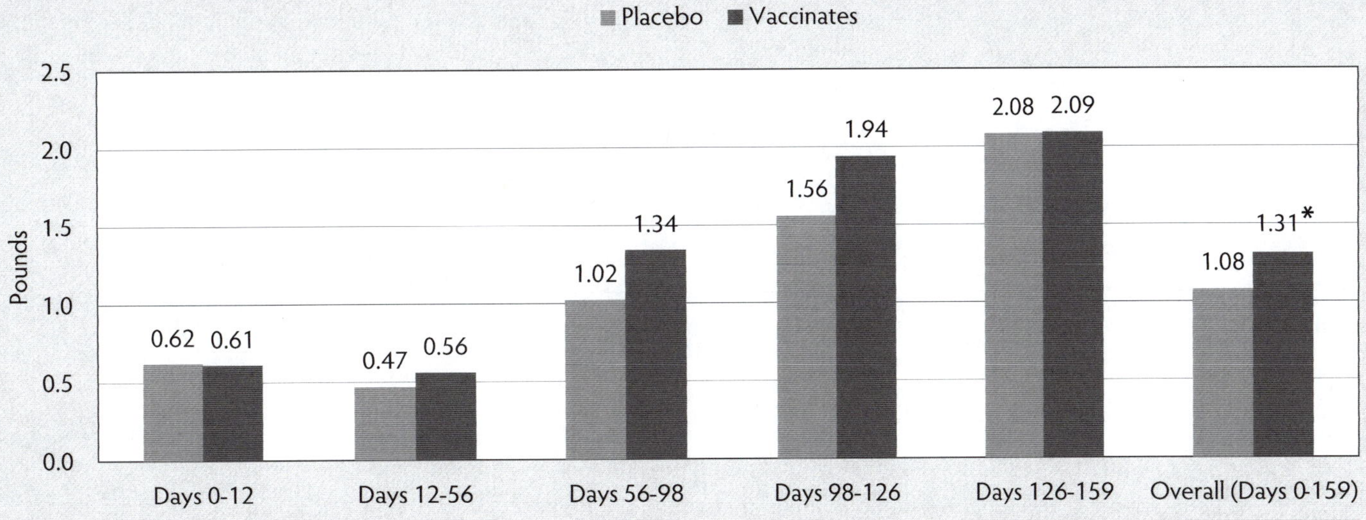

* $P < 0.0001$

Increased growth rate and livability resulted in 61.4 pounds more pork being marketed/head enrolled in this study in the vaccinated pigs as compared to the nonvaccinated controls. At a conservative $50/live cwt, this would be an additional $30.70/head. There were also labor and medication cost savings due to a reduction in the number of spot treatments administered, 41 vs. 60.

Conclusions

A single dose of Circo/MycoGard administered as early as 10 days of age, even in the presence of maternal antibodies, and severe health challenges is protective against disease caused by Mhp, PCV2a, PCV2b, and PCV2d.

References

1. Pointon AM, Davies PR, Bahnson PB. Disease Surveillance at Slaughter. In: Straw B, D'Allaire S, Mengeling W, Taylor D, eds. Diseases of Swine. Ames, Iowa: Iowa State Univ Pr, 1999:1111–1132.

2. Personal Communication, Dr. Darin Madsen, Iowa State University, 2017.

3. Personal Communication, Dr. Kent Schwartz, Iowa State University, 2017.

4. Segales J., Domingo M. Porcine circovirus type 2 infection: postweaning multisystemic wasting syndrome and other conditions. IPVS. 2002;35-42.

Determination of the effectiveness of Aivlosin® Water Soluble Granules (62.5% w/w tylvalosin) in drinking water of pigs for the control of swine respiratory disease (SRD)

R. Kaptur[1], DVM; D. Rosener[1], DVM; E. Abbott[2], BVMS, PhD, MRCVS; C. Eliopoulos[3], DVM; K. Rooney[4], MS, DVM; R. Domangue[5], PhD

[1]Pharmgate Animal Health, Wilmington, North Carolina; [2]ECO Animal Health, London, United Kingdom; [3]ECO LLC, Princeton, New Jersey; [4]The Veterinary Consultancy, LLC, Montvale, New Jersey; [5]Statistical Consulting Services, Broadway, Virginia

Introduction

In the majority of swine producing regions, pigs are housed in large groups, often times in areas with dense pig populations.[1] High stocking density in a closed environment facilitates transmission of airborne pathogens within a herd.[2]

Respiratory disease is one of the most significant health challenges facing the swine industry and is the leading cause of mortality.[3] Throughout nursery and grow/finish operations, a combination of infectious bacterial agents and environmental stressors often leads to SRD. The health challenges and subsequent financial losses caused by SRD include fever, depression, dyspnea, increased mortality, decreased weight gain, increased feed costs, increased condemnation at slaughter and increased costs for treatments.[1]

SRD is defined as the component of acute respiratory disease in swine associated with bacterial pathogens such as *Actinobacillus pleuropneumoniae* (APP), *Bordetella bronchiseptica, Haemophilus parasuis, Pasteurella multocida,* or *Streptococcus suis* (FDA CVM Guidance 178). Many of these pathogens are naturally found in the upper respiratory tract and can be out competed by non-pathogenic commensal flora; thus unable to infect an immunocompetent pig. Primary pathogens, such as viruses that damage the respiratory epithelium (e.g., swine influenza virus, porcine reproductive and respiratory syndrome virus) and mycoplasmas predispose the pig to secondary bacterial invaders by lowering the local and systemic defense mechanisms of the host.[1] When the pigs' pulmonary defenses are compromised, these pathogens become secondary invaders and multiply resulting in severe systemic disease such as polyserositis, polysynovitis, and meningitis.[4]

Aivlosin® (62.5% w/w tylvalosin) Water Soluble Granules (WSG) has recently been approved by the United States Food and Drug Administration (FDA) for the control of SRD associated with *Bordetella bronchiseptica, Haemophilus parasuis, Pasteurella multocida* and *Streptococcus suis*, in groups of swine in buildings experiencing an outbreak of SRD. Approval was based upon the results of a multi-site natural infection study. The study objective was to determine the effectiveness of Aivlosin® administered continuously for 5 consecutive days at an inclusion rate of 50 parts per million (ppm) tylvalosin in drinking water for the control of SRD. In order for a pathogen to be considered for the label claim, at least 30 isolates were required from a minimum of 30 pigs.

Materials and methods

This was a multi-site masked study with 980 commercial crossbred pigs sourced from farms with a history of naturally acquired bacterial respiratory infections. The four study sites were in Nebraska, Minnesota, Iowa, and Saskatchewan, Canada. Pens of 14 pigs were randomly assigned to one of two groups: treated (50 ppm tylvalosin) or untreated (0 ppm tylvalosin). The study was conducted as a randomized complete block design with the investigational staff masked to treatment groups. Pigs had *ad libitum* access to an age appropriate non-medicated, commercial ration from arrival to the end of the study. General health status was assessed daily and any health abnormalities were evaluated by the Clinical Investigator (CI). Pigs were scored twice daily for clinical signs of SRD from arrival through Day 0. When at least 15% of the pigs were judged clinically affected at each site, treatment was initiated (Day = 0). A clinically affected pig met the following criteria: either 1) had all three of the following signs: moderate or severe depression score, moderate or severe respiratory score and a rectal temperature of 104.0° F or greater OR 2) SRD was confirmed at necropsy. After treatment began, pigs were clinically scored once daily from Day 1 through the end of the study. Pen feed consumption was measured from Day 0 through Day 8. Individual body weights of all study animals were measured on Day 0 and prior to necropsy, which included any animal removed during the study period. Due to the large number of pigs at each farm, half of the pigs in each pen were randomly assigned to

Aivlosin® is a Registered trademark of ECO Animal Health, Ltd., London, UK necropsy on either Day 7 or Day 8. All animals remaining on the assigned necropsy day (Study Day 7/8) were clinically scored, euthanized, and necropsied. At necropsy, gross lung lesions were scored and lung tissue samples collected. Lung tissue samples were cultured to determine the bacterial pathogens present.

The primary variable, treatment success was defined as a pig with normal or mild depression scores, and normal or mild respiratory scores, and a rectal temperature less than 104.0°F on the Day 7 clinical scores (Table 1). The secondary variables average daily gain and feed efficiency were calculated from the pen feed consumption and body weight data. The lung lesion score of each pig was calculated by adding the percent area of all the lung lobes that contained pneumonic lesions relative to the entire lung mass. Right cranial, left cranial, right middle, left middle and accessory lobes were each 10% of the entire lung. Right caudal and left caudal lobes were each 25% of the entire lung.

Though not part of the study protocol, the treatment success rate of only the clinically affected (CA) pigs was also calculated though not statistically analyzed. This would show any improvement in success rate for the tylvalosin-treated pigs in this study that were sickest at the start of treatment.

Results and discussion

A relatively large number of naturally infected SRD pigs were used in this study. Pigs were housed in different geographic locations across North America (NA). This helped provide an accurate representation of the NA swine market within the study population. Furthermore, by waiting to initiate treatment at each site until at least 15% of the pigs were clinically affected, the trial successfully simulated the conditions around an acute outbreak of respiratory disease.

A statistically significant improvement was seen in the Aivlosin® treated group compared to the non-treated control group in regards to the primary variable, treatment success. See the treatment success rate (%) in Table 1 ($P = 0.0353$). The secondary variables, average daily gain (ADG) ($P = 0.0297$) and pen feed efficiency (FE) ($P < 0.0001$), seen in Table 1 represented as kg/pig/day and weight gain/feed intake, respectively, were also statistically significantly different in favor of the Aivlosin® treated group vs the non-treated controls.

Table 1: Summary of variables

Treatment group	Aivlosin® WSG	Non-treated controls
Treatment success rate(%)	48.5[a]	41.6[b]
Pen average daily gain (ADG, kg/pig/day)	0.46[a]	0.39[b]
Pen feed efficiency (kg feed/kg gain)	0.67[a]	0.60[b]
Mortality %	3.9*	6.5*
Pen mean lung lesion scores (%)	12.95	15.65

[ab] Rows with different letters are significantly different at $P < 0.05$

* Statistical analysis was not applicable for this variable due to low counts.

Mortality in the treated group was reduced 40% compared to the control group, 3.9 and 6.5 respectively (Table 1) but this was not analyzed statistically because overall mortality was low (< 10% as specified in the protocol). Additionally, pen mean lung lesion scores were also numerically improved in the treated group (Table 1). Treatment success was analyzed using a generalized linear mixed model with response being pen proportion of successes and logit link function. Pen mean lung lesion score, pen average daily weight gain and pen feed efficiency were analyzed with a normal-based linear mixed model with a specification for heterogeneous variances if appropriate. All statistical tests were conducted at the 0.05 level of significance.

For the CA subpopulation, all sites combined and for three of the four individual sites, the treatment success rate was higher in the tylvalosin-treated pigs than in the untreated pigs. For all sites combined the clinical outcome for the tylvalosin-treated pigs was 30.4% Success whereas for the untreated pigs it was 19.4% Success, a 56.7% improvement (Figure 1).

To ensure that treatment success due to Aivlosin® is correlated to successfully treating the SRD pathogens listed on the label claim, the FDA CVM has very stringent requirements. Each pathogen listed on the label had to have a minimum of 30 isolates obtained from at least 30 different pigs. *Bordetella bronchiseptica, Haemophilus parasuis, Pasteurella multocida,* and *Streptococcus suis* were isolated from a total of 107, 166, 124, and 115 pigs, respectively (Table 2).

References

1. *Diseases of Swine,* Tenth Edition. Edited by Zimmerman et al. 2012. 348-353.
2. Buddle et al. 1997. *Aust Vet J* 75: 274-281.
3. National Animal Health Monitoring Service (NAHMS) *Swine 2012 Part III: Changes in the U.S. Swine Industry, 1995–2012.* p. 40, 45.
4. Nielsen and Danielsen. 1975. *Nord Vet Med* 27: 20-25.

Figure 1: Success rates in pigs that were Clinically Affected (CA) at enrollment and in all enrolled pigs

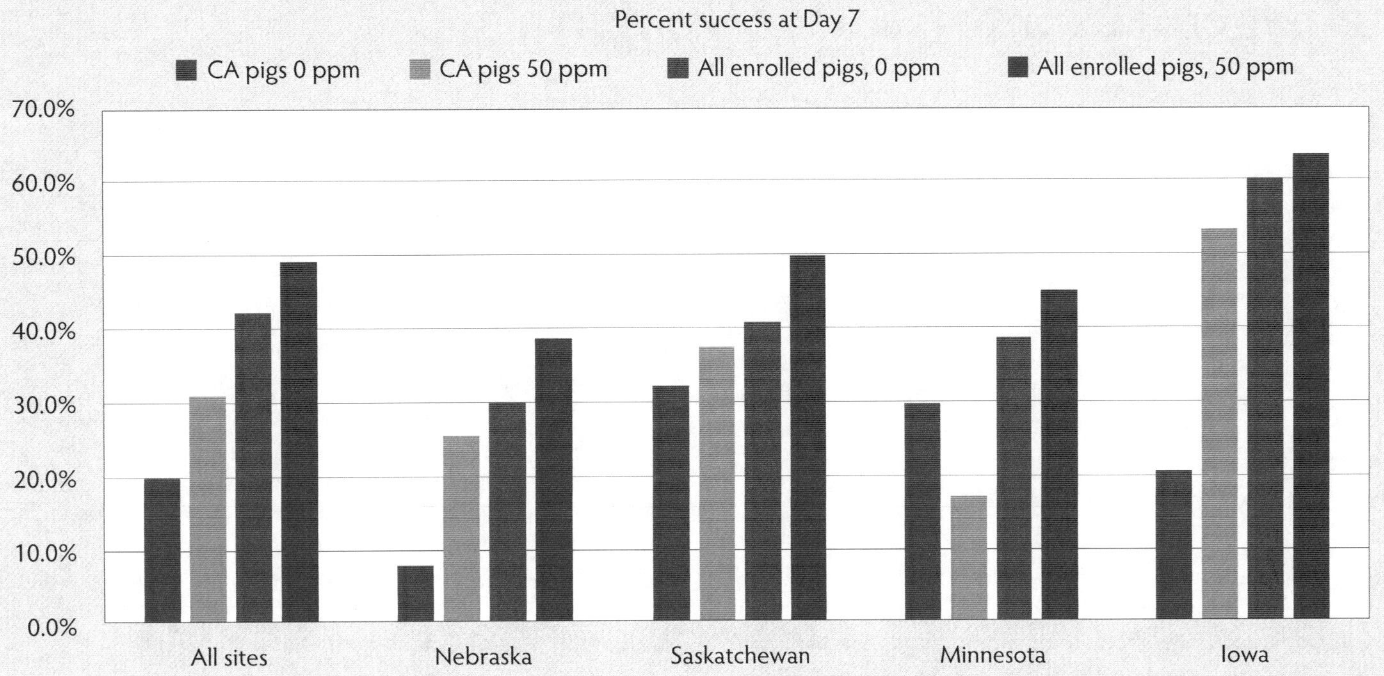

Table 2: Summary of pathogens isolated

Pathogen	Aivlosin® WSG	Non-treated controls
Bordetella bronchiseptica	61*	46*
Haemophilus parasuis	81*	85*
Pasteurella multocida	58*	66*
Streptococcus suis	57*	58*

* Represents isolate frequency across all 4 study sites.

Maximus Sow™ next-generation sow ESF and management information system

Thomas Stein, DVM, MS, PhD
Maximus Systems & Maximum Ag Technologies; St Hubert, Quebec, Canada Jenison, Michigan

In response to consumer demand and retail customer requirements, pork producers are converting sow farms from stall to pen gestation systems. As producers either remodel or build new barns, they are also installing static or dynamic ESF (electronic sow feeding) systems. Sow identification with RFID tags is a key part of making an ESF system work. It won't be long before RFID tags are present on almost all sows and gilts in commercial sow farms. Omnipresent RFID tags provide a real opportunity to advance in-barn, sow-side data collection technology which will improve the speed and accuracy of farm data collection as well as reduce the overall cost of data collection and data management.

The biggest challenge in using software effectively has been data acquisition and data entry. Many pork producers continue to send hand-written barn sheets to a central data office where clerical staff enter data by hand. Proprietary hand-held data collection devices have not gained much market share because the economics and form factor haven't worked (small screen, small keyboards, proprietary and expense devices leading to vendor lock-in).

Maximus Systems in collaboration with Maximum Ag Technologies have introduced Maximus Sow™, a next-generation sow management system. It includes: (1) a seamless connection to data streaming from Maximus controllers used for running sow farm buildings as well as sow feeding and behavior data streaming from ESF systems including the Maximus static ESF feed stations; (2) in-barn data collection system called WANDA™, which uses an RFID tag reader on a wand-like device linked via Bluetooth to a hand-held computer tablet. WANDA runs on either Android or iOS (Apple) tablet computers; (3) customer-specific private cloud database that stores and integrates all these data; and, (4) full sow management reporting, data analysis, custom report building and database applications for *ad hoc* data queries.

The tablet computer contains a simple and straightforward sow record software application which synchronizes with a cloud database. With WANDA, barn workers can read a sow's RFID tag, bring up her history on the tablet, review her sow card, see her lifetime performance, and view actions needed or alerts for that sow. They can also enter basic sow data such as breeding, pregnancy check, farrowing, weaning, movements, etc. These data immediately update a sow's record on the tablet computer as well as updating the cloud database. Importantly, WANDA's cloud database immediately synchronizes with the Maximus Sow™ system.

WANDA provide several advantages: larger screen, larger keyboard, full sow histories, easier data entry, real-time data collection, and better ergonomics by using a wand-like device to read RFID ear tags. More importantly, there is no vendor lock-in because WANDA can send data to any sow software system.

WANDA also links to and synchronizes with the Maximus barn controller system and the Maximus ESF system from Maximum Ag Technologies, which means that each sow's daily feed intake in gestation automatically connects to their history. This also gives barn workers the ability to manage or change a sow's intake on-the-fly and in real-time as they do their daily check on sow body condition in gestation.

Development of Ingelvac Provenza™: A new tool against IAV-S

Marc A. Eichmeyer[1]; Wesley Johnson[1]; Eric Vaughn[1]; Michael B. Roof[1]; Philip W. Hayes[2]
[1]Boehringer Ingelheim Vetmedica, Inc., Ames, Iowa; [2]Boehringer Ingelheim Vetmedica, Inc., St. Joseph, Missouri

Introduction

Influenza A virus of swine (IAV-S) has long been a source of economic loss for swine producers. It has been estimated that IAV-S results in a loss of $3.23 per pig through closeout.[1] Given the economics of swine production, it is essential to minimize diseases in an effort to maximize profitability. IAV-S infection can result in decreased fed conversion efficiency and increased time to market, thereby having a negative impact on swine production.

Currently, three major subtypes of IAV-S circulating in US swine populations, H1N1, H1N2 and H3N2 IAV-S. Disease is characterized by a rapid onset of high fever, lethargy, loss of appetite, labored abdominal breathing and coughing.[2] Duration is typically 2-6 days and while mortality is low with most animals recovering, weight loss can be severe. IAV-S is also a contributing factor in chronic respiratory disease problems and Porcine Respiratory Disease Complex.[3]

Vaccines for IAV-S disease have been limited to inactivated products containing various isolates and/or hemagglutinin types. These vaccines have struggled to overcome maternal-derived antibodies and induce immunity prior to the incidence of disease. With concerns around piglet vaccine efficacy, the historical focus has been to vaccinate gilts and sows to increase antibody levels conveyed to the piglets. This concept centered on the theory that the higher maternal-derived antibody levels would delay disease until the animals are more capable of enduring disease and minimizing loss. In reality, this delays the onset of disease until more extensive investments have been made.

Materials and methods

The vaccine is based on a truncation of the NS1 gene (Figure 1).[4] This was developed in collaboration with Mt. Sinai School of Medicine (MSSN) and the USDA-ARS. Using reverse genetics, H1N1 and H3N2 viruses encoding a carboxy-truncated NS1 were generated and tested for safety and efficacy. A bivalent H1N1/H3N2 vaccine was developed for intranasal administration to piglets as young as 1 day of age.

Results and discussion

When a wild-type influenza strain infects a cell, there is normally an intact NS1 protein generated, which in turn suppresses the host cell's natural interferon defense system. With a supression of the host cell's interferon system, the virus can escape a main defense mechanism. As a result, the host develops a poor cell-medicated immunity to the virus. However, when an NS1-truncated influenza virus infects a cell, the NS1 protein generation is not only diminished in quality, but the truncated NS1 gene product can no longer act as an interferon antagonist. Because the interferon system is not shut down, the host cell can develop a more complete immune response.

The bivalent vaccine was evaluated by the USDA through both an environmental and general risk assessment and determined to be safe for use. It was also tested and demonstrated host and alternate animal safety, as well as host animal efficacy, in support of a USDA license.

Ingelvac Provenza™ not only provides significant protection against IAV-S, it also eliminates the use of needles through intranasal administration, minimizes stress on the animals, reduces labor required for vaccination of older animals and the need for a second injection.

References

1.Dykhuis Haden, C., Painter, T., Fangman, T., Holtkamp, D. Assessing production parameters and econimic impact of Swine Inflenza, PRRS and Mycoplasma hyopneumoniae on finishing pigs in a large production system. Proceedigns 2012 AASV Annual Meeting. P75-76.

2. Van Reeth, K., Brown, I.H., Olsen, C.W., 2012. Swine Influenza. In: Zimmerman J.J. (Ed.), Diseases of Swine. Wiley-Balckwell, Chichester West Sussex.

3. Vincent, A.L., Perez, D.R., Rajao, D., Anderson, T.K., Abente, E.J., Walia, R.R., Lewis, N.S. Influenza A virus vaccines for Swine. Vet. Micro. 206 (2017): 35-44.

4. Solórzano, A., Webby, R.J., Lager, K.M., Janke, B.H., García-Sastre, G., Richt, J.A. Mutations in the NS1 Protein of Swine Influenza Virus Impair Anti-Interferon Activity and Confer Attenuation in Pigs. JOURNAL OF VIROLOGY, June 2005, p. 7535–7543.

Figure 1: Truncation of the NS1 gene

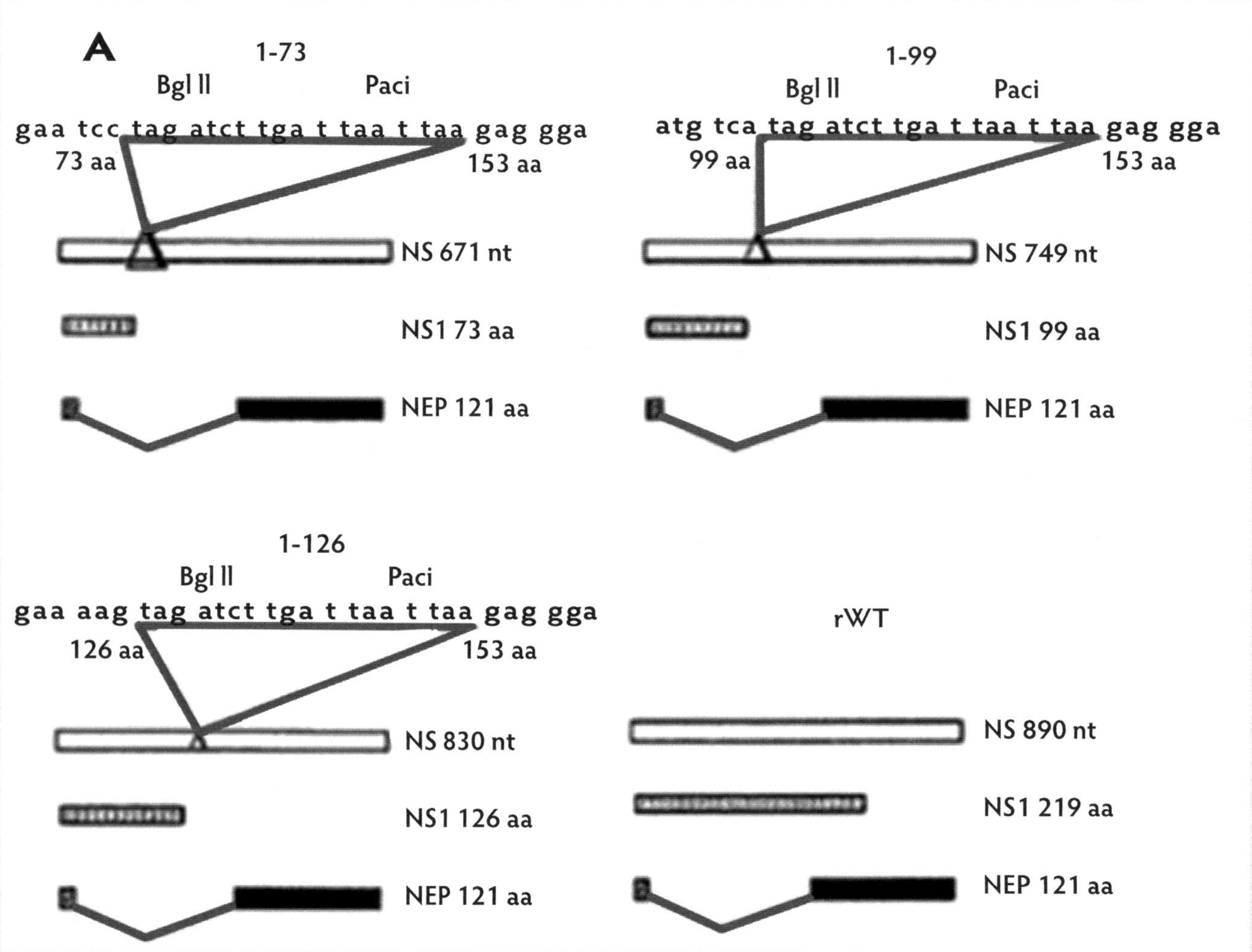

Laboratory efficacy evaluation of intranasal vaccination with Ingelvac Provenza™ in newborn piglets

R. A. Smiley[1], MS; W. Johnson[2], MS; B. Fergen[2]; PhD; M. Eichmeyer[2]; T. J. Kaiser[1], DVM MS
[1]Boehringer Ingelheim Vetmedica, Inc., St. Joseph, Missouri; [2]Boehringer Ingelheim Vetmedica, Inc., Ames, Iowa

Introduction

Influenza A virus in swine (IAV-S) is an acute respiratory disease which causes substantial economic losses in pig production. One major control strategy includes the use of killed vaccines in breeding females to confer passive immunity to their offspring. Recently, USDA-APHIS has licensed an attenuated live viral vaccine (Ingelvac Provenza™) for use in pigs as young as one day of age. The bivalent vaccine (H1N1 and H3N2) uses NS1-truncated technology. Two laboratory studies evaluated efficacy after challenge with either a heterologous H1N2 virus or a heterologous H3N2 virus administered intratracheally. Primary parameters were differences in clinical signs and virus shedding during a 5-day post-challenge phase and lung lesions.

Materials and methods

Between 1 and 5 days of age, entire litters of pigs born to IAV-S seronegative dams were intranasally administered vaccine or saline. Approximately 4 weeks post-vaccination, pigs were weaned and controls and vaccinates were commingled within a pen for the challenge phase. A/Swine/North Carolina/001169/2006 H1N2 challenge strain was used for Study #1, and A/Swine/Nebraska/97901-10/2008 H3N2 was used for Study #2. Pigs were restrained manually, and the challenge material was placed intratracheally just prior to the tracheal bifurcation. Daily for five days, pigs were observed for clinical signs of IAV-S infection, and nasal swabs were collected daily for evaluation of viable virus by virus isolation. On the fifth day, pigs were humanely euthanized and necropsied, and lungs were evaluated for lesions consistent with influenza infection. Both studies were performed under the supervision of the Boehringer Ingelheim Vetmedica Inc. (BIVI) Institutional Animal Care and Use Committee. A schematic of the study design is shown in Table 1.

Results and discussion

Both studies showed significant reduction in nasal virus shedding, lung lesions and clinical signs in pigs vaccinated with Ingelvac Provenza™ compared to controls. When challenged with H1N2, comparisons of the controls to vaccinates are presented in Figure 1 for lung lesions and clinical signs and in Figure 2 for nasal virus shedding. When challenged with H3N2, comparisons of the controls to vaccinates are presented in Figure 3 for lung lesions and clinical signs and in Figure 4 for nasal virus shedding. These two studies demonstrated a 1-mL intranasal dose of Ingelvac Provenza™ provides effective protection for pigs vaccinated as early as 1 day of age against both a heterologous H1N2 strain and a heterologous H3N2 strain particularly when animals are susceptible to stress or IAV-S exposure such as weaning and commingling events.

References

Data on file: BIVI Study 2013200 and Study 2013232.

Table 1: Study design

Study	Group	#	Treatment	Commingle/ challenge	Necropsy
#1	Vaccine	26*	1-5 days of age	**H1N2 challenge** 26-30 days of age	5DPC 31-35 days of age
	Control (saline)	31			
#2	Vaccine	30	1-4 days of age	**H3N2 challenge** 31-34 days of age	5DPC 36-39 days of age
	Control (saline)	26			

* One pig died 1 day following challenge due to an unrelated cause.
DPC = days post-challenge

Figure 1: H1N2 Challenge: Median lung lesion scores & percentage of pigs with clinical signs

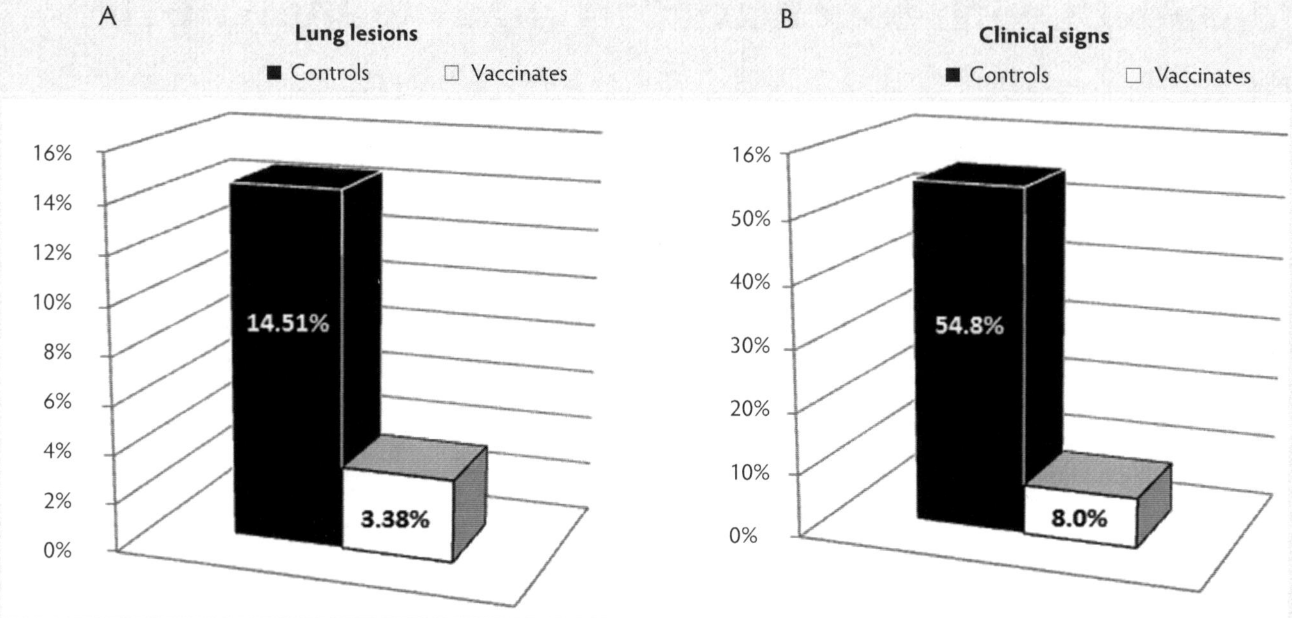

* A pig was considered affected if it had any clinical sign indictive of influenza (sneezing, coughing, nasal/ocular discharge, respiratory signs of behaviral signs) on at least two days.

Figure 2: H1N2 Challenge: Percent of pigs with virus isolation positive nasal swabs

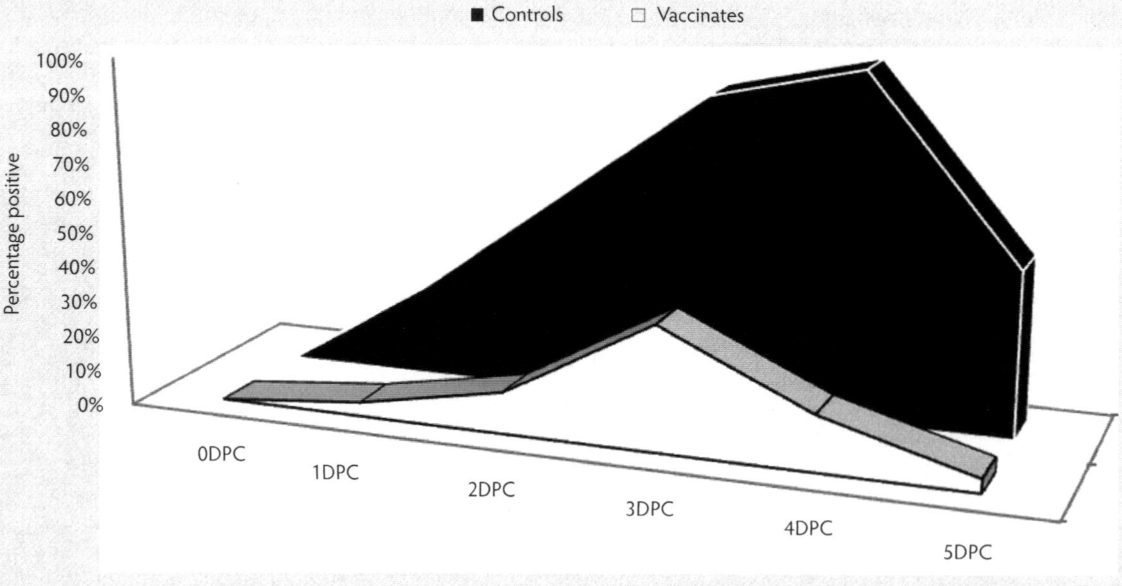

	0DPC	1DPC	2DPC	3DPC	4DPC	5DPC
□ Vaccinates	0%	4%	12%	36%	16%	4%
■ Controls	0%	26%	58%	90%	100%	48%

DPC = days post-challenge

Figure 3: H3N2 Challenge: Median lung lesion scores & percentage of pigs with clinical signs

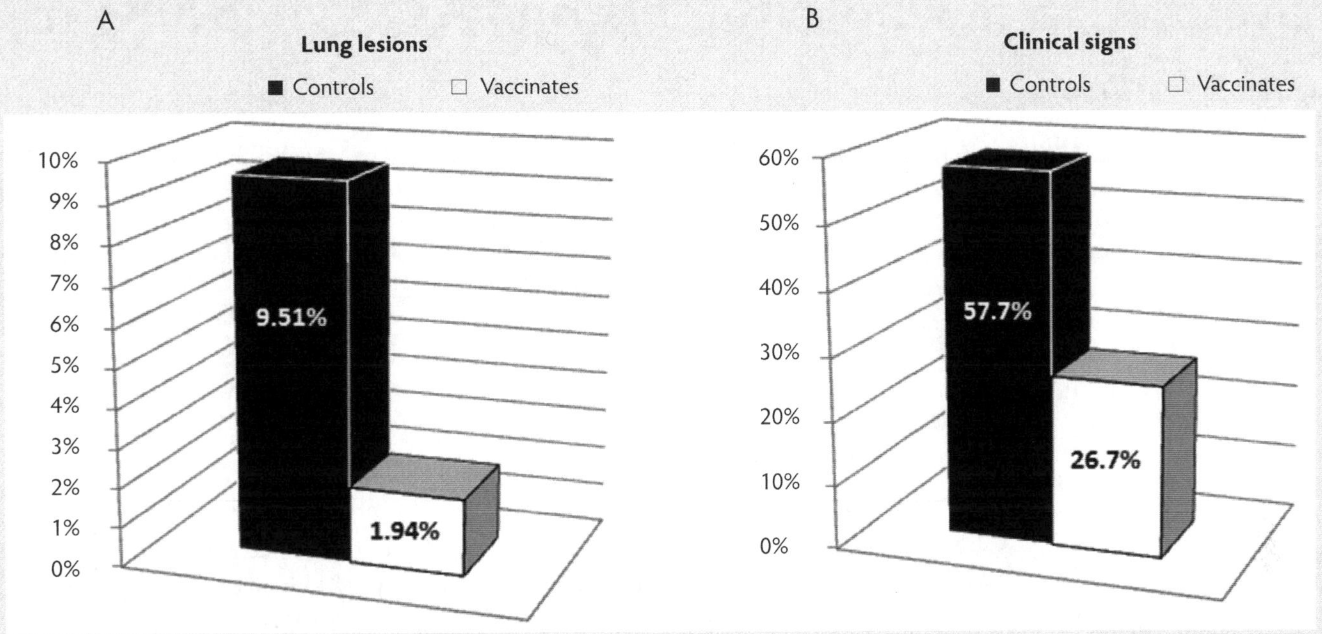

* A pig was considered affected if it had any clinical sign indictive of influenza (sneezing, coughing, nasal/ocular discharge, respiratory signs of behaviral signs) on at least two days.

Figure 4: H3N2 Challenge: Percent of pigs with virus isolation positive nasal swabs

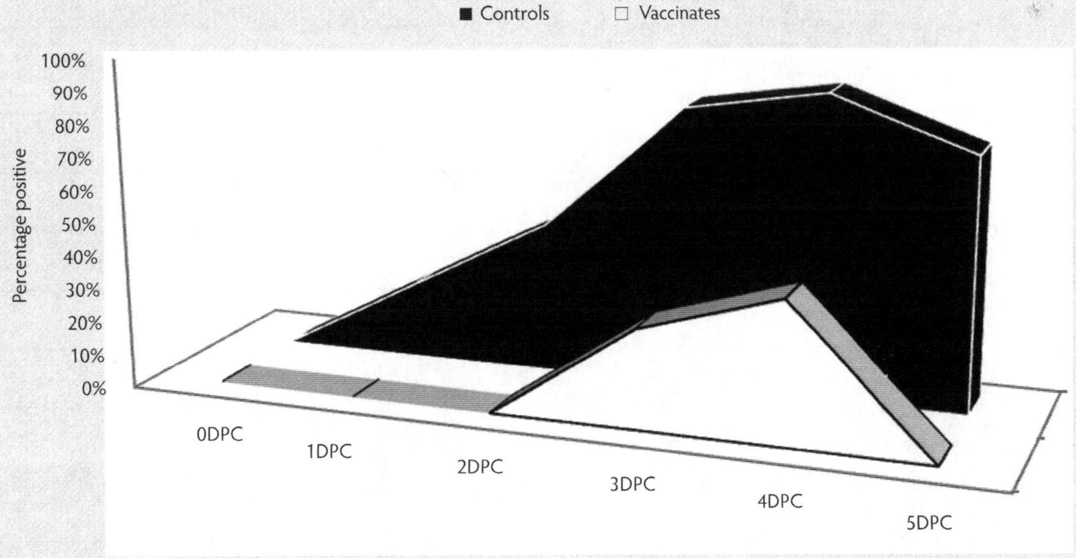

	0DPC	1DPC	2DPC	3DPC	4DPC	5DPC
□ Vaccinates	0%	0%	0%	30%	43%	0%
■ Controls	0%	23%	46%	85%	92%	77%

DPC = days post-challenge

The effect of intranasal vaccination with Ingelvac Provenza™ in newborn piglets on IAV-S shedding in nasal swabs

Marika Genzow[1]; Christa Goodell[2]; Troy J. Kaiser[2]; Wesley Johnson[2]; Marc Eichmeyer[2]
[1]Boehringer Ingelheim Vetmedica GmbH, Ingelheim, Germany, [2]Boehringer Ingelheim Vetmedica, St. Joseph, Missouri

Influenza A in swine is an acute respiratory disease of swine caused by type A influenza viruses, which cause great economic losses in pig production. Major control strategies include the use of killed vaccines (KV) in breeding females to confer passive immunity to their offspring. An H1N1 and H3N2 NS1-truncated modified live influenza virus vaccine (Ingelvac Provenza™) has been recently licensed by USDA APHIS. The aim of this study was to investigate the effect of an intranasal vaccination of newborn pigs with or without maternally derived antibodies (MDA) on virus shedding in nasal swabs (shown by virus isolation) through intratracheal challenge experiments with either a heterologous H1N2 or H3N2 virus.

To evaluate protection of the pigs in the presence of maternal antibodies over time, three groups were administered intratracheal challenges at 4 weeks (H1N2), 5 weeks (H3N2) or 9 weeks (H1N2) post-vaccination, respectively. Non vaccinated challenged seropositive and seronegative groups were included as controls. The challenges were successful, as evidenced by increase in body temperature in all groups and respiratory signs following intratracheal administration of either IAV-S virus in the non-vaccinates.

The results of this study showed a significant decrease in the incidence and duration (Figure 1) of shedding viable virus in nasal swabs for vaccinated newborn piglets with or without MDA.

This study indicates that intranasal vaccination with Ingelvac Provenza™ of newborn piglets with maternal antibodies can be a valuable tool for reducing the prevalence of heterologous H1N2 and H3N2 IAV-S in pig herds by reduction of virus shedding.

Figure 1: Median duration of shedding over time

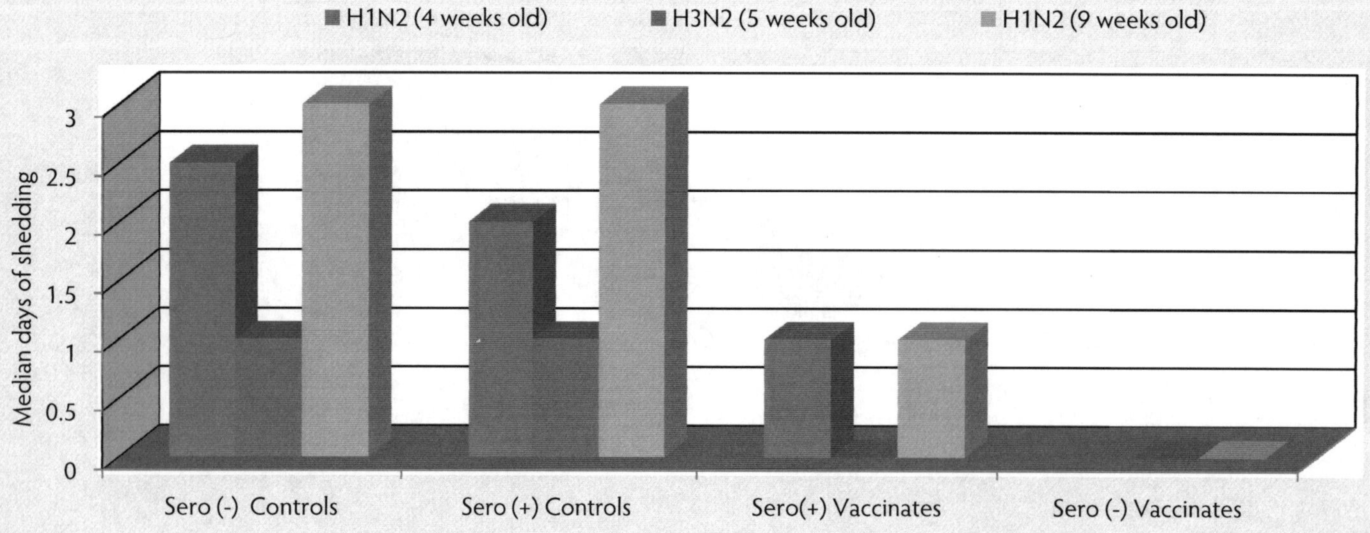

Understanding factors influencing pig quality measured using pig weaning weight

C. Bruns, PhD; R. Noel; B. McNeil; J. Sonderman, PhD; T. Rathje, PhD
DNA Genetics, Columbus, Nebraska

Introduction

Maintaining pig quality with increasing litter size is an issue currently facing the swine industry. This is in part due to the negative relationship existing between litter size and pig weaning weight.[1] One management practice that has been implemented to manage large litter sizes is cross-fostering, where some pigs are placed on nurse sows when the birth sow may be unable to nurse her pigs. How this protocol and other management practices are executed may impact wean pig quality. Weaning weight is an indicator trait for pig quality. Heavier weaning weight has been associated with faster growth rate and fewer days to market.[2,3] Understanding the factors that affect pig weaning weight can aid producers in enhancing pig quality. The objective of this study was to examine factors that influence pig weaning weight.

Materials and methods

The data used for this study were recorded from February to October 2017 on 40,616 purebred Landrace and Yorkshire pigs from four sow farms with similar management protocols. All pigs born alive were weighed individually at birth and all foster movements were recorded for each individual pig. When weighing pigs at weaning, it was known which sow the pig nursed and how long the pig nursed. Pig weaning weight was analyzed using a model with fixed effects of nurse sow parity, number of foster events for the pig, and sex, covariates of birth weight and wean age, and random effects of nurse litter, birth dam, and wean week within site.

Results and discussion

An increase in birth weight of 0.1 kg was related to an increase in weaning weight of 0.28 and 0.25 kg for the Yorkshire and Landrace pigs, respectively ($P < 0.05$). In both breeds, females weighed less than males at weaning by 0.1 kg ($P < 0.05$). Each day increase in weaning age was related to 0.2 kg ($P < 0.05$) increase in weaning weight for both breeds. Yorkshire and Landrace pigs that were not fostered from their birth dam weighed 0.2 and 0.1 kg more ($P < 0.05$) than pigs that were cross-fostered, respectively. The effect was magnified to -0.5 and -0.4 kg for Yorkshire and Landrace pigs that were fostered to multiple nurse sows prior to weaning ($P < 0.05$). All factors were related to weaning weight in a similar manner for both breeds, except nurse sow parity. Interestingly, higher parity nurse sows weaned significantly lighter ($P < 0.05$) Yorkshire pigs; however higher parity nurse sows weaned significantly heavier ($P < 0.05$) Landrace pigs. When examining pigs that were cross-fostered to one nurse sow, each day after birth before the pig was fostered resulted in 0.07 and 0.03 kg ($P < 0.05$) lighter weaning weights for Yorkshire and Landrace pigs, respectively. Knowing how certain management practices can influence pig quality as measured by weaning weight can aid producers in recognizing the impact on quality associated with management decisions and system limitations.

References

1. Kaufmann et al. Genetic parameters for individual birth and weaning weight and for litter size of Large White pigs. *J Anim Breed Genet* 2000;117:121.

2. Wolter and Ellis. The effects of weaning weight and rate of growth immediately after weaning on subsequent pig growth performance and carcass characteristics. *Can J Anim Sci* 2001;81:363.

3. Klindt. Influence of litter size and creep feeding on preweaning gain and influence of preweaning growth on growth to slaughter in barrows. *J Anim Sci* 2003;81:2434.

eYeGrow - weight monitor for finisher

Guido Klement, Dr. agr; Simon Lague; Patrick van Dijk, B.Eng., B.ICT
Fancom BV, Panningen, Netherlands

Do you ever wonder if your finisher are growing enough? And can you predict the end weight early on? Or would you rather be surprised at the end?

eYeGrow (Figure 1) is a 3D infrared camera system that offers permanent stress-free weighing. Making use of machine learning and AI to reach an accuracy within 3%, from 7,500 weightings per day, 40 pigs. The system measures daily, uniformity and number of weightings, the latter being a useful welfare indicator in terms of pen activity. The system works on straw or slats, is cloud ready and therefore can integrate into any system. Due to infrared technology it operates 24/7 and does not need extra light (Figure 2). Benefits include no stress to your pigs, training required or manual intervention. The system is easy to install, can retrofit to existing pens and connects via PC or mobile device. Upgrade path to individual pig, due out 2018. Beside average weight of the pigs inside the pen, the system provides average daily weight gain, number of readings and information about uniformity of the group (Figure 3). Permanent growth monitoring enables better forecasting of the delivery weights and early detection of problems. Also optimization of feed programs, meaning a reduction of nutrient excess, nitrogen and phosphorus excretion.

Time saving

The eYeGrow is an automatic weight monitor for finisher. A 3D camera installed in the ridge of the section continually monitors and registers the animals so the computer can accurately calculate their weight. This is an entirely automatic and stress-free process for the animals. Manual weighing causes stress which will be reflected by a growth dip. eYeGrow prevents this

Figure 1: eYeGrow growth sensor installed under the ceiling

Figure 2: 3D infrared picture of a hog

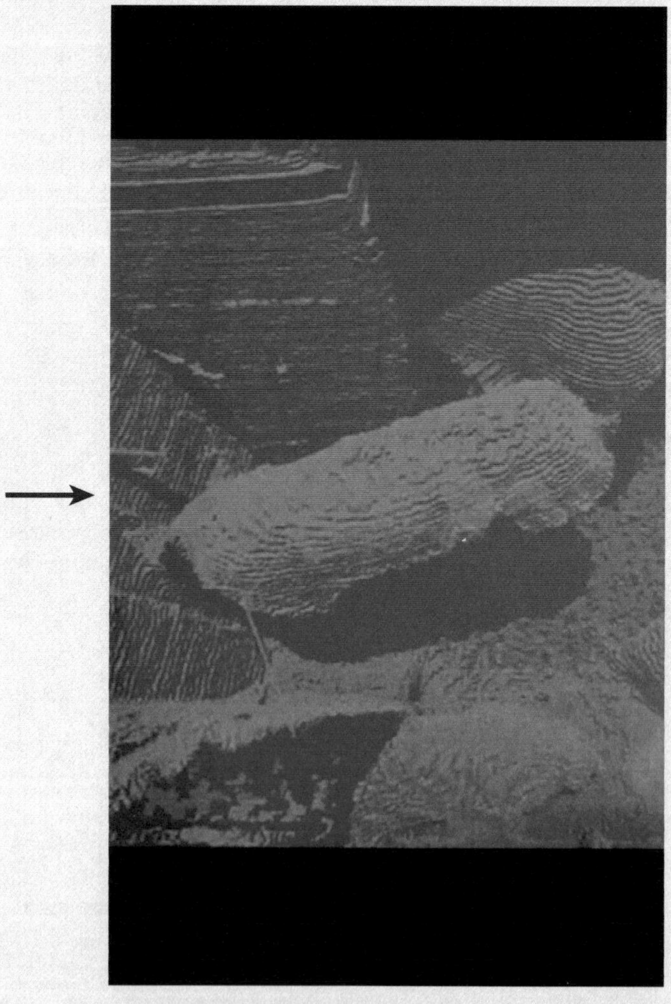

Figure 3: Information overview on a screen

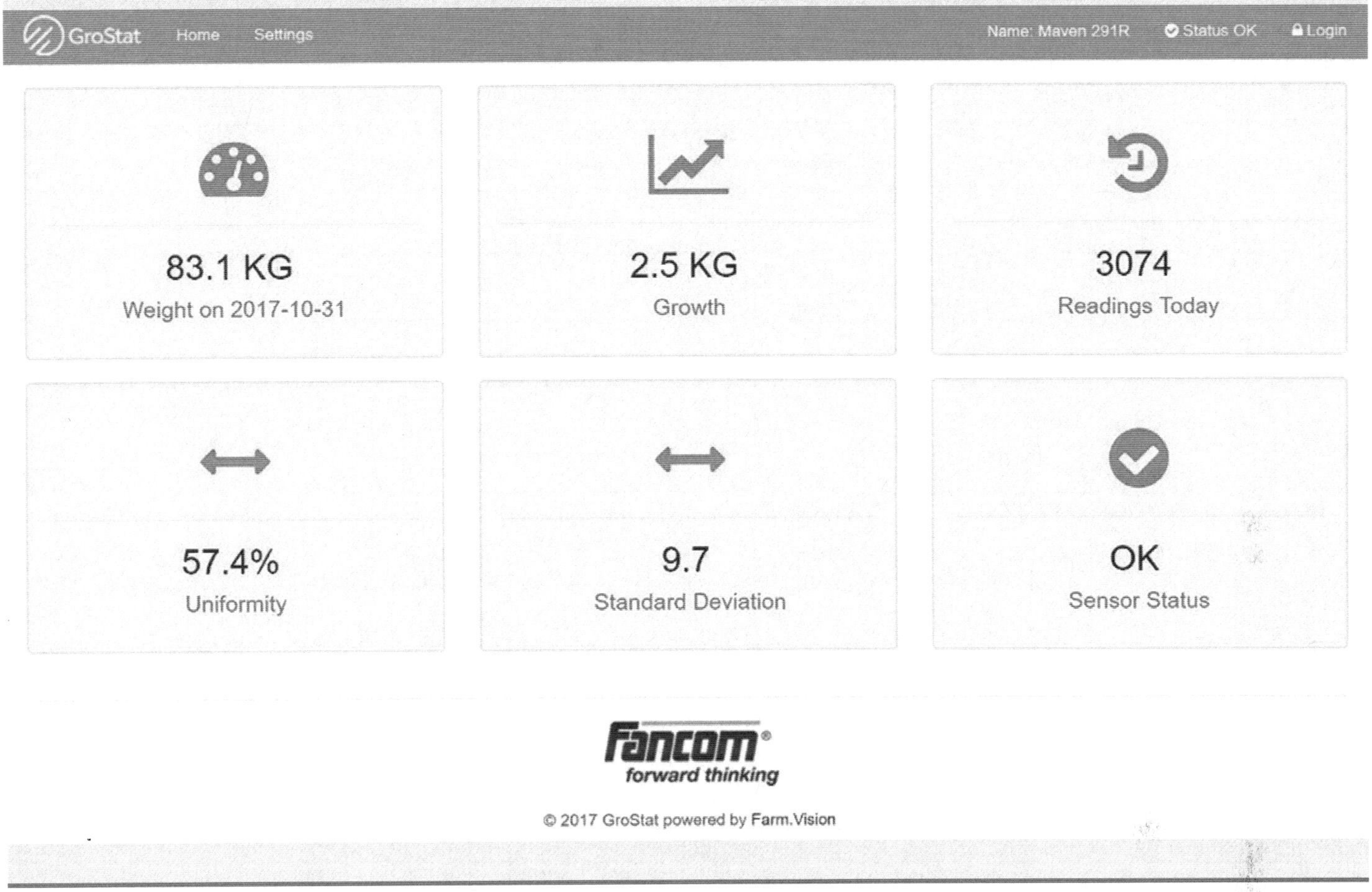

growth dip and requires no additional labour input.

Save on feed costs

eYeGrow allows the development of a group of finisher to be followed from day to day. You can easily adapt your operational management towards an optimal feed strategy and save feed costs.

Improved delivery strategy

You can also predict the delivery weight of the group far more accurately and achieve a higher price for the supplied finisher.

Insight into growth

The measurement data is stored in the cloud and important management information such as average weight, growth per day, number of measurements and uniformity is available 24/7 no matter where you are. The system can also be linked to FarmManager.

Features

- More than 97% accuracy (Figure 4)
- Easy installation without calibration
- Fits perfectly in existing houses
- Does not take up space in the pen or affect the number of animals
- Low maintenance
- No additional lighting needed
- Suitable for finisher > 40kg

Benefits

- Automatic weighing saves labour
- Stress-free weighing prevents growth loss
- Helps improve feed strategy to reduce feed cost
- Improved delivery strategy prevents weight reductions

Figure 4: Data from study showing accuracy of the system

Use of the swine bioassay test to evaluate the infectivity of UV-irradiated commercial liquid porcine plasma

E. Blázquez[1,2]; J. Ródenas[1]; C. Rodríguez[1]; J. M. Campbell[3]; J. Segalés[2,4]; J. Pujols[2]; J. Polo[1,3]
[1]APC Europe, S.A., Granollers, Spain; [2]IRTA-CReSA, Bellaterra, Spain; [3]APC Inc, Ankeny, Iowa; [4]Facultat de Veterinària, Universitat Autònoma de Barcelona (UAB), Bellaterra, Spain

Introduction

Spray dried plasma (SDP) is a functional protein source used in pig diets due to its beneficial effects on post-weaning performance and survival. Although the manufacturing process of SDP involves several safety features, additional safety steps may be investigated. Ultraviolet light at 254 nm wavelength (UV-C) is a non-thermal process that disrupts cellular transcription and replication, leading to death of bacteria, viruses and molds. The objective of the present study was to assess the risk of virus transmission to piglets using both liquid native plasma and UV-C irradiated plasma at different UV-C doses. Biosafety tests evaluated Porcine circovirus type 2 (PCV2) and Porcine reproductive and respiratory syndrome virus (PRRSV) as usual contaminant indicators because their importance as pathogens of concern for the global swine industry. Furthermore, PCV2 is one of the most heat and chemical resistant viruses.

Materials and methods

Ten 8-L batches of liquid plasma collected from a commercial abattoir (each batch of plasma was collected from a group of more than 10,000 pigs) were analyzed for presence of PCV2 and PRRSV. One 8-L batch that contained a high number of PCV2 DNA copies measured by real-time quantitative PCR (qPCR) was selected as source material for the study. Presence of PRRSV was also verified by a reverse transcriptase qPCR (qRT-PCR) method. The selected plasma batch was divided in 3 sub-batches: non-irradiated, irradiated at 3000 J/L, and irradiated at 9000 J/L.

A total of 37 piglets (17 ± 3 days of age) were randomly divided into 5 groups. At approximately 14 days of age blood samples collected from these piglets were verified negative for PRRSV antibodies and genome; in addition, these animals had low PCV2 antibody ELISA S/P values (of maternal origin) and were negative for PCV2 qPCR. Experimental groups were established once piglets were seronegative by ELISA against PCV2, at the age of 50 days.

Group 1 (n = 7) represented a negative control and was injected with 10 mL of phosphate buffer saline (PBS) solution. Group 2 (n = 8) received 10 mL of plasma UV-C irradiated at 9000 J/L. Group 3 (n = 8) received 10 mL of plasma UV-C treated at 3000 J/L. Group 4 (n = 8) was injected with 10 mL of native non-UV-C irradiated plasma (raw native liquid plasma). Finally, group 5 (n = 6) was the positive control, and was inoculated with10 mL of PCV2 at a dose of 100 $TCID_{50}$/mL. The products were administrated by intra-peritoneal route to all groups. All studied groups were located in different isolation rooms.

Biosafety of treated plasma was evaluated in pigs in terms of clinical signs, viremia, and antibody seroconversion against PRRSV and PCV2. Piglets were observed daily during a 50-day study period. Blood samples were collected on days 0, 15, 30 and 50 post-injection. Serum samples were analyzed for the presence of antibodies against PCV2 (Ingezim Circo IgG, 11.PCV.K.1/5 ELISA, INGENASA, Madrid) and PRRSV (IDEXX PRRS X3 Ab Test ELISA, IDEXX). Presence of virus was measured by qPCR for PCV2 (LSI VetMAX™ Porcine Circovirus Type 2 Quantification) and qRT-PCR for PRRSV (LSI VetMAX™ PRRSV EU/NA Real-Time PCR Kit), both test from Thermo Fisher Scientific.

Results and discussion

The selected plasma batch used in this study contained 2.71×10^4 PCV2 DNA copies/mL (Ct = 30,96) and a Ct value of 35.35 for PRRSV (European strain); moreover, it had an ELISA S/P ratio for PCV2 of 0,80 (low to moderate) and 2,99 (high) for PRRSV antibodies.

The biosafety test was validated by the negative control group (group 1) that remained free of PRRSV and PCV2 and showed absence of seroconversion against those viruses during the whole study period. In contrast, all piglets from group 5 (inoculated with PCV2) got infection in the first 15 days post-inoculation as measured by qPCR, indicating their susceptibility to PCV2 infection. These animals already seroconverted against PCV2 by day 30 post-inoculation.

Piglets injected intraperitoneally with UV-C irradiated plasma at both doses (group 2: 3000 J/L, and group 3: 9000 J/L) did not get infection by PCV2 or PRRSV and no antibodies were detected against both viruses. However, piglets injected with raw native plasma (group 4) were found PRRSV positive by qRT-PCR starting at day 15 after injection; moreover, PRRSV antibodies were first detected also at 15 days of the experiment until the end of the study. No evidence of PCV2 infection by qPCR or ELISA was detected in group 4 piglets.

These results demonstrate that UV-C treatment is a useful step to reduce load of viruses (at least PRRSV in the present study) in liquid plasma. Therefore, it should be considered as an intermediate additional safety feature for the manufacturing process of spray-dried plasma. Interestingly, although the raw native plasma contained PCV2 to a level of more than 10^4 DNA copies/mL, it resulted non-infective in intraperitoneally injected, susceptible pigs.

High quality semen program – HQS

P. Quadros[1], DVM, MS; K. Calderam[1], DVM, MS; J. Rocha[1], DVM; J. Quadros[2], DVM; R. Ausejo[3], DVM
[1]Bretanha Importação e Exportação Ltda. Passo Fundo – RS, Brazil; [2]MB Swine Reproduction, LLC, Charlotte, North Carolina; [3]Magapor Inc. Ejea de Los Caballeros, Zaragoza, Spain

Introduction

Ten years ago, in Brazil, our sales team figured out that the quality control made for semen doses produced was very poor – I could say that nothing was done. Why? Maybe Brazilian producers did not know the main importance of semen doses. Or maybe they did not know what the damage that one semen dose with low quality can do at reproductive rates, especially to the farrow rate and total born. Of course, before the indicator of farrow rate gets worse, some symptoms can be observed at the farm. The photo in Figure 1 shows a serious problem that happened in a farm in the south of Brazil. This farm received semen from a boar stud located 3 hours away from the farm. What was unbelievable was that the discharge began just 4 hours after the sows had been inseminated - around 80% of the sows had the problem. The bacteria found inside the semen doses was *Serratia Marcescens*. (Figure 2). We believe more in prevention than in treatment. After the diseases are inside the herd, it is much more expensive to eliminate or control than it is to prevent. For these motives, keeping a strong and standardized program to control the quality of semen doses before they are delivered to the farms, can make the difference between the success of the boar stud and facing big trouble.

Figure 1: Sow with discharge caused by contaminated semen

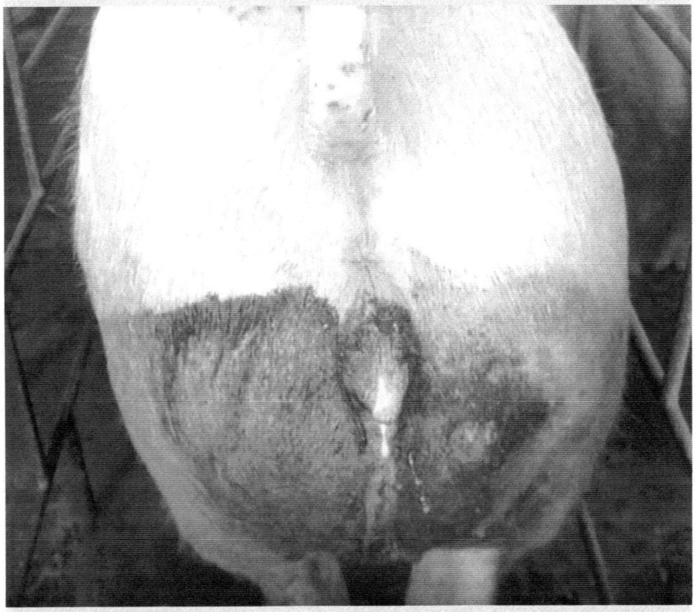

Description of the HQS – High quality semen program

The unique objective of the one dose of semen is to carry and split the genetic of the best boars to as many sows as possible. Not less important, is to be sure that the breeding sow will produce a good number of piglets. At the end of the day, semen doses are a very simple product: it is the result of a mix of the: ejaculated plus treated water and one preservative – all these raw materials are processed and mixed to result in one semen dose to move the genetics from a place to another - However, to be sure of the reproductive results and dissemination of the genetics, the semen dose must meet 4 assumptions of quality. 1) Effectively done with the right boar, to carry the right genetics – 2) The contamination inside the doses must be controlled – 3) The number of spermatozoa useful inside of the dose must be as expected, with the right concentration – 4) as in the last assumption of quality, the doses must have a shelf life in accordance with the extender used. Furthermore, based on our beliefs, thinking at the doses components, we believe that everything inside a semen dose – after the bag has been closed – is what was put inside. I mean, if you find *Serratia spp* inside a semen dose, it means in some moment during the flow of production you either put it inside or allowed it to go inside. For this reason, the most important principle of the HQS – High Quality Semen Program – is prevention – it is based in considering measures and taking actions to make sure that the semen doses will comply with their 4 assumptions of quality written above.

Execution and methodology used

The execution and methodology used to make the HQS program are based on: a) a technical audit made by a veterinary; b) exams and quality monitoring; c) reports and action plans. Everything is thought based on the flow of production shown in Figure 3; All steps of production are evaluated and measured. Keep in mind that the final objective is prevention. The program starts with a visit to audit the boar stud. In this visit, our veterinaries whom are specialized in reproduction issues, do an evaluation of the production flow considering 75 points. Each point is given a score (excellent, good, regular or bad). The 75 points evaluated can affect the final quality of the dose or likewise impact the productivity of the boar stud. These visits will occur every two or three months depending on the agreement made with the owner of the boar stud. In each auditing visit, a report and an action plan are generated based on the points that must be improved. The action plan is monitored with the help of the boar stud's team in between the each of the visits. Also, the evolution of

Figure 2: Laboratory report where isolated *Serratia marcescens* occurs

ISOLAMENTO BACTERIANO

TÉCNICA: Isolamento em meios de cultura

As três (03) amostras de sêmen submetidas a cultivo bacteriológico apresentaram os seguintes resultados:

Amostra 37640 diluído: incontáveis;
Amostra 37689 diluído: 1.400 UFC/mL;
Amostra 897 diluído: 5.700 UFC/mL.

Na amostra de sêmen (37640 diluído) foram isoladas colônias bacterianas identificadas através de provas bioquímicas como *Serratia marcescens*.

Metodologia Utilizada: Metodologia interna

the boar stud is monitored by a graphic as shown in Figure 4. The monitoring of the HQS – Program is made based on exams and tests done throughout the production flow. In terms of contamination, we usually test it at a different point of the production flow – for example, we systematically test for water contamination in: the source of water, the water box, the water going in to the reverse osmosis equipment, the water leaving the OR, the water stored to be used, and finally the water mixed with the extender. The same flow analysis is made for the semen, from the moment it leaves the boar until it is mixed with the ready extender (water plus extender). By performing all of these analysis throughout the production flow we are able to accurately identify the critical point where the system is being contaminated. Another important assumption of quality for semen doses is the "concentration". People usually measure the average concentration, of course, and also identify the standard deviation and coefficient of variation. Our program's focus is to ensure that the average is correct, but we must also guarantee that the coefficient of variation is below 15% - It is very important to look for a decrease in the average concentration, nowadays we work with 1.25 billion of useful cells in doses of semen, because I want to get more doses from my best boars to maximize the use of my better genetic. For this reason, when I decrease the average concentration, I must have a low coefficient of variation to be sure that the doses I am sending to farms are able to result in a sow's pregnancy. Figure 5 generically explains what can happen when a semen dose produced has a high coefficient of variation. Figure 5 was created with the use of the statistical tool "student T Table". Bearing in mind shelf life, which is the last assumption of quality of a semen dose, its measure is done inside the lab by an operational production team. The main goal is to know every boar - We know that the durability of semen has a high correlation with the boar's individuality – clearly we want to know for how long the samples kept in the lab will be useful, in order to warn farmers when we find doses with problems. Controlling an individual semen dose's shelf life – knowing every boar individually will allow us classify the boars keeping in mind his productivity and his semen's durability. Which we can then put in a graph like the one showed in Figure 6.

Conclusion

The HQS Program was created to help our clients to be more thoughtful regarding the production flow of semen doses. It is based on preventive actions, and working before trouble arises. It is not a program to compete with a veterinary consultant, labs or other professionals. Our goal is help our clients produce more semen doses per boar per year, produce more piglets (fathered by the best boars) per sow per year, and therefore help our clients save money.

Figure 3: Production flow for a boar stud

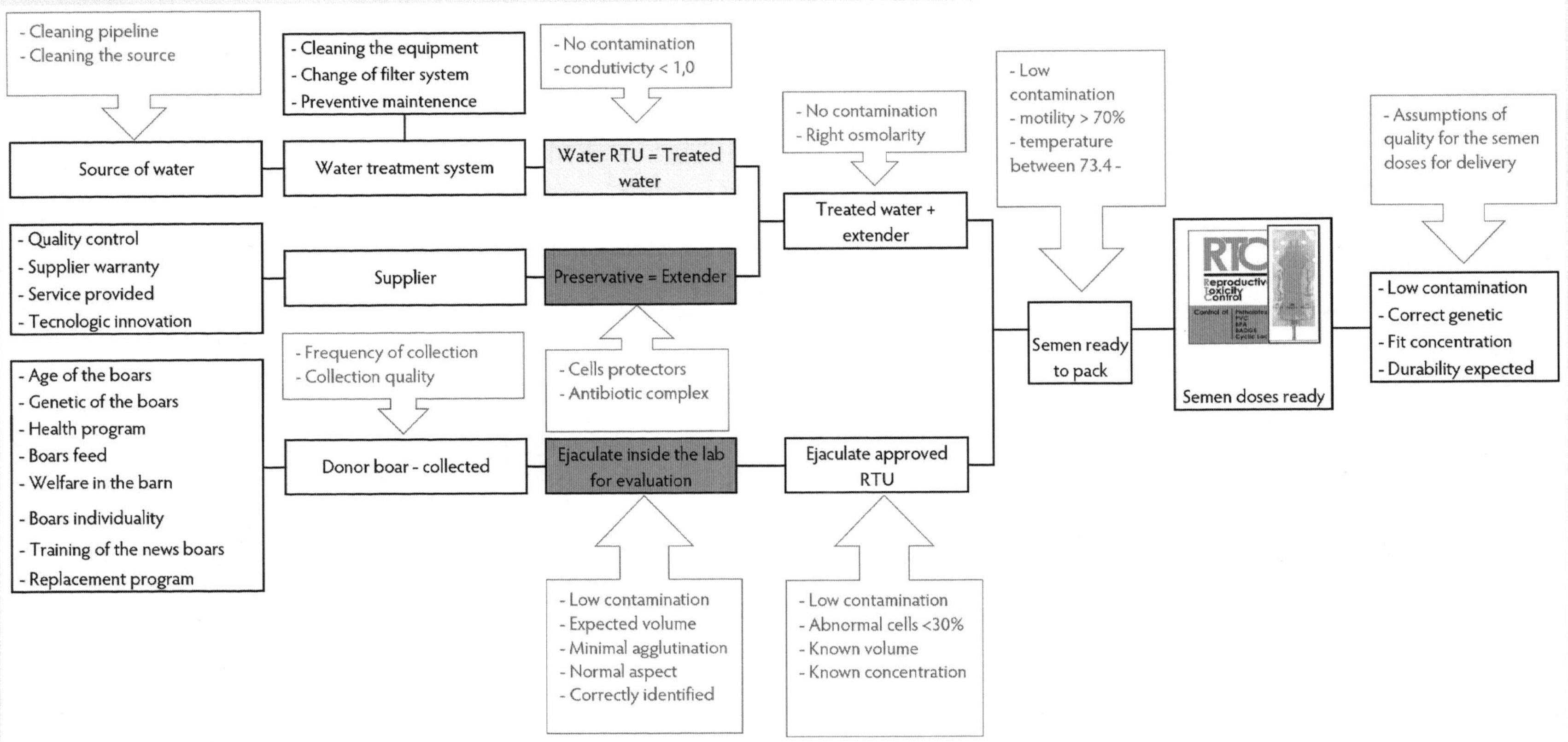

Figure 4: Evolution of evaluation from auditing visits

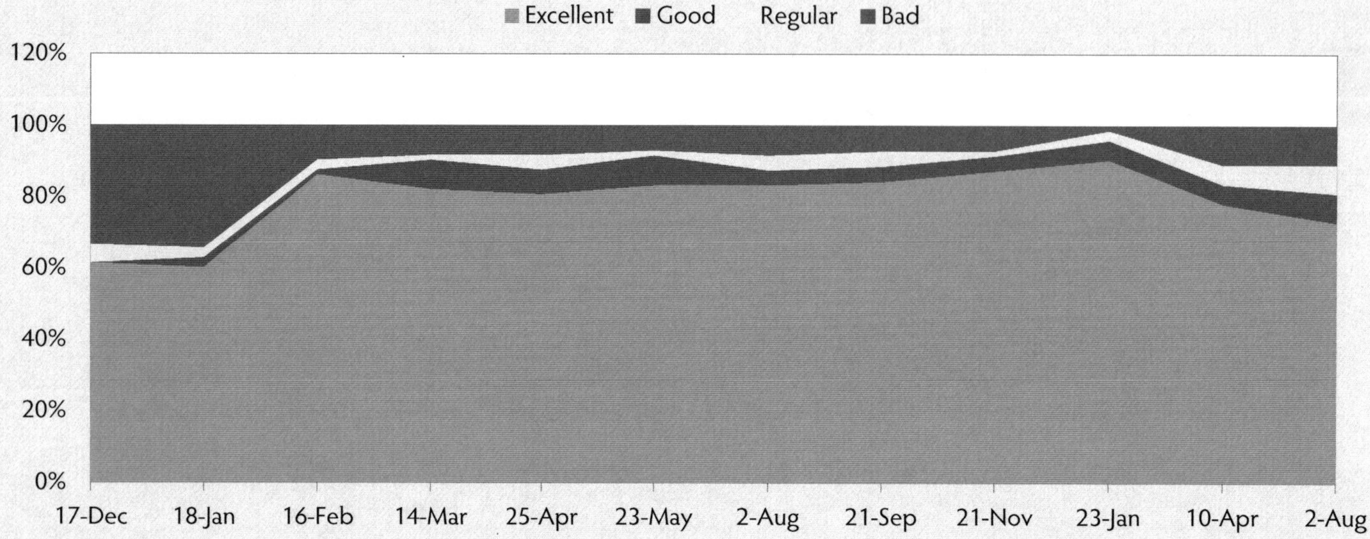

Figure 5: Example of how the coefficient of variation can impact the percentage of doses with low concentration sent to the farms

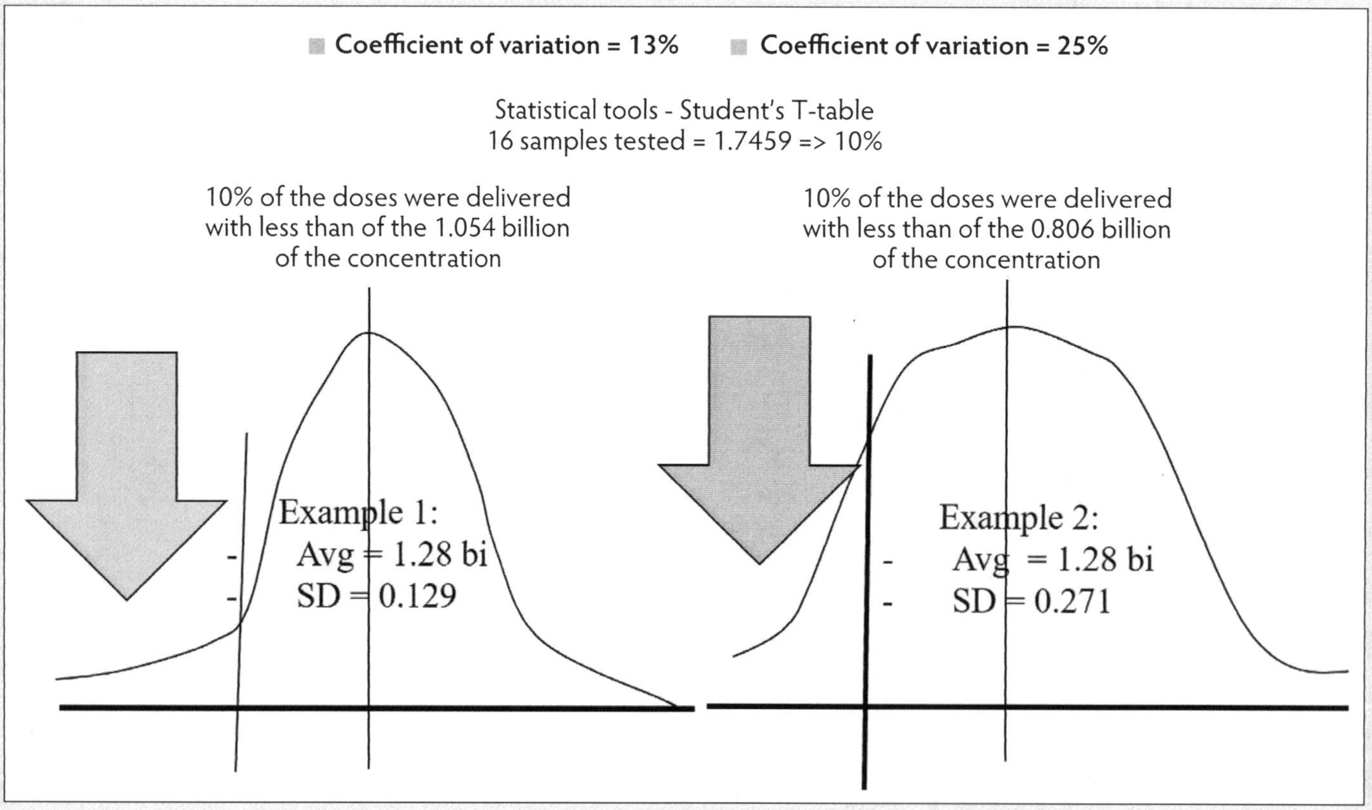

Figure 6: Illustration of boars classified by the durability of their semen doses

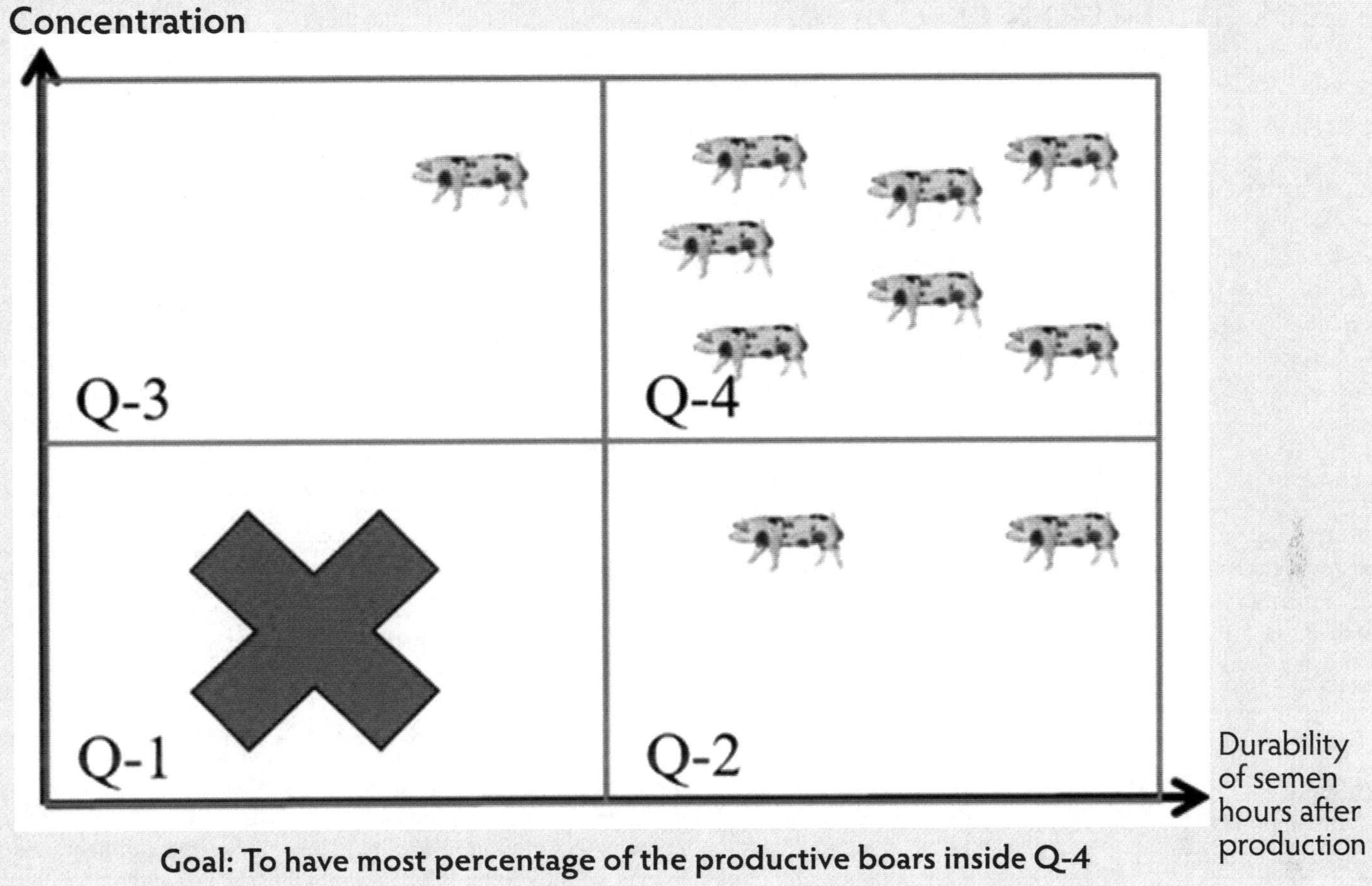

Cleaning and disinfecting critical surfaces: Farrowing crates and trailers

Dr. Ricardo Muñoz
Neogen

Introduction

The biggest challenge to cleaning and disinfecting a surface is the presence of biofilm and organic matter, which can easily become the substrate for bacterial growth and the development of other microbiological challenges. Biofilm is the number one biosecurity risk simply because it cannot be seen. Biofilm accumulation becomes critical when it limits the effect that disinfectants have on a surface.

Biofilm is a surface-coating layer made of a polymeric extracellular secretions and microorganisms. It occurs in an animal production environment as secretions accumulate on a matrix over time. Biofilms act as a mechanical barrier that lessens the effectiveness of disinfectants. Acidic cleaners can prepare a surface for disinfection by penetrating the slimy layer of biofilm so that disinfectants can be used more effectively and efficiently.

Any cleaner program should be based on results from tools that provide numerical results, such as adenosine triphosphate (ATP) testing, which offer more specific, timely information which producers can use to adjust their cleaning processes before problems become visible.

When it comes to battling biofilms, dry cleaning removes most dust and gross organic matter, but then a wet cleaning truly removes the biofilm and any additional organic matter. If a biofilm is allowed to remain, it negatively impacts the performance of the disinfectants used after the cleaning stage of the cleaning and disinfecting (C&D) process. This allows surfaces, such as farrowing units floor panels and/or aluminum-trailer to become challenging to the C&D process.

Materials and methods

Floor panels in farrowing crates: These are critical surfaces to clean and disinfect because of the potential amount of slime caused by secretions that accumulate on surfaces, resulting from exposure to placenta, blood and other secretions. A field observation was performed in four different sow farms in Brazil. The selection of these four facilities was done based on the recent historical records of high incidence of enteric-diarrhea problems right after the weaning period or early in the nursery time. Readings of ATP luminescence were done using AccuPoint Advance® after the regular cleaning process and after verification that those readings were very high (more than 300 relative light unites (RLU)). The same floor panels were cleaned using a more appropriate cleaner, dilution rate and after improving the mechanical application of the detergent.

Results and discussion

ATP readings performed right after the improved cleaning resulted in numbers lower than 300 RLU. Different parameters were collected right after the herds left the nursery room (weeks 10-12 according with the farm flow) The results of these four field observations show a trend that the lower the ATP readings in particular floor panel surfaces were, the better the performance of the herds in terms of enteric problems and body weight before the starting of the finishing stage (Table 1). The benefit of this type of hygiene verification tool is the timing and accuracy of obtaining a reference number, which can help a producer visualize objectively how good or poor the cleaning was done on critical surfaces before placing sows for the farrowing process. The value could help producers decide on corrective actions regarding how cleaning crews should handle dilution, mechanical action or detergents (acid/alkaline).

Aluminum-based surface from hog trailers: Eight trailers were tested — two swabs per trailer — to detect the level of cleanness of their internal surfaces. Trailers 1 and 5 are in the red zone, so they need to be re-cleaned and disinfected. Trailers 2 and 4 are in the yellow zone, meaning there is room for improvement, and the rest of the trailers showed acceptable cleaned and disinfected conditions (Figure 1).

Summary

For best biosecurity practices, cleaning is a very important step. Before disinfection can take place, producers must do the best dry clean and wet clean they can in order to reduce 90% or more of organic material and biofilms that act as barriers to disinfectants.

A quick hygiene evaluation of any critical surfaces, with the purpose of identifying how well the cleaning process was performed, is possible with the use of AccuPoint Advanced® to measure ATP. Before disinfectant is applied, AccuPoint Advanced® can be used to evaluate how well the surface has been prepared for disinfection, allowing for the maximum virus and bacteria elimination – in other words, the efficacy of cleaning.

Monitoring and/or verification: this has become one of the key elements to move biosecurity programs from a descriptive format to a more systematic program. In such a program, protocols should generate metrics rather than just qualitative-descriptive assessments. As other farm processes, such as vaccinations and intestinal integrity, are evaluated, it is necessary and critical to start implementing numbers behind biosecurity protocols that could be evaluated in a continuous way.

Table 1: Summary of key performance indicators

Name	ATP	ATP	Enteric issue	Enteric issues	% Antibiotics	% Antibiotics	Vaccines - Rota	Vaccines - Rota	Other enteric	Other enteric	(+) weight
	Before	After	Before	After	Before	After	Before	After	Vaccine-Before	Vaccine - After	After
Farm 1	8905	121	yes	Sharp decrease	7%	5%	na	na	10%	10%	
Farm 2	1295	20	yes	Sharp decrease	5%	4%	50%	0%	10%	0%	
Farm 3	6574	56	yes	Sharp decrease	5%	3%	na	na	na	na	200g
Farm 4	54231	34	yes	Sharp decrease	6%	3%	50%	0%	0%	0%	230g

Figure 1: Results from eight hog trailers

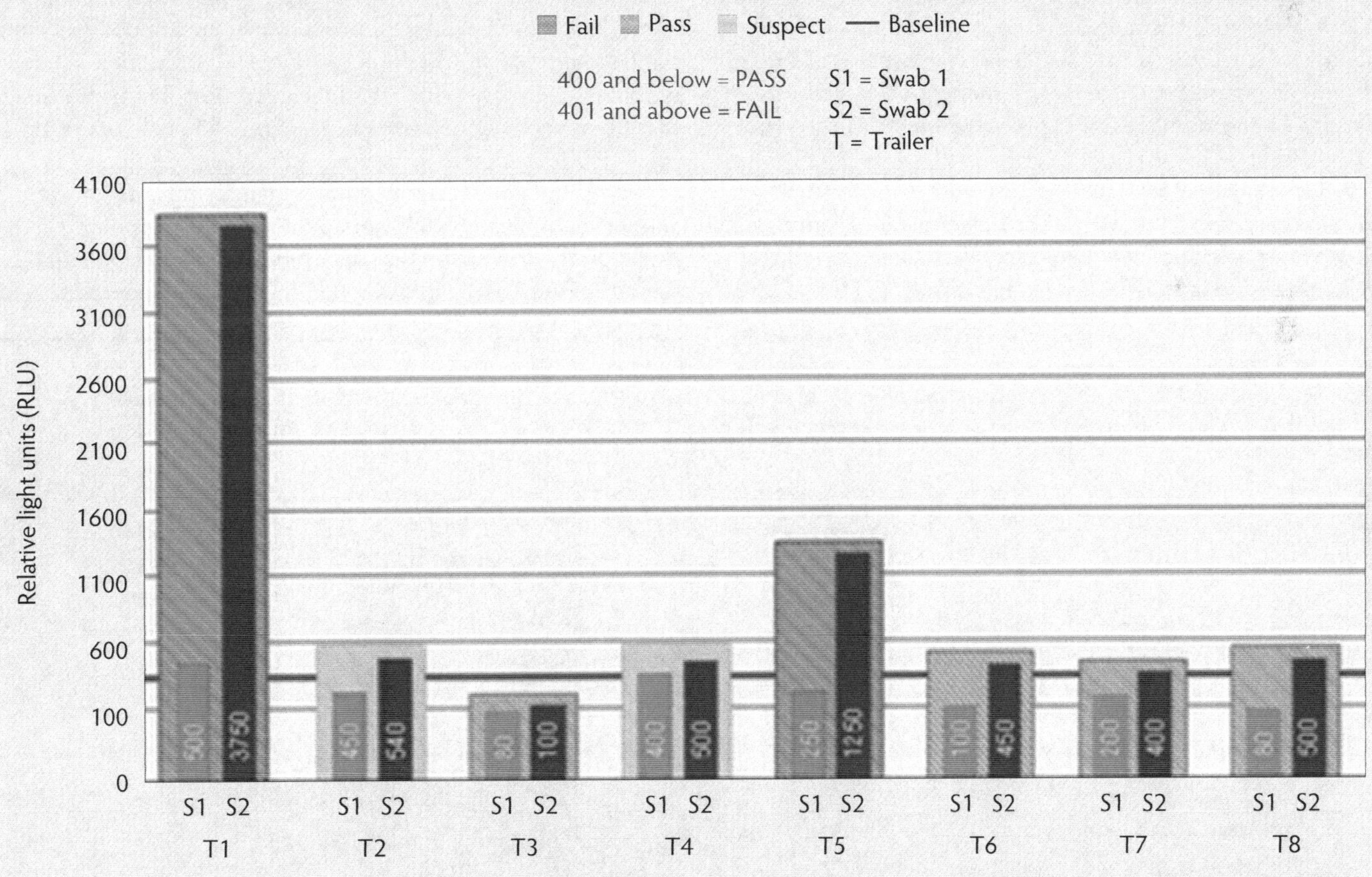

Protecting animal health and profitability via air filtration solutions, a front line defense against PRRSV, PEDV, and influenza

J. McGuire; M. Osborne, PhD
AAF Flanders, Louisville, Kentucky

Introduction

While virus outbreaks are not the only risk sow farm facility managers have to consider when allocating capital for operations, it is one that should be given serious consideration, based on its potential to significantly impact production and costs. Research findings on the impact of these diseases clarify the risk of substantial losses due to an outbreak. A 2013 study published in *the Journal of Swine Health and Production* estimated that the annual cost of productivity losses due to PRRSV in US national breeding and growing pig herds to be $664 million, up from $560 million in 2005. This equates to a loss of $1.8 million per day by the US pork industry (Figure 1). An additional $477.8 million is estimated to be lost each year on outbreak-related costs, including animal care and biosecurity. As an example, a feeder-pig operation with an endemic PRRSV infection in the nursery reported a 70% loss in profits due to a reduction of over $5.00 per pig attributed to the nursery stage alone, based on decreased growth rates, increased feed conversion, and increased mortality.[1] The likelihood of sustaining losses is increased if the operation is located within a five mile radius of other sow farms. The PRRSV virus can travel airborne for five miles or more, and its ability to constantly change creates the potential for genetic evolution of the strain.

Role of air filtration

Trials conducted by the University of Minnesota Swine Disease Eradication Center (SDEC) found that the risk of the indirect spread of these diseases can be reduced with a comprehensive biosecurity program that includes air filtration. Some ventilation systems in swine facilities are designed to supply fresh air and control indoor temperature. However, they are not optimally designed to provide air filtration. An effective air filtration system traps airborne viruses and their contaminants, preventing the viruses from entering and spreading throughout a facility. Disease can be transmitted by a variety of hosts in addition to air, including humans, supplies, equipment, manure, water, insects, and semen. A biosecurity program that addresses these other methods of transmission is critical, but is not a substitute for the significant impact of air filtration in preventing the introduction

Figure 1: Annual cost of productivity losses due to PRRSV in the US

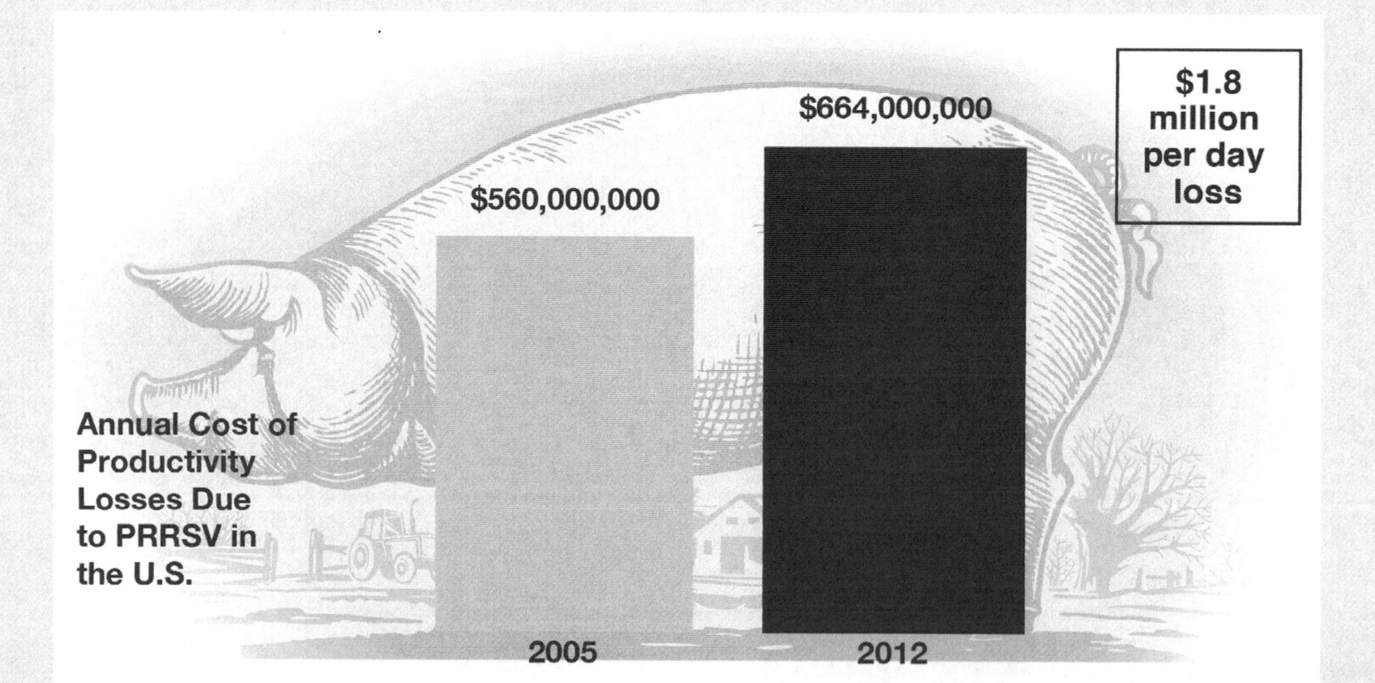

of the virus to the herd. Facilities without proper air filtration are eight times more likely to have an outbreak than facilities with sufficient filtration.[2]

When investing in an air filtration system for both herd protection and profitability, the following guidelines should be followed, in order to ensure maximum protection from the ravages of these viruses: (1) Choose filters with a low resistance to airflow, which reduces the number of filters needed, eliminating or decreasing the need for additional filter housings and building extensions; this lowers energy consumption and labor costs, while at the same time reducing waste; (2) Ensure buildings are airtight; (3) Monitor accurately the differential pressure between the interior and exterior of your building; (4) Ensure filter brackets are airtight upon installation; (5) Choose filters with the recommended level of filtration for your facility and level of risk; (6) Follow manufacturer recommendations for replacing filters; (7) Install an efficient backdraft damper.

While cost is always a significant factor, there should also be consideration regarding whether an operation can afford to not invest in an air filtration system. For example, a single outbreak of PRRS can cost two times more than investing in a filtration system, or the equivalent of operating the system for four to five years. The bottom line is that if the decision prevents one severe outbreak, it has paid for itself.[3] The cost of equipping a facility with filters varies, based upon the configuration, age, and maintenance history of the building and equipment, with newer facilities generally requiring less of an investment to filter.

Air filtration and airborne pathogens

Viruses such as PRRSV are a major concern for sow farms, boar studs, and growing herds (Table 1). Air filtration prevents airborne pathogens from entering and spreading throughout a farm, preventing costly outbreaks of a broad spectrum of diseases that impact swine health, production, and operating costs. An air filtration system implemented to protect against the virulent PRRS virus also prevents airborne transmission of endemic *Mycoplasma hyopneumoniae* (M hyo), the cause of Enzootic Pneumonia and Porcine Epidemic Diarrhea (PEDv), which is currently a significant threat to European farms and is being monitored closely by the US Department of Agriculture.

In order to prevent airborne transmission of pathogens and therefore protect animal health and profitability, air filtration solutions must produce and maintain positive indoor air quality. Filters should be utilized that have high minimum efficiency reporting value (MERV) ratings due to their efficiency in removing dust particles and airborne contaminants from the air, including disease-causing viruses and bacteria. Additionally, moisture-resistant filters, frames, and components that prevent degradation and clogging in high moisture environments, such as sow farms that use evaporating cooling systems for temperature control, should be considered. Extremely durable, longer lifecycle filters help reduce maintenance costs and produce less waste.

A filter framing system should be used that is designed specifically for the swine environment and includes an interlocking design that allows a combination of frames to form a grid. This provides the ability to build modular filter banks of prefilters and final filters that will hold in place in horizontal applications. The frames should be durable and corrosion resistant, creating a lasting, airtight seal to prevent the entry of pathogens into the facility.

Table 1: The devastating effects of PRRSV

While the severity and long-term effects of PRRS may vary, the potential for devastating losses remains consistent.
With an acute outbreak, more than half of a herd's litters may be affected.

Effects on breeding herd	Effects on growing herd
Anorexia	Increased mortality in conjunction with increases in other respiratory infections
Fever	Slower growth rates
Cyanosis	Non-uniform performance
1% – 2% increase in mortality	Potential for persistent respiratory disease with secondary infections
1% – 2% increase in late-term abortions	
Up to 20% increase in premature farrowing	
Dead and mummified piglets in farrowed litters	
Variations in breeding and farrowing intervals	
Temporary decline of semen quality	
10% – 40% percent increase in pre-weaning piglet mortality	
Increased prevalence of secondary infections	

Additional air quality precautions

Comprehensive air filtration tools play a key role in the optimization for clean indoor air in the swine environment. A handheld reporting tool that displays data from real time monitoring of the filter system's performance and automatically tracks unit location, filter type, filter size, pressure drop, MERV rating, and particulate matter levels of 1, 2.5, and 10, gives the user the ability to proactively resolve potential issues in air quality before they arise (Figure 2, 3). Such a program can optimize preventive maintenance scheduling, decrease deferred maintenance, save money, reduce risk, and gain time. Additionally, by monitoring pressure drop, facility managers can determine the changeout point for the swine barns' filters that offers the best value and energy efficiency.

Protecting animal health and profitability can be achieved through optimal air filter performance and maintenance. Combining the selection of the proper filter, the use of compatible filter frames, and the utilization of a reporting tool with the ability to monitor real time air quality, can definitively accomplish these goals.

References

1. Kerkaert BR, Pijoan C, Dial G. Financial impact of chronic PRRS. *Proc Allen D. Leman Swine Conf.* 1994; 217-218.

2. Dee, S., Cano, J.P., Spronk, G., Reicks, D., Ruen, P., Pitkin, A. and D. Polson. 2012. Evaluation of the long-term effect of air filtration on the occurrence of new PRRSV infections in large breeding herds in swine-dense regions. *Viruses,* 4: 654-6620.

3. Ricard, M. and Pouliot, F. 2013. Air Filtration in Swine Buildings. Centre de développement du porc du Québec inc., 10.

Figure 2: Real-time monitoring and automatic tracking

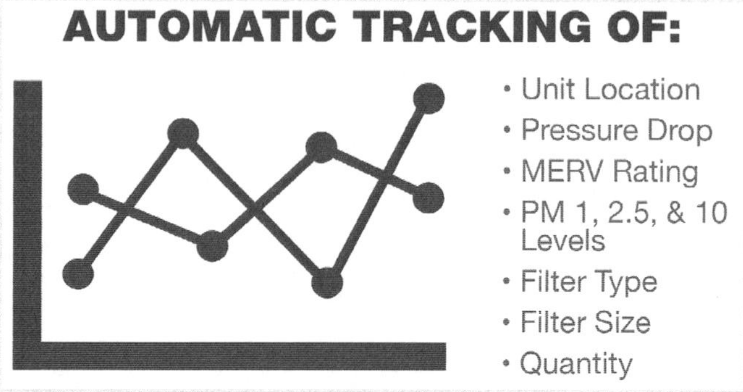

Figure 3: On-demand air quality analytics

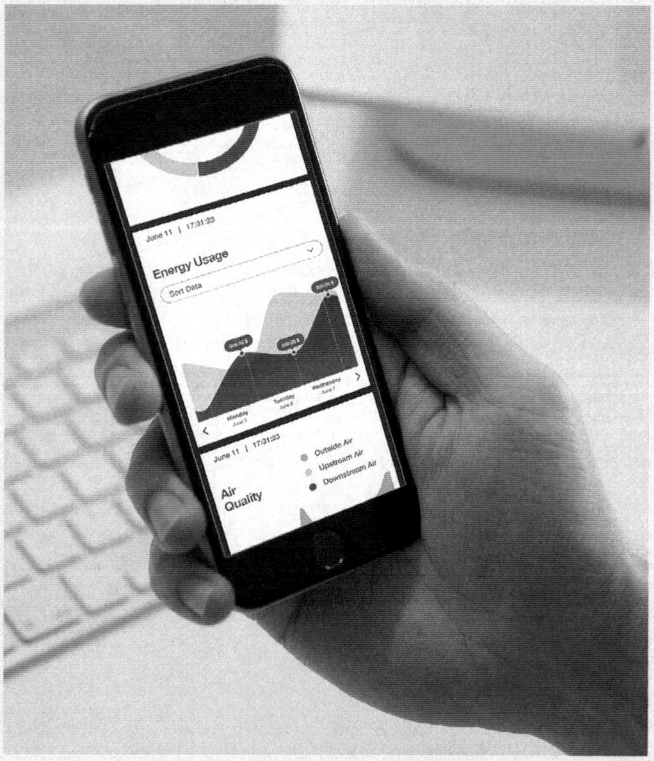

Field safety study of a vaginal mucosal vaccination for porcine reproductive and respiratory syndrome virus (PRRSV) in sows

Glen Almond[1]; Cary Sexton[2]; Juliana Bonin Ferreira[1]; Andy Smythe[3]; Jessica Seate[3];
[1]College of Veterinary Medicine, North Carolina State University, Raleigh, North Carolina;
[2] Livestock Veterinary Services, Kinston, North Carolina; [3]Aptimmune Biologics

Introduction

Porcine reproductive and respiratory virus (PRRSV) represents the most significant pathogen in the US pork industry. Current vaccines often are beneficial; however, swine veterinarians and producers recognize the fact that heterologous strains of the virus often overwhelm the immunological protection provided by the commercial vaccines. Mucosal vaccines, specifically ones produced as autogenous vaccines, offer a new method of immunization against PRRSV. For piglets and recently weaned pigs, the administration of an intranasal, mucosal vaccine is relatively easy to achieve on commercial farms. In contrast, intranasal vaccination of sows represents a major obstacle for producers. Therefore, the objective of this study was to evaluate the clinical safety of a mucosal PRRS vaccine administered intravaginally to sows in various stages of their reproductive cycle.

Materials and methods

The study was conducted on a 2000-head, commercial sow farm. Forty sows (mixed parity) were selected, based on the stage of reproduction. The Treatment (TMT) Group included 10 pregnant sows (mid-gestation) and 10 open sows. Each sow received the killed vaccine (2 ml dose; Strain 1-7-3 [killed], Barricade PRRS®) as a small volume infusion via the OvuGel gun into the anterior vagina. The Control (CON) Group included 10 pregnant sows (mid-gestation) and 10 open sows (in same breeding groups as the treatment animals). The CON group received a placebo intravaginal infusion. All sows were monitored for 7 days post vaccination. Blood samples were collected from sow at day 0 (prior to vaccination or placebo), day 21, and day 50. This sampling scheme was based on the temporal sequence of events, ie antibody production, as previously described.[1] In addition, oral fluids were collected from each sow at the time of blood sampling on days 0 and 21. The blood was transported to the laboratory and centrifuged to collect serum All serum samples were tested for ELISA and neutralizing antibody responses (Fluorescent Foci Neutralization Test) at Iowa State University Diagnostic Laboratory. The IgA levels in oral fluid samples also were evaluated with ELISA testing at the ISU Diagnostic Lab. The test results were analyzed by an ANOVA with treatment, reproductive stage, day and interactions as the independent variables. Means were compared with Tukey's test.

Since a critical component of the study was to evaluate the clinical safety of the intravaginal, mucosal PRRS vaccine, each animal was assessed by clinical scoring and feed consumption for 7 days after the administration of the vaccine. Subsequent reproductive parameters also were evaluated.

Results and discussion

As shown in Table 1, many of the sows had existing ELISA and FFN titers prior to the vaccine and placebo infusions. In fact many of the CON sows had greater ($P < 0.05$) FFN titers at the initiation of the study. Few other differences between TMT and CON sows were evident in the open or pregnant sows. One open CON sow died prior to Day 21 and her data was excluded from the study. All other open sows (both CON and TMT) conceived at a subsequent mating (prior to Day 21), and all pregnant sows remained pregnant throughout the course of the study. At no time did the TMT sows become inappetant nor lose their pregnancies. The typical route of natural PRRSV infection is through mucosal membranes. Therefore, this method of vaccination appears to be a logical and valid approach. To conduct whole-herd vaccinations, producers would have the option of vaccinating sows without the use of needles for IM injections. Studies with PRRSV-naïve sows are needed to show antigenic stimulation.

References:

1. Lopez OJ, Osorio FA. Role of neutralizing antibodies in PRRSV protective immunity. *Veterinary Immunology and Immunopathology* 2004; 102:155–163.

Table 1: PRRSV ELISA (S/P Ratio) and FFN titers in open and pregnant sows following the administration of an intravaginal infusion of killed Strain 1-7-3, Barricade® vaccine or placebo. The IgA ELISA was performed on oral fluid samples collected on Day 0 and Day 21. Due to the large variation, no differences in FFN nor IgA were detected. ND = not determined.

Status		Day	PRRS X3	SE	IgA	SE	FFN	SE
Open	Placebo	0	1.02[a]	0.13	1.35	0.84	91.56	47.00
		21	0.98[ab]	0.13	2.46	0.77	161.33	47.00
		50	0.88[abc]	0.13	ND		225.8	47.00
	Vaccine	0	0.51[abc]	0.13	1.65	0.87	63.00	47.00
		21	0.36[c]	0.13	1.00	0.82	54.20	46.42
		50	0.37[bc]	0.13	ND		15.60	47.00
Mid-gestation	Placebo	0	0.35[c]	0.13	0.84	0.84	202.67	47.00
		21	0.46[abc]	0.13	1.45	1.11	224.44	47.00
		50	0.34[c]	0.13	ND		38.00	47.00
	Vaccine	0	0.40[bc]	0.13	0.86	0.89	157.60	164.52
		21	0.54[abc]	0.13	1.41	0.94	74.40	44.60
		50	0.50[abc]	**0.13**	ND		42.00	44.60

[abc] Within column, means with different superscripts differ (*P* < 0.05).

The effects of Defusion® products on the performance of nursery pigs fed diets contaminated with vomitoxin

Kari Saddoris-Clemons, PhD; Sara Ebarb, MS; Colleen Fowler, MS; Sabrina Williams, MS; Don Giesting, PhD
Provimi, Brookville, Ohio

Introduction

Vomitoxin, or deoxynivalenol (*DON*) is a mycotoxin, produced by *Fusarium graminearum* that is commonly found in corn, other grains, and their processed co-products. Vomitoxin is most prevalent when wet conditions are present before and during harvest. *Fusarium graminearum* (Gibberella zeae) produces both DON and zearalenone, and contaminated feedstuffs may contain both of these mycotoxins. Vomitoxin contamination of grains can lead to reductions in feed intake and performance when fed to swine. Levels as low as 1 ppm of DON negatively impact feed intake (ADFI) and daily gain(ADG), and levels of 5 ppm can cause severe reductions in feed intake (Diekman et al., 2005). In addition, DON negatively impacts intestinal barrier function, impairs the immune system, and increases oxidative stress (Pierron et al, 2016; Weaver et al., 2014b). In the 2017 corn crop, most corn from parts of Ontario, Indiana, Illinois, and Ohio had elevated levels of DON.

Defusion® is a complex product formulated to support animal health and growth. A decade of internal and external research have been conducted to establish efficacy of this product in swine diets containing quality challenged corn and dried distillers grains and solubles (DDGS), specifically when DON is measured as an indicator of quality. Defusion has consistently shown improvements in ADG, ADFI, and feed efficiency in diets with quality challenged corn with DON as high as 5 ppm. Recently, efforts have been focused on enhancing Defusion to provide additional support to swine during periods of DON exposure. The objective of the trial evaluated the impact of a formulation update to Defusion on ADG, ADFI, gain/feed (GF), and return over feed costs (ROFC) in diets with quality challenged corn and DDgs with 3.2 ppm DON in 25-48 lb pigs.

Materials and methods

The study was conducted utilizing 2 barns, for a total of 1822 pigs, with 12 pens per treatment. The study was a completely randomized block design. Pigs were split sex housed and pens contained equal numbers of heavy, medium, and light weight pigs, to allow for similar initial body weight for all pens. Barns were blocked by location and pens were randomly allocated to treatment groups within block. Treatment groups were the following: 1) Positive control (PC) with clean corn and DDGS, 2) Quality challenged (QC) corn and DDGS + 0.25% of a Sulfur (S)-based ingredient (SBI), 3) QC corn and DDGS + 0.25% of Defusion, 4) QC corn and DDGS + 0.25% Defusion Prime, 5) QC corn and DDGS + 0.25% of a SBI + 0.01% Yeast derivative, 6) QC corn and DDGS + 0.25% Defusion + 0.01% Yeast derivative, and 7) QC corn and DDGS + 0.25% Defusion Prime 0.01% Yeast derivative. A description of the dietary treatments can be seen in Table 1.

Pigs were assigned to treatment diets at 23.5 days post-weaning (~41 days of age). Pen weights were taken on D0 (23.5 post-weaning) and again at the end of the study on D18 (41.5 post-weaning). All diets were analyzed for DON content at Trilogy Laboratory (Washington, MO) by HPLC. Feed intake was measured and mortality and removals were recorded. Average daily gain, ADFI, FG, return over feed costs (ROFC) and the probability of mortality and removals occurring were determined.. Statistics were analyzed using PROC GLIMMIX in SAS 9.3 (SAS, Cary, NC). Differences were considered significant a $P < 0.05$, and trends were reported at $P < 0.10$. Pre-planned contrasts included : 1) Clean corn vs S containing compound, 2) S containing compound vs Defusion and Defusion Prime, 3) Defusion vs Defusion Prime and 4) No yeast derivative vs diets containing the yeast derivative. Only results for the Defusion formulation aspect will be discussed in this abstract.

Results and discussion

Dietary analysis indicated the PC diet contained 1.1 ppm of DON and the average of the treatment diets contained 3.2 ppm DON. Table 2 shows the impact of S compound, Defusion, and Defusion Prime on pig performance. Feeding quality challenged corn and DDGS + SBI significantly reduced ($P < 0.01$) overall ADG, ADFI, final BW, and negatively impacted ($P < 0.10$) feed efficiency and cost/lb of gain compared to pigs fed the clean diets. The inclusion of either Defusion or Defusion Prime significantly improved ($P < 0.01$) overall ADG, ADFI, final BW, feed efficiency, and reduced ($P < 0.01$) cost/lb of gain compared to pigs just fed the SBI. Pigs fed Defusion Prime had significantly ($P < 0.05$) higher overall ADG and tended ($P < 0.10$) to have higher overall ADFI and final BW compared to pigs fed Defusion.

Return over feed costs (ROFC) was calculated assuming gain was worth $1.00/lb and accounted for any differences in mortalities and removals and is shown in Table 3. Pigs fed the quality challenged diets + SBI had lower ($P < 0.01$) ROFC compared to pigs fed the clean diets. This is not unexpected, as no product on the market is known to consistently mitigate 100% of the losses incurred in pigs fed diets contaminated with moderate to high levels of DON. The inclusion of either Defusion or Defusion Prime

Table 1: Description of dietary treatments

Trt Number	Description	DON contaminated corn and DDGS	SBI*	Defusion	Defusion Prime	Yeast Derivative
1	PC	No	No	No	No	No
2	S based compound	Yes	Yes	No	No	No
3	Defusion	Yes	No	Yes	No	No
4	Defusion Prime	Yes	No	No	Yes	No
5	S +Yeast	Yes	Yes	No	No	Yes
6	Defusion + Yeast	Yes	No	Yes	No	Yes
7	Defusion Prime + Yeast	Yes	No	No	Yes	Yes

* SBI= sulfur based ingredient

Table 2: The effect of sulfur based ingredient (SBI), Defusion, and Defusion Prime on ADG, ADFI, and feed efficiency in late nursery pigs fed diets contaminated with DON

	Treatment					Contrast *P*-value		
Variable	PC	SBI	Defusion	Defusion Prime	SEM	PC vs SBI	SBI vs Defusion and Defusion Exra	Defusion vs Defusion Prime
Initial BW, lbs	27.5	27.5	27.5	27.5	0.926			
ADG, lbs/d	1.146	0.953	1.010	1.044	0.033	< 0.01	< 0.01	< 0.05
ADFI, lb/d	1.528	1.325	1.372	1.404	0.037	< 0.01	< 0.01	< 0.10
Feed efficiency lb/lb	1.319	1.439	1.370	1.351		< 0.01	< 0.01	NS
Final BW, lbs	48.7	45.0	45.8	46.1	0.683	< 0.01	< 0.01	< 0.10

had a significantly higher ($P < 0.01$) ROFC compared to pigs fed diets with only a SBI. ROFC was $0. 0.48 higher for pigs fed Defusion Prime compared to pigs fed Defusion, but this difference was not significant ($P < 0.10$).

The probability of a mortality and removal occurring is shown in Table 4. The probability of a mortality and removal occurring tended to be higher ($P < 0.10$) in pigs fed the quality challenged corn + SBI compared to pigs fed the clean corn. The inclusion of either Defusion or Defusion Prime significantly reduced ($P < 0.05$) the probability of a mortality or removal occurring compared to pigs fed diets with only SBI. No differences ($P > 0.10$) were detected between the probability of a mortality and removal occurring between pigs fed diets with Defusion compared to pigs fed diets with Defusion Prime.

Under the conditions of this study, both the original Defusion product and Defusion Prime were effective in mitigating more of the negative performance effects and reducing mortality and removals observed in pigs fed diets containing quality challenged corn and DDGS than just a single SBI. The Defusion Prime formulation was more effective at mitigating some of the negative impacts of feeding quality challenged corn and DDGS on ADG and ADFI compared to the original Defusion formulation, without increasing cost/lb of gain and numerically improving ROFC. Pigs fed Defusion Prime had a 9.5% higher ADG, a 6% higher ADFI, improved feed efficiency by 6%, lowered cost/lb of gain by over 5% and improved ROFC by $1.54 compared to pigs fed only a SBI. Defusion Prime will replace Defusion in the marketplace due to the additional benefits observed with the updated formulation.

References

1. Diekman, M., T. Coffey, R. Hill, E. D. Purkhiser, D. Reeves, and L. Young. Mycotoxins and swine performance. Pork Information Gateway. PIH-129. 07-06-05.

2. Pierron, A., I. Alassane-Kpembi, and I. Oswald. 2016. Impact of mycotoxin on immune response and consequences for pig health. Animal Nutrition. 2: 63-68.

3. Weaver, A. C., M. T. See, and S. W. Kim. 2014b. Protective effect of two yeast based feed additives on pigs chronically exposed to deoxynivalenol and zearalenone. Toxins 6:3336-3353.

Table 3: The effect of sulfur based ingredient (SBI), Defusion, and Defusion Prime on cost/lb of gain and closeout ROFC in late nursery pigs fed diets contaminated with DON

	Treatment					Contrast *P*-value		
Variable	PC	SBI	Defusion	Defusion Prime	SEM	PC vs SBI	SBI vs Defusion and Defusion Exra	Defusion vs Defusion Prime
Cost/lb gain	0.202	0.221	0.212	0.209	0.013	< 0.01	< 0.01	NS
ROFC[a], $	16.44	12.95	14.01	14.49	0.52	< 0.01	< 0.01	NS

[a] Calculation assumed 1 lb of gain was worth $1.00

Table 4: The effect of a sulfur based ingredient (SBI), Defusion, and Defusion Prime on mortality and removals in late nursery pigs fed diets contaminated with DON

	Treatment				Contrast *P*-value		
Variable	PC	SBI	Defusion	Defusion Prime	PC vs SBI	BI Defusion and Defusion Prime	Defusion vs Defusion Prime
Mortality and Removals,% (lower and upper limit)	1.9 (1.22, 3.29)	3.7 (2.04, 4.34)	1.7 (1.09, 3.10)	2.2 (1.36, 3.44)	< 0.10	< 0.05	NS

Effects of an antioxidant support technology on sows and piglets

K. Price[1], PhD; C. Fowler[1], MS; D. Giesting[1], PhD; M. Van Den Bosch[2], PhD; J. Zhao[3], PhD; B. Humphrey[1], PhD
[1]Provimi, Brookville, Ohio [2]Cargill Animal Nutrition Innovation Center, Velddriel, The Netherlands
[3]Cargill Animal Nutrition Innovation Center, Elk River, Minnesota

Introduction

Throughout pregnancy and lactation, sows experience physiological stress that can cause an imbalance in their oxidative state. In late gestation and lactation, sows are in catabolic metabolism. The catabolic condition causes tissue mobilization, increased free radical production, and thus elevated oxidative stress.[1] Additionally, the evolution of genetic selection for hyper prolific sows has resulted in increased litter size.[2] Unfortunately, this improvement can be accompanied by lower piglet livability, including increased incidence of early mortality and stillbirth. The decrease in livability and higher early mortality may be associated with oxidative stress. Litter size and litter weight gain have been shown to be strongly negatively correlated with increased free radicals in the plasma of piglets.[3] Provimi provides technology that has been shown to support antioxidant status in multiple species through the Promote®AOX™ products. For sows, this support for the antioxidant system may improve the performance of highly prolific animals and their litters. Two studies were conducted to determine if providing additional antioxidant support with AOX could improve antioxidant status of sows and piglets, piglet livability, and litter performance.

Materials and methods

The first study was conducted in Olszty, Poland with the University of Warmia and Mazury at a facility typical of EU swine operations with continuous farrowing. Sows were housed in fully enclosed buildings with ventilated and heated rooms to maintain animal comfort. The study utilized 52 breeding females (PIC; all parities) divided into four treatments. Sows were blocked by parity, and treatments were applied from day 1 of pregnancy until weaning. The treatments were as follows; 1) Negative control with no supplemental vitamin E (NC); 2) 50 ppm supplemental vitamin E in gestation and 75 ppm supplemental vitamin E in lactation (Medium E); 3) 100 ppm supplemental vitamin E in gestation and 150 ppm supplemental vitamin E in lactation (High E); and 4) treatment 2 with 100 ppm AOX in gestation and 150 ppm AOX lactation (Medium E+AOX). Treatments are shown in Table 1. Typical European barley-wheat-triticale diets were used. Sow reproductive performance and litter performance were collected. Blood samples from the sow at farrowing and weaning, as well as piglets at 21 days of age were collected for oxidative status analyses. Serum samples were tested for Vitamin A and E concentrations, total antioxidant status (TAS), and concentrations of antioxidant enzymes (superoxide dismutase, (SOD), and glutathione peroxidase (GPx)). Data were analyzed using one-way ANOVA and Duncan's test with STATISTICA 10 (Palo Alto, CA).

The second study was conducted at a commercial research facility with 540 breeding females (Genesus KS-11 gilts, and PIC C-29 sows; all parities). Each treatment utilized 90 females. The experimental design consisted of a 2 x 3 factorial to determine the effect of AOX inclusion, vitamin E level, and its interaction (Table 2). Gilts and sows were moved to the group housing section of gestation at approximately 40 days post-breeding and fed a common diet. One of two gestation treatments was fed from day 85 of gestation to farrow loading, which occurred at day 109 to 112 gestation. The treatments were; 1) 60 ppm vitamin E (Vitamin E, G1), and 2) 60 ppm vitamin E +120 ppm AOX (AOX, G2). Predicted farrowing dates were 114 d post-breeding. One of three lactation treatments was fed from farrowing loading to weaning. The lactation treatments were; 1) 60 ppm vitamin E (Vit E, L1), 2) 60 ppm vitamin E+ 120 ppm AOX (AOX, L2), and 3) 120 ppm vitamin E (High E, L3). Sows were limit fed 6 lb of lactation treatment per day from farrow loading until farrowing, and then the lactation diet was provided ad libitum. Piglet weight and feed intake was determined on day 0, 7 and 20 or 21. Probability data for piglet mortality and livability is presented as a value between 0 and 1. The observations were analyzed using a binomial model. Average mortality was calculated with the following formula: number died / (total born alive + piglets added or taken away at cross fostering)*100. Average livability was calculated with the following formula: 1 - (number stillborn + number died/ (total born + piglets added or taken away at cross fostering))*100. Data were analyzed using the MIXED procedure of SAS version 9.3 (Cary, NC). Differences were defined as trends at $0.10 > P > 0.05$ and considered significant at $P < 0.05$.

Results and discussion

Litter performance for the first study is summarized in Table 3. Compared to NC, the number of weaned piglets per litter tended to increase ($P < 0.06$) in sows fed medium E, high E, or medium E+AOX (11.73, 11.82, 11.55 vs 9.80). Also, sows fed medium E, high E, or medium E+AOX had significantly ($P < 0.05$) higher litter weights at weaning (185.10, 185.92, 189.55 lb) compared to NC (149.58 lb). Average weaning weights of piglets from sows fed medium E, high E, or medium E+AOX were 15.79, 16.12, and 16.49 lb, respectively, whereas piglets from NC weighed 15.30 lb; however, the difference was not statistically significant.

Table 1: Treatment structure for Study 1

Treatment	Gestation		Lactation	
	Vitamin E (ppm)	AOX (ppm)	Vitamin E (ppm)	AOX (ppm)
1 Negative control	-	-	-	-
2 Medium E	50	-	75	-
3 High E	100	-	150	-
4 Medium E+AOX	50	100	75	150

Table 2: Treatment structure for Study 2

Treatment	Gestation		Lactation	
	Vitamin E (ppm)	AOX (ppm)	Vitamin E (ppm)	AOX (ppm)
1 (G1+L1)	60	-	60	-
2 (G1+L2)	60	-	60	120
3 (G1+L3)	60	-	120	-
4 (G2+L1)	60	120	60	-
5 (G2+L2)	60	120	60	120
6 (G2+L3)	60	120	120	-

Table 3: Effects of supplemental vitamin E and AOX on litter performance

Treatment	NC	Medium E	High E	Medium E +AOX	SEM	*P*-value
Born alive	11.50	12.82	13.18	13.00	0.308	0.224
Stillborn	1.70	1.36	1.09	1.27	0.173	0.677
Mummies	0.20	0.18	0.20	0	0.055	0.500
Litter birth weight, lb	34.37^{b}	40.79^{a}	42.04^{a}	42.64^{a}	1.151	0.037
Piglet birth weight, lb	3.00	3.17	3.20	3.28	0.051	0.223
Number of weaned per litter	9.80	11.73	11.82	11.55	0.173	0.055
Litter weaning weight, lb	149.58^{b}	185.10^{a}	185.92^{a}	189.55^{a}	5.135	0.014
Piglet weaning weight, kg	15.30	15.79	16.12	16.49	0.335	0.657
Pre-weaning mortality, %	13.25	8.37	10.64	11.06	0.852	0.261

^{a,b} $P \leq 0.05$

The heavier weaning weights may be partially explained by the numerically heavier birth weights for pigs from sows fed medium E, high E, or medium E+AOX, which were 3.17, 3.20, and 3.28 lb, respectively, compared to 3.00 lb for NC. In this study, adding AOX to a nutritional level of vitamin E in sow diets increased litter birth weights and weaning weights compared to the NC and was numerically greatest for both.

The effects of vitamin E and AOX supplementation on antioxidant status of sows and piglets is summarized in Table 4. Blood samples were taken from sows at farrowing and weaning. Vitamin E concentration in serum was significantly higher ($P < 0.001$) in sows fed high E and medium E+AOX (1.16 and 1.14 mg/ml) compared to sows fed medium E or NC (0.74 and 0.42 mg/ml) at farrowing. At weaning, vitamin E concentration remained significantly higher ($P < 0.001$) for high E and medium E+AOX compared to NC (1.13 and 1.06 mg/ml). Similarly, vitamin A concentration in serum was significantly higher ($P < 0.001$) in sows fed medium E+AOX (0.17 mg/ml) than sows fed medium E or NC (0.11 and 0.07 mg/ml) at farrowing. At weaning, vitamin A concentration in serum was similar across treatments. At farrowing and weaning, sows fed medium E+AOX had greater ($P < 0.001$) TAS than NC sows (1.20 vs 1.14 and 1.29 vs 1.16 mmol/L). Also at farrowing and weaning, SOD and GPx was greatest ($P < 0.001$) in sows fed medium E+AOX compared to NC and medium E. Sows fed high E had intermediate levels of SOD and GPx.

Blood samples were taken from piglets at weaning. Concentrations of vitamin E and vitamin A were greater ($P < 0.001$) in piglets from sows fed medium E+AOX or high E compared to medium E or NC. Also, TAS, SOD, GPx was higher ($P < 0.001$) in piglets from sows fed medium E+AOX or high E compared to medium E or NC. In this study, antioxidant status of sows and piglets was improved by adding AOX to a nutritional level of vitamin E or feeding a high level of vitamin E to sows through an increase in Vitamin E, Vitamin A, TAS, SOD, and GPx level.

The effects of vitamin E and AOX supplementation for the second study are summarized in Tables 5 and 6. When AOX was supplemented in the gestation diet, the probabilities for pigs born alive and weaned were numerically increased compared to the control (0.94 vs 0.93 and 0.89 vs 0.87, respectively). The probabilities for stillborn and mummies were numerically lower for the AOX treatment in gestation compared to the control, but the differences were not statistically significant. Average birth weight tended to be higher ($P = 0.09$) when AOX was fed during gestation (Table 5). The probability of piglet livability was numerically increased when sows were fed AOX in gestation compared to the control (Table 6). Relatedly, the converse, probability for piglet mortality, was numerically decreased for sows fed AOX. In this study, the inclusion of AOX in the gestation diet tended to increase piglet birth weight. The data also suggest that livability may be positively affected by the addition of AOX to the

Table 4: Effects of supplemental vitamin E and AOX on antioxidant status of sows and piglets

Treatment	Time	NC	Medium E	High E	Medium E +AOX	SEM	*P*-value
Sows							
Vitamin E*	Farrowing	0.42[c]	0.74[b]	1.16[a]	1.14[a]	0.050	<0.001
	Weaning	0.75[c]	1.02[b]	1.13[a]	1.06[ab]	0.026	<0.001
Vitamin A*	Farrowing	0.07[c]	0.11[b]	0.15[ab]	0.17[a]	0.007	<0.001
	Weaning	0.16	0.15	0.15	0.16	0.002	0.091
TAS**	Farrowing	1.14[b]	1.18[a]	1.19[a]	1.20[a]	0.007	0.007
	Weaning	1.16[b]	1.18[b]	1.20[ab]	1.29[a]	0.010	<0.001
SOD***	Farrowing	61.13[c]	77.14[b]	79.60[ab]	82.82[a]	1.458	<0.001
	Weaning	61.88[d]	79.50[c]	90.93[b]	140.27[a]	4.715	<0.001
GPx***	Farrowing	27.00[d]	32.77[c]	37.64[b]	40.11[a]	0.835	<0.001
	Weaning	27.23[d]	34.78[c]	36.48[b]	40.27[a]	0.766	<0.001
Piglets							
Vitamin E*	Day 21	0.52[d]	1.41[c]	1.61[a]	1.48[b]	0.070	<0.001
Vitamin A*	Day 21	0.078[d]	0.16[c]	0.20[b]	0.25[a]	0.011	<0.001
TAS**	Day 21	0.86[c]	1.16[b]	1.18[ab]	1.24[a]	0.026	<0.001
SOD***	Day 21	98.01[b]	100.61[b]	129.54[ab]	134.80[a]	2.743	<0.001
GPx***	Day 21	27.40[c]	29.12[b]	29.91[b]	34.58[a]	0.457	<0.001

[a,b,c] $P \leq 0.05$
* In mg/ml serum; **mmol/L serum; ***U/ml of blood

Table 5: Main effects of vitamin E or AOX in gestation on reproductive and piglet performance

Gestation (d 85 to d 111)	Vitamin E	AOX	SEM	*P*-value
Probability[1] born alive	0.93	0.94	0.008	0.368
Probability still born	0.07	0.06	0.007	0.455
Probability mummified	0.002	0.001	0.001	0.527
Probability weaned	0.87	0.89	0.012	0.136
Litter birth weight (lb)	37.41	38.21	0.783	0.160
Average birth weight (lb/pig)	3.11	3.17	0.049	0.094

1 Probability: between 0 and 1. Statistically analyzed by binomial model

Table 6: Main effects of vitamin E or AOX in gestation on piglet mortality and livability (farrowing to weaning)

Gestation (d 85 to d 111)	Vitamin E	AOX	SEM	*P*-value
Average total mortality (%)[1]	12.64	10.93		
Average total livability (%)[2]	82.39	84.39		
Probability[3] mortality	0.14	0.13	0.012	0.207
Probability[3] livability	0.80	0.82	0.014	0.119

1 Average total mortality per sow is calculated by; = #died/ (total born alive + piglets added or taken away at cross fostering)*100.
2 Average total livability per sow is calculated by; 1-(# stillborn + # died/ (total born + piglets added or taken away at cross fostering))*100
3 Probability: between 0-1. Statistically analyzed by binominal model

gestation diet. Though increased probability of livability was not statistically significant in this study, the data from both studies suggest that AOX may support litter performance. The benefits of technology that supports the antioxidant system in sows and piglets, like AOX, may be greatest when sows are under increased stress, especially highly prolific sows struggling with low birth weights and increased preweaning mortality.

References

1. Bernardi et al. Plasma nitric oxide, endothelin-1, arginase and superoxide dismutase in pre-eclamptic women. *J. Obstet. Gynecol. Res.* 2008;34:957–963

2. Marantidis et al. Association of BF gene polymorphism with litter size in a commercial pig cross population. *Animal Reproduction Science.* 2013;141:75-79

3. Zhao, Y. Oxidative Stress Status and Reproductive Performance of Sows. Ph.D Thesis. North Carolina State University. 2011

Impact of two training implementation plans on post-training competency expression of personnel on farm

S. Probst Miller, DVM
AgCreate Solutions, Monticello, Illinois

Introduction

Training personnel is an essential part of ensuring competency expression, engaging employees, creating a safe workplace, and developing a workplace that people want to stay.[1,2,8,9] Proper implementation of health and welfare programs by production managers and veterinarians is achieved when integrated with a training program embraced top down by the Human Resources and production management team.[6,7] In this training implementation case study, a traditional style of training program (Training Program A) was compared against a new training program (Training Program B). Impact on personnel competencies, engagement scores and safety were measured. These results were completed by the system implementing the program to determine which training program to implement across their system.

Materials and methods

Training Program A was the traditional program utilized in this pork business. In this program farm managers were required to use the system's existing Standard Operating Procedure (SOP) booklet to train staff. Training Program A had a 12-week training guide for each area: Breeding, Farrowing, Nursery, Finishing. Employees were trained in the area they worked, and if they stayed in that area, cross-training did not occur. Each week of the training had specific tasks that needed to be taught according to the SOP, but teaching method was left up to the manger. Typically, managers would read or talk through the SOP's verbally; and as needed, demonstrate in barn, and provide follow up specific to them or customized to the situation. A check in with each employee was done after 12 weeks to see if things were going ok.

Training Program B used online learning management software, Pork Avenue Training Portal (delivered by AgriSchool Party Lmt and created by AgCreate Solutions)[5]. This program follows a measured "See it, Do it, Teach it" philosophy. Assets consist of videos followed by simulations of desired end skills, graphical SOP's and In Barn Verification Experiences. All learning experiences are assigned on online learning management software which records video and simulation completion as well as completion of the in-barn verification experience to an individual's transcript.

For Training Program B, the system's hog training manager created a custom 26-week orientation curriculum within Pork Avenue choosing 40 of the 120 learning experiences to support and help drive understanding of the system's existing SOPs. He designated and trained 2 service people outside of the farms on how to do the in-barn verification of competencies. Lessons were assigned and when completed, the service people trained to verify skills went to the barn and made sure employees knew how to physically do the task in a safe manner and verify employee understanding of the why behind the process. Additional lessons in the program were subsequently assigned depending on the employee's role in the program. For example, farrowing personnel were assigned more farrowing lessons and wean to market personnel were assigned more wean to market lessons, etc.

Over 100 employees were enrolled in Training Program B for an average 8.6 months. The employees enrolled were new employees in the system and competency expression was compared to over 500 employees of a wide variety of experience levels in the system.

For Training Program B, on average each employee was assigned 59 learning experiences on average over an 8.6-month period. On average, 46 lessons were started and 43.8 were completed yielding a 95% completion rate once lessons were initiated. Over this period 5.1 lessons were completed per employee per month on average. In addition to the learning experiences online, each employee over the first 26 weeks, received over 6 hours of one on one verification following the in-barn verification checklists.

To measure impact in this case study, verifiers compared testing results and observed for physical expression of competencies on farm. To compare the programs, verifiers chose 70 key competencies they wanted to see in action on farm. In Training Program B, employees saw these competencies listed on their In-Barn Verification Checklists, but did not know which competencies were being measured in the Training Program efficacy evaluations. For each competency chosen, correct expression was either "right" or "wrong." The farm's total competency ranking was listed as a percent of competencies being implemented physically on farm.

Engagement scores were measured using Gallop's Q12 program in Q1 of 2017. Re-measurement is scheduled in Q4 of 2017, but not completed across all farms at the time of this publication. Reported injuries was the measurement of safety impact. Research shows that businesses with higher engagement scores are on average 18% more productive, 12% more profitable, and experience a 62% reduction in work related accidents. More engaged employees also have 27% lower absenteeism and are 31% less likely to leave the organization. "Gallup's employee engagement work is based on more than 30 years of in-depth behavioral economic research involving more than 17 million employees."[2,9]

Safety impact was measured by recording reported injury incidents across all employees.

Results

For expressed competency, Training Program A achieved a 59% level of competency. Training Program B achieved over 85% on competency. This was a 26-percentage point increase over Training Program A.

Engagement scores were trending upwards at the time of this paper's publication. At the time of presentation, more information will be available.

Safety has been a long-term focus of this company. At the time of measurement, 6 months post implementation of the training program, zero injuries had occurred across the whole system.

Discussion and conclusion

The science of how adults learn suggests that a "See it, Do it, Teach it" program is necessary to achieve physical expression of competencies. Research shows that, on average, employees can express competencies on about 10% of what they read and 20% of what they hear.[6,7] The managers efforts on Training Program A did increase competency expression from these levels to a 59% competency expression on farm. However, this study shows that utilizing programs like Pork Avenue that offer and measure the "See it, Do it, Teach it" experiences had significant impact on competency expression on farm raising it 26 percentage points above the farm's traditional level. This system used the study to determine whether to enroll the rest of their employees in the program. Post study, the system has decided to enroll all 600 employees in Training Program B.

Of note, this system decided to have people outside of the farms to do In Barn Verifications. They felt that managers were often anxious to get people moving on tasks. The 3rd party eye, helped them make sure the employee was truly able to physically do the task on farm according to their desired competency expression. Other systems implementing Training Program B utilize designate supervisors or managers for In Barn Verification experiences with good success.

References

1. Collins J. Built to Last, Successful Habits of Visionary Companies, 1994.

2. Gallup Research

a.) https://q12.gallup.com/public/en-us/Features

b.) http://news.gallup.com/poll/180404/gallup-daily-employee-engagement.aspx

c.) http://www.goalbusters.net/uploads/2/2/0/4/22040464/gallup_q12.pdf.

3. Firkins, Larry, Getting there from here, AASV 2009.

4. Kliebenstein et al, Employment in the US Pork Industry, http://lib.dr.iastate.edu/cgi/viewcontent.cgi?article=1205&context=ans_air, 2006.

5. Pork Avenue Training Portal, www.porkavenuetraining.com.

6. Probst Miller, Sarah, Applying the Science and Physiology of How Adults Learn to Our On-farm Approach as Practitioners, AASV Proceedings, 2016.

7. Probst Miller, Sarah, AASV, The Impact of Trained and Untrained Personnel on Product Implementation and Subsequent Pig Health, AASV Proceedings 2016.

8. Probst Miller, Sarah, et al, Employee Engagement on Sow Farms, EVP 2009.

9. Wagner R. et al, 12 The Elements of Great Managing, 2006.

The impact of Swine Awake®, a new, non-antimicrobial protein, in natural disease conditions in newly weaned pigs

J. Tyler Holck[1], DVM, MS, MBA; Rob Musser[2], PhD; Ran Song[2], PhD; Kim Friesen[2], PhD
[1]Feed His People, LLC, Gilbert, Iowa; [2]NutriQuest, Mason City, Iowa

Introduction

Swine Awake® is a novel, bio-active, non-antimicrobial small protein derived from direct feed microbial technology. In pigs, Swine Awake has been associated with the induction of immunomodulatory cytokines thus providing improved immunity and performance in the health challenges.[1-3] Response to Swine Awake was initially observed to mitigate health challenges in commercial sow herds experiencing PEDv outbreaks.[4] Swine Awake has also been shown to improve ADG, reduce lung lesions and mediate the immune response in a PRRSv challenge model [5] and is being evaluated in an SIV challenge model as well.

The objective of this study was to assess the impact of Swine Awake on wean-to-finish cull, mortality, and individual animal treatments when administered in the water for 21 days post-placement.

Materials and methods

A single, high-health sow flow from Southeast US was selected as the source herd. Each week approximately 5200 pigs were transported to central Iowa and placed in wean-to-finish barns at three weeks of age. Barn was the experimental unit and a total of twenty barns across four sites were selected based upon a history of respiratory disease in the early nursery phase. Sites ranged from two barn to ten barn in size. Placements occurred over a 5 week period and barns were paired within sites by placement dates with one receiving Swine Awake treatment and the other serving as a non-treated control. Two Swine Awake-treated and two non-treated control barns were placed weekly for five (5) weeks providing ten (10) replicates for comparison. Each barn placed an average of just over 1300 pigs (Table 1) with the treated barns receiving Swine Awake in the water for 21 consecutive days at an estimated dosage of 5 mg/pig/day.

The primary parameter of measure was mortality as recorded on the weekly barn sheets. Individual animal treatments were also recorded by barn on the weekly barn sheets and final cull data was also captured as defined as pigs not sold to the full value market. A student t-test was performed for these parameters. Oral fluid samples were collected approximately 8 weeks post-placement or at 11 weeks of age to characterize disease status of each site. All oral fluid samples were tested for PRRSv ELISA and PCR, SIV PCR, and PED PCR at Iowa State University Veterinary Diagnostic Laboratory.

Results and discussion

All sites were positive for PRRSV, SIV, and PED with the exception of a two-barn site, which was negative for both SIV

Table 1: Summary data

Parameter	Control group	Swine awake	Difference	*P*-value
Total barns placed	10	10	0	
Total pigs placed	13173	13274	+101	
Average pigs per barn	1317	1327	+10	
Total individual treatments	3029	2607	-422	
% Treatments (of pigs placed)	23.0%	19.6%	-3.4%	< 0.0001
Total culls	81	74	-7	
% Culls (of pigs placed)	0.61%	0.56%	-0.05%	0.54
Total mortality (thru 22 wks)	815	700	-115	
% Mortality (of pigs placed)	6.2%	5.3%	-0.9%	0.002
Total cull & mortality	896	774	-122	
% Cull & mortality	6.8%	5.8%	-1.0%	0.002

and PED. A PRRSv isolate from each site was selected for sequencing and all were characterized as RFLP pattern 1-3-2 or "Fostera-like", consistent with PRRSv vaccine administered approximately 6 weeks earlier. The exception was the isolate from the ten-barn site which was characterized as wild-type with an RFLP pattern of 1-7-4 even though these pigs had been vaccinated as well. Clinical observations during the treatment period of the first 3 weeks post-placement included moderate to severe diarrhea at three of the four sites excluding the 2 barn site. Mass medication with either gentamycin in the water or chlortetracycline/tiamulin in the feed was provided at each of affected sites. Clinical signs at the two barn site were less severe during the first 3 weeks post-placement and the overall health of the pigs at that site was very good compared to the other sites during the entire study.

The overall performance data is summarized in Table 1 including numeric differences between treatment groups and statistical analysis of chi-square test. Pigs treated with Swine Awake had fewer individual animal treatments ($P < 0.0001$), less culls at final cut ($P = 0.54$), lower mortality ($P = 0.002$), and subsequently lower total cull/mortality ($P = 0.002$) compared to non-treated controls. The reduction in mortality was observed already during the first three weeks post-placement with 122 mortalities (0.93%) in the control group compared to only 79 (0.59) in the Swine Awake treated barns. Of more importance is the reduction in mortality continued throughout the entire grow-out period as shown in Figure 1, resulting in a reduction of nearly a full percent in mortality at 22 weeks post-placement. When cull pigs were factored in with mortality, Swine Awake improved full value marketable pigs by a full percentage. The number and percentage of pigs requiring individual antibiotic treatments were also dramatically reduced in the Swine Awake barns with 422 less pigs treated compared to control barns.

Growth parameters such as average daily gain (ADG), average daily feed intake (ADFI), and feed efficiency (FE) were not able to be collected due to marketing procedures by site within this system. Given the differences in mortality and culls, it would be expected that these parameters would likely be improved with the treatment of Swine Awake. Pig ADG and feed efficiency have been shown to be improved in the PRRSv challenge study [5] conducted with Swine Awake by 21% and 13%, respectively, and are currently being assessed in other commercial field settings.

Conclusions

Under the conditions of this field trial utilizing over 26,000 pigs, Swine Awake provided in the water for 21 consecutive days post-placement reduced total cull and mortality by 1.0 percent compared to non-treated controls. In addition, Swine Awake treated pigs required less individual animal treatments compared to non-treated controls. The improvements in total cull/mortality and individual treatments were more dramatic at the sites which experienced more severe health challenges, both enteric and respiratory.

References

1. Brennan K et al. Interleukin 8. *xPharm: The Comprehensive Pharmacology Reference*, 1-4, 2007

2. Lopez-Castejon G et al. Understanding the mechanism of IL-1β secretion. *Cytokine & Growth Factor Reviews*, 2011;22(4), 189–195.

3. Kajdaniuk D et al. Transforming growth factor β1 (TGFβ1) in physiology and pathology. *Endokrynol Pol*, 2013;64(5), 384–396.

4. Data upon request, NutriQuest

5. Song, R et al. A novel, non-antimicrobial protein improves growth performance and immune responses of pigs experimentally infected with porcine reproductive and respiratory syndrome virus. *J. Anim. Sci.*, paper in press

Figure 1: Cumulative mortality by week post-placement

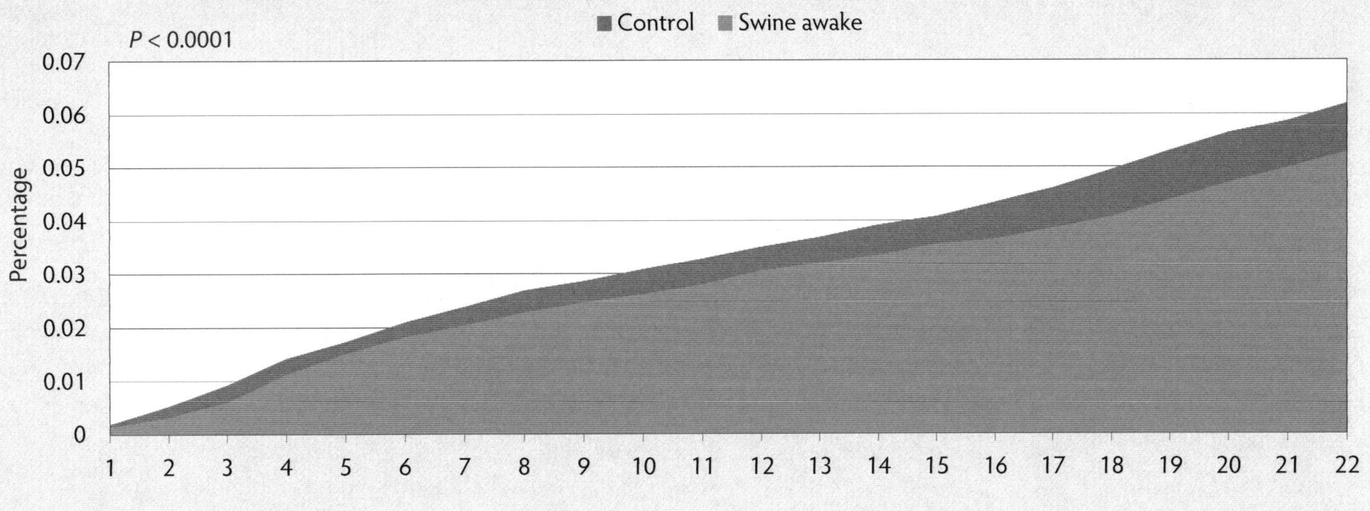

Next generation sequencing for veterinary diagnostics and vaccinology

B. Hause, MS, PhD
Cambridge Technologies, Worthington, Minnesota

DNA sequencing has been widely utilized by swine veterinarians and diagnosticians since the early 2000's to study the genetics and molecular epidemiology of swine viruses. The focus of these investigations mainly involved glycoprotein 5 (GP5) of porcine reproductive and respiratory syndrome virus (PRRSV) and the hemagglutinin (HA) gene of influenza A virus (IAV). This testing has become commonplace and is offered by most veterinary diagnostic laboratories for $100-150 per sample.

Advances in DNA sequencing technology are rapidly expanding its use for veterinary diagnostics. Next generation sequencing (NGS) technology has become increasingly reachable for diagnostic laboratories owing to decreased equipment and consumable costs and increased ease of operation and data analysis. The principle benefit of NGS as compared to Sanger sequencing is that NGS generates orders of magnitude more sequence data from a sample. This ability can be leveraged to analyze larger regions or complete viral and bacterial genomes. Additionally, the shear amount of DNA sequence generated for each sample allows for non-targeted sequencing of samples, enabling unbiased pathogen detection without *a priori* knowledge of their presence or sequence. This latter ability is utilized for metagenomic sequencing, the direct analysis of genetic material from clinical samples.

Cambridge Technologies uses NGS for diagnostic testing and genetic analysis of isolated bacteria and viruses. For isolated viruses, NGS is used to assemble complete genome sequences. Genetic analysis of complete genomes provides the greatest resolution of viral phylogeny and provides the most complete picture of virus mutations which affect antigenicity and pathogenicity.[1] NGS of bacteria utilizes multi locus sequence typing (MLST). A number of genes (typically 6-8) are amplified from the bacterial genome, sequenced and the concatenated sequences are analyzed, providing fine genetic resolution of bacterial lineage.

In addition to NGS of isolated bacteria and viruses for strain selection, metagenomic sequencing is a diagnostic offering at Cambridge Technologies. This powerful technology is commonly used in cases with unusual clinical presentations, in cases where the expected pathogen based on clinical disease is not identified and for herd or inoculum profiling.[2-4] Metagenomic sequencing is poised to replace targeted PCR's for pathogen detection and has the benefits of lower overall cost and the ability to detect novel and emerging viruses. This overview of NGS for veterinary diagnostics and vaccinology aims to illustrate use of this emerging technology by swine veterinarians.

References

1. Chen Z et al. Genetic diversity in envelope genes of contemporary US porcine reproductive and respiratory syndrome virus strains influenza viral antigenicity. Res Vet Sci. 2017;115:432.

2. Arruda PH et al. Detection of a novel sapelovirus in central nervous tissue of pigs with polioencephalomyelitis in the USA. Transboud Emerg Dis. 2017;64:311.

3. Palinski R et al. A novel porcine circovirus distantly related to known circoviruses is associated with porcine dermatitis and nephropathy syndrome and reproductive failure. J Virol. 2016;91:e01879-16.

4. Hause BM et al. Virus detection using metagenomic sequencing of swine nasal and rectal swabs. J Swine Health and Prod. 2016;24:304.

Pooling sensitivity of oropharyngeal swabs when sampling for influenza A and porcine reproductive and respiratory syndrome viruses

Rachel Schulte[1]; Thomas Wetzell[2], DVM; Ethan Schmaling[2]; Adam Schelkopf[3], DVM
[1]Iowa State University, Ames, Iowa; [2]Boehringer Ingelheim Vetmedica, St. Joseph, Missouri;
[3]Pipestone Veterinary Services, Pipestone, Minnesota

Introduction

Sow farm monitoring for porcine reproductive and respiratory syndrome virus (PRRSV) focuses on weaned pig diagnostic testing of various sample types, the most popular of which is serum. Recently, oral pharyngeal swabs have been successfully used to detect several swine pathogens including influenza A virus of swine (IAV-S),[1] as well as to detect PRRSV in wean-age pigs, especially when the prevalence of infection is high.[2] A number of sow farms have implemented oral pharyngeal (OP) pooling surveillance protocols. Still, there has been no research done to determine pooling sensitivity and therefore this practice may not reliably detect the viruses.

Objective

The objective of the project is to compare the sensitivity of PCR results from 6 pools of 5 oral pharyngeal samples for IAV-S and PRRS in ready-to-wean pig populations versus individual samples.

Materials and methods

200 individuals were selected for IAV-S and 200 different individuals were selected for PRRSV PCR tests for a total of 400 ready to wean pigs from 2 sow farms with expected IAV-S and PRRS positive status. 100 randomized pooled samples of 5 were then tested for IAV-S and 100 randomized pooled samples of 5 were tested for PRRS PCR where 1 known positive was mixed with 4 known negatives based on the pigs' individual test results. Negative samples were re-used in subsequent pools when required.

These 400 pigs were sampled by inserting a BD BBL Culture-Swab in the back of the oral cavity and swabbing the soft palate vigorously per standardized production system protocol.

Results

The individual IAV-S PCR results correlated to the pooled results at a level of .957, while the PRRS PCR individual results had a correlation of .905 in relationship to the pooled results.

For IAV, 100% (53/53) were pooled positive with initial individual CT under 33, yet only 17.8% (8/45) were positive over 33. For PRRSV 98.0% (100/102) were pooled positive with initial individual Ct under 34 yet only 33.3% (2/6) were positive for Ct's over 34.

Conclusions

The OP-swabbing technique is easily implemented and will detect IAV-S and PRRSV. However, there may be value in increasing number of pools tested for routine surveillance, especially when prevalence of infection is low and sensitivity is diminished.

References

1. Strobel, R, et al. Evaluation of diagnostic sampling techniques used in the farrowing house to increase sensitivity of detection and sequencing of influenza A virus in swine (IAV-S). AASV. 2016;77.

2. Glisson K, et al. An evaluation of oropharyngeal swabbing as a diagnostic technique for the detection of porcine reproductive and respiratory syndrome virus (PRRSV) in weaned pigs. In: Proceedings of the 48th Annual Meeting of the American Association of Swine Veterinarians (February 25-28, 2017). Page 65.

Medium chain fatty acids improve growth and alter fecal microbial populations in nursery pigs

Jordan Gebhardt[1], BS; Steve Dritz[1], DVM, PhD; Jason Woodworth[1], PhD; Mike Tokach[1], PhD; Joel DeRouchey[1], PhD; Bob Goodband[1], PhD; Megan Niederwerder[1], DVM, PhD; Samodha Fernando[2], PhD; Waseem Abbas[2], BS; Tom Burkey[2], PhD
[1]Kansas State University, Manhattan, Kansas; [2]University of Nebraska, Lincoln, Nebraska

Introduction

The lag in performance suffered by pigs immediately after weaning appears to have intensified in recent years with the rapid increase in litter size. Reduction in use of growth-promoting antibiotics has increased the need for dietary strategies to aid pigs through this challenging period. Medium chain fatty acids (MCFA) are natural compounds with anti-viral and anti-bacterial characteristics against pathogens which can be found in livestock feed and pet food. Previous research from our group has demonstrated that adding MCFA to feed can prevent PEDV infection in pigs when the feed is inoculated with PEDV after processing. Some individual MCFA appear to be more effective than others for this purpose. The antiviral and antibacterial properties of MCFA led us to postulate whether MCFA have the potential to improve growth of nursery pigs. Therefore, our objective was to determine the impact of MCFA in nursery pig diets on growth performance, fecal microbial populations, and fecal consistency.

Materials and methods

A total of 360 pigs [DNA (Columbus, NE) 400 × 200; initial BW = 6.7 ± 0.07 kg] were randomized to pens (5 pigs/pen) with pens blocked by BW and randomized to dietary treatment (9 pens/treatment). All MCFA used (Sigma Aldrich, St. Louis, MO) were guaranteed ≥ 98% purity. Treatment diets were fed in two dietary phases from d 0 to 14 (7 to 11 kg) and d 14 to 35 (11 to 23 kg). Eight dietary treatments included increasing levels (0, 0.25, 0.5, 1.0, and 1.5%) of a MCFA blend (1:1:1 ratio of C6, C8, and C10) as well as treatments with 0.5% of either C6, C8, or C10. Fecal microbiome analysis was performed using 16s rDNA sequencing on 0 and 1.5% MCFA blend samples collected on d 0 and 14. Data were analyzed as a randomized complete block design using a generalized linear mixed model (PROC GLIMMIX; SAS Institute, Inc., Cary, NC) with pen as the experimental unit.

Results and discussion

From d 0 to 14, increasing levels of MCFA increased (Table 1; linear, $P < 0.001$) ADG by 8, 15, 20, and 26% for pigs fed 0.25, 0.5, 1.0, and 1.5% MCFA, respectively, compared with the control. The improvements in growth were due to improved ADFI (linear, $P = 0.003$) and F/G (quadratic, $P = 0.012$). Pigs fed 0.5% C6 and 0.5% C8 had greater ($P \leq 0.018$) ADG than pigs fed the control diet without MCFA. Pigs fed 0.5% C8 had greater ADFI than control fed pigs ($P = 0.023$), and pigs fed 0.5% C6, 0.5% C8, or 0.5% C10 had improved F/G ($P \leq 0.005$) compared to control fed pigs. Pigs fed 0.5% C8 had a marginally significant increase ($P = 0.094$) in ADFI compared to pigs fed 0.5% blended MCFA. Pigs fed 0.5% C10 tended to have poorer ($P = 0.060$) F/G compared to pigs fed the 0.5% MCFA blend diet. The growth benefits to the MCFA continued into the second phase, such that final BW was improved ($P < 0.05$) by dietary addition of C6 (0.9 kg), C8 (1.2 kg) or increasing levels of the MCFA blend (0.5, 1.2, 1.3, and 1.7 kg, respectively) compared to control. Pigs fed 0.5% C6 or 0.5% C8 had greater ($P \leq 0.027$) ADG compared to pigs fed the control diet, and F/G was improved ($P \leq 0.003$) when pigs were fed 0.5% C6, 0.5% C8, or 0.5% C10 compared to control. Increasing dietary inclusion of MCFA resulted in firmer stools as evidenced by improved fecal consistency (linear, $P = 0.010$). No evidence ($P > 0.05$) of a treatment or day effect in fecal microbial alpha and beta diversity was observed. Fecal family relative abundance of Streptococcaceae increased (treatment × day, $P = 0.013$) in control fed pigs over time compared to 1.5% MCFA blend fed pigs, whereas relative abundance of Turicibacteraceae and Clostridiales increased (treatment × day, $P \leq 0.048$) over time in 1.5% MCFA blend fed pigs compared to control. There was no evidence of a difference detected among treatments over time ($P > 0.05$) for other bacterial families (n = 77) or phyla (n = 18) examined. In summary, the use of MCFA, particularly C6 and C8, in nursery pig diets offers significant potential to improve growth performance and alter microbiota during the challenging time immediately after weaning.

Table 1: Influence of medium chain fatty acid (MCFA) on nursery pig growth performance[1]

	Added MCFA, %[2]								
		C6:C8:C10[3]				C6	C8	C10	
	0	0.25	0.5	1.0	1.5	0.5	0.5	0.5	SEM
BW, kg									
d 0	6.7	6.7	6.7	6.7	6.7	6.7	6.7	6.7	0.07
d 35	21.5	22.0	22.7	22.8	23.2	22.5	22.8	22.2	0.34
d 0 to 14									
ADG, g	255	275	294	305	323	297	314	269	12.9
ADFI, g	331	333	343	365	376	353	373	325	13.9
F/G	1.30	1.21	1.17	1.20	1.17	1.19	1.19	1.22	0.020
d 0 to 35									
ADG, g	423	436	457	461	472	451	457	441	9.1
ADFI, g	630	621	642	667	664	644	649	630	14.1
F/G	1.49	1.42	1.41	1.45	1.41	1.43	1.42	1.43	0.013

1 A total of 360 pigs [DNA (Columbus, NE) 400 × 200; initial BW = 6.7 ± 0.07 kg] were used in a 35-d experiment with 5 pigs per pen and 9 pens per treatment.
2 Sigma Aldrich (St. Louis, MO).
3 Consisted of a 1:1:1 blend of C6, C8, and C10.

An assessment of hemoglobin concentrations in relation to reproductive stage and parity in sows

Kayla Castevens; Glen Almond, DVM; Juliana Bonin Ferreira, DVM

Introduction

Iron is important for enzyme metabolization and oxygen transport. It is typically bound to proteins, such as hemoglobin (Hb), with 65% in red blood cells and the rest stored as ferritin. In sows, iron requirements increase with gestation and lactation due to utero-transfer, hemorrhage, and milk excretion. If demands exceed intake, erythropoiesis and ferritin mobilization are initiated, and non-degenerative anemia or ferriprive anemia will occur when reserves are depleted. Hb measurement typically is a reliable indicator of anemia.[1] To prevent anemia in piglets, routine iron injections are given shortly after birth. In contrast, the iron requirements of sows are ignored despite the high reproductive demands in the modern pork industry. Therefore, this study was designed to evaluate Hb concentrations at various stages of gestation and lactation and at multiple parities. The overall goal was to determine if sows are anemic at any particular stage or parity and if there is there is an association between low Hb concentrations and number of stillbirths/mummies.

Materials and methods

The study was a cross-sectional design including 1730 sows from 7 different farms (2400-4000 sows/farm). Approximately 250 samples were collected from each farm with 10 sows/parity/stage. Stages were defined as early (~25-35 days), mid (~50-70 days), late (~100+ days) gestation, and early (< 7 days postpartum) and late (> 14 days postpartum) lactation. Parity groups were 0 (gilts), 1, 2, 3, and 4+ according to the parity of the sow when she was bred. The Hb concentrations were measured on the farm using a HemoCue Hb 201+ ™. A blood sample was taken from the ear and loaded (10μL) into a disposable cuvette via capillary action. The cuvette was placed in the HemoCue and the resulting Hb concentration displayed and recorded within 60 seconds. Data was analyzed using an analysis of variance, with farm, parity, and stage as the independent variables. Means were compared with Tukey's HSD test.

Results and discussion

Results revealed significant differences in Hb concentrations according to reproductive stage with mid gestation having the highest ($P < .05$) at 10.5 ± .07 g/dL, decreasing in late gestation (9.5 ± .06 g/dL) to the lowest in early lactation (9.0 ± .06 g/dL), and rising over late lactation (9.3 ± .06 g/dL) and early gestation (10.1 ± .06 g/dL). This cycle was consistent among parities and farms. There also was a significant effect of parity, with Hb concentrations decreasing with increasing parities: gilts (10.1 ± .08 g/dL), parity 2 (9.6 ± .07 g/dL), parity 4+ (9.4 ± .06 g/dL). A comparison of mean Hb from the seven farms according to reproduction stage and parity can be found in Figure 1. Analysis failed to reveal an association between Hb concentration and number of stillbirths, contrary to previous studies. This method of Hb assessment provided a quick and inexpensive method for on-farm use. Veterinarians and personnel can easily use it when anemia is suspected. If a sow is anemic in mid-gestation when Hb levels are expected to be high, it is likely that she would be severely anemic by early lactation and will not recover from the anemic state without intervention. Severe anemia, although not seen overall in this study could lead to reproductive issues or sow death, whereas an iron injection of 500 mg or oral administration of 100 ppm was shown to correct sow anemia in previous studies.[2,3] Another study on diets in piglets revealed that iron was better absorbed when given with a diet that contained resistant starches versus fully digestible starch diets due to the phytates. Phytates form insoluble complexes with several minerals, including iron, which interferes with absorption. The resistant starches reform hydrogen bonds during retrograde amylose formation possibly decreasing the number of potential binding sites for iron leaving more molecules free for absorption.[4] For digestible diets, it has been suggested that phytase enzymes can be given to neutralize the phytate therefore causing it to release the minerals with which it has formed complexes. While further study is necessary to evaluate more parameters of Hb and reproductive performance, iron supplementation or phytase added to diets at critical periods or onset of anemia may sustain a sow at higher reproductive performance levels and in production longer at relatively low cost. It may also serve as a means to help access sow health and welfare overall.

Reference

1. Normand, V., Perrin, H., Auvigne, V., Robert, N., Laval, A. (2012) Anaemia in the sow: a cohort study to assess factors with an impact on haemoglobin concentration, and the influence of haemoglobin concentration on the reproductive performance. *Veterinary Record* 171, 350

2. Lamana García-Fuente, J. (1996) Effect of amino acid-chelated minerals on swine reproduction: a practical case. Proceedings of 14th IPVS Congress, Bologna, Italy. p 450.

3. Moore, R W., Redmond, Hlivingston, C. W. (1965) Iron deficency anemia as a cause of stillbirths in swine. *Journal of American Veterinary Medical Association* 147, 746-8

4. Morais, Mauro B, et al. "Effect of Resistant and Digestible Starch on Intestinal Absorption of Calcium, Iron, and Zinc in Infant Pigs." *Pediatric Research*, vol. 39, no. 5, 1996, pp. 872–876.

Figure 1: Comparison of mean Hb values at different stages of reproduction at various parities. A clear pattern can be seen with the reproductive stages as well as on overall decrease with increasing parity. The standard deviations (for parity) ranged from 1.5 (Parity 0) to 1.2 (all other parities).

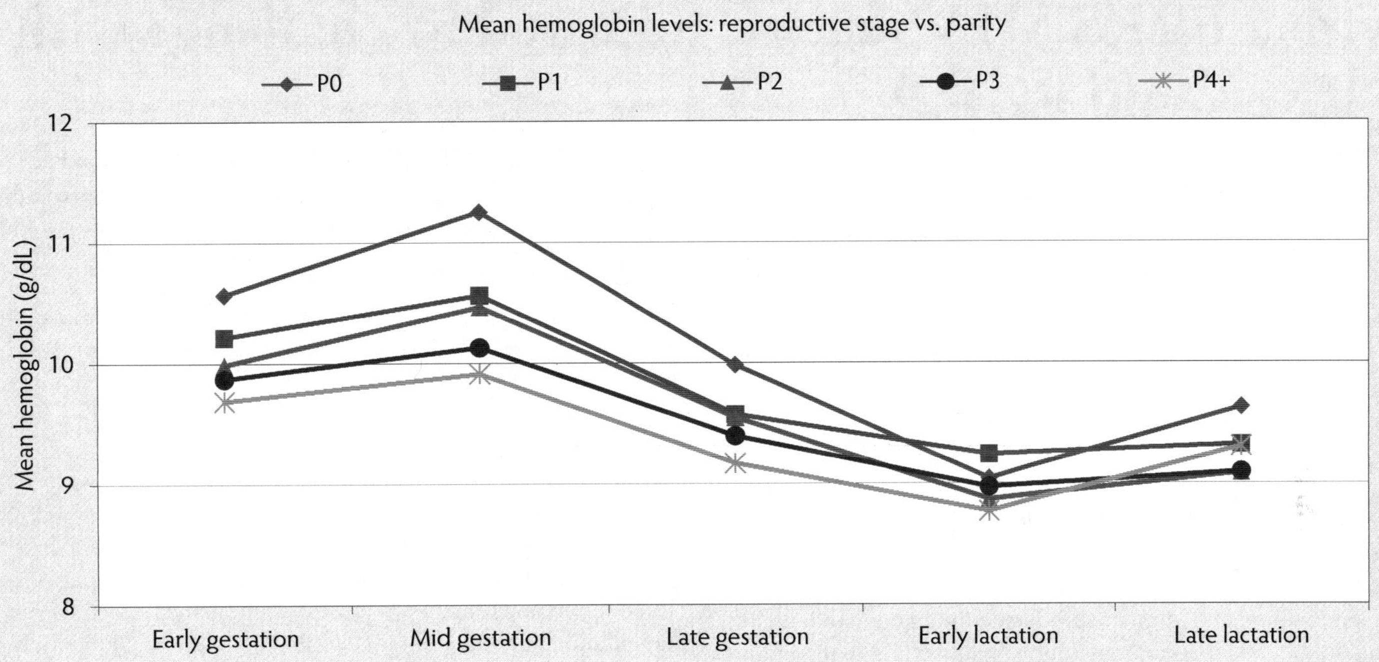

Pre-challenge microbiome composition is associated with improved weight gain in pigs after vaccination with a PRRS MLV vaccine followed by challenge with PRRSV and PCV2b

L.A. Constance[1], J.B. Thissen[2], C.J. Jaing[2], K.S. McLoughlin[3] A.G. Cino-Ozuna[1,4], R.R.R. Rowland[1], M.C. Niederwerder[1,4]

[1]Department of Diagnostic Medicine/Pathobiology, College of Veterinary Medicine, Kansas State University, Manhattan, Kansas; [2]Physical & Life Sciences Directorate, [3]Computations Directorate, Lawrence Livermore National Laboratory, Livermore, California; [4]Kansas State Veterinary Diagnostic Laboratory, Kansas State University, Manhattan, Kansas

Introduction

PRRSV and PCV2b are two of the most significant pathogens affecting the swine industry. Worldwide, co-infections are common, resulting in respiratory disease and reduced growth rates in nursery pigs.1,2 PRRS modified live virus (MLV) vaccines are widely used to reduce PRRS-associated loses; however, vaccination does not provide sterilizing immunity or prevent weight gain variation after PRRSV challenge.3 Our previous research demonstrated that the pre-infection gastrointestinal microbiome is associated with weight gain following co-infection with PRRSV and PCV2b.4 Still, it remained unknown if similar effects would be seen in pigs vaccinated with PRRS MLV. The objective of the current study was to determine if pre-infection microbiome profiles were associated with weight gain in pigs vaccinated with a PRRS MLV and subsequently challenged with PRRSV and PCV2b.

Materials and methods

Twenty-eight days post-vaccination (dpv) and at the time of PRRSV/PCV2b challenge, fecal samples were collected from an experimental population of 50 nursery pigs. At 42 days post-infection (dpi), 20 pigs were retrospectively identified as having high or low growth rates during the post-challenge period.

Results and discussion

Average daily gain post-challenge was 0.94 ± 0.06 kg and 0.71 ± 0.11 kg in high and low growers, respectively ($P < 0.001$; unpaired t-test). Comparing virus load and histopathology of the two groups revealed no significant differences in the detection of PRRSV and PCV2 in the serum or in the severity of lymphoid depletion in the tonsils. Using the Lawrence Livermore Microbial Detection Array (LLMDA), the microbiome profiles of the two growth rate groups were characterized. Overall, the microbial family and species diversity were similar between the high and low growth rate groups; no significant difference was present in the mean number of families or species detected. At the level of microbial composition, several significant differences were detected. In the high growth rate pigs, increased species were detected in the genera *Megasphaera* and *Spirochaeta* ($P = 0.03$; Mann-Whitney U test). Species in the *Prevotella* genus also had a trend towards increased detection in high growth rate pigs ($P = 0.07$; Mann-Whitney U test). An unclassified Clostridiales bacterium (GenBank accession no. BAIB00000000.2) was found in a significantly higher proportion of high growth rate pigs ($P = 0.02$; Fisher's exact test). Although microbiome diversity did not appear to predispose outcome in vaccinated pigs, several shifts in pre-challenge microbiome composition were associated with improved growth rates in pigs after vaccination and co-infection. This data suggests that microbiome composition may play a role in predisposing outcome following PRRS MLV vaccination and subsequent co-challenge with PRRSV and PCV2b. Overall, modulating the microbiomes of nursery pigs to have one or more of these characteristics may provide a beneficial growth effect in the presence of viral infection.

Acknowledgements

This work was supported by the USDA NIFA Award #2013-68004-20362, and the State of Kansas National Bio and Agro-Defense Facility Fund.

References

1. Chand RJ, Trible BR, Rowland RR (2012). *Curr Opin Birol.* 2:256-263.
2. Segales J (2012). *Virus Res.* 164:10-19.
3. Han K., Seo H.W., Shin J.H., et al. (2011) *Clin Vaccine Immunol.* 18:1600-1607.
4. Ober RA, Thissen JB, Jaing CJ, et al. (2017). *Vet Microbiol.* 208:203-211.

Investigations of prevalence and causes of lameness in various age groups of gilts

Shelby Perkins[1], BS; Emily Severt[2], DVM; Darin Madson[3], DVM; Doug King[4], DVM; Pete Lasley[2], DVM
[1]University of Missouri College of Veterinary Medicine, Columbia, Missouri; [2]Smithfield Hog Production Missouri, Princeton, Missouri; [3]Iowa State University Diagnostic Laboratory, Ames Iowa; [4]Merck Animal Health, Columbia, Missouri

Introduction

Lameness is cited as a top reason for culling replacement gilts and sows within the breeding herd. Early culling results in significant economic losses throughout the industry. Within a Midwest production system, thirty-five percent of the gilts culled or euthanized in the last year were due to lameness. This high prevalence of lameness significantly impacts selection rates. The purpose of this descriptive observational study was to determine the prevalence and causes of lameness in individual gilts.

Methods

The study was performed in a production system during the summer of 2017 in two different flows of pigs. To identify which age was most affected, five age groups were selected: 10 to 15 weeks of age, 16 to 20 weeks of age, 21 to 25 weeks of age, 26 to 35 weeks of age, and mature gilts. A random number generator was used to select the pens within each barn to be evaluated and all gilts in each pen were given individual gait scores. Scoring was based on a zero to four scale where zero is normal, one is stiff, two has obvious gait abnormalities, three has one or more non-weight bearing limbs, and four is unable to rise. Oral fluid (OF) samples were collected in each flow and tested for *Mycoplasma hyorhinis, Mycoplama hyosynoviae, Erysipelothrix rhusiopathiae, Erysipelothrix tonsillarum*, and *Erysipelothrix sp2* by PCR. These samples were used as a screening test to estimate prevalence in the general population since the animals to be necropsied were used to provide individual specific diagnostic etiologies. Necropsies were performed on non-select animals within the age group exhibiting highest prevalence of lameness. Prior to euthanasia, each animal was scored for gait and body condition on a three-point scale. Genetic line, age, visible lesions, and inflammation were also noted for each individual animal. The second rib from each animal was used to measure bone quality and quantity.

Results and discussion

The largest number of lame animals found during gait scoring were 26 to 35 weeks of age with over forty-one percent lame followed by thirty-three percent lame in the 21 to 25 weeks of age group as shown in Table 1. OF sample results showed no occurrence of any *Erysipelothrix*. Both *M hyorhinis* and *M hyosynoviae* were present in multiple age groups with some groups approaching 100%. For individual diagnostic work-up, thirty-seven gilts were chosen and scored for necropsies. Of these thirty-seven, fourteen were purebred, twenty-one were F1 generation, and two were a cross between these genetics with nineteen percent of the animals being overweight in body condition. Necropsy results indicated lameness noted was primarily due to a non-infectious cause. Osteochondrosis lesions were present in over fifty-one percent of the animals necropsied. Over thirty-one percent of necropsied animals had proliferative synovitis. Sampling of bone ash, calcium, and phosphorus of the necropsied animals were found to be within normal limits. The swine industry has selected animals for rapid growth with increasing weights taxing joints. The Midwest production system involved is testing nutritional changes in gilt diets at younger ages in an attempt to slow the growth rate and reduce the incidence of lameness. Future studies will be needed to determine if interventions decrease the incidence of lameness and, more importantly, to see if these improvements translate into less culling in the breeding herd.

Table 1: The distribution of lameness by age in weeks is displayed. Gilts were scored from 0 to 4 based on normal gait (% 0s), stiffness (% 1s), obvious gait abnormalities (% 2s), one or more non-weight bearing limbs (% 3s), or inability to rise (% 4s). The final column represents the total percent of lame gilts in each age group among all stages.

Age	Percent of animals scoring 0	Percent of animals scoring 1	Percent of animals scoring 2	Percent of animals scoring 3	Percent of animals scoring 4	Total percent of lame animals
Mature gilts	76.86%	20.34%	2.81%	0.00%	0.00%	23.14%
26-35	58.90%	29.91%	9.59%	1.60%	0.00%	41.10%
20-25	67.30%	26.46%	5.94%	0.30%	0.00%	32.70%
15-20	77.97%	17.80%	3.83%	0.40%	0.00%	22.03%
10-15	89.86%	8.91%	1.02%	0.20%	0.00%	10.14%

The cost of porcine reproductive and respiratory syndrome virus and porcine epidemic diarrhea virus in finishing pigs

K. Henness[1], BS; K. Dion[2], DVM; D. Holtkamp[3], DVM, MS; R. Swalla[4], DVM
[1]University of Illinois, Urbana, Illinois; [2]The HANOR Company, Webster City, Iowa;
[3]Iowa State University, Ames, IA; [4]Zoetis, Parsippany, New Jersey

Introduction

Porcine reproductive and respiratory syndrome virus (PRRSV) in growing pigs has been estimated to cost the US swine industry $361.85 million annually.[1] Porcine epidemic diarrhea virus (PEDV) emerged in the US in 2013 and has also had substantial financial impacts on the swine industry.[2] It is currently unknown what cost PEDV, individually and combined with PRRSV, imposes on infected lots of growing pigs. The objective of this study was to determine the financial and production impacts of PRRSV and PEDV on finishing pig performance.

Materials and methods

Data from diagnostic reports, clinical cases, surveillance programs, and closeout reports were collected from 2014 to 2017 for a large Midwest production system. Information obtained included average daily gain (ADG), feed conversion ratio (F/G), percent mortality, percent culls, and veterinary costs for each lot. A PRRSV and PEDV disease status was assigned to each lot of pigs according to the disease status. Each lot received a separate classification for PRRSV and PEDV. Disease status A lots were negative for virus at weaning and remained negative throughout the growth period, including pigs vaccinated with commercial modified live PRRSV vaccine that tested negative at weaning. Disease status B lots were negative for virus at weaning and virus challenge occurred during the growth period. Disease status C lots were positive for virus at weaning. Status A/A was considered baseline for this study. Data were analyzed using PROC MIXED in SAS (Cary, NC). Costs per pig was determined by using a wean to finish budget model (D. Holtkamp, Iowa State University).

Results

The results are shown in Table 1. All parameters of interest proved to be statistically significant ($P < 0.01$) among the various disease statuses.

Discussion

This study confirms that raising pigs infected with PRRSV or PEDV will cost more than healthy baseline pigs. However, substantial differences are seen when pigs are infected with both viruses. Timing of infection may be important, as pigs that were initive for PRRSV at weaning had substantially increased cost/pig for both those infected only with PRRSV and both PRRSV and PEDV. Pigs that were infected with PRRSV and PEDV during the growing period also had substantially increased cost over baseline. As indicated, both viruses cause significant production losses and this information can be used to make financially conscious decisions about vaccination protocols, pig movement, veterinary intervention costs as well as implementing preventative measures such as biosecurity changes and maintaining surveillance initiatives.

References

1. Holtkamp DJ, Kliebenstein JB, Neumann EJ, et al. Assessment of the economic impact of porcine reproductive and respiratory syndrome virus on United States pork producers. *J Swine Health Prod.* 2013;21(2):72-84.

2. Schulz, Lee L., and Glynn T. Tonsor. "Assessment of the economic impacts of porcine epidemic diarrhea virus in the United States." *Journal of Animal Science.* 93.11 (2015): 5111.

Table 1: Results of PRRSV and PEDV disease status with associated productivity and economic significance

PRRSV/PEDV	A/A	A/B	B/A	B/B	C/A	C/B	C/C
Number of lots analyzed	647	252	125	55	559	276	5
ADG, lbs.	1.68	1.68	1.67	1.62	1.66	1.66	1.77
F/G	2.73	2.75	2.71	2.83	2.68	2.69	2.76
Mortality, %	8.31	8.17	8.73	9.44	9.96	12.81	14.47
Culls, %	3.81	3.99	3.63	4.44	3.83	4.00	3.47
Vet costs/pig, $	5.18	5.00	5.32	5.55	6.13	6.46	8.92
Difference from baseline, $	--	- 0.17	- 0.59	- 7.08	- 2.50	- 6.10	- 6.79

Immuno-modulatory influence of antibiotic therapy on the humoral immune response to vaccination in weaned pigs

J. Scott, BS; G. Almond, DVM, PhD; J. Fogle, DVM, PhD, DACVIM
College of Veterinary Medicine, North Carolina State University, Raleigh, North Carolina

Introduction

In the commercial swine industry, pigs are commonly vaccinated and administered long-acting antibiotics simultaneously, especially around the time of weaning.[1,2] With this production practice, it is assumed that no interactions are taking place between the vaccinations and antibiotic therapy. Recent reports, however, demonstrate that some antibiotics, including enrofloxacin and ceftiofur, modulate the immune response, thus hindering the response to vaccinations.[1,2] Given swine production practices, this immuno-modulatory action of antibiotics needs to be investigated in pigs vaccinated and receiving antibiotics at weaning.

Objective

Investigate the influence of the long acting antibiotics enrofloxacin (Baytril®100), tulathromycin (Draxxin®25) and ceftiofur (Excede®) on antibody and immune cell response in weaned pigs vaccinated with a porcine circovirus type 2 (PCV-2)/*Mycoplasma hyopneumoniae* (M hyo) vaccine.

Materials and methods

Pigs (n = 64) were selected at weaning (day 0) and assigned to 8 treatment groups (n = 8/group). Treatments included control (CON), a PCV-2/M hyo vaccine (Fostera®PCV MH) only (VAX), Draxxin only (TUL), Excede only (CEFT), Baytril only (ENRO) and the combination of the PCV-2/M hyo vaccine with the antibiotics (VAX+TUL, VAC+CEFT, VAC+ENRO). Vaccinations and antibiotics were administered on day 0. Body weights and blood samples were collected on days 0, 14, 21 and 35. Complete blood counts, number of immune cells (lymphocytes, T-cells, B-cells) and antibody titers against M hyo and PCV-2 were determined. Results were analyzed using repeated measures ANOVA with treatment, time (age of pig) and treatment*time interaction used as factors. Sow was included as a covariate and random effect while measurement of response at time 0 was a fixed effect. Least square means were estimated, and unadjusted *P*-values and adjusted *P*-values (Tukey's method) were determined.

Results and discussion

Antibody titer results are shown in Figure 1. In all treatment groups, Mhyo antibody titers were highest at day 0 and lowest at day 35, indicating no apparent antibody response to the vaccine. This may be due to maternal antibodies interfering with vaccine response, lack of sufficient antigen in the vaccine to induce an antibody response, or a delay in antibody response beyond the 35 day sampling period. The vaccine only group and groups receiving vaccine and antibiotic combinations had increased PCV-2 antibody titers at day 35, indicating a response to the vaccine. Contrarily, the control treatment group and groups receiving antibiotics only had highest PCV-2 antibody titers at day 0 and lowest titers at day 35. Immune cell populations were significantly impacted by time (age of pig) relative to treatment administration. However, immune cell populations did not appear to be significantly affected by any of the treatments. Furthermore, there was no interaction between response to the PCV-2 vaccine and the three antibiotics. This lack of interaction does not support previous reports that long-acting antibiotics modulate the immune system and impact vaccination responses.[1,2] The previous studies were conducted in pigs between 10 and 12 weeks of age using different vaccines. This may explain the discrepancies between the current study and other reports. Further studies need to be conducted to determine whether vaccination and antibiotic therapy practices around weaning should be altered.

References

1. Pomorska-Mól, M et al. Ceftiofur hydrochloride affects the humoral and cellular immune response in pigs after vaccination against swine influenza and pseudorabies. *BMC Vet Res.* 2015;11:268.

2. Pomorska-Mól, M et al. Immune response in pigs treated with therapeutic doses of enrofloxacin at the time of vaccination against Aujeszky's disease. *Res Vet Sci.* 2015;68:74.

Figure 1: Antibody titers (expressed as S/P ratio) against porcine circovirus type 2 (PCV-2) and *Mycoplasma hyopneumoniae* (M hyo) during different days of the study. CON pigs received neither vaccine nor antibiotic while VAX pigs received the PCV-2/M hyo vaccine but no antibiotics. ENRO pigs received enrofloxacin (Baytril®100) but no vaccine, and TUL pigs received tulathromycin (Draxxin®25) but no vaccine. CEFT pigs received ceftiofur (Exede®) but no vaccine. VAX+ENRO pigs received the vaccine and enrofloxacin. VAX+TUL pigs received the vaccine and tulathromycin, and VAX+CEFT pigs received the vaccine and ceftiofur.

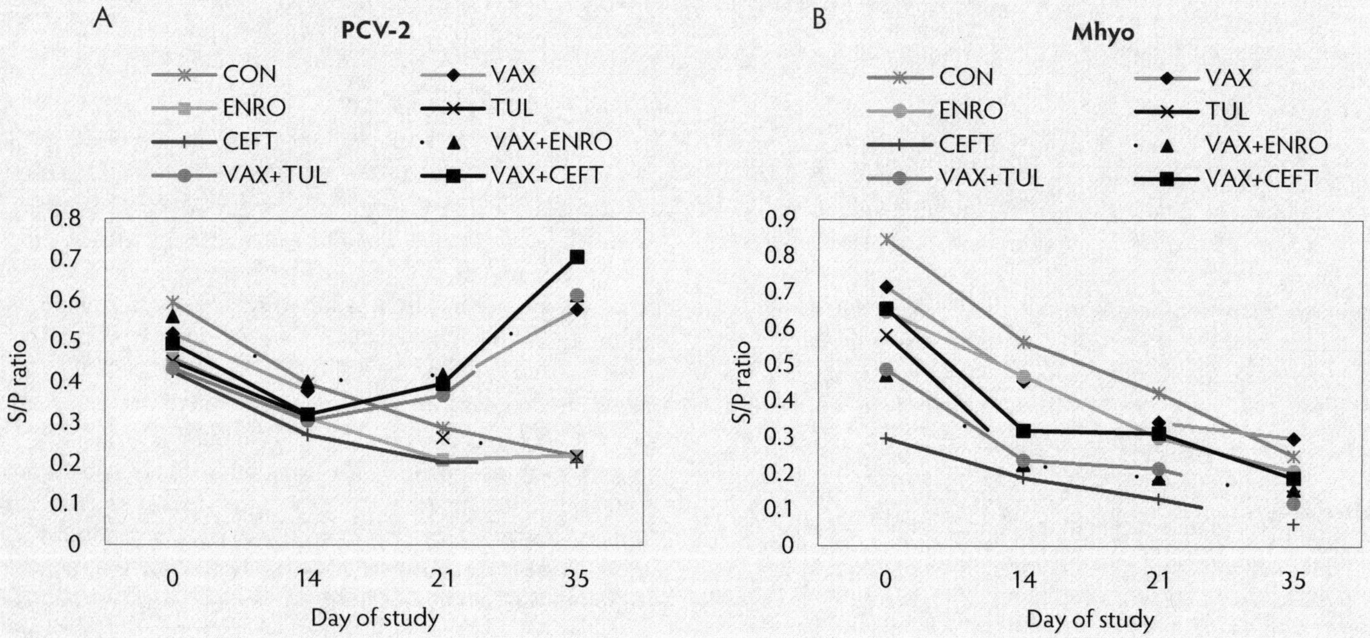

Comparison of serological responses in traditional and parity segregated flows for *Mycoplasma hyopneumoniae* and *Lawsonia intracellularis*

Elizabeth Noblett[1], MAS; Joshua Duff[2], DVM; Samuel Copeland[2], DVM; Eugene Nemechek[3], DVM
[1]North Carolina State University, Raleigh, North Carolina; [2]Maxwell Foods LLC, Goldsboro, North Carolina; [3]Zoetis, Parsippany, New Jersey

Introduction

Should the timing of vaccination and medication for *Mycoplasma hyopneumoniae* (M hyo) and *Lawsonia intracellularis* (Li) change depending upon the makeup of the breeding herd from which the pigs are sourced? Maxwell Foods uses a modified parity-segregation plan allowing gilts to farrow twice before being dispersed to older parity farms, and downstream pig flows are segregated based upon parity status of sow farm of origin, as compared to a traditional mixed-parity farm. The objective of this study was to investigate exposure to M hyo and Li within modified parity-segregated and traditional flows through antibody detection.

Materials and methods

Four pig flows were tested from separate breeding herds: a gilt farm (P0→P1), a P2 farm, a P3+ farm, and a traditional farm with monthly gilt introductions (TRAD). Clinical signs within the flows were not observed or documented. From each of these four pig flows, 20 sera samples were collected from pigs at 3, 6, 9, 12, 15, 18, and 21 weeks of age. A total of 560 sera were submitted to Iowa State Veterinary Diagnostic Laboratory for ELISA for Li (Svanova, Uppsala, Sweden) and M hyo (Idexx, Westbrook, ME).

Results and discussion

M hyo antibodies were observed across all flows at 3, 9 and 12 weeks of age (Table 1). Antibody presence at 3 weeks was considered maternal antibody and at 9 and 12 weeks was attributed to vaccination, since piglets were vaccinated at 3 and 6 weeks for M hyo. Animals positive for Li antibodies were observed at 3 weeks of age in the traditional and P0→P1 flow and were attributed to maternal antibodies. Only the dams in the P0→P1 and TRAD flows received vaccinations for Li. Across all four flows, animals were positive for Li antibodies at 21 weeks of age, indicative of exposure. The results were notably different when comparing the P3+ the number of animals positive for Li versus the other three flows tested. The results were not pertinent for the presence of M hyo across the flows, due to the inability to distinguish maternal antibodies from vaccinations. The P3+ flow had the fewest number of Li positive animals, validating the absence of a medicated feed pulse to control Li in the finisher. As no other differences were noted within flows tested, this provided support to the production system decisions that existing medication and vaccination programs adequately control M hyo and Li disease in finishing. This type of cross-sectional testing is a potential method for other producers to validate their disease control programs, and may be used with additional pathogens.

Table 1: Percent seropositive animals by flow for M hyo and Li

Age (weeks)	3	6	9	12	15	18	21
M hyo							
P0->P1	10%	0%	85%	90%	10%	0%	0%
P2	65%	25%	32%	20%	0%	0%	28%
P3+	15%	10%	89%	60%	24%	5%	21%
TRAD	40%	70%	45%	35%	0%	5%	0%
Li							
P0->P1	40%	10%	0%	0%	5%	80%	15%
P2	0%	0%	0%	0%	5%	10%	44%
P3+	0%	0%	0%	0%	0%	0%	5%
TRAD	10%	0%	0%	0%	5%	20%	10%

Utilizing serum antibody levels in post vaccinated piglets to evaluate maternal antibody interference and vaccination compliance

J. Tubbs[1], MS; E. Byers[2], DVM
[1]Auburn University CVM, Auburn, Alabama; [2]Prestage Farms, Clinton, North Carolina

Introduction and objective

Understanding maternally derived antibody (MDA) levels in peri-weaned pigs is critical in determining appropriate vaccination timing within the herd to avoid potential vaccination interference by MDAs. Measuring MDA levels is especially important in herds that utilize pre-farrowing vaccination in sows for determining when levels are low enough not to interfere. The study objective was to evaluate serological responses to a porcine circovirus type 2 (PCV2) and *Mycoplasma hyopneumoniae* (Mhp) combination vaccine in weaned pigs following two, one mL injections at three and six weeks of age (WOA). In addition, this study analyzed if MDA levels at the time of the first vaccination influenced post-vaccination antibody levels. The overall purpose was to determine if post-vaccination serology results are useful in determining if pigs received both vaccinations.

Materials and methods

Comparison of maternally derived antibodies from eight sow farms determined site inclusion. Four sites administered a pre-breeding vaccination (PBV) while four had no pre-breeding vaccination (N-PBV) protocol. Based on MDA, one sow farm from each category were selected for the study. Thirty litters from both the PBV and N-PBV sites were selected and categorized by parity (P) into one of three groups of 10 sows: P1, P2-3, and P4+. Blood sample collection and vaccination was completed one day before weaning and prior to when the farm staff routinely vaccinate litters. Within each litter, one of the biggest pigs was selected for blood sample collection and tagging, followed by administration of one mL of vaccine to the full litter. After weaning, the pigs from each sow farm were placed at two separate nurseries. At three weeks of placement in the nursery (six WOA), a second one mL vaccination was given. Three weeks following the second vaccination (nine WOA), a blood sample was collected for comparison and analysis.

Results and discussion

Study results presented in Table 1 are arranged by PBV site and N-PBV site as well as by PCV2 and Mhp testing. A cut-off of ≥ 1280 was used for PCV2 antibody titer threshold for high MDA levels at first vaccination and a positive vaccination response while a. cut-off of > 0.400 S/P ratio was used for Mhp. PCV2 IFA results show positives at three WOA of 3 of 30 for the N-PBV group and 28 of 30 for the PBV groups. S/P ratios for Mhp show 13 of 30 samples positive (0.428 avg.) for the N-PBV group and 24 of 30 samples positive for the PBV group (0.918 avg.). Following both vaccination events, serotesting yielded the following results at nine WOA for PCV2: N-PBV 19 of 21 responded, 2 did not; PBV 7 of 30 responded, 7 did not, 16 were unknown. Unknown response was categorized for being above the threshold for response, but having the same or higher than titers at weaning, which could be persisting MDA. For Mhp results at nine WOA, S/P ratios showing response to vaccination (> 0.400) were as follows: N-PBV group 18 of 21 responded; PBV group 25 of 30 responded. These results demonstrated MDA interference for the PCV2 vaccination while Mhp vaccination was able to induce an antibody response despite high levels of MDAs at weaning. This is significant when considering timing and monitoring of both PCV2 and Mhp vaccination. Also, these results indicate that post vaccination monitoring is better accomplished through measuring Mhp antibodies, as it is difficult to monitor post vaccination antibodies for PCV2 when pigs have high levels of MDA at the time of vaccination.

References

1.Jayappa H, Davis R, Rapp-Gabrielson V, Wasmoen T, Thacker E, Thacker B. Evaluation of the efficacy of *Mycoplasma hyopneumoniae* bacterin following immunization of young pigs in the presence of varying levels of maternal antibodies. *Proc AASV*. Nashville, TN. 2001. 237-241.

2.Woolums AR. Vaccinating calves: New information on the effects of maternal immunity. *Proc. AABP*. Vancouver, British Columbia. 2007. 10-17.

Table 1: Pre and post vaccination diagnostic results for PCV2 and Mhp

Laboratory test	Pig age/ vacc. status	Variable	Unit of measure	Results by farm MDA status	
				N-PBV	PBV
PCV2 IFA	3 wk/ Pre-vacc.	Titer	Geomean	100.8	3880.2
		MDA (> 1280)	Pos/tested	3/30	28/30
	9 wk/ Post-vacc.	Titer	Geomean	2921.3	1403.9
		Response (> = 1280)	No	2	7
			Yes	19	7
			Unknown	0	16
Mhp ELISA	3 wk/ Pre-vacc.	Serostatus	Pos/tested	13/30	24/30
		S/P ratio	Avg	0.428	0.918
		MDA (> .400)	Pos/tested	13/30	24/30
	9 wk/ Post-vacc.	Serostatus	Pos/tested	21/21	29/30
		S/P ratio	Avg	1.273	1.264
		Response (S/P > 0.400)	No	0	1
			Yes	18	25
			Unknown	3	4

Assessment of the association between respiratory pathogen burden and the productivity of growing pigs

R. Stika[1]; D. Holtkamp[1], MS, DVM; C. Goodell[2], MS, DVM, PhD; E. Lowe[2], DVM; E. Kluber[3], MS, DVM, MBA; D. Baum[1], MS, DVM, PhD; J. Maurer[3], VMD; R. Bi[1], MS; C. Wang[1], PhD; J. Zimmerman[1], DVM, PhD, DACVPM; L. Gimenez-Lirola[1], MS, PhD; C. Rademacher[1], DVM; Seth Playter[2], BS; Ethan Schmaling[2], BS

[1]Iowa State University, Ames, Iowa; [2]Boehringer Ingelheim Animal Health, Duluth, Georgia; [3]Smithfield, Milford, Utah

Introduction

Growing pigs in the US face the threat of numerous and concurrent respiratory pathogens including porcine reproductive and respiratory syndrome virus (PRRSV), influenza A virus of swine (IAV-S), *Mycoplasma hyopneumoniae* (Mhp) and porcine circovirus type 2 (PCV2). Efforts to estimate the production and economic impact of these pathogens is confounded by the multitude of factors that impact productivity and the inability to attribute losses specifically to disease. The concept of pathogen burden has been applied in human medicine to study the association between infectious pathogens and various diseases. The objective of this study was to develop a way to characterize respiratory pathogen burden in pigs and to estimate the association between pathogen burden and the productivity of growing pigs.

Materials and methods

A lot was defined as a group of pigs that were placed and closed out together. Eleven bi-weekly pen based oral fluid (OF) samplings were collected from each lot of pigs starting on date of placement. Each sampling had 6 pen-based OF samples tested per lot. Samples were tested by real time polymerase chain reaction (PCR) for PRRSV (US and EU strains), IAV-S, PCV2, and Mhp to determine levels of pathogen burden. Pathogen burden was measured as percent of OF samples positive by PCR for each pathogen. Data was organized into four phases to determine phase specific pathogen trends: 1) placement phase (weeks 1-2); 2) nursery phase (weeks 4-8); 3) early finisher phase (weeks 10-16); and 4) late finisher phase (weeks 18-24). Close out data, stocking density and bi-weekly mortality were collected. For the statistical analysis, K-means clustering method was applied to partition the lots into K groups based on the percentage of positive PCR results for each pathogen in each phase of production.

Results and discussion

Twenty lots of pigs from a single production system were enrolled. The clusters of pathogen burden were associated with mortality for several pathogens and combinations of pathogens. For example, the cluster analysis resulted in two distinct clusters of burden for Mhp: Lots in cluster 1 (10 lots) had low *Mhp* burden throughout the 4 phases (Figure 1A) and lots in cluster 2 (10 lots) had high Mhp burden in the last 2 phases (Figure 1B). Lots in cluster 2 experienced an increase in mortality in the early and late finishing phases. This study demonstrates that pen based, bi-weekly OF collections can be used to identify patterns of respiratory pathogen burden in groups of growing pigs that are associated with phase-specific mortality. Measuring pathogen burden to estimate more precisely the production and economic impact of respiratory disease in growing pigs will help veterinarians and producers make better resource allocation decisions to improve animal health.

Fig. 1A: Cluster 1 of Mhp burden vs. phase mortality; **Fig. 1B:** Cluster 2 of Mhp burden vs. phase mortality

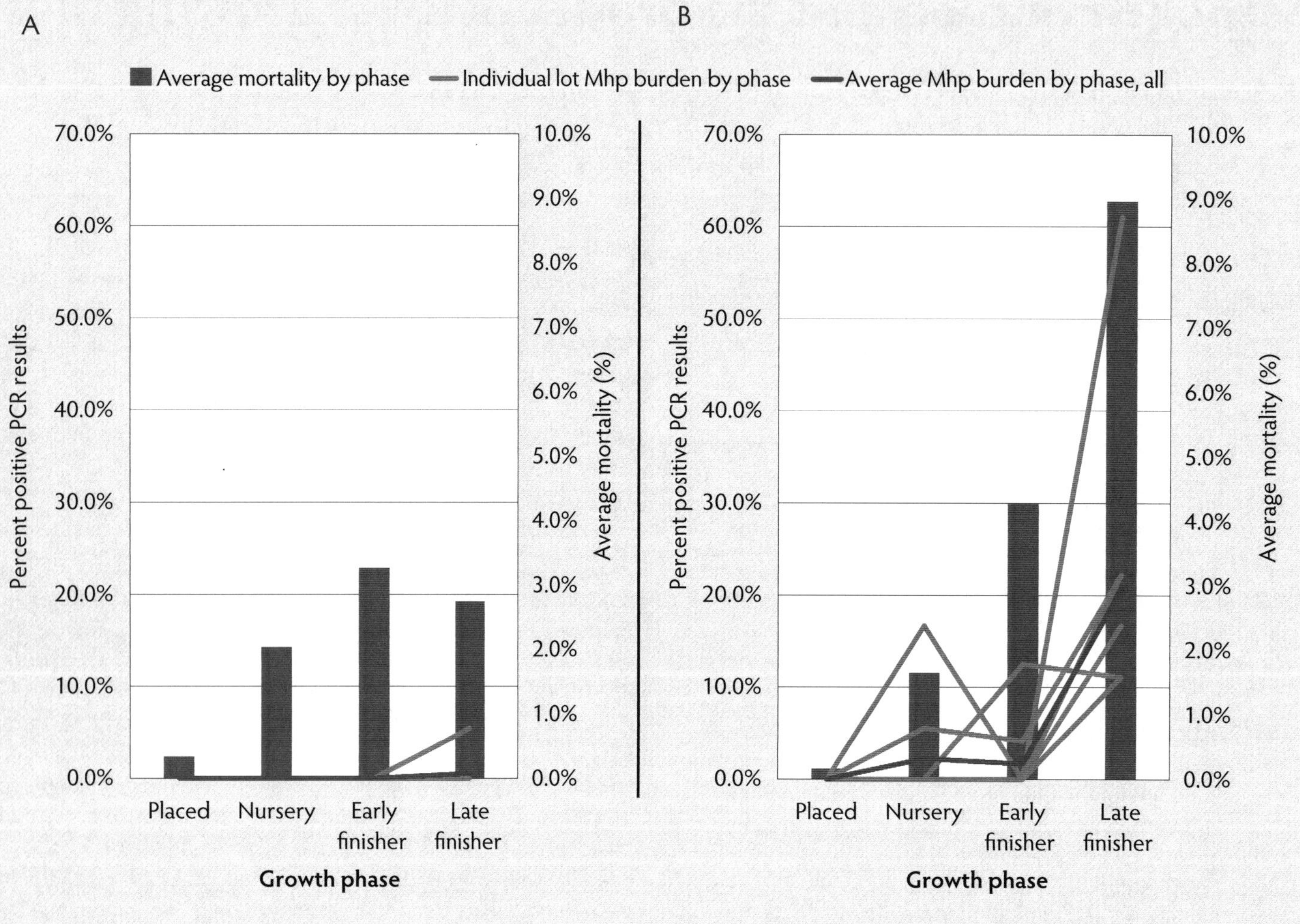

A novel approach to water medication calculations

Brent Sexton, BS; Alejandro Ramirez, DVM, MPH, PhD
Iowa State University, College of Veterinary Medicine, Ames, Iowa

Introduction

Many modern swine finishing facilities utilize water medicating systems to rapidly mass treat pigs. Conventional methods of calculating water medication rates typically do not account for variations in either water consumption or water wastage. It is not uncommon for water wastage to exceed 50% of the total water disappearance.[1] Drinker style and the weight of the hogs can also greatly impact water disappearance.[2] This has the potential to both waste medication and under-medicate hogs. This study aims to develop a method to better calculate water medications, accounting for changes in water consumption and waste.

Materials and methods

Before a new method to calculate water medications could be developed, a comprehensive review of recent literature was performed in order to understand factors affecting water consumption and waste. A retrospective study consisting of 7 wean-to-finisher groups using nipple drinkers and dry feeders was then performed to better understand weekly water disappearance through the entire wean-to-finish process. The data was compiled in Excel and evaluated for abnormal values. Age, temperature, and water usage were evaluated for the entire finishing periods, with periodic weights and average daily feed intake (ADFI) available from 5 of the 7 (71%) groups. The compiled data was used to determine an estimated weight based on age, as well as to estimate ADFI and water usage at weekly intervals. These estimates were made by plotting the parameters relative to each other in Excel and applying a trendline. The most appropriate trendline and formula was determined by comparing R^2 values and visual fit. These values were compiled in a table, which was expanded to include estimated water usage for various drinker styles based on published values.

Results

The results of the literature review identified age, weight, ambient temperature, ADFI, and drinker style as factors of primary concern. Other factors were identified as having the potential to impact water disappearance, but only when far outside normal ranges. After evaluating the data, temperature did not have a substantial impact on water consumption in this study, likely due to the relative consistency of the barn temperatures. Age, weight, and ADFI had strong relationships with water usage and exhibited substantial collinearity. This indicates that two parameters, drinker style and either age, weight, or ADFI are needed to estimate water usage. The change in water usage as a function of weight, seen in Figure 1, indicates that water usage does not exhibit a linear relationship with weight ($R^2 = 0.3949$, trendline not shown). The equations used to estimate water usage can be easily updated as new data is acquired.

Figure 1: Relationship of weight vs water usage

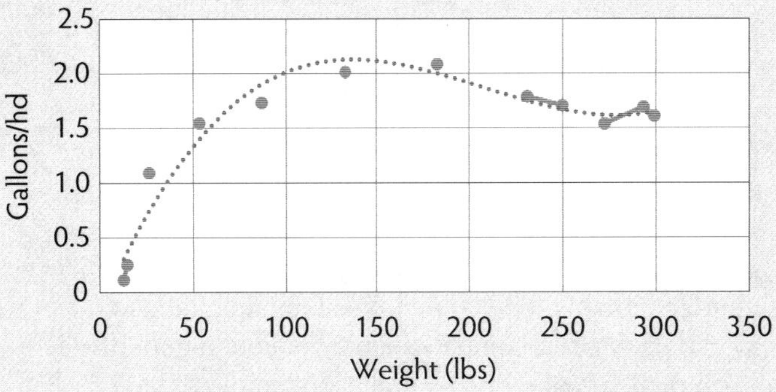

Discussion

Because of the intricacies in making these estimations, a spreadsheet (and potentially a future mobile app) has been developed that can automatically calculate the necessary gallons of stock solution and packets of medication needed, based on the pigs' age or weight. The calculations are based directly on the results from this study. The spreadsheet also accounts for water wastage based on drinker design, to ensure more accurate dosing. This information will be made available to the public and updated periodically, allowing producers and practitioners to quickly and accurately calculate the medication rates and total medication needed to appropriately treat their hogs.

References

1. S. Torrey, E.L.M. Toth Tamminga, T.M. Widowski. Effect of drinker type on water intake and waste in newly weaned piglets. *Journal of Animal Science.* 2008;86:1439-1445. doi:10.2527/jas.2007-0632

2. R.V. Muhlbauer, L.B. Moody, R.T. Burns, J. Harmon, K. Stalder. Water Consumption and Conservation Techniques Currently Available for Swine Production. *National Pork Board Research Report.* 2010;NPB#09-128:1-32

Comparison of individual oral fluids, pooled oral fluid samples and Swiffer™ environmental samples of drinkers for the detection of IAV-S and PRRSV by PCR

Taylor Homann[1]; Chris Sievers[2]; Jana Morgan[3]; Brad Leuwerke[2]
[1] University of Minnesota, CVM; [2] Swine Vet Center; [3] Boehringer Ingelheim

Introduction

Oral fluids (OF) are a valuable sampling tool for swine veterinarians. One drawback to OF is that young and/or diseased pigs commonly do not chew ropes enough to obtain a sufficient volume for testing. Swiffer™ environmental sampling (ES) of water cups may give a comparable sample. Pooling OF through multiple pens may also be a technique to sample more pigs with the same diagnostic resources. The objective of this project was to compare the diagnostic sensitivity of pooled pen OF and ES (using Swiffer™ kits) using individual pen OF as the standard.

Materials and methods

For two barns, 15 ES and 15 individual pen OF, paired, were obtained for each time point. After individual samples were collected, pen OF were pooled by 3 for 5 total pooled samples. 35 total samples per time point. Samples were collected at days 0, 4, 7, 14, 21 and 28, with sampling starting at 3 days placed into nursery. Ropes were hung from the front gate of each pen to sample only OF from that pen. ES were taken using Swiffer™ cloths to sample the bottom of water cups (both pans and bowls), focusing around nipples. Gloves were changed between each sample taken. Individual pen OF were the standard for true positives and true negatives. Sensitivity is true positives over true positives plus false negatives. Specificity is true negatives over true negatives plus false positives.

Discussion

Results are reported in Table 1. Environmental samples of water cups had a sensitivity of 71% (IAV) and 60% (PRRS). Pooled OF samples had a sensitivity of 78% (IAV) and 91% (PRRSv) relative to individual pen OF samples. Decreases in sensitivity and specificity may be offset by increases in other efficiencies like time and number of pigs sampled per diagnostic test.

Table 1: Percent positive and average Ct values for all sample types (IAV-S and PRRSv). ES and OF pooled were compared to OF individual as the standard for sensitivity and specificity (IAV and PRRSv). Each sample has a total n of 134.

		OF	ES	OF Pooled
IAV-S	% Pos	16%	14%	16%
	Avg Ct	32.41	30.27	33.15
	Sensitivity	-	71%	78%
	Specificity	-	94%	97%
PRRSv	% Pos	64%	34%	77 %
	Avg Ct	33.32	35.50	33.41
	Sensitivity	-	60%	91%
	Specificity	-	68%	87%
Time		44.76 min	5.21min	-

Evaluation of serological responses to Circumvent PCV M G2 in sows vaccinated pre-farrowing and in pigs with maternally-derived antibodies at time of first vaccination

Jamie Madigan[1], BS; Amanda Anderson[2]; Jeremy S Pittman[3], DVM, MS, DABVP
[1]North Carolina State University, Raleigh, North Carolina; [2]Iowa State University, Ames, Iowa;
[3]Smithfield Hog Production- North Region, Waverly, Virginia

Introduction

Porcine circovirus type 2 associated disease (PCVAD) and *Mycoplasma hyopneumoniae* (Mhp) are important systemic and respiratory diseases of swine. For the most part, PCVAD and Mhp are controlled by vaccinating pigs at an early age. In addition, breeding herds may use these vaccines to stabilize the herd and booster maternal immunity to protect offspring. It is unclear if sow vaccination influences the vaccination response in young pigs. Accordingly, the objectives of the experiments reported here were to evaluate dam serological responses after vaccination and the effect of maternally-derived antibody (MDA) on pig serological responses following two vaccinations with 1 mL of Circumvent PCV M G2 (G2) at weaning and four weeks later.

Materials and methods

The study was conducted on four commercial breed-to-wean farms. Each farm was vaccinating breeding animals with 2 mL of G2 two weeks prior to farrowing (2 WPF). To evaluate serological responses in sows vaccinated prior to farrowing, on one farm (Farm A), 60 gestating gilts were randomly allocated to receive 2 mL G2 2 WPF (VX) or no vaccine (NVX). Gilts were bled prior to and 2 weeks post-vaccination (at farrowing). All piglets remained on their birth dam for the duration of lactation and one pig from each litter was ear tagged, blood sampled, and vaccinated with 1mL of G2 at weaning. In the other three farms (B, C and D), all sows were vaccinated prior to farrowing and thirty to be weaned litters; 10 each from 3 parity groupings (1, 2-3, and 4+) were selected and 1 piglet from each litter was ear tagged, blood sampled and vaccinated with 1 mL of G2. All pigs from all four farms were revaccinated in the nursery with 1mL G2 28 days later, after their first vaccination. Blood samples were collected 21 days following the second vaccination. All samples were tested individually by PCV2 ELISA, PCV2 IFA, and Mhp ELISA. Where appropriate, results were analyzed by the Kruskal-Wallis non-parametric ANOVA, and Pearson correlation.

Results and discussion

In Farm A VX gilts, the geomean PCV2 IFA pre- and post-vaccination titers were 3,989 and 3,774, respectively, while the titers in NVX gilts declined from 4,101 to 347 ($P < 0.05$). In VX gilts, the Mhp ELISA average S/P ratios at pre- and post-vaccination increased from 0.739 to 1.735, while titers declined in NVX from 0.767 to 0.480 ($P < 0.05$). The titers of the dams in both groups at farrowing for both PCV2 and Mhp were highly correlated with the titers of their pigs at weaning ($P < 0.05$). In pigs from VX gilts, PCV2 IFA titers decreased from 2560, to 609, while in pigs from NVX gilts, titers increased from 188 to 2808. For PCV2 ELISA 30/30 and 26/28 of the pigs from VX gilts were positive before and after vaccination, respectively. The corresponding geomean titers were 5668 and 757 respectively.

In vaccinated pigs from Farms B, C, and D, the geomean PCV2 IFA titers before and after vaccination for all 3 farms were 4,740 and 492, respectively. For PCV2 ELISA, 89/90 and 65/87 of the pigs were positive before and after vaccination and the corresponding geomean titers were 3,384 and 413. For Mhp ELISA, 87/90 and 66/87 of pigs were positive before and after vaccination and corresponding S/P ratios were 2.056 and 0.649. Post-vaccination PCV2 and Mhp antibody levels were higher in the pigs from one of the three farms ($P < 0.05$) but the reason for this difference was not readily apparent. Table 1 presents all of the results in detail.

In summary, vaccination of sows prior to farrowing increased the sows PCV2 and Mhp titer levels and the titer levels of the pigs at weaning were highly correlated with their dam's titers at farrowing. The high levels of maternal antibodies to both PCV2 and Mhp made it difficult to determine how well the pigs "responded", however the majority of pigs did have measureable antibody titers beyond what is expected for persistence of MDAs, suggesting they did respond to vaccination. This data only represents the humoral immune response to the vaccination, as measured by antibody titers. Overall, the data suggests that adjustment of the timing of the sow and/or piglet vaccines may be warranted, and more work should be done to determine how these adjustments influence response to vaccination of pigs.

Table 1: Summary of all data collected from all four farms

Farm	Animal/timing	Test	Value	Averages, geomeans, rates: Control	Averages, geomeans, rates: Vaccinated
FARM A DATA	Sows two weeks pre-farrowing	PCV IFA	Titer	4102	3774
		MH ELISA	S/P Ratio	0.767	0.739
			No. Positive	26/30	26/30
	Sows at farrowing	PCV IFA	Titer	347	3792
		MH ELISA	S/P Ratio	0.480	1.735
			No. Positive	22/30	30/30
	Pigs at first vaccination	PCV ELISA	No. Positive	27/30	30/30
			S/P Ratio	0.818	1.460
			Actual Titer	727	5668
			Adjusted Titer	520	2502
		PCV IFA	Titer	188	2560
		MH ELISA	No. Positive	8/30	27/30
			S/P Ratio	0.353	1.117
	Pigs three weeks after second vaccination	PCV ELISA	No. Positive	30/30	26/28
			S/P Ratio	1.253	0.831
			Actual Titer	2923	757
			Adjusted Titer	1505	538
		PCV IFA	Titer	2808	609
		MH ELISA	No. Positive	29/30	23/28
			S/P Ratio	1.059	0.782
FARMS B, C, D DATA	Pigs at first vaccination	PCV ELISA	No. Positive		89/90
			S/P Ratio		1.299
			Actual Titer		3384
			Adjusted Titer		1552
		PCV IFA	Titer		4740
		MH ELISA	No. Positive		87/90
			S/P Ratio		2.056
	Pigs three weeks after second vaccination	PCV ELISA	No. Positive		65/87
			S/P Ratio		0.642
			Actual Titer		413
			Adjusted Titer		296
		PCV IFA	Titer		492
		MH ELISA	No. Positive		66/87
			S/P Ratio		0.649

Validation of a urine dipstick as a tool to assess the potential impact of urinary tract health on reproductive performance and sow mortality

J. Bremer[1], BS; B. Peters[2], DVM; M. Jansen[3], DVM; E. Toohill[2], DVM; D. Amodie[3], MS
[1]North Carolina State University, Raleigh, North Carolina; [2]The Maschhoffs LLC, Pittsfield, Illinois;
[3]Zoetis, Parsippany, New Jersey

Introduction

In the past, effects of urinary tract infections on a sow's overall productivity have been researched.[1] Complications include increased wean-to-estrus interval, dysuria, periparturient complications and general discomfort for sows in farrowing. At present, assessments of urine characteristics are done infrequently. This study compared conventional lab-based urinalysis (UA) with an on-farm dipstick test to evaluate the reliability of the dipstick test and determine the impact of urinary tract health on sow longevity and reproductive performance.

Materials and methods

Urine samples from 292 late-term sows of varying age and parity were obtained from four commercial farms (Illinois). Data recorded consisted of sow identification, parity, breed date, UA, and dipstick results. The first urine samples (n = 141) were obtained via free catch and with tampons as described previously.[2] Post collection, an aliquot from each urine sample was pipetted onto each of 10 pads on a Health Wiser Urinalysis Dipstick (Davao City, Philippines). The remainder of the sample was submitted to either Iowa State University or University of Tennessee Veterinary Clinical Pathology Lab for UA and urine sediment evaluation. Urine characteristics evaluated included: glucose, blood, ketones, protein, nitrites, leukocytes, bilirubin, urobilinogen, specific gravity, and pH. Sediment was also analyzed on UA only. SAS (9.4, Cary, NC) was used to analyze the accuracy between the two urine testing methods (dipstick vs UA) for 141 samples. Results from the dipstick and UA were categorized as either normal or abnormal and analyzed using a Generalized Linear Mixed Model (GLMM).

Upon validation, the remaining 151 animals were tested solely with the dipstick. The sow's enrollment in the trial terminated upon her successful weaning, return to estrus, and re-breeding. Failure to accomplish any of these classified this animal as experiencing an "abnormal event" and it was noted when assessing urine dipstick results. Positive reaction for a pad indicated an abnormality, including pH outside of the range 5.5-7.0 and specific gravity below 1.010.

Results and discussion

Urine dipsticks were shown to be a reliable tool that can be used on-farm to assess sow kidney function and hydration status quickly and accurately. The dipsticks were consistent at finding abnormalities within the urinary profile of late term sows. No significant differences ($P < 0.05$) for glucose, ketones, protein, and specific gravity results were observed between the dipstick and lab UA values (Table 1). 25% of sows not re-bred post weaning and 60% with increased stillborn or mummified piglets had protein in the urine pre-farrow. Of the 43 animals whose abnormal event was death or culling 40% also had protein in the urine and 25% were hyposthenuric or unable to concentrate their urine. These values show that insufficient urinary tract health could have contributed to the sow's diminished performance and/or ultimate fate. A dipstick exam costs 13cents/sow compared to UA alternatives priced at $15.50/sow. By performing a urinalysis prior to clinical presentations, the dipstick can provide an opportunity for intervention, resulting in improved sow health, breeding efficiency and enhanced productivity.

References

1. Dial GD, MacLachlan NJ. Urogenital infections of swine. Part 1. Clinical manifestations and pathogenesis. *Compendium on continuing education for the practicing veterinarian*. 1988; 10: 63-71.

2. Nickel M, Skoland K, Forseth A, Brown J, Thompson V, Breuer M; Warner R, O'Brien K, Canning P, Thomas P, Hoogland M, Lauterbach N, Karriker L. Development and validation of ante-mortem urine collection techniques for gilts and sows. *Journal of Swine Health and Production*. 2017.

Table 1: Summary of in-lab UA vs. on farm urine dipstick results confirming consistency.

Variable	% Matches	*P*-value	Result
Glucose	98.58%	0.970	No significant difference
Ketones	94.33%	0.074	No significant difference
Protein	53.90%	0.103	No significant difference
Specific gravity	85.82%	0.607	No significant difference

Evaluation of antibody titer of the weaned pig following vaccination with a killed influenza A virus vaccine in the presence of maternal antibodies both pre and post heterologous influenza A virus challenge

Abigail Ruane[1], BS; Mark Wagner[2], DVM; Abigail Redalen[3], DVM; Deborah Amodie[3], MS
[1]Iowa State University, Ames, Iowa; [2]Fairmont Veterinary Clinic, Fairmont, Minnesota; [3]Zoetis, Parsippany, New Jersey

Introduction

Influenza A virus of swine (IAV-S) causes an acute respiratory disease resulting in significant productivity losses in all stages of production.[1,2] Natural exposure to IAV-S in the early nursery stage is common and can predispose pigs to various production setbacks through the finisher stage. The presence of maternal antibodies (MA) in weaned pigs makes the utilization of vaccine challenging. Conflicting evidence exists regarding the value of vaccinating in the presence of MA in order to provide sufficient immunity.[3,-6] The objective of this study was to evaluate serological antibody response and weight gain of weaned pigs following vaccination with a commercially available killed IAV-S vaccine (FluSure XP®, Parsippany, NJ) in the presence of MA, both pre and post IAV-S challenge with heterologous virus.

Materials and methods

Fifteen day old piglets (n = 101) from an IAV-S positive sow herd were screened using a Swine-IAV-S Gamma hemagglutination inhibition (HI) Zoetis XP-012 HI serological panel at University of Minnesota Veterinary Diagnostic Laboratory (UMN VDL). At weaning, 42 pigs that tested positive for MA presence were identified as being either high or low antibody titer status (Low 10-20; High ≥ 40). Pigs were placed in a nursery at weaning and randomly allocated to treatment groups by titer status and weight, with 21 pigs per pen. Each pig was assigned to one of two treatments: vaccinated with FluSure XP® vaccine (T01) or unvaccinated (T02). The pig was the experimental unit and pigs from both treatment groups were comingled. T01 pigs were vaccinated on day -42 (d-42) and d-21 of the trial. Both treatment groups were challenged with a heterologous (93.3% homologous) gamma H1N1 influenza strain on d0. An additional 10 of those initially screened nursery pigs with no measurable MA presence were randomly selected as negative controls (T03) and housed in the same nursery as T01 and T02 pigs pre-challenge. On d0, T03 pigs were relocated to a new nursery to prevent IAV-S exposure. Blood samples from each pig were collected (days -42, -21, 0, 7, 14) and submitted to the UMN VDL for HI to compare differences in titers throughout the trial. The primary outcome variable was post-challenge titer responses between T01 and T02 pigs. Weight gain was analyzed as a secondary outcome of the study. Individual weights were recorded on d-42, d0 and d14 and compared between the three groups. Results were analyzed using ANOVA (SAS 9.4, Cary, NC).

Results and discussion

T01 demonstrated significantly higher ($P < 0.05$) antibody titers on d14 than T02 in pigs with low and high levels of MA (Table 1). At time of challenge (d0), the low MA group in T01 had significantly higher ($P = 0.0446$) titer when compared to T02. There were no significant differences in weight at d-42 or at time of challenge (d0). T02 was significantly heavier on d14 than T01 ($P = 0.0332$). Results from this study suggest vaccinating recently weaned pigs in the presence of low or high levels of MA primes the immune system with a stronger and faster serological antibody response to a natural IAV-S virus challenge as compared to unvaccinated pigs with similar MA profiles. Vaccinating at weaning can thus potentially reduce the negative health and productivity impact IAV-S has on the growing pig.

Table 1: Geometric mean of antibody titers by day of trial

Category	High	High	High	High	High	Low	Low	Low	Low	Low
Day	-42	-21	0	7	14	-42	-21	0	7	14
T02	47.57[a]	7.94[a]	5[a]	5[a]	5[b]	14.7[a]	5.4[a]	5[b]	5[a]	5[b]
T01	44.9[a]	6.4[a]	5.69[a]	5.69[a]	13.85[a]	15.87[a]	5.83[a]	9.26[a]	7.94[a]	20[a]

All titers that were recorded as "negative" were given a value of 5 prior to statistical analysis.
[a,b] Values with different superscripts within each column are statistically different ($P > 0.05$).

References

1. Questions & Answers: Influenza A Virus in Swine (IAV-S). American Association of Swine Veterinarians. 2016.

2. H. Kothalawala, M.J.M. Toussaint, E. Gruys. An overview of swine influenza. Veterinary Quarterly. 2006;28:2, 45-53, DOI: 10.1080/01652176.2006.9695207.

3. M. Gramer, S. Detmer. Brief overview of influenza immunology and vaccinology. American Association of Swine Veterinarians. 2012.

4. P. Kittikoon, E. Thacker, D. Nilubol, S. Yu, B. Erickson, B. Janke, T. Hoover, S. Sornsen. Immune response and effect of maternal antibody interference on vaccination with a bivalent swine influenza vaccine. American Association of Swine Veterinarians. 2005.

5. P. Kitikoon, D. Nilubol, B. Erickson, B. Janke, T. Hoover, S. Sornsen, E. Thacker. The immune response and maternal antibody interference to a heterologous H1N1 swine influenza virus infection following vaccination. Veterinary Immunology and Immunopathology. 2006;112.3-4: 117-28.

6. A. Scheidt, M. Ficken. Effect of swine influenza virus maternal antibody on active immunization (maternal antibody override). American Association of Swine Veterinarians. 2003.

Characterization of skin variations that may influence transdermal absorption of topical drugs

S. Anthony[1], BS; J. Brown[1], DVM; D. Madson[2], DVM, PhD, DACVP; K. Klemesrud[1]; M. Rohde[1], BS; E. Akin[1], BS; S. Radke[2], DVM; K. Skoland[1], BS; M. Breuer[1], BS; A. Forseth[1], DVM; P. Canning[1], DVM, PhD; K. Hayman[1], DVM; H. Kittrell[1], DVM; L. Karriker[1], DVM, MS, DACVPM

[1]Swine Medicine Education Center, Iowa State University, Ames, Iowa; [2]Veterinary Diagnostic & Production Animal Medicine, Iowa State University, Ames, Iowa

Introduction

Topical application of pharmaceuticals has proven to be a viable, alternative route of drug administration within veterinary medicine. Implementation of topical drug application at piglet processing foregoes the extra stress, number of injection sites and needle sticks of standard drug administration. For topical application to become a viable option in swine medicine, an objective evaluation of skin characteristics must occur. The objective of this study was to characterize various skin samples across the body of processing age piglets to aid in determining the anatomical location(s) that will most likely lead to transdermal absorption and systemic circulation of a drug.

Materials and methods

In total, ten, 4-5 day old piglets were obtained from a commercial sow farm in central Iowa and brought to Iowa State University's Livestock Infectious Disease Isolation Facility (LIDIF). Upon arrival to LIDIF, the piglets were euthanized via blunt force trauma. Using a sample site diagram and key, 32 anatomical locations were marked on each piglet by placing a dot via permanent marker on the skin for identification. Four methods were used to collect skin samples from the piglets. Method one involved tenting the skin with a pair of forceps and using a scalpel blade to remove the skin above the underlying subcutaneous fat and muscle. Method two utilized a 3-millimeter punch biopsy to remove two samples from each site including the snout and between the digits. Method three used tissue scissors to cut out a middle section of the ear. Orientation (dorsal and ventral surface) was determined upon microscopic evaluation of the histology slide. Method four included using a scalpel blade to incise above the lateral aspect of the coronary band and slice a sagittal section from the fore and hind limb including the coronary band and hoof wall. All samples were placed in individually labeled cassettes and stored in jars containing 10% formalin for fixation. Samples were submitted to Iowa State University's Veterinary Diagnostic Lab and histologically evaluated to determine characteristics of average stratum corneum, epidermal, dermal, and subcutaneous fat thickness in micrometers and the average number of hair follicles. To determine the average thickness of each layer in all 340 samples, three measurements in micrometers were taken, one from the middle and from each end of the tissue section or where there were distinct visible differences and an average computed. Similarly, to determine average number of hair follicles a count was taken from three areas of each sample site and the average computed as a count. Statistical analysis was performed using a generalized linear mixed model using SAS software, version 9.4 (SAS Foundation, Cary, NC).

Results and discussion

Statistical analysis of the data discovered significant differences ($P < 0.0092$) between the highest and lowest average sample site measurement for each skin characteristic evaluated (Table 1). These results reflect the importance of site selection for topical pharmaceutical application. Variations in skin thickness and hair follicle density across the body of a processing aged piglet reflect key factors that may influence transdermal absorption and pharmacokinetic principles of a drug. This can assist in targeted site selection for drug application that can take into consideration practical factors such as ease of administration, animal and administrator safety, as well as maximizing time and efficiency. This research can be used as a foundation in pharmacokinetic studies to evaluate bioavailability of pharmaceuticals when administered topically and how systemic drug levels may differ based on site skin characteristics as measured in this study.

Table 1: Average sample site measurement for each skin characteristic evaluated

	Stratum corneum thickness (μm)	Epidermal thickness (μm)	Dermal thickness (μm)	Subcutaneous thickness (μm)	Average number of hair follicles*
Highest value	154.96 Right, front coronary band of the lateral digit	466.99 Right, front coronary band of the lateral digit	824.90 Right, perianal region	1893.90 Left, ventral pinnae, cervical attachment	14.5017 Right, lateral carpus
Lowest value	11.5730 Vulva	34.5640 Vulva	232.37 Vulva	306.92 Vulva	0.3 Vulva
Average	37.17	91.5	563.54	787.599	4.3767

* Average number of hair follicles reported as a count

Monitoring mature sows for *Mycoplasma hyopneumoniae* (Mhp) pre-and post-Mhp inoculated gilt entry

Megan E. Bloemer[1], BS; Thomas Painter[2], DVM; Adam Schelkopf[3], DVM
[1]University of Illinois College of Veterinary Medicine, Urbana, Illinois; [2]Zoetis, Parsippany, New Jersey;
[3]Pipestone Veterinary Services, Pipestone, Minnesota

Introduction

Mycoplasma hyopneumoniae (Mhp) is a well-known swine respiratory pathogen that affects up to 80% of pigs worldwide, and is a primary cause of enzootic pneumonia. A dry, non-productive cough, labored breathing, and pneumonia characterize enzootic pneumonia. Research has shown that Mhp infection can last as long as 240 days, evidenced by positive polymerase chain reaction (PCR) results.[1] The economic consequences of this include chronic reduction in feed efficiency and average daily gain as well as an increased co-infection with other respiratory pathogens. Recent studies have shown, when Mhp is identified in a herd along with Influenza A virus-swine (IAV-S) and/or porcine reproductive and respiratory syndrome virus (PRRSV), the additive effect has a drastic impact on mortality.[2] The objective of this study was to determine if the Mhp health status of parity 2 and greater (P2+) sows changes when gilts previously inoculated with Mhp in the gilt development unit (GDU) are introduced into the gestation barn. We hypothesized when inoculated gilts are introduced, P2+ sows will become re-infected with *Mhp* and shed for an additional 240 days. This will confound the timeline for the herd to become Mhp negative.

Materials and methods

The selected sow farm has an on-site wean-to-finish GDU connected to the gestation barn by an enclosed hallway. The farm was selected because clinical signs of Mhp were seen on farm and downstream, and Mhp was confirmed by diagnostic testing. The decision was made to purposefully expose gilts to Mhp in the GDU to begin herd closure for Mhp. Gilts 8 to 15 weeks old in the GDU were inoculated with Mhp on June 27th. They started entering the sow farm in late August, 9 weeks post inoculation (PI). Older gilts in the GDU (16 to 25 weeks of age) were not directly inoculated with Mhp, but did have exposure to Mhp inoculated gilts and continued to enter the sow farm uninterrupted. On July 6th, 9 days PI- but prior to entry of inoculated gilts, 12 gilts, 12 first parity (P1) animals, and 30 P2+ animals were deep tracheal swabbed to determine their Mhp status by PCR. The same 30 P2+ sows were retested on August 8th (42 days PI) and September 12th (77 days PI) along with 30 additional P2+ sows. For sampling, 29.5" EQUIVET Uterine Culture Swabs and 5ml Becton Dickenson Falcon tubes- Polystyrene Round Bottom Tube were used. Gilts were preferentially selected based upon if they were bred and in the gestation barn for greater than 60 days, P1 animals were randomly selected, and P2+ sows were selected based upon if they were housed next to a P1 sow or gilt. Samples were pooled by three within parity group and submitted for Mhp PCR at the University of Minnesota Veterinary Diagnostic Laboratory. Positive pools were individually tested and sequenced.

Results and discussion

All three sampling timepoints yielded no positive results. The first two sow farm samplings in July and August have established a herd baseline of Mhp status of the P2+ sows which is negative. The older sows in the herd will continue to be monitored periodically by deep tracheal swabs until Mhp naïve gilts enter the herd in February 2018. Pending further herd testing results conducted in November and January, we can determine if the current herd closure timeline many producers use of 240 days after 80% of Mhp inoculated gilts are positive needs to be extended.

References

1. Pieters M., et al. An assessment of the duration of *Mycoplasma hyopneumoniae* infection in experimentally infected population of pigs. *Veterinary Microbiology* Aug 2008.

2. Haden C., Painter T., et al. Assessing production parameters and economic impact of swine influenza, PRRS and *Mycoplasma hyopneumoniae* on finishing pigs in a large production system. *Proc AASV*, 2012, 75-76.

Testing of the efficacy of various methods of urine collection for detection of flunixin meglumine (Banamine) in gilts and barrows

C. Brooks[1], BS; T. Mays[2], MS; B. Dominguez[1], DVM, MS
[1]College of Veterinary Medicine, Texas A&M University, College Station, Texas,
[2]Texas A&M Veterinary Medical Diagnostic Laboratory, College Station, Texas

Introduction

According to Livestock Slaughter Annual Summary, USDA, NASS, April 2015, Texas was 14th in a ranking of states by hogs and pigs inventory. A large part of the swine industry in Texas is show pigs, with over 48,000 hogs raised each year for show. Show pigs are a significant source of education regarding the swine industry for students and the public. One issue facing the show pig industry is drug residues. Drug residues can negatively affect public perception of the swine industry. At present, winning show pigs are drug tested at each show using free catch urine collection. The tampon method is proposed to be quicker and easier for those procuring the sample.[1] The objective of this study was to test the efficacy of two methods of urine collection, free catch and tampon methods, in the detection of flunixin meglumine (Banamine). The drug flunixin meglumine (Banamine) was chosen because it is one of the common residue violations in show pigs. The withdrawal period for flunixin meglumine (Banamine) is 12 days.

Materials and methods

Twelve 4- month old pigs, six barrows and six gilts, were selected for urine collection and drug testing. The pigs were divided into two groups consisting of 3 barrows and 3 gilts in each. Urine was obtained at 6 (+/- 2 hours) and 12 (+/- 2 hours) hours post administration of a 6 mL dose of intramuscular flunixin meglumine (Banamine) via tampon method and free catch. Sample collection from the groups alternated methods at 6 and 12 hours. Males 4-6 and Females 4-6 samples were collected with tampons at 6 hours and Males 1-3 and Females 1-3 were collected with tampons at 12 hours (Figure 1). The tampon method consisted of using a supersized, unscented tampon with a plastic applicator. The cost of 36 Equate Super tampons was less than $6. For gilts, the tampon was inserted into the vestibule. For barrows, the tampon was inserted into the preputial sac. Tampons were transferred to a Ziploc® sandwich bag and then the urine was pressed from the tampon into a 50 mL sample collection tube. Free catch samples were obtained with a 50 mL sample collection tube. Paired (6 and 12 hour) urine samples from eleven of twelve pigs (one barrow did not urinate for 14 hours, so no sample was available for collection) were submitted to the Texas A&M Veterinary Medical Diagnostic Laboratory. The data was analyzed using Analysis of Variance (ANOVA) to determine the significance of pig, pig gender, free catch vs tampon, and concentration.

Results and discussion

Testing and analysis showed that both the free catch and tampon methods were effective in finding flunixin in the urine. Analyses conclude that the concentration of the drug obtained by tampons is significantly higher, except for gilts at 6 hours. The tampon samples obtained greater concentrations of the drug at both 6 hours and 12 hours and was found to be significant. However, the data is not normally distributed. The highest concentration of flunixin in a free catch sample was 3 ng/mL, as compared to 38 ng/mL in a tampon sample (both at 6 hours). Possible reasons for this may be that with tampons a more concentrated sample is obtained, there may be some component in the tampons that binds to the drug better, or that the tampon absorbs drug while it is in the vestibule or preputial sac prior to urination. More research is necessary to determine the cause of the variability in data and the real world implications of using tampons in drug testing of show pigs. This study look to the feasibility of using tampons to collect urine for drug residue testing, as residues often occur before the established withdrawal time, we elected to test within hours of administration, future studies will look at the ability to detect the drug after the established withdrawal time.

Reference

1. Nickel, M, et al. Development and Validation of Ante-Mortem Urine Collection Techniques for Gilts and Sows. 48th Annual Meeting of the American Association of Swine Veterinarians, Feb. 2017.

Figure 1: Comparison of the concentration of flunixin meglumine (Banamine) in the urine of swine using free catch versus tampon collection methods. Note: No urine was obtained from Male 2 at 6 hours and Male 4 and 12 hours.

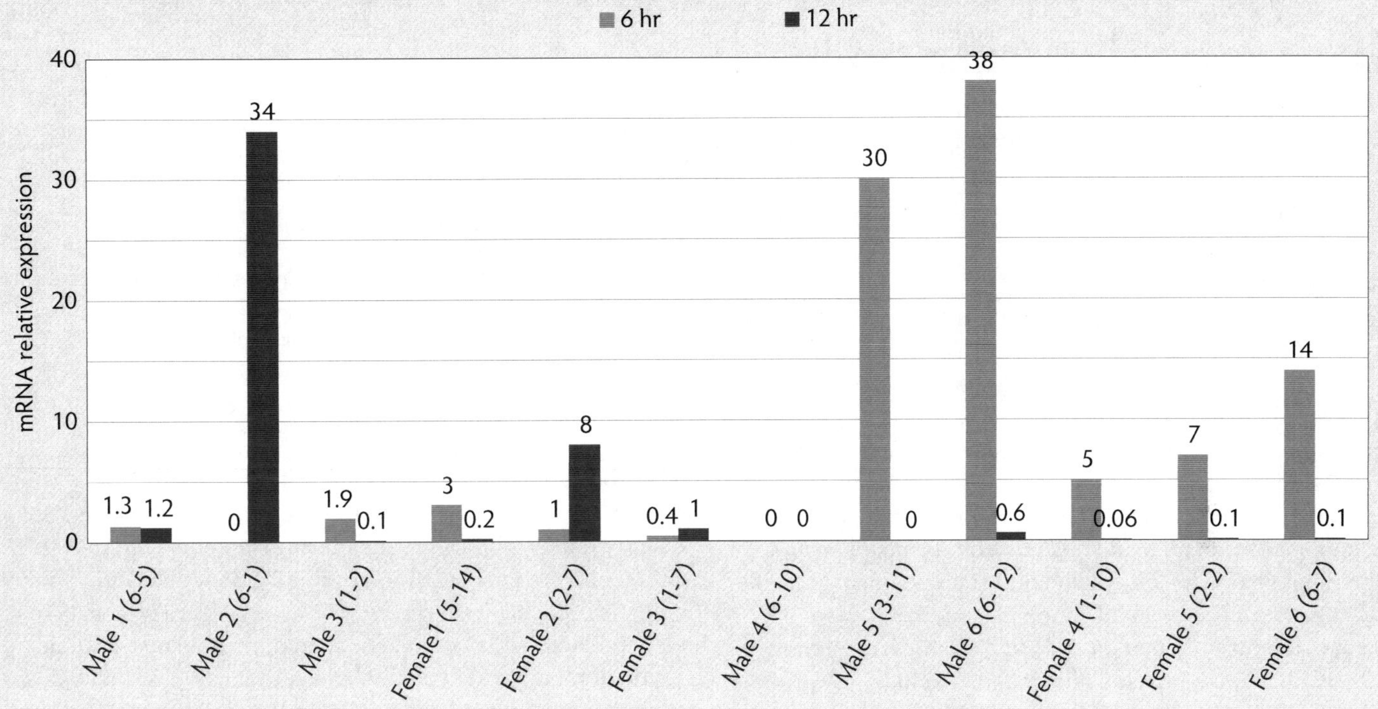

Effects of Improvest,® a gonadotropin releasing factor analog-diphtheria toxoid conjugate, on boar behavior during trailer transport

Alexandria D. Bufford[1]; Noel Garbes[2], DVM; Wesley Lyons[3], DVM
[1]Tuskegee University, Tuskegee, Alabama; [2]Zoetis, Parsippany, New Jersey;
[3]Pipestone Veterinary Services, Sycamore, Illinois

Introduction

Transport of cull boars is a growing welfare concern and new methods of management are needed. The transport of unfamiliar boars in open trailers can lead to displays of dominance and aggressive fighting, which can lead to severe injuries or death upon arrival (DOA). Common techniques to manage this are to segregate, detusk, or simply euthanize boars on farm. Lame or badly injured boars cannot be harvested causing a loss for both the farmer and the cull market. An effective boar culling program should be a part of any good breeding herd management program. Immunological castration of cull boars may reduce boar aggression and provide a safe way to transport cull boars to market. The objective of this study was to investigate the use of Improvest® as a temporary immunological castration in reducing displays of aggression in the boars during shipment.

Materials and methods

This field study began on May 20, 2017 when the first group of boars receiving their first injection of Improvest. Treatment groups were based on trailer load with the composition of treated and non-treated boars varying within loads. Culled boars were treated with two injections, one injection, or no injection of Improvest. An average of 12 boars were culled at a time from the stud, boars were typically culled every 4-5 weeks to make room for replacement boars in the herd, and a total of 2 trailer loads were evaluated during transport to market. Group A all boars were treated with two doses of Improvest, while Group B consisted of a mixed group of boars with no dose, one dose, and two doses. The boars' initial interactions were observed in the holding area where they were first commingled by the use of crates. Each boar was individually identified and assessed for lameness, lesions, and wounds prior to loading on the trailer. Boars were transported from the stud to the cull market in a trailer with two compartments. The total transport time was approximately 2 hours. Two GoPro cameras captured boars' interaction with one another during transport. Boars were then assessed again during unloading of the trailer at market for wounds, lesions, lameness, and DOA. Video data was then analyzed and compared between each treatment group to assess differences.

Results

Boars treated with two doses of Improvest, as per the label, had decreased displays of aggression during transport. Group A had a total of 7 bouts of aggression during a 2 hour duration of trailering with 0% change in lameness or wound scores post trailering. Group B had 165 bouts of aggression during the two-hour duration of transport. Post transport lameness for Group B had no change, while 58% of boars in Group B had a change in wound scoring. Results from this study showed aggression levels can be reduced in cull boars with proper implementation of Improvest as a means of humane transport procedures.

Discussion

The industry today is constantly looking for new ways to improve how we care for pigs, making it better for the pig and public perception of the industry. Giving two doses of Improvest to cull boars prior to transport may lower aggression and provide an option to improve welfare conditions during transport. The information provided in this study highlights the need for further research in regards to welfare and boar transport.

References

1. Hook, Trevor J., Joseph M. Stookey, and Harvey Wagner. *Animal Welfare Rethinking Cull Boar Transport*. N.p.: CVMA = ACMV, 1997. Web.

2. Marchant-Forde, J.N., J.P. Garner, D.C. Lay, Jr., A.K. Johnson, and R.M. Marchant-Forde. "Understanding and Reducing Aggression Using Pre-exposure, When Sows Are Mixed in a Grouped Gestation System." N.p., n.d. Web.

Assessing the impact of perinatal antibiotics on performance and antibiotic resistant bacteria in pre-weaned pigs

B. Burton[1]; M. Zeineldin[1]; A. Megahed[1]; A. Harding[2]; J. Lowe[1,2]
[1]University of Illinois College of Veterinary Medicine, Urbana, Illinois; [2]Lowe Consulting Ltd., Mahomet, Illinois

Introduction

It is important to understand how common management practices, such as antimicrobial administration, might affect the complex microbial ecosystem of the gut and ultimately impact animal health. In addition, the emergence of antibiotic-resistant bacteria is an urgent public health issue creating pressure to reduce the use of antimicrobials. Understanding the impact of antibiotic administration on host welfare and performance is crucial in the development of strategies to utilize antibiotics while reducing some of the possible adverse consequences. The benefits of utilizing perinatal antibiotics in pigs have been inadequately documented. The objective of this study was to understand the impact of perinatal antibiotics on pig growth, mortality, and the prevalence on both antimicrobial resistance genes (ARGs) and culturable antibiotic resistant bacteria (ARB) in pre-weaned pigs.

Materials and methods

A randomized complete block design was used for this study. Forty-eight litters were blocked to one of six treatments (N = 8) by farrowing day, dam parity group, and litters with a minimum of 11 pigs. Within litter, all pigs received the same treatment. Pigs were weighed, and treatments administered within 24 hours of age, after litters were reduced to an appropriate size when deemed necessary. The allocated treatments were as follows: control (saline 1cc), tulathromycin (2.5 mg/kg IM, TUL), ceftiofur crystalline free acid (5 mg/kg IM, CCFA), ceftiofur hydrochloride (5 mg/kg IM, CHC), oxytetracycline (22 mg/kg IM, OTC) and Procaine Penicillin G (33,000 units/kg IM, PPG). Two pigs per litter were individually identified, and weights and deep fecal swabs were collected at days 0 (pre-treatment), 5, 10, 15 and 20. Once collected, all samples were placed on dry ice and held at -80°C until analysis. The weights of all piglets in each litter were recorded on study day 0 and at weaning (day 20). Mortalities were recorded throughout the study. High-throughput next-generation sequencing was used to assess the presence of ARGs. Briefly, samples were processed for DNA extraction using MO-Bio extraction kit (Qiagen), following manufacturer instructions. The extracted DNA was then subjected to Fluidigm Access Array Amplification to amplify and enrich specific ARGs (Tet O, Tet W, Tet C, Sul 1, Sul 11, bla CTX (lactamase) and erm (B) (macrolide). The final pooled Fluidigm libraries were then sequenced using the Illumina MiSeq platform. Sequencing data were processed and analyzed with the QIIME algorithms, using an operational taxonomic unit approach. The occurrence, diversity, and abundance of each ARG were assessed using a repeated-measures ANOVA model to determine if the treatment or the time point were significantly related to the presence of ARGs. In addition, supernatant from swabs was plated on blood agar plates containing 1 of 5 antibiotics, which corresponded to treatment groups (n = 600). MALDI-Tof Biotyper System (Bruker) was used for viable bacterial identification.

Results and discussion

Preliminary analysis of the data shows that while antimicrobial treatment had no effect on individual weight gain, litter weight gain, or mortality, it was associated with significant changes in the abundance of ARGs and on the prevalence of ARB. The timing of the changes appear to be antibiotic dependent. Although antibiotic usage has been documented to increase performance in nursery and finishing pigs,[1,2] the results specifically from this study indicate that there is no performance benefit in preweaned pigs. At the same time, there were changes in the prevalence of both ARG and ARB suggesting that there is potential for adverse effects later in the growing period.

References

1. Tangen, J. et al. A comparison of the effect of ceftiofur crystalline free acid and procaine penicillin-G administered at either birth or processing on suckling piglet health and performance. AASV Annual Meeting. 2012:109-112.

2. Zhou, C et al. A global multilocation study to evaluate the effect of ceftiofur sodium on preweaning mortality and growth performance. Swine Health and Production. 2000; volume 8, number 3:113-117.

The impact of observer experience and training methods on behavioral scores in piglets

E. DeCaluwe-Tulk, HBSc; H. Golightly, HBSc; K. Reynolds, BSc, MSc, DVM; K. Richardson; T. L. O'Sullivan, DVM, PhD
Department of Population Medicine, Ontario Veterinary College, University of Guelph

Introduction

Behavioral observations are useful tools in helping to understand pain experienced by animals.[1] However, the use of observers may introduce a source of error due to the inherent subjectivity associated with individual observer interpretation of behavior.[1,2] The training and validation of observers should be considered to improve the validity and hence utility of behavioral observations when conducting on-farm clinical trials or when training new production staff in animal husbandry.[1,3]

Objectives

The objectives of this study were to determine the impact of training method and observer experience on behavior scores (BSs) and specifically: to compare BSs assigned by inexperienced observers trained by two different methods; to assess methods of observer training; and to compare BSs assigned by trained observers without prior experience observing piglets to those assigned by an experienced observer.

Methods

Six observers without prior experience observing piglets were assigned to the inexperienced (IE) group. One previously trained, experienced (E) pig observer was recruited. The IE (n = 6) were randomly assigned to one of two training methods 1) discussion training (DT), where participants (n = 3) engaged in group discussion of the ethogram definitions; and 2) visual training (VT) where participants (n = 3) viewed a PowerPoint™ presentation containing photos and videos demonstrating the behaviors and ethogram definitions. The same ethogram which included behavioral indicators of pain such as back-arching, tail-wagging, rump scratching, orbital tightening, and ear position was used for both of the IE groups as well as for the E observer.[4,5] Each observer (n = 7) participated in 4 video sessions which involved viewing 1-minute video clips of 48 unique piglets in a systematically randomized order to which they were blinded. Two-way random effects models were used, and intraclass correlation coefficients (ICC) were calculated to assess the agreement between the IE observers and between the IE group and E (Stata Statistics/Data Analysis® v 14.0).

Results and discussion

Regardless of training method, poor correlation (ICC < 0.5, range 0.16-0.42) was observed between the IE group and E for all behaviors except for tail-wagging which had a ICC = 0.58 and 0.65 for the visual and discussion groups, respectively.[6] Similarly, poor correlation was also observed between IE groups for all behaviors (ICC range: 0.24-0.44) except for tail-wagging (ICC = 0.56). This suggests observer experience is an important factor influencing the reliability of behavior scores. The results of this study suggest that other training methods should be considered and explored for the training of inexperienced observers such as new production staff.

References

1. Meagher (2009). Appl. Anim. Behav. Sci., 119(1), 1-14.
2. Landis & Koch (1977). Biometrics, 159-174.
3. Hannaford et al. (1997). J. N. Am. Benthol. S. 16(4), 853-860.
4. Di Giminiani et al. (2016). Front. Vet. Sci., 3.
5. Hay et al. (2003). Appl. Anim. Behav. Sci., 82(3), 201-218.
6. Koo & Li (2016). J. Chiropr. Med., 15(2), 155-163.

Table 1: Inter-observer agreement of piglet behaviour scores assigned by inexperienced observers (n = 6) compared to an experienced observer (n = 1)

		Intraclass correlation coefficient*		
		IE (VT, n = 3) compared to IE (DT, n = 3) [95% CI]	IE (VT, n = 3) compared to E (n = 1) [95% CI]	IE (DT, n = 3) compared to E (n = 1) [95% CI]
Behavior	Back arching	0.281 [0.189-0.397]	0.319 [0.211-0.451]	0.420 [0.298-0.552]
	Tail wagging	0.569 [0.458-0.675]	0.579 [0.461-0.688]	0.651 [0.539-0.749]
	Rump scratching	0.244 [0.160-0.353]	0.159 [0.082-0.284]	0.372 [0.258-0.502]
	Orbital tightening	0.438 [0.324-0.560]	0.539 [0.411-0.661]	0.374 [0.242-0.527]
	Ear position	0.260 [0.173-0.372]	0.199 [0.115-0.322]	0.542 [0.424-0.656]

* two-way random-effect logistic regression models
IE = inexperienced observer group
E = experienced observer
VT = visual training participant viewed a PowerPoint™ presentation containing photos and videos demonstrating the behaviors of interest and ethogram definitions.
DT= discussion training where participants engaged in group discussion regarding the behaviours of interest and ethogram definition.

Validation and application of accelerometers to sow behavior and management

Mary Jane Drake, PhD; Thomas D. Parsons, VMD, PhD, DACAW
School of Veterinary Medicine, University of Pennsylvania, Philadelphia, Pennsylvania

Introduction

Observation of animal behavior is implicit in the management of sows. Current methods for the study of sow behavior, including activity levels and postural changes, are usually limited to video capture and analysis. This approach is time consuming and can be subjective. The aim of this study was to assess the use of accelerometers on sows to monitor activity levels and postures quantitatively and efficiently. More specifically, this study sought to validate a commercially available accelerometer for cows in sows for automated step counting and posture determination. A secondary objective was to use this accelerometer to study sow behavior in farrowing pens in the peri-parturient period.

Materials and methods

Yorkshire-Landrace sows (DNA line 241) parity 2-5 were included in the study. IceTag accelerometers (IceRobotics, UK) were attached to the distal limb of sows along the metacarpus or metatarsus by securing the sensor to a Velcro strap around the circumference of the limb and applying a top layer of VetWrap to prevent rotation of the sensor. The accelerometers logged movement in three axes 16 times per second and a proprietary algorithm was used to automatically calculate motion index, steps taken, and time spent lying or standing. For validation studies, 6 sows were observed for a total of 328 mins either for short intervals (20-45 min) in gestation pens using a handheld camcorder or for longer intervals in farrowing pens (1-2 hours) using overhead video surveillance cameras. Recorded video was compared to posture and step data output from the accelerometers in order to determine their accuracy and sensitivity. Following validation, the IceTags were used to monitor the behavior of 11 sows across 3 different farrowing groups in the peri-parturient period from the time sows entered the farrowing room until 48-72 hours post-farrowing.

Results and discussion

Comparison of video and accelerometer data revealed that the latter were able to accurately capture sow locomotion in both open housing situations (gestation) as well as in more confined farrowing pens. For 95% of the validation observation period, the accelerometer was within 3 steps of the analyzed video. A regression line fitted to the scatter plot of step counts using the two methods had a high degree of correlation ($F(1,318) = 25840$, $P < 0.001$, $m = 0.965$, $R^2 = 0.983$) (Figure 1). As the accelerometers are able to detect posture (lying vs. standing), we affixed accelerometers to both the forelimb and hindlimb of sows in order to assess additional postures (sitting, kneeling) by comparing the reported posture of the two limbs. This also provided more resolution in analyzing the frequency of postural changes. Using this configuration, an increase in the activity levels of sows was observed beginning approximately 24 hours before farrowing. Activity levels peaked at 4-6 hours before farrowing as the frequency of postural changes increased to 15-25 changes/hr, compared with 0-5 changes/hr at baseline (> 48hr before farrowing). This was related to a ~4-fold increase in mean posture changes ($t(11) = 1.72$, $P < 0.001$) in the 12 hours prior to farrowing compared to baseline (> 48hr). These activity increases are consistent with behavior changes observed in free-ranging and enclosed sows during the nest building phase of labor.[1,2]

The IceTag technology is used for heat detection and health monitoring in cows. Preliminary results from the current investigation indicate that these accelerometers can also be reliably used for monitoring sow locomotion and posture. The availability of an efficient and objective method for studying sow behavior expands our ability to assess sow husbandry and welfare. For producers, application of this technology promises to serve as an alert system for farrowing (e.g. activity spikes), lameness, or injury.

Figure 1: Comparison of device and video counted steps with regression analysis

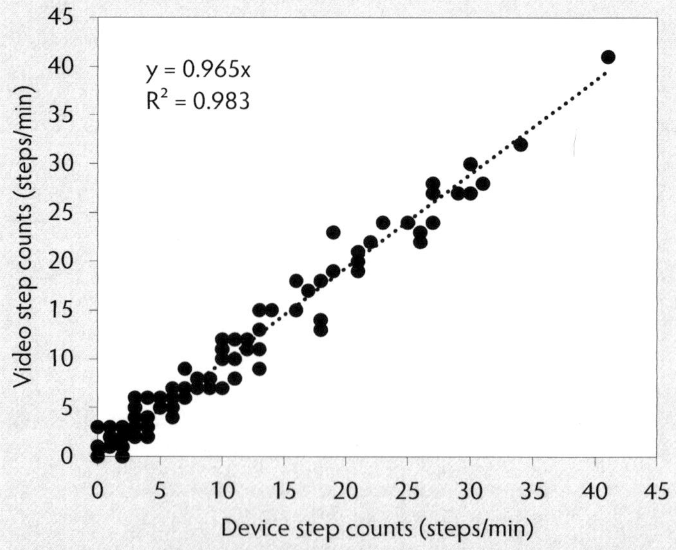

References

1. Jensen, Per, K. Floren, and B. Hobroh. "Peri-parturient changes in behaviour in free-ranging domestic pigs." *Applied Animal Behaviour Science* 17.1-2 (1987): 69-76.

2. Hansen, Krista E., and Stanley E. Curtis. "Prepartal activity of sows in stall or pen." *Journal of Animal Science* 51.2 (1980): 456-460.

Multiplexed phylogenetic comparison of PRRSv sequences using the MinION nanopore sequencer

T. Engle, BS, MS; S. M. Todd, BS, MS; K. K. Lahmers, BS, DVM, PhD, DACVP
[1]Virginia-Maryland College of Veterinary Medicine, Virginia Polytechnic Institute and State University, Blacksburg, Virginia

Introduction

Porcine reproductive and respiratory syndrome is considered one of the most devastating swine diseases affecting commercial swine production worldwide. Challenges with prevention and immune protection involve the modulated immune response and high genetic diversity. Fortunately, today, phylogenetic analyses have been heavily studied due to the significant global economic impact of PRRSv. The objective of this study was to investigative new sequencing technologies to further advance diagnostic medicine. In this study, we compare phylogenetic analyses of PRRSv sequences from accredited veterinary diagnostic laboratories with binned sequences from the MinION nanopore sequencer.

Materials and methods

Serum samples were obtained from PRRSv positive sow farms and RNA was extracted using Qiagen QIAamp Viral RNA Mini kit (Qiagen, Germantown, MD) and converted to cDNA using random hexamers and SuperScript IV First-Strand Synthesis System (Invitrogen, Carlsbad, CA). PRRSv viral genomes were quantified in triplicates using SYBR Green-based quantitative PCR for the ORF7 reading frame.[1] For phylogenic comparison, primers were designed to flank the entire ORF5 region and so that PRRSV-specific primer sequence makes up > 50% of the primer sequence with barcoding sequence flanking. Both primers are based on VR2385 sequence in the areas of ORF4 (forward) and ORF6 (reverse) and contain overhangs (italics) for barcoding.

ORF4 ONTbar F: 5' *TTTCTGTTGGTGCTGATATTG-C*AGACCATGAGGTGGGCAACTGTTTTAGCCTGTC 3'

ORF6 ONTbar R: 5' *ACTTGCCTGTCGCTCTATCTTC-*CCAAGAGCACCTTTTGTGGAGCCGTGCTATCAT 3'

Then each amplicon was barcoded using a commercially available kit (Oxford Nanopore) with known sequences to serve as an identifiable tag to distinguish each sample. Barcoded samples were mixed together in groups of 12 for library preparation. Stages of library preparation include ligation of adapter sequences and library quantification. Sequences were run on a version R9.5 flowcell and a MinION Mark II with MinKNOW software version 1.7.14 for 4 hours. Sequences were determined using Albacore version 2.0.2 and all of the samples are binned based on barcode sequences using Porechop version 0.2.3.[2] Finally, consensus sequences were determined for each binned sequence and bootstrap analysis performed using Genenious version10.2.3. Synthetic mixed isolates were prepared by mixing reads from two separate isolates at varying ratios and determining if separate consensus sequences were identified.

Results and discussion

Consensus sequences from nanopore sequencing were compared to traditional Sanger sequencing. Consensus sequence shared 99.1-100% identity with sequences of published isolates and when compared to traditional sequencing. Synthetic mixed isolates with 95% sequence identity were detected at a ratio of 2:1. Currently, we have barcoded 12 samples at a time, but the ability to barcode 96 is currently available. This would reduce the cost of sequencing to under $4 per sample while allowing for rapid turnaround time (< 2 hours for library preparation) and the ability to detect mixed infections in a sample such as oral fluids. As the swine industry continues to grow, viral sequencing technology will remain an integral role in phylogenetic analyses and diagnostic medicine. The results from this study demonstrate that MinION nanopore sequencing is as an alternative sequencing strategy for rapid phylogenetic analysis. This approach is comparable in cost per sequence to contract services, has shorter turnaround time, and, with minimal startup cost, the financial barrier is eliminated with this approach.

References

1. Ni YY et al. Attenuation of Porcine Reproductive and Respiratory Syndrome Virus by Molecular Breeding of Virus Envelop Genes for Genetically Divergent Strains. Journal of Virology. 2013. 87:1;304.
2. R. Wick. Porechop repository. 2017. Available at **https://github.com/rrwick/Porechop**. Accessed 29 October 2017.

Evaluation of the impact of iron dosage on post-weaning weight gain, and mortality

L. Fredericks[1], BS; C. Olsen[2], DVM; A. Maschhoff[3], DVM; C. Shull[3], PhD
[1]Iowa State University College of Veterinary Medicine, Ames, Iowa; [2]Pharmacosmos Inc., Watchung, New Jersey;
[3]The Maschhoffs LLC, Carlyle, Illinois

Introduction

Prior studies have shown that modern genetics leading to rapid growth rates potentiate an increased requirement for iron, in order to maintain adequate hemoglobin production.[1] If the iron requirement of modern pigs exceeds that provided by a standard 200 mg dose, health and performance may be negatively affected.[2] While studies have shown an improvement in weight gain in the first 3-weeks post weaning, no studies have followed such improvements through the wean to market period.[1] Therefore, the objective of this study was to determine the impact of iron dosage on wean-to-market weight gain and wean to 8-weeks post-wean (WPW) mortality.

Materials and methods

In one commercial sow farm, 16,551 piglets were sorted by birthweight within each litter and assigned to one of three treatment groups within 24 hours of life. Group A (n = 5,494) received 200 mg of Uniferon® iron dextran at the time of processing (day 5-7 of life). Group B (n = 5,538) received 100 mg of iron within 24 hours of life and 100 mg of iron at processing. Group C (n = 5,519) received 200 mg of iron within 24 hours of life and 200 mg of iron at processing. Mortality data was collected from all pigs through 8-WPW. A subset of piglets within each treatment group (n = 1,400) were weighed individually at birth, weaning, 8-WPW, and 18-WPW. Hemoglobin (Hb) levels were measured at weaning and 3-WPW in a subset of piglets (n = 100) within each group of piglets to be weighed.

Results and discussion

Mortality, weight, and hemoglobin data up to 8-WPW were analyzed using a one-way analysis of variance (ANOVA) model. No significant differences in post-weaning mortality were identified between treatment groups. At weaning, Group C had a significantly higher average hemoglobin (Hb) concentration (111 g/L) compared to groups A (94.2 g/L) and B (98.2 g/L), and was the only group with a mean in the optimal range (>110g/L). At 8-WPW, Group A had a significantly lower average Hb concentration (95 g/L), while Groups B and C averaged in the sub-clinical range (101 g/L and 102 g/L, respectively). In addition to group means, it is important to evaluate the distribution of hemoglobin values within each treatment group. As shown in Table 1, the majority of piglets tested in Group C had optimal Hb at weaning while only 8 and 15 piglets in Groups A and B respectively had the same or greater Hb concentrations at that time. Irrespective of iron dosage, piglets with optimal Hb at weaning had significantly greater 8-WPW body weight compared to piglets with clinical (< 90 g/L) or sub-clinical (90-110 g/L) iron deficiency anemia at weaning (Table 1). The results of this study indicate that maintaining optimal hemoglobin levels through weaning support improved post-weaning weight gain compared to piglets with clinical or sub-clinical iron deficiency anemia at the time of weaning.

References

1. Jolliff JS, Mahan DC. Effect of injected and dietary iron in young pigs on blood hematology and postnatal pig growth performance. *J Anim Sci*. 2011;89(12):4068-4080.

2. Perri AM, Friendship RM, Harding JCS, O'Sullivan TL. An investigation of iron deficiency and anemia in piglets and the effect of iron status at weaning on post-weaning performance. *J Swine Heal Prod*. 2016;24:10-20.

Table 1: Impact of blood hemoglobin concentration at weaning on 8-week post weaning bodyweight in piglets

Hemoglobin status	Piglets per treatment group				8-WPW weight (lbs)
	A	B	C	Total	
Clinical anemia (< 90 g/L)	26	16	5	47	64.56[b]
Sub-clinical anemia (90-110 g/L)	66	69	31	166	67.76[b]
Optimal (> 110)	8	15	64	87	73.72[a]

[1] Values within columns not connected by the same letter superscript are statistically different ($P < 0.05$)

Assessment of intra- and interrater agreement of pain behavior scores in castrated piglets

H. Golightly, HBSc; E. DeCaluwe-Tulk, HBSc; K. Richardson; K. Reynolds, DVM; T. L. O'Sullivan, DVM, PhD
Department of Population Medicine, Ontario Veterinary College, University of Guelph

Introduction

On-farm clinical trials using observer-based behavioral measurements may be scrutinized if observer agreement assessment is not included in the trial design. How reliably a category of behavior is measured is influenced by observer practice and experience, frequency of occurrence of the behavior, observer fatigue and the adequacy of the behavior definition.[1] The objective of this study was to evaluate the agreement within and between two observers assigning scores related to piglet pain behavior indicators. This study was a component of a larger on-farm clinical trial.

Materials and methods

Piglets enrolled in a larger ongoing clinical trail using live-scan behavior scoring to identify levels of pain in piglets post-castration, were randomly selected and individually video-taped. Forty-eight unique, one-minute long video clips were systematically randomized into three sets. Two observers, with extensive knowledge and experience of swine behavior independently viewed each video three times, with randomization of sets, and a wash-out period, taking place in between each viewing. Observers scored observed behavior according to ethograms that were validated for the larger trial. One and two-way random effect models and intraclass correlation coefficients (ICCs) were evaluated to interpret intra- and interrater reliability of the different behavior scores by behavior category (Stata Statistics/Data Analysis® v 14.0).

Results and discussion

The results demonstrate that experienced observers who are practiced with an ethogram can achieve moderate (ICC = 0.5-0.75) to good (ICC = 0.75-0.9) intrarater agreement however, they may not achieve comparable interrater agreement (Table 1).[2] These results also emphasize that some behaviors may be more reliably measured by multiple observers than others. For example, back arch and ear position had lower interrater ICCs and hence appear to be more difficult to be measured reliably between observers (Table 1). It has become increasingly common for swine practitioners to conduct on-farm trials, some of which may utilize behavioral indicators of pain such as those presented here. Practitioners advising on such trials should consider implementing a suitable observer training method and assessment of observer agreement throughout the trial to ensure that the data collected can be interpreted in a meaningful way.

References

1. Martin P, Bateson P. (2007). *Measuring behaviour: an introductory guide.* 3rd ed. New York: Cambridge University Press; 2007: 80-81.

2. Koo TH, Li MY, A guideline of selecting and reporting intraclass correlation coefficients for reliability research. *J Chiropr Med.* 2016; 15:155-163.

Table 1: Intraclass correlation coefficients[a] of intra- and interrater agreement of two observers scoring piglet pain behavior following castration.

Behavior	Back arch [ICC 95% CI]	Tail wag [ICC 95% CI]	Rump scratch [ICC 95% CI]	Orbital tightening [ICC 95% CI]	Ear position [ICC 95% CI]
Intra-rater[b]	A: 0.59 [0.44, 0.73] B: 0.67 [0.54, 0.79]	A: 0.71 [0.59, 0.81] B: 0.86 [0.78, 0.91]	A: 0.71 [0.58, 0.81] B: 0.75 [0.64, 0.84]	A: 0.77 [0.66, 0.85] B: 0.75 [0.64, 0.84]	A: 0.70 [0.57, 0.81] B: 0.62 [0.48, 0.75]
Inter- rater[c]	0.36 [0.25, 0.50]	0.73 [0.64, 0.82]	0.66 [0.54, 0.76]	0.68 [0.57, 0.77]	0.38 [0.26,0.52]

[a] one and two-way random effect models.
[b] comparison between 3 viewings of 48 unique piglets (video-taped) by the same observer.
[c] comparison between observers rating the same 48 piglets 3 times each.
ICC: Intraclass correlation coefficient.
CI: confidence interval.

In vitro culture of porcine B cells

Kevin L. Gustafson; Michael P. Murtaugh

Introduction

B cells are a critical component of the adaptive immune system in pigs. Following activation by an antigen, B cells will go down one of two pathways. Many B cells become activated and produce antibodies that are specific for a single epitope on a specific pathogen. A small number of B cells will differentiate into long-lived memory B cells that can be quickly activated to produce antibodies when the antigen is encountered a second time. This memory and specificity is the basis behind vaccination. A virulent antigen of a specific pathogen is introduced to the pig, then the pig will produce memory B cells that are activated if and when naturally exposed to the virulent form of the pathogen, resulting in protection against disease. Our lab has performed studies to determine a method to isolate these rare antigen specific memory B cells and to determine a culture method to maximize bulk expansion of porcine B cells (1). However, no studies have been performed trying to grow antigen specific memory B cells in vitro following isolation. The objective of this study is to produce a lymphocyte culture system that maximally expands antigen specific memory B cells by adding porcine reproductive and respiratory syndrome virus (PRRSv) non-structural protein 7 (nsp7), along with various cytokines and receptor agonists to an in vitro culture system. This culture system will allow for the characterization of key protective memory B cells.

Materials and methods

B cells were isolated from porcine spleen. IL-2, a cytokine produced by T cells in vivo and known to cause B cell proliferation, was added to the B cells in culture along with various other proliferation and differentiation factors. A Toll Like Receptor 9 (TLR9) agonist was also tested to determine its ability to stimulate porcine B cell proliferation, viability, and antibody production. B cells were cultured in plates with round bottom or flat bottom wells, and proliferation and functional antibody secretion were measured after 4 days.

Results

When the cytokine IL-2 was added to the culture, B cell viability increased 2-fold compared with the current culture systems (Fig. 1). IL-2 also increased both IgM and IgG antibody isotype production from B cells, indicating this cytokine has a potent activating effect. B cells incubated with the addition of the TLR9 agonist produced higher amounts of antibodies as compared to cells that were incubated in the current culture systems.

Figure 1: Cultures with IL-2 showed a higher level of viability over 4 days compared with cultures that lacked IL-2

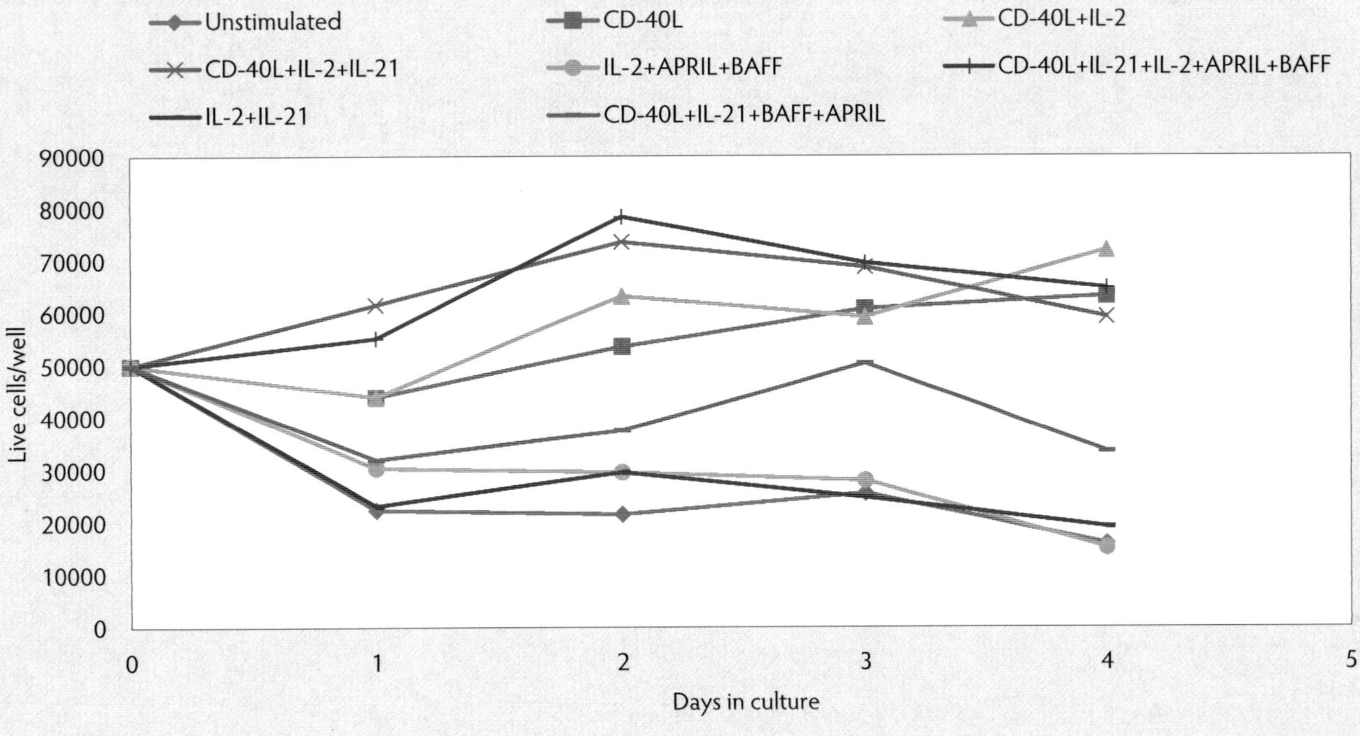

Discussion

The addition of IL-2 to B cell culture systems increases B cell viability and antibody production. Addition of a TLR9 agonist increased antibody production of B cells, indicating TL9 agonists can be used to stimulate B cells to produce antibodies that can be analyzed. IL-2 will be added to the culture system with antigen to activate and analyze nsp7 specific memory B cells in our in vitro system. The findings from this study will be applied to culturing and cloning of memory B cells for receptor and RNA analysis. This research allows for the study of rare memory B cells that are important for protecting the animal against PRRSv infection.

References

1. Rahe, M., Gustafson, K., Murtaugh, M. B cell tetramer development for veterinary vaccinology. Viral Immunology. 2017, Jul 31.

Understanding the impact of farrowing induction on pigs born alive

Olivia Hardebeck, BS; Luis Ochoa, PhD; Laura Greiner, PhD; Joe Connor, DVM
University of Illinois CVM; Carthage Veterinary Service, Carthage, Illinois

Introduction

Induction of parturition can be used to synchronize when most sows give birth in a farrowing room. Benefits of induction may include increased time of staff supervision of farrowing which (i.e. reducing the numbers of farrows occurring overnight) and improving the efficiency of production practices such as cross-fostering.[1] In addition, increased supervision may result in fewer stillborns per litter than animals that farrowed naturally with minimal supervision.[2] However, for every day of gestation, a piglet grows on average rate increased 26 grams per day.[1] Therefore, it is expected for piglets from induced sows to be lighter at birth compared to piglets born without induction. The objective of this study was to determine if the timing of induction affected preweaning mortality, piglet morbidity, the number of stillborns in a litter, and birth weight of pigs.

Materials and methods

This study was conducted using 68 pregnant gilts (first parity) and 164 pregnant sows (second or more parity). Within parity (i.e., gilts or sows), gestating animals were randomly assigned to one of the following treatments: 1) induction on day 114 of gestation; 2) induction on day 115 of gestation; and 3) induction on day 116 of gestation. Sows were induced according to treatment using a sterile solution with naturally occurring prostaglandin F_2 alpha. The control group consisted of animals that farrowed naturally before their induction or had milk on the date of induction date. At farrowing, treatment, crate, room, gestation length, number of live, stillborn, and mummies were recorded. The live piglets, stillborns, and mummies were weighed and ear notched according to treatment. Stillborns were categorized as attended (born during working hours of 6:00 AM to 3:00 PM) or unattended (born outside working hours). Cross-fostering did not take place until the piglets were weighed. At weaning, piglets were weighed and their treatment notch was recorded.

Results and discussion

Summary of results is presented in Table 1. As expected, gestation length was significantly different between treatment groups and controls ($P < 0.001$). These gestation lengths are consistent with the timing of their induction. The control group and treatment 1 had shorter gestations than treatments 2 and 3 with gestation lengths of 115.76 and 115.26 respectively. Treatment 2 which were induced on day 115 of gestation had an average gestation length of 116.19 days. Finally, treatment 3 had the longest gestation length of 117.05 days. Timing of induction had little or no effect ($P > 0.05$) in birth weight of pigs born alive, but significantly ($P = 0.035$) reduced the number of pigs born alive. Sows induced on days 115 and 116 of gestation had a lower number of born alive compared to animals induced on day 114 or not induced at all. The number of stillborn pigs seem to be not affected by induction ($P = 0.25$). Interestingly, while the number of pigs born alive between the control and sows induced on day 114 of gestation was similar, the variation in farrowing length of the control group was near double that of treatment 2 (0.731 vs. 0.431 days, respectively). This data suggest that inducing sows at 114 days of gestation could be used to reduce the variability on the start of farrowings without reducing the number of pigs born alive.

References

1. Nguyen, K., G. Cassar, and RM Friendship. "Stillbirth and preweaning mortality in litters of sows induced to farrow with supervision compared to litters of naturally farrowing sows with minimal supervision." *Swine Health Production* 19, no. 4 (2011): 214-17. https://www.aasv.org/shap/issues/v19n4/v19n4p214.pdf.

2. Gunvaldsen, RE, C. Waldner, and JC Harding. "Effects of farrowing induction on suckling piglet performance." *Swine Health Production* 15, no. 2 (2007): 84-91. https://www.aasv.org/shap/issues/v15n2/v15n2p84.pdf.

Table 1: Least square means for the effect of day of farrowing induction on the length of gestation, total pigs born, pigs born alive and stillborn rate

Item	Treatment				SEM	*P*-value
	Control	1	2	3		
Day of induction[1]	N/A	114	115	116	-	-
Number of sows	44	38	41	30	-	-
Average sow parity	2.39	2.84	2.32	2.60	1.265	0.162
Gestation length, days	115.76[a]	115.26[a]	116.19[b]	117.05[c]	0.100	< 0.001
Std. Dev. of gestation length, days	0.731	0.431	0.446	0.186	-	-
Total pigs born	15.42	15.58	15.25	13.84	0.051	0.191
Pigs born alive	14.45[a]	14.28[a]	13.28[ab]	12.35[b]	0.482	0.035
Born alive birth weight, kg	1.38	1.37	1.30	1.47	0.245	0.177
Stillborn per litter[2]	0.7	0.68	1.17	0.87	-	0.25
Average stillborn weight, kg	1.35	1.28	1.19	1.18	0.117	0.51

[1] Days of gestation.
[2] Data analyzed using χ^2
[abc] Means with different superscript within a row differ ($P \leq 0.05$).

Physical and behavioral indicators of animal welfare on farms using electronic sow feeders

J. Hurst; T. D. Parsons, VMD, PhD., DACAW; M. K. Pierdon, VMD, DACAW
Swine Teaching and Research Center, University of Pennsylvania,
School of Veterinary Medicine, Philadelphia, Pennsylvania

Introduction

With increasing consumer demand for crate-free pork, our industry is transitioning towards pen gestation systems yet many remain concerned for the welfare of sows outside of individual housing. Within electronic sow feeding (ESF) systems, various housing and mixing strategies exist, in part, to limit social aggression. Group composition may be kept constant throughout gestation or animals may be added and removed regularly (static vs. dynamic). Farmers may also choose to retain sows in breeding stalls until the embryo is implanted rather than move sows into groups immediately following estrus (post-implantation vs. preimplantation). We collected physical and behavioral observations of pens and individual sows to compare the effects of static vs. dynamic housing and preimplantation vs. post-implantation mixing on sow welfare in ESF systems.

Materials and methods

Observations were made of 695 individual sows, in 185 open pens, on eleven ESF farms. Sows' physical well-being was assessed by quantification of lesions per PQA Plus Site Assessment Guide 3.0. A subset (n = 399) were scored on their response to a standardized human approach paradigm with sows categorized as retreating from contact, initiating contact, or neutral. Pens were scanned for activity level and expression of stereotypic behaviors (e.g. sham chewing and bar biting) prior to introduction of a novel stimulus. Pen responsiveness to the novel object was assessed via a modification of Smulders et. al. protocol wherein we recorded the time to first snout–ball contact and the occurrence of contacts during the two-minute observation. For statistical analysis via STATA v. 15.1 (STATA Corp LLC, College Station, TX, USA), we used Spearman rank sum correlations, non-parametric Mann Whitney U-tests and mixed effect models to compare metrics across farms, pens and individuals.

Results and discussion

Breeding and farrowing performance was similar for all farms: Farrowing Rate 83.0 ± 4.0, Totalborn 14.8 ± 0.9, Pigs Weaned/Sow/Year 24.0 ± 3.5. Overall, we observed scratch counts to decrease with increasing gestation day and parity ($P < 0.001$), with fewer scratches on sows in static pens than sows in dynamic pens (39.5 ± 5.1 vs. 58.2 ± 6.7, $P < 0.05$). Behaviorally, sows housed in static pens were faster to touch the novel object (14.6 ± 28.4 vs. 51.8 ± 98.0, $P < 0.001$) and did so more frequently (0.09 ± 0.046 vs. 0.02 ± 0.013, $P < 0.001$) and responded more positively to human approach (0.38 ± 0.78 vs. 0.21 ± 0.61, $P < 0.01$) than sows in dynamic pens. Static pens were more active, expressed as a proportion of sows standing versus sitting or lying down, than sows in dynamic pens (0.33 ± 0.27 vs. 0.19 ± 0.20, $P < 0.01$). A comparison of mixing strategies showed preimplantation sows had fewer scratches (41.9 ± 3.7 vs. 59.2 ± 6.6, $P < 0.05$) and responded more negatively to human approach than post-implantation sows (0.29 ± 0.72 vs. 0.51 ± 0.73, $P < 0.05$). Those sows with vulva bites were further along in gestation than those without (57.9 days ± 30.9 vs. 64.1 days ± 29.2, $P < 0.05$). Vulva bites also increased on farms where sows were given less time in the ESF station ($R = -0.78$, $P < 0.01$). The relationship between greater scratches and more tentative behavior may indicate social inhibition of sows in dynamic pens, where social hierarchy is less stable. For producers currently employing or transitioning to ESF, our results suggest an effect of static housing on reducing the negative consequences of social aggression in pen gestation. Where vulva biting is a problem, producers should see that their ESF parameters allow sows adequate time in the station. Finally, farms in our study demonstrate how ESF systems can be successfully implemented to meet consumer demand for crate-free pork yet remain competitive with industry benchmarks.

Efficacy of interferon-inducing PRRSV vaccine candidate A2P90 against challenge with atypical pathogenic strain MN184

R. Kass[1], BS; Y. Zhang[2], PhD; Z. Ma[2], PhD; A. Karuppannan[1], DVM, PhD;
P. Halbur[1], DVM, MS, PhD; T. Opriessnig[1,3]; Dr med vet, PhD
[1]Department of Veterinary Diagnostic and Production Animal Medicine, College of Veterinary Medicine, Iowa State University, Ames, Iowa; [2]VA-MD College of Veterinary Medicine and Maryland Pathogen Research Institute, University of Maryland, College Park, Maryland; [3]The Roslin Institute, University of Edinburgh, Midlothian, Scotland, United Kingdom

Introduction

Effective control of porcine reproductive and respiratory syndrome virus (PRRSV) through vaccination remains difficult due to the high genetic variability and a high mutation rate of the virus. Recently A2MC2, a PRRSV strain which induces interferon (IFN) type I activity, pivotal for triggering innate and adaptive immunity, was isolated.[1] A2MC2 was passaged 90 times and the resulting attenuated vaccine candidate strain A2P90[2] protected pigs against homologous challenge.[3] The objective of this study was to further evaluate the efficacy of this novel PRRSV vaccine candidate using intramuscular or intranasal vaccination routes against challenge with atypical, heterologous PRRSV MN184.

Materials and methods

The experimental protocol was approved by an Institutional Animal Care and Use committee (IACUC No 3-16-8220-S). Forty-two 3-week old PRRSV free pigs were randomly assigned to five groups including NEG-CONTROL (no vaccination, non-challenged), POS-CONTROL (no vaccination, MN184 challenge), A2P90-IN (A2P90 intranasal vaccination, MN184 challenge), A2P90-IM (A2P90 intramuscular vaccination, MN184 challenge), and COM-MLV (commercial PRRSV vaccine, MN184 challenge). Vaccination was done at 3 weeks of age, challenge was at 8 weeks of age, and all pigs were euthanized 10 days post challenge (dpc). Serum samples and nasal swabs were collected at regular intervals and analyzed by a real-time RT-PCR (viral load), ELISA (IgG antibodies) and a fluorescent focus neutralization assay (neutralizing antibodies).[3] Blood monocytes at challenge were tested by ELISpot (cell mediated immunity). The model used to analyze continuous data collected over time was a repeated measures analysis of variance. A *P*-value of less than 0.05 was considered significant.

Results and discussion

At the time of challenge 8/8 COM-MLV, 4/9A2P90-IN, and 6/9 A2P90-IM pigs had developed PRRSV-specific antibodies. Pigs in both A2P90 groups had significantly higher MN184 specific neutralizing antibody levels interferon gamma secreting mononuclear cells compared to COM-MLV. Of the four pigs which seroconverted in the A2P90-IN group, three had no detectable PRRSV nasal shedding on dpc 3 and one pig showed complete absence of systemic viremia or nasal shedding through the entire period after challenge. The obtained results confirm that A2P90 is a promising vaccine candidate. If further developed, the A2P90 vaccine candidate has immense practical value, especially as a mucosal vaccine, to protect pigs against heterologous and highly virulent PRRSV strains in the field.

References

1. Wang et al. Enhancing neutralizing antibody production by an interferon-inducing porcine reproductive and respiratory syndrome virus strain. *Vaccine*.2013;31:5537-5543.

2. Ma et al. Sustaining interferon induction by a high-passage atypical porcine reproductive and respiratory syndrome virus strain. *Sci Rep*.2016;6:36312.

3. Fontanella E. et al. An interferon inducing porcine reproductive and respiratory syndrome virus vaccine candidate elicits protection against challenge with the heterologous virulent type 2 strain VR-2385 in pigs. *Vaccine*.2017;35:125-131.

Determining the necessity of peri-processing antimicrobial use in piglets

K. Klemesrud[1]; A. Forseth[1], DVM; E. Akin[1], BS; S. Anthony[1], BS; M. Rohde[1], BS; M. Breuer[1], BS; K. Hayman[1], DVM; K. Skoland,[1] BS; J. Brown,[1] DVM; H. Kittrell,[1] DVM; P. Canning,[1] DVM, PhD; P. Thomas,[2] DVM, MS; Christopher Rademacher[3], DVM; L. Karriker,[1] DVM, MS, DACVPM

[1]Swine Medicine Education Center, Iowa State University, Ames, Iowa; [2]Audubon-Manning Veterinary Clinic, Audubon, Iowa; [3]Veterinary Diagnostic & Production Animal Medicine, Iowa State University, Ames, Iowa

Introduction

As the swine industry continues to strive toward more judicious use of antimicrobials, further research is needed to demonstrate their efficacy and value. Administration of antimicrobials at processing (tail-docking and castration) is a common industry practice. However, there is no current, scientifically validated data indicating that antimicrobial administration at processing improves animal health and well-being. The objective of this study was to evaluate the benefit of peri-processing antimicrobial use compared to no antimicrobial administration at processing.

Materials and methods

Nine hundred sixty piglets from four commercial sow farms were selected for the study. Enrollment criteria included (a) piglets from unprocessed litters consisting of either four boars and four gilts or four piglets of one gender, (b) piglets born to non-gilt litters, and (c) piglets not exposed to antimicrobials post-farrowing. Litters meeting the selection criteria were randomized and selected using Microsoft Excel. Four gilts and/or four barrows within a litter were individually tagged with one of four colored ear tags. Each color represented a treatment group: beta-lactam (15,000 IU/lb), macrolide (2.5 mg/kg), tetracycline (19.8 mg/kg), or saline control (2cc/piglet). At the time of processing, antimicrobials were administered via intramuscular injection in the post-auricular region using a 20 gauge x 0.5" needle. A 4-pound dose was used for all piglets.

Outcomes measured from processing to weaning included percent mortality, wound healing, complete blood cell counts (CBC), average daily gain (ADG), navel size, and percent culls at weaning. Scoring castration wounds and tail-docking wounds occurred two days post-processing, seven days post-processing and at weaning. Wounds were evaluated on a 3-point scale: 1-exudate or blood, 2-swollen/reddened, 3-no swelling, redness, exudate, or blood. Complete blood cell counts were analyzed 2 days post-processing. The piglets were weighed to the nearest tenth of a pound at processing and weaning and ADG was calculated for each piglet. Piglet navels were measured to the nearest millimeter at weaning. Health observations were recorded daily to monitor for clinical signs of disease, including lameness, abscesses, hernias, scour, and cough.

Results and discussion

Preliminary data analysis did not reveal statistically significant ($P < 0.05$) differences between navel size of treated and non-treated animals, between castration wound scores of treated and non-treated animals, nor between castration wound scores of different antimicrobial classes. Data analysis did not reveal statistically significant ($P < 0.05$) growth (ADG) or mortality differences between treated and non-treated animals and there were no differences observed between the three antimicrobial classes evaluated during this study. The results from this study suggest more research is needed to better understand the effect of peri-processing antimicrobial use.

Dynamics of porcine parainfluenza virus shedding in sows and their piglets

Allison Knox, BS[1]; Amber Stricker, DVM, MS[2]
[1]University of Illinois, Urbana, Illinois; [2]Suidae Health & Production, Algona, Iowa

Introduction

Porcine parainfluenza virus 1 (PPIV1) has become more prevalent in commercial swine production since its discovery in 2008.[1] Similar to other Respirovirus species, PPIV1 has been identified in the upper respiratory tract and has been implicated as a contributor to the porcine respiratory disease complex, which can be detrimental to productivity and animal welfare.[2] Previous studies indicate that clinical signs are typically observed in piglets 10 to 21 days of age; however, there is limited data regarding clinical signs and shedding in pigs outside that age range.[3] The objective of this study was to characterize the dynamics of PPIV1 shedding in sows and their piglets from a clinical sow herd.

Materials and methods

A 6,200 head commercial sow farm endemic with PPIV1 for 6 months was utilized. The observational unit of this study was the sow and litter. Sows and piglets were enrolled in the study based on parity of the sow. 10 parity 0 gilts, 10 parity 1 or 2 sows, and 10 sows parity 3 or higher were enrolled. Oropharyngeal (OP) swabs were collected from these sows at the time of farrowing and tested for the presence of PPIV1 by PCR at the Iowa State University Veterinary Diagnostic Laboratory. In addition, 33 pigs were also enrolled at one week of age (WOA) based on parity of the dam: 11 piglets born to parity 0 sows, 11 piglets born to parity 1 or 2 sows, and 11 piglets born to parity 3 or higher sows. The pigs were ear-tagged to identify them through the duration of the study. Rectal temperatures and OP swabs were collected from the pigs beginning at 1 WOA and were continued once weekly through 7 WOA. Following confirmation of viral shedding, a one-time nasal swab sampling was performed at 4 WOA in conjunction with the OP swabs to compare the Ct scores between the two sampling methods. OP swabs and nasal swabs from the piglets were also evaluated for PPIV1 by PCR assays.

Results and discussion

None of the 30 sows sampled tested positive for PPIV1 on OP swabs. The lack of positive oropharyngeal PCR assays in these sows indicate that sows may not be directly shedding the virus at the time of farrowing, suggesting an alternative method of viral transfer within the farm that continues to infect the suckling pigs. The earliest positive OP swab from the piglets was at 2 weeks of age from one pig (Table 1). By 3 WOA, all piglets tested positive for PPIV1 via OP swab. The piglets began testing negative at 5 WOA, however, some individuals were still shedding virus at the last sampling time at 7 WOA. This data indicates that the ideal time for PPIV1 sampling in this farm is at weaning and the two weeks post weaning. At the week 4 sampling period when both nasal and OP swabs were collected, nasal swabs were shown to have significantly lower Ct scores compared to the OP swabs. The average difference between the two sampling methods was 2.7 cycle times, which corresponds to a 7.3-fold difference. This data indicates that nasal swabs are more likely to yield higher virus levels and may be a preferred sampling method to diagnose PPIV1. No correlation was identified between pig gender or sow parity and PPIV1 Ct scores of the OP swabs. However, a negative correlation was identified at week 3 between rectal temperatures and OP swab Ct scores, indicating that pigs with lower Ct scores or greater virus levels had higher rectal temperatures. This correlation was not found in the following weeks. Continued studies of PPIV1 in endemic herds are indicated to identify viral transmission routes and better understand the impact of PPIV1 on swine health and production.

References

1.Lau SK, Woo PC, Wu Y, et al. Identification and characterization of a novel paramyxovirus, porcine parainfluenza virus 1, from deceased pigs. *Journal of General Virology* 2013; 94:2184–2190.

2.Gauger PE, Lin S, Bade S, Park J, et al. Porcine Parainfluenza Virus Type 1 (PPIV-1) Disease, Diagnostics, and Clinical Significance. *ISU James D. McKean Swine Disease Conference Proceedings, November 3-4, 2016*;106-114.

3.Palinski RM, Fang Y, Lang Y, et. al. Widespread detection and characterization of porcine parainfluenza virus 1 in pigs in the USA. *Journal of General Virology*. 2016; 97(2):281-286. doi:10.1099/jgv.0.000343.

Table 1: Piglet oropharyngeal and nasal swab PCR assay results with rectal temperature data, and the timeline of viral shedding in PPIV1 positive piglets.

	PCR results by pig age and swab type						
	Oropharyngeal			Nasal			
Pig age	Avg. CT	No. positive	Lowest CT	Avg. CT	No. positive	Lowest CT	Avg. rectal temp (F)
Week 1	38.1	0/33	38.1		Not done		102.8
Week 2	37.9	1/33	32.6		Not done		103.1
Week 3	31.0	32/32	19.5		Not done		102.5
Week 4	30.6	32/32	22.6	27.9	32/32	20.9	102.9
Week 5	34.5	28/32	21.0		Not done		103.4
Week 6	37.1	16/32	33.2		Not done		103.8
Week 7	37.6	5/32	31.7		Not done		103.3

Onset and duration of positive PCR results

	No. and percent positive pigs by week of age or no. of weeks						
Parameter	1	2	3	4	5	6	7
Week at first positive	None	1/32 (3.1%)	31/32 (96.9%)	None	None	None	None
Week at last positive	None	None	None	3/32 (9.4%)	12/32 (37.5%)	12/32 (37.5%)	5/32 (15.6%)
No. of weeks positive	2/32 (6.3%)	13/32 (40.6%)	12/32 (37.5%)	5/32 (15.6%)	None	None	None

Comparison of pathogenic *Salmonella* spp. detection from porcine oral fluids, over-the-shoe booties, and fecal samples

S. Naberhaus[1]; A. Krull[2], DVM, PhD; B. Arruda[2], DVM, PhD; P. Arruda[2], DVM, MS, PhD; D. Magstadt[2], DVM, MS; F. Matias Ferreyra[2], DVM; I. Honorato Gatto[2], DVM; H. Meiroz de Souza Almeida[2], DVM; O. Sahin[2], DVM, PhD; A. Kreuder[1,2], DVM, PhD, DACVIM

[1]Veterinary Microbiology and Preventive Medicine, Iowa State University, Ames, Iowa; [2]Veterinary Diagnostic and Production Animal Medicine, Iowa State University, Ames, Iowa

Introduction

Salmonellosis in pigs, which is typically seen during the weaning or finishing phase, may result in clinical disease or in establishment of a carrier state.[1,2] In a carrier state, *Salmonella* is shed intermittently in the feces, creating a public health concern as a result of potential pork contamination.[1] Accurate and timely detection of *Salmonella* is vital to the initiation of appropriate treatment protocols and prevention strategies. Veterinarians commonly question which sample type is best for herd level surveillance of *Salmonella* in swine herds. Oral fluids, currently widely used for the detection of PRRSV and SIV, are quick and inexpensive to collect. Over-the- shoe booties, while easy to collect, are not currently used as frequently for surveillance of pathogens in the swine industry. Fecal samples are the most commonly used sample for detection of many viruses and bacteria in swine but are more time consuming to collect than the previous two sample types. The goal of this project was to compare the accuracy of oral-fluids and over-the-shoe booties to the current gold standard of fecal sampling for identification of *Salmonella* using pigs experimentally inoculated with various serotypes of *Salmonella*.

Materials and methods

Five-week old pigs were experimentally infected with 10^9 CFU of *Salmonella* via oral gavage and tonsil swab using isolates obtained from clinical samples submitted to the Iowa State University Veterinary Diagnostic Laboratory. Twenty pigs were infected with *Salmonella* Typhimurium and 20 with *Salmonella* I 4,[5],12:i:- (2 separate groups with 5 pens, 4 pigs in each pen). Rectal temperatures were taken on all pigs daily until 7 days post-inoculation (PI) then biweekly for the remaining 3 weeks. Samples were collected on days 2, 4, 7, 14, 21 and 28 post-inoculation. Over-the-shoe booties were collected during pen walk-through at each time point, with one pair of booties worn through all pens per infecting serotype. Oral fluids were collected by hanging a rope in each pen of 4 pigs for 30-60 minutes. Fecal samples were collected via digital rectal palpation of each pig. Samples were stored at -80°C from the time of collection until quantification of *Salmonella* was performed. *Salmonella* quantification was completed using the standard plate count method on Xylose-Lysine-Tergitol-4 (XLT4) agar. All samples were also enriched in Buffered Peptone Water for 18-24 hours and were then plated to Brilliant Green (BG) agar and XLT4 agar. A subset of colonies for each sample characteristic of *Salmonella*, black on XLT4 agar and red on BG agar, were confirmed as *Salmonella* using MALDI-TOF-MS.

Results and discussion

The median temperatures of pigs peaked on day two post-inoculation at 102.8° F and returned to near baseline levels by day five PI at 101.3° F. Average *Salmonella* quantities in the feces peaked at day two PI at 8.04×10^4 CFU/mL. *Salmonella* was still detectable in the feces from enrichment at 28 days PI. From oral fluids, average *Salmonella* quantities peaked at day 3 PI at 5.6×10^2 CFU/mL. *Salmonella* was no longer detectable from oral fluids past day seven PI. Over-the-shoe booties were only sporadically positive throughout the testing period, with the highest average level of *Salmonella* detected being only 100 CFU/mL at day 4 PI. *Salmonella* was only detectable from the booties through day four PI.

Selection of the most appropriate sample type is important for proper surveillance of pathogens. Over-the-shoe booties proved to be an unacceptable sample type to detect *Salmonella* given that booties positive for *Salmonella* were not reflective of the level of disease nor were they consistently positive. Oral fluids would be an acceptable sample type for detection but only during the acute phase of a *Salmonella* outbreak when *Salmonella* levels are the highest. Enrichment of those oral fluid samples would be recommended to ensure there is adequate *Salmonella* for detection through culture. Although feces is the most time-consuming to collect, it should remain the gold standard for *Salmonella* detection as the fecal samples were most reflective of the level of disease as indicated by the rectal temperatures in addition to having the highest levels of *Salmonella* for the longest amount of time post-infection.

References

1. Alsop, J. "An outbreak of salmonellosis in a swine finishing barn". J. Swine Health Prod. 2005:I3(5): 265-268.

2. Harris, DLH. "Intestinal Salmonellosis in Pigs". Merck Veterinary Manual. Available at: https://www.merckvetmanual.com/digestive-system/intestinal-diseases-in-pigs/intestinal-salmonellosis-in-pigs. Accessed 3 November 2017.

Worker satisfaction and animal empathy on sow farms using electronic sow feeders

A. Palmieri; T. Parsons, VMD, PhD, DACAW; M. Pierdon, VMD, DACAW

Introduction

Caretakers are essential to a successful swine farm. It appears that no research has been done to assess the reaction of workers to changes in group housing of sows. Sows may be grouped using different strategies. Dynamic pens place animals from multiple breeding groups together, adding or removing animals from pens each week. Static pens contain one, or a portion of one breeding group per pen. Farms may also mix sows at various times. Pre-implantation mixing is placing sows into group pens immediately after breeding. Post-implantation mixing is the placement of sows into group pens after pregnancy confirmation. Surveys of worker attitudes in dairy cows showed milk yield was correlated with worker attitudes.[1] It was hypothesized that animal empathy might have a similar correlation in swine caretakers and sow production.

The study was designed to use two survey instruments. First, to determine whether a relationship exists between job satisfaction and grouping strategy or mixing strategy on farms that implement electronic sow feeders. Second, to examine the relationship between animal empathy and productivity measures.

Materials and methods

Ten farms that met the study criteria, utilizing the same electronic sow feeders (Schauer Agrotronic Compident 7) and management by an integrator, participated in the study. Farm size ranged from 1435 to 5900 sows. Five farms used static pens and 5 used dynamic pens. Two farms used post-implantation mixing and 8 used pre-implantation mixing. Five members per farm (N = 48) were selected via random number to participate in a survey encompassing both job satisfaction[2] and animal empathy.[3] Participants responded to statements on a scale from 1 (strongly disagree) to 7 (strongly agree) with 4 being a neutral response. The survey responses were collected, and total scores were calculated. STATA v 15.1 was used for statistical analysis. Non-parametric Mann Whitney U tests were performed to compare average farm level job satisfaction by grouping and mixing strategies. Mann Whitney U test was used to compare animal empathy between male and female respondents. Spearman rank sum correlation was done for average farm level animal empathy score and farrowing rate, as well as pre-weaning mortality, for a period of 6 months prior to the survey.

Results

There was no significant difference between job satisfaction scores when farms with dynamic pens (223.7+/-17.2) and static pens (216.8+/-10.2) ($P > 0.05$) were compared. There was no significant difference in job satisfaction scores between pre-implantation mixing (220.0+/-13.7) and post-implantation mixing (218.4 +/-17). Animal empathy score was not significantly correlated with farrowing rate ($R = -0.27$, $P > 0.05$) or preweaning mortality ($R = -0.36$, $P > 0.05$). Females (114.2+/-14.3) had higher animal empathy scores than males (100 +/-18.9) ($P < 0.05$).

Discussion

Interestingly, no difference was found in job satisfaction depending on housing type or mixing strategy. Dynamic pens represent a significant departure from facilities that use gestation stalls to farms where animals from multiple breeding groups are spread across several pens. In static pens, the breeding groups are kept intact, and animals are easier to find. Despite these differences, there was no impact on job satisfaction scores. Total empathy score, as well as gender differences in the animal empathy scores, were consistent with results of lay people that took the same survey.[3] The survey instrument used in this study was oriented toward small animals. Future studies should use questions more applicable to food animals to determine if such correlations exist for US swine caretakers. If an appropriate survey instrument were to be developed, scores could be used to place employees in suitable roles on the farm, or perhaps, to screen employees prior to offering employment.

References

1. Waiblinger, S., Menke, C., & Coleman, G. (2002). The relationship between attitudes, personal characteristics and behaviour of stockpeople and subsequent behaviour and production of dairy cows. *Applied Animal Behaviour Science*, *79*(3), 195-219.

2. Schmidt, S. W. (2007). The relationship between satisfaction with workplace training and overall job satisfaction. *Human Resource Development Quarterly*, *18*(4), 481-498.

3. Paul, E. S. (2000). Empathy with animals and with humans: Are they linked? *Anthrozoös*, *13*(4), 194-202.

Evaluation of analgesics for castration and tail docking in 3-5 day old piglets

Andrea Patterson, BScH; Karine Talbot, DVM

Statement of the problem

The National Farm Animal Care Council of Canada published the Code of Practice for the Care and Handling of Pigs in 2014. The Code of Practice requirements state that:

- As of July 1st 2016 castration performed at any age must be done with analgesics to help control post-procedure pain
- As of July 1st 2016 tail-docking performed at any age must be done with analgesics to help control post-procedure pain

Along with the requirements for castration and tail-docking it is recommended practice that analgesics are used to control pain when ear tattooing, notching, or punching. There are multiple drugs available for use during these procedures, however little research has been conducted to determine which is most efficacious. This trial therefore will give industry partners a critical evaluation of the currently available analgesics, including price consideration, so that scientifically informed decisions can be made in the processing of piglets.

Objective

The objective of this trial is to evaluate the effectiveness of different pain medications licenced for use in food animals in Canada on post-operative pain associated with castration and tail docking.

Materials and methods

A total of 212 piglets between the ages of 3 and 5 days, were included in this trial. Seven treatment groups were used with three sub-treatments per group for a total of 21 groups. Piglets were randomly assigned a treatment, which was blind to the person administering the dosage. A total of 10 navigation times (NT) were taken for each piglet. Piglets were first trained to navigate through the chute with hurdles, then baseline assessments were taken, a total of 4 baseline measurements were taken and averaged to account for individual variation. Piglets were assessed at -15 minutes, 0 minute, 15 minutes, 30 minutes, 60 minutes and 120 minutes. Assessment was done through a chute, with 2 12cm hurdles, and time was counted from 4 feet on the ground to nose through the exit, and both feet over the second hurdle. Each piglet was assigned to a treatment group; castrated and tail docked, just tail docked and nothing at time 0 and received their drug treatment at this time.

Statistical analysis was done using STATA 13, outliers were determined as anything above or below 3X the standard deviation.

Results

Statistical analysis is ongoing.

Conclusions and industry implications

Next steps include testing analgesics in combination with iron to determine if there are any pharmaceutical interactions as well as if mixing effects efficacy.

Using estrone sulfate urine test to determine pregnancy in sows and gilts

Meagan Rennie, BS; Brandon J. Dominguez, DVM MS

Background

Pregnancy detection in sows and gilts is critical to the efficiency of commercial swine operations. Waiting for return to estrous is unpredictable, and the use of ultrasound can be time consuming and involve costly equipment. Several studies have targeted hormones produced during pregnancy as urinary indicators of continued gestation.[1] To our knowledge, only estrone sulfate has a cheap, quick, and publicly available urine test claiming to detect levels significant for pregnancy in swine. A pack of fifty tests comes out to only two dollars per test, relatively inexpensive given the cost of production, and requires no animal restraint in contrast to ultrasound procedures. In this study, we hope to evaluate the efficacy of Emlab Genetics' P-TEST[2] in detecting estrone sulfate levels and thus the pregnancy status of sows compared to transabdominal ultrasound.

Materials and methods

Sows had ultrasound examination for pregnancy diagnosis between days 28 and 49 of pregnancy. Urine was collected according to the method described by Nickel, et al.[3] An ultra-size tampon was inserted vaginally and secured to each animal via tape. The tampon was expelled during urination and at least 1.5 mL of urine introduced into one P-TEST vial per pig. If any color change away from amber occurred, this was indicative of pregnancy. Results of the P-TEST were compared with those gained from ultrasound performed at the same time.

Results

Of 25 tampons retrieved, 6 yielded sufficient urine for testing. Results of the P-TEST are compared with those gained from ultrasound performed at the same time (Table 1).

Sensitivity (compared to ultrasound) = 50%

Specificity (compared to ultrasound) = 50%

Positive Predictive Value = 66.7%

Negative Predictive Value = 33.3%

Discussion

Our results indicate that the P-test does not offer a competitive alternative to the use of ultrasound in determining the pregnancy status of sows. Follow through until farrowing will provide the gold standard of pregnancy detection. The sample size is small due to the time available to collect urine on the farm. Due to unforeseen weather phenomena, we were unable to attempt replication of this experiment. A larger sample size and further studies may help further elucidate the benefit of using the P-Test for on farm pregnancy diagnosis.

References

1.Pork Checkoff. Pregnancy Diagnosis in Swine. Available at: http://old.pork.org/filelibrary/factsheets/pigfactsheets/newfactsheets/08-04-03g.pdf. Accessed Jul 26, 2017.

2.Emlab Genetics. P-TEST. Available at: http://www.emlabgenetics.com/Documents/PTEST-%201X%20(2016).pdf. Accessed Jul 26, 2017.

3.Nickel, M; Skoland, K; Forseth, A; Brown, J; Thompson, V; Breuer, M; Warner, R; O'Brien, K; Canning, P; Thomas, P; Hoogland, M, Lauterbach, N, Karriker, L. Development and validation of ante-mortem urine collection techniques for gilts and sows. Proceedings of the 48th Annual Meeting of the American Association of Swine Veterinarians, 2017. Denver, CO.

Table 1: Ultrasound and P-Test results

	Ultrasound positive	Ultrasound negative	
P-Test positive	2	1	3
P-Test negative	2	1	3
	4	2	6

Evaluation of BoarBetter® pheromone in optimizing swine reproductive efficiencies in two commercial breeding farms

M. Rohde[1], BS; J. Brown[1], DVM; S. Anthony[1], BS; K. Hayman[1], DVM; P. Canning[1], DVM, PhD; K. Klemesrud[1]; E. Akin[1], BS; A. Forseth[1], DVM; H. Kittrell[1], DVM; M. Breuer[1], BS; K. Skoland[1], BS; C. Rademacher[2], DVM; L. Karriker[1], DVM, MS, DACVPM

[1]Swine Medicine Education Center, Iowa State University, Ames, Iowa;
[2]Veterinary Diagnostic and Production Animal Medicine, Iowa State University, Ames, Iowa

Introduction

One of the goals of agriculture is to produce larger quantities of food for safe consumption with fewer animals. While swine litter sizes have risen in recent years, farrowing rates remain variable and below the biological potential of the sow. Animal Biotech has developed a synthetic triple pheromone product, BoarBetter®, with the potential to enhance breeding and farrowing rates. It is hypothesized that more pheromone stimulation could lead to improved estrus behavior, ovulation, pregnancy, total born, and number born alive on commercial swine operations. The goal of this study was to evaluate breeding, conception, and farrowing successes and rates as well as total born and number born alive in sows exposed to the BoarBetter® product.

Materials and methods

Two commercial sow farms in central Iowa participated in the study. Each farm weaned three days every week. All qualified sows from the first two weaning days of each week were enrolled in the study for seven weeks. Sows from the third weaning of each week were not included in the study; serving as a buffer between treatment and control groups. Sows were selected using three enrollment criteria: a body condition score of 2/5 or greater, parity of two through seven, and no culling designation at the time of weaning. There were seven groups of sows in the study, three treatment groups and four control groups, alternating each week during the study. Control groups served as a negative control. The study began and ended with control groups to monitor seasonal variation of sow estrus behavior. Treatment groups participated in the following protocol: on days four through eight post weaning, one hour prior to heat checking and immediately after post-cervical artificial insemination, four to five milliliters of BoarBetter® was applied to participants' snouts until each sow received two breeding services. After breeding, both treatment and control groups followed unmodified farm protocols through farrowing.

Results and discussion

Between the two farms, 1,627 sows were enrolled in the study. Table 1 depicts the results from each farm as well as a weighted total for the farms. Breeding success was calculated by dividing the number of sows that were bred by day seven post weaning by the number of sows enrolled into the study. Conception success was calculated by dividing the number of sows confirmed pregnant on transabdominal ultrasound by the total number of sows enrolled in the study. Conception rate was calculated by dividing the number of sows confirmed pregnant on transabdominal ultrasound by the number of sows that were bred day four through seven post weaning. Weighted totals for both treatment and control groups were calculated using proportions based on enrollment numbers at each farm. A 3.01% increase in breeding success conveys that BoarBetter® could be an effective supplementation to teaser boars. Remaining data collection is ongoing and will conclude during the third week of December 2017. Pending continued success, BoarBetter® may contribute to progress in improving pigs per sow per year. Final data will be analyzed to determine if the production differences are of significant statistical value.

Table 1: This table demonstrates the initial BoarBetter® farm data

	Farm A treatment	Farm A control	Farm B treatment	Farm B control	Treatment total	Control total
Total sows enrolled (A)	272	357	428	570	700	927
Enrolled sows bred day 4-7 post wean (B)	223	299	341	420	564	719
Enrolled sows preg. check positive (C)	208	287	316	383	524	670
Breeding success*	81.99%	83.75%	79.67%	73.68%	80.57%	77.56%
Conception success†	76.47%	80.39%	73.83%	67.19%	74.86%	72.28%
Conception rate‡	93.27%	95.99%	92.67%	91.19%	92.90%	93.04%

* Breeding success = # sows bred day 4-7 post wean / Total sows enrolled = B/A
† Conception success = # sows preg. check positive / Total sows enrolled = C/A
‡ Conception rate = # sows preg. check positive / Enrolled sows bred day 4-7 post wean = C/B

Evaluation of farrowing stall sanitation protocols with ATP bioluminescence

Marjorie Schleper[1], BS; Laura Dalquist[2], DVM; Tyler Te Grotenhuis[3], DVM; Steven J. P Tousignant[4], DVM, PhD
[1]University of Minnesota, College of Veterinary Medicine, St. Paul, Minnesota, [2]Swine Vet Center, St. Peter, Minnesota, [3]Christensen Farms, Sleepy Eye, Minnesota, [4]Boehringer Ingelheim Vetmedica, Inc., Duluth, Georgia

Introduction

Farrowing stall cleanliness is important for maintaining the health of neonatal piglets. Currently there is no rapid, stall-side method of detecting organic material left after washing. Adenosine triphosphate (ATP) bioluminescence is used to determine the level of organic material, expressed as relative light units (RLU), on cleaned surfaces. This investigation utilized ATP bioluminescence, conventional bacterial culturing and polymerase chain reaction (PCR) analysis to assess cleanliness of stalls washed with four different protocols.

Materials and methods

A preliminary study was conducted to test for potential interference between Synergize disinfectant (Preserve International, Reno, Nevada) and the ATP device (Neogen Food Safety, Lansing, Michigan). Samples of Tri-Bar, cast iron, and plastic flooring were cleaned and sampled via the ATP device, then sampled again after application of the disinfectant. It was determined that there was no interference, and this system was used on-farm to assess washing protocols. Four washing treatments (assigned by stall row) were tested: hot water without foaming of the disinfectant (HNF), hot water with foaming (HF), cold water without foaming (CNF), and cold water with foaming (CF). Utilizing the ATP device's randomize function, one stall in each treatment was selected for pre-wash sampling. After washing, each stall was inspected by farm staff and re-washed if organic material was observed. Then the originally selected stalls and four additional stalls in each row were sampled. Then disinfectant was applied and stalls were resampled. Each stall was sampled in three areas; the back and front of the piglet creep area and the back of the sow area. Two 4" diameter rings were utilized for sampling, one for ATP and one for culture and rotavirus A, B and C PCR (Stuart's Media swabs). The Stuart's swabs were cultured on blood agar and microbial growth was scored 0-4. Each phase of the sanitation procedure was timed.

Results and discussion

In this trial, all treatments were successful in reducing ATP, bacteria, and rotavirus C levels, although these data suggest cold water treatments had a greater reduction in bacterial growth and rotavirus positive samples than hot water treatments (Table 1). Although a rotavirus A, B, and C panel was conducted, only rotavirus C was found in this study. The mere presence of rotavirus nuclear material doesn't necessarily indicate that it is infectious, but it does mean that there was incomplete removal from the stalls. This may mean that other microbes were still present, some of which may be viable. Foaming had a greater reduction in bacterial growth than non-foaming treatments. This may be due to more even and complete application of the detergent and the disinfectant. The HF row was likely contaminated after the clean stage samples were collected before final sampling, leading to an elevated RLU level and number of PCR positive samples. This confounding event highlights potential sources of contamination that can occur due to time restraints in completing the sanitation of the room. This also highlights the importance of washing order of structures within the room. There were no appreciable differences between treatments for washing time, contradicting the theory that hot water decreases washing time. It was also found that the ATP device was user friendly and the RLU levels tended to agree with culture and PCR results. Based on these findings, this device could effectively be used to evaluate current on-farm cleaning protocols and to aid in training programs for new hires. Finally, the ATP bioluminescence device shows potential as a rapid, stall-side method for determining levels of organic material remaining after washing, as RLU values generally decreased throughout the cleaning process.

Acknowledgements

The authors would like to thank Drs. Mike Eisenmenger and Daniel Boykin as well as Seth Playter for their contributions to this project.

Table 1: Mean relative light units, mean culture scores, and percent PCR positive results for the four treatments at each stage of testing, hot water no foaming (HNF), cold water no foaming (CNF), hot water foaming (HF), and cold water foaming (CF).

	Mean RLU				Mean culture score				% PCR positive			
	HNF	CNF	HF	CF	HNF	CNF	HF	CF	HNF	CNF	HF	CF
Dirty	7480	6631	8251	5891	3.3	2.7	3.7	3.3	100	100	100	100
Cleaned	1066	194	954	639	2.7	2.3	2.5	2.5	93	73	93	100
Disinfected	6	111	1571	350	1.5	0.6	1.1	0.4	53	53	100	53

Effects of mechanical lead system for boar control during heat detection on libido and salivary androstenone

N. Sugai[1]; S. Probst Miller[2], DVM; R. V. Knox[1], PhD
[1]University of Illinois College of Veterinary Medicine, Urbana-Champaign, Illinois;
[2]AgCreate Solutions, Monticello, Illinois

Introduction

Daily boar exposure for weaned sows shortens weaning-to-estrus interval.[6,13] Boar libido impacts a technician's ability to correctly identify heat.[6] Mechanical leads are used for human safety and to reduce labor for boar movement and heat detection. To date, we are unware of other studies that have measured impact of mechanical lead on libido and salivary androstenone level. This study compared two options for mechanical leads, the Boar Bot (BB) and Contact-O-Max (CM) on measures of boar libido during heat detection and effects on salivary androstenone as a potential indicator for libido.

Materials and methods

Twelve boars from 5 different farms were observed during morning heat detection from June through July 2017. Farms ranged in size from 2700 to 6200 sows. Boars observed were 50% Meishan and commercial breed. Since different treatments were located on different farms, for BB; 5 boars were used to observe libido in response to 2759 sows over 9 days. For CM, libido of 7 boars to 2840 sows was observed over a 9-day period. Boar libido behaviors were ranked (1- 4) with 1 = not interested and up to 4 = strong activity. Scores on a per sow basis included contact time (head toward sow), chomping/saliva production, urination, vocalization, and the time for the boar to be moved from snout to tail while in front of a sow stall. Saliva samples were obtained from boars via ropes before and after a heat check shift. Samples were extracted and stored at -80°C and later processed for gas chromatography analysis of androstenone.[1]

Results

Statistics were performed using SAS for the main effects of mechanical lead, and boar and sow order of exposure (1-200, and 201-400 sows/day). BB boars showed doubled contact time ($P < 0.0001$), increased chomping/saliva production ($P < 0.0001$), and increased urination ($P < 0.02$), with females over CM boars. There was no significant difference for vocalization (Figure 1). It was common to observe boar libido waning over time for both lead systems. CM libido waned more than BB ($P < 0.05$). There was a difference between pre- and post- androstenone for all boars, but no significant difference between BB and CM boars.

Discussion and conclusion

Of note to veterinary practitioners providing reproductive consulting, boars in CM tended to lay down more frequently, especially as heat detection shifts progressed. By choice, BB farms rotated boars more frequently than CM farms, resulting in better libido scores. In fact, all CM farms used the same boar for the

Figure 1: Overall means for behavior signals for mechanical lead systems

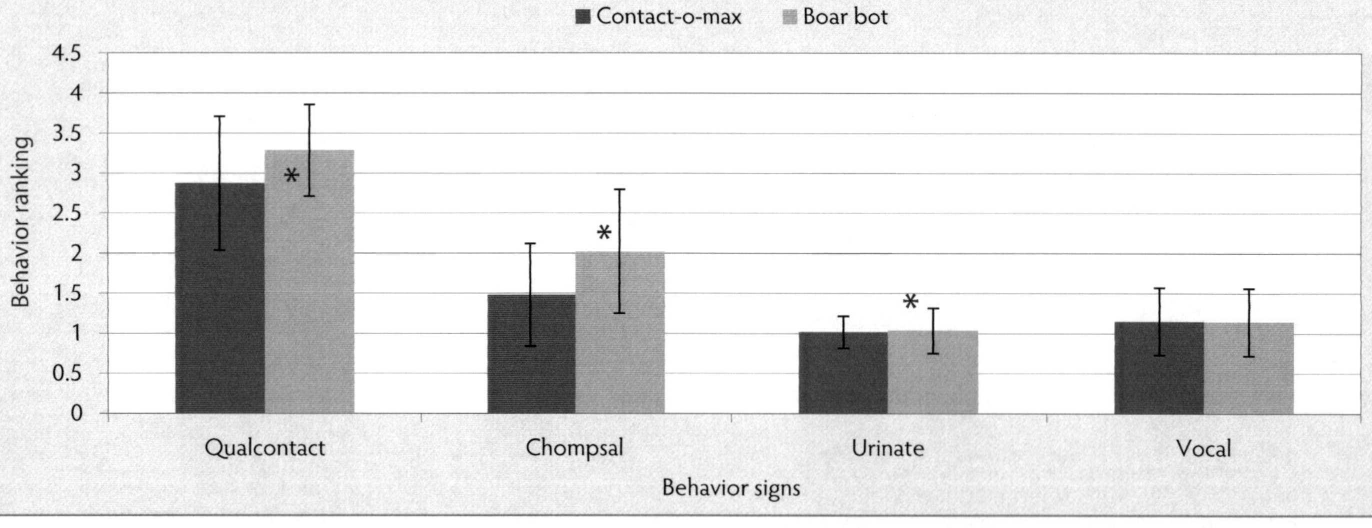

entire morning. We observed that some farms had a favorite boar (usually elderly) they chose to use every day. Farms using the same boar daily had lower libido scores. In addition, psychological attachment may make staff reluctant to cull boars viewed as "pets." Training on why a boar replacement program is important to impact boar libido and human safety could be valuable to the industry. Of note is that the farm with highest post heat checking salivary androsteneone did the best job of rotating, culling old, and training new boars. This infers that testing at height of boar excitement or just after, could be a better timing to measure boar libido. As previous work on androstenol levels are looking at boar stud boars, this study was the first to extrapolate the procedure for libido. Therefore further studies will be necessary to compare if these results are more correlated to other factors (eg. age and genetics) compared to boar libido. Overall this study shows boar handling for heat checking has a significant difference on the boars' overall libido and performance. Investigating further on the effects of age and training for boars can help formulate better heat checking boar protocols for improving heat checking efficiency on sow farms.

References

1. Booth, W.D., 1987, J. Reprod. Fert. 81, 427-431.

2. Cassar G, Kirkwood RN, Seguin MJ, et al. Influence of stage of gestation at grouping and presence of boars on farrowing rate and litter size of group-housed sows. J Swine Health Prod. 2008;16(2):81 85.

3. Castaneda, J., Orihuela, A., Rocha, G., 2002, Poster, Production and Management, IPVS, 649

4. East, M.L., Dehnhard, M., Rohrmann, H., and Kauffold, J., 2013, Chemical Signals in Vertebrates 12, 381-390. (ch 30 measure pheromones)

5. Hemsworth, P.H., Hansen, C., 1990. Applied Animal Behavior Science 28, 281-285.

6. Kemp, B., Soede, N.M., Langgendijk, P., 2005. Theriogenology 63, 643-656.

7. Kirkwood, R., 2009, American Association of Swine Veterinarians (AASV) AAS-VPCS, 9-14.

8. Knox, R.V., (year) American Association of Swine Veterinarians (AASV), 1-10.

9. Knox, R.V., 2004 Advances in Pork Production 15, 307-313.

10. Knox, R.V., Miller, G.M., Willenburg, K.L., Rodriguez-Zas, S.L., 2002 J Amin Sci 80, 892-899.

11. Knox, R.V., Rodriguez-Zas, S.L., Sloter, N.L., McNamara, K.A., Gall, T.J., Levis, D.G., Safranski, T.J., Singleton, W.L., 2013. J Anim Sci. 91, 433-445.

12. Knox, R.V., Willenburg,K., Miller, G., 2001. American Association of Swine Veterinarians (AASV) 541-544

13. Knox, R.V., Willenburg, K., Miller, G., 2002, Poster Production and Management IPVS, 659

14. Langendijk,P., Soede, N.M., Kemp, B., 2000. J. Anim. Sci. 78, 871-878.

15. Marchant-Forde, J.N., 2010. Proceedings of the 21st IPVA Congress, 36-39.

16. Marchese, S., Pes, D., Scaloni, A., Carbone, V., Pelosi, P., 1998 Eur J. Biochem. 252, 563-568. (Lipocalins of boar saliva...)

17. Penaflor-Siller, J., Sosa-Ferreyra, C.F., 2002, Poster Production and Management IPVS, 683

18. Petznick, T., 2013 American Association of Swine Veterinarians (AASV) Seminar topics, 23-24.

19. Soede, N.M., Langendijk, P., Kemp, B., 2002, Proc. 17th IPVS 1:15-24.

20. Sonderman, J.P., 2016 American Association of Swine Veterinarians (AASV) Annual Meeting (Proceedings), 3-5.

21. Thacker,B., 2001, Iowa State University Swine Disease Conference for Swine Practitioners, 55-60.

Comparison of oral fluids versus tonsil swabs to diagnose *Mycoplasma hyosynoviae* infection prior to onset of clinical signs

Grzegorz Tarasiuk[1]; Luis Ochoa[2], PhD; Micah Jansen[3], DVM; Laura Greiner[4], PhD; Attila Farkas[4], DVM
[1]University Centre of the Veterinary Medicine UJ-UR Cracow, Poland, [2]Carthage Innovation Swine Solution, Carthage, Illinois; [3]Zoetis, Parsippany, New Jersey; [4]Carthage Veterinary Service, Carthage, Illinois

Introduction

Mycoplasma hyosynoviae (*Mhs*) is a commensal organism of upper portions of the respiratory tract of pigs. Clinical lameness associated with *Mhs* typically occurs in three to five month old pigs.[1] However, it is unclear if *Mhs* can be found in the upper respiratory tract by diagnostic methods such as tonsil swabs or oral fluids, and how they correspond to clinical signs of arthritis related to *Mhs*.

Objective

The objective of this project was to determine which diagnostic method, oral fluids or tonsil swabs, is more accurate in predicting the incidence and timing of *Mhs* clinical signs.

Materials and methods

The study began upon arrival of gilts (approximately 21 days of age) to the isolation unit of a 6,000 head commercial sow farm with a history of *Mhs*. Fifty pigs were randomly selected, tagged, and housed in one pen. Oral fluid was collected every week by hanging ropes in each pen (six pens) for 30 minutes. Tonsil swabs were collected from tagged pigs at the beginning of the study and at four and eight weeks post-arrival into isolation. Once moved into the gilt developer unit (GDU), all tagged pigs were kept in the same pen and they were not treated with any antibiotics. Lameness scores were recorded three times per week for all tagged pigs based on Zinpro's locomotion scale (score of 0 to 3 based on severity of lameness, with 0 being normal and 3 having reluctance to bear weight on one or more legs). Nontagged pigs were also assessed for lameness three times per week and the number of scores greater than or equal to 1 was recorded by pen. Oral fluid samples were collected from each pen weekly and tested for *Mhs* via PCR. Tonsil swabs were collected weekly from all tagged pigs and frozen. Once a pig showed signs of lameness, no more samples were collected from that individual and they became eligible for antibiotic treatment. *Mhs* PCR was then run on all previous tonsil swabs collected from pigs showing signs of lameness. Tonsil swabs from seven non-clinical pigs were tested as negative controls.

Results and discussion

At the end of the study, 15 of 50 tagged pigs showed signs of lameness (score ≥ 1). PCR analysis was performed on 198 tonsil swabs; all of them were negative to *Mhs*. PCR results obtained from oral fluids revealed that pigs from one of the six pens were positive for *Mhs* on weeks 2, 3, and 5 post-arrival in the GDU. One additional pen was positive for *Mhs* on week 5. For both diagnostic methods, incidence of lameness scores ≥ 1 did not differ between positives and negatives. In conclusion, in the constraints of this study, oral fluids appeared to be more sensitive than tonsil swabs in the detection of *Mhs*. Additionally, detection of *Mhs* did not correlate with incidence or timing of clinical signs.

References

1. Jeffrey J. Zimmerman, Locke A. Karriker, Alejandro Ramirez, Kent J. Schwartz, Gregory W. Stevenson et al. Diseases of swine 10th edition p.789.

Medicated starter feed in weaned piglets and ensuing effects on overall health

Bex Weaver, BS; Owen Rae, DVM; Brian Payne, DVM

Introduction

Since the 2017 Veterinary Feed Directive (VFD) guidelines were instituted, many producers have reevaluated their use of feed-grade antibiotics and must strive for limited and judicious use. Legitimate concerns for antibiotic resistance will continue to socially push the industry in the same direction.[3,4] The fear of development of resistant pathogenic bacterial strains and antibiotic meat residues has been an increasing point of public scrutiny. With stricter VFD guidelines including prevention of antibiotic use for growth promotion, some swine producers are being forced to change the way they have approached herd health. The question remains, do piglets really need antibiotics to remain healthy and grow to be a similar weight? It must be determined if there is a quantifiable difference in health status between the two groups, fed a medicated or non-medicated feed. The objective of this study was to assess growth and health parameters in newly weaned piglets fed medicated and non-medicated starter feed.

Materials and methods

A single, blinded pilot study was conducted with 100 weaned piglets from 10 litters. Piglets were weaned at 21 ± 3 days of age and randomly divided into two groups. The treatment group was given CTC medicated feed for seven days, and the control group was given non-medicated feed for seven days. On day eight, both groups were switched to the standard non-medicated cornmeal based diet. Individual piglet weights were recorded at weaning (day 0), and every seven days for six weeks, and then a final weight when they exited the nursery 8 weeks later. Serial blood sample collection was conducted on two piglets from each group, the initial heaviest and lightest pig from each litter, once per week to monitor each piglet's health status (via PCV and TP). Piglets were observed daily, and fecal score was assigned to each pen based on presence (1) or absence (2) of diarrhea.

Results and discussion

We found that the pigs in both control and experimental groups had similar weight gain and health parameters, as measured in PCV and TP, signifying that the health of piglets in both studies was similar. The lab work was not meant to be a diagnostic tool, but to highlight any sick animals to differentiate a disease status in each group. Weights of pigs in both groups were similar throughout the duration of the study, except in week six when the treatment group (19kg) weighed significantly more than the control group (16.6kg), signifying a health challenge. The overall weight gain in the treatment group (21.5kg) was only slightly higher than that of the control group (19.26kg), and the average daily gain was also higher in the treatment group (0.38kg/day) as compared with the control (0.36 kg/day). Unno reported similar results of no weight difference, using additional antibiotics to CTC (chlortetracycline, sulfathiazole, penicillin; 2:2:1).[2] The PCV in both groups remained similar during the study, except on day seven when the control group (35.6%) was slightly higher as compared with that of the treatment group (34.8%). According to a study conducted by Kairie, an elevated PCV is seen with presence of enterotoxgenic *E coli*,[1] and the control group most likely had higher pathogenic bacterial load. The total protein was higher in the control group (6.1mg/L) than that in the treatment group (5.8mg/L) on day fifty-six only, suggesting possible illness or dehydration. Finally, the overall morbidity, associated with diarrhea and other disease, was significantly higher in the control group (57.7%) than in the treatment group (38.5%). Morbidity including only diarrhea was similar in both groups, suggesting that prophylactic use of CTC does not significantly prevent diarrhea. According to van der Fels-Klerx et al, variation in results of similar studies have been attributed to differences in hygiene status, prophylactic antibiotic use, and treatment of sick piglets.[3]

References

1. Kairie, E., et al. "Acute Phase Response of Piglets Fed Diets Containing Non-Starch Polysaccharide Hydrolysis Products and Egg Yolk Antibodies Following an Oral Challenge with *Escherichia Coli* (K88)." Canadian Journal of Animal Science 89.3 (2009): 353-60. Web.

2. Unno, Tatsuya, et al. "Effects of Antibiotic Growth Promoter and Characterization of Ecological Succession in Swine Gut Microbiota." *Journal of Microbiology and Biotechnology* 25.4 (2015): 431-8. Web.

3. van der Fels-Klerx, H. J., et al. "Farm Factors Associated with the use of Antibiotics in Pig Production." *Journal of animal science* 89.6 (2011): 1922-9. Web.

4. Weber, Nicolai, et al. "Occurrence of Diarrhoea and Intestinal Pathogens in Non-Medicated Nursery Pigs." *Acta Veterinaria Scandinavica* 57 (2015): 64. Web.

Figure 1: Weight gain is shown on a weekly basis, revealing a significant difference in mean weight on week six of the study.

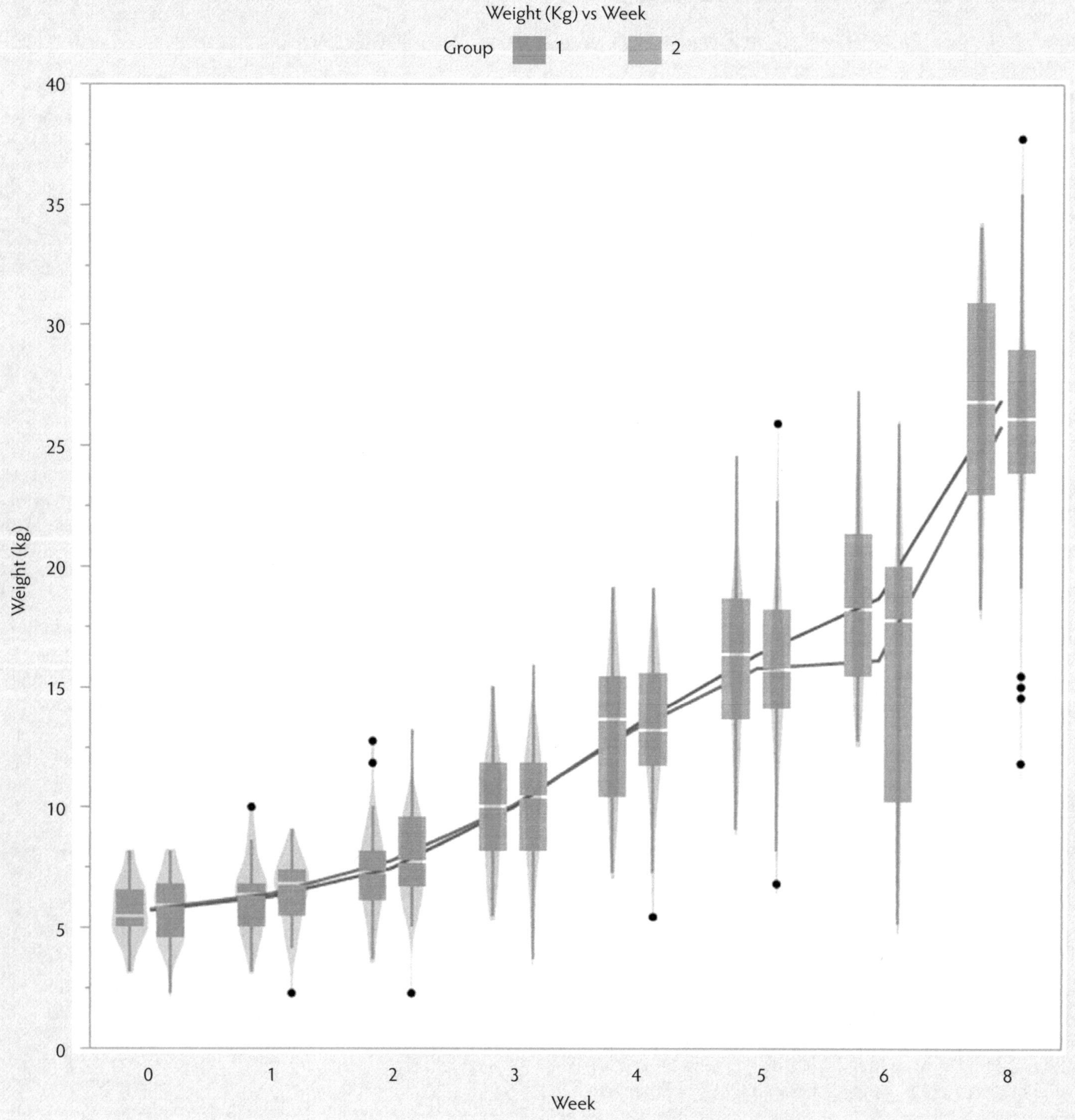

Comparative evaluation of family oral fluids and piglet sera to detect PRRSV RNA by PCR

Marcelo N. de Almeida[1]; Jeffrey J. Zimmerman[1]; Derald J. Holtkamp[1]; Chris Rademacher[1]; Hans Rotto[2]; Peter Schneider[2]; Daniel C. L. Linhares[1]

[1]Iowa State University, Ames, Iowa; [2]Innovative Agriculture Solutions, LLC, Ames, Iowa

Background

The conventional strategy for monitoring porcine reproductive and respiratory syndrome virus (PRRSV) in breeding herds relies on testing serum samples from suckling piglets (usually 30-60 piglets). This approach is time consuming, costly, and, due to the small sample size, probably misses the detection of PRRSV infection at low prevalence. For these reasons, practical, affordable and more effective monitoring and surveillance systems are needed. Oral fluids offer specific advantages for surveillance of swine populations: (a) they are easily collected by a single person; (b) can be collected without stress or risk to pigs or people; (c) at the population level they provide a higher probability of analyte detection than serum, and (d) can be used to screen for nucleic acid and/or antibodies for a growing number of pathogens. Thus, the aims of this study were: 1) to assess sampling methods to obtain oral fluids from due-to-wean pigs; and 2) to compare oral fluids to individual serum samples for detection of PRRSv RNA by real time reverse transcriptase polymerase chain reaction (rRT-PCR).

Methods

Aim 1: "Litter samples" (oral fluids from piglets within a litter; n = 170 litters) and "family samples" (oral fluids from a sow and her respective piglets; n = 120 families) were compared in 5 field experiments. The rate of "successful sampling", defined as obtaining ≥ 0.5 ml of oral fluids, was used to compare the two sampling approaches. Data recorded at sampling included time of the day, age of piglets, rope height, training before sampling (Y/N), and use attractants (Y/N).

Aim 2: PRRSV detection in individual piglet serum versus family oral fluids (FOF) was compared in 72 litters. Serum samples were collected from all piglets in litters from which oral fluids were collected. Samples were tested by PRRSV rRT-PCR and the results compared between FOF and serum.

Results and discussion

FOF samples (73.3%, range 56% to 100%) were collected more successfully than litter samples. The highest success rate in litter samples (43.5%, range 0% to 88%) was achieved in piglets ≥ 18 days of age, with collection started around 6:00 AM and the ropes hung so that they nearly touched the floor. Pre-soaking ropes with peanut butter, and training piglets by exposing them to a rope the day before the actual collection day slightly improved oral fluid collection.

PRRSV rRT-PCR was conducted on 72 FOF and on matching sera from 699 piglets, which was all piglets in the litters from which FOF was obtained. In serum, 24.6% (172/699) of samples were positive. Thirty-two litters (44.4%) had ≥ 1 piglet positive for PRRSV by rRT-PCR. Among these 32 litters, 27 (84.4%) of the FOF were PRRSV rRT-PCR positive. All litters with ≥ 3 viremic piglets were positive for PRRSV by rRT-PCR on FOF, but only 50% of litters (5/10) with 1 or 2 viremic piglets were positive on FOF. All litters with negative serum results (n = 40) were also negative by FOF.

Conclusion and implications

The success rate for collection of family oral fluids (FOF) was greater than for litter oral fluid samples. PRRSV nucleic acid was detected in FOF from naturally infected litters under field conditions, and can be used to significantly improve monitoring and surveillance systems for pathogen detection in due-to-wean pig populations by allowing cheaper, more frequent, and more comprehensive sampling of suckling pigs and sows.

Assessment of slaughter surveillance based on oral fluid samples

Marcelo N. de Almeida[1]; Jeffrey J. Zimmerman[1]; Fabio Vannucci[2]; Daniel C. L. Linhares[1]
[1]Iowa State University, Ames, Iowa; [2]University of Minnesota, St. Paul, Minnesota

Background

The development of practical, affordable, and effective monitoring and surveillance systems (MOSS) for tracking pathogens (endemic and/or exotic) in swine populations over time and space is crucial to the future of the industry. Although serum is the traditional surveillance sample, oral fluid specimens are increasingly recognized as a bona fide alternative. The objective of this study was to determine whether surveillance could be done using oral fluid samples collected in a US abattoir. Porcine reproductive and respiratory syndrome virus (PRRSV) and Seneca Valley virus (SVV) were used to represent endemic and emerging pathogens, respectively.

Methods

Oral fluid (n = 10) and serum (n = 10) samples were collected from 36 groups of finishing pigs within 48 hours of transport from the farm to the abattoir. At the abattoir, oral fluids (n = 3) were collected from the same groups while held in lairage. Samples were tested for the presence of PRRSV or SVV RNA by rRT-PCRs and PRRSV or SVV antibody by ELISA. If one sample from the group tested positive by a test, the group was considered positive. The frequency of detection was analyzed both on the basis of individual samples and by group results. Group results were used to evaluate the agreement between farm and abattoir results.

Results

Oral fluids were successfully collected from 32 of the 36 (89%) groups at the abattoir. A total of 316 serum, 319 on-farm oral fluids, and 96 abattoir oral fluids were available for testing (Table 1).

Conclusion and implications

Obtaining swine oral fluids at the abattoir was feasible, allowing detection of anti-PRRSV antibodies, and PRRSV and SVV RNA. PRRSV and SVV ELISA group results were the same for farm and abattoir samples. More studies are needed to further investigate how to improve the agreement of molecular testing between locations. Validating oral fluids-based abattoir testing would allow screening a significant number of pig populations on a much simpler, faster and cost-effective fashion, as compared to sampling individual pig sites in the field.

Table 1: Frequency of PRRSV and SVV detection by ELISA and rRT-PCR in serum and oral fluids collected on-farm, and oral fluids and environmental samples collected at abattoir.

	PRRSV		SVV	
	RNA (+)	ELISA (+)	RNA (+)	ELISA (+)
On-farm sampling				
Serum[1]	6 (1.9%)	295 (93.4%)	-	0 (0%)
Oral fluid – individual results[2]	55 (17.2%)	319 (100%)	0 (0%)	0 (0%)
Oral fluids – group results[3]	16 (50%)	32 (100%)	0 (0%)	0 (0%)
Abattoir sampling				
Oral fluid[4]	17 (17.7%)	96 (100%)	55 (57.3%)	0 (0%)
Oral fluids – group results[5]	9 (28.1%)	32 (100%)	21 (65.6%)	0 (0%)

1 n = 316
2 n = 319
3 Farm results based on rRT-PCR results from on-farm oral fluids
4 n = 96
5 Farm result based on rRT-PCR results from abattoir oral fluids

Human-like H3N2 and H3N1 swine influenza A viruses are resistant to neutralization due to antigenic variation

Srivishnupriya Anbalagan, M.Sc., Ph.D; Joshua D Elston; Jessica L Peterson
Newport Laboratories Inc., Worthington, Minnesota

Introduction

Pigs play a major role in the evolution of influenza A viruses (IAV). There is evidence for swine-to-human and human-to-swine transmission of IAV leading to considerable genetic and antigenic diversity in both H1 and H3 IAVs circulating in swine and humans. At Newport Laboratories (NPL), routine antigenic characterization of swine IAVs using a high throughput serum neutralization (HTSN) assay identified that approximately 20% of field isolates in 2014 were not neutralized by a panel of reference antisera (H3 clusters I, III, IV, and IVA) representing circulating subtypes and clusters of swine IAV in North America. The objective of this study was to genetically and antigenically characterize the neutralization-resistant isolates.

Materials and methods

Virus isolation was conducted in swine testicle (ST) cells grown in Dulbecco's minimum essential medium (DMEM) containing 5% fetal bovine serum at 37°C with 5% CO_2. HTSN was conducted as previously described.[1] Whole genomes were amplified using the method as previously described.[2] Amplified products were sequenced using the Ion Personal Genome Machine™ (PGM™) sequencing platform (Life Technologies, Grand Island, NY) as previously described.[3] Sequence reads were assembled into contigs using the SeqMan NGen program (DNAstar, Madison, WI). Sequences were analyzed using MEGA (version 7) and phylogenetic analysis was conducted using maximum likelihood analysis with 1000 bootstrap replicates.

Results and discussion

Whole genome sequencing and further phylogenetic analysis identified that the neutralization-resistant isolates had a human-like HA gene. 67% of the human-like swine IAV isolates were H3N2 and the rest were H3N1 subtypes. 50% of the human-like H3N2 isolates had human origin HA and NA. The rest of the isolates had a swine origin NA gene. To show that the resistance to neutralization is due to antigenic diversity between different clusters, a representative isolate from the human-like swine H3 subcluster was picked, hyperimmune antisera was generated, and the HTSN was conducted using new antisera. The results indicate that all the isolates that carried human-like HA were neutralized by the new antisera and no cross reactivity was observed with reference antisera from other clusters suggesting that the isolates with human-like H3 were antigenically distinct from the other circulating isolates. Since 2014, 20% of swine IAV isolated in our laboratory have carried a human-like HA gene. A majority (88%) of human-like IAV isolates had a H3N2 subtype, and only 12% of the isolates had a H3N1 subtype. Except for one isolate (H3N2) that had human origin HA and NA genes, all of the other neutralization-resistant isolates had human-like HA and classical swine lineage N1 or N2. Similar to our previous observation, all of the human-like swine IAV were neutralized by new antisera. Significant genetic and antigenic heterogeneity of this human-like swine IAV from other swine IAVs suggest that isolates from this cluster may be required in commercial and autogenous vaccines to protect swine from human-like swine IAV infection.

References

1. Hause BM et al. Antigenic categorization of contemporary H3N2 Swine influenza virus isolates using a high-throughput serum neutralization assay. *J Vet Diagn Invest.* 2010; 22:352-359.

2. Zhou B et al. Single-reaction genomic amplification accelerates sequencing and vaccine production for classical and Swine origin human influenza a viruses. *J Virol.* 2009;83:10309-10313.

3. Anbalagan S et al. Whole genome analysis of epizootic hemorrhagic disease virus identified limited genome constellations and preferential reassortment. *J Gen Virol.* 2014;95:434-441.

Minimal *Mycoplasma hyopneumoniae* genetic variability within production flows

A. Anderson Betlach,[1] BS; E. Fano,[2] DVM, MS, PhD; A. Sponheim,[1,2] DVM; L. Dalquist,[3] DVM; M. Pieters,[1] DVM, PhD
[1]College of Veterinary Medicine, University of Minnesota, St. Paul, Minnesota; [2]Boehringer Ingelheim, Duluth, Georgia; [3]Swine Vet Center, St. Peter, Minnesota

Introduction

Mycoplasma hyopneumoniae (*M hyopneumoniae*) is the etiologic agent of enzootic pneumonia, a prevalent respiratory disease known to economically impact swine herds.[1,2] Diagnostic tools, such as Multiple Locus Variable number tandem repeat Analysis (MLVA) and P146 sequencing have been used to characterize this microorganism at the molecular level.[3,4] The implementation of those tools has allowed veterinarians to gain insight regarding the potential origin and transmission of *M hyopneumoniae* variants in the field. Multiple variants have been identified within and among swine farms using MLVA; however, the extent of genetic variability within various production stages in swine flows has not been evaluated.[5] Therefore, the objective of this study was to assess the genetic variability of *M hyopneumoniae* within swine production flows.

Materials and methods

Four *M hyopneumoniae* positive production flows across various states were selected for this investigation. Within each flow, a maximum of ten laryngeal swabs were obtained from 1-3 farms per production stage (i.e. gilt developer unit, sow farm, nursery, and finisher) for up to three years. Laryngeal swabs were collected from pigs showing clinical signs suggestive of *M hyopneumoniae* infection or from randomly selected pigs in the case that no clinical signs were observed. All samples were tested by real-time PCR for *M hyopneumoniae* detection. MLVA typing based on characterizing Variable Number Tandem Repeats (VNTRs) of two adhesion loci, namely p97 and p146, and p146 sequencing were performed using the DNA from 1-5 *M hyopneumoniae* positive samples per farm. Genetic variability within and across flows was assessed by evaluating the two molecular characterization tools and geographical distribution using BioNumerics, Geneious,™ MEGA, and a geospatial analysis software.

Results and discussion

In one production flow, a single *M hyopneumoniae* variant (100% p146 sequence similarity and one MLVA type) was identified across all four production stages. In the remaining three flows, 1-4 MLVA types and a 98-100% sequence similarity were identified. Nevertheless, MLVA types varied by 1-2 VNTRs for either locus (Figure 1). From a geographical standpoint, there appeared to be no relationship between location and variant distribution. Therefore, the genetic variability of *M hyopneumoniae* within swine production flows was minimal over time and was flow specific. From this study, it can be hypothesized that evolutional pressures associated with farm management practices may not have a significant effect on *M hyopneumoniae* genetic variability. In addition, *M hyopneumoniae* variants across all production stages appeared to originate from vertical transmission. Further research should be performed to determine the effect of co-mingling flows with different *M hyopneumoniae* genetic diversity on swine production and clinical severity.

References

1. Thacker EL, Minion C. Mycoplasmsosis: Zimmerman JJ, et al., eds. Diseases of Swine. 10th ed. Ames, IA. 2012; 779-797.

2. Holtkamp D, et al. Assessing production parameters and economic impact of swine influenza, PRRS, and *Mycoplasma hyopneumoniae* on finishing pigs in a large production system. AASV. 2012; 75-76.

3. Vranchx K, et al. Multiple-Locus Variable-Number Tandem-Repeat Analysis is a suitable tool for differentiation of *Mycoplasma hyopneumoniae* strains without cultivation. J Clin Microbiol. 2011; 49:2020-2023.

4. Bogema DR, et al. Characterization of cleavage events in the multifunctional cilium adhesion Mhp684 (P146) reveals a mechanism by which *Mycoplasma hyopneumoniae* regulates surface topography. J Amer Society Microbiol. 2012; 3:1-11.

5. Pantoja LG, et al. *Mycoplasma hyopneumoniae* genetic variability within a swine operation. J Vet Diag Inves. 2016; 28:175-179.

Figure 1: *M. hyopneumoniae* MLVA variants based on flow & production stage. A) Flow 1; B) Flow 2; C) Flow 3; D) Flow 4

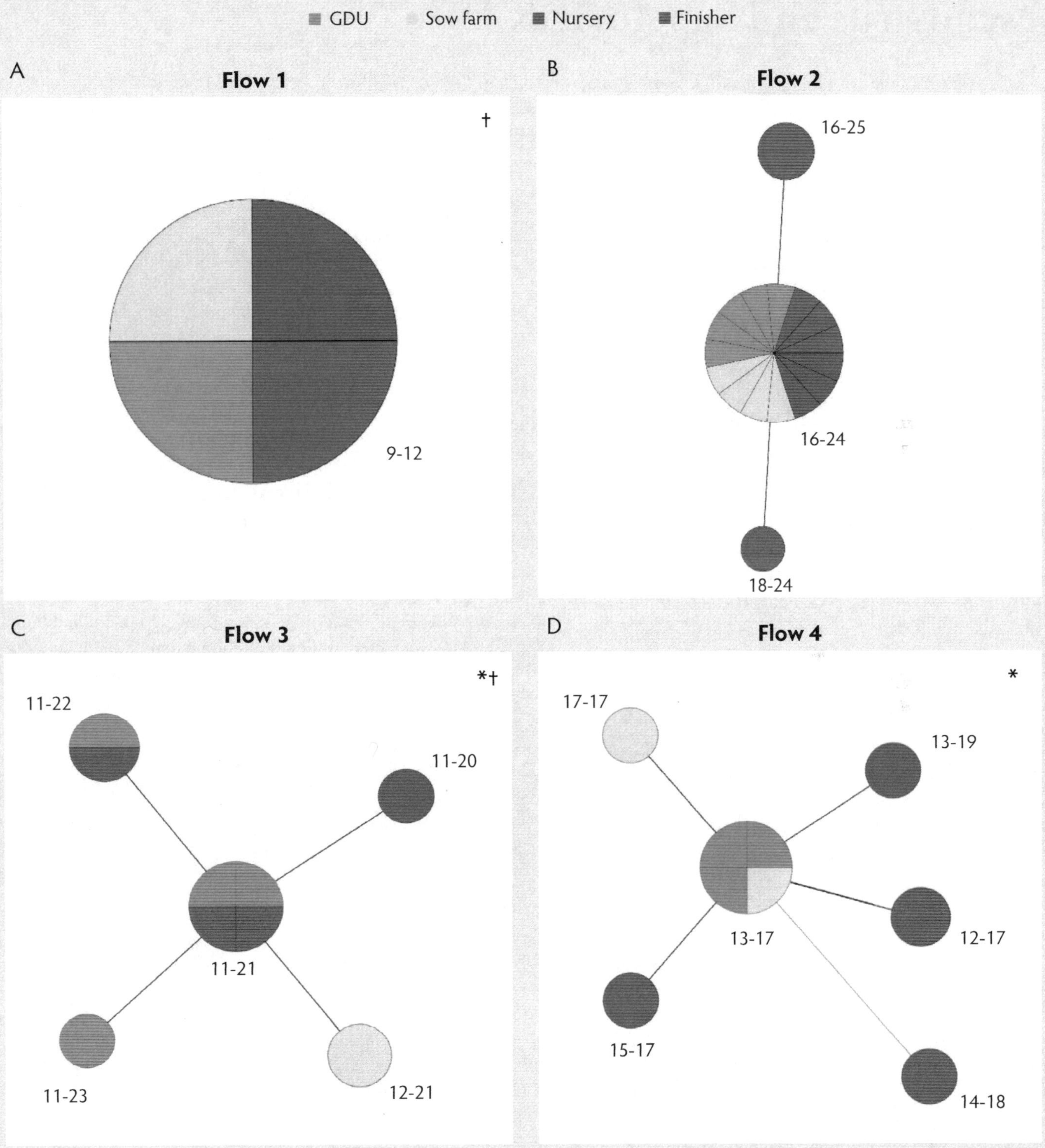

* Samples collected over two years
† 99-7-100% sequence similarity

New duplex real time RT-PCR to detect porcine Teschovirus and Sapelovirus

M. Barrera[1]; F. Chamba[2]; T. Knutson[1]; Y. Jiang[1]; S. Gresch[1]; F. Vannucci[1,2]; D. Marthaler[1-3]
[1]Veterinary Diagnostic Laboratory, University of Minnesota, St. Paul, Minnesota; [2]Veterinary Population Medicine Department, University of Minnesota, St. Paul, Minnesota; [3]College of Veterinary Medicine, Kansas State University, Manhattan, Kansas

Introduction

Porcine Teschovirus (PTV) and porcine Sapelovirus (PSV) are enteric picornaviruses that can cause neurological disorders in pigs.[1] With the increase of neurological cases associated with these viruses, improved diagnostic capabilities are required to detect these viruses in United States (US) pig populations since current conventional RT-PCR tests were developed in 2002 based on a limited number of isolates.[2] In this study, we developed, optimized and validated a new duplex real time RT-PCR to detect PTV and PSV, and assessed the geographic distribution of these two viruses in diagnostic samples.

Materials and methods

Multiple primer and probe sets were designed for each virus using complete genome sequences (18 PTV and 13 PSV strains) from GenBank, including recent US sequences. Analytical sensitivity was assessed using 10-fold serial dilutions of PTV and PSV isolates. Analytical specificity was assessed using a panel of 54 porcine pathogens, including Porcine Enterovirus G. Finally, inter- and intra-assay repeatability was evaluated testing 10-fold serial dilutions 5 and 10 times respectively, for both viruses. Diagnostic sensitivity and specificity was evaluated using a conventional RT-PCR[1]. Geographic distribution of PTV and PSV in diagnostic samples was assessed using 231 samples within 108 submissions from the University of Minnesota Veterinary Diagnostic Laboratory (UMN-VDL). Submissions included environmental, fecal and intestinal samples from pigs of different ages and US states.

Results

In the optimization phase, we tested 7 set of primers and probes, 6 RT-PCR kits, 4 primers and probes concentrations, 2 sample volumes, 2 final volumes, 2 internal controls concentrations and 4 fluorescent dyes, yielding 94 combinations for testing. The final PTV and PSV primer and probe sets target the 5'UTR region of each virus. Limit of detection was 1.0×10^{-6} and 1.0×10^{-8} 10-fold serial dilutions for PTV and PSV respectively. Coefficient of variation for both viruses was less than 9.2% and 1.9% for inter- and intra-assay repeatability. While the diagnostic sensitivity was 100% for both viruses, the diagnostic specificity was 88% and 60% for PTV and PSV respectively. The diagnostic specificity is low because our new duplex real time RT-PCR can detect lower levels of PTV and PSV in diagnostic samples compared to the conventional RT-PCR. Finally, we assessed the geographic distribution of PTV and PSV in 108 submissions from 15 US states, which identified PTV and PSV in 82% (89/108) and 72% (78/108) of the submissions. Interestingly, 70% (76/108) of submissions tested positive for both viruses.

Conclusions and discussion

A new duplex real time RT-PCR has been developed, optimized and validated based on the most current sequences and can be used to investigate neurological disorders in pigs. The new duplex RT-PCR is specific to both viruses and more sensitive than the reported conventional RT-PCR. We found that PTV and PSV were widely detected in the tested submissions. Finally, co-infections of PTV and PSV were common in this subset of submissions. This study highlights the need for more research to further elucidate the geographic distribution, prevalence and genetic diversity of these viruses in US pig populations.

Acknowledgments

This study was funded by the Swine Health Information Center (SHIC).

References

1. Arruda, P. H. et al. Detection of a novel sapelovirus in central nervous tissue of pigs with polioencephalomyelitis in the USA. Transbound Emerg Dis 64, 311-315, doi:10.1111/tbed.12621 (2017).

2. Palmquist, J. M. et al. Detection of porcine teschovirus and enterovirus type II by reverse transcription-polymerase chain reaction. J Vet Diagn Invest 14, 476-480, doi:10.1177/104063870201400605 (2002).

Potential use of outer membrane proteins as subunit vaccines against *Streptococcus suis* and *Haemophilus parasuis*

Susan L. Brockmeier[1], DVM, PhD; Kirsten Eberle[1], PhD; Samantha Hau[1], DVM, PhD; Kathy Mou[1], PhD; Crystal L. Loving[1], PhD; Tracy L. Nicholson[1], PhD; Jinhong Wang[2], PhD; Sarah E. Peters[2], PhD; Lucy Weinert[2], PhD; Duncan J. Maskell[2], PhD; Alexander W. Tucker[2], DVM, PhD
[1]USDA, ARS, National Animal Disease Center, Ames, Iowa;
[2]Department of Veterinary Medicine, University of Cambridge, Cambridge, United Kingdom

Introduction

Both *Streptococcus suis* and *Haemophilus parasuis* are important invasive bacterial pathogens of swine, commonly causing meningitis, arthritis, polyserositis, and septicemia. Due to the presence of many serotypes and high genotypic variability, broadly protective efficacious vaccines are not readily available. We have used functional genomic and proteomic methods to select protein candidates for evaluation in vaccine trials testing both homologous and heterologous protection against these pathogens.

Materials and methods

The selection of *S suis* protein candidates for inclusion in a vaccine was accomplished by identifying fitness genes through a functional genomics screen using transposon mutagenesis with *in vitro* organ culture using swine turbinate and tracheal tissue. The list of fitness genes was then reduced by selection of predicted surface-associated proteins and identification of proteins conserved across a wide range of isolates to enhance the prospect of cross-protection. Candidate proteins were evaluated in a vaccine trial. Six-week-old, Caesarian-derived, colostrum-deprived piglets (8 per group) were vaccinated with the proteins either intramuscularly, intranasally, or by both routes, twice 2 weeks apart and subsequently challenged with a highly virulent type 2 strain of *S suis*.

In addition, using an immunoproteomic strategy based on 2-D gel electrophoresis, we screened for possible cross protective factors to identify potential vaccine candidates against *H parasuis*. Caesarian-derived, colostrum-deprived piglets (5 per group) were vaccinated (twice, 3 weeks apart) with bacterins made from one of two different serovar 5 strains of *H parasuis* and then challenged with the homologous strain or a heterologous serovar 1 strain to determine cross-protection. Sera from both vaccinated groups were used to probe western blots from 2-D gels of outer membrane protein preparations form each of the strains of *H parasuis* to identify immunoreactive proteins. Mass spectrometry was then used to identify proteins of interest.

Results and discussion

Five *S suis* proteins were identified via the above methods, and intramuscular vaccination with a combination of these proteins induced IgG antibody titers to each individual protein and provided substantial protection from challenge with virulent *S suis* (Figure 1A). Both the immune response elicited and degree of protection were dependent on the parenteral route of delivery. The reactivity of the sera from vaccinated pigs against several diverse *S suis* strains indicates a potential for cross-protection.

Pigs that were vaccinated with either of the two *H parasuis* serovar 5 bacterins were protected against homologous challenge (Figure 1B and 1C). However, the same protection was not observed with a heterologous serovar 1challenge as protection was observed only in pigs vaccinated with one of the bacterins (Figure 1B and 1C). Serum from both the heterologous protected and non-protected groups was used to probe western blots from 2-D gels of outer membrane proteins of the three strains of *H parasuis*. Three outer membrane proteins of interest that reacted with serum from animals protected from heterologous challenge but not serum from non-protected pigs were identified by mass spectrometry. A vaccine trial to determine the efficacy of vaccination with these 3 proteins against challenge with multiple strains of *H parasuis* is underway.

Figure 1: Survival curves of (A) pigs vaccinated with a *S. suis* subunit vaccine intranasally (IN), intramuscularly (IM) or by both routes (IN/IM) after challenged with virulent *S. suis* and (B and C) pigs vaccinated with one of two *H. parasuis* bacterins (SV5 Bacterin 1 or SV5 Bacterin 2) after challenge with the homologous serovar (SV) 5 strain or a heterologous SV1 strain.

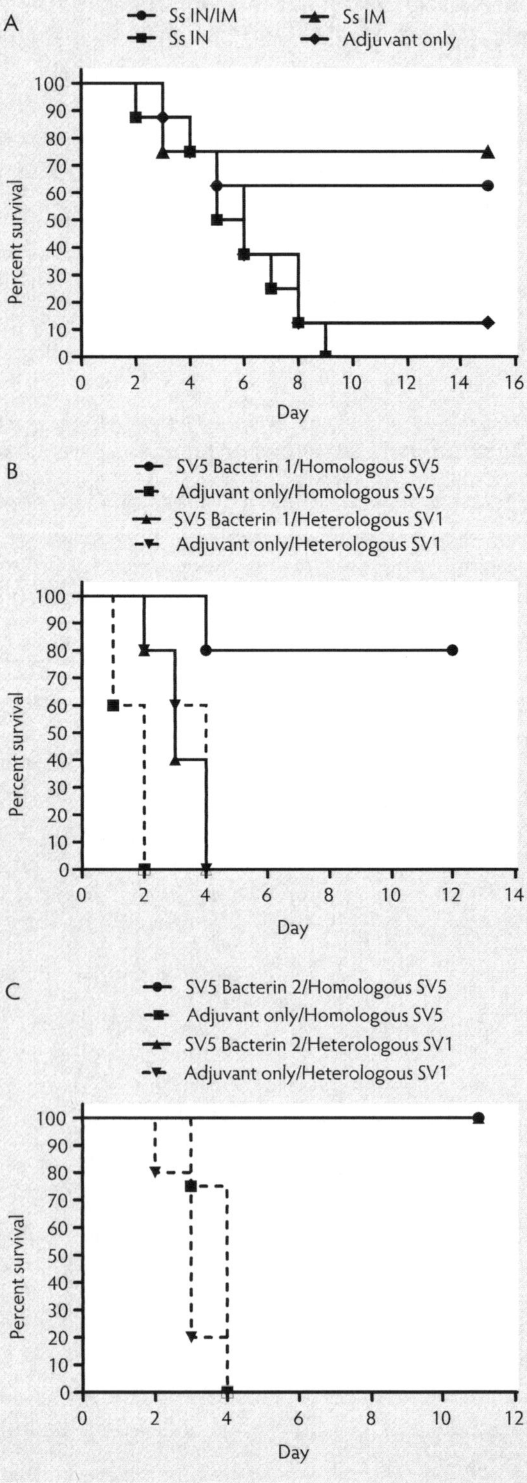

Comparison of historic and contemporary strains of Senecavirus A

Alexandra Buckley[1]; Baoqing Guo[2]; Vikas Kulshreshtha[1]; Albert van Geelen[1]; Kyoung-Jin Yoon[2]; Kelly Lager[3]
[1]Oak Ridge Institute for Science and Education and National Animal Disease Center, US Department of Agriculture, Ames, Iowa; [2]College of Veterinary Medicine, Iowa State University, Ames, Iowa; [3]National Animal Disease Center, US Department of Agriculture, Ames, Iowa

Introduction

In the past, detection of Senecavirus A (SVA) from swine samples in the United States was sporadic and originated from cases presenting with various clinical signs. Starting in July of 2015, there was a significant increase in the detection of SVA often in association with farms experiencing vesicular disease and/or an increase in neonatal mortality. After Koch's postulates were fulfilled and SVA was proven to be a causative agent for vesicular disease, speculation remained surrounding the sudden increase in SVA infections and why attempts to reproduce disease had failed in the past. One hypothesis was that contemporary strains were more pathogenic than historical strains. Our objective was to study disease progression of historical and contemporary SVA isolates in growing pigs.

Materials and methods

Commercial swine aged 16-20 weeks old (n = 54) were split into 6 challenge groups (n = 8) and 1 control group (n = 6). Three historical isolates and three contemporary isolates with titers ranging from $10^{5.1}$ – $10^{6.8}$ TCID50/mL were given intranasally on 0 days post inoculation (dpi) to each of the challenge groups. Animals were regularly bled, rectal swabbed, and oral swabbed through 14 dpi. Animals were observed daily for any clinical signs of vesicular disease. Serum and swabs were tested by real-time PCR for SVA nucleic acid detection. Serum was also tested for neutralizing antibody response to the homologous challenge virus by virus neutralization assay and cross neutralizing antibodies to heterologous challenge isolates.

Results

All isolates used in the study were able to induce clinical disease with the development of vesicles either on the coronary bands or the snout. The number of pigs presenting with clinical signs in each challenge group ranged from 5/8-8/8. All animals in each challenge group replicated virus and had a least one PCR positive rectal swab for SVA. There were slight differences in onset and duration of shedding among the six different isolates, but overall most pigs were positive for SVA in oral and/or rectal swabs by 4 dpi and were still positive for virus in rectal swabs at 14 dpi. Cross neutralizing antibody titers against the oldest isolate, SVV001/2002, and HI/2012 were low for all challenge groups. The group challenged with SVV001/2002 had low cross neutralizing titers against all other isolates. The groups challenged with contemporary isolates had good neutralizing antibody responses to the contemporary isolates and Can/2011.

Conclusions

This study demonstrated that vesicular disease can be experimentally reproduced in growing pigs with both historic and contemporary isolates of SVA. In addition, the results suggested there were not large differences in clinical presentation between strains, which was contrary to our hypothesis. As expected, higher neutralizing antibody titers were observed against homologous and more closely related strains. Further research will be needed to help determine the cause of the sudden increase in vesicular disease due to SVA infection in the United States swine population.

Serological evaluation of influenza A virus vaccine antibody responses in replacement gilts on twelve breeding farms in the United States

J. Carlson[1], MS; A. L. Vincent[2], DVM, PhD; J. Zhang[1], PhD; P. C. Gauger[1], DVM, PhD
[1]Iowa State University, Ames, Iowa; [2]United States Department of Agriculture/Agricultural Research Service, Ames, Iowa

Introduction

Influenza A virus (IAV) is a component of the porcine respiratory disease complex, causing significant economic losses for United States (US) swine producers. As a result, efforts have been made to monitor the presence and distribution of IAV in US swine. However, serological evaluation to detect IAV antibodies and determine immune status in replacement gilts is lacking. This has important implications for vaccination and acclimation protocols for incoming gilts in sow farms. The objective of this study was to assess the presence of IAV antibody in non-vaccinated, new replacement gilts prior to vaccination and their antibody status post entry into the breeding herd.

Materials and methods

A total of 504 sera were collected from breeding-age gilts prior to initial vaccination (Pre-Vac) from 12 swine breeding farms (11 states) representing 4 geographical regions (3 farms per region). A total of 789 sera were collected post vaccination and after gilts were integrated into the breeding herd. Post vaccination samples were collected over three time periods over one year; Post-Vaccination (Post-Vac), Post-Integration 1 (P-Int 1) corresponding to parity 1, and Post-Integration 2 (P-Int 2) corresponding to parity 2. Serum was evaluated with a nucleoprotein (NP) ELISA (IDEXX AI MultiS-Screen AB, Westbrook, ME) and a hemagglutination inhibition (HI) assay using homologous vaccine antigens (commercial or farm-specific) based on farm vaccination protocols. The HI assay was also used to evaluate antibody against three H1 phylogenetic clusters (γ, δ-1b, δ-2) and three H3 antigenic clades (red, green, and human-lineage) representing the predominant IAV currently circulating in swine.

Results

Collectively, 38.3% (193/504) of the gilts were positive for NP ELISA antibodies, and 50% (6/12) of the farms had at least one NP ELISA positive gilt at Pre-Vac. Influenza HI antibodies to at least one representative antigen were detected in 16.9% (85/504) of the gilts and 41.7% (5/12) farms at Pre-Vac. The farms receiving IAV antibody positive gilts at the Pre-Vac time period included KS, MO, ND, NE, PA, and TX. At Post-Vac, 86.5% (281/325) of the gilts were positive for NP ELISA antibodies excluding the only non-vaccinated ND farm. Additionally, 71.5% (213/289) of the gilts were HI antibody positive to one or more of their vaccine-specific homologous antigens. The percent of gilts with positive homologous HI antibody titers varied widely across all farms based on vaccine type, individual vaccine antigen, and time period. Replacement gilts lacking Pre-Vac IAV antibody demonstrated more uniform vaccine-induced HI antibody responses such as the NC-2 farm which demonstrated > 94.3% of gilts with positive homologous HI antibody titers to all 4 antigens immediately after vaccination (Post-Vac). In contrast, the MO farm demonstrating 87.5% of their gilts with NP ELISA antibody at Pre-Vac had 0.0% of the gilts demonstrating Post-Vac HI antibodies to 3 of 4 homologous vaccine antigens. The TX farm also had 96.0% of gilts NP ELISA positive at Pre-Vac; however, less than 9.1% of gilts demonstrated positive HI antibody titer to all 4 homologous vaccine antigens at Post-Vac. These data suggest new replacement gilts with IAV antibody may induce a non-uniform, inadequate post-vaccination immune response compared to gilts without prior IAV antibody.

Conclusions

These data demonstrate diversity in the IAV antibody status and their IAV vaccine antibody response in replacement gilts prior to and post vaccination. Approximately 50% of the farms lacked IAV NP ELISA antibody positive gilts suggesting the age, multiplier, and geographical location may influence the IAV antibody status of replacement gilts. More importantly, our data suggests naïve gilts demonstrate a more uniform HI antibody response against vaccine antigens and a higher percentage of gilts with HI antibody positive titers compared to IAV antibody positive gilts. In summary, the IAV antibody status of replacement gilts should be monitored using appropriate serological assays pre- and post-vaccination to help monitor IAV immune status and vaccine antibody responses that may suggest changes in vaccination strategies such as the appropriate time of administration, or if vaccine antigen updates are necessary to match IAV strains circulating in the farm and reduce the propensity for endemic infections.

Evaluation of chlorine dioxide virucidal efficacy against porcine epidemic diarrhea virus, porcine reproductive and respiratory syndrome virus, Senecavirus A and influenza A virus

Qi Chen[1]; Julian N. Rosenberg[2]; Douwe F. Mason[2]; Phil C. Gauger[1]
[1]Department of Veterinary Diagnostic and Production Animal Medicine, College of Veterinary Medicine, Iowa State University, Ames, Iowa; [2]The Sabre Companies, Slingerlands, New York

Introduction

Chlorine dioxide (ClO_2) has been proven to be a safe environmental disinfectant to both human and animals, as the residual product lacks any known negative health effects. For the decontamination or disinfection of commercial farms and food processing facilities, DiKlor™ ClO_2 can be generated in high quantities and applied as a liquid or gas treatment. In this *in vitro* study, virucidal efficacy of liquid DiKlor™ ClO_2 was evaluated at different concentrations against four virus species responsible for some of the major porcine diseases. These included porcine epidemic diarrhea virus (PEDV), porcine reproductive and respiratory syndrome virus (PRRSV), Senecavirus A (SVA) and influenza A virus (IAV).

Materials and methods

PEDV was propagated in Vero cells and diluted to 3×10^5 $TCID_{50}$/ml. PRRSV virus was propagated on M145 cells to a concentration of 5.6×10^5 $TCID_{50}$/ml. SVA was propagated on H1299 cells and diluted to a concentration of 5.6×10^7 $TCID_{50}$/ml. IAV was propagated on MDCK cells with a final concentration of 5.6×10^4 $TCID_{50}$/ml. Working stocks of ClO_2 were prepared immediately prior to administering the treatment to avoid prolonged exposure to light. Each concentration of ClO_2 was mixed with the different virus isolates or negative cell culture media at 1:1 volume ratio to achieve final concentrations of 5 ppm, 100 ppm, or 200 ppm CLO_2 within 5 seconds. The mixtures were then incubated at room temperature for either 30 or 60 seconds (s) prior to adding Dey/Engley neutralizing (D/E) broth into the mixture at 1:1 v:v ClO_2:D/E ratio to neutralize the ClO_2 from further reaction with the viruses and prevent cell cytotoxicity. The mixture of virus, ClO_2, and D/E broth was subsequently titrated on cells appropriate for each virus propagation. The mixture of negative culture media, ClO_2, and D/E broth, and mixture of virus and D/E broth without ClO_2 treatment were also titrated on each cell line to evaluate non-specific cytopathic effects (negative control and no-treatment virus control, respectively). Viral specific fluorescent antibody staining assays were performed to confirm virus infection and calculate the virus titrations.

Results

The 100 ppm ClO_2 dose achieved a 1.5-log_{10} reduction in PEDV at both 30s and 60s, whereas 200 ppm ClO_2 reduced PEDV titer 3.4-log_{10} in 30s and 4.5-log_{10} at 60s (Figure 1A). Both 100 and 200 ppm ClO_2 inactivated at least 2.8-log_{10} PRRSV in 60s, which was the limit of detection for PRRSV titration (Figure 1B). Interestingly, non-specific viral inhibition was observed in PRRSV titration at 10-fold dilutions of virus-D/E broth mixture. Thus, PRRSV titer limit of detection was 3×10^2 $TCID_{50}$/ml, and the virucidal efficacy of ClO_2 against PRRSV could only be assessed to 2.8-log_{10} reduction. As for SVA, 100 and 200 ppm ClO_2 inactivated 4.8-log_{10} and at least 6.6-log_{10} SVA in 60s, respectively (Figure 1C). A 2.1-log_{10} reduction was observed with IAV at 100 ppm, and at least a 3.8-log_{10} reduction was achieved with IAV at 200 ppm after 60s (Figure 1D). The virucidal efficacy of ClO_2 against SVA and IAV could only be assessed to 6.6-log_{10} and 3.8 log_{10}, respectively, due to the limit of detection of virus titration. In the 5 ppm viral suspension challenges, ClO_2 treatment did not exceed a 1.1-log_{10} reduction in virus titers at either 30s or 60s contact times.

Discussion

In summary, these data indicate that in vitro liquid treatment with DiKlor™ ClO_2 at 200 ppm can achieve virucidal efficacy against various porcine viruses (PEDV, PRRSV, SVA, and IAV) with a relatively short 60-second contact time. Indeed, notable virus inactivation was also observed with 100 ppm ClO_2 after 60s. Together, these results suggest that ClO_2 is an effective environmental disinfectant for swine facilities and production systems with proven efficacy against the major porcine viruses responsible for disease currently facing the swine industry.

Figure 1: Virus titers reported for A) PEDV, B) PRRSV, C) SVA and D) IAV after in vitro treatment with 5ppm, 100 ppm and 200 ppm of ClO2 for 60 seconds compared to non-treatment controls.

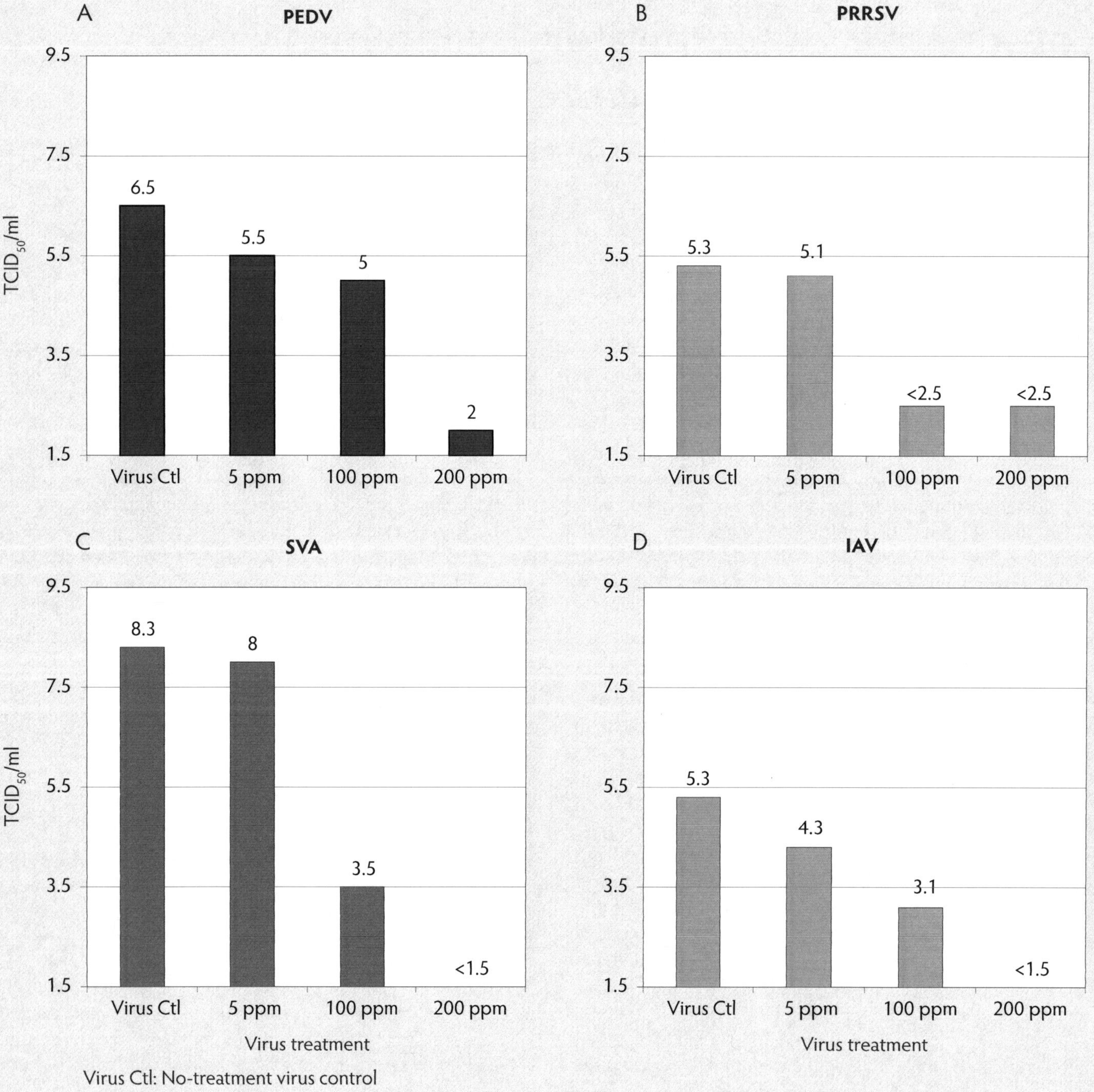

Identification and genetic characterization of a novel deltacoronavirus

Qi Chen[1]; Chenghuai Yang[1]; Ying Zheng[1]; Phil Gauger[1]; Tavis Anderson[2]; Karen Harmon[1]; Jianqiang Zhang[1]; Kyoung-Jin Yoon[1]; Rodger Main[1]; Ganwu Li[1]
[1]Veterinary Diagnostic and Production Animal Medicine, College of Veterinary Medicine, Iowa State University;
[2]National Animal Disease Center, USDA-ARS, Ames, Iowa

Case description and methods

During routine health surveillance testing at a grow-finish swine farm in Illinois, porcine DCoV (PDCoV) was detected weakly positive (Ct 34.2-39) using the ISUVDL real-time RT-PCR (rRT-PCR) targeting the nucleoprotein gene in oral fluid samples from clinically healthy pigs. Additional testing was conducted on pig rectal swabs, feed delivered to the farm, and feces collected from pigs and terrestrial wild birds located inside a hoop facility. Interestingly, several avian fecal samples were tested positive (Ct 26.3-37.3) using the ISUVDL PDCoV rRT-PCR, but pig rectal swabs and feed samples were negative. Full-length sequencing was attempted on the positive avian fecal samples for further characterization of the virus using NGS technology.[1]

Results

A genome of 25,842 nucleotides was *de novo* assembled, and the genome queried against the GenBank database using the basic local alignment search tool (BLAST). This revealed the genome to be most closely related to viruses in the genus *Deltacoronavirus*. The virus, designated DCoV USA/IL/ISU690/2017 (hereafter ISU690), shared 87% and 88% identity with PDCoV HKU15/155 (JQ065043) and a sparrow DCoV HKU17-6124 strain (JQ065045), respectively. The genome shared limited identity with other DCoVs: magpie-robin coronavirus (79%), bulbul coronavirus (80%), white-eye coronavirus (80%), and thrush coronavirus (80%). Phylogenetic analysis based on the whole genome sequences of alpha-, beta-, gamma-, and deltacoronaviruses yielded a topology consistent with established relationships in the *Coronavirinae* subfamily. This dendogram revealed that the ISU690 strain shared a common ancestry with both PDCoV and avian DCoVs, but the ISU690 strain was on an independent branch, suggesting it is a novel DCoV different from currently described porcine and avian DCoV strains (Figure 1). Several genes (S, E, M, and NS6) of ISU690 showed the highest identity to PDCoV (72%-93%) while other genes (ORF1ab, N, and NS7) showed the highest identity to sparrow DCoV (87%- 99%): this pattern of discordant similarity suggests that this novel DCoV may have emerged through a recombination event. Since DCoV in pig samples had high Ct values and sequencing was unsuccessful, it is unknown whether the DCoV detected in pig samples was derived from the avian DCoV ISU690 strain or not. Sequence alignment revealed that PDCoV rRT-PCR primers/probe are highly homologous to the corresponding sequence of the ISU690, so it is plausible that the positive PDCoV rRT-PCR results from sparrows were due to the DCoV ISU690 cross-reacting with PDCoV rRT-PCR. This incidence raised a concern that the specificity of diagnostic assays for PDCoV can be compromised using environmental samples such as feces collected from floors and oral fluids, which can be contaminated by other DCoVs.

Conclusions

We identified a novel deltacoronavirus (DCoV) using NGS. More surveillance and sequence data are needed in order to understand the potential interspecies transmission of DCoVs. In light of our observation, further follow-up may be desired when unexpected PCR results are observed or when clinical signs are inconsistent with clinical signs.

References

1. Q Chen et al. (2016). Pathogenesis comparison between the United States porcine epidemic diarrhoea virus prototype and S-INDEL-variant strains in conventional neonatal piglets. Journal of General Virology (2016), 97, 1107–1121.

Figure 1: Phylogenetic tree based on 58 representative coronavirus complete genomes.

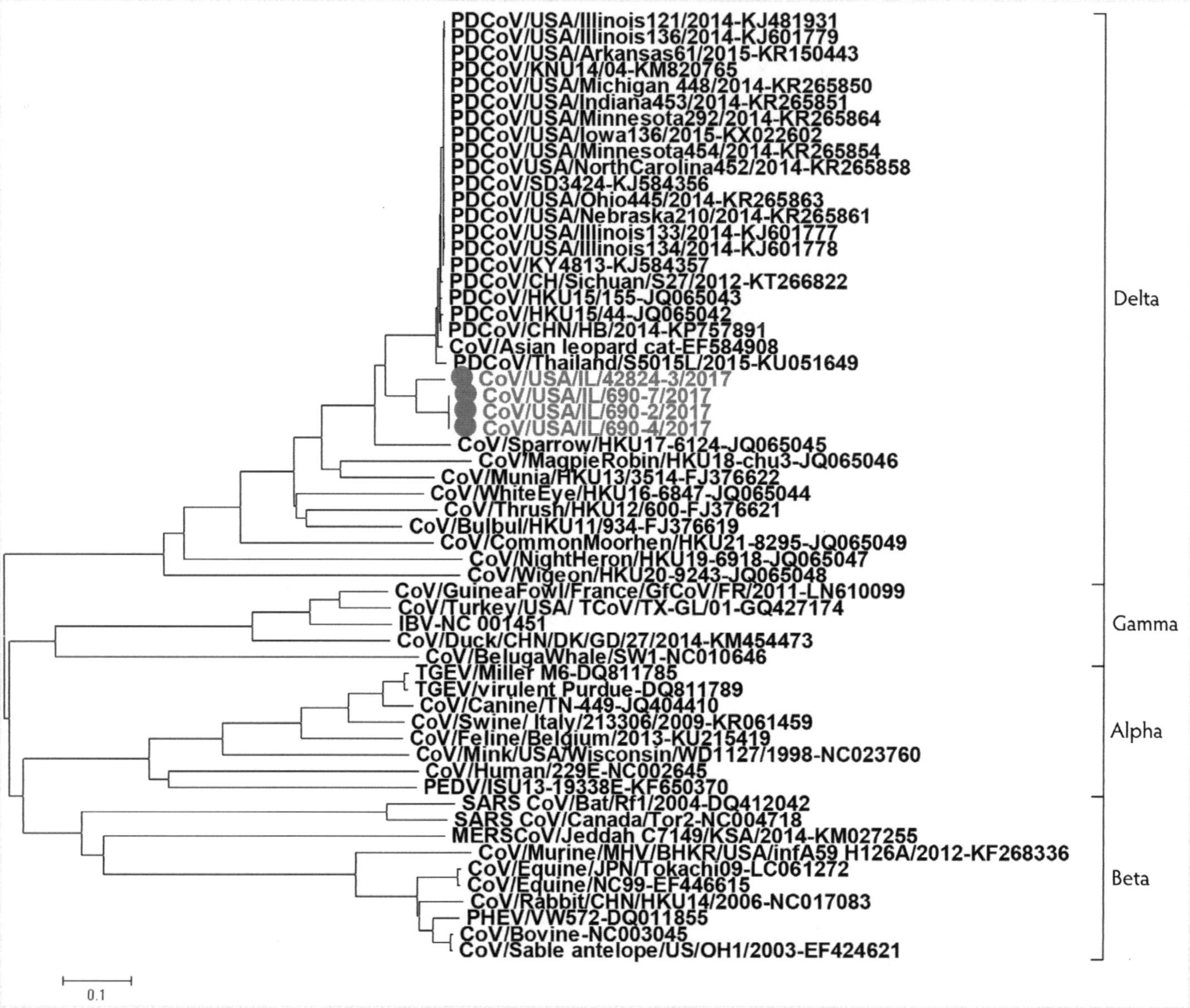

Effect of chemical clarification of oral fluids on the detection of PRRSV IgG

Y. A. Henao-Díaz[1], DVM; L. Giménez-Lirola[1], PhD; K. Poonsuk[1], DVM, PhD; C. Wang[2], PhD; J. Ji[2], MS; T. Y. Cheng[1], DVM, MS; R. Main[1], DVM, PhD; J. Zimmerman[1], DVM, PhD
[1]College of Veterinary Medicine, Iowa State University, Ames, Iowa;
[2]College of Liberal Arts and Sciences, Iowa State University, Ames, Iowa

Introduction

Oral fluid antibody ELISAs combine the advantage of a welfare-friendly, aggregate specimen with results highly correlated with serum antibody responses.[1] However, oral fluid samples often contain feed, feces, and other environmental contaminants that could affect pipetting accuracy and/or test performance. Therefore, the objective of this study was to improve the handling characteristics of oral fluid specimens by removing particulates suspended in the sample using chemical clarification (chitosan). In addition, the impact of chemical clarification on antibody diagnostic stability over time was evaluated using a PRRSV oral fluid ELISA.

Methods

Oral fluid samples of known status were generated by vaccinating PRRSV-negative pigs (n = 17) with PRRSV MLV vaccine (Ingelvac®, Boehringer Ingelheim Vetmedica, GA). Oral fluid specimens were collected from individual pigs on day post vaccination (DPV) -7, -5, -3, 0, 3, 6, 7, 8, 9, 10, 14, 21, 28, 35 and 42, and then subdivided into 4 aliquots. Each aliquot was treated with one of 3 lyophilized chitosan-based "clarifying" formulations (A, B, and C), with the 4th aliquot serving as an untreated control. Aliquots were tested using a commercial PRRSV oral fluid ELISA (IDEXX Laboratories, Inc.) immediately after treatment (day post treatment 0). Thereafter, samples were held at 4°C and re-tested on DPTs 2, 4, 6, and 14. Both immediate and temporal treatment effects of oral fluid treatment were analyzed separately for oral fluid samples collected prior to the expected appearance of detectable antibody (≤ DPV 7) and oral fluids collected after the expected appearance of antibody (≥ DPV 14). Quantitative (sample-to-positive ratio) and qualitative (positive vs negative) effects on the PRRSV oral fluid ELISA results were evaluated using nonparametric statistical analyses.

Results

Statistical analyses (Kruskal-Wallis test and Cochran's Q test) detected neither immediate (DPT 0) nor residual effects (DPT 2, 4, 6, and 14) of treatments on PRRSV oral fluid ELISA quantitative or qualitative results.

Conclusions

The chitosan-based formulations evaluated in this experiment visually reduced the level of particulates in oral fluid samples, but did not affect the PRRSV oral fluid antibody ELISA results. The results suggest that this, or similar, treatments could be used in the field to improve oral fluid handling characteristics without affecting testing results.

References

1. Kuiek AM, el at. Comparison of serum and oral fluid antibody responses after vaccination with a modified live (MLV) porcine reproductive and respiratory syndrome virus (PPRSV) vaccine in PRRS endemic farms. *Trop Anim Health Prod.* 2015;47:1337.

Novel behavioral analysis of piglets treated with Metacam®thirty minutes before and at time of castration

J. Law[1], DVM; G. Cunningham[2], BSc, DVM, MVetSc; N.J. Cook[3], PhD
[1]Prairie Swine Health Services, Red Deer, Alberta, Canada; [2]Boehringer Ingelheim Canada Ltd.;
[3]Alberta Agriculture and Forestry, Lacombe Research and Development Centre, Lacombe, Alberta, Canada

Introduction

The use of pain control in swine is a key tool as our industry continues to maximize pig welfare and comfort. Research regarding pain control and mitigation in piglets during surgical procedures is ongoing and there is a need for objective behavioral tools to further analyze the effectiveness of available options. A proof of concept was published using a handling chute as a novel behavioral tool to analyze castration pain in piglets.1 Our objective was to determine if Metacam®(5 mg/ml) is effective for post-operative pain control when given at castration, or 30-minutes prior to castration using the novel behavioral test validated at the Prairie Swine Centre.

Materials and methods

The study was approved by the Animal Care Committee of the Lacombe Research Centre and animals were cared for in accordance with the guidelines set out by the Canadian Council on Animal Care in Science.[2] Ninety-five piglets 6-8 days of age from the Lacombe Research and Development Centre were randomly placed into six treatment groups. Uncastrated groups: A (saline at sham castration), B (Metacam® at sham castration) and E (Metacam® 30 minutes prior to sham castration). Castrated groups: C (Saline at castration), D (Metacam® at castration) and F (Metacam® 30-minutes prior to castration). Piglets were trained prior to the trial in the handling chute which was 1.5 meters long and contained two hurdles that were each 10 centimeters tall. Four baseline runs were timed to establish individual baseline averages. Ninety minutes after the baselines were completed, the piglets were collected, and groups E and F were treated with Metacam®. All piglets were timed through the handling chute 15 minutes prior to castration. At castration or sham-castration the remaining groups were treated with either 0.08 ml/kg of saline or Metacam® (5mg/ml). Castration was performed in accordance with standard commercial practice using one horizontal cut in the scrotum and tearing the spermatic cord.[3] Immediately after castration (T0) the piglets were run through the chute, timed and repeated at 15 minutes (T15), 30 min (T30), 1hr (T60), 2hrs (T120) and 24hrs (T1440). The results were analyzed using the time recorded at each chute run minus the baseline average. A positive result means the piglet/group was slower than the baseline average, whereas a negative result means the piglet/group was faster than the baseline average. The model analyzed included the fixed effects of treatment (castration, sham, Metacam® and Metacam® 30 and their interactions) and the random effects of piglet and litter. Statistics completed in IBM® SPSS® Statistics version 22, 2013.

Results and discussion

The piglets given Metacam® at castration (group D) and the piglets given Metacam® 30-minutes prior to castration (group F) showed no significant differences between group at any time at, or after castration ($P > 0.10$) (Figure 1). Group C (castrated, saline) was significantly slower ($P < 0.05$) when compared to all other groups at T15 and T30.

There were no significant differences between group D (Metacam® at castration, castrated) and group F (Metacam® 30-minutes before castration, castrated). This indicates that regardless of when Metacam® is administered as an analgesic for castration, it provides measurably improved pain control. Group C was the positive control group and the significant increase in time through the chute for the time points measured after castration indicates the piglets in treatment group C were experiencing significantly more pain according to this novel behavioral test. This research contributes to the data validating the handling chute model for behavioral analysis of castration pain.

Metacam® treatment at the time of castration or 30-minutes prior provided increased piglet comfort according to this model.

References

1. Bilsborrow K et al. An investigation of a novel behavioral test to assess pain in piglets following castration. Can. Vet. J. 2016; 96: 376-385.

2. Canadian Council on Animal Care in Science 2009. CCAC guidelines on: the care and use of farm animals in research, teaching and testing. Canadian Council on Animal Care, Ottawa, ON, Canada.

3. Rault JL et al. Castration induced pain in pigs and other livestock. Anim Behav Sci. 2011;135(3):214-225.

Figure 1: Mean time (time minus baseline average) at each time point for groups E and F. There were no significant differences between groups E and F at or after T0 ($P > 0.10$). A positive result means the piglet was slower than baseline average, whereas a negative result means the group was faster than their baseline average.

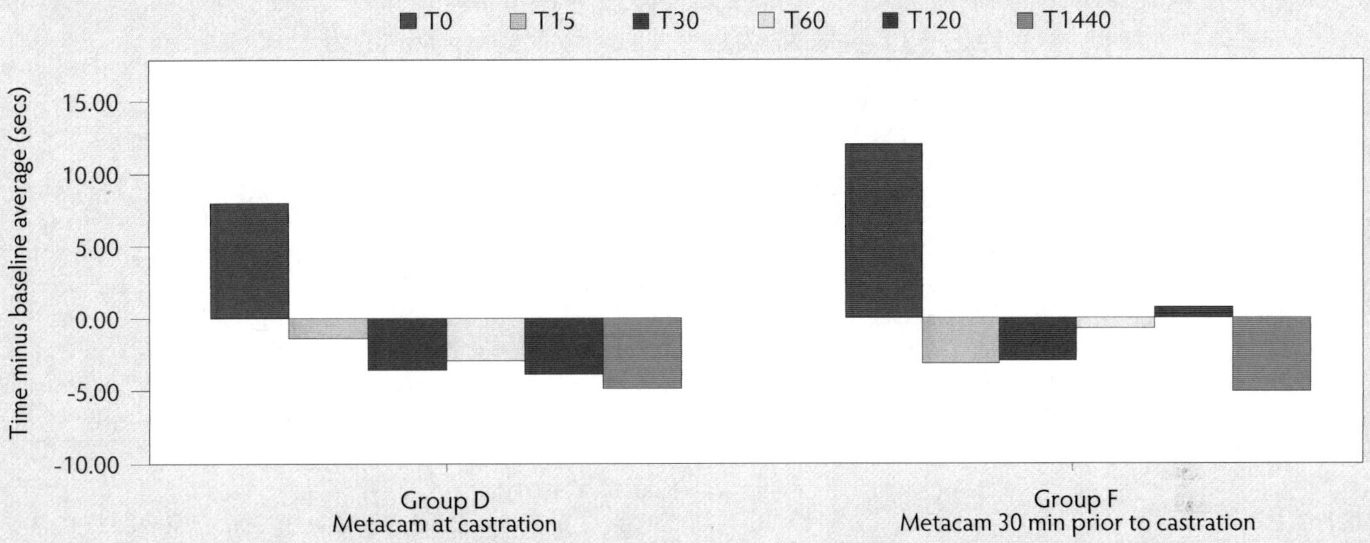

Porcine parvovirus type 2 associated with lymphocytic perivascular lesions in nursery and finishing pigs

Lacey Marshall Lund; Stephanie A Rossow; Fabio A Vannucci
University of Minnesota, St. Paul, Minnesota

Introduction

PPV2 has been identified in various countries with high prevalence in the US swine population, however its clinical relevance remains to be elucidated.[1,2] Epidemiological studies have largely focused on the detection of PPV2 by PCR. Additionally, the interpretation of PCR-positive results becomes more challenging with the presence of co-infections which may contribute to the clinical presentation. The objective of this study was to investigate the presence of PPV2 mRNA and its association with perivascular lesions using *in situ* hybridization (ISH).

Methods

An antisense probe was chosen to target PPV2 ORF mRNA (2801-3986nt) representing transcriptionally active virus. Formalin-fixed paraffin embedded (FFPE) tissues from cases submitted to the University of Minnesota Veterinary Diagnostic Lab (UMN-VDL) were retrospectively selected for ISH testing based on their PPV2 PCR results. FFPE samples that were PCR-negative for PPV2 but positive for other agents (including PPV1, PPV4, PPV6 and PPV7) were also tested to evaluate probe specificity.

Results and discussion

PPV2 PCR revealed lower Ct values ranging from 15 to 22 in samples with lymphocytic perivasculitis. The distribution of PPV2 mRNA in PCR-positive samples showed an association with the presence of lymphocytic perivasculitis and was located in the lymphocytic infiltrates (lung, joint and subcutaneous tissues) and within the cytoplasm of endothelial cells (Figure 1). Interestingly, PPV2 was also observed in lymphoid follicles of the lymph nodes and broncho-associated lymphoid tissues.

Conclusions

The present study is the first to describe the presence of PPV2 in association with systemic perivascular inflammation. This finding may represent an important advance for understanding the potential role of PPV2 in systemic syndromes in nursery and finishing pigs.

References

1. Hijikata M, Abe K, Win KM, Shimizu YK, Keicho N, Yoshikura H. Identification of new parvovirus DNA sequence in swine sera from Myanmar. Jpn J Infect Dis. 2001;54(6):244-245.

2. Xiao CT, Gerber PF, Giménez-Lirola LG, Halbur PG, Opriessnig T. Characterization of porcine parvovirus type 2 (PPV2) which is highly prevalent in the USA. Vet Microbiol. 2013;161(3-4):325-330.

Figure 1: A) Lung, perivascular lymphoplasmocytic infiltrate, B) Lung, PPV2 mRNA (red) peribronchiolar and perivascular, C) Subcutaneous tissue, PPV2 mRNA (red) within endothelial cells

Age susceptibility of pigs to *Mycoplasma hyorhinis* challenge

Brian Martinson[1,2]; F. C. Minio[1]; Jeremy Kroll[2]; Joseph Hermann[2]
[1]Interdepartmental Microbiology, Veterinary Microbiology and Preventive Medicine, Iowa State University, Ames, Iowa;
[2]Swine Research & Development, Boehringer Ingelheim Animal Health, Ames, Iowa

Introduction

Mycoplasma hyorhinis (MHR) causes disease in three- to ten-week old animals but most commonly causes polyserositis and polyarthritis in pigs around seven weeks of age. Economic losses due to MHR can be attributed to reduced performance (weight gain and feed conversion) and culling. Literature has suggested an age limitation to the susceptibility of MHR. A robust experimental infection protocol for MHR has been established in seven-week old caesarian derived-colostrum deprived (CDCD) pigs which results in severe pericarditis and lameness.[2] In this study, we evaluated this MHR challenge model in the same litters of animals at seven, ten, thirteen, and sixteen weeks of age.

Materials and methods

One hundred-twenty CDCD pig were blocked by litter and randomly assigned to one of four groups (n = 25 each) for experimental challenge or one of four groups (n = 5 each) as non-infected, age-matched controls. Pigs in Group 1 were inoculated with MHR at 20 mL/intraperitoneal injection (IP) on Day 0 (at 7 weeks of age), 10 mL/intravenous injection (IV) on Day 1, and 10 mL/intranasal application (IN; 5 mL/nostril) on Day 2, for a total administered dose of 5.73 × 10^9 CCU/pig. Pigs in Groups 2, 3 and 4 were inoculated with MHR in identical fashion beginning on Day 21 (at 10 weeks of age; 2.55 × 10^9 CCU/pig), Day 42 (at 13 weeks of age; 3.72 × 10^9 CCU/pig), and Day 63 (at 16 weeks of age; 4.26 × 10^9 CCU/pig), respectively. Pigs in Groups 5-8 remained uninoculated. Pigs were observed daily for lameness during their respective challenge periods, from the first challenge day (0dpc) through 21 days post challenge (21dpc). Weight measurements were taken just prior to challenge on 0dpc and just prior to euthanasia on 21 dpc. At 21dpc, challenged and age-matched negative control pigs were euthanized and a gross pathological examination was performed. The thoracic and abdominal cavities were opened and examined for evidence of polyserositis. Pericardial swabs (eSwab™, Copan; Murrieta, CA) were taken from all pigs. Pleural swabs (eSwab™) were taken from only those animals exhibiting pleuritis. Serosal swabs (eSwab™) were taken from only those animals exhibiting peritonitis. For each pig, a single swab (eSwab™) was used to sample the articular surfaces of both elbows and both stifles, and another swab for the carpi and tarsi. All swabs were tested for MHR by real-time PCR.

Results and discussion

Lameness was observed in over 60% of the animals in the first three age groups but fell to 33% in the oldest group. The number of animals with arthritis fell from 100% at seven weeks, to 56% at ten weeks and approximately 25% at both thirteen and sixteen weeks of age. Pericarditis was observed in 87% of the seven week challenge group, 28% in the ten week challenge group, 8% in the thirteen week challenge group and 4% in the sixteen week challenge group. Over 80% of the joint swabs were PCR positive at seven weeks of age but fell to less than 40% in subsequent challenges. Less than 30% of the pericardial swabs from all challenged groups were PCR positive. All samples collected from each non-challenged group were negative for MHR by PCR. All challenged groups showed an average daily gain (ADG) of at least 0.6 lbs/day less than their age-matched non-challenged control group, with the largest margin noted at thirteen weeks of age with a 1.2 lbs/day difference.

Results of this study show that animals were susceptible to MHR-associated lameness through sixteen weeks of age while susceptibility to MHR-associated polyserositis appeared to peak at seven weeks of age.

References

1. Martinson et al. Age susceptibility of caesarian derived colostrum deprived pigs to *Mycoplasma hyorhinis* challenge. *Vet Microbiol.* 2017; 210:147-152.

2. Martinson et al. Development and optimization of a cell-associated challenge model for *Mycoplasma hyorhinis* in seven-week old caesarean derived-colostrum deprived pigs. *Can J Vet Res.* 2017; in press.

Fecal microbiota transplantation improves outcome in pigs following co-infection with porcine reproductive and respiratory syndrome virus and porcine circovirus type 2d

M. C. Niederwerder[1,2], DVM, PhD; R. R. R. Rowland[1], PhD; L. A. Constance[1], MS; S. C. Fernando[3], PhD; M. L. Potter[4], DVM, PhD; M. A. Sheahan[1], MS; T. Burkey[3], PhD; R. A. Hesse[1,2], PhD; A. G. Cino-Ozuna[1,2], DVM, PhD, DACVP

[1]Department of Diagnostic Medicine/Pathobiology; [2]Kansas State Veterinary Diagnostic Laboratory, College of Veterinary Medicine, Kansas State University, Manhattan, Kansas; [3]Department of Animal Science, University of Nebraska, Lincoln, Nebraska; [4]Abilene Animal Hospital, Abilene, Kansas

Introduction

Co-infections involving porcine reproductive and respiratory syndrome virus (PRRSV) and porcine circovirus type 2 (PCV2d) are common on a global scale and contribute to significant losses in swine production due to poor weight gain and respiratory disease in growing pigs. The diversity and composition of the gut microbiome, or collection of microorganisms living in the gastrointestinal tract, has been shown to impact outcome following PRRSV/PCV2 co-infection in nursery pigs. Previous work by our group has demonstrated a consistent association between increased microbiome diversity and improved outcome characteristics following co-infection, including reduced virus replication, improved weight gain, and decreased clinical disease. In the current study, we sought to determine the effects of modulating the microbiome through fecal microbiota transplantation (FMT) on outcome following co-infection.

Materials and methods

Sibling pairs from a high health commercial source were divided into two groups at weaning and housed separately. The FMT group (n = 10) was administered the transplant for 7 consecutive days prior to co-infection and the control group (n = 10) was administered a mock-transplant with sterile saline over the same time period. All pigs were co-infected with PRRSV and PCV2d and followed for 42 days post-infection (dpi).

Results and discussion

Clinical signs of porcine circovirus associated disease (PCVAD), such as muscle wasting, dyspnea, tachypnea, pallor, lethargy, rough hair coat and nasal discharge, peaked between 18 and 30 dpi. Morbidity and mortality due to PCVAD was significantly higher in the control group compared to the transplanted group; 70% mortality compared to 20% mortality was seen in the control and transplanted groups, respectively ($P = 0.0447$, Mantel-Cox test). The majority of control pigs lost weight (70%) whereas only 30% of FMT pigs had evidence of weight loss at any time during the study. On 21 and 28 dpi, increased PCV2d replication and increased PRRSV replication, respectively, were detected in the serum of control pigs. Seroconversion for PRRSV and PCV2 was detected with the FMT pigs having higher antibody titers for both viruses. Upon examination at necropsy, the FMT group had reduced severity of lesions characteristic of PRRSV/PCV2 co-infection. Overall, this study provides evidence that microbiome modulation through FMT prior to co-infection with PRRSV and PCV2d improves outcome and suggests that a microbiome therapeutic may be an alternative method for disease control.

Acknowledgements

This work was supported by the Kansas State University College of Veterinary Medicine Success for Young Investigators Grant Program, start-up funds provided by the Kansas State University College of Veterinary Medicine, and the State of Kansas National Bio and Agro-Defense Facility Fund.

Increased microbiome diversity at the time of infection is associated with improved growth rates of pigs after co-infection with porcine reproductive and respiratory syndrome virus (PRRSV) and porcine circovirus type 2 (PCV2)

Rebecca A. Ober; James B. Thissen; Crystal J. Jaing; Ada G. Cino-Ozuna; Raymond R. R. Rowland; Megan C. Niederwerder

Introduction

Pork is the most widely consumed protein around the world and global pork production is forecast to reach a record of 111 million tons in 2017 (USDA 2016). Worldwide, porcine reproductive and respiratory syndrome virus (PRRSV) and porcine circovirus type 2 (PCV2) are two of the most significant pathogens affecting the swine industry, costing billions in lost production over the last 3 decades. The microbiome, or collection of microorganisms within the gastrointestinal tract, has been associated with outcome during systemic viral infections in both humans and mouse models. The role of the microbiome and its impact on the response to systemic infections in swine is an emerging area of study and the focus of our work. To investigate the porcine microbiome, we utilized the Lawrence Livermore Microbial Detection Array (LLMDA), which was developed to detect all known microbes for which whole genome sequences are available.

In a recent study, we confirmed a potential role for the microbiome in outcome following PRRSV/PCV2 co-infection, important gaps in our knowledge remained, such as microbiome characteristics that predispose outcome and the role of the microbiome in subclinical infections. In the current study, we investigated the early microbiome properties that predispose high and low growth rates after co-infection in subclinically affected pigs. Our results demonstrate that both microbiome diversity and composition prior to infection play a role in weight gain following PRRSV/PCV2 co-infection.

Materials and methods

At approximately 8 weeks of age (average age 54.5 ± 2.6 days), 50 pigs were infected with PRRSV and PCV2b. The viruses were recombined to yield a 2-mL dose consisting of 103.6 TCID50 PCV2b and 105 TCID50 PRRSV in MEM. Body weights of individual pigs were collected upon arrival and on 0, 7, 14, 21, 28, 35, and 42 days post-infection (dpi). Fecal samples were collected from all 50 pigs during the week prior to co-infection. At 35 dpi, 20 pigs were selected to represent high growth rate pigs (n = 10) and low growth rate pigs (n = 10) based on average daily gain. Pigs in the high growth rate group had the highest ADG and pigs in the low growth rate group had the lowest ADG.

Results and discussion

Three families had notable differences in the number of species detected between the two groups. First, there was a trend towards greater species diversity of the high growth rate pigs in the *Streptococcaceae* family; 4.2 species compared to 2.8 species detected in the low growth rate group ($P = 0.0998$, Mann-Whitney U test). Overall, 12 Streptococcus species were detected in the feces of these pigs. Second, there was a significant decrease in the species diversity of the *Methanobacteriaceae* family in the high growth rate group. An average of 1.4 *Methanobacteriaceae* species were detected in the high growth rate pigs versus an average of 2.5 species detected in the low growth rate pigs ($P = 0.0086$, Mann-Whitney U test). Third, an increased number of *Ruminococcaceae* species were detected in the high growth rate group; 1.5 species versus 1.1 species in the low growth rate group ($P = 0.0573$, Mann-Whitney U test). This data suggests that microbiome composition at the time of virus exposure may play a role in determining outcome during co-infection. Microbiome diversity was determined by calculating the number of families and species detected in the feces of each pig. The mean number of families detected in the low growth rate group was 13.1 with a range of 9-16 compared to a mean of 15.1 with a range of 10-18 for the high growth rate group. The difference in family diversity between the two groups was significant ($P = 0.0385$, Mann-Whitney U test). Species diversity between the two groups followed a similar trend. The mean number of species detected in the low growth rate group was 32.7 with a range of 29-37 whereas the high growth rate group had a mean of 36.9 species with a range of 29-42. The difference in species diversity was also significant ($P = 0.0328$, Mann-Whitney U test). Overall, high growth rate pigs had greater microbiome diversity, on both a family and species level, than low growth rate pigs. This data suggests that increased microbiome diversity may contribute to improved weight gain in pigs after co-infection.

Figure 1: Fecal microbiome diversity in pigs with high and low growth rates after co-infection with PCV2 and PRRSV. Data is shown as the total number of microbial families **(A)** and microbial species **(B)** detected by DNA microarray on 0 days post-infection (dpi) for individual pigs. Group means and standard deviations are represented by horizontal lines. Significantly greater microbiome diversity was detected on both a family and species level in high growth rate pigs (*$P = 0.0385$ and $P = 0.0328$, respectively; Mann-Whitney U test).

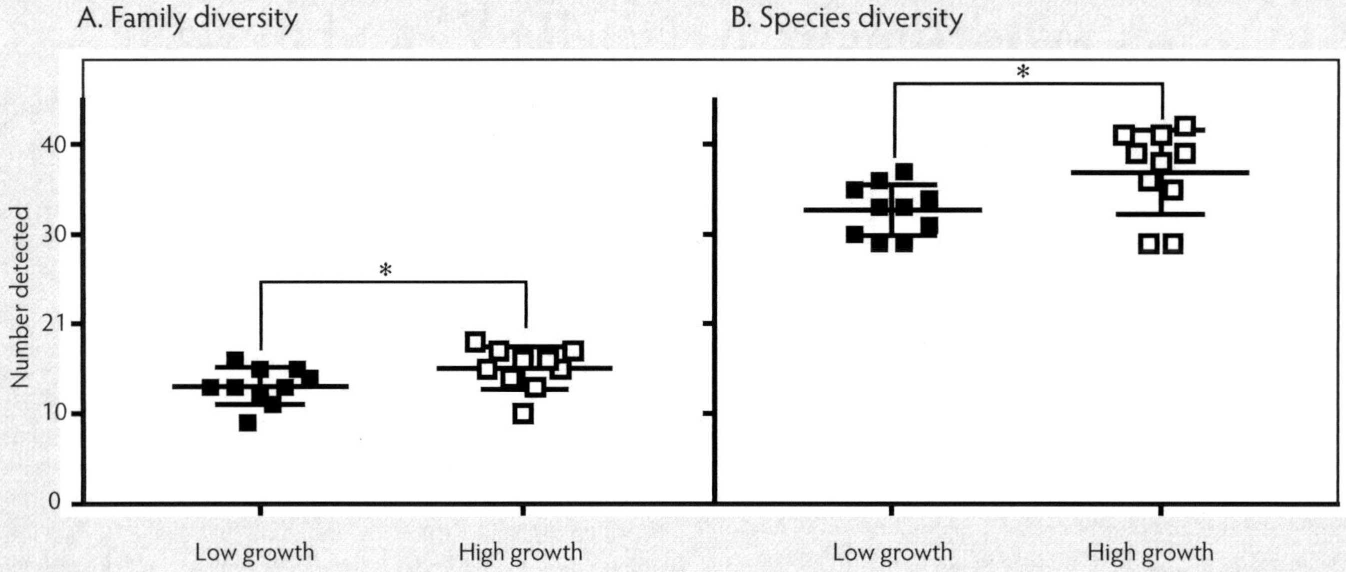

(Ober et al.2017)

Short summary on the low prevalence of PCV2 on pigs at birth

L. Ochoa[1], PhD; L. Greiner[1], PhD; J. Connor[1], DVM; J. Waddell[2], DVM; E. Fano[2], DVM
[1]Carthage Innovative Swine Solutions, LLC, Carthage, Illinois; [2]Boehringer Ingelheim Vetmedica Inc.

Introduction

During the summer of 2016, several commercial sow farms located in the US Midwest were sampled in search of pigs born viremic to PCV2 to be used for a study. The lower than expected PCV2 prevalence observed in this sampling motivated us to report the following summary.

Materials and methods

The collection of samples was conducted in 2016 between June 7 and July 22nd as part of a selection process to identify a farm to source pigs from for a study looking at different effects of PCV2 viremia at birth on the performance of pigs. Based on PCV2 history from PUCS and other tissues (i.e. fetal tissues), farms were expected to have a high (> 30%) prevalence of pigs being born positive to PCV2. Farms were screened one farm at the time, moving to the next farm if the PCV2 results did not meet the requirements for the study. A total of 12 farms were used in the first screening. Farms varied on purpose (i.e., commercial or multiplication sites) and health status as follows: 1) PRRS status: 7 farms were negative, 3 PRRS MLV vaccinated, and 2 positive (wild type); 2) Purpose: 8 commercial sites and 4 multiplication sites. Size of farms ranged from 2,500 to 6,000 sows and with all farms being PEDV negative. For each farm, sampling was conducted on a single day (i.e., 1 day/farm), using litters from sows actively farrowing, which resulted in different number of litters being used between farms (6 to 19 litters per farm). Blood samples were collected from piglets immediately after birth (before colostrum intake) via puncture of the jugular vein (25 to 78 pigs per farm). At time of birth, pigs were bled immediately or briefly placed in plastic totes until they were bled to avoid colostrum intake. After each day of collection, samples were submitted to BI HMC lab and individually analyzed for PCV2 via PCR.

Results and discussion

The number total number of samples collected from each farm as well as the number of positive and negative results are presented in Table 1. From a total of 469 samples submitted for analysis, only 1 sample was PCV2 positive. These results were contrary to expectations and indicated an extremely low prevalence of PCV2 viremic newborn pigs within the tested production system. Interestingly, similar results were obtained by Knox (2017) who showed a 0% PCV2 prevalence from newborn piglet serum samples (pre-colostrum intake; pooled per litter; N = 60 litters) in a study conducted in 2016 in a different farm within the same production system. Moreover, a study conducted by Schulte (2017) on a different production system also showed a 0% prevalence from piglet serum samples (≤ 4 h post-birth; pooled per litter; N = 78 litters).

Conclusion

Results suggest that the current prevalence of PCV2 viremia in pigs at birth is potentially low across production systems in the US Midwest. Further studies will be needed to better understand the apparent reduction in the prevalence of PCV2 in piglets at birth.

References

1. Knox A. et al: Correlating sampling methods to porcine circovirus type 2 (PCV2) viremia status of sows and piglets in a sow herd with or without PCV2 vaccination during gestation. AASV. 2017;69.

2. Schulte R. et al: Sample comparison for detecting porcine circovirus type 2 in breeding herds. AASV. 2017;72.

Table 1: Description of farms (i.e., health status and type of farm), number of litters and pigs sampled, and results of PCR PCV2 analysis of piglets sampled.[1]

Farm ID	Herd PRRS status	Farm type	Litters	Pigs	PCV2 qPCR result	
					Positive	Negative
1	Negative	Commercial	12	34	0	34
2	Negative	Commercial	10	40	1 (CT 21)	39
3	Negative	Commercial	18	41	0	41
4	Negative	Multiplication	17	39	0	39
5	Negative	Multiplication	11	38	0	38
6	Negative	Multiplication	2	5	0	5
7	Negative	Multiplication	12	78	0	78
8	MLV	Commercial	14	42	0	42
9	MLV	Commercial	6	25	0	25
10	MLV	Commercial	19	45	0	45
11	Positive	Commercial	15	41	0	41
12	Positive	Commercial	14	41	0	41
		Total	150	469	1	468

[1] Blood samples were collected from newborn pigs, immediately after birth and before suckling colostrum.

Effects of PCV2-viremia level of piglets at birth on the efficacy of PCV2 vaccine against exposure to PCV2-b determined by viremia levels and growth rate of pigs

L. Ochoa, PhD; L. Greiner, PhD; J. Connor, DVM; T. Pacion
Carthage Innovative Swine Solutions, LLC, Carthage, Illinois

Introduction

The study was carried out over a 12-week period to look at the effects of PCV2-viremia level of piglets at birth on the efficacy of a PCV2 vaccine against exposure to PCV2-b determined by viremia levels and overall growth rate of pigs.

Materials and methods

A total of 50 barrows and gilts (average BW = 8.4, SD 0.37 kg). The study was conducted as a RCBD (Block = Body weight) with a 3 × 2 factorial of the following treatments: 1) qP-CR-PCV2 viremia levels at birth: Negative vs. Low (≤ 103 DNA copies of PCV2) vs. High (> 103 DNA copies of PCV2); and 2) PCV2 vaccination: Placebo vs. Vaccine (CircoFLEX®). Pigs were sourced from a commercial sow farm. Pigs were ear tagged and bled via jugular puncture immediately after birth (before colostrum intake) to determine their PCV2 status via qPCR. Based on PCV2 status at birth (identified as PCV2a by ORF2 sequencing), viremia in pigs was classified as Negative, Low or High. At approximately 18 days of age, pigs were weighed (wean weight). Two days after weighing, 48 pigs of similar BW were selected (16 per PCV2 class) and were bled and transported to the Carthage Veterinary Research Facility (BSL2 facility; Champaign, IL) where the study was conducted. Upon arrival, 6 pigs of similar BW (2 per PCV2 class), were randomly assigned to a pen, and vaccine treatment was randomly assigned within pen to each pair of pigs of the same PCV2 class. One day after arrival, PCV2 vaccine or placebo was administered to pigs intramuscularly according to treatment. Pigs were individually weighed on days 23, 41, 63 and 84 after arrival. Blood samples were collected at start and end of study and every time pigs were weighed and were submitted to BI HMC lab and individually analyzed for PCV2 via qPCR. On day 23 post-arrival, all pigs were given an intranasal and intramuscular dose (1 ml each) of PCV2-b isolate (104.5 TCID50). Pigs were fed standard diets formulated to meet or exceed NRC (2012). Feed and water were offered ad libitum throughout the study period. Growth data was analyzed using the PROC MIX procedure of SAS with pigs as the experimental unit.

Results and discussion

Results are presented in Table 1. High birth viremia pigs tended ($P = 0.09$) to be heavier at birth (1.56 kg) than pigs born negative or with low levels of PCV2 (1.41 and 1.38 kg, respectively). Post-wean growth data was adjusted to a common wean weight of 6.22 kg. There were no significant ($P < 0.10$) treatment interactions. However, the positive effects of vaccination tended to be bigger ($P = 0.6$ and $P = 0.12$) within Negative pigs for ADG d23 to 84 and final BW, respectively, compared to vaccine effects in Low and High viremic pigs. Body weight at end of study (d 84) for Negative (72.4 kg), Low (70.4 kg), and High (71.1 kg) viremic pigs was not statistically different ($P = 0.50$). There was a numeric trend ($P = 0.15$) for Vaccinated pigs to be heavier (72.3 kg) than Placebo pigs (70.3 kg) at the end of the study. There were not statistical differences or trends ($P \leq 0.05$ and $P \leq 0.10$, respectively) for average daily gain for any of the interim periods between viremia class treatments. Vaccinated pigs tended ($P = 0.106$) to grow faster (0.92 kg/d) than Placebo pigs (0.88 kg/d). All pigs were PCV2 negative at time of weaning regardless of viremia levels at birth. The PCV2 challenge was successful based on PCV2 viremia levels in samples collected post-exposure on d 24. In general, both the number of PCV2 positive animals and their qPCR values were numerically lower for vaccinated pigs, which support the trends observed in growth. Overall, despite the small pathogen load in the of research facility, the present data suggest benefits in pig performance from PCV2 vaccination in the presence of a PCV2 challenge. Current data suggest that viremia status of pigs at birth may impact the effects of PCV2 exposure in later life stages of the pig. Moreover, under the conditions of this study, a greater effect of PCV2 vaccination could be expected on pigs born PCV2 negative, which is of relevance given the low incidence of PCV2 viremia in pigs at birth (close to 0% in over 12 farms pre-screened for this study; data not shown).

References

1. National Research Council. 2012. *Nutrient Requirements of Swine: Eleventh Revised Edition*. Washington, DC: The National Academies Press.

® Registered trademark of Boehringer Ingelheim Vetmedica, Inc.

Table 1: Least square means for the effect of PCV2 viremia levels at birth and PCV2 vaccine at weaning on the body weight and growth rate of pigs[1,2]

Item	Viremia at birth (Vir)			SEM	Vaccine (Vac)[3]		SEM	*P*-value		
	Negative	Low	High		No	Yes		Vir	Vac	Vir × Vac
Body weight, kg										
Birth	1.41	1.38	1.56	0.076	1.41	1.49	0.067	0.09	0.23	0.37
Weaning (d0)	6.29	5.93	6.45	0.271	6.19	6.25	0.244	0.19	0.79	0.22
Day 23[4]	18.06	17.65	18.16	0.530	17.72	18.19	0.449	0.74	0.39	0.76
Day 42	32.99	31.82	32.29	0.551	32.02	32.72	0.446	0.33	0.27	0.55
Day 63	52.9	51.6	52.4	0.90	51.7	52.8	0.73	0.56	0.29	0.26
Day 85	72.4	70.4	71.0	1.31	70.3	72.3	1.11	0.50	0.15	0.44
ADG, kg										
Birth to wean	0.257	0.261	0.255	0.004	0.259	0.256	0.003	0.56	0.48	0.99
Wean (d0) to day 23[4]	0.47	0.46	0.48	0.021	0.46	0.48	0.018	0.73	0.40	0.78
Day 23 to 42[4]	0.83	0.79	0.78	0.027	0.79	0.81	0.023	0.40	0.68	0.81
Day 42 to 63	0.91	0.90	0.91	0.027	0.90	0.91	0.024	0.87	0.42	0.11
Day 63 to 85	0.93	0.90	0.89	0.003	0.88	0.93	0.033	0.49	0.11	0.12
Birth to day 85	0.68	0.66	0.67	0.013	0.66	0.68	0.107	0.60	0.17	0.40
Wean (d 0) to day 85	0.77	0.75	0.75	0.015	0.75	0.77	0.013	0.05	0.15	0.46
Day 23 to 85[4]	0.89	0.87	0.87	0.021	0.86	0.89	0.018	0.52	0.23	0.55

1 Post-weaning data was adjusted to a common wean weight of 6.22 kg.
2 Low viremia ≤ 103 DNA copies of PCV2; High viremia > 10^3 DNA copies of PCV2.
3 CircoFLEX®, Boehringer Ingelheim Vetmedica, Inc.
4 On day 23 of study, all pigs were given an intranasal and intramuscular dose (1 ml each) of PCV2-b isolate ($10^{4.5}$ $TCID_{50}$).

Measuring air filter efficiencies in the discrete PRRS virus range

M. Osborne, PhD; D. Sharpley
AAF Flanders, Louisville, Kentucky

Introduction

A new evaluation procedure using Optical Particle Sizer/Scanning Mobility Particle Sizer/Condensation Nuclei Counter (OPS/SMPS/CNC) particle size measurement has been developed to study the suitability of air filters for PRRS virus reduction in the range where possible individual and agglomerated virus freely exist, in the E00 range of 0.03 - 0.1 µm and the E0 range of 0.1 - 0.3 µm. This procedure demonstrates that air filters exhibit a 10 - 30% reduced filtration efficiency for particles in the E00 and E0 range over efficiency for the previously measured American Society of Heating, Refrigerating and Air-Conditioning Engineers (ASHRAE) ranges of E1 0.3 - 1.0 µm, E2 1.0 - 3.0 µm, and E3 3.0 - 10 µm. This 10 - 30% reduced filtration efficiency represents a greater biosecurity risk that the PRRS virus can infiltrate a barn.

Materials and methods

The new procedure has been applied to both microglass media filters and to synthetic fiber filters. The microglass media filters, which work on the principle of mechanical filtration, show an approximate 10% and 15% lower filtration efficiency for the E00 and E0 range versus the E1 range. Synthetic filters, which work on the principle of both mechanical and electrostatic filtration, were tested with the media as normally received, i.e. with the electrostatic charge as normally installed, and also tested with the media in a discharged state, whereby the media has lost its charge (electrostatic charge often discharges over the lifetime of the filter).

Results and discussion

The new procedure shows that the synthetic media filters' E00 and E0 range efficiencies are approximately 15% and 18% lower than the filter's corresponding E1 efficiency, respectively. In the discharged state, the synthetic media filters showed even greater reduced efficiency: a reduced E00 efficiency by 25%, and a reduced E0 efficiency by 30%, versus the synthetic media filter's charged E1 efficiency. This implies that there is a significantly greater risk in the use of synthetic media filters (either electrostatically charged or in a discharged state) versus microglass media filters for PRRS virus reduction. The OPS/SMPS/CNC procedure utilizes a TSI Model 3787 General Purpose Water-Based Condensation Particle Counter (GP-WCPC), which can detect airborne particles down to 5 nm (0.005 µm) in diameter using state-of-the-art, water-based condensation particle counting technology.

Analysis of the importation of feed ingredients, live animals and genetic material into the United States to reduce the risk of foreign swine disease and health hazard introduction

Gilbert Patterson, VMD, MPH, DACVPM
Center for Animal and Human Health in Appalachia, Lincoln Memorial University-College of Veterinary Medicine, Harrogate, Tennessee

Introduction

Amid a growing body of evidence suggesting that porcine epidemic diarrhea virus (PEDV) may have been first transmitted into the US swine herd from contaminated feed ingredients imported from abroad, the need for a better understanding of the quantity, quality, and process of livestock feed importation to the US has been recognized.[1] This study quantified the magnitude of international trade and its contribution to the US swine industry, with the purpose of identifying potential animal health hazards represented in swine feed ingredients that are sourced internationally. Current knowledge of the quantity and source country of specific ingredients is critical to understanding the risk of disease introduction from swine feed ingredients, as well as preserving the health and safety of the food supply.

Method

International trade data was organized and analyzed from the United States International Trade Commission (USITC) website. The USITC database contains information on imported products, categorized into a list of codes known as the Harmonized Tariff Schedule (HTS). Queries were designed to create a comprehensive listing of specific products that are imported into the US with potential to be included in swine diets and may contribute to the transmission of pathogens which are foreign to the US swine industry. The database includes information on country of origin, quantity of product, date of entry, and port of entry into the US The database also features a flexible interface which allows specific types of products to be viewed in greater or lesser detail and ranked in comparison to other importing countries.

Results

423 specific HTS product codes were identified as likely inputs to the swine supply chain. These specific inputs were divided into two broad categories: 1) live animals and viable genetic material (8 codes), and 2) products that may contribute to disease transmission through the intentional or non-intentional inclusion in the swine feed supply chain. Item 2 was further sub-divided into three categories: a) animal-based ingredients (82 codes), b) plant-based ingredients (233 codes), and c) other miscellaneous trace ingredients (100 codes).

The result is a dynamic database that allows the user to rapidly examine either broad categories of feed inputs or very specific ingredients of interest. This document is not meant to be a finished product; rather it is built in a way so that it can be easily amended and refined by the author to provide a current overview of global feed ingredient sourcing. As an example, 7.1 million kilograms of soybean meal, a common ingredient in swine feed that has experimentally been shown harbor viable PEDV for extended time periods,[1] was imported into the United States from 13 foreign countries in 2016. The largest volume of this product (59.66%) of this product originated from China (Table 1).

Discussion

This work represents an initial step towards building a comprehensive listing of imported products that may be introduced

Table 1: Soybean meal* imported into the USA from foreign countries in 2016

Exporting country	Sum of 2016	% of total 2016
China	4,272,903	59.66%
Canada	2,078,293	29.02%
Denmark	241,577	3.37%
India	187,497	2.62%
Germany	162,073	2.26%
Netherlands	115,808	1.62%
Mexico	63,750	0.89%
Japan	18,278	0.26%
Ecuador	9,000	0.13%
Korea	6,242	0.09%
Thailand	3,250	0.05%
Malaysia	2,700	0.04%
Taiwan	648	0.01%
Grand total (Kilograms)	**7,162,019**	**100.00%**

* Adapted from HTS Code 1208100000, "FLOURS AND MEALS OF SOYBEANS" in USITC database.

into the pork supply chain and contribute to the transmission of pathogens that are foreign to the US swine industry. While a growing body of evidence implicates the importance of feed ingredient sourcing and quality control in preventing disease transmission to susceptible species, this information is critical to understanding the risks involved in global livestock feed ingredient sourcing. Additionally, this information provides a basis to identify gaps in border security against foreign pathogens, and may be used to help prevent future transmission events as diseases emerge internationally.

References

1. Dee, S., Neill, C., Singrey, A., Clement, T., Cochrane, R., Jones, C., Patterson, G., Spronk, G., Christopher-Hennings, J., Nelson, E., 2016. Modeling the transboundary risk of feed ingredients contaminated with porcine epidemic diarrhea virus. BMC Veterinary Research, Mar 12;12(1):51.

Oral fluid specimens can be clarified ("cleaned up") without affecting porcine epidemic diarrhea virus (PEDV) isotype-specific (IgG, IgA) ELISA responses

K. Poonsuk; L. Giménez-Lirola; R. Magtoto; D. Baum; C. Rademacher; J. Brown; J. Ji; J. Q. Zhang; C. Wang; R. Main; J. Zimmerman
Iowa State University, Ames, Iowa

Introduction

Oral fluid samples often contain particles of feed, feces, and other materials from the environment that might affect pipetting and/or test performance. Removal of particulates by centrifugation or filtration is not practical because these procedures require too much time to be performed routinely in high through-put diagnostic laboratories. The objective of this study was to reduce the suspended particulate matter in oral fluid specimens using chemical clarification without adversely affecting the detection of PEDV.

Methods

Aliquots of PED oral fluid samples collected under experimental conditions (Study One) and under field conditions (Study Two) underwent one of 3 chemical treatments (A, B, C) and were then tested by IgG and IgA whole virus PEDV ELISAs. Treatments were evaluated in terms of 1) effect on the diagnostic performance of PEDV IgG and IgA ELISAs and 2) the stability of PEDV-specific IgG and IgA in samples held at 4°C over the course of 7 days, i.e., the residual effect of treatment.

In Study One, oral fluid samples from pigs inoculated with PEDV were collected on day post inoculation (DPI) -3, 0, 5, 10, 15, 20, 25, 30, 35, and 42. In Study Two, oral fluid samples were collected on DPIs -4, 0, 3, 7, 10, 14, 17, 21, 24, 28, 31, 35, 38, and 42. Each oral fluid sample was split into 4 aliquots and treated with one of 3 chemical treatments. The 4th aliquot served as an untreated control. Treatments consisted of adding the oral fluid sample to a 5 ml tube containing the lyophilized chemical treatment, vortexing briefly, then centrifuging for 3 minutes.

All samples were tested by PEDV IgG and IgA ELISAs immediately following treatment (day post treatment (DPT) 0). Thereafter, treated and control oral fluid samples were stored at 4°C and tested again on DPT 2, 4, and 6 to evaluate the residual effect of the treatment. Serum samples were tested by PEDV IgG and IgA ELISAs on day 0, then held at 4°C and tested again on 2, 4, and 6 to be compared with oral fluid antibody responses.

Results and conclusions

Treatments were easily administered, i.e., the lyophilized chemicals were easily resuspended in oral fluid samples and the clarifying effect was immediate. Statistical analysis (nonparametric ANOVA) of oral fluid IgA and IgG S/Ps found that neither treatment nor time affected the ELISA S/P results ($P > 0.05$). That is, pairwise comparisons of DPT 2, 4, and 6 results to DPT 0 IgA and IgG S/Ps detected no significant differences ($P > 0.05$). Thus, chemical treatment removed particulates from oral fluids without affecting test performance. Given the low cost and ease of use, this approach could be used either in the field or in veterinary diagnostic laboratories to improve the characteristics of oral fluids tested for PEDV antibody.

Work in progress: Development of an ELISA for the detection of FMDV 3ABC antibody in swine oral fluid specimens

K. Poonsuk[1]; L. Giménez-Lirola[1]; J. Chiotti[2]; K. Kamrud[3]; J. Zimmerman[1]
[1]Iowa State University; Ames; Iowa; [2]VMRD Inc.; Pullman; Washington;
[3]Synthetic Genomics Vaccines Inc.; La Jolla, California

Introduction

FMDV is difficult to control because it is highly contagious, widely geographically distributed, has a broad host range, is able to establish persistent infection, is genetically high heterogeneous, and invokes weak protective immunity. For these reasons and for its impact on trade, FMDV remains one of the most economically important infectious agents of livestock, e.g., the predicted cost of an incursion into the US is $260 to $3,270 million (Schoenbaum, M.A. and Disney. 2003. *Preventive veterinary medicine*, *58*(1), pp.25-52).

An effective response to the appearance of FMDV in North America will require rapid detection of foci of infection. Nucleic acid-based assays are useful for this purpose, but ELISAs based on pan-FMDV non-structural proteins, e.g., 3ABC, also have advantages: 1) cheap, fast, and easy to perform using common technology, 2) compatible with DIVA-compatible vaccines, 3) detection of antibody regardless of FMDV serotype, and 4) well-established concept.

Serum-based FMDV surveillance using 3ABC ELISAs have been utilized in countries around the world. Beside the advantages of a welfare-friendly, aggregate mucosal specimen, oral fluids are the sample of choice for the detection of both antibodies and pathogens for several diseases. Therefore, the objective of this study is to develop a FMD 3ABC ELISA capable of detecting FMDV antibody in oral fluid specimens.

Methods

1. **FMD 3ABC antibody negative and positive serum and oral fluid samples**
   Samples of known 3ABC antibody status are required for ELISA development. Pigs (n = 20) will be randomly assigned to two groups, i.e., vaccinated (n = 15) and negative control (n = 5) group. Vaccinated pigs will receive a RNA vaccine designed to stimulate the production of antibody against 3ABC (Synthetic Genomics®, La Jolla, CA) on days post-vaccination (DPV) 0 and 21. Serum samples will be collected from individual pigs on DPV -7, 0, 5, 7, 10, 14, 17, 21, 24, 28, 35, 42, 49, and 56. Pen-based oral fluid samples will be collected twice daily throughout the collection period.
2. **Develop prototype FMDV 3ABC antibody ELISAs**
   Blocking and/or indirect FMDV 3ABC ELISAs will be optimized using the set of specimens generated in Part 1 by adjusting reagent concentrations (3ABC antigen, secondary antibody, and sample dilution), incubation time, and incubation temperatures.
3. **Assay evaluation**
   ELISA format that is most appropriate will be used as prototype ELISA to test oral fluid samples from Senecavirus A infected pigs and oral fluid samples submitted to the Iowa State University Diagnostic Laboratory to evaluate the diagnostic specificity of the assay. Future work will include evaluation of test performance using known FMDV-positive samples.

General trends of *Mycoplasma hyopneumoniae* detection by PCR in cases submitted to the Iowa State University Veterinary Diagnostic Laboratory from 2004 - 2016

G. Rawal DVM; P. Arruda MS, PhD; C. J. Rademacher DVM, BS; Ran Bi MS; D. C. L. Linhares DVM, MBA, PhD
College of Veterinary Medicine, Iowa State University

Background

Mycoplasma hyopneumoniae (Mhp) is an economically important concern to swine producers as it increased the cost of production by $ 2.50 per pig in grow-finish phase with additional cost of therapeutics by $ 0.75-$ 0.90 per pig at weaning (Schwartz, 2015). The isolation of Mhp by culture is a method for the detection of Mhp. However, the low success rate to isolate Mhp, limits the use of bacteriology on Mhp diagnosis of field cases. With the development of PCR, the molecular diagnostics has become a common practice. This method is suitable for fast, sensitive, and specific detection of Mhp in field samples.

Objective

To describe the general diagnostic trend of Mhp detection using quantitative PCR (qPCR) of cases submitted to the Iowa State University Veterinary Diagnostic Laboratory (ISU-VDL) from 2004 - 2016 by selected variables including year, sample type, age, season and geographic location.

Methods

Diagnostic data was gathered from the ISU-VDL database from 2004-2016 in which Mhp qPCR was performed. The data was cleaned by excluding cases with non-conventional sample types for Mhp diagnostics (e.g., abomasum, conjunctival swab, and liver), and cases with inconclusive, suspect, blank, or pending results for Mhp qPCR. The data description was done by using SAS and Tableau. Cases were categorized by season (December through February as winter, March through May as spring, June through August as summer, and September through November as fall), by age group, (cases from pigs with < 21 days of age were defined as suckling, 21 days to 6 weeks of age as nursery, 6-11 weeks of age as growing, 11 weeks to 200 days as finishing, and > 200 days as adult) and by sample type. Saliva was considered as oral fluid, and assorted, lung swab, serum, pleural swab, pharyngeal swab, oropharyngeal swab and tonsil swab were considered as other. Infrequent sample types such as fluid, fibrin, lymph node, tonsil scrapings, tonsil, fresh tissue, fibrin swab were excluded as they comprised less than one percent of submissions.

Results and discussion

There were 93,680 cases of Mhp, from which 25,339 tested qPCR-positive (27.04%), with 22.4%, 28.3%, 28.1%, 28.2%, 33.6%, 18.9%, 32.8%, 31.2%, 40.7%, 23.6%, 27.4%, 27.6%, and 17.9% respectively from 2004 to 2016. The fall seasons received the highest number of cases per year, and highest prevalence of Mhp detection. The finishing age group had highest number of Mhp cases. However, there was a trend of increasing the number of cases from suckling and adult groups in recent years, which may be associated with monitoring of Mhp in herds undergoing control or elimination projects. Mhp detection rate was 66% in lung homogenate, 64% in bronchial swab, 49% in broncho-alveolar lavage, 34% in tracheal swab, 33% in lung, 27% in other, 18% in laryngeal swab, 18% in swab, 10% in nasal swab, and 8% in oral fluids. The relatively low positivity rate in 2016 (17.9 %) was associated with increased testing from adults, suckling, and from oral fluids.

Conclusions

This study describes a general trend on Mhp diagnostics from 2004-2016. Oral fluid was the sample type with lower prevalence of positive results, and lung homogenate the highest. Increasing number of Mhp cases in suckling and adult pigs may indicate efforts to monitor progress towards Mhp control and/or elimination project.

References

1. Goodwin et al. Production of enzootic pneumonia in pigs with a mycoplasma. Vet Rec.1956;77:1247–1249.

2. Kobisch et al. Swine mycoplasmoses. Rev Sci Tech. 1996;15:1569-1605.

Figure 1: *Mycoplasma hyopneumoniae* detection by PCR from 2004 to 2016

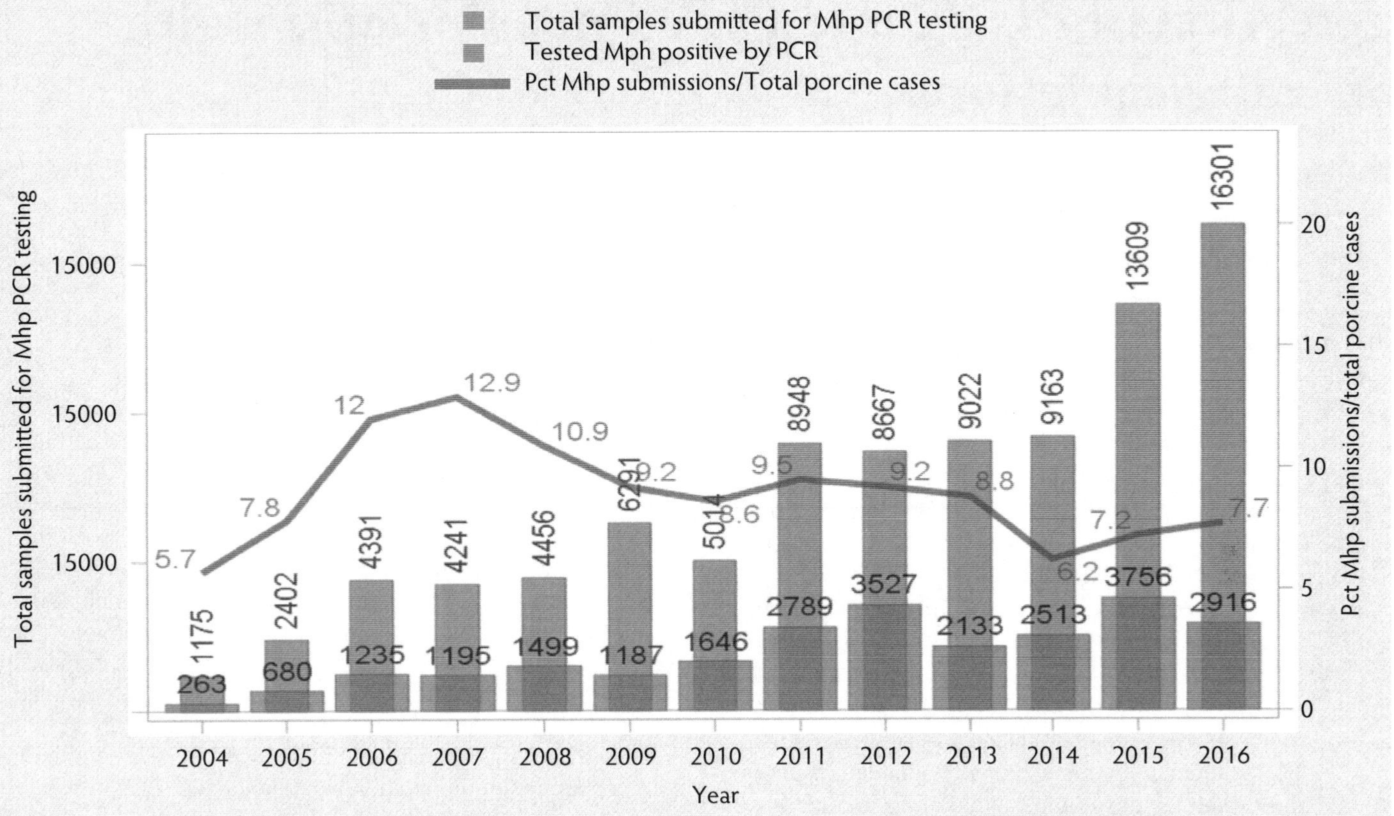

Development of a duplex assay for simultaneous detection of PCV2 and PCV3 in association with histological lesions

TP Resende[1], DVM, MS, PhD student; Benjamin Miller[1], BS; Todd Knutson[1], BS, PhD; Douglas Marthaler[2], BS, PhD; FA Vannucci[2], DVM, MS, PhD
[1]Department of Veterinary and Biomedical Sciences, University of Minnesota, St. Paul, Minnesota
[2]University of Minnesota Veterinary Diagnostic Laboratory, University of Minnesota, St. Paul, Minnesota

Introduction

A novel type of porcine circovirus, porcine circovirus type 3 (PCV3), has been recently identified in sows with dermatitis and nephropathy syndrome as well as in post-weaning pigs with multisystemic perivascular inflammation.[1] PCV2 is also produces similar lesions in pigs but due to its high prevalence in swine populations worldwide, it is necessary to associate clinical, gross and microscopic lesions to the diagnostic tests (PCR and immunohistochemistry) to have a definitive diagnosis. Considering that PCV2 and PCV3 can cause similar clinical signs and lesions, it is crucial to determine the involvement of each virus within the lesions, and then determine the best practices for its prevention and control in affected herds. The objective of this study was to develop a duplex assay for simultaneous *in situ* detection of PCV2 and PCV3 in association with histological lesions.

Materials and methods

Retrospective PCR screening: a PCR assay targeting the capsid gene of PCV3 was previously designed to investigate the presence of PCV3 DNA from 477 tissue homogenate samples submitted to the University of Minnesota Veterinary Diagnostic Lab (UMN-VDL) in 2017. This sample type is routinely tested for PCV2 and PRRSV and contains a pool of tissues including, lung, heart, kidney and lymphoid tissues.

Duplex assay for in situ detection of PCV2 and PCV3: Based on the results from the PCR screening, formalin-fixed paraffin embedded tissues were retrospectively selected based on clinical signs, histological lesions and detection of PCV2 and PCV3 by PCR. An *in situ* hybridization (ISH) duplex assay was developed targeting mRNA of the *rep* gene of PCV2 (Genbank: KX298474.1) and PCV3 (Genbank: HQ839721.1). ISH was performed on tissue slides according to a previously described protocol.[2] PCV2 and PCV3 probes were hybridized, amplified and stained using alkaline phosphatase (green) and horseradish peroxidase (red) labelling systems, respectively. Positive hybridization signals represent transcriptionally-active virus in association with the histological lesions.

Results and discussion

From a total of 477 tissue samples tested by PCR, 27.2% (n = 130) were positive for PCV3, 20.7% (n = 99) positive for PCV2 and 9% (n = 43) positive for both viruses. The PCV2 probe was tested in samples only positive for PCV3 and vice-versa to evaluate probe specificity. There was no evidence of cross-reaction between the probes. This finding supports the high degree of difference between PCV3 and PCV2 replicase protein sequences, approximately 55% identity. The analytical sensitivity of the duplex assay compared with PCR demonstrated the ability to detect PCV2 at 31ct and PCV3 at 30Ct. Strikingly, both viruses were predominantly observed in lymph nodes, spleen (white pulp), perivascular lymphohistiocytic infiltrates in the heart and peribronchiolar areas. Occasionally, while PCV2 was observed surrounding the central arteriole in the white pulp of the spleen, the dual detection of the PCV3 in the same tissue section revealed its presence in the marginal zone between the white and red pulp (Figure 1). This finding characterizes the ability of both viruses to replicate in the same tissue but with potential tropism to different cell types during concurrent infections.

Figure 1: Spleen. Co-infection PCV2 (green) and PCV3 (red). PCV2 surrounding the central arteriole of the white pulp and PCV3 in the marginal zone between the white and red pulp.

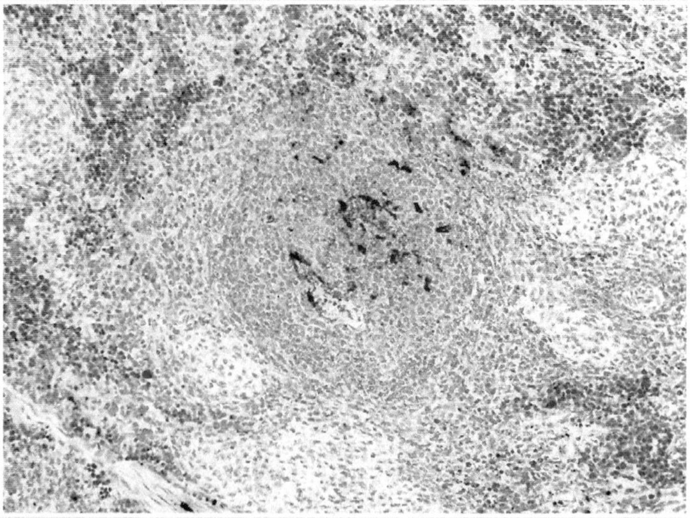

Conclusions

This study has demonstrated the endemic distribution of PCV3 in the US swine population. Therefore, similarly from what is observed in PCV2, the diagnosis of PCV3-associated syndromes requires: a comprehensive clinical evaluation of the herd, gross and histological lesions, and detection of the virus within the lesions. This study describes the development of a novel diagnostic tool for dual detection of PCV2 and PCV3 in association with the histological lesions. This technology allows an accurate diagnosis focusing on the specific contribution of PCV2 and PCV3 to clinical disease.

References

1. Phan TG et al. Detection of a novel circovirus PCV3 in pigs with cardiac and multi-systemic inflammation. *Virol J.* 2016 Nov 11;13(1):184.

2. Wang F, et al. RNAscope: a novel in situ RNA analysis platform formalin-fixed, paraffin-embedded tissues. *J Mol Diagn*. 2012;14:22–29.

Factors that impact PRRS time to stability

Juan Sanhueza[1], DVM, MS, PhD; Carles Vilalta[1], DVM, PhD; Andreia G. Arruda[2], DVM, MS, PhD
[1]Veterinary Population Medicine Department, College of Veterinary Medicine, University of Minnesota, Saint Paul, Minnesota; [2]The Ohio State University, Department of Preventive Veterinary Medicine, Columbus, Ohio

Introduction

The AASV developed a terminology to classify breeding herds according their porcine reproductive and respiratory syndrome virus (PRRSV) status. Briefly, farms are classified as positive unstable, positive stable, provisionally negative and negative according to PCR and serology testing of piglets and gilts.[1] The time between the outbreak and 4 monthly negative PCR results in weaned piglets is called time to stability (TTS). It was previously observed that farms using field virus or that had experienced a previous outbreak during the last three years achieved stability sooner.[2] However, there is a lack of knowledge on what other factors could influence PRRS TTS. Therefore, the aim of this study was to analyze how different factors impact TTS.

Material and methods

The TTS from sow farms that had PRRS outbreaks from 2011 to March, 2017 and participate in the Morrison's Swine Health Monitoring Project (MSHMP) were assessed. These farms belonged to six US swine production companies compliant with the AASV guidelines for PRRS status classification at all times. A Cox proportional hazards model was built to examine variables potentially associated with TTS. Variables investigated included system, herd size, outbreak starting season, year of the outbreak, whether the farm had an outbreak with porcine epidemic diarrhea (PED) during the stability process, and whether the farm had a PRRS outbreak during the previous year. Model selection was guided by the Akaike Information Criterion (AIC) and Wald-test p-values of candidate variables. Farm and system were included in the model as random effects (i.e. shared frailty) since the same farm could have had multiple outbreaks and farms are clustered within systems.

Results and discussion

A total of 172 outbreak events in 87 farms from six systems was used for the analysis. The median TTS was 41.5 weeks (1st Q: 31 weeks and 3rd Q: 55 weeks). This is higher compared to the 26 weeks previously described.[2] The voluntary reporting nature of data coming for the present study, as well as differences in demographics and location of farms used herein might partially explain this difference. Variables retained in the final model were outbreak starting season, year of the outbreak, herd size, system and whether the farm had a PRRS outbreak during the previous year. Outbreaks that occurred during autumn or winter were more likely to achieve stability sooner (shorter TTS) than outbreaks that occurred during summer or spring (Figure 1). Additionally, TTS was shorter in farms that had a PRRS outbreak in the previous year and in farms that had a herd size of more than 2500 sows. Finally, having a PED outbreak during stability process was not associated with the TTS. These results suggest that few farms are able to achieve stability during winter months and therefore outbreaks occurring during spring or summer months would last longer on average than the ones occurring during winter or autumn.

References

1. Holtkamp, D.J., D.D. Polson, and M. Torremorell, *Terminology for classifying swine herds by porcine reproductive and respiratory syndrome virus status.* Journal of Swine Health and Production, 2011. 19(1): p. 44-56.

2. Linhares, D.C., et al., *Comparison of time to PRRSv-stability and production losses between two exposure programs tocontrol PRRSv in sow herds.* Prev Vet Med, 2014. 116(1-2): p. 111-119.

Figure 1: Kaplan-Meier curves survival estimates (i.e. time to stability) by PRRS outbreak starting season

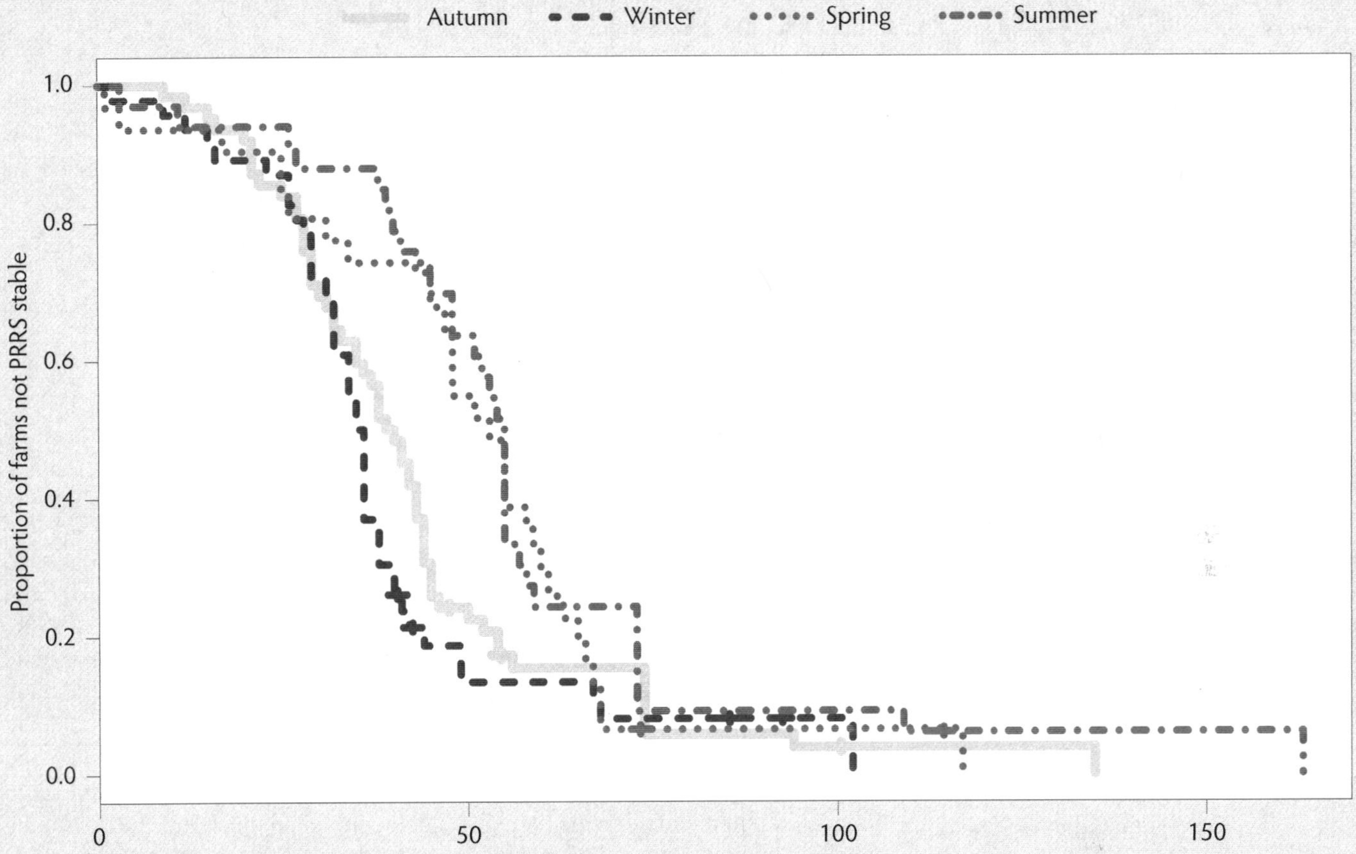

Phenotypic relationship and amino acid variability of the VP7 gene in porcine rotavirus B and C

F.K. Shepherd[1], BS; F. Chen[2], BS; T.P. Knutson[3], PhD; M.R. Culhane[4], DVM, PhD; M.P. Murtaugh[1], PhD; D.G. Marthaler[5], PhD
[1]Department of Veterinary and Biomedical Sciences, University of Minnesota, St. Paul, Minnesota;
[2]State Key Laboratory of Agricultural Microbiology, Huazhong Agricultural University, Wuhan, China;
[3]Minnesota Supercomputing Institute, University of Minnesota, Minneapolis, Minnesota;
[4]Department of Veterinary Population Medicine, University of Minnesota, St. Paul, Minnesota;
[5]Department of Diagnostic Medicine/Pathobiology, Kansas State University, Manhattan, Kansas

Introduction

Rotavirus B and C (RVB and RVC) are causes of acute diarrheal disease in swine throughout the world.[1] Vaccines for RVB and RVC are lacking because they are difficult to propagate in cell culture. Subunit vaccines expressing antigenic epitopes of RVB and RVC provide an alternative strategy for preventing rotavirus disease in swine. Using major genotypes of the RVB and RVC VP7 outer capsid protein from strains circulating in the United States and Canada, this study investigates amino acid variability and bioinformatically predicts the neutralizing epitopes to identify subunit vaccine candidates to prevent enteric disease due to RV infections.

Material and methods

Between 2010 and 2017, porcine intestinal and fecal samples were submitted to the University of Minnesota Veterinary Diagnostic Laboratory from the United States and Canada. Using RT-qPCR,[2] samples were determined positive for RVB (n = 179) or RVC (n = 369) and the VP7 genes were Sanger sequenced upon client request. The G genotypes were assigned based on 80% or 85% nucleotide identity cutoffs for RVB and RVC, respectively.[3,4] MUSCLE alignments of translated sequences were used to calculate mean pairwise identity and the number of amino acid (aa) functional groups at each residue location. Antigenic epitope prediction of predominant genotypes used EPCES (`http://sysbio.unl.edu/EPCES`). Analysis of positive and negative selection (dN-dS) was performed using SLAC (Single-Likelihood Ancestor Counting), FUBAR (Fast, Unconstrained Bayesian Approximation), FEL (Fixed Effects Likelihood), and MEME (Mixed Effects Model of Evolution) methods in DataMonkey (`www.datamonkey.org`). Selection was inferred if identified by at least 3 methods.

Results and discussion

Sequencing revealed G8 (n = 1), G11 (n = 2), G12 (n = 15), G14 (n = 11), G16 (n = 68), G17 (n = 3), G18 (n = 17), and G20 (n = 52) G genotypes within the RVB dataset. Five strains also represent 4 tentative new genotypes of RVB (G22-G25). RVC G genotypes G1 (n = 27), G3 (n = 4), G5 (n = 46), G6 (n = 280), and G9 (n = 12) were identified. Potential antibody binding sites were predicted for the predominant genotypes of RVB and RVC in the dataset (Table 1). A 4-aa insertion in RVC G6 strains (sites 245-248) was highly antigenic. The aa functional diversity at several antigenic sites (n = 7 for RVB and n = 13 for RVC) was highly conserved. Identification of negative selection at antigenic epitopes was surprising, but provides an opportunity to target these residue in a subunit vaccine due to their tendency to resist diversifying nucleotide substitutions. This study provides the framework for protein modeling of hypervariable regions and antigenic epitopes for the VP7 proteins. Genotypic analysis of the VP4, the other outer capsid protein and target of neutralization, may enhance our knowledge to design a subunit vaccine to prevent the spread and control clinical disease of RVB and RVC in swine populations.

References

1. Chang K, et al. "Reoviruses (rotaviruses and reoviruses)." *Diseases of Swine*. Eds. Jeffrey J. Zimmerman, et al. Ames, Iowa: Wiley and Sons, 2012. 621-632.

2. Marthaler D, et al. Rapid detection and high occurrence of porcine rotavirus A, B, and C by RT-qPCR in diagnostic samples. *J Virol Methods*. 2014; 209:30-34.

3. Marthaler D, et al. Detection of substantial porcine group B rotavirus genetic diversity in the United States, resulting in a modified classification proposal for G genotypes. *Virology*. 2012:85-96.

4. Marthaler D, et al. Identification, phylogenetic analysis and classification of porcine group C rotavirus VP7 sequences from the United States and Canada. *Virology*. 2013: 189-198.

Table 1: Amino acid conservation and evolutionary selection properties of predicted antigenic sites of porcine rotavirus B and C VP7.

Genotype (Predicted number of antigenic residues)	Number of conserved epitopes	Number of positively selected residues	Number of negatively selected residues	Locations of negatively selected and conserved residues
RVB strains				
G12 (n = 21)	6	0	5	None
G14 (n = 15)	5	0	4	67, 161
G16 (n = 21)	5	1	6	160
G18 (n = 19)	5	0	4	158
G20 (n = 16)	3	0	8	61, 158, 179
RVC strains				
G1 (n = 15)	7	0	5	86, 87, 128
G5 (n = 25)	10	0	5	90, 128, 227
G6 (n = 25)	6	0	17	90, 92, 193, 195
G9 (n = 23)	10	0	0	None

Benchmarking pen-side sequencing for infectious disease diagnosis

Shaoyuan Tan; Michael P. Murtaugh
Department of Veterinary and Biomedical Sciences University of Minnesota

Introduction

On-farm, pen-side molecular disease diagnosis is increasingly needed and valuable for swine producers. It would provide faster results and facilitate rapid decision-making, an important consideration in an industry in which delays in moving animals due to unknown health status can disrupt flow patterns and schedules, or cause disease outbreaks with great economic losses even in the midst of regional control programs.

Here, we evaluated the capacity of the MinION portable sequencer for detection of bacterial and viral pathogens. The MinION sequencer (Oxford Nanopore Technologies) is a handheld device that can read sequences >100,000 bases in length.[1] Nanopore sequencing has been shown to be able to discriminate individual nucleotides by measuring the change in electrical conductivity as DNA molecules pass through the pore.[2] At this time, it only requires about 2 h for sample preparation, and generates data in real time at a rate of ~400 bases per second per pore for 450-500 pores continuously for minutes to hours.[3] The sequencing run can be stopped anytime when enough data has been generated.

Materials and methods

We used PRRSV as a model for virus detection by MinION sequencing. PRRSV genomic RNA was purified by Qiagen extraction and cDNA was synthesized using Superscript II (Thermo Fisher Scientific). The MinION strand-switching cDNA synthesis protocol was used to create the PRRSV MN184 cDNA library, which was loaded on an R9 flow cell. Sequencing data were generated using MinION and MinKNOW, uploaded to Metrichor for real-time basecalling, and then assembled by Canu.

Similarly, we used *Escherichia.coli* as a model for bacterial detection by MinION sequencing, at the same time, we compared MinION with PacBio (Pacific Biosciences), which is a competing single molecule, long-read sequencing platform. Identical *E coli* DNA samples were submitted to the UMN Biomedical Genomics Center core facility for SMRT sequencing on the PacBio instrument, and for in-lab MinION sequencing using R9 flow cell and chemistry and the 2D ligation genomic DNA protocol.

Results and discussion

From PRRSV sequencing, a total of 9405 reads (mean read length 1130 bp) and a contig of approximately 7.5 kb with a raw accuracy of 92% and a 94.1% consensus accuracy was obtained. This quality level is more than sufficient to discriminate virus types and even different strains. But in theory we should get the entire ~15 kb genome instead of only a 7.5 kb contig. Alignment of reads and assembly of a contig showed predominate alignment to the 3' end of the genome, indicating that the large majority of cDNA synthesis was primed from the polyA tail, with a minor proportion of internal priming. This indicates that the read length was limited by the reverse transcription step. Optimization of reverse transcription and primer design is a way to achieve complete genome coverage in single reads. Another way to overcome the limit of reverse transcription is to directly sequence the RNA, which Oxford Nanopore has now made a reality. Even though direct RNA sequencing is still very early on in its development cycle, Oxford Nanopore will continue to improve key properties such as throughput and per read accuracy.

For the *E coli* sequencing, Table 1 shows the comparison results of MinION and PacBio. Notably, raw read accuracy is equivalent, but the consensus accuracy of MinION is considerably lower than PacBio due to the read throughput and depth of coverage. This result shows that MinION is a suitable tool for rapid molecular diagnosis at present, and with improved yield, it is possible to achieve a consensus comparable to PacBio. In this study MinION output was a function of sample input amount and running time, neither of which were optimized for this comparison. Optimization of input DNA quality, quantity and running time will greatly improve MinION results.

Table 1: Comparison of *E coli* sequencing on MinION and PacBio instruments

	MinION R9	PacBio
Accuracy	0.85	0.86
Total yield	65,506,364 b	918,848,964 b
Total reads	4,286	65,041
Mean read length	5,350 b	14,100 b
Contigs	Number: 72 Largest: 55,213 b	Number: 1 Largest: 4,592,923 b
Width of coverage	~29%	100%
Depth of coverage	~7	~126
Consensus accuracy	92.3%	99.99%

References

1.Loman, N.J., J. Quick, and J.T. Simpson, A complete bacterial genome assembled de novo using only nanopore sequencing data. Nat Methods, 2015. 12(8): p. 733-5.

2.Olasagasti, F., et al., Replication of individual DNA molecules under electronic control using a protein nanopore. Nat Nanotechnol, 2010. 5(11): p. 798-806.

3.Quick, J., et al., Real-time, portable genome sequencing for Ebola surveillance. Nature, 2016. 530(7589): p. 228-32.

Associations between PRRS farm status and productive parameters in Spanish sow farms

D. Torrents[1]; J. Miranda[1]; A. Puigredon[1]; A. Ramirez[2]; D. Linhares[2]
[1]Laboratorios HIPRA S. A, Amer, Spain; [2] Iowa State University, Ames, Iowa

Introduction

PRRS is an endemic swine disease causing significant productive and economic loses in pig farms. Spain is one of the main global pig producers. However, data on the impact of PRRS virus (PRRSV) infection on Spanish, or other European pig farms are still scarce. Likewise, there is a lack of systematic and periodic monitoring of PRRSV in Europe, which limits the knowledge on PRRS incidence or prevalence over time. The aim of this study was to establish, for the first time, a systematic monitoring system for PRRS in Spanish sow farms, to classify farms according the PRRS status using terminology currently adopted in the US swine industry.[1] Furthermore, we investigated associations between PRRS status of breeding herds, and selected production parameters, allowing the characterization, for the first time, of production and economic impact of PRRSV infection under Spanish field conditions.

Materials and methods

Between Feb to Mar 2017, 40 breeding herds (total of 93,800 sows) belonging to one large integrated group located in Spain enrolled voluntarily in a one-year program for the PRRSV monitoring project. The PRRSV systematic monitoring project was designed based on the AASV guidelines. More specifically, study farms adopted a diagnostic monitoring protocol, which consisted of monthly individual due-to-wean blood testing of 30 piglets. Serum from individual blood were pooled (5 pools of 6), and tested for PRRSV RNA by RT-PCR.[2] Farms were classified as: Negative (N), Provisional negative (PN), Positive Unstable (PU) and Positive Stable (PS). Farms were considered N when all ELISA test were negative and no clinical evidence of PRRS were observed. Starting under the assumption that all positive farms were PU, farms reached PS status after 4 consecutive negative PCR test for all tested pools. When at least one pool was PCR-positive, farms are kept in the PU status. For the evaluation of production data, weekly data were gathered from each farm including the following weekly parameters: average born alive per litter, average of stillbirths and mummies per litter, average of total born per litter, and total pigs weaned per week .

Results and discussion

At the beginning of the study, in February 5 farms (12.5%) were classified as N, and remained N all along the study period. From February 2017 until October 2017, after 8 consecutive monthly samplings (Figure 1A), 5 farms (12.5%) have been classified as PS for the whole study period, and 9 (22.5%) as PU for the whole study period as well. Twenty-one farms (52.5%) presented alternating status, 10 farms shifting from PS to PU, 6 farms from PU to PS, and 5 farms showed a double shifting from PU to PS to PU. Among farms classified as PU during the whole study period, just 1 presented positive results in all samplings, suggesting 'high' PRRSV circulation. Meanwhile, the other 9 farms presented different patterns of alternating positive and negative results. This was in agreement with data from literature,[3] demonstrating intermittent pattern of PRRS RNA detection by RT-PCR in herds undergoing PRRSV control/elimination.

Percentage of PS farms (Figure 1B) constantly increased from March (40%) to June (69%) but later it constantly decreased again from July to October (37%). In order to elucidate the possible causes of this decrease in the percentage of PS farms, epidemiologic investigations will be performed.

Conclusions

We have established for the first time in Spain the basis for a systematic PRRS monitoring protocol in accordance with the current US standards. Our data is preliminary as of the time of submission of this poster but we will have 4 more months of data for all 40 farms by March 2018 allowing to have enough data to completely classify all enrolled farms and better understanding the dynamics of PRRS stability. At that time will have 13 herd samplings, and will also have PRRS ORF-5 sequence information, which will better characterize the PRRSV strains currently circulating in these Spanish farms.

References

1. Holkamp et al. Terminology for classifying swine herds by porcine reproductive and respiratory syndrome virus status. *J Swine Health and Prod.* 2011;19(1):44-56.

2. Martinez et al. Simultaneous detection and genotyping of porcine reproductive and respiratory syndrome virus (PRRSV) by real-time RT-PCR and amplicon melting curve analysis using SYBR Green. *Res Vet Sci.* 2008;185(1):184-93.

3. Linhares, DCL; Torremorell, M; Cano, JP; Morrison, RB. 2014. Comparison of time to PRRSv-stability and production losses between two exposure programs to control PRRSv in sow herds. *Preventive Veterinary Medicine.* 2014 Sep 1;116 (1-2):111-9.

Figure 1A: Results of PCR of positive farms from Feb to Oct 2017.

Farm\Week	1	2	3	4	5	6	7	8	9	10	11	12	13	14	15	16	17	18	19	20	21	22	23	24	25	26	27	28	29	30	31	32	33	34
230	P				N*				N*				N*						N*				N*				N*				N*			
359	N				P				N*				N*						N*				N*				N				P			
772	N*				N*				N*				N*						P				N				N				N			
449	N*				N*				N*				N*						N*				N*				N*				N*			
450	P				N				P				N*						N*				N*				N*				N*			
830	N*				N*				N*				N*						N*				N*				N*				N*			
127	N				P				P				P						N				N				P				P			
54	N				P				N				P						N*				N*				N*				N*			
74	P				P				P				P						N*				N*				N*				N*			
40	P				P				N*				N*						N*				N*				P				P			
58	P				N				N				N						P				P				P				P			
125	P				P				N				N						P				N				N				N			
682		N				P				N*					N*					N*				N*				N*				N*		
281		N*				N*				N*					N*					N*				N*				N*				N*		
383		N*				N*				N*					N*					N*				N*				N*				P		
332		N*				N*				N*					N*					N*				P				N				N		
45		P				P				P					N					P				N				P				P		
351		P				P				P					P					P				N				N				P		
868			P				P				P					P					P				P				P				P	
355			N				N				N					P					P				P				P				P	
138							N*				N*				N*							N*				P				P				P
232			N*				N*				N*					N*					N*				P				N				N	
243			N*				N*				N*					N*					N*				N*				N*				N*	
145			P				P				P					P					P				N				P				N	
174			N				P				P					N					N				N				N*				P	
159			N*				N*				N*					N*					N*				N*				N*				P	
147			N*				N*				N*					N*					N*				N*				N*				P	
417				N*				N*				N*					N*					N*				N*				N*				N*
419				N*				N*				N*					N*					N*				N*				P				N
857				N*				N*				N*					N*					N*				N*				N*				P
136				N				N				P					N					N				N				P				N
137				P				N				N					N*					N*				N*				N*				N*
747				N				N				P				N*					N*				N*				N*					P
128				N				N				N					N*					N*				N*				P				N
92				P				P				P					N*					N*				N*				N*				P

P: PCR positive; N: PCR negative; N*: 4 consecutive PCR negative results

Figure 1B: Percentage of PS and PU farms. Feb to Oct 2017.

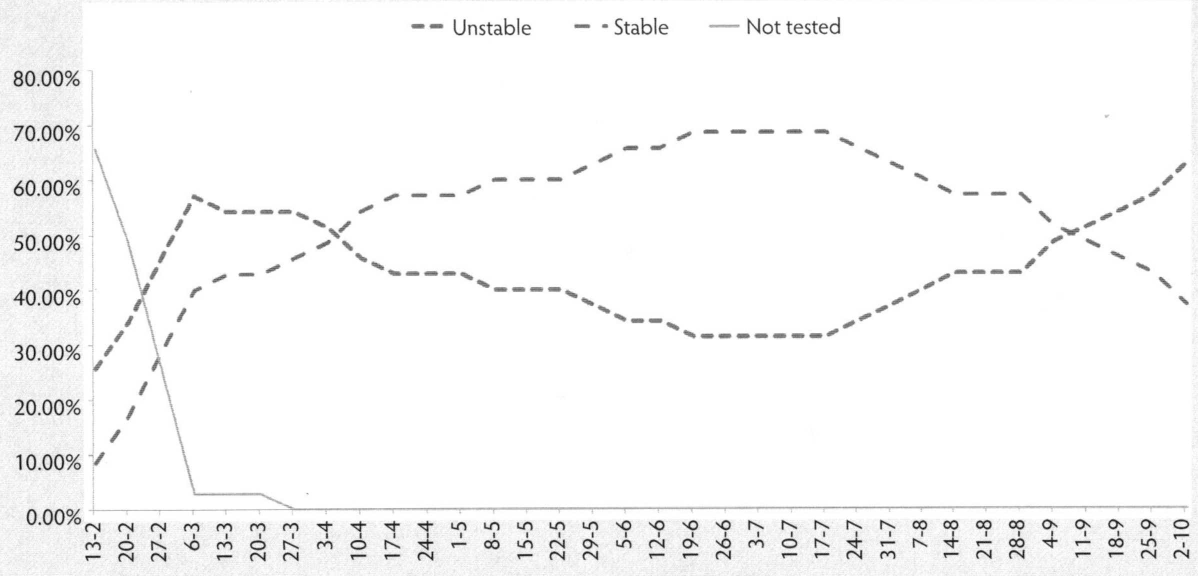

Live virus immunization (LVI) with a recent 1-7-4 PRRSV isolate elicits broad protection against PRRSV challenge in finishing age swine

Albert G. M. van Geelen[2]; Alexandra Buckley[2]; Vikas Kulshreshtha[3]; Susan Brockmeier[1]; Laura Miller[1]; Damarius Fleming[2]; Crystal Loving[4]; Kay Faaberg[1]; Kelly Lager[1]
[1]US Department of Agriculture (USDA), Agricultural Research Service (ARS), Virus and Prion Research Unit; [2]Oak Ridge Institute for Science and Education; [3]Iowa State College of Veterinary Medicine; [4]US Department of Agriculture (USDA), Agricultural Research Service (ARS), Food Safety and Enteric Pathogens

Introduction

Porcine respiratory and reproductive syndrome (PRRS) continues to be the most economically important disease affecting commercial pig production in the United States and worldwide. Its considerable sequence and antigenic variation, coupled with the limited protection offered by current vaccine options impedes our abilities to control PRRS virus (PRRSV). PRRSV isolates of 1-7-4 lineage have been associated with severe clinical cases in the Mid-Western US. Commercially available vaccines are often based on older viral strains and offer limited efficacy against heterologous challenge. Live virus immunization (LVI), a form of autologous vaccine, is based on immunizing healthy swine with serum from a diseased pig infected with a recent local isolate affecting that farm or area. Since the emergence of PRRS, this practice has been frequently utilized with reported success in immunizing replacement gilts, suggesting it does confer better protection than commercial vaccines. However, recently there have been field reports that LVI is not as efficacious as it once was, implying there may have been a change in how PRRSV interacts with the pig. We decided to test LVI efficacy with a recent 1-7-4 PRRS isolate[2].

Materials and methods

To evaluate the efficacy of LVI with contemporary 1-7-4 strains, 60 finishing age swine[3] were used in this study and divided into 7 groups. Three groups were challenged with 2ml serum I.M. from an NADC34 infected pig, a current 1-7-4 strain.[2] Each of these LVI groups were again challenged after six weeks intranasally with 5 × 10E4 TCID50 of either NADC34 (homologous challenge), NADC36, a close relative (98.4% homologous whole genome), or SDSU73, a known moderately pathogenic 1-4-4 strain that is 83.2% homologous with NADC34. Each of these groups were compared with an equivalent but naïve group and evaluated for pyrexia, ADG, viral load, lung lesions and virus neutralization titer.

Results and discussion

LVI with NADC34 protected all three LVI groups against homologous or heterologous challenge by lowering viral load by almost three logs, by eliminating negative effects on ADG of all three viruses and by eliciting broadly neutralizing calculated antibody titers of greater than 100 at 50% inhibition in 5 out of 8 pigs in the homologous challenge group, 5 out 8 in the NADC36 group, and 7 out of 9 in the SDSU73 group. The heterologously challenged group with SDSU had significantly higher PRRS ELISA titers than the other two groups.

Conclusions

LVI with NADC34 induces equivalent neutralization and protection per group from challenge with the NADC34, NADC36 or heterologous virus SDSU73. When sera of individual pigs were examined by western blotting, individually distinct reactivity patterns against viral proteins were observed between pigs in a group, suggesting individually distinct anti-PRRSV antibody responses. Future work aims to elucidate if a predictive value can be gleaned from these patterns for efficacy of protection against future challenge.

References

1. Workman AM, Smith TP, Osorio FA, et al. Complete Genome Sequence of Highly Virulent Porcine Reproductive and Respiratory Syndrome Virus Variants That Recently Emerged in the United States. *Genome Announc* 2016;4.

2. van Geelen AGM, Anderson TK, Lager KM, Das PB, Otis NJ, Montiel NA, Miller LC, Kulshreshtha V, Buckley AC, Brockmeier SL, Zhang J, Gauger PC, Harmon KM, Faaberg KS: Porcine reproductive and respiratory disease virus: Evolution and recombination yields distinct ORF5 RFLP 1-7-4 viruses with individual pathogenicity. *Virology*. 2017 Oct 30;513:168-179

3.: At necropsy their BWT distribution was: 172lb < 10 <200lb, 200lb < 19 < 250lb, 250lb < 23 < 300lb, and 300lb < 8 < 374lb.

Good Vaccination Practice: It all starts with a good vaccine storage temperature

F. Vangroenweghe[1]; K. Kelderman[1]; F. Meulders[2]; R. Sol[3]; D. Hendrickx[4]
[1]Elanco Animal Health, BU Swine & Poultry, Antwerpen, Belgium; [2]DAP Varkensdierenartsen, Netersel, The Netherlands; [3]DAP VUG, Voorthuizen, The Netherlands; [4]DAP Zuid-Oost, Deurne, The Netherlands

Introduction

Recent introduction of strategies to reduce antibiotic use in food animal production implies an increased use of vaccines in order to prevent the economic impact of several important diseases in swine.[1] Good Vaccination Practice (GVP) is an overall approach on the swine farm aiming to obtain maximal efficacy of vaccination through good storage, preparation and finally application to the target animals. In order to have a better insight into the vaccine storage conditions, the temperatures in the vaccine storage refrigerators were measured during farm visits over a period of 1 year.

Materials and methods

The actual refrigerator temperature was measured in 126 swine farm vaccine storage refrigerators on the occasion of a consultative farm visit in Belgium and The Netherlands. The digital thermometer sensor (MOXX Thermometer; TFA® Dostmann GmbH & Co., Wertheim, Germany) was installed in the refrigerator in a standardized way: 1) The sensor was inserted into a cardboard packaging box of a veterinary medicinal product (VMP) present in the refrigerator, 2) The VMP package was positioned in the central part of the body of the refrigerator (not the door shelf, not the upper nor lower shelf), 3) Temperature measurement was allowed for at least 45 min and 4) The actual refrigerator temperature was noted including date, season and country of measurement. Additionally, it was registered if a thermometer or other temperature monitoring system was already present in the vaccine storage refrigerators included in the study. Measured temperatures were categorized based on the season they were measured: S1 (winter; 21/12 – 20/3), S2 (spring; 21/3 – 20/6), S3 (summer; 21/6 – 20/9) and S4 (autumn; 21/9 – 20/12). Differences among seasons were analyzed using oneway analysis of variance (ANOVA) of temperature by season. Significant differences were detected using a Tukey-Kramer test ($P < 0.05$). The distribution of the vaccine storage refrigerator temperatures over the year was plotted in a histogram with intervals of 3° C.

Results

Less than 10% of the farms had a thermometer in their vaccine storage refrigerator at the moment of the visit. Temperature measurement revealed that only 71% of the measured refrigerators were in line with the recommended temperature range of +2° C to +8° C (Table 1). Both below +2° C and above +8° C temperatures could be registered during all seasons of the year. Compliance was significantly ($P < 0.05$) lower during summer with an average temperature of 9.2° C and only 43% of the measured temperatures within the recommended range (Table 1).

Discussion

The present study clearly showed the need for continuous education on vaccine storage management. In our study, there were few (< 10%) on-farm vaccine refrigerators which had a thermometer present at the moment of the farm visit. This is somehow in accordance with another study in human medicine where the presence of a thermometer was 11%.[2] In veterinary medicine, the correct storage of vaccines is crucial since both too low and too high temperatures can provoke damage to specific vaccine types.[2] Adjuvanted killed or subunit vaccines can be damaged (e.g. structure of aluminium hydroxide in adjuvans) by too low temperatures (below 0° C), whereas lyophilized live vaccines are susceptible (e.g. loss of vaccine potency) to heat damage by temperatures above +8° C.

Conclusions

In conclusion, knowledge and awareness of vaccine storage conditions are crucial under practical field conditions in swine herds. Focus on a correct on-farm vaccine storage is part of the responsible veterinarians' guidance in order to obtain the required vaccine efficacy.

References

1. Speksnijder DC, Mevius DJ, Bruschke CJM, Wagenaar JA. Reduction of veterinary antimicrobial use in the Netherlands – the Dutch success model. Zoonoses & Public Health 2015;62(Supp. 1):79-87.

2. Yuan L, Daniels S, Naus M, Brcic B. Vaccine storage and handling – Knowledge and practice in primary care physician's offices. Canadian Family Physicians 1995;41:1169-1176.

Table 1: Distribution (%) of on-farm refrigerator temperatures in specific temperature category: below +2° C, between +2° C and +8° C, above +8° C

	Below +2° C	Between +2° C and +8° C	Above +8° C
S1, winter	4	77	19
S2, spring	0	81	19
S3, summer	4	43	52
S4, autumn	6	71	24
Summary	3	71	26

Cup waterer:pig ratio on growth performance of growing and finishing pigs

C. M. Vier[1], DVM; S. S. Dritz[1], DVM, PhD; M. D. Tokach[1], PhD; M. A. D Gonçalves[2], DVM, PhD; F. Gomez[2], Msc; D. Hamilton[2], PhD; J. C. Woodworth[1], PhD; R. D. Goodband[1], PhD; J. M. DeRouchey[1], PhD
[1]Kansas State University, Manhattan, Kansas; [2]Genus PIC, Hendersonville, Tennessee

Introduction

Water is an essential nutrient that fulfills many physiological functions,[1] yet guidelines for water delivery equipment are based on limited data. Water helps to regulate body temperature, aids in the digestion of feed, transports nutrients, and removes toxins from the body. Given these critical roles, water access and availability on pig farms may impact a pigs' feed intake, health, and performance. General recommendations regarding the number of pigs per drinking device were proposed more than 30 years ago,[2] but limited data is available in modern production systems stocked with modern genetics to support these recommendations. Therefore, the objective of this trial is to determine the effects of increasing cup waterer to pig ratios on growth performance of growing and finishing pigs housed under commercial conditions.

Materials and methods

A total of 1,134 pigs (PIC 359 × Camborough; initial pen average body weight 35.7 ± 1.17 lb) were used in a growth trial with 11 replicate pens per treatment and 27 pigs per pen. Pigs were housed in pens that provided 6.85 square feet per pig, and the typical ratio of cup waterer to pigs in this system is 1:27. The trial was conducted during summer months from mid-May to mid-September. Treatments consisted of 1, 2, or 3 cup waterers per pen resulting in 27, 13.5, or 9 pigs per cup waterer. Treatments were randomly assigned to pens at placement based on location in the barn in a randomized complete block design. Pigs were fed a common corn-DDGS-soybean meal-based diet throughout the duration of the trial. Pens of pigs were weighed and feed delivery and disappearance were determined approximately every 14 days from the start to conclusion of the trial for determination of average daily gain (ADG), average daily feed intake (ADFI), and feed efficiency (F/G). Cups with fecal material were cleaned daily, and the number of cups cleaned was tracked for each pen in two periods from d 0 to 45, and d 80 to the end of the trial. Data were analyzed using the GLIMMIX procedure of SAS 9.4 (SAS Foundation, Cary, NC). Polynomial contrasts were implemented to assess the functional form of increasing the cup waterer:pig ratio with pen as the experimental unit. Differences in the percentage of days that had waterers cleaned per total days observed was analyzed using a binomial distribution. Growth performance was analyzed as normally distributed. Results were considered significant $P \leq 0.05$, and marginally significant at $P \leq 0.10$.

Results and discussion

Increasing the number of cups per pen resulted in a quadratic ($P < 0.001$) increase in the percentage of days that cups needed to have fecal material removed from d 0 to 45 and a linear increase ($P < 0.001$) from d 80 to 113 (Table 1). The lower percentages of cups needed cleaning during the second period for the 2 and 3 cup treatments suggests pigs are using the multiple cup treatments more frequently. From d 0 to 45, there was no evidence increasing the number of cups improved ADG, ADFI, and F/G. Similarly, no evidence for differences in ADG and F/G were observed from d 45 to 74, but there was a marginal evidence for a linear ($P < 0.074$) increase in ADFI as the number of waterers increased from d 45 to 74. From d 74 to 113, a linear ($P < 0.05$) improvement in ADG and body weight (BW) was observed as the number of cup waterers increased. Contrarily, no evidence of differences was detected for ADFI and F/G from d 74 to 113. Overall, ADG improved linearly ($P < 0.038$) as the number of cup waterers increased, with no evidence for differences in ADFI and F/G. Results from this study indicate the optimal water cup to pig ratio changes as pigs grow. Water availability appears to be more critical at heavier weights as seen by the linear improvement in growth performance as the number of drinking devices increased. However, increasing the number of cups increased management associated with cleaning cups. Further characterization of whether water cup cleanliness is associated with disease transmission and growth performance is needed.

References

1. Roubicek, D. Water metabolism. Animal Growth and Nutrition, H. Hafez and I. Dyer, eds. Philadelphia: Lea and Febiger.1969:353-373.

2. MWPS. Midwest Plan Service. Swine housing and equipment handbook. Publication n. MWPS-8. Iowa State University, Ames; 1983.

Table 1: Effects of increasing the cup waterer to pig ratio on growth performance of growing and finishing pigs[1]

Item[2]	Cups: Pig no.	Cup waterers 1	2	3	SEM	Probability, P <	
		1:27	1:13.5	1:9		Linear	Quadratic
d 0 to 45							
ADG, lb		1.64	1.65	1.67	0.013	0.136	0.771
ADFI, lb		3.06	3.04	3.06	0.028	0.945	0.462
F/G		1.87	1.85	1.85	0.012	0.174	0.510
d 45 to 74							
ADG, lb		1.91	1.91	1.93	0.016	0.431	0.507
ADFI, lb		4.71	4.74	4.81	0.039	0.074	0.623
F/G		2.46	2.48	2.49	0.015	0.108	0.687
d 74 to 113							
ADG, lb		2.10	2.19	2.21	0.031	0.026	0.129
ADFI, lb		5.58	5.63	5.70	0.057	0.155	0.904
F/G		2.66	2.59	2.59	0.033	0.157	0.283
d 0 to 113							
ADG, lb		1.84	1.87	1.88	0.013	0.038	0.646
ADFI, lb		4.44	4.45	4.51	0.033	0.198	0.571
F/G		2.41	2.39	2.40	0.016	0.527	0.259
BW, lb							
d 0		35.6	35.9	35.7	1.17	0.479	0.084
d 45		110.1	110.5	111.3	0.55	0.095	0.699
d 74		165.6	165.9	167.3	0.83	0.158	0.594
d 113[3]		244.4	247.5	248.4	1.48	0.044	0.504
Days with cups cleaned, %							
d 0 to 45		0.1	16.3	36.3	-[4]	0.001	0.001
d 80 to 113		0.5	13.3	29.4	-[5]	0.001	0.581

[1] A total of 1,134 pigs (PIC 359 × Camborough; initial pen average body weight 35.7 ± 1.17 lb) were used in a growth trial with 11 replicate pens per treatment and 27 pigs per pen. Treatments consisted of 1, 2, or 3 cup waterers per pen resulting in 27, 13.5, or 9 pigs per cup waterer. Treatments were randomly assigned to pens at placement based on location in the barn in a randomized complete block design.
[2] Initial body weight was used as a covariate.
[3] Topping of 3 pigs per pen occurred on d 113.
[4] SEM were 0.14, 2.89, and 4.63 for treatments 1, 2, and 3, respectively.
[5] SEM were 0.46, 1.33, and 2.94 for treatments 1, 2, and 3, respectively.

Model to estimate PRRSV introduction in filtered farms with negative pressure

C. Vilalta[1], DVM, PhD; K. Janni[2], PhD; C. Alonso[1], DVM, MSc, PhD; M. Torremorell [1], DVM, PhD
[1]College of Veterinary Medicine, University of Minnesota, Saint Paul, Minnesota; [2]Department of Bioproducts and Biosystems Engineering, University of Minnesota, Saint Paul, Minnesota

Introduction

In order to protect swine farms from airborne porcine reproductive and respiratory syndrome virus (PRRSV) introductions, producers invest in enhanced biosecurity practices including air filtration.[1] Unfortunately, PRRSV introduction in filtered farms can still happen if there is significant building or air filter leakage or if filters have insufficient particulate matter removal efficiency. Understanding the risks associated with virus introduction in filtered farms is important to help producers decide how to best protect their animals. Here we present a filtered sow barn model to estimate in-barn airborne virus concentrations and we assessed the relative importance of virus introduction through the filters and farm leaks. We estimated virus concentrations in barns with filters of Minimum efficiency reporting value (MERV) 14 and 16 ratings, conditions of winter and summer ventilation, used filters, and different distribution of virus concentration by particle size.

Materials and methods

A deterministic model was developed with adjustable inputs including barn size, number of filters, filter removal efficiency, ventilation flow, fan capacity, leakage rate, particle size distribution and viral concentration among some of the most important inputs.[2] Scenarios were evaluated for a 3,000 sow herd with 0.5 filters (2 ft^2) per sow and conditions of winter air flow (20 cubic feet minute (cfm)), summer (200 cfm) and filter particle removal efficiencies of MERV 14 and MERV 16 filters, and a MERV14 used filter (45%, 81% and 90% removal efficiency for particle size ranges of 0.3-1, 1-3 and > 3 microns, respectively). Total virus concentration per cubic meter (RNA virus copies/m^3) were assumed to be100 distributed as follows, 10%, 10%, and 80% for particle size ranges of 0.3-1, 1-3 and > 3 microns, respectively.[2]

Results

The model estimated higher viral concentration inside the barn in winter compared to summer for both, MERV 14 (19.1 vs 13.3 RNA copies/m^3 of air) and MERV 16 (13.9 vs 7.4) filters, although the total amount of virus entering a barn (virus copies per second) were higher in summer (3,250 for MERV 14 and 1,655 for MERV 16) than in winter (536 for MERV 14 and 386 for MERV 16). However, virus introduction through leakage was more important in the winter than summer for both types of filters (Table 1).

As expected, virus distribution by particle size affected the amount of virus introduction through the filter but not the amount entering through leaks. Finally, effect of pressure drop in used filter efficiency indicated increased risk of virus introduction through the filter but leakage rate was not affected.

Table 1: Estimated PRRSV RNA copies in barn concentration per volume of air (RNA copies/m^3) and % of the total shown within brackets

	Winter		Summer	
	Filter	Leak	Filter	Leak
MERV 14	10.5 (55)	8.6 (45)	11.3 (84.5)	2.1 (15.5)
MERV 16	4.5 (33)	9.3 (67)	4.9 (65.6)	2.6 (34.4)

Discussion

The model developed allows users to compare quantitatively different scenarios and assesses the risk of virus introduction into filtered barns. Under the conditions tested, the variables that affected the model the most were filter efficiency, amount of virus present in the air outside the barn and virus distribution by particle size. Model results also indicated that higher barn virus concentrations were obtained with lower mechanical ventilating rates and higher leakage rates. Barn static pressure was depended on type of ventilating system and filter conditions and impacted barn virus concentrations. Higher filter MERV rating indicated higher virus removal by the filter, however risk of leakage increased for MERV 16 filters in particular in the winter.

Although the model was parameterized using data obtained from field measurements and experimental studies,[3] it may not fully represent field conditions but it helps to point out high risk areas and areas for which data is limited and that need further research.

References

1. Dee et al. Further assessment of air filtration for preventing PRRSV infection in large breeding pig herds. *Vet Rec.* 2010 Dec 18;167(25):976-7. doi: 10.1136/vr.c6788.

2. Janni et al. Modeling airborne virus concentrations in filtered swine barns with negative-pressure ventilating systems. ASABE 2017 Paper No. 170058. St. Joseph, MI: ASABE. doi.org/10.13031/aim.201700558.

3. Alonso et al. Concentration, Size Distribution, and Infectivity of Airborne Particles Carrying Swine Viruses. *PLoS One.* 2015 Aug 19;10(8):e0135675. doi: 10.1371/journal.pone.0135675.

Use of tylvalosin (Aivlosin®) water medication for *Mycoplasma hyopneumoniae* elimination

P. Yeske[1], DVM, MS; R Kaptur[2], DVM; D. Rosener[2], DVM; Maria Pieters[3], DVM, PHD; Chris Sievers[1], DVM, MS
[1]Swine Vet Center St. Peter, Minnesota; [2]Pharmgate Animal Health, Wilmington, North Carolina;
[3]University of Minnesota St. Paul, Minnesota

Introduction

There are many examples of herd elimination *Mycoplasma hyopneumoniae* that can be done.[1] Previous *in vitro* studies indicate Tylvalosin (Aivlosin®) is effective against *Mycoplasma hyopneumoniae*[2] and has been used for elimination programs. The decision to embark on an elimination program was made on a swine herd with an acute outbreak of PRRSv and PEDv. With the closure in place for the two viral pathogens, the project was extended to include *Mycoplasma hyopneumoniae.*

Materials and methods

The selected herd was a 1350-sow farrow-to-wean farm. It was closed to new animal introduction for 240 days. Since the sow herd was doing internal replacement of gilts, a negative source was identified to use as a source of replacement gilts until the elimination was completed and then would go back to being closed and internally producing its own replacement gilts. The sow herd was vaccinated with M+Pac Merck® at 30 days prior to starting medication in the water and 78 days prior to completion to booster herd immunity prior to elimination. Just prior to the medication program, the entire gestation facility was washed down, as farrowing rooms are washed with each turn, no extra washing was performed.

Tylvalosin (Aivlosin®) at 25ppm was used as water medication for 28 days to the entire sow herd for elimination. Aivlosin was mixed fresh daily and administered with a proportioner water medicator. All piglets were treated with a dose of Tulathromycin (Draxxin) at the beginning of the water medication period, and again 14 days after the initial injection. After the initial day of the medication program all pigs were treated at birth and again 14 days later. This protocol was performed for 4 weeks as well. A statistical sample of sows were tested using laryngeal swabs for detection of *Mycoplasma hyopneumoniae* genetic material by PCR.

Results and discussion

One hundred eighty sow samples (laryngeal swabs) were collected following the medication of the sow herd and piglets. Samples were collected in duplicate and pooled in 5s. All pools were negative except for 1 pool that resulted suspect. The duplicate swabs were run individually in only the suspect pool. The herd was declared negative based on PCR testing and sentinel gilts and finishing pigs will be followed downstream 6 months following the elimination. Based on these results, Tylvalosin (Aivlosin®) at 25 ppm in water medication to the sow herd and piglet injection with Draxxin® are another tool that can be used for Mycoplasma elimination in farrow-to-wean herds.

References

1. Holst S, Yeske P, Pieters M; Elimination of *Mycoplasma hyopneumoniae* from breed-to-wean farms: A review of current protocols with emphasis on herd closure and medication; AASV/ SHAP / v23n6p321.

2. Tavio M, et al; *In Vitro* activity of Tylvalosin against Spanish field strains of *Mycoplasma hyopneumoniae*; Veterinary Record, published online September 2, 2014, doi: 10.1136/vr.102458.

Preventing PRRS outbreaks in swine environments: Monitoring air filtration for airborne diseases

N. Woolard; M. Osborne, PhD
AAF Flanders, Louisville, Kentucky

Understanding the threat that swine operations face from the virulent and costly porcine reproductive and respiratory syndrome (PRRS) virus, as well as other pathogens with the potential to have a significant impact on herds, production levels, and operating costs, is at the forefront of swine producers' minds. Air filtration is a front-line defense against airborne transmission of PRRS and other pathogens, and comprehensive, real-time information to monitor the biosecurity risk associated with air filtration investment strategies is available to help swine facility managers reduce the potential for PRRS outbreaks. This information is generated by a next-generation monitoring and management program that directly correlates air filtration system performance and Indoor Air Quality (IAQ) by tracking particulate penetration levels in real-time and alerting swine facility managers to indoor air quality threats. An app-driven alert is sent to a mobile device, which signals that particulate levels are at an unacceptable concentration, enabling swine facility managers to resolve issues before animal health is negatively impacted. The program also tracks the life cycle status of air filters and their associated total cost of ownership, in order to optimize change-out points. Effective air filtration and science-based monitoring of the swine barn environment reduces biosecurity risk, lowers cost, and reduces piglet mortality rate.

Pursuing American Board of Veterinary Practitioners (ABVP) certification in Swine Health Management

Jeffrey A. Husa, DVM, Diplomate ABVP

Introduction

How might ABVP credentials benefit you? In a 2017 survey, Swine Health Management (SHM) Diplomates responded with "challenged me to broaden expertise for my producers", "opened doors for career advancement", and "made me a better practitioner." No matter what sparked your interest, we encourage you to come join us!

Applicant requirements

A. Graduate of AVMA accredited / approved college of veterinary medicine, or holder of ECFVG certificate.
B. Legally qualified to practice in the US or other country.
C. Demonstrate unquestionable moral character and ethical professional behavior.
D. Background in swine medicine and production in one of the following ways:
   a. At least 5 years of clinical practice experience prior to credentials submission, and at least 6 years in practice prior to the credentialing exam. The first year need not be in swine practice, however must have worked in a predominantly swine-oriented practice during the 5 years prior to credentialing exam, OR:
   b. Complete an ABVP-approved Residency plus at least 1 year of swine practice.

Entrance examination

Board certification in SHM begins by successful completion of an entrance exam, offered each year at the AASV Annual Meeting. This assesses your basic knowledge of disease, economics, epidemiology and more. To sit the exam, submit an application before midnight on January 15th. Additional details can be found at `www.abvp.com`. The Applicant Handbook is also available at this address, and is an excellent guide throughout the entrance and credentialing processes. Applicants who fail may retake this exam in succeeding years through the same process.

The certification process*

After passing the entrance exam, select an ABVP advisor, subject to approval by the SHM Regent. Then complete a plan of study which you develop in collaboration with your advisor. We recommend that you submit your credentials application within 34 months of your entry exam. Credentials documents include: veterinary diploma, curriculum vitae, synopsis of veterinary practice, self-report of job experience, CE records, three applicant evaluation forms, and two publications: either two case reports, or one peer-reviewed article and one case report. You will also begin studying for the certification exam during this time. Once your credentials are accepted, it's time to study, study, study for the next year. The certification exam is administered during our AASV Annual Meeting. It includes written and oral evaluations with short answer and problem solving components. The credentialing process will push, stretch, and energize you! As with the entrance exam, applicants whose credentials are not accepted or who fail portions of the certification exam may reapply the following year.

Maintenance of certification*

Your credentials remain valid for 10 years. Between the 8th and 10th years after you initially credential, you will recertify through the accumulation of points or by reexamination. Complete details about either pathway can be found in the Maintenance of Certification Handbook available at the ABVP website under Forms and Documents. ABVP urges all Diplomates to submit in their 8th year to best ensure continued success.

Conclusion

Certification in Swine Health Management through ABVP will stretch you! The journey will hone new skills and enhance old ones. You will demonstrate knowledge and your ability to solve critical problems...problems which may make all the difference for you, your producers, their pigs, and consumers world-wide. Come join us!

*Resources

For most current application, credentialing, and maintenance requirements, visit `www.abvp.com`.

Identification of the complete swine reproductive pheromone

John J. McGlone[1,2], PhD; Garrett Thompson[2], MS; Michael Kuhn[2,3], DVM, MBA
[1]Texas Tech University, Lubbock, Texas, [2]Animal Biotech, Dallas, Texas, [3]Westminster, Colorado

Introduction

Litter sizes have steadily increased on commercial farms. Farrowing rates have largely not changed in recent years. Opportunities to improve farrowing rate include better application of boar pheromones that stimulate sow behavior and physiology. Our goals were (1) to identify boar-unique molecules from saliva, (2) test each molecule alone and in combination to determine the most potent pheromone, (3) to apply that behaviorally-active pheromone to weaned sows to determine if it might stimulate LH secretion.

Materials and methods

Three studies were conducted to 1) identify unique boar pheromones, 2) test the behavior elicited by these compounds alone or in combination and 3) evaluate serum LH levels following administration of the compounds.

In study 1[1] saliva was collected from boars and sows at the Texas Tech university farm. SPME-GC-MS was utilized to assay for common and unique volatile molecules. A comparison was made between compounds found in boar saliva compared to those found in sow saliva.

In study 2[2] the three compounds identified in study 1 were administered to sows on a commercial farm either alone or in one of various combinations. All investigational compounds were formulated in alcohol base and administered intra-nasally on days 4 and 5 post-weaning. In each case, sows in different treatment groups were housed in different rooms or air-spaces to avoid cross contamination. Investigators blinded to treatment then performed evaluations on each animal to determine signs of estrus including lordosis, pricked ears, vocalization and standing still. A numerical score was assigned based on intensity of these signs.

In study 3[3] the triple combination of compounds was administered to 3 sows. An indwelling catheter was utilized to collect blood from sows beginning 10 minutes prior to administration of the triple compound. Blood then was then collected at time of administration and continuing every minute for 10 minutes and again at 20 and 30 minutes. Blood was processed to serum and analyzed for presence of Luteinizing Hormone (LH) by ELISA. All results were compared to the T-10 sample.

Results and discussion

Three boar-unique bio-active molecules were identified in saliva: Androstenone, Androstenol, and Quinoline. These same compounds were confirmed in boar saliva collected from intact males at a commercial farm and confirmed the presence of these three molecules. This combination of compounds has been patented as Boar Saliva Analog (BSA).

When administered to 947 sows on a commercial farm, we found that BSA induced the full complement of sexual behaviors – an effect not elicited by any single or paired combination of molecules. Androstenone and androstenol more frequently caused sows to stand still and arch their back (lordosis). Quinoline more frequently caused sows to prick their ears. Only when BSA was present was the full sow sexual behavioral score increased compared with the control.

Analysis of serum LH levels shows an elevation of LH beginning immediately after intra-nasal administration of BSA and continuing for 30 minutes. (Figure 1) Additional studies have shown that BSA increased breeding, conception and farrowing rates in field trials on commercial farms.[4]

BSA is the active ingredient in a new product called BOARBETTER® to be sold in 2018. Use of pheromone molecules provides a natural way to improve reproductive efficiency on commercial farms.

References

1. Study #AB2015-1
2. Study #AB2015-2
3. Study #AB2017-7
4. Kuhn, M., J. J. McGlone, G. Thompson. 2018. Field studies to evaluate the efficacy of BOARBETTER®, a new pheromone spray for sows. AASV industrial partners' presentation.

Figure 1: Serum LH levels of sows in response to Boar Saliva Analog (BSA) administration

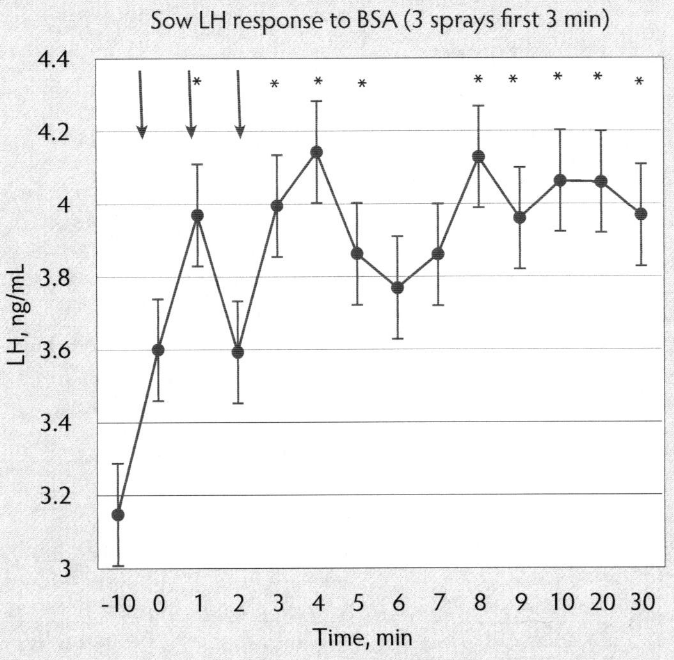

* Indicates significant difference ($P \leq 0.05$) from timepoint -10 minute. Sows (N = 3) on post-weaning day 4 were sprayed with BSA immediately after blood draws at min 0, 1 and 2 (indicated by downward arrows). Serum LH levels were significantly increased ($P > 0.05$, SEp = 0.14) compared to baseline (-10) during min 1 and 3 and post-BSA treatment, min 3, 4, 5, and 8 to 30 as indicated by the asterisk (*). All means were reported using PROC GLM Least Square means ± pooled standard error.

Exploring profitability differences when average wean-finish production metrics show minor numerical differences between groups

Dennis DiPietre[1], PhD; Lance Mulberry[1], BS; Eduardo Fano[2], DVM MC PhD; Mitchell Johnson[2], BS MBA
[1]KnowledgeVentures, LLC$_{USA}$; [2]Boehringer-Ingelheim Vetmedica, Inc

Introduction

It is common in the protocols devised for comparisons of treatments between two products to only use physical production metrics such as comparative finished animal weights, growth rates, feed conversions and mortality. It is also common for all the animals to be sold at once either on a single predetermined date or at the attainment of a predetermined average group weight. While this type of protocol assures fair comparative production metrics it does not match the way US producers market animals and does not measure comparative profitability differences. We illustrate why it is important to consider profit differences and typical producer marketing strategies even when common physical production metrics in comparative groups are essentially equal.

Materials and methods

We use the raw data from a trial comparing the efficacy of Ingelvac FLEXCombo (one dose) and Circumvent PCVM G2 (two doses). In this trial, a total of 704 pigs were randomly assigned to two treatment groups FLEXCombo (hereafter T1) = 331 pigs; Circumvent PCVM G2 (hereafter T2) = 333 pigs and a non-vaccinate control group of 40 pigs. Animals were weaned at 21 days of age and placed in a 1,200-head wean-to-finish barn under similar conditions. On trial day 1, group T1 and T2 were administered vaccine and on trial day 22, T2 received its second dose. All pigs were weighed on trial day 71 and trial day 154 at which time they were taken off test. In the original published research, the authors report essentially equal average growth and equal average finished weight outcomes for both T1 and T2 treatment protocol groups. In addition, the same number of animals died in each treatment group by the completion of the study (17 head) and the culls differed by one animal resulting in statistically indistinguishable cull rates.

We create a stochastic, bio-economic profit optimization model to test for profitability differences in the trial groups by simulating the typical marketing patterns of US producers and optimizing the sale dates for three marketing groups to maximize return over feed costs. We determine the optimal marketing day for a topping load (first marketing; heaviest 16.67% of total group), a closeout load (middle group; next heaviest 66.67%) and a tail-end load, final 16.67%, the lightest weight animals. Using actual individual animal weights from the trial, we fit weight distributions for T1 and T2 at day 71 and day 154 along with estimating four polynomial average growth curves (T1 Barrows, T1 Gilts, T2 Barrows, T2 Gilts) between days 71 and passing through day 154 weights. From these average growth curves, we fit unique individual animal growth curves and forced their collective daily gains between day 71 and the optimal marketing day to be in correlation with the actual trial data. The diet feed ingredients and cost supported an average daily feed intake consistent with an average daily gain of 1.65lbs and average feed conversion ratio of 2.51 from 13 to 290 lb W-F. Feed ingredient prices were randomly selected from distributions fit to recent historical data with means corn \$3.40/bushel; soybean meal/ton \$313.00; and DDGS/ton \$114.63. The animals were priced using a base price randomly selected from a distribution fit to recent historical hog prices with mean \$64.03/cwt in the carcass. Premiums to base were determined using an actual US pricing grid (Schedule 111) anonymized and made publicly available through USDA Packers and Stockyards division.

Results

To avoid calculating the difference associated with different turn lengths, we forced the sale of both marketing groups when the first one reached optimal weight and assume finishing space was not competed for under 175 days. The optimal marketing days were day 141 (tops), day 157 (closeout), day 175 tail enders. Return over feed cost was net -\$0.16/head for T1 vs T2 for the topping load, net \$0.59/head for T1 vs T2 for the closeout loads, and net \$2.70/head for T1 vs T2 in the closeout. Weighted by animals sold per load, total advantage net \$0.814/head for T1 vs T2 (Figure 1.). We did not increase mortalities beyond marketing day 154 as they were identical up to day 154 for each group in the original data. Differences in standard deviation of marketing weights, which are influenced by disease and many other factors, possibly vaccination choice, may result in different profit outcomes even when average weights are not different.

References

1. Eduardo Fano, et al, Comparison of efficacy between two PCV2 vaccination protocols under PCV2d field exposure. AASV. 2017;95-97.

2. Robert Morrison, Making decisions based on economics, not only on performance. London Swine Conference. 2004;119-133.

Figure 1: Three marketing groups live weight lbs

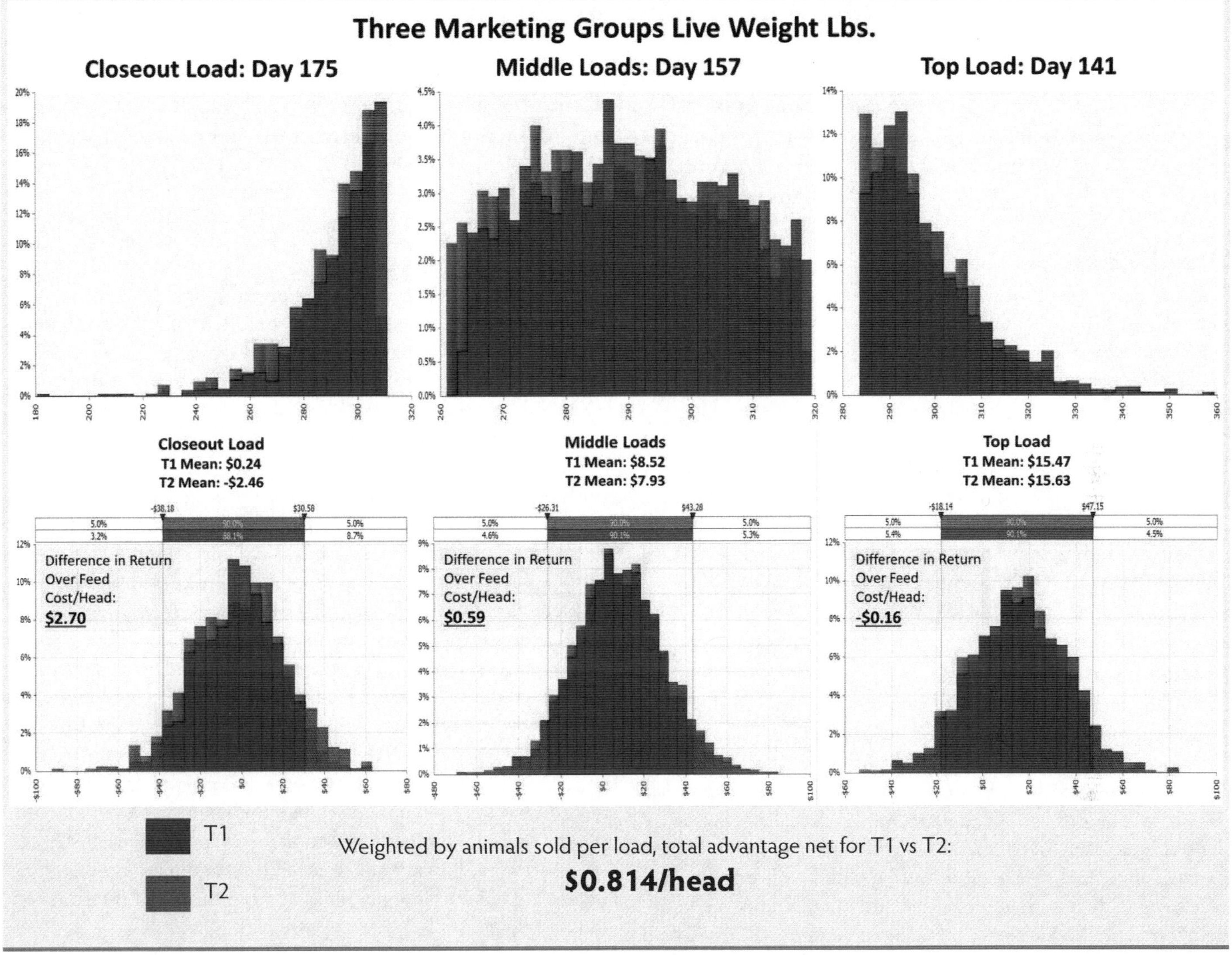

PEDV, PRRSV, and Seneca Valley virus PCR positive flies (*Musca domestica*). What now?

Grant Allison[1], DVM; Phil Gauger[2], DVM, MS, PhD; Jianqiang Zhang[2], MD, MS, PhD; Mark Engesser[3], DVM; Gene Spellman[4], BS

[1]Walcott Veterinary Clinic, Walcott, Iowa; [2]Vet Diagnostic and Production Animal Medicine, Veterinary Diagnostic Laboratory, Iowa State University, Ames, Iowa; [3]ME Swine Service, Rose Creek, Minnesota; [4]Central Life Sciences, Schaumburg, Illinois

Introduction

Despite extensive investments in biosecurity measures and protocols, farms still have unexplainable cases of viral diseases caused by porcine epidemic diarrhea virus (PEDV), porcine reproductive and respiratory syndrome virus (PRRSV), and more recently Seneca Valley virus (SVV). Many potential sources of viruses for these diseases have been studied in the past. Insect vectors such as flies have been investigated as a possible source of PRRSV.[1] An unexplored and possibly overlooked area is the ability of flies to serve as a source (vector) of PEDV and SVV. The purpose of this preliminary study is to determine if flies can harbor PEDV, SVV and/or PRRSV and serve as a vector in the transmission of these economically important viral diseases.

Materials and methods

Oral fluid samples were collected from a 2,400 head finisher in Eastern Iowa that was PCR positive for PEDV 30 days post diagnosis. The 4,800 head finisher in Southern Minnesota was PCR positive for both SVV and PRRSV 24 days post diagnosis. Flies were collected with Captivator® jug liquid traps (Starbar®, Central Life Sciences) to reduce any cross-contamination from the environment that may occur during sample collection. Traps were hung six to a room with a negative control that prevented flies from entering the device but allowed room air to circulate in the traps. Flies randomly enter traps by their own volition. Sterile water was used to solubilize trap attractant. Flies were netted out of the traps into numbered zip-lock bags, frozen and shipped overnight (PEDV) or delivered same-day (PRRSV/SVV) to the Iowa State University Veterinary Diagnostic Lab (ISU VDL). A minimum of 100 flies per trap/bag was targeted. Liquid from the trap was also collected for subsequent PCR testing. Flies were processed as a 10% homogenate solution using phosphate buffered saline. Homogenates were clarified by centrifugation and supernatants were analyzed by real-time reverse transcription polymerase chain reaction (rRT-PCR) per routine protocols at the ISU VDL for PRRSV, PEDV and SVV. The remaining supernatants were stored at – 80°C. Seneca Valley virus has a propensity to grow in cell culture and virus isolation was conducted in H1299 cells (ATCC CRL-5803) per routine procedures at the ISU VDL.

Results and discussion

Iowa PEDV herd: Flies were PCR positive for PEDV with cycle threshold of 34.9 (Ct 34.9). Minnesota PRRSV/SVV herd: one out of six traps were PCR positive for PRRSV (Ct 31.6). Two out of six traps were PCR positive for SVV (Ct 30.5 and Ct 30.4). All fly attractant liquid samples were rRT-PCR negative. Virus isolation attempts were conducted on the two PCR positive SVV fly samples. The Ct 30.5 sample was VI positive and confirmed by PCR. The initial passage of cell culture material (P0) had a Ct of 20, and the subsequent passage (P1) had a Ct of 17. These results indicate that the fly sample contained viable SVV (Table 1). Flies may bypass all currently known biosecurity measures and have multiple routes of entry into swine farms. Flies are ubiquitous, can fly up to four miles,[2] live and breed in close proximity to swine, and can be transported by farm vehicles. There also is a proven model for transmission of a corona virus by a flying vector.[3] Are flies a potential risk factor in the transmission of PED, PRRS and/or SVV? At present, there is no standard protocol for fly control in the industry. There are steps and strategies that can be easily taken to lessen this risk from flies, both in a convenient and economical, e.g. feed-thru, method. The next step is to confirm that flies are positive via PCR for one or more of these agents and conduct a piglet bioassay to confirm that transmission to susceptible swine via flies is possible.

References

1. Otake S, Dee SA et al. Transmission of porcine reproductive and respiratory syndrome virus by houseflies, (Musca domestica Linnaeus). *Vet Rec* 2003. 152: 73-76.

2. Nazni WA, Luke H et al. Determination of the flight range and dispersal of the house fly, Musca domestica (L.) using mark release recapture technique. *Trop Biomed* 2005 Jun;22(1):53-61.

3. *Diseases of Swine*. 9th Edition 2006 pg 494-495.

Table 1: Diagnostics

Virus	Date	Sample	Result	Comment
PED	5/15/17	Rectal swabs	positive (Ct 18.9)	Pool of 5; site status
PED	6/15/17	Oral fluid	positive (Ct 29.6 – 32.8)	4 of 6 OF samples positive
PED	6/15/17	Flies	positive (Ct 34.9)	30 days positive PI site
PRRS	8/04/17	Oral fluid	positive	
PRRS	8/28/17	Flies	positive (Ct 31.6)	Trap (1 of 6 positive)
PRRS	8/28/17	Liquid	negative	Liquid attractant in traps
SVV	8/04/17	Oral swab	positive	
SVV	8/28/17	Flies	positive (Ct 30.5 & 30.4)	Trap (2 of 6 positive)
SVV	8/28/17	Liquid	negative	Liquid attractant in traps
SVV	8/28/17	Flies (Ct 30.5)	positive (Ct 20.0)	Virus Isolation passage "0"
SVV	8/28/17	Flies	positive (Ct 17.0)	Virus Isolation passage "1"

A *Brachyspira* challenge study to evaluate medication options

Chris Puls[1], PhD; Tom Weber[1], PhD; Matt Ritter[1], PhD; Lyle Kesl[2], DVM, PhD
[1]Elanco Animal Health, Greenfield, Indiana; [2]Veterinary Resources, Inc., Ames, Iowa

Introduction

Previous research (Elanco, data on file) suggested that narasin may have *in vitro* efficacy against *Brachyspira hyodysenteriae*. The primary objective of this study was to evaluate the effect of feeding 27.2 g/ton narasin on the severity of enteric disease *in vivo* in pigs challenged with *Brachyspira hyodysenteriae* in comparison to untreated challenged controls or those treated with 35 g/ton tiamulin in the feed.

Materials and methods

This study was carried out over a fixed-time of 28 days using a randomized complete block design (blocking factor = weight) with 3 dietary treatments: 1) Control, 2) 27.2 g/ton narasin, and 3) 35 g/ton tiamulin. Ninety (90) pigs (30/treatment) housed in mixed-sex groups of 5 (6 pens/treatment) were intragastrically challenged over 3 days with a field-isolate of *Brachyspira hyodysenteriae* (Iowa State University, Ames, IA). Individual pig was the experimental unit for all enteric disease parameters (diarrhea, depression, gauntness, and colon lesion scores) and for animal body weights and the incidence of morbidity and mortality. Pen was the experimental unit for growth performance outcomes.

Feed was withheld for approximately 4 hours prior to and 1-2 hours after the challenge procedure, but was otherwise offered *ad libitum* throughout the study. Diarrhea (4-point scale), depression (4-point scale) and gauntness (3-point scale) scores were collected daily by a masked observer. Body weights were recorded on Days 0, 7, 14, 21, and 28, and colon lesions (4-point scale) were scored on Day 28 or at the time of necropsy if the animal died or was euthanized.

Data were analyzed using JMP12.1 (SAS Inst. Inc., Cary, NC). Mean daily depression, diarrhea and guantness scores, along with the colonic lesion scores and growth performance data were analyzed using mixed model methodology in the Fit Model function. When ANOVA indicated a significant difference ($P < 0.05$), means were separated using Tukey's HSD test. Diarrhea incidence and animal morbidity and mortality data were analyzed via logistic regression using the Fit Model function. Means were compared using odds ratios when the overall model was significant ($P < 0.05$).

Results and discussion

Feeding tiamulin at 35 g/ton resulted in lower ($P < 0.05$) colon lesion score, incidence of diarrhea, and mean diarrhea, depression, and gauntness scores compared to the other treatments. Compared to the negative control, feeding narasin at 27.2 g/ton resulted in lower ($P < 0.05$) incidence of diarrhea and lower ($P < 0.05$) mean diarrhea, depression, and gauntness scores. However, feeding narasin had no effect ($P > 0.05$) on colon lesion score.

Pigs fed tiamulin were heavier ($P < 0.05$) than pigs on the negative control, with pigs fed narasin being intermediate ($P < 0.05$). Additionally, pigs fed tiamulin had the lowest ($P < 0.05$) incidence of morbidity and mortality (0%) of the three treatments; pigs fed narasin had lower ($P < 0.05$) incidence of morbidity and mortality than those on the negative control.

For the overall growth period (Day 0 to 28), pigs fed tiamulin at 35 g/ton had greater ($P < 0.05$) ADG and ADFI than the other treatments. Pigs fed narasin at 27.2 g/ton had improved ($P < 0.05$) ADG and ADFI compared to the negative control. Pigs fed tiamulin at 35 g/ton had improved ($P < 0.05$) G:F compared to the negative control, with those fed narasin being intermediate and different ($P < 0.05$) than the other two treatments for this trait.

The results of this study suggest that feeding narasin to pigs challenged with *Brachyspira hyodysenteriae* may lower the incidence of clinical signs of enteric disease and improves growth rate, feed intake, and gain efficiency. However, the lack of improvement in colon lesion score compared to the negative control is of concern and minimizes the clinical significance of these improvements. In addition, these data validate the high efficacy of tiamulin against *Brachyspira hyodysenteriae*.

Table 1: Effects of dietary treatment on enteric disease outcomes.

Item	Dietary treatment: Negative control	Narasin 27.2 g/ton	Tiamulin 35 g/ton	SEM	*P*-value
No. of pigs	30	30	30	-	-
Disease parameters					
Colon lesion score[1]	2.69^{a}	2.40a	0.00^{b}	0.13	< 0.0001
Diarrhea incidence, %	100^{a}	83.33^{b}	6.67^{c}	--	< 0.0001
Mean diarrhea score	1.35^{a}	1.02^{b}	0.02^{c}	0.04	< 0.0001
Mean depression score	0.69^{a}	0.45^{b}	0.001^{c}	0.02	< 0.0001
Mean gauntness score	0.73^{a}	0.53^{b}	0.005^{c}	0.02	< 0.0001
Growth performance					
Individual animal data					
No. of pigs	30	30	30	-	-
Body weight, kg					
Initial	11.2	11.2	11.3	0.3	0.98
Final	15.4^{c}	23.1^{b}	29.8^{a}	1.2	< 0.0001
Morbidity and mortality, %	60.0^{a}	36.7^{b}	0.0^{c}	--	< 0.0001
Pen performance data					
No. of pens	6	6	6	-	-
Overall ADG, kg/d	-0.026^{c}	0.234^{b}	0.661^{a}	0.036	< 0.0001
Overall ADFI, kg/d	0.403^{c}	0.646^{b}	0.978^{a}	0.042	< 0.0001
Overall G:F	-0.065^{c}	0.345^{b}	0.677^{a}	0.057	< 0.0001

^{a,b,c} Means within a row with different superscripts differ ($P \leq 0.05$).

Key VFD learnings from 2017

B. Cowan[1], MS; J.T. Holck[2], DVM, MS, MBA; C. J. Rademacher[3], DVM
[1]GlobalVetLINK, Ames, Iowa; [2]Feed His People, LLC, Gilbert, Iowa;
[3]Iowa State University Swine Extension, Ames, Iowa

Introduction

Implementation of the Veterinary Feed Directive (VFD) Final Rule created several challenges for stakeholders in the livestock industry in 2017. The intent of this paper is to share key learnings to minimize issues in the future.

Critical issues

Stakeholder education

A primary challenge to implementation was stakeholder knowledge of the new VFD rule including the legal scope of a VFD. A "One" approach was utilized by GlobalVetLINK (GVL), Iowa State University Swine Extension, and others to define the VFD scope as one veterinarian, one client, one feed distributor, one (or more) production sites, and one VFD drug (along with any FDA approved combination drugs). Defining the number of head as the total number of animals treated for the entire length of the VFD (animal spaces X turns of a barn) and clarifying that production classes, indications of use, dose, and duration were taken directly from FDA approved drug labels were common stakeholder questions. No extra-label use in feed medications was emphasized for all stakeholders.

Record keeping

The two year VFD retention requirement for veterinarians, clients, and feed distributors increased the need for collaboration and communication among these stakeholders. Electronic solutions such as GVL FeedLINK®, which enables compliance and meets retention requirements, and PrairiESystems Feed Allocation System, which confirms a VFD is present prior to allowing a feed order, are valuable tools for stakeholder collaboration and compliance.

Learnings

General impressions

Primary concerns recorded by GVL during 2017 included: **(1)** whether to write a VFD for a group of animals, an entire site, or an entire flow of animals – by group allows for better tracking but requires more VFDs to be written while by writing by flow requires fewer VFDs but also provides less detailed oversight. Of the approximately 50,000 FeedLINK swine VFDs written in 2017, 73% were written for groups of less than 5,000 pigs (Figure 1); **(2)** educating producers to label use of medicated feeds; and **(3)** awareness of which drugs are now VFD status.

Ask CVM

FDA Center for Veterinary Medicine established an "Ask CVM" process to respond in a timely manner with answers to industry questions related to the VFD ruling. A comprehensive list of these questions and answers is referenced below.1 The following is an example of a commonly asked question:

Q: Producer A wants to feed 2 pulses of chlortetracycline (CTC) in the nursery phase (8 week duration of growth). In a single group of a 1,000 pigs, they get 2 weeks of CTC at the beginning (weeks 0-2 at 15 lbs. of weight) and again the last 2 weeks of the group (weeks 7-8 at 60 lbs. of weight). How many VFD's are required for this group?

A: A veterinarian cannot issue a VFD that authorizes a duration of use that is inconsistent with the directions for use described on the product labeling. In the example provided, if the approval limits the treatment to 14 days, the VFD can only authorize that approved duration. Issuing a VFD that authorizes a 14-day course to be repeated for the same animals would be considered an illegal extra-label use. However, if the veterinarian reassesses the animals involved after a single course of therapy (i.e., drug administered according to the labeled dose and duration), the veterinarian may decide that additional therapy is warranted.

Summary

Several challenges related to implementation of the VFD ruling were experienced in 2017. GVL, Iowa State University, FDA CVM, and others have helped educate veterinarians, producers, and feed distributors for a successful transition. Keep current with recent Ask CVM questions and answers at the reference below.

References

1. Iowa Pork Industry Center, Veterinary Feed Directive Questions and Answers Supplied by FDA Center for Veterinary Medicine. 2017. Available at: `www.ipic.iastate.edu/info/VFDfaq.pdf`. Accessed 13 November 2017.

Figure 1: GVL FeedLINK 2017 Swine VFDs

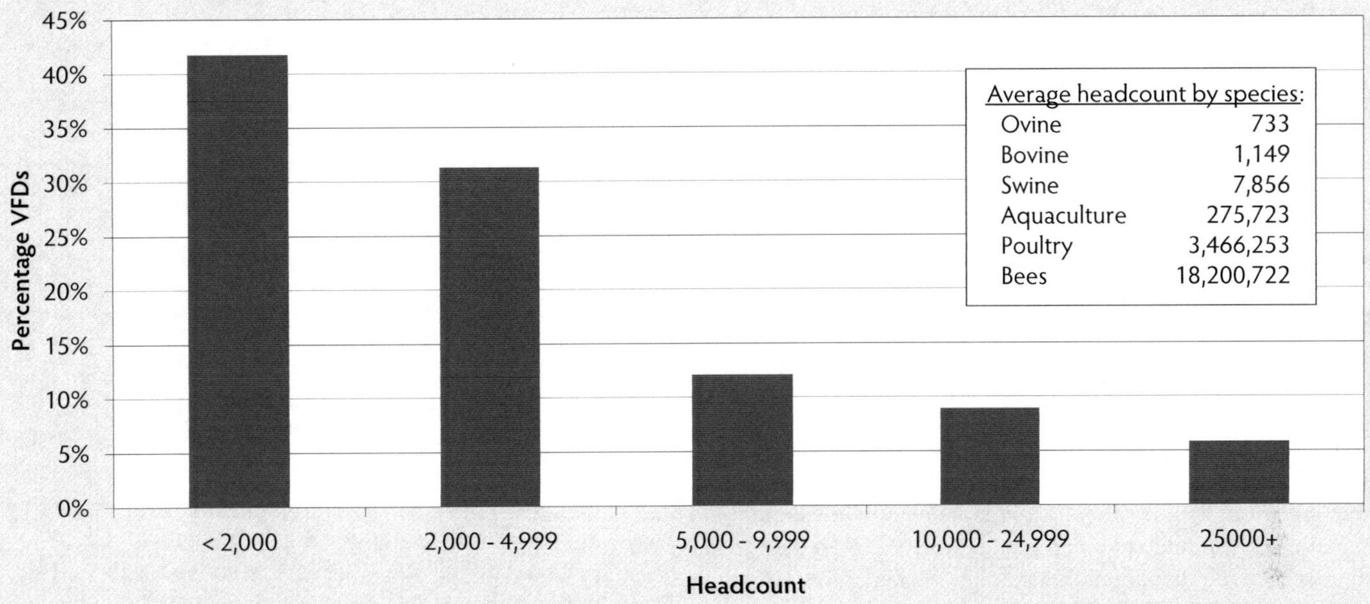

Efficacy of Porcilis™ Ileitis against *Lawsonia intracelluaris* challenge and post-vaccination seroresponse kinetics

Brad Thacker[1], DVM, PhD; Greg Armbruster[1], DVM; Brett O'Brien[1], VMD, MS; Robyn Fleck[1], DVM; Nathan Winkelman[2], DVM; Adam Mueller[2], DVM

[1]Merck Animal Health, Madison, New Jersey; [2]Swine Services Unlimited, Inc., Rice Minnesota

Introduction

Porcilis™ Ileitis (Merck Animal Health - MAH, Madison, New Jersey, USA) was introduced to the US market in late 2015 as an aid in the control of ileitis caused by *Lawsonia intracellularis,* as an aid in the reduction of colonization by *Lawsonia* and as an aid in the reduction of duration of fecal shedding.[1] The main objective of this study was to evaluate the efficacy of Porcilis™ Ileitis using a challenge model developed and performed by an independent investigator under laboratory conditions similar to licensing studies performed by MAH Research and Development.[2] In addition, this challenge model is used to inoculate by oral gavage seeder pigs for field studies conducted by MAH and we wanted to better characterize the disease induced by the direct challenge.[3] A secondary objective was to measure weekly serum antibody responses following vaccination to determine if and when serotesting could be used for compliance monitoring.[1]

Materials and methods

Seventy, three-week-old, cross bred, *Lawsonia* seronegative pigs were used in the study. On study day -1, pigs were tagged, and their gender and litter were recorded at the sow farm. Pigs were then weaned and moved to an isolation facility for the remainder of the study. At arrival, pigs were weighed, bled and randomly allocated to treatment within litter with blocking by start weight and gender. The next day (study 0), 40 pigs were vaccinated per label directions (2 mL, IM) with Porcilis™ Ileitis and 30 pigs were not vaccinated (controls). Ten of the vaccinated pigs were randomly assigned to be bled weekly. All pigs were weighed and bled on study days -1, 40 and 64. On study day 40, all pigs except for the 10 weekly bled pigs were challenged by oral gavage with 50 mL of a gut mucosal tissue homogenate containing approximately 1.03×10^9 of *Lawsonia*. Pigs were then observed every second day for general demeanor, abdominal distension and fecal consistency and color. Fecal samples for qPCR testing were collected on 0, 15 and 22 days post challenge (DPC). On 24 DPC, pigs were necropsied and the intestinal tracts were removed for macroscopic lesion scoring and sample collection. For macroscopic lesion scoring, jejunum, ileum, cecum and colon were scored for lesion severity and length. These values were used to calculate a total score that represented the sum of lesion severity (0-3 scale) times the lesion length (cm) for the four segments. For histopath (0-2 scale) and immunochemistry (IHC; 0-3 scale) evaluations, ileum was collected 1 inch proximal to the ileocecal junction. For tissue PCR, ileal mucosa was scraped 2 inches proximal to the ileocecal junction. All laboratory testing except for serology was done at the Iowa State University Diagnostic Laboratory using standard methods and previously reported scoring criteria.[2] For serum antibody testing, PCV2 ELISA (bioScreen Ileitis Antibody ELISA, Svanova, Uppsala Sweden) was done by MAH. Statistical analysis was done by Kruskal-Wallis One-Way AOV, General One-way AOV, Chi Square or Fisher Exact Test depending on data type and distribution (Statistix™ 10, Analytic Software, Tallahassee, FL).

Results and discussion

The results are presented in Table 1. Compared to control pigs, Porcilis™ Ileitis vaccinated pigs had significantly less macroscopic and microscopic lesions and lower levels of tissue infection, fecal shedding and feces discoloration while weight gain from challenge to necropsy trended to be greater. The percent reduction by parameter in vaccinated pigs compared to control pigs was: total lesion score, 94.4%; histopath, 69.1%; IHC, 69.4%; mucosal tissue infection, 99.0%; and fecal shedding, 99.3% and 99.8% at 15 and 22 DPC, respectively. This level of protection was similar to what we have reported in previous studies.[1,2] Likewise, these results characterized the severity of this challenge model and demonstrated the efficacy of Porcilis™ Ileitis against this model. For the weekly bled pigs, seroconversion (6/10, avg. % inhibition = 32.9%) was first detected 2 weeks post vaccination and thereafter the rates of seropositive pigs each week varied between 8 to10 out of 10 while the average percent inhibition by week varied between 51.8% and 70.0%.

References

1. Thacker B et al. Field experiences with Porcilis™ Ileitis: The newest tool for controlling *Lawsonia intracellularis*. AASV. 2017;191.

2. Roerink F et al. Porcilis™ Ileitis: 20-week duration of immunity against *Lawsonia intracellularis* challenge. AASV. 2016;144.

3. Winkelman N et al. Validation of an ileitis mucosal homogenate seeder pig model to study the natural transmission of *Lawsonia intracellularis* in grow-finish pigs. AASV. 2017;170.

Table 1: Results of vaccination – challenge study comparing Porcilis™ Ileitis vaccinated pigs to non-vaccinated control pigs

Parameter	Variable		Vaccine group: Porcilis	Vaccine group: Control	*P* Value
Group size	**No. pigs**		**29**	**30**	
Macroscopic lesions	% pigs with score	0	58.6	16.7	0.0003
		1	24.1	36.7	
		2	17.2	10.0	
		3	0.0	36.7	
	Avg. lesion length		4.6	51.5	0.0002
	Avg. total lesion score		7.4	132.1	0.0002
Microscopic lesions	Avg. score		0.34	1.10	0.0011
	% pigs with score	0	82.8	40.0	0.0025
		1	0.0	10.0	
		2	17.2	50.0	
Immunochemistry	Avg. score		0.48	1.57	0.0008
	% pigs with score	0	82.8	40.0	0.0057
		1	0.0	10.0	
		2	3.4	3.3	
		3	13.8	46.7	
Tissue infection levels	Copies per mL (median)		2,238,324	232,352,150	<0.0001
Fecal shedding	Copies per mL (median)	15 DPC	585,106	89,785,520	0.0011
		22 DPC	60,068	30,307,582	0.0004
Weight gain challenge to necropsy	Pounds per day		1.94	1.79	0.0738
Lawsonia ELISA	Percent inhibition	At 1st Vac.	-11.5	-12.2	0.6431
		At Chall.	53.5	0.1	0.0000
		At Necr.	86.0	77.2	0.0014
	Percent seropositive	At 1st Vac.	0.0%	0.0%	NC*
		At Chall.	89.7%	0.0%	0.0000
		At Necr.	100.0%	100.0%	NC

* NC = not calculated

Streptoccus suis: Research overview

Brittanny Wiener, BS; Eric Bumgardner, MS; Paulraj Lawrence, PhD
Biological Development and Research Department, Newport Laboratories Inc., Worthington, Minnesota

Introduction

Streptococcus suis is a gram positive bacterium that causes septicemia, bronchopneumonia, arthritis, and fatal meningitis in young piglets. Piglets are usually infected through exposure to asymptomatic sows. Thirty-five different serotypes of *S suis* have been identified,[1] however, recent work has suggested that only 29 of these are *S suis* with the remaining six originating from different *Streptococcus* species.[2,3] In addition, many field isolates remain non-typeable using traditional methods. Of the 35 serotypes described, serotype 2 is most commonly associated with disease, but serotypes 1, 1/2, 3, 4, 5, 7, 8, 9 and 16 are also commonly isolated from diseased swine populations in the United States.[1] Newport Laboratories has developed a comprehensive, unambiguous method of capsular typing based on whole genome sequencing and bioinformatics.[4] This capsular typing approach can accurately identify serotypes even when strains possess novel capsular loci and can be used to determine which strains to include when formulating herd specific vaccines. However, it is essential to assess the performance of vaccines formulated using this approach in a vaccination-challenge model in order to determine whether these vaccines might benefit swine producers. This study was designed to evaluate the use of two different types of growth media, two inactivation methods, and two adjuvants in order to identify the best method for producing vaccines against *S suis*. The results indicate that in addition to including the correct serotype strain in the vaccine, the selection of the appropriate adjuvant is essential for producing a vaccine that is able to protect pigs from challenge. The optimal vaccine formulation was able to provide complete protection against a homologous lethal challenge in pigs.

Materials and methods

A group of naïve pigs was divided into nine treatment groups and a control group. Treatment groups were designed to test the role of the media used to produce the antigen, the method used for inactivation, and the adjuvant used in formulation in inducing a protective response against a homologous challenge. Animals were vaccinated two times at a three week interval and were challenged approximately two weeks after their last vaccination. Clinical signs and mortality were evaluated post-challenge (Figure 1). Injection site reactions were evaluated post-vaccination.

Results and discussion

Vaccines formulated using both inactivation methods afforded the same level of protection regardless of which method was used. Vaccines formulated using antigen grown in Gopher Gold medium did have slightly less morbidity and mortality than those formulated with antigen grown in tryptic soy broth (TSB). However, the difference was only measurable when vaccines were formulated with adjuvant B. Vaccines formulated in adjuvant A afforded substantially better protection than vaccines formulated with adjuvant B. The vaccine formulated with antigen grown in TSB, inactivated with formalin, and adjuvanted in adjuvant A provided complete protection from a lethal homologous challenge. No morbidity or mortality was observed post challenge in this group. There were mild injection site reactions, mostly redness and swelling, which subsided with no blistering/granuloma or tissue damage. Overall, it appears that when used in conjunction with a robust serotyping method, inactivated vaccines can provide complete protection against a lethal *S suis* challenge.

References

1. Higgins R et al. Description of six new capsular types (29-34) of *Streptococcus suis*. J Vet Diagn Invest. 1995;7:405.

2. Hill JE et al. Biochemical analysis, cpn60 and 16S rDNA sequence data indicate that Streptococcus suis serotypes 32 and 34, isolated from pigs, are *Streptococcus orisratti*. Vet Microbiol. 2005; 107:63.

3. Tien LHT et al. Reappraisal of the taxonomy of *Streptococcus suis* serotypes 20, 22, 26, and 33 based on DNA-DNA homology and sodA and recN phylogenies. Vet Microbiol. 2013;162:842.

4. Lawrence PK et al. Map-based serotyping of *Streptococcus suis* field isolates for vaccine formulation. Proc. AAVLD Ann Meet. 2015.164.

Figure 1: Survivability post-challenge

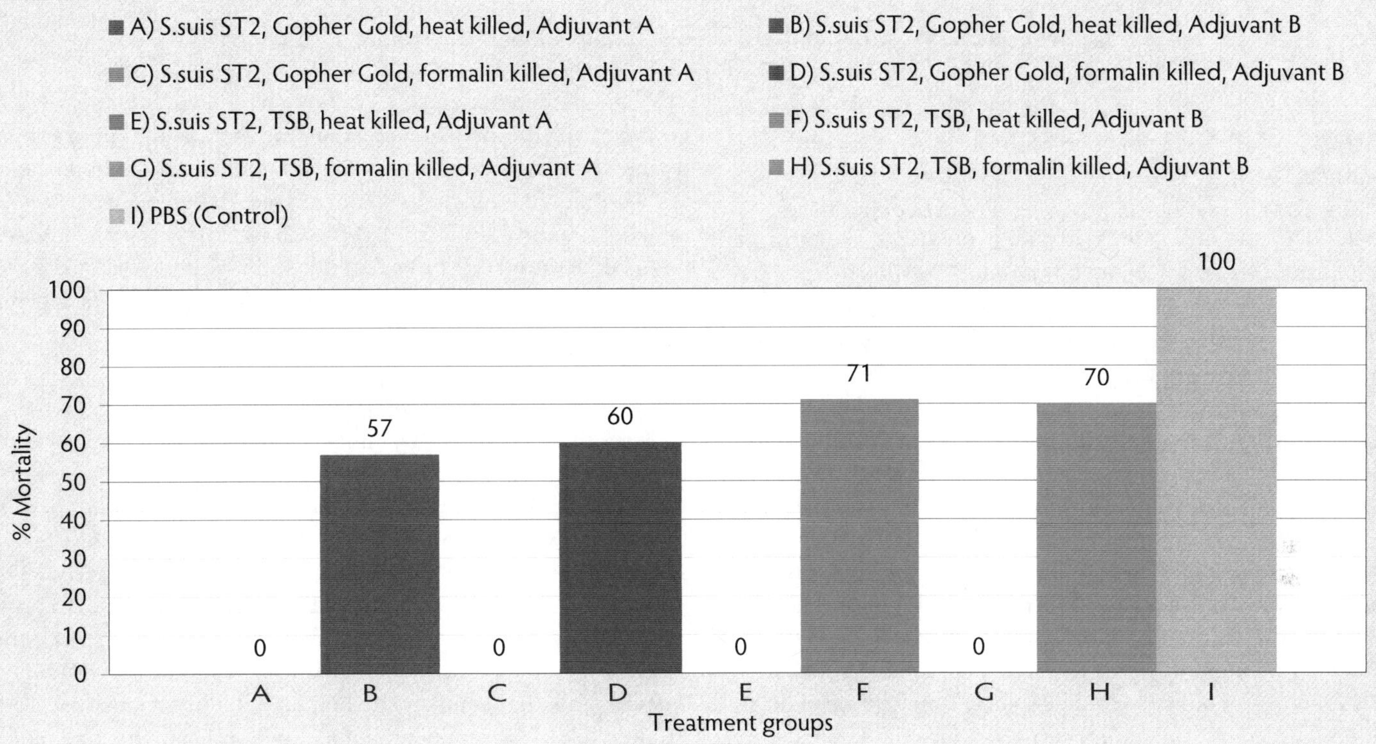

Mineral source matters conception to market

B. Lawrence[1], PhD; T. Loeffler[1], DVM; Y. Ma[1], PhD; L. Greiner[2], PhD
[1]NOVUS International Inc., St. Charles, Missouri; [2]Carthage Veterinary Services, Carthage, Illinois

Introduction

Zinc (Zn), copper (Cu), and manganese (Mn) function in numerous physiological processes. Pharmacological Zn (1,500 – 3,000 ppm) and Cu (125 – 250 ppm) have also been used for diarrhea prevention. Clinical mineral deficiencies are rare. Mineral inadequacy, however, may manifest as reduced appetite, dermatitis, reduced birth weight, anemia, low fat soluble vitamin status, structural abnormalities or compromised locomotion, osteochondrosis, slipped tendons, reduced reproductive efficiency/low fertility, and increased morbidity/mortality. Supplying trace minerals from the most available source to the sow and offspring from conception to finish is critical for optimizing health and performance while minimizing mineral interactions that reduce mineral availability. Supplying trace minerals from the most available sources is also important for reducing environmental footprint. Metal (Zn, Cu, or Mn) Methionine Hydroxy Analogue Chelate (MMHAC) is an American Association of Feed Control Officials recognized class of chelated organic trace that meets these needs.

Materials and methods

Mineral source impact on pigs from conception to market was evaluated in two trials. Trial 1 was a 3 year sow trial using sister sow farms in which 50% of the total trace mineral (165 ppm Zn, 16 ppm Cu, and 38 ppm Mn) was supplied as MMHAC to one farm. The other sow farm received 100% of Zn, Cu and Mn as sulfate. Near the end of the sow trial, individual pig weights from 130 litters, comprising approximately 1,690 pigs per dam mineral source, were evaluated for birth weight according to mineral source. Also near the end of the sow trial, 1,200 pigs from dams fed each mineral program were weaned at approximately 20-d of age and 13.4 pounds. Pigs were assigned to common diets at 25 pigs per pen with pen gain and feed/gain measured at days 0, 10, 21, 42 and 161. Trial 2 was performed wean-finish in which Cu was supplied as either MMHAC or Tribasic Copper Chloride (TBCC). Approximately 1,725 pigs (14.6 pounds and 21-d of age) were housed at 25 pigs per pen and 23 pens per treatment in a wean-to-finish facility. Dietary treatments were Cu as MMHAC or TBCC and a negative control (0 ppm supplemental Cu). TBCC was fed at 150 ppm Cu throughout the study. Pigs fed Cu as MMHAC received 150 ppm Cu day 0 – 7, 80 ppm Cu day 7 – 42, and 60 ppm Cu day 42 post-wean to market. Initial pig marketing occurred at 153 days post-weaning (4 pigs/pen) , a second marketing occurred at d-163 (14 pigs/pen), with final marketing at d-179.

Results and discussion

Substitution of 50% of sow mineral with MMHAC increased piglet birth weight by 0.24 pounds per pig ($P < 0.001$) across parities. Statistical improvements of 0.24 and 0.32 pounds per pig in birth weight were observed with MMHAC fed sows for Parity 2 ($P < 0.05$) and Parity 3+4 ($P < 0.001$) respectively. MMHAC substitution for inorganic mineral also improved post-wean offspring performance. Pigs from MMHAC fed dams were heavier ($P < 0.01$) at days 10, 21, and 42 post-wean as well as heavier ($P < 0.02$) at market (1.85, 2.56, 5.47, and 4.30 lbs/pig respectively). The results from these trials indicate that although no deficiency symptoms where present, and total mineral supplementation met or exceeded recognized requirements, mineral source was important for improving piglet birth weight as well as wean-finish weight gain, particularly during the nursery period. In the wean-finish Cu trial, weight of pigs marketed was greater (8.1 lbs/pig) for the MMHAC group ($P < 0.01$) compared to 0 supplemental Cu (Table 1). Gain between the first and second marketing periods was also greatest ($P < 0.05$) for the MMHAC group. Across all pigs marketed, supplying Cu as MMHAC wean-finish resulted in 4.5 lbs/pig more liveweight. Feed conversion was improved with MMHAC Cu compared to 0 supplemental Cu during days 0 – 7 ($P < 0.10$), 7 – 21 ($P < 0.05$), 42 – 153 ($P < 0.01$), and days 153 – 165 ($P < 0.05$). These results highlight the importance of Cu supplementation for gain and efficiency wean-finish and the ability of Cu as MMHAC to be used at a lower level to achieve the same or better performance as a higher levels of inorganic Cu. The results of these trials also support the importance of minerals source for optimal pig gain and efficiency from conception to market.

Table 1: Impact of Cu source on wean-finish growing pig performance

	Control	MHAC-Cu	TBCC	*P* <
Gain, lbs/d				
Day 0-7	0.282	0.306	0.288	0.09
Day 7-21	0.780[a]	0.815a[b]	0.818[b]	0.02
Day 21 - 42	1.468[a]	1.479[ab]	1.523[b]	0.04
Day 42 - 153	1.966	1.968	1.999	0.12
Day 153 - 165	2.319[a]	2.482[b]	2.321[a]	0.05
Day 165 - 179	1.689	1.759	1.561	0.12
Feed /gain				
Day 0-7	1.477	1.356	1.393	0.09
Day 7-21	1.285[b]	1.245[a]	1.253[ab]	0.01
Day 21 - 42	1.582	1.592	1.574	0.13
Day 42 - 153	2.710[b]	2.652[a]	2.688[ab]	< 0.01
Day 153 - 165	2.631[b]	2.475[a]	2.666[b]	0.5
Day 165 - 179	4.566	4.310	4.854	0.19
Top pigs wt., lbs				
Day 153	286.81[a]	294.95[b]	290.94[ab]	< 0.01
Day 165	300.02	303.35	303.21	0.24
Closeout d-179	301.39	304.36	302.16	0.34

How accurate and precise are current options for testing deoxynivalenol?

Sarah Weiland
NutriQuest, LLC

Introduction

Mycotoxin levels in grains can increase animal feeding costs by decreasing use of distillers dried grains with solubles (DDGS), driving up the price of uncontaminated grains, and increasing the need for additives or binders to mitigate the toxin impact. To aggressively use DDGS and make informed decisions regarding other grains, it is crucial to accurately and precisely test grains for mycotoxins. With many lab and quick test options currently on the market, it can be difficult to determine which is the most reliable. With this issue in mind, NutriQuest performed a proficiency test to determine the variability and accuracy of labs and quick tests currently on the market for the analysis of DON in corn and DDGS.

Materials and methods

This proficiency test was completed to evaluate the accuracy and precision of seven labs and five quick tests currently testing deoxynivalenol (DON) levels in animal feed ingredients. The following labs were analyzed in this proficiency test: Midwest Labs, NDSU, SGS, Trilogy Labs, MVTL, Romer Labs, and Neogen. Quick tests analyzed were: Charm ROSA® DONQ2, Envirologix QuickScan DON FLEX, Neogen Reveal® Q+, Romer AgraStrip® DON WATEX, and Vicam DON-V®. Two subsamples were pulled from eight corn and eight corn DDGS samples to be analyzed in duplicate by each lab for a total of four observations per initial sample. The same subsampling procedure was used for quick test analysis with eight corn samples and seven corn DDGS samples, again tested in duplicate for a total of four observations per initial sample. All quick tests were completed following test protocol by the same technician trained in each test. Samples were also analyzed in quadruplicate by the Schmale Lab at Virginia Tech which was used as a control. All deoxynivalenol levels were reported in parts per million (ppm). Accuracy was measured calculating z-scores using the Virginia Tech sample mean and standard error as the population mean and standard error. Variability within lab and quick test was measured by calculating coefficient of variation (CV) among the four readings per sample.

Results

For the seven labs analyzed, average CV ranged from 16-60% in corn and 2-10% in corn DDGS. Lab z-scores ranged from 5.04-9.16 in corn and 1.11-16.39 in corn DDGS. It was common for the same corn sample to yield DON results varying 2 ppm and the same corn DDGS to yield DON results varying up to 6 ppm among the different labs. For the five quick tests analyzed, average CV ranged from 7-22% in corn samples, and 3-20% in corn DDGS; only two quick tests had CV below 10% for both commodities. Average z-scores in corn ranged from 6.97-42.80, with only two quick tests below 10. Average z-scores for corn DDGS ranged from 6.67-11.51, with three quick tests below 10. It was common for the same corn and corn DDGS to yield DON results ranging 3 ppm among the different quick tests.

Conclusion

This proficiency test identified two labs and two quick tests that perform far superior to their competition. The analysis also demonstrated that the reliability of mycotoxin testing data is largely impacted by the labs or quick tests used.

Comparing the immunogenicity of Circo/MycoGard™ with other combination vaccines used in the field

A. Hintz[1], DVM; R. Kaptur[1], DVM; D. Rosener[1], DVM; T. Schwartz[2], DVM
[1]Pharmgate Animal Health, Wilmington, North Carolina; [2]Suidae Health and Production, Algona, Iowa

Introduction

Enzootic pneumonia caused by *Mycoplasma hyopneumoniae* (M hyo) is one of the most economically significant respiratory pathogens in swine,[1] while porcine circovirus type 2 (PCV2) and PCVAD remains an issue in commercial herds throughout the swine industry.[2] Vaccination with a combination product including PCV2 and M hyo is utilized successfully to minimize the effects of the two pathogens. There are multiple commercial products available with different dosing regimen. Circo/MycoGard™ (CMG) is the latest PCV2/M hyo combination vaccine to the market, with a ready-to-use 1 dose, 1 mL delivery method that is approved for administration to pigs as young as 10 days of age. The objective of this study was to determine the immunogenicity of CMG compared to current combination vaccines, along with a comparison of different dosing strategies using CMG.

Materials and methods

One hundred-seventy-five (175), 3-week old, weaned crossbred pigs were tagged, weighed and randomly allocated to 5 pens with 35 pigs/pen across 7 treatment groups, totaling 150 pigs in vaccinated groups and 25 pigs in the control group. These animals were sourced from a M hyo negative flow of pigs. Each pig was randomly assigned to one of seven treatments: 1. Circo/MycoGard (1 dose, 1 mL) 2. Circo/MycoGard (2 dose – 0.5 mL given twice – this is an extra-label dosing protocol) 3. Circo/MycoGard + MycoGard-1™ (2 dose – 1 mL of each mixed together with 1 mL given twice – this is an extra label dosing protocol) 4. FLEXCombo® (1 dose – 2 mL) 5. Circumvent® PCV M G2 (1 dose – 2 mL) 6. Circumvent® PCV M G2 (2 dose 1 mL given twice) 7. Non-vaccinated control (Sterile saline – 1 mL). The pigs were weaned at 3 weeks of age and given the first dose of vaccine. Those receiving a 2nd dose were vaccinated again at 6 weeks of age, and the study terminated at 10 weeks of age. All pigs were bled at 3, 6, and 10 weeks of age as well and samples were ELISA antibody tested for PCV2 and M hyo at the Iowa State University Veterinary Diagnostic Lab. Blood samples were collected prior to the vaccine being administered on the weeks of sampling.

Results and discussion

Pigs that received 1 dose of CMG had a higher percentage of positive M hyo ELISA results (24%), than those that received one dose of FLEXCombo® (0%) or Circumvent® PCV M G2 (16%) at 10 weeks of age. Those pigs that received two doses of CMG (groups 2 and 3) had at least 75% positive M hyo ELISA results at 10 weeks of age. The pigs that received 1 dose of CMG had 48% positive titers for PCV2 at 10 weeks of age, with higher percentages for those pigs receiving multiple doses. There also was no titer response in the control group for PCV2 nor M hyo, indicating these ELISA positive animals were due to vaccination (Table 1). These results indicate that Circo/MycoGard produces at least as good of a humoral immune response against PCV2 and M hyo as the current combination vaccines on the market with a smaller dose volume, along with fewer adverse reactions noted by field veterinarians.[3]

References

1. Holst S et al. Herd closure and medication protocols for elimination of *M hyopneumoniae*. *J Swine Health and Prod.* 2015;23:321-330.

2. Madson D, PCV2 infection/disease in the breeding herd: An update. AASV. 2012; 445-448.

3. Pharmgate Animal Health feedback from field veterinarians.

™ Circo/MycoGard and MycoGard-1 are trademarks of Pharmgate Animal Health, Wilmington, NC.

Table 1: PCV2 and M hyo ELISA results

Group	PCV2 Dilution > = 160 3 weeks of age	PCV2 Dilution > = 160 6 weeks of age	PCV2 Dilution > = 160 10 weeks of age	M hyo ELISA Positive at 3 weeks of age	M hyo ELISA Positive at 6 weeks of age	M hyo ELISA Positive at 10 weeks of age
1) Circo/MycoGard 1 Dose	4/25*	6/25	12/25	1/25*	2/25	6/25
2) Circo/MycoGard 2 dose^	10/25*	4/25	21/25	1/25*	0/25	22/25
3) Circo/MycoGard + MycoGard 2 dose^	12/25*	6/25	15/24	0/25*	0/25	18/24
4) FLEXCombo 1 dose	8/25*	8/25	20/23	2/25*	0/25	0/23
5) Circumvent PCV M G2 1 dose	13/25*	22/25	17/24	3/25*	8/25	4/24
6) Circumvent PCV M G2 2 dose	14/25*	24/24	24/24	0/25*	3/24	23/24
7) Non-vaccinated control group	9/25*	7/25*	2/25*	1/25*	0/25	0/25

* False positives or maternal antibody detection
^ Extra-label dose

Effects of level of Defusion® Prime on performance of growing pigs fed diets containing vomitoxin

C. Fowler, MS; S. Ebarb, MS; P. Xue, PhD; K. Saddoris-Clemons, PhD; S. Williams, MS; D. Giesting, PhD
Provimi, Brookville, Ohio

Introduction

Deoxynivalenol (DON), commonly called vomitoxin, is a mycotoxin that was detected in much of the corn harvested in the US in 2016.[1] Provimi research has shown that pigs fed DON-contaminated grains can exhibit decreased average daily feed intake (ADFI) and average daily gain (ADG) and poorer feed efficiency (FG) and health. These physiological effects often decrease profitability.[2,3] Promote® Defusion® Prime is a proprietary product developed by Provimi and formulated with multiple components including feed preservatives to support animal health and growth. In the study described below, Defusion Prime (DP) was fed to growing pigs at 0, 0.25%, 0.375%, and 0.50% in diets that contained measured levels of DON.

Materials and methods

The study was conducted in a commercial grower-finisher barn with 1,080 pigs in pens with 24 pigs/pen and a stocking density of 6.7 ft^2/pig. The design was a completely randomized block design where pigs were penned by sex with animals of similar body weight. Barn was blocked by initial body weight and pens were randomly allocated to treatment groups within weight block. A corn and soybean meal-based diet with less than 1.5 ppm DON was included as a "low DON" control. A basal diet was formulated with corn and corn DDGS containing DON. The treatments were 1) Control diet with low DON; 2) Basal diet with no diet preservatives; 3) Basal diet with 0.25% DP; 4) Basal diet with 0.375% DP; and 5) Basal diet with 0.50% DP. For the first half of the trial, the basal diet contained 2.5 to 3.3 ppm DON. In the second half of the trial, the basal diet contained 4.7 to 5.7 ppm DON. Diets with DP often have analyzed values of DON that are lower than the same diet without DP. This observation indicates DP may have an effect on DON in the feed prior to consumption. Pigs were on test from 64 to 200 pounds. The effects on ADG, ADFI, FG, and return over feed cost (ROFC) were analyzed using the PROC GLIMIX procedure in SAS 9.3 (Cary, NC).

Results and discussion

The results from the study are summarized in Table 1. The control diet with less than 1.5 ppm DON was compared to the basal diet with no feed preservatives. The higher level of DON in the basal diet decreased ADG ($P < 0.01$), ADFI ($P < 0.01$), and final body weight ($P < 0.01$) compared to the control diet. Feed efficiency was not affected by DON level.

In the basal diet, increasing the level of DP from 0 to 0.50% linearly ($P < 0.01$) and quadratically ($P < 0.01$) affected ADG, but did not affect ADFI ($P > 0.10$). The greatest ADG was observed in pigs fed the 0.25% DP diet. Increasing the level of DP had a quadratic effect on FG ($P < 0.01$). The pigs exhibited the best FG when fed 0.25% DP. Final body weight was linearly ($P < 0.01$) and quadratically ($P < 0.05$) affected by the inclusion level of DP. The heaviest pigs fed the basal diet at the end of the study also received 0.25% DP. They were 12.1 lb heavier than pigs fed the basal diet with no feed preservative (199.0 lb vs 186.9 lb).

The cost of feed for each treatment and a value of one US dollar ($) per pound of live body weight was used to calculated ROFC. Increasing the level of DON in the diet decreased ROFC by $14.64 per pig ($108.69 vs $94.05). Increasing the level of DP in the basal diet linearly and quadratically affected ROFC. The best ROFC was achieved when pigs were fed 0.25% DP at $10.36 more than pigs fed the basal diet with no feed preservative.

References

1. Provimi Corn Harvest Report, unpublished data. 2016.

2. Williams et al. Provimi unpublished data. 2009.

3. Ebarb et al. Provimi unpublished data. 2017.

Table 1: Effects of Defusion® Prime and DON on growth performance of finishing pigs and economic return

	Control	Defusion prime† 0	0.25%	0.375%	0.50%	SEM	Linear	Quadratic	Effect of DON††
Initial BW, lb	63.93	64.23	63.88	64	64.59	3.25	0.731	0.347	0.689
Final BW, lb	206.5	186.85	198.99	197.17	195.66	5.61	0.006	0.011	< 0.001
ADG, lb/d	2.01	1.73	1.9	1.88	1.85	0.04	0.002	0.002	< 0.001
ADFI, lb/d	4.99	4.29	4.54	4.52	4.5	0.15	0.113	0.256	< 0.001
F:G, lb/lb	2.48	2.47	2.38	2.42	2.43	-	0.105	0.005	0.772
ROFC, $	108.69	94.05	104.41	102.32	100.1	1.735	0.001	< 0.001	< 0.001

† Level of Defusion Prime added to the basal diet that contained 2.5 to 5.7 ppm DON
†† Control vs basal diet with no preservatives

Effect of diet and sex on feed efficiency and the fecal microbiome at slaughter weight

L. M. G. Verschuren[1,2], MSc; M. P. L. Calus[1], PhD; A. J. M. Jansman[1], PhD; R. Bergsma[2], PhD; J. R. Dunkelberger[3], PhD; E. F. Knol[2], PhD; H. Gilbert[4], PhD; O. Zemb[4], PhD
[1]Wageningen University & Research, Wageningen Livestock Research, Wageningen, The Netherlands; [2]Topigs Norsvin Research Center B.V., Beuningen, The Netherlands; [3]Topigs Norsvin USA; Burnsville, Minnesota; [4]INRA-INPT-ENSAT-Université de Toulouse, GenPhySE, Castanet-Tolosan, France

Introduction

Feed is a major cost of pig production. Therefore, feed efficiency (FE) is a critical trait for the swine industry. Availability in the market and cost of feed ingredients dictate changes in feed composition. As a result, fiber level and composition can vary between pig diets. Microbiota in the gastrointestinal tract play an important role in fiber digestion because they produce enzymes that break down fiber structures and deliver volatile fatty acids to the pig. These volatile fatty acids can be used as metabolic energy sources.[1] As such, microbial fermentation could influence FE in pigs. The aim of this study was to investigate the association between FE and fecal microbiome in commercial grower-finisher pigs. In addition, the effects of diet composition and sex on fecal microbiome composition were studied.

Materials and methods

A total of 154 three-way crossbred grower-finisher pigs (Synthetic boar × (Large White × Landrace)) were either fed a diet based on corn/soybean meal (CS) or a diet based on wheat/barley (WB). Littermates were split randomly over the two diets and sexes were housed in separate pens (10 animals per pen) equipped with individual feed intake registration feeders. Fecal samples were collected on the day before slaughter (mean body weight 122 kg) and DNA was extracted and sequenced for the V3V4 16S ribosomal DNA regions. Sequences were clustered according to operational taxonomic units (OTU) for each individual, forming a table of OTU abundancy. A discriminant analysis using principal components was applied to the dataset to test separation for diet, sex and FE extreme groups (10 high and 10 low FE animals for each diet by sex combination).

Results and discussion

Pigs on different diets had very distinct microbiome composition, needing only two OTUs to separate the diets ($P = 0.018$). This finding is consistent with results reported in the literature, indicating that diet composition affects gut microbial composition.[2] In addition, males and females could be separated based on 18 OTUs ($P = 0.002$). Previous studies in weaners did not detect a difference between sexes in fecal microbiome,[3] but in agreement with our study, did detect a difference at slaughter age,[4] which is most likely due to sex steroid hormones.[5] Results also show that microbiome composition significantly differed between sexes for pigs fed a (fiber-rich) WB diet, but not for pigs fed the CS diet. In conclusion, our results show a diet and sex dependent relationship between fecal microbial composition and FE in pigs at slaughter weight. These results demonstrate the potential to modulate microbiome composition, and therefore feed efficiency, via changes in diet formulation.

This study is part of the Feed-a-Gene Project, funded from the European Union's H2020 Program under grant agreement no 633531.

References

1. Ingerslev, A. K., P. K. Theil, M. S. Hedemann, H. N. Lærke, and K. E. B. Knudsen. 2014. Resistant starch and arabinoxylan augment SCFA absorption, but affect postprandial glucose and insulin responses differently. Br. J. Nutr. 111: 1564-1576.

2. Bauer, E., B. A. Williams, H. Smidt, R. Mosenthin, and M. W. Verstegen. 2006. Influence of dietary components on development of the microbiota in single-stomached species. Nutr. Res. Rev. 19: 63-78.

3. Mach, N., M. Berri, J. Estelle, F. Levenez, G. Lemonnier, C. Denis, J. J. Leplat, C. Chevaleyre, Y. Billon, J. Dore, C. Rogel-Gaillard, and P. Lepage. 2015. Early-life establishment of the swine gut microbiome and impact on host phenotypes. Environ Microbiol Rep 7: 554-569.

4. Xiao, L., J. Estellé, P. Kiilerich, Y. Ramayo-Caldas, Z. Xia, Q. Feng, S. Liang, A. Ø. Pedersen, N. J. Kjeldsen, and C. Liu. 2016. A reference gene catalogue of the pig gut microbiome. Nat Microbiol 1: 16161.

5. Org, E., M. Mehrabian, B. W. Parks, P. Shipkova, X. Liu, T. A. Drake, A. J. Lusis. 2016. Sex differences and hormonal effects on gut microbiota composition in mice. Gut Microbes 7: 313-322.

Organic acids modes of action and impacts on intestinal microbiota and performance in swine

J.A. Mesonero Escuredo[1,2], DVM, ECPHM Resident; P. Bruinenberg[2], PhD; N. de Groot[2], DVM; P. Roubos[2], PhD; N. Ferguson[2], PhD; J.R.E. del Castillo[3], DMV MSc, PhD; J. Carr[4], BVSc, PhD, MRCVS, Dipl. ECPHM; D. Maes[1], DVM, PhD, Dipl. ECPHM, Dipl. ECVPH
[1]University of Ghent, Belgium; [2]Trouw Nutrition (Nutreco); [3]Québec Animal Pharmacology Research Group, University of Montreal, Canada; [4]www.carrsconsulting.com, Australia;

Introduction

Organic acids (OA) and their salts are globally used in animal nutrition for microbial preservation and supporting animal health. They have been proven effective in maintaining growth performance (GP) since the EU banned antibiotic growth promoters (AGP) in 2006. The use of OA in animal nutrition can increase body weight (BW), improve feed conversion ratio (FCR) and reduce colonization of pathogens in the intestine.[1,2] The aim of this study is first to review the mode of action and to investigate the impact of additional designed blends of OA on GP and health in pigs.

Mode of action of OA (MofA)

The antimicrobial effect of an OA is dependent on its concentration, molecular size, pKa value, pH of the environment, relative lipophilicity as indicated by its Log P_{ow}, and the bacterial surface area, permeability and thickness of membrane and bacterial wall. In addition, for solute ionizing substances, the membrane charge, transmembrane ionic gradient and possibility that the solute may be a substrate of a carrier protein must be considered. Many OA are substrates of the monocarboxylate transporters.[3] Recent studies show that the MofA in bacteria, yeast and moulds is related to dissipation of the proton gradient ("classical" uncoupling effect), acidification of the cytosol, membrane disruption, osmotic stress by anion accumulation and metabolic disturbance by inactivating specific enzymes. Application of salts of OA (buffered, BOA) can deliver an anion for OA production at a low pH, such as that present in the stomach.

Materials and methods

The minimally inhibitory concentration (MIC) of various weak organic acids was determined by incubation with growing cells of relevant (pathogenic) gram-positive and gram-negative bacteria in a microtiter plate format. Based on the acquired MIC data, a drinking water additive (DWA), being a blend of free and BOA, was designed. The effect of the DWA (0.02%, targeting pH 3.8) on GP and microbiota of piglets was evaluated. In study 1 (S1) (water pH 7.9; DWA pH 3.8), 280 weaned piglets were divided into 2 randomized groups (10 replicates of 14 piglets): group 1 (G1) compound feed with 400 ppm Amoxicilline + 300 ppm Neomycin and group 2 (G2), compound feed with 400 ppm Amoxicilline + DWA (AB). The piglets were monitored on GP during the first month after weaning. S2 (water pH 7.4; DWA pH 3.8), followed 576 weaned piglets (12 replicates of 16 piglets) randomly allotted to 3 groups: G1, control, G2, control diet with chlortetracycline (phase1 1100, phase2 440, phase 3 550 ppm) and G3, DWA. The animals were monitored on GP during 42 days. S3 (water pH 8.0; DWA 3.7) followed 120 piglets (20 replicates of 3 piglets): G1, control and G2, DWA. The piglets were monitored for GP and 10 animals per treatment were sacrificed at day 42. DNA was isolated from ileum and jejunum samples and sequenced using the Illumina HiSeq. Statistical method used for S1 and S2 was ANOVA t-test, $P < 0,05$; for S3 mixed procedure SAS, $P < 0,05$.

Results and discussion

S1 showed a significantly improved ($P \le 0.0001$) BW, ADG and FCR for DWA compared to control. S2 showed that piglets receiving DWA had a higher feed intake (FI) and BW compared to control ($P < 0.05$), with comparable results when using either an AB or a DWA. S3 showed significant improvements in BW gain, ADG and FI for the group receiving the DWA ($P < 0.05$). Besides the improvement in GP, a reduction of pH in the stomach was observed for the DWA group compared to the control, and microbiota analysis showed a significant reduction of genus streptococcus in the ileum and jejunum ($P < 0.05$)(Figure 1). Further pharmacology studies should be performed on the MofA. However the results on GP and microbiota analysis indicate that OA can be used as a valid alternative for antimicrobial medication, as well to improve production in swine.

References

1. F. Van Immerseel et al. The use of organic acids to combat Salmonella in poultry: a mechanistic explanation of the efficacy, Avian Pathology 2006, 35:3, 182-188C.

2. C. Lückstädt, S. Mellor. The use of organic acids in animal nutrition, with special focus on dietary potassium diformate under European and Austral-Asian conditions, Recent Advances in Animal Nutrition 2011 – Australia 18

3. Sepponen K, et al. Expression of CD147 and monocarboxylate transporters MCT1, MCT2 and MCT4 in porcine small intestine and colon. Vet J. 2007; 174:122-8

Figure 1: Microbiota analysis showed a significant reduction of the genus *Streptococcus* in the ileum of the DWA group vs control ($P < 0.05$)

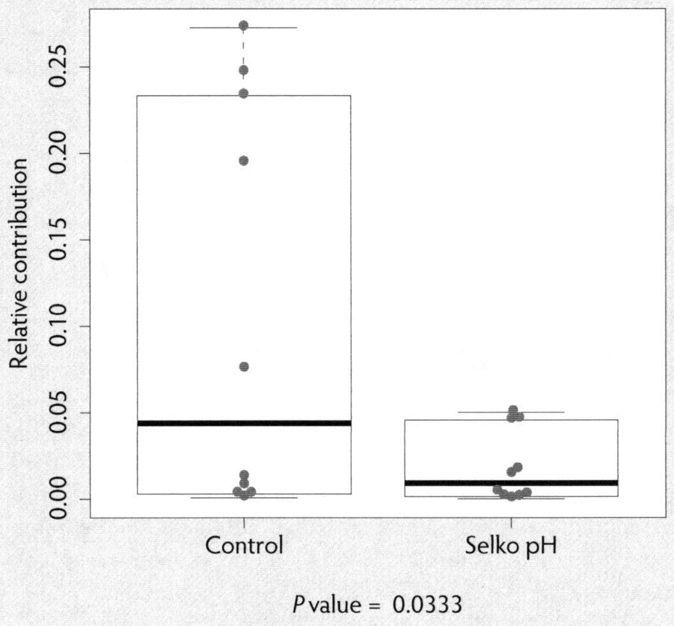

A dose confirmation study evaluating 100 g/ton of LINCOMIX® (lincomycin hydrochloride) for the reduction of the severity of effects of respiratory disease associated with *Mycoplasma hyopneumoniae*

Michael Senn[1], DVM, MS; Raksha Tiwari[2], BVSc & AH, MS, PhD; Tom Painter[1], DVM
[1]Zoetis, Parsippany, New Jersey; [2]Zoetis, Kalamazoo, Michigan

Introduction

The objective of the study was to evaluate the efficacy of LINCOMIX® (lincomycin hydrochloride) to support the label indication, "For the reduction in severity of effects of respiratory disease associated with *Mycoplasma hyopneumoniae* (M hyo) when fed at 100 grams of lincomycin per ton of complete feed for 21 days".

Materials and methods

Eight locations with a history of respiratory disease associated with *Mycoplasma hyopneumoniae* (M hyo), and confirmed to have a minimum of 20% incidence of M hyo determined by PCR testing of deep tracheal/bronchial swabs were enrolled in the study (n = 1,719 pigs).[1] Pigs were assigned to treatment by pen according to a randomized complete block design. Pigs were randomized to pen based on weight. Treatment groups were T01: Non-medicated fed continuously for 21 days, and T02: Lincomix®, 100 grams / ton fed continuously for 21 days. Groups exhibiting acute respiratory disease, concurrent disease, or other physical conditions that interfere with progression of respiratory disease, or response to therapy were excluded from the study. All pigs were weighed on study days 0 and 21 as well as all mortalities. Personnel involved in treatment administration were not involved in data collection. Once a day each pen was evaluated for pigs with clinical signs associated with acute swine respiratory disease (SRD). Animals with Respiratory Scores (RS) or Depression Score (DS) of ≥ 2 with a scale of 0–3 had their scores recorded. Animals with RS > 2 and DS > 2 had their rectal temperatures recorded. Animals that met the minimum criteria for acute SRD (RS = 2, DS = 2 and rectal temperature ≥ 104° F) or with RS or DS equal to 3 were identified, rectal temperature taken, weighed, euthanized and necropsied. On study day 14 and 21 all animals were evaluated for coughing and scored on a scale of 0-2. Lung lesions were scored at necropsy for all enrolled animals. Lung lesions and average daily gain were analyzed using a general linear mixed model while acute SRD was analyzed using a generalized linear mixed model.

Results and discussion

Gross lung lesions in the T02 (Lincomix) group were significantly reduced compared to T01 (controls) ($P = 0.001$). The T02 (Lincomix) group demonstrated significantly improved final weights (P 0.011), average daily gain ($P = 0.001$), average daily feed intake ($P = 0.0021$), and gain to feed ratio ($P = 0.0040$) compared to T01 (controls). The T02 (Lincomix) group demonstrated a significant reduction in coughing scores ($P = 0.0014$) on day 21 of the study, compared to T01 (controls) (Table 1). A total of 892 lung or bronchi samples were tested via PCR for M hyo 756 samples (85%) were verified as positive. In this study, Lincomix was effective in the reduction in effects of respiratory disease associated with *Mycoplasma hyopneumoniae* in swine at a dose of 100 grams of lincomycin per ton of complete feed fed for 21 days.

References

1. Data on file, Zoetis Inc, Study A121C-US-14-150.

Table 1: Results

	T01 (Controls) Least squares means	T02 Lincomix® (100 grams/ton) Least squares means
Final weight (kg)	65.51[a]	67.40[b]
Average daily gain (kg)	0.64[a]	0.73[b]
Average daily feed intake (kg)	1.74[a]	1.84[b]
Gain : Feed	0.36[a]	0.40[b]
Day 21 cough (% Abnormal)	17.19[a]	10.01[b]

[ab] Means in rows not sharing the same superscript are significantly different ($P < 0.05$).

Evaluating data: What do we really know?

Eric R. Burrough, DVM, PhD, DACVP
Veterinary Diagnostic Laboratory, Iowa State University, Ames, Iowa

Introduction

The German philosopher Immanuel Kant founded the philosophical doctrine of transcendental idealism.[1] At its core, this doctrine holds that there are noumena, things that exist regardless of our interpretation (the thing itself), and phenomena, which are the apparent realities interpreted by our minds. In this branch of epistemology (*episteme* = knowledge), Kant argues that we can never truly know the thing itself only the mental picture developed by our minds and therefore our minds contextualize and limit reality. Essentially, the world as we know it is what we believe it to be and it is therefore imperative that we understand the justification of our beliefs and our susceptibility to biases.

When an animal or population of animals is exhibiting clinical signs of disease, a veterinarian typically uses *a priori* knowledge,[1] often gained from textbooks and scientific journals, to formulate a prioritized differential diagnosis and likely has a strong belief as to which differential is most likely. While this belief may well be true, it cannot be considered knowledge without proper evidence or justification.[2] If confirmed, this belief becomes *a posteriori* knowledge, gained through experience, and serves as a framework to guide future clinical investigations. Accordingly, it is critical that *a posteriori* knowledge be acquired and validated through a purposeful, repeatable process.

Here it bears noting that all true beliefs are not necessarily knowledge if they are the result of luck, chance, or random events. In other words, the justification process is equally important as the truth of the belief itself as poorly justified true beliefs do not contribute to actual knowledge (Figure 1). For example, if two diagnostic submissions from a population of pigs with salmonellosis yield a positive PCR for *Salmonella* Typhimurium, and one was collected from a sacrificed diarrheic pig with observed gross lesions and the other from a random environmental swab, the sample from the purposely collected lesioned tissue would be more complete knowledge than the random sample that produced a positive result because the process that selected the lesioned tissue is more likely to produce future disease confirmations. The random method in this scenario is thereby potentially dangerous in that it may be interpreted that such a process will be successful in the future. Our cumulative knowledge about these processes and the predictive value of associated diagnostic results determines the diagnostic sensitivity and specificity of various tests and approaches.[3]

Another potential confounding variable in our human data acquisition process is our susceptibility to inattentional blindness.[4] Inattentional blindness, also known as perceptual blindness, is a psychological lack of attention whereby an individual fails to perceive something that is in plain sight. Human brain capacity is essentially a zero-sum game and when an individual focuses their attention intently on a specific task or thing it becomes impossible for the individual to fully account for everything occurring around them and a temporary blindness to events outside of their focus ensues. This is an important consideration when evaluating data generated in the field as it is truly possible for an observer to lose sight of the forest for the trees.

Accordingly, when examining a population of animals for specific signs of an assumed disease it can be possible to overlook and ignore other issues in the population and may lead to conformational bias where you simply find what you are looking for. This occurs through reductionist functions in the left brain that break complex subjects into distinct separate components rather than employing more integrated, holistic approaches associated with right side of the brain.[5] The four circles approach to swine herd evaluation,[6] which begins with a big picture view of the site and ends with individual animal examinations, is an effective method to help minimize bias and incorporates both holistic and reductionist functions of the brain.

Lastly, there is often an illusion of cause in data collection that occurs through overzealous pattern detection.[4] In this form of inductive reasoning, one can infer cause from coincidence creating inherent biases that are taken as evidence and ultimately treated as fact. Even a statistically significant association between two things does not imply cause, only association, and one must avoid the leap from correlation (belief) to causation (truth) without proper evidence and justification (knowledge). This type of

Figure 1: Venn diagram illustrating the relationship between truths, beliefs, and knowledge

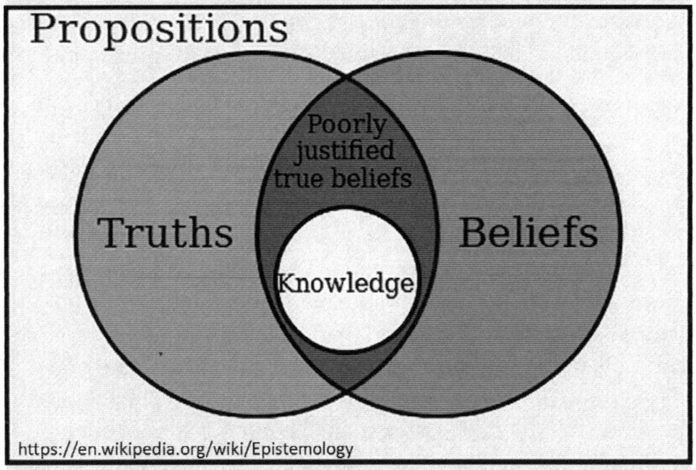

https://en.wikipedia.org/wiki/Epistemology

induction, particularly related to the detection of potential novel pathogens, is common in the literature and must be interpreted with caution.

Evaluating data

Diagnostic data

Cumulative diagnostic data offers the possibility that disease occurrences, interventions, and outcomes in animal populations can be analyzed over time with the aim of continual improvement. However, such an approach to disease control will only be valid if the original diagnoses and assumptions of causation are properly justified. Databases must have accurate inputs and must be continuously monitored and improved through appropriate controls and feedback loops.[7] Without such controls, databases may actually create erroneous assumptions and should be carefully interpreted in the context of their creation.

The most critical step in the veterinary diagnostic process, and thereby the ultimate validity of the diagnostic data created, is the examination of the animal(s) in their environment and the selection of which animal(s) to sample and which samples to submit for diagnostic testing. It is indisputable that diagnostic accuracy in a population depends heavily on obtaining the right sample from the right animal at the right time. Increasingly in modern swine medicine this critical step is being performed by field personnel who lack the level of training and experience necessary to perform this task. Field samples are then sent to the laboratory often without sufficient clinical context or communication of specific diagnostic questions and testing is left to the discretion of the diagnostician. When these data are returned to the supervising veterinarian they must be aligned with the clinical context and herd history such that treatment and intervention strategies can applied. The veterinarian must use these data in an effort to determine proximate and ultimate causes of disease. Given the myriad potential sources of bias and error in a diagnostic investigation, the swine practitioner is well-served to be actively involved in sample acquisition and test selection as no diagnostic test can overcome the limitations of inaccurate sampling. Simply sending more samples or doing more testing does not necessarily increase the odds of finding the right answer or increase the value of a diagnostic database and may actually only add confusion.

Diagnostic databases are important to modern swine practice but there are clear limitations, not the least of which is that data struggles with context. A diagnostic test result by itself is often a small part of the picture as many animals selected for diagnostic testing suffer from cumulative insults that may have been ignored. The individual interpreting diagnostic data must fully understand the context of the case including the case definition and population history. As more data is acquired from a population it is easy to find many statistically significant correlations; however, these are often spurious and can be misleading without proper context. Even raw data can be misleading as it was collected and structured in accordance with someone's perceptions and personal bias.

Peer-reviewed data

While the information age has brought forth many blessings, it has inadvertently given access to an unfettered stream of scientific information and nearly real-time sharing of data. Much of this scientific data appears in sources that are labeled as peer-reviewed, but the value of this designation is becoming somewhat grey as the rigor of the peer-review process is highly variable. Most traditional peer-reviewed print journals continue to maintain high standards as their printed product is provided as a service to their subscribers who pay a fee. The advent of open-access publishing, which had the original intent of bringing scholarly research to the masses, flipped the script and placed the onus of publishing fees on the contributing scientist. In the academic reality of publish or perish, many academic scientists are willing to fund publication to bring their work into the public eye. Not surprisingly, this new opportunity to generate large publication fees, typically more then one thousand dollars per article, and low overhead associated with online publishing gave rise to predatory open-access publishers who are willing to publish almost anything if the submitter is willing to pay. Adding to the peer-review quagmire is the shrinking body of academic faculty with adequate time to devote to proper peer-review which forces journals to accept less qualified reviewers and more superficial reviews than what was historically accepted.

With this in mind, now more than ever, any consumer of scientific literature is wise to thoroughly evaluate the data provided in a scientific manuscript and to decide for themselves if the claims made are valid. Simply reading the abstract of a peer-reviewed article can be dangerously misleading, particularly if you are not familiar with the publisher or the source. High throughput molecular testing has made finding new potential pathogens easier than ever and in the rush to be first to publish there are many peer-reviewed papers that use statistical associations and inductive reasoning to infer cause. Accepting something as knowledge without fully understanding its justification is a risky proposition. There is still some truth to the adage that as a consumer you often get what you pay for.

Summary

As veterinarians, it is essential that we understand the genesis of what we consider knowledge. What do we think we know and why do we believe it to be true? To quote the British philosopher Bertrand Russell, "It is not what the man of science believes that distinguishes him, but how and why he believes it." A certain amount of scientific due diligence should be applied to all diagnostic and peer-reviewed data before it is accepted as fact. Inattentional blindness, inductive reasoning, and conformational bias are realities that must be considered and can impact all datasets. The more personally involved a clinician is in the acquisition of diagnostic data the more reliable their conclusions are in the context of the case. Inarguably you can only find what you look for, and you only recognize what you know, but how much data is either inappropriately interpreted or not even accounted for because the collector of the data simply did not know?

References

1. Kleinman P. *Philosophy 101: From Plato and Socrates to Ethics and Metaphysics, an Essential Primer on the History of Thought.* Avon, MA: Simon & Schuster, Adams Media; 2013.

2. Schick T, Vaughn L. Knowledge, Belief, and Evidence. In: *How to Think About Weird Things: Critical Thinking for a New Age.* 4th ed. New York, NY: McGraw-Hill, 2005; pp. 114-149.

3. Saah AJ, Hoover DR. "Sensitivity" and "specificity" reconsidered: The meaning of these terms in analytical and diagnostic settings. *Ann Intern Med* 1997;126:91-94.

4. Chablis C, Simons D. *The Invisible Gorilla: How Our Intuitions Deceive Us.* New York, NY: Crown Publishers / Random House, Inc.; 2009.

5. Newberg A, Waldman MR. *Why We Believe What We Believe.* New York, NY: Simon & Schuster, Free Press; 2006.

6. Ramirez A, Karriker LA. Herd Evaluation. In: Zimmerman JJ, Karriker LA, Ramirez A, Schwartz KJ, Stevenson

GW, eds. *Diseases of Swine.* 10th ed. Ames, IA: Wiley-Blackwell, 2012; pp. 5-17.

7. Nielsen AC. Data warehouse for assessing animal health, welfare, risk management and communication. *Acta Vet Scand* 2011;53(Suppl 1):S3.

Portraying the industry in a positive light

Erin Brenneman

We have all heard the statistics. Less than 2% of the population of the United States farms. But less than 2% of that 2% dedicates time and efforts towards actively sharing what we do so that the 98% might have an accurate glimpse of what it is really like, and how it has changed from what they thought to be true.

With the social media revolution, the way we converse and get information has changed forever, and it is only going to change faster and more drastically moving forward. With such a small amount of people participating in that conversation, can we in agriculture possibly keep up?

I was born and raised in a suburb of Chicago and never even had to drive past a farm for the first 18 years of my life. Traffic was a way of life, distance was judged in time instead of miles. And a second thought was never given to the farmer who raised your food, because food simply came from the grocery store. I began my strong love for animals through horses. Throughout high school I worked and rode at a horseback riding lesson program in the city. It was here I purchased my first horse with the money I made working at the barn and paid off his board each month by exercising horses and working with the lesson horses.

When it came to making my college decision I easily chose Iowa State for its amazing agriculture program and beautiful campus. Not only did I graduate there with a degree in animal science, but I also met my husband, a pig farmer from southeast Iowa.

We married in 2004 and moved back to his home to farm with his family. When I got there, I was amazed by the amount of people, effort, and care that it takes to bring food to our table. I tried everything there was to try on the farm and my heart settled in the farrowing house. There I learned and practiced every day to become the best at my new-found passion, day one care.

Since then my roles have shifted slightly. I am now a mother to two farm boys, and an advocate for sharing the farm story. I believe my outsider perspective brings a refreshing look on what people find fascinating and wonderful about our everyday world in agriculture. Known for wearing my fun scrub caps and using snapchat to share my farrowing days, I have created a brand "sowmomma" that shows the fun, and hard-working life of a pig farmer.

Sharing about farming can be a full-time job, or just a few minutes a day. But one thing is for sure, every little bit helps! Try to touch on small things, often those are the most fascinating to people on the "outside". Keep it light, keep it fun. We are people, not fact spouting robots. Be yourself. Find a voice that fits you and stick with it.

The platforms

Facebook

Facebook is a wonderful, very private platform. It can be considered one of the most capable of private settings of them all. It is a good place to share with your family and close friends. You are able to select who sees what you post. It can be public too. Facebook applies an algorithm that, if you like numbers, can be fun to play with. It provides some of the best analytics for tracking progress and following. Facebook is a place where you can share stories with pictures and albums. It is not really a "news based" platform but rather a storytelling place. It is an active platform, but definitely fading out with the younger crowd. If you want to follow our farm on Facebook, our name is Brenneman Pork Inc.

Twitter

Twitter is a very fast, news/event oriented platform. If you are tweeting about something that happened yesterday, you're too late! Uses hashtags to group common interests. You are limited in your characters (280) so get your point across quick! Twitter is a very open platform that allows you to gain a following quickly. However, remember that because of this there can be a lot of "trolls" and people hiding behind fake accounts to attack. Some people welcome the challenge, some prefer to have a less committed online presence. If you want to follow me on Twitter, my username is @sowmomma

Instagram

Instagram is a very friendly, non-conversational platform. It is the place for pretty pictures, pet accounts, and food! Instagram uses a lot of hashtags too. Equipped with its own photo-editing, Instagram encourages the abstract and pretty sides of interests. This is a more preferred platform for many food bloggers. You can provide links, but Instagram is mainly about visual rather than informational. Which is nice for all of the amazing things we see every day as farmers and people in the agricultural community. If you want to follow me on Instagram, my user name is @sowmomma

Snapchat

My favorite. Fast, easy, and to the point. You are able to show bits of your day as you move through it in a format that people will take (albeit very short) time and attention to watch. Snapchat provides filters for certain days, silly faces, and locations. It is a great way to tell your story in a fun and interesting manner! Gaining a following is hard in snapchat, but it is fun. If you want to follow me on snapchat, my username is "spookgal".

As you move through your social media journey you will find what works and what doesn't work. What makes people tick? What do they find interesting, and what falls flat? I have found that it is the small things that we do in our everyday routine that are most fascinating to people outside of our industry. Taking a shower, changing your boots, towel drying pigs are all examples of such wonderful small things that we do every day to ensure our pigs are healthy. There is no need to overthink and overdo anything on social media. Nobody wants to be bombarded with data and details (ok, maybe except for swine veterinarians...). People relish in discovering the human element of our business. The pigs and the information about them are just an added bonus. Time after time my most viewed and viral posts have nothing to do with pigs, but rather to do with the world around us. The community, the country, our staff. Never underestimate the power of a beautiful country sunset, or that amazing team member. One important tip I always recommend to people who are just getting started on social is to take plenty of pictures. Driving down the road, walking through the barn, visiting a client, seeing a perfect sunrise. Snap pictures of all of it. You don't have to use them all, but you will have them in your "arsenal" when it is a day that you can't come across any content to use. Staying frequent, or at least regular on social will gain you "points" with the algorithms, and consider you more interesting to look at.

Find one platform, and stick with it, don't try to be a hero and do them all, unless you want to, then by all means do! Find your passion, take lots of pictures, and get out there and share! Thank you!

Agriculture 2025: Global, local and high tech

Lowell B. Catlett, PhD
Regents Professor and Dean, Emeritus, New Mexico State University

The generally accepted estimate for world population growth over the next 25 years is for an additional 2 billion pushing the total to approximately 9 billion people that agriculture will have to support. The additional 2 billion people will reside in countries that have the fastest growing middle classes and personal incomes thus the expectation that per capita meat consumption will increase[1] The need to increase meat production over the next several years will be impacted by several factors such as climate change, animal welfare, consumer expectations and technology. How the meat industry responds to the increased demand given many constraints should be foremost in the strategic plans of both companies and organizations that want to make sure consumers that want meat in their diets can have meat in their diets.

Global demand for meat

Economic theory states that when the price of a product increases, the quantity demanded of the product will decrease. This assumption is based on the "law of diminishing utility" which states that when a consumer uses a product they get a certain amount of satisfaction (utility) but an additional item of the same product would not yield as much satisfaction as the first one, thus additional items "diminish in utility". A pork chop dinner at a restaurant offers a certain amount of satisfaction to a patron, but when offered a second chop at the same price the patron will likely already be sated and not be inclined to pay the same price for a second chop. To induce the patron to consume a second chop which clearly will be less satisfying the restaurant would have to reduce the price. This concept is at the heart of markets. When the price of pork increases consumers will generally buy less and vice versa. How responsive price and quantity are to each other is called price elasticity of demand. If the price of pork increases 10 percent, will consumers reduce their quantity demanded of pork by more than 10% or less? If the quantity demanded response is more than 10% the product is said to be elastic and if it less than 10% it is said to inelastic.

How responsive quantity demanded is to price changes is largely determined by the amount and availability of substitutes. In most developed countries, various types of meat are substitutes for one another. If poultry increases in price, most consumers will not only reduce the amount of chicken they demand, but will often increase the consumption of another meat. They may, of course, not increase the consumption of another meat, but may consume more pasta. The relationship between the price of one product and how it relates to the demand for another product is called cross price elasticity. If one meat is a substitute for another then the relationship between the price of one meat and the quantity consumed of another meat is inverse. If one meat product, say ground beef is a complement to another, ground pork, then the relationship between the price of one meat and the quantity consumed of another is positive. For example, a consumer may make their own sausage which includes both ground beef and pork. If beef prices increase then the quantity demanded of beef by the consumer will fall and likewise so will the demand for ground pork because they use them together.

Price elasticity of demand and cross price elasticity are the two most important economic drivers in determining meat consumption in developed economies. Consumption of meat is fairly stable and changes as the relative prices of various types and varieties change. Per capita consumption of meat in the US in 1981 was 193.7 pounds and increased to 214.5 pounds in 2016. A modest 10% growth over 35 years. However, beef consumption fell from 77.2 pounds in 1981 to 56.5 pounds in 2016, pork fell from 54.2 pounds to 50.1 pounds but the broad category of poultry rose from 59.3 pounds to 107.6 pounds.[2] This is not unimportant, but certainly is not the major reason meat production will change globally during the next 25years. This new change is coming from developing countries.

The demand for meat will increase if one of four things occur—population increases, consumers get more money, consumers change their tastes and preferences and the price of substitutes for meat increase. The last two are, as discussed previously, the reasons for changes in meat consumption in developed countries. The first two—population growth and income growth are driving the growth worldwide in meat consumption. Since 1981 the world's population has increased 60% to its current level of 7.5 billion people and at the same time the number of people living on $1.90 per day has decreased 41% since 1990.[3] Clearly the very poor as a percentage of the population has dropped dramatically and as they got more money, one of the first things they did was change their diets to include more meat protein. A similar trend has occurred worldwide as countries have experienced major economic growth thus increases in per capita income. For example, in 1981 China had a Gross Domestic Product (GDP) of 281 billion US dollars, Brazil had a GDP of 183 billion US dollars and India had a GDP of 197 billion US dollars for a combined total for the three countries of 661 billion US dollars. In 1981 the GDP of Japan was larger than China, Brazil and India combined—as were the economies of United Kingdom, Germany and France. Fast forward to 2016 and the GDP of China is 11.2 trillion US dollars, Brazil GDP is 1.8 trillion US dollars and India is 2.3 trillion US dollars for a combined total of 15.3 trillion which is more than three times as large as Japan's GDP of 5 trillion US dollars.[3]

Since 1981 the combined economies of China, Brazil and India have increased 23-fold while their populations have increased only 1.6-fold for a per capita gain in economic output of 14-fold. Add the other 190 plus countries in the world where economic output has increased faster in the last few years than population growth and it is easy to see why the growth in the demand for meat will more than likely double in the next 25 years. As the economic wellbeing of people continues to grow they will also demand more consumer goods of all kinds as well as other types and varieties of food. Witness the growth in consumers in developing countries wanting access to electric power, automobiles and restaurants.

Local ways to meet the global demand for meat: The case for intensive animal production

During the last 50 years the success of intensive animal production is readily observed in feed conversions, animal physical health, containment of diseases and overall productivity in milk, beef, pork and poultry. Milk production per cow has doubled in the last 30 years and feed efficiency for beef, pork and poultry has likewise doubled. Where once (1980) it took on average approximately 10 pounds of feed to produce a pound of beef it now takes less than 5 pounds, pork production when from 6 pounds to less than 3 with many efficient operations now at 1.6 pounds and poultry dropped from 3 pounds to 1.3. All of these gains have occurred in intensive type operations, not in open pasture systems. To be sure open pasture systems have reaped some gains during the last several years using improved genetics and supplemental nutrition. Additionally, the concentrations of animal wastes have led to many secondary by products that are economically viable such as compost, sterilized bedding, fertilizer and energy. In economic terminology, these waste items would once have been called an externality meaning the market place did not put a value on them, that is, they were external to the market system. This is one of the major problems of pastoral type agriculture in that there is no direct economic value only indirect in the form of manure providing fertilizer value.

Pastoral type animal production will continue and perhaps even provide an extra economic incentive to produces as they capture the market segment that will pay for type of production, but it simply cannot provide for the doubling of meat consumption during the next 25 years. Intensive animal production provides the necessary set up for the continued application of technologies both in production and in handling of the waste. One future area that needs scientific study is the impact to the environment when wastes can be turned into valuable products and the impact to carbon sequestration.

As animal production increases via intensive operations, animal welfare will be foremost in the scientific arena as the balance between production efficiency and animal mental health will dominate new designs for concentrated operations. Additionally, the animal human bond will be part of the new designs to incorporate another externality—the value of animals to human health. The new research dairy design at Cornell University in Ithaca, New York, is an example of both function and form in new animal facilities.

High tech

To achieve a large increase in meat production will require continued applications of new emerging technologies. Many of the technologies are based on "hard" science in the form of sensors, robotics and genetics but some are "soft" in the area of human-animal bonds and health. The following list is not exhaustive but is provided only as a point of stimulating discussions in the meat industry.

Sensors

Almost every day a new app for smart phones is released that provide everything from heart monitoring to car locations. Replacing the current camera lens with a molecular lens that electric impulses can change the focus of 60 times per second is currently being developed for instant blood tests, physical heart health and metabolic rates. Infrared lasers can provide even more surface details and this whole field will revolutionize animal health. Qualcomm announced in 2012 the Tricorder XPrize for the best hand-held devices to measure health and awarded the first-place prize of $2.6 million dollars to Final Frontier Medical Devices and the second-place prize of $1 million to Dynamical Biomarkers Group in 2017.

New microprocessors that cost about 1000 US dollars can complete DNA profiles in less than two hours are currently available. This changes everything about both human health and animal health. Some interesting spins include prescription diets for both humans and animals—an individualized diet for each animal can best be performed in concentrated local operations that monitor animals 24 hours a day.

Robotics and drones

Great strides have been made during the last decade with robotics that are more sensitive and dexterous that can provide assistance in animal care and monitoring. Robotic milkers and animal handling robots are the two fastest growing areas of robots in agriculture. Commercial small-scale drones are available for less than 1000 US dollars that can carry up to 5 pound payloads for up to 6 miles. Autonomous driving vehicles are moving rapidly towards commercial reality as both Ford and GM have stated the desire to have autonomous driving vehicles available for sale as early as 2020.

3-D Printing

Perhaps no new technology offers a greater upending of manufacturing worldwide than the ability to now "print" new products. 2017 is already emerging as the year of bio printing as human livers, kidneys and most importantly new vaccines and medicines have been partially printed. 3-D printing of hard devices using plastics and metals is well developed and now the biggest movement is to use the same technology to print the building blocks of biology to rapidly produce vaccines, medicines and body organs.

Human-animal bonds

New medical research is now showing how both human and animals change physically when exposed to each other (known collectively as "bonding chemicals") and this will be another major contribution to new designs for intensive animal production facilities. Furthermore, a study in Canada found that early life household exposure to pets reduces the risk of children being overweight and having allergic reactions.[4]

Big data/cloud

Being able to assemble large amounts of data from seemingly unrelated areas will forever change human and animal health when a small outbreak of a disease is known and within hours a vaccine is "printed" and the outbreak controlled. This is an emerging technology but one that offers great promise in animal and human welfare issues.

Summary

As the speed and capacity of computers continues to increase, the ability to gather and use information on all aspects of production agriculture will explode. Indeed, this is happening in all aspects of our lives and has caused a lot of thinking about when (or if) artificial intelligence will exceed humans. Regardless of when, or if, that event occurs it is useful today to understand that information is vitally important in all aspects of agriculture and has tremendous value. The mechanical, chemical and biological revolutions that in the past have and continue to have major impacts will no doubt be dwarfed by the information revolution. The value of knowing when, where, how and all aspects of the life of either a crop or an animal will exceed the actual value of the product itself. The information revolution ties global and local producers and consumers together in ways not possible just a decade ago.

The next 25 years offers animal producers, the meat industry and researchers a golden period when more people than ever before can experience meat in their diets as the demand for meat worldwide doubles. New technologies and new attitudes about animals will allow for intensive animal production that captures efficiencies, increases the physical and mental health of both humans and animals.

References

1. Catlett, L.B., "Consumer Demand and Regional Preferences for Meat," in Processed Meats, Woodhead Publishing, 2011, J.P. Kerry and J.F. Kerry, Editors

2. United States Department of Agriculture Statistics, 2016. Available at: usda.gov. Accessed September 30, 2017.

3. International Monetary Fund Statistics, 2016. Available at: IMF.org. Accessed August 15, 2017.

4. Hein. M. Tun, et.al. Exposure to household furry pets influences the gut microbiota of infants at 3-4 months following various birth scenarios, Microbiome, 2017. Available at: microbiomejournal.biomedcentral.com. Accessed November 8, 2017.

Experiences with new sow farm technologies

N. H. Williams[1], MS, PhD; H. Williams[2], MS
[1]Iowa Select Farms, Iowa Falls, Iowa; [2]Department of Animal Science, Kansas State University, Manhattan, Kansas

Introduction

Iowa Select Farms (Iowa Falls, IA) is a 25-year-old company owned farrow-to-finish system located primarily in Iowa. The majority of Iowa Select facilities were built in the early- to mid-1990's with an average sow farm size of approximately 4,200 sows. Facility footprint consisted of stall gestation with 4 gestation barns and 2 farrowing barns that housed 32 sows per room. Gilt entry consisted of either mature, heat exposed gilts or bred gilts at approximately 10 weeks of gestation. Gilts were housed in retrofitted finishing facilities located primarily in central Iowa. Historically, the majority of gilt development and sow farms experienced field strain exposure to PRRSV on an annual basis, which limited productivity and economic performance.

Over the past two years, efforts have been made to convert to open pen gestation and implement positive pressure filtration in these facilities. To date, 65% of the sow system has been converted to open pen gestation and 56% of the system has been retrofitted to positive pressure ventilation systems. Additionally, we have constructed 9 positive pressure filtered gilt development units for production of bred gilts for gilt entry into our sow farms. Furthermore, we have constructed 4 new positive pressure filtered 7,200 open pen gestation sow farms.

This paper will review our progress to date and lessons learned throughout the implementation of these new technologies.

Transition to open pen gestation

There are many factors to consider when transitioning to open pen gestation. These include: feeding system type, when sows are mixed into groups, optimal stocking density, grouping strategy, and number of sows per pen. Based on barn footprint, age of facilities, cost, and input from our customer, our decision was to implement a floor fed small pen system with groups made post-second pregnancy confirmation in our retrofitted facilities. Part of this decision was based on our gilt development system. Our gilts are housed in groups of 12 to 16 from the nursery through selection and breeding. They are gestated off-site in floor fed systems in groups of 12. Based on personal experience we believed maintaining this pen size would minimize production loss during this transition. Moreover, we transition these facilities by removing bred inventory to off-site finishing facilities fitted with drop feeding systems to minimize production loss during this transition.

Our open pen gestation system consists of breeding and housing in stalls until second pregnancy check. Sows are then moved to small pens of 12 to 14 with partial stanchions and are fed with a floor feeding system. Sows are penned by breed date, parity, and body condition score. Opportunity sows, which are sows with potential foot and leg or body condition issues, are penned together. Feeding regimen consists of 1.5× maintenance for the first 5 days and then maintenance for the duration of pregnancy. Our experience demonstrates that this increased feeding level minimizes aggression and reduces fallout. Daily observation is conducted for non-competitive sows and they are removed to individual stalls. These are primarily sows with foot and leg issues but also includes low body condition sows. Approximately, 3 to 4% of sows are removed from pens.

Sow farm staff transition training is the most critical component to make this transition successful. Sow farm managers receive training in existing open pen gestation facilities prior to implementation of open pen gestation in their farms. Strict operating procedures are developed and implemented across the system.

Implementation of filtration systems

Prior to 2016, Iowa Select Farms had 1 negative filtered farm. This was a multiplier that was retrofitted in 2012. This farm was successful in mitigating disease risks with no new introduction of field strain PRRSV. However, this farm required extensive time and cost to maintain the facility for negative filtration. In 2016, Iowa Select Farms made the decision to retrofit older sow farm facilities at the same time it was transitioning to open pen gestation. After extensive research into filtration type and technology and based on the footprint of existing facilities, the decision was to implement positive filtration. To date, we have remolded 18 existing and filter constructed 4 new sow farms and 9 gilt development units with positive pressure filtration.

The filtration concept is similar in all facilities with a filter room consisting of an outside cool cell approximately 8 ft. from an interior filter bank. The filter bank consists of 2 in. thick 20 × 20 in. pre-filter with a MERV rating of 8. After the pre-filter, there is a V-bank filter with a MERV rating of 15. Air is then pressurized into the facility with high efficiency direct-drive fans which provide positive ventilation. Size and number of filter rooms depends on facility size and farm footprint. Ventilation is controlled by maintaining minimum pressure to mitigate the risk of disease entry.

In addition to filtration system, implementation of several other biosecurity measures have been added including Danish entry systems, ultraviolet chambers, transition rooms, and supply entry rooms. Biosecurity processes are monitored through video surveillance at all entry points. Key lessons learned include: filter type, filter testing, filter installation and inspection, and employee training. While only 2 years in to this process, we have seen success in disease reduction.

Table 1: Reproductive performance trends

	2016	2017	Change
Total born	14.1	14.6	+ 0.5
Born alive	12.6	13.2	+ 0.6
Pigs weaned/sow	10.5	10.8	+ 0.3
PSY	25.6	27.0	+ 1.4
Sow mortality (%)	11.2	12.1	+ 0.9

Managing highly prolific sows

Deborah Murray, DVM
New Fashion Pork, Jackson, Minnesota

Introduction

Gilt development breaks down into three main buckets; 1) Health, 2) Nutrition 3) Management (environment). All three of these categories come together as a piglet on the sow. How is colostrum intake being managed? Are piglets being moved or do they stay with the sow they were born to? Do we have vaccine programs in place for the sows prior to farrowing to enhance her immunity and reduce sow shedding of pathogenic organisms when she is moved into farrowing? Are our farrowing crates adequately sized for larger litters? Is sanitation as good as it could be, can it be improved? All of these factors as well as others come together to dictate our piglet wean weight and subsequent nursery and gilt development health and production. So where do we start? We start with the sow and her care during gestation.

Sow conditioning

Sow conditioning is challenging due to the subjectivity of the evaluator and individual manager/supervisor's personal preferences. To help standardize this we have recently adopted a standardized measurement tool for the farm managers to use and a corresponding feed box adjustment for that sow. Far too often we see fat sows which are not adjusted early in gestation and then allowed to come into farrowing still overweight. We know these sows do not milk as well and have more difficulty farrowing, and then have a subsequent breed back difficulty. These challenges are predicated from sows going into lactation in a catabolic state, and with more body weight stress to joints and feet is also increased. Under conditioning leads to its own problems including lower total born and difficulty breeding after weaning sow's next litter. To head this off, sows are measured at breeding and her feed is adjusted accordingly, she is then measured again at pregnancy check with a spot check on day 70 to keep her on the right track and make further adjustments if needed. Typically, the day 70 adjustment would never be made to an overweight sow but would be made to an underweight sow at this point in her gestation. This process places the highest percentage of sows in the "happy medium" that optimizes health and performance. On a 1-3 body condition scale, our goal is 90% to fall in the middle or ideal category. Since we also know that feed ingredients change and corn ages, we have also implemented feed box measurement monthly with 20 boxes measured per barn per box type along with the sow measurement. This has driven objectivity to our sow conditioning and also has given us a better position going into lactation.

Pre-farrow and gestation vaccination strategy

Although our vaccination strategies are farm specific they have many commonalities throughout the system. As farms have removed feed medications as well as Penicillin as a treatment option we have seen higher incidence of lameness in sows and by also removing most of the downstream feed medications it became more apparent what our sows are shedding and what is leaking from our sow farms. All of our farms are vaccinated pre-breeding for PCV2, in addition sows would also receive pre-farrow vaccinations for *E coli/Clostridium/Rota, Ery*, and *M hyorhinis*. We see very early colonization of *Ery* and *M hyorhinis* with subsequent clinical disease and without pre-farrow vaccination booster it becomes very difficult to get in front of piglet clinical disease and allow time for piglet vaccination. We also will feedback specifically for Rota but actively try to avoid other pathogens being inadvertently feedback as well. In addition, whole herd vaccinations are also performed with salmonella as well as additional *Ery* as the vaccines do not offer complete protection but do help to reduce shedding and clinical disease.

Piglet care and lactation feeding

Gilt development starts at birth with split suckling and colostrum management. On our multiplication farms we want to cater to the gilts are they are the future of our herds. We know lifetime sow performance is largely impacted from very early in life. They therefore become the first priority in terms of colostrum intake, as their litters tend to be large farrowing crate space can be a limitation resulting in higher pre-wean mortality and poorer colostral immunity. We have replaced smaller farrowing crates with larger crates to take better care of the piglets we have in them. We also identify larger gilts at birth in conjunction with a larger litter size to target selection of those gilts later in the nursery. All piglets receive supplemental vitamin D at birth as our sows tend to run deficient and our piglets will also be deficient when entering the nursery if not supplemented (below rickets levels). Without supplementation we will see a higher incidence of kyphosis in the nursery and move a gilt from being selectable to a non-select. The majority of vaccinations piglets receive are in the sow farm as compliance tends to be much better and this allows us to get in front of clinical disease in the nursery. The sow farms which we have seen the highest production on full feed when sows enter into farrowing and continue to have access to clean fresh feed at all times.

Nursery management

As gilts enter the nursery we don't want to limit their productive potential and is a critical transition period and space can become crucial l. The target stocking density of these animals at 2.5 sq ft/head on a six week nursery turn. A dedicated individual preselects gilts while in the nursery with a final selection prior to sow farm shipment (either late nursery or early in the gilt grower facility). By adding a dedicated individual for selection the gilt retention rate was increased from by over 10%. All gilts for the system are raised on rural water in the nursery which has further helped to reduce fall out due to poor starters and loose stools associated with poor well water.

Gilt development

Gilts enter into the sow farm from 50-130 lbs and are batched in by birth week. Deliveries are no fewer than once per month with weekly entries for farms closer to multiplication and shorter transport times. Most sow-farms would have on site GDU's that are continuous flow. This allows for good exposure to farm specific pathogens. Boar exposure would begin at approximately 175lbs and is continuous with boar pens placed strategically in the GDU. Age and weight of breeding is farm specific *(M hyopneumoniae* positive farms tend to be a bit later as growth is impeded due to the Mycoplasma exposure and infection earlier in life). Most farms optimize subsequence performance when bred at 31 weeks of age and at 315lbs with a minimum of 300lbs. As lameness can be and has historically been very challenging in the growing gilts it becomes evident if the vaccine program compliance was good throughout this time. Gilts would not typically receive any feed grade medications and rarely necessitate water medications either. The primary drivers for lameness in our system include *M hyorhinis*, *E rhusiopathiae*. If this lameness is not prevented we have historically seen a dramatic decrease in conception rates on gilts and will also have significant fall out and removal rates despite antibiotic interventions.

Staff training

Staff training is always continuous and ongoing. For example, farrowing staff tends to like to move pigs and will gravitate toward that regardless of the set protocol or training in place so constant review of procedures and follow up is critical. Very clear written guidelines in multiple languages are also critical to understanding when coupled with one on one training starting with the farm manager and team lead then working toward individual employees in the farm. Some things that are often obvious to experienced personnel is not obvious to a new employee especially when they do not come from a farm or livestock background. Some staff may not learn most effectively by reading and require primarily hands on training or picture type instructions. It is often an eye opener, even for a farm that is good about minimizing pig movement, when the farm faces a disease challenge and full McREBEL is implemented. The farrowing staff is generally shocked to see how many good quality pigs the sows and gilts are capable of weaning. Inevitably after the disease is eliminated and the farm resumes normal pig movement this lesson is often quickly forgotten and re-training is many times necessary

Leading change to improve piglet survival

Larry L. Coleman, DVM

In recent years, US swine production has seen gains in sow productivity as measured by piglets weaned per sow per litter farrowed. In fact, one of the largest swine record databases in the US, Swine Management Services (SMS) from Fremont, NE, reveals piglets weaned per litter have moved from approximately 9.17 pigs weaned per litter to 10.96 pigs over the last twelve years. A few of the reasons for this increase are: improved genetics, improved nutrition, and improved management. Unfortunately, these increases in sow productivity have not always been accompanied by an increase in the percentage of total born piglets that are alive at weaning. In fact, the same SMS database indicates that piglet survival at weaning today is at 78.8% while 12 years ago, it was at 79.2%. This is somewhat surprising considering in many cases there have been investments; not only in better facilities but also in improved management techniques and employee training. Certainly, the tendency for a sow to have more piglets can add to the challenges faced by each piglet, but this does not entirely explain why the modern US swine industry does not have a positive trend in the percentage of total pigs born alive at weaning.

Many techniques have been developed over the years to save a higher percentage of piglets born, and often, when well implemented, they have resulted in certain farms being well above the norm in terms of piglet survival. Unfortunately, many farms/systems still struggle with implementing strategies to prevent stillbirths and reduce preweaning mortalities. The following is a brief list of the techniques that have become proven strategies to save piglets:

Farrowing assistance

Sows often need help farrowing or weak/stillborn piglets will result. Interventions cannot wait if piglets are to be saved. Many US farms have moved to practicing extended farrowing care in the farrowing house, even moving to monitoring sows 24 hours/7 days a week. This practice can be highly successful if implemented correctly. Unfortunately, many farms have tried to extend hours of care and have eventually returned to normal one-shift care because of lack of improvement or difficulty in implementing the new schedule.

Keeping newborn piglets warm

Postpartum, newborn piglets chill quickly. Techniques to prevent chilling include: drying of piglets, hot boxing, drying powders, moving piglets, and double heat zones for piglets to stay warm. Unfortunately, initiatives to make improvements in preventing piglet hypothermia have often been met with frustration.

Colostrum intake

Unless newborns receive colostrum (first milk) as close to immediately after birth as possible, their chances of surviving in the farrowing house or even in the nursery are drastically reduced. Techniques to improve uniform colostrum intakes across all piglets born include split suckling, cross fostering, and other efforts to ensure that no piglet is left behind. Unfortunately, these management efforts are often only average and therefore lead to only average results.

Fallback piglets

Large litters are a challenge for many individual pigs. It is very common for pigs to suffer from suboptimal milk intake and eventually become a "fallback." Techniques to spot these piglets and intervene on their behalf are often successful, but even then, when measured across all farms, results can be disappointing. Initiatives to improve this area are often implemented but prove to be unsuccessful.

Since all of these practices are widely known and vast sums of money are spent on caretaker training, why has the US swine industry not seen a corresponding decrease in the number of piglets that are lost per litter? In my opinion as a practicing veterinarian, these strategies listed above: while well documented, well thought out, and well understood, are often not practiced in the concise and consistent manner which facilitates improved piglet survival. Few farms seem to be able to put the entire playbook into everyday practice. Why? I would suggest we need to look away from the science of pig farming and pay more attention to the psychology of people involved in the swine industry. We need to determine why stakeholders (owners/supervisors/caretakers/veterinarians) are often satisfied with the status quo of ever increasing weaning numbers but also with the higher percentage of deaths. For information on why people are often satisfied with the status quo (and as a result often resist positive change) I would suggest we look at information from John P. Kotter in his many books and materials on creating change. Mr. Kotter lists 3 major initiatives which must be in place before positive change can occur. I believe they apply to this problem and are worth exploring. These initiatives are as follows:

A climate for change

What describes a climate for change? First and most importantly, there must be a sense of urgency. Someone has to awaken out of complacency and recognize that their system/farm has a problem with piglet survival percentage, as compared to current and future competitors - some of whom are able to have piglet survival rates of 90%. This will take leadership and could come from the ownership group, the supervisors, the farm manager, the farrowing crew, or the attending veterinarian. Unfortunately, most organizations suffer from a shortage of leaders and unless there is a leader concerned about excessive piglet losses, no change will occur. A sense of urgency to correct the problem must spread throughout all stakeholders in the operation and the resistance to change must be overcome. This is often very difficult and to make matters worse, the difficulty is often underestimated. Most change initiatives fail badly because of this misunderstood level of difficulty.

Engaging the organization

This step involves getting everyone on the farm to buy in to the vision of improved piglet care. It can be difficult at first, but will be facilitated when the owners take steps to show the importance of the change initiatives. This might include remodeling facilities and/or providing extra expenditures for labor. Also, offsite trips to farms that are successful in this area can be a powerful motivating tool for stakeholders. Possibly in a larger system, a farm can be developed that can become a model for other farms in the system. It is imperative, however, that caretakers truly believe that those in higher management have made sacrifices that demonstrate the desired change is important. Unfortunately, often caretakers do not sense that upper management is willing to sacrifice to make changes so, why should they? As a result, complacency reigns.

Implementing and sustaining change

Survivability of piglets is an everchanging parameter. Disease outbreaks and personnel changes can occur. For change to stick, leaders must recognize that a new improved culture of high expectations in piglet care must be fully cemented into place. This effort may well take years. This difficulty is also usually underestimated.

Unfortunately, as long as the swine business is profitable and sow productivity is increasing, there is little urgency for owners, caretakers, or veterinarians to be seriously concerned about this problem. Hence, the current state of affairs when it comes to piglet survival trends.

So, what is the role of swine veterinarians when it comes to leading change in preventing piglet losses?

First: They themselves must not be complacent about piglet care; and they must model this sense of urgency to all stakeholders. They need to realize that in many cases of substandard piglet care complacency is the real problem rather than a lack of training. Often training will occur almost spontaneously when there is a true sense of urgency about this problem among all stakeholders.

Second: The swine veterinarian should be the most knowledgeable person on the farm to accurately describe the physiological needs of the unborn and born piglets. Who has better training to understand the needs of a young fragile piglet than the veterinarian? Therefore, the physiological needs of the piglet must be vigorously advocated for by the veterinarian on a regular basis. Veterinarians must not become discouraged and accept the status quo as unchangeable.

Third: They must realize that an inspirational leader is needed that is able to create a sense of urgency. If the herd veterinarian is not a naturally inspirational leader, they must be willing to bring in outside consultants who are inspirational and knowledgeable to help dismantle the complacency and the resistance to change. Certainly, there is still much work to do after the outside consultants leave and the real work of building a culture of high expectations in piglet care begins. Changing the culture can be very difficult, but it is achievable.

Lastly: The swine veterinarian should be a tireless advocate of what the farm culture could be when it comes to meeting the needs of young piglets. In many cases, this will require determination, persistence, and a willingness to learn themselves. Most of all, maintaining an unwavering belief that with time, change can occur. As always, everything rises and falls on leadership.

For more information on creating change, I would encourage those who are interested to read John P. Kotter's materials.

Update on feeding strategies for the highly prolific sow

M. B. Menegat[1], DVM, MS; S. S. Dritz[1], DVM, PhD; C. M. Vier[1], DVM; M. D. Tokach[2], PhD; J. M. DeRouchey[2], PhD; J. C. Woodworth[2], PhD; R. D. Goodband[2], PhD
[1]College of Veterinary Medicine, Department of Diagnostic Medicine and Pathobiology, Kansas State University, Manhattan, Kansas; [2]Department of Animal Sciences and Industry, Kansas State University, Manhattan, Kansas

Introduction

The improvements in prolificacy in modern sows have markedly increased the number of piglets gestated, nursed, and weaned per sow. The ability of sows to provide nutrients to support fetal growth and to produce milk to nourish nursing piglets has, therefore, been enhanced. These remarkable improvements in productivity have prompted the reassessment of proper nutritional programs for the highly prolific sow. Additionally, as larger litter sizes have led to a decline in piglet birth weight, efforts have been focused on nutritional strategies to increase piglet birth weight. The purpose of this review is to provide an update on recent nutrition research and feeding strategies for the highly prolific sow during the course of gestation and lactation.

Gestation

The goals of the nutrition program for gestating sows consist of meeting the nutrient requirements for maintenance and growth of the breeding female and for adequate development of the conceptus, while managing body condition. The NRC[1] divides the amino acid and energy requirements for gestating sows into requirements for maintenance, growth, fetus, mammary tissue, uterus, and placenta and fluids. Noticeably, the greatest estimated protein deposition and energy requirement during gestation are associated with maintenance and weight gain of the sows[1] (Figures 1 and 2). Early gestation represents the best opportunity for replenishing body reserves, whereas in late gestation, both estimated protein deposition and energy requirement are exponentially increased and directed towards fetal growth and mammary development. Since maintenance and growth account for the majority of the requirements during gestation, an estimate of body weight is extremely important to accurately feed the gilt and sow.

Nutrition after breeding

The purpose of sow nutrition after breeding is the maintenance of pregnancy and embryo survival, which is mostly ensured by adequate progesterone levels. An increased feed intake after breeding has been conventionally presumed to be detrimental to embryo survival because of increased hepatic metabolism of systemic progesterone.[2] However, sows also have a local supply of progesterone through a direct countercurrent mechanism from the ovaries to the oviducts and uterine horns. This mechanism provides much higher concentrations of progesterone in a pulsatile fashion and is not subject to hepatic metabolism.[3] In support of this theory, recent studies found that an increased feeding level (2.8 vs. 1.5 kg/d) in the immediate period after breeding resulted in increased progesterone secretion and embryo survival in early pregnancy in gilts.[4] Moreover, severe undernutrition at days 10-11 of pregnancy, which coincides with the period of embryo elongation, reduced embryo survival and total number of piglets born in gilts.[5] Therefore, it would be recommended to avoid feed deprivation in the immediate period after breeding and to feed thin sows to recover body condition as quickly as possible.

Nutrition in early-mid gestation

The early-mid gestation period comprises most of the gestation course. The purpose is to maintain well-conditioned sows to improve reproductive efficiency, sow longevity, and welfare. Under-conditioned females are predisposed to have reproductive failure, development of sores, and higher risk of mortality; whereas over-conditioned females are prone to dystocia, reduced feed intake in lactation, and higher risk of culling due to locomotion problems. The impact of undernutrition throughout gestation on birth weight of piglets is modest because sows have high capacity to mobilize maternal reserves to support placental and fetal development under prolonged nutritional deficiency.[6] However, restricted feeding levels (1.8 kg/d) to thin gilts (12 mm backfat) reduced birth weight by 100 g/piglet.[7] The effect of inadequate nutrition on fetal growth in immature gilts is related to a greater nutrient requirement for own body growth and development compared to mature sows, leading immature gilts to compete with the fetus for nutrients when in deficiency.[6] Therefore, it is important to adjust feed allowance of thin gilts during gestation to avoid compromising piglet birth weight.

On the other hand, feeding well-conditioned gilts and sows above the requirements in gestation accrues unnecessary feed expenses and leads to an over-conditioned breeding herd. It has been recognized that increasing feed allowance during most of the gestational period (early and mid-gestation) provides no benefit to piglet birth weight or litter size.[7,8] In fact, sows that accrue excessive backfat deposition during gestation show reduction in feed intake during lactation, excessive body weight loss at weaning, and difficulties to maintain milk production and litter growth.[9] It was estimated that for every 1 mm increase in backfat at the end of gestation there is a reduction of 60-120 g of daily voluntary feed intake in lactation, with a noticeable impact on reproductive performance when backfat is greater than > 20 mm.[9] In a recent yet-to-be-published study, the daily requirement of SID lysine in gestation was studied in a large number of gilts and sows (971 females) individually fed in a group-housing system.[10]

The findings appear to suggest that increasing dietary SID lysine (11.0, 13.5, 16.0, and 18.5 g/d) during the course of gestation is associated with maternal body protein deposition, supported by increased body weight gain and unchanged backfat thickness (Table 1). However, increasing dietary SID lysine did not seem to have an impact in piglet birth weight.

Body condition

The major goal of body condition scoring remains unchanged: identify thin sows to bring them back to an adequate body condition and prevent over-conditioned sows at farrowing. Excessive weight gain during gestation has a negative impact on stillborn rate[11,12] and well-documented negatives on lactation feed intake, as well as a significant economic impact. The traditional methods to estimate body condition of gilts and sows are visual scoring, flank measurement, and backfat measurement with ultrasound.

The visual scoring is usually performed on a 5-point scale by evaluating the bone prominence and fat cover on shoulders, ribs, backbone, and hips. The target is to have 90% of the sows in ideal condition score at farrowing. Although visual scoring is an inexpensive and low-labor method to assess body condition, it is a subjective method. The flank measurement is an objective method to estimate body weight from a derived allometric equation. It is obtained using a cloth tape to measure from the point of one flank over the back of the sow to the point of the other flank immediately in front of the back legs. The flank measurement provides a simple, yet more accurate, method of categorizing sows into weight categories, and coupled with ultrasound measurement of backfat is a more objective measure of body condition.[13] The backfat measurement is an objective method to estimate body condition, although it is poorly correlated with muscling. The backfat thickness is measured at the last rib approximately 10 cm off the midline using an A-mode or real-time ultrasound. The target is to have more than 75% of the sows between 12 and 22 mm of backfat depth at farrowing.

Recently, another method to determine body condition of gilts and sows was developed to standardize body condition for modern genetic lines.[14] The sow body condition caliper describes substantial variation in sow weight, backfat, loin depth, and body condition score. The caliper quantifies the angularity from the spinous process to the transverse process of the back at the last rib of sows, and it is based on the premise that the back becomes more angular as sows lose weight, fat, and muscle. The ideal caliper angle is between 12 to 15-unit range.

Nutrition in late gestation

While little evidence supports an increase in feed intake in early and mid-gestation to improve birthweight, increasing feed intake in late gestation seems to improve piglet birth weight modestly.[11] Estimated protein deposition greatly increases from day 60 of gestation onwards[1] (Figure 1) due to rapid fetal growth and mammary development. Amino acid requirements exponentially increase in late gestation and are influenced by larger litter sizes[1] (Table 2). Interestingly, a recent study with high-performing gilts and sows revealed that energy intake rather than amino acid intake in late gestation provides a response in piglet birth weight.[11] Higher energy intake in late gestation increased piglet birth weight by 30 g per piglet in gilts and sows. However, increasing energy intake also increased stillborn rate by 2.1% in sows, but not in gilts. A meta-analysis reviewing the existing literature on bump feeding in late gestation showed only a modest increase on piglet birth weight by an average of 28 g per piglet in gilts and sows.[15] Also, increasing both energy and amino acid intake in late gestation resulted in increased body weight gain. The meta-analysis estimated that each 1 kg increase in daily feed allowance from d 90 to farrowing is associated with approximately 7 kg increase in body weight of gilts and sows.[15] Also, increasing SID lysine levels during the course of gestation is associated with an increase in body weight of gilts and sows by approximately 6 kg.[10] Altogether, these recent findings in gestating sows suggest that the characteristics of modern genotypes seems to consist of improved feed efficiency and propensity for growth. Therefore, increasing energy intake during late gestation seems to have a modest effect on individual piglet birth weight and a negative effect on stillborn rate in sows probably due to the increase in body weight gain.[11]

Table 1: Effects of increasing dietary standardized ileal digestible lysine (SID Lys) in gestation on sow growth and piglet birth weight.[1]

	SID Lys, g/d					Probability, P <	
	11.0	13.5	16.0	18.5	SEM	Linear	Quadratic
Initial BW, kg	187.3	185.8	187.2	187.0	1.92	0.777	0.304
Final BW, kg	232.9	233.4	236.8	239.1	0.76	0.001	0.205
Initial backfat, mm	14.7	14.7	14.8	14.8	0.26	0.650	0.824
Final backfat, mm	17.3	17.5	17.5	17.5	0.25	0.612	0.657
Total born, n	15.5	15.7	15.5	15.6	0.209	0.993	0.858
Piglet BW, kg	1.28	1.27	1.29	1.28	0.014	0.574	0.972

[1] Study with a total of 971 group-housed and individually-fed sows and gilts from d 5 to 112 of gestation. Gilts and sows received 2.1 and 2.3 kg/d of feed (5.3 and 5.7 Mcal NE/d), respectively.[10]

Figure 1: Estimated total protein deposition (Pd) of second-parity sows during gestation (adapted from NRC[1])

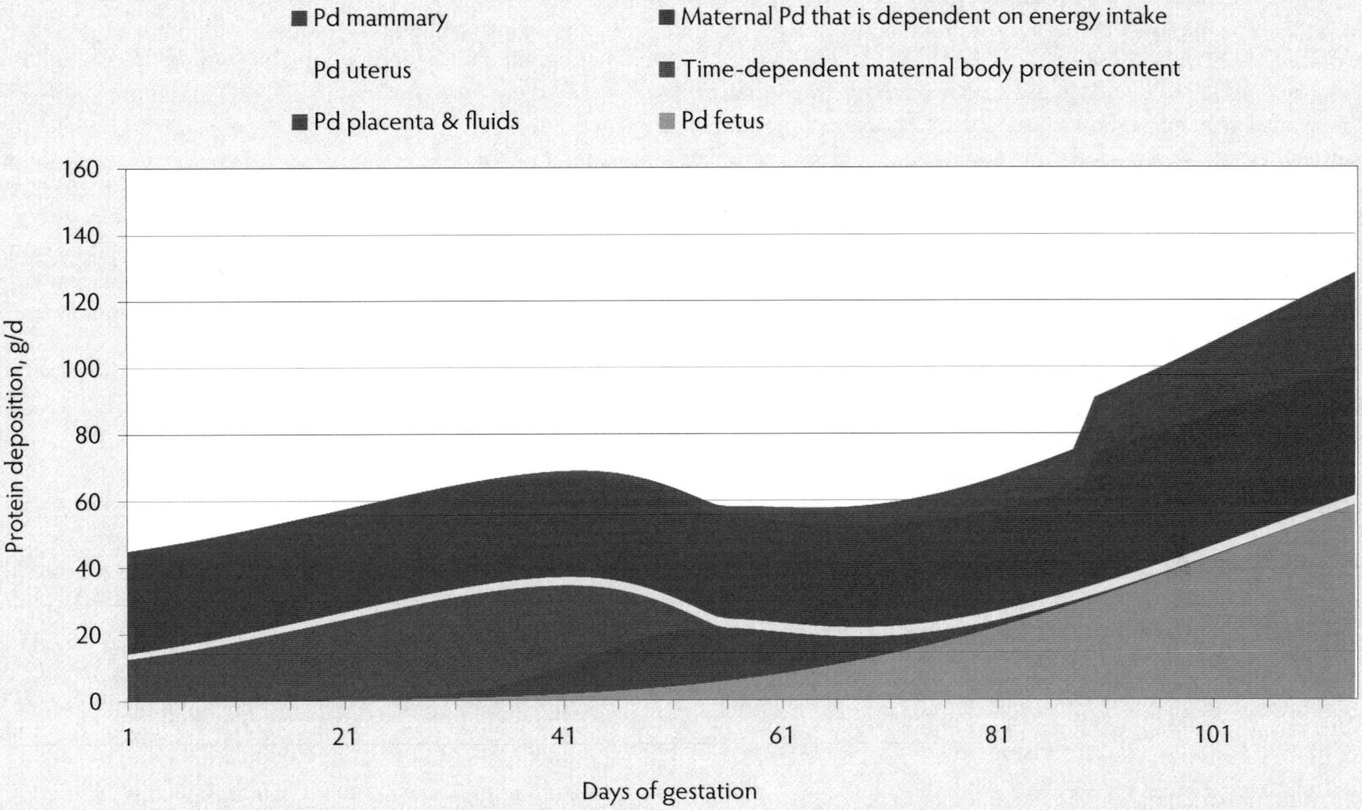

Table 2: Estimated daily standardized ileal digestible lysine requirement (g/d) of gilts during gestation.[1]

	Days of gestation		
Total born	**1 to 45**	**45 to 90**	**90 to 114**
10	9.3	9.2	16.4
12	9.5	9.6	17.1
14	9.4	9.9	17.7
16	9.4	10.2	18.4

[1] Feeding level of 1.83 kg/d and 2.40 kg/d from d 1 to 90 and from d 90 to 114, respectively, of a diet containing 2,530 kcal NE/kg (NRC[1])

Gestation feeding systems

World-wide housing systems of gestating sows are gradually shifting from individual stalls to group housing systems. In the US, this change is driven by animal welfare concerns, state legislatures, and food companies' demands. Group housing systems generally allow a similar nutritional program to individual housing, but taken into account some particularities regarding social interaction, activity level, environmental temperature, feed wastage, and adjustment of body condition score.

Activity level and environmental temperature influence energy requirements of sows.[1] Group-housed sows have increased activity level compared to individually-housed sows, as they have more social interaction and spend more time standing and walking and less time lying.[16] The energy requirements of sows are also increased to maintain body temperature when environmental temperature is below the thermoneutral zone[1]. The lower critical temperature for individually-housed gestating sows is 20°C, but because group-housed sows are able to exchange body heat with each other, the lower critical temperature is decreased to 16°C. This indicates that sows in groups can be housed at lower environmental temperatures before the energy requirements begin to increase.[17]

Group housing systems have characteristic feeding systems that fall into the category of simultaneous or sequential feeding systems. Simultaneous feeding systems are characterized by once or twice a day feed allowance for all sows simultaneously, as in the case of free access stalls, floor feeding, and long trough feeding systems. Sequential feeding systems are characterized by continuous feed allowance throughout the day and sows are fed one after the other, as in the case of electronic sow feeding (ESF) systems.[18] Group housing allows for a natural competition for feed that is typically imposed by dominant sows over the timid ones. Competitive feeding is particularly observed in floor feeding, trough feeding, and non-gated stalls systems. However, it is important to notice that feeding strategy, feeder design, and pen layout can ameliorate the dispute over access to feed.[19] Another reasonable way to ameliorate aggression in these systems appears to be to increase feed intake.[20] Thus, many group housing systems create variation in sow weight and limit the ability to evaluate and control sow body condition.[19] In most group housing systems, it is difficult to adjust individual feeding levels to get sows in an adequate body condition. For example, in floor feeding or trough feeding the overall feeding level should be adjusted to prevent sows from becoming over- or under-conditioned. However, this situation can create feed wastage and variation in body condition. In fact, observations in the field indicate feed wastage is higher in these systems. On the other hand, in ESF systems the individual feeding level can be adjusted for each sow. However, it requires extra time and labor to assess individual body condition within the group of sows.[17]

Observations in large scale US productions systems have indicated that there are not adequate guidelines for optimizing feeding at the individual sow level. Although the feeding level can be adjusted for the individual sow in ESF systems, so much of the nutrient requirement is dependent on sow body weight, which is unknown in group-housed sows. Successful implementation of ESF program systems are focusing on feeding a constant level with the ESF system and simplifying the management of the system. Also, acquainting sows to the feeding system and maintaining full feed intake is often a challenge, especially in gilts and parity 1 sows.[21] Training of replacement gilts is becoming a priority in farms with ESF systems[22], because gilts unfamiliar with the system are reluctant to enter the feeding station to eat. Even with adequate training, parity 1 sows are reluctant to consume their daily feed allowance throughout the course of gestation. Parity 2 and 3+ sows seem to be able to maintain adequate feed intake throughout gestation, but even sows accustomed to the feeding station do not consume the allotted feed allowance for 7 days after placement.[21]

Gestation feeding recommendations

Our practical experience and observations from around the world indicate that the strategies provided herein can be used successfully. As the modern sow is very resilient, simplification of gestation diets to a single phase throughout pregnancy seems to be an effective feeding strategy. General recommendations would be to provide 2.1 to 2.3 kg/d of a gestation diet containing 3,300 kcal/kg ME (2,560 kcal/kg NE) and 0.60% SID lysine. Feeding levels can be adjusted for gilts and sows considering their higher requirements for growth and maintenance, respectively.

The nutritional impact on piglet birth weight is modest. Genetic selection can have a much higher impact in this regard.

Figure 2: Estimated daily metabolizable energy requirements of gilts during gestation (adapted from NRC[1])

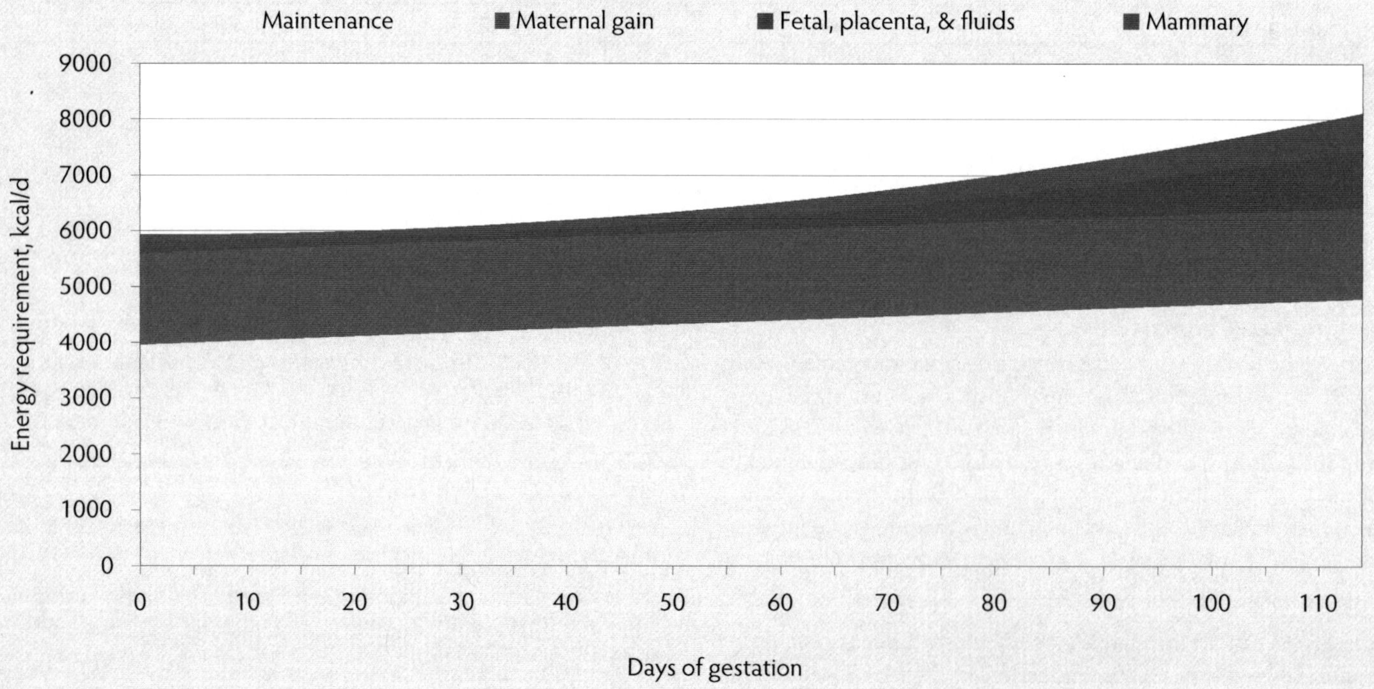

Incorporation of individual piglet birth weight in the genetic selection index was able to recover, within 2 years, 60 g of the 100 g decrease in birth weight in the last decade.[23] Also, inclusion of litter size at day 5 after birth as an index of genetic selection was able to improve neonatal survivability even in larger litters.[24]

Peripartum

Feed allowance has been traditionally limited in the peripartal period to prevent impaired lactational feed intake.[25] However, recent findings demonstrated that sows fed *ad libitum* in the peripartum are able to voluntarily reduce feed intake around farrowing and avoid hypophagia during lactation.[26] Considering that mobilization of maternal body reserves is actively occurring around the peripartal period and progressing throughout lactation, the practice of limiting feed allowance in the peripartum could have an impact on the magnitude of body weight and backfat loss during lactation.[27] Also, many farms have transitioned from hand feeding to feeding systems that allow *ad libitum* intake during lactation. Sows provided extra feed or *ad libitum* feed intake in the peripartum lost less backfat during lactation compared to restricted-fed sows.[26,27] The *ad libitum* intake resulted in improvement in piglet growth rate and weaning weight in well-conditioned sows, but in impaired piglet growth rate and weaning weight in over-conditioned sows (> 22 mm backfat).[26]

Feed allowance increase in the peripartal period can also influence yield and nutritional composition of colostrum.[27] Colostrum is produced during the last month of gestation but mainly during the last week before farrowing.[28] Sows provided extra feed for a week prior to farrowing improved colostrum yield and secretion of lactose and fatty acids through colostrum while maintaining immunoglobulin levels.[27] As a result, piglet weight gain was improved during the first day of life. Body condition in the peripartal period can also influence colostrum production and mammary gland development. Sows with 17 to 23 mm backfat in the peripartum were found to have the highest colostrum yield and nutrient concentration in colostrum compared to thin (< 17 mm) or fat (> 23 mm) sows.[27] Gilts at farrowing with an average of 23 mm of backfat in the peripartum have shown optimal mammary development compared to an average of 14 mm or 18 mm of backfat deposition, with a positive impact on subsequent litter performance.[29]

Therefore, providing *ad libitum* feeding in the peripartum increases total feed consumption prior to weaning and reduces mobilization of body reserves.[26,27] A practical strategy is to allow *ad libitum* feed intake upon placement in the farrowing room (i.e. within one week prior to farrowing). The benefits are noticed as improvements in piglet growth rate and weaning weight, as well as colostrum yield and nutritional composition. However, *ad libitum* feeding is advantageous for sow performance provided that backfat thickness is adequate, emphasizing the importance of not having over-conditioned sows and gilts (> 22 mm) at farrowing.

Lactation

The main goal of the nutrition program for lactating sows is to maximize feed intake to sustain milk production without excessive mobilization of body reserves. The estimated requirements for energy and amino acids are mainly determined by maintenance and milk production, with the latter driven by litter size and litter growth rate[1] (Table 3). However, if intake is insufficient to support the requirements, both fat and protein stores are mobilized from the lactating sow. Ideally, the curve of feed intake should mimic the curve of requirements during lactation (Figure 3), which coincides with the curve of milk production. However, it may not reflect the variability within the population of lactating sows. Therefore, *ad libitum* feeding during lactation is recommended to maximize milk production, minimize mobilization of body reserves, increase litter growth rate, and improve subsequent reproductive performance.[25]

The genetic selection for hyperprolific sows increased the demand for milk production to support a larger litter size and growth rate. However, feed intake has not increased to the same extent as milk production (Figure 3), resulting in a negative energy balance during most of lactation. Additionally, greater mobilization of protein stores is observed in lean genotype sows. Therefore, nutrient requirement estimates and feeding strategies have been reevaluated to ensure optimum performance of modern sows.

Nutrition in lactation

Dietary energy is important to alleviate sow body weight loss during lactation. Feeding a lactation diet with an energy concentration of 3.25 Mcal ME/kg improved body condition and

Table 3: Estimated daily standardized ileal digestible lysine requirement (g/d) of first-parity sows during lactation.[1]

	Weaning weight, kg		
Total born	**5.8**	**6.0**	**6.2**
10	51.3	52.1	52.8
12	55.5	56.4	57.3
14	59.7	60.8	61.8
16	64.0	65.3	66.5

[1] Feeding level of 6.0 kg/d of a diet containing 2,565 kcal NE/kg in a 21-d lactation (NRC[1])

Figure 3: Estimated daily metabolizable energy requirements and intake of multiparous sows during lactation (adapted from NRC[1])

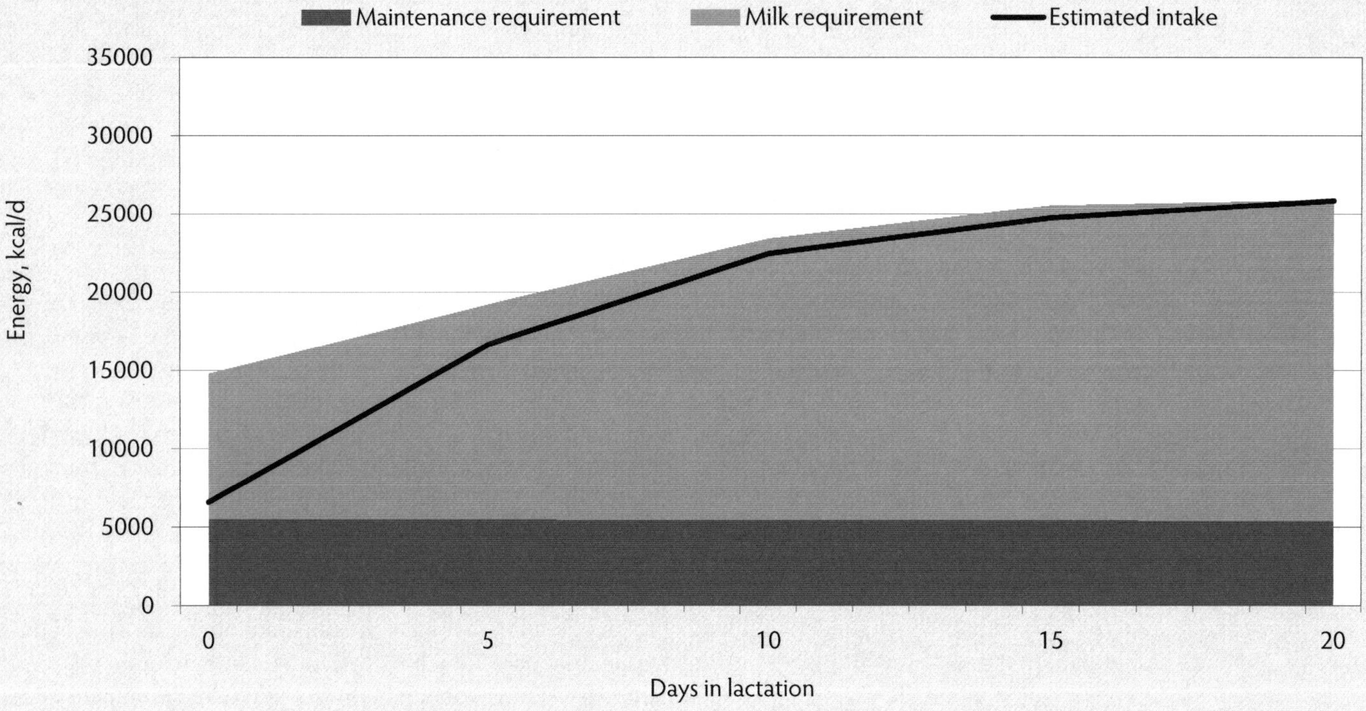

voluntary feed intake than feeding a diet with less than 3.2 Mcal ME/kg.[30] Feeding a diet with more than 3.4 Mcal ME/kg resulted in reduced feed intake.[30] In addition to energy concentration, increased dietary amino acid levels reduce body weight loss[30] and protein loss during lactation,[31] suggesting that body protein catabolism likely occurs to support both insufficient energy and amino acid intake. Dietary amino acid levels also have the potential to improve litter growth rate in high-producing sows.[30,31] Litter growth rate was optimized at an estimated SID lysine:ME ratio of 3.05 g/Mcal, which corresponds to approximately 60 g/d of SID lysine.[30] Furthermore, it has been recently suggested that dietary intake of both single essential amino acids and balanced protein have an influence on sow and litter performance of modern genotype sows.[32] For high-producing sows nursing and weaning 13-14 piglets, the optimal litter body weight gain was observed up till 13.5% SID CP in the diet. This improvement in litter performance was driven by increased milk yield and daily protein output in milk, emphasizing the importance of milk production and composition to support the requirements of nursing piglets.[32]

Traditionally, lactating sows are fed with a single diet until weaning. Recently, a feeding regimen based on individual energy and amino acids requirements for maintenance and milk production was proposed.[33] The 2-diet feeding regimen provided a basal diet and a lactation supplement to adequately support the dynamic requirements for milk production of individual sows during lactation. This feeding regimen demonstrated a potential to reduce nutrient deficiency throughout lactation, but it needs to be further developed into a feeding strategy with applicability in commercial settings.

Thermal stress in lactation

Sows are increasingly susceptible to thermal stress during lactation, especially high-producing sows with larger litter size and greater milk production.[34] In response to thermal stress, sows reduce voluntary feed intake and milk production, with further impact on body reserves mobilization, litter growth performance, and subsequent reproductive performance. Daily feed intake is estimated to be decreased by 3.7% for each 1°C rise in ambient temperature above 25°C.[1] A nutritional strategy to mitigate the effects of thermal stress is the addition of fat in lactation diets with the purpose of increasing energy density and decreasing heat increment in sows.[35,36] In addition, water availability is of utmost importance during lactation, particularly when the ambient temperature is above the thermoneutral zone of sows (18-20°C). Daily water consumption of a lactating sow averages 27.5 L and, as feed intake, is negatively correlated with sow body weight loss and positively correlated with piglet weaning weight.[37]

Subsequent reproductive performance

Ultimately, lactation catabolism impacts subsequent reproductive performance of sows, particularly in the case of first-parity sows.

It is well recognized that excessive weight loss and mobilization of body reserves during lactation is associated with prolonged wean-to-estrus interval, high incidence of anestrus and return to estrus, low farrowing rate, less piglets born in the subsequent farrowing, as well as high culling rate due to reproductive failure.[25,38] However, the reproductive performance of contemporary sows is increasingly resistant to the negative effects of lactational catabolism.[39] This could be related to changes in biology of modern genetic sows, as well as increase in lactation length, use of advanced feeding systems in lactation, and application of proper breeding techniques.[19,38]

Wean-to-estrus interval

The feeding strategy of sows during wean-to-estrus interval is focused on recovery of body reserves lost during lactation and maximization of ovulation rate to ensure a large litter size in the subsequent farrowing. It is well stablished that increasing feed intake of weaned sows in poor body condition improves reproductive performance.[40] However, sows in good body condition do not seem to benefit from increasing feeding allowance during wean-to-estrus interval. Graham et al.[41] found similar wean-to-estrus interval (5.1, 5.0, and 5.0 d), conception rate (95.6, 95.6, and 94.7%), and subsequent total born (14.3, 13.9, and 13.9 pigs/sow) in sows with a body condition score > 2.75 (5-point scale) fed 2.7, 3.6, or 5.4 kg/d, respectively, from weaning to estrus. Considering the requirements of lysine and energy during the wean-to-estrus interval[1], feeding 2.5 kg/d of a conventional gestation diet after weaning seems to be sufficient to meet the requirements and provide additional support for recovery of body reserves (Figure 4). This suggests that *ad libitum* feeding and offering lactation diets during the wean-to-estrus interval is not needed.

References

1. NRC. Nutrient requirements of swine. 11th rev. ed. National Academic Press, Washington, DC. 2012.
2. Jindal R, Cosgrove JR, Aherne FX et al. Effect of nutrition on embryonal mortality in gilts: association with progesterone. *J Anim Sci.* 1996;74:620.
3. Krzymowski T, Kotwica J, and Stefanczyk-Krzymowska S. Uterine and ovarian countercurrent pathways in the control of ovarian function in the pig. *J Reprod Fertil Suppl.* 1990;40:179.
4. Athorn RZ, Stott P, Bouwman EG et al. Effect of feeding level on luteal function and progesterone concentration in the vena cava during early pregnancy in gilts. *Reprod Fertil Dev.* 2013;25:531.
5. Langendijk P, Bouwman EG, Chen TY et al. Temporary undernutrition during early gestation, corpora lutea morphometrics, ovarian progesterone secretion and embryo survival in gilts. *Reprod Fertil Dev.* 2016;29:1349.
6. Campos PHRF, Silvva BAN, Donzele JL et al. Effects of sow nutrition during gestation on within-litter birth weight variation: a review. *Animal.* 2012;6:797.
7. Amdi C, Giblin L, Ryan T et al. Maternal backfat depth in gestating sows has a greater influence on offspring growth and carcass lean yield than maternal feed allocation during gestation. *Animal.* 2014;8:236.
8. Nissen PM, Danielsen VO, Jorgensen PF et al. Increased maternal nutrition of sows has no beneficial effects on muscle fiber number or postnatal growth and has no impact on the meat quality of the offspring. *J Anim Sci.* 2003;81:3018.
9. Kim JS, Yang X, Pangeni D et al. Relationship between backfat thickness of sows during late gestation and reproductive efficiency at different parities. *Acta Agric Scand.* 2015;65:1.
10. Thomas LL. Kansas State University. Unpublished. 2017.
11. Gonçalves MAD, Gourley KM, Dritz SS et al. Effects of amino acids and energy intake during late gestation of high- performing gilts and sows on litter and reproductive performance under commercial conditions. *J Anim Sci.* 2016;94:1993.
12. Mallmann AL. Setor de Suínos da UFRGS. Unpublished. 2017.

Figure 4: Estimated daily standardized ileal digestible lysine requirements (g/d), metabolizable energy requirements (kcal/d) and intake of multiparous sows during wean-to-estrus interval (adapted from NRC[1]).[1]

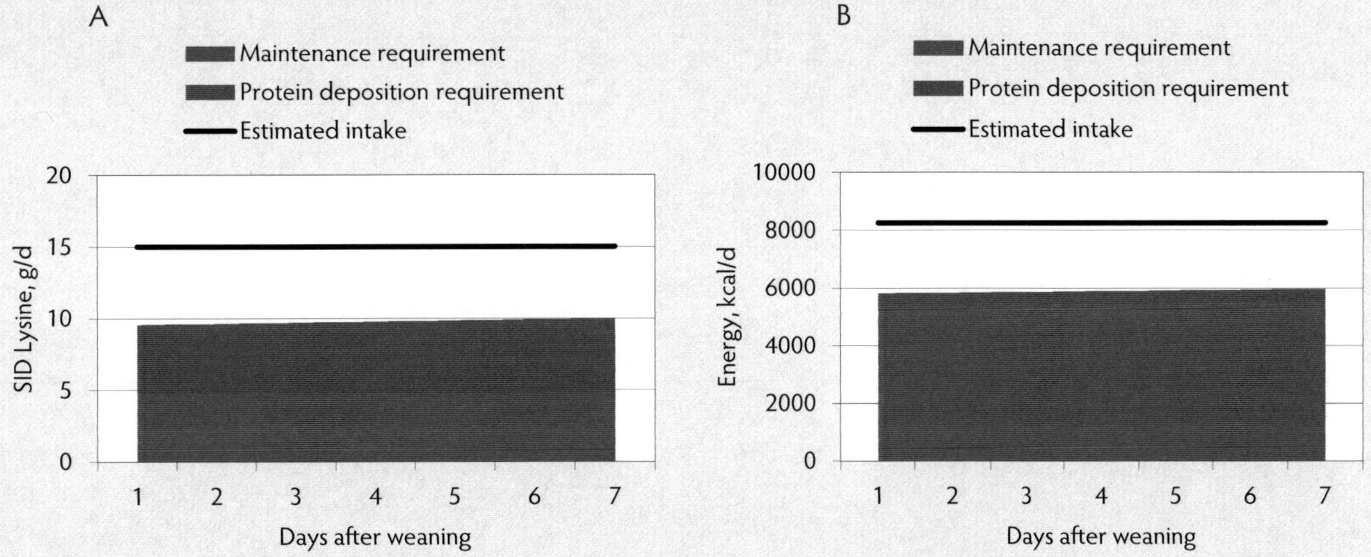

[1] Assuming 1 kg/d gain and feeding level of 2.5 kg/d from d 1 to 7 after weaning of a gestation diet containing 3,300 kcal ME/kg and 0.60% SID lysine.

13. Young MG, Tokach MD, Aherne FX et al. Comparison of three methods of feeding sows in gestation and the subsequent effects on lactation performance. *J Anim* Sci. 2004;82:3058.

14. Knauer MT and Baitinger DJ. The sow body condition caliper. *Appl Eng Agric.* 2015;31:175.

15. Gonçalves MAD. Late gestation lysine and energy effects in sows and dose-responses to tryptophan and valine in nishing pigs. 2015. PhD Dissertation. Kansas State University, Manhattan, KS.

16. Salak-Johnson JL, DeDecker AE, Horsman MJ et al. Space allowance for gestating sows in pens: behavior and immunity. *J Anim Sci.* 2012;90:3232.

17. DeRouchey J and Tokach M. Group housing systems: nutritional considerations. *National Pork Board Factsheet* #03647-3/13. 2013.

18. Spoolder HAM and Vermeer HM. Gestation group housing of sows. In: Farmer C (ed), *The Gestating and Lactating Sow*. Wageningen Academic Publishers, The Netherlands. 2015.

19. Solà-Oriol D and Gasa J. Feeding strategies in pig production: sows and their piglets. *Anim Feed Sci Tech.* 2016; http://dx.doi.org/10.1016/j.anifeedsci.2016.07.018

20. Muller TL. Nutritional Management Strategies to reduce aggression at mixing of unfamiliar sows. *Aust Pork CRC Research Report* #1C-115. 2015.

21. Thomas LL, Dritz SS, Tokach MD et al. Lessons learned from managing electronic sow feeders and sow body weight data. *Kansas Agricultural Experiment Station Research Reports.* 2016;2:8.

22. Vier CM, Gonçalves MA, Thomas LL et al. Gilt training for electronic sow feeding systems in gestation. *Kansas Agricultural Experiment Station Research Reports.* 2016;2:8.

23. Gonçalves MAD, Vier C, Dritz S et al. Impacto da nutrição da porca gestante no peso dos leitões ao nascimento. In: Barcellos DE et al. (eds), *Avanços em sanidade, produção e reprodução de suínos*. Universidade Federal do Rio Grande do Sul, Porto Alegre, Brazil. 2015.

24. Nielsen B, Su G, Lund MS et al. Selection for increased number of piglets at d 5 after farrowing has increased litter size and reduced piglet mortality. *J Anim Sci.* 2013;91:2575.

25. Koketsu Y, Dial GD, Pettigrew JE et al. Characterization of feed intake patterns during lactation in commercial swine herds. *J Anim Sci.* 1996;74:1202.

26. Cools A, Maes D, Decaluwé R et al. Ad libitum feeding during the peripartal period affects body condition, reproduction results and metabolism of sows. *Anim Reprod Sci.* 2014;145:130.

27. Decaluwé R, Maes D, Cools A et al. Effect of peripartal feeding strategy on colostrum yield and composition in sows. *J Anim Sci.* 2014;92:3557.

28. Devillers N, Le Dividich J, and Prunier A. Physiologie de la production de colostrum chez la truie. *INRA Prod Anim.* 2006;19:29.

29. Farmer C, Martineau J-P, Méthot S et al. Comparative study on the relations between backfat thickness in late-pregnant gilts, mammary development and piglet growth. *Transl Anim Sci.* 2017;1:154.

30. Xue L, Piao X, Li D et al. The effect of the ratio of standardized ileal digestible lysine to metabolizable energy on growth performance, blood metabolites and hormones of lactating sows. *J Anim Sci Technol.* 2012;3:11.

31. Gourley KM, Nichols GE, Sonderman JA et al. Determining the impact of increasing standardized ileal digestible lysine for primiparous and multiparous sows during lactation. *Transl Anim Sci.* 2017;doi:10.2527/tas2017.0043

32. Strathe AV, Bruun TS, Geertsen N et al. Increased dietary protein levels during lactation improved sow and litter performance. *Anim Feed Sci Technol.* 2017;232:169.

33. Pedersen TF, Bruun TS, Feyera T et al. A two-diet feeding regime for lactating sows reduced nutrient deficiency in early lactation and improved milk yield. *Livest Sci.* 2016;191:165.

34. Cabezón FA, Schinckel AP, Richert BT et al. Analysis of lactation feed intakes for sows including data on environmental temperatures and humidity. *PAS.* 2016;32:333.

35. Rosero DS, van Heugten E, Odle J et al. Response of the modern lactating sow and progeny to source and level of supplemental dietary fat during high ambient temperatures. *J Anim Sci.* 2012;90:2609.

36. Choi Y, Hosseindoust A, Shim Y et al. Evaluation of high nutrient diets on litter performance of heat-stressed lactating sows. *Asian-Australas J Anim Sci.* 2017;11:1598.

37. Kruse S, Traulsen I, and Krieter J. Analysis of water, feed intake and performance of lactating sows. *Livest Sci.* 2011;135:177.

38. Koketsu Y, Tani S, and Iida R. Factors for improving reproductive performance of sows and herd productivity in commercial breeding herds. *PHM.* 2017;3:1.

39. Patterson JL, Smit MN, Novak S et al. Restricted feed intake in lactating primiparous sows I: effects on sow metabolic state and subsequent reproductive performance. *Reprod Fertil Dev.* 2011;23:889.

40. Baidoo SK, Aherne FX, Kirkwood RN et al. Effect of feed intake during lactation and after weaning on sow reproductive performance. *Can J Anim Sci.* 1992;72:911.

41. Graham A, Touchette KJ, Jungst S. et al. Impact of feeding level post-weaning on wean to estrus interval, conception and farrowing rate, and subsequent farrowing performance. *J Anim Sci.* (Abstr). 2015;316:145.

Sow mortality: Impact on performance and root causes

Clayton L. Johnson, DVM
Carthage Veterinary Service, Carthage, Illinois

Introduction

Sow mortality is a frustrating problem for United States pork producers and veterinarians. Dead sows cause lost production and lost revenue while potentially implying an animal welfare challenge. All of these factors contribute to decrease farm staff morale. Frustrations have increased since 2014 as sow mortality benchmarks consistently trend upward. Figure 1 below reflects weekly sow death loss in our managed sow system over the last 3 years. Similar to others in the United States we see a continued increasing trend. Our sow mortality challenge appears to be uniquely American with most international pork production partners failing to report noticeable increases in sow mortality over the same time frame. Anecdotally the increasing sow mortality appears to be independent of disease status, genetics, nutrition, environment and any other individual risk factor. Simply put, whatever your sow mortality level a couple of years ago, it is probably increased today.

Clear and concise problem statements are required for veterinarians to accurately diagnose the root cause of a health problem. Quality problem statements always start with records reviews. Fortunately, we have a tremendous amount of sow mortality information in the form of reason codes. Regardless of the producer's record system, sows removed from the herd are assigned a reason code. This allows for basic statistical analysis to determine which reason codes are predominately associated with the increase in sow mortality. Looking across numerous records platforms it is clear that increasing sow mortality has been driven by a combination of locomotive challenges, prolapses and sudden death/unknown. We must acknowledge the accuracy of these sow removal codes is in many cases questionable.

Impact of sow mortality

In addition to the challenge of record accuracy, producers and veterinarians must overcome the fact that while sow mortality is frustrating it is not considered to be a primary economic driver of farm or production system profitability. Said another way, producers can often gain a greater economic benefit by improvements in pre-wean mortality rate, stillborn rate, conception rate, and numerous other performance metrics versus similar improvements and sow mortality. The combination of low to moderate economic impact with a moderate to high difficulty to improve means sow mortality simply doesn't receive the time, diagnostic dollars or intervention efforts that both producers and veterinarians would ideally like to apply.

Figure 1: Weekly sow death loss

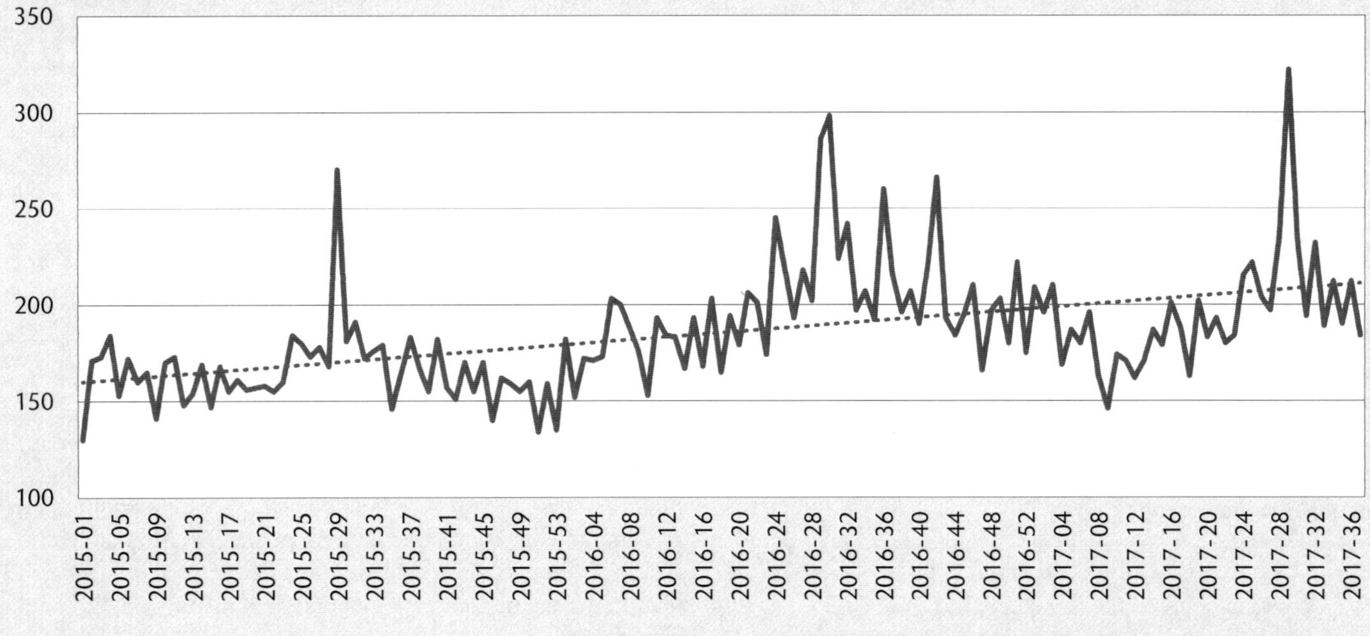

The following factors should be considered when quantifying the impact of sow mortality:

- Sow Cull Value
- Unborn piglet inventory value

Both values have tremendous and predictable seasonal swings. As such using annual averages is appropriate when looking at impact over time. Cull value can be determined using an average weight per parity and calculating your sow mortality as a percentage by parity. This allows you to estimate an average sow weight at the time of death to which you can apply an average annual cull value to derive the final Sow Cull Value for your herd. Unborn piglet inventory is highly dependent on the piglet sale or transfer price as well as the gestation stage at time of death. Where available, producers should use wean pig revenue values that are specific to their herd. This captures the genetic merit of multiplication herds as well as any negotiated premium making this estimate as specific as possible to the herd a question. Similar to using parity profiling to determine Cull Sow Value, we can use duration of gestation at the time of death to determine Unborn Piglet Inventory Value. Unborn piglet inventory can be assigned a value by week post breeding linearly correlated to the duration of gestation. Producers can assess the percentage of sow death loss at each week of gestation to calculate a final value for their herd. The combined Cull Sow Value and Unborn Piglet Inventory Value provides the producer with the real economic loss valuation for sow mortality in their herd or production system.

While the "hard" costs of sow mortality are quantifiable the "soft" costs of sow mortality should also be considered. It takes a tremendous amount of labor to remove a dead sow from the facility in a bio secure manner. This process results in labor and equipment costs as well as the inevitable worker safety costs associated with such manual labor. Further, the frustration associated with giving this effort on a regular basis should not be overlooked. Farm staff morale can be impacted, particularly given this effort often falls on the shoulders of the same animal care takers every day. These animal care takers are trained and skilled in keeping animals alive and dealing with the consequences of a dead sow has an unquantifiable impact on motivation and psyche of our employees. Metrics regarding sound mortality serve another tool to demotivate employees. These metrics are not only in the form of a common production management reports but also now include various Producer audit programs including the Common Swine Industry Audit as well as packer specific audits. Farm managers and employees are aware that sow mortality is an industry problem and frustrated by the fact that while we don't have solutions, the industry continues to apply pressure through the form of measurement and benchmarking.

Causes of sow mortality

Cause of death is most often assessed by farm staff based on an assessment performed at the time of death. This assessment generally includes gross observations of the carcass, treatment history and current stage of production. Cause of death is occasionally determined by necropsy, which provides a better diagnosis. While necropsy provides superior evidence we lack personnel trained in this skill on all farms. Even if necropsy skills were present on farm, committing the time and considerable manual labor required to necropsy all dead sows would generally not be practical or possible. Regardless of how the cause of death is assessed, all dead sows are assigned a reason code.

Many removal codes represent extremely broad categories with numerous subjective interpreters assigning the code. Fortunately, several causes of sow death are so easy to appreciate we can feel comfortable in the accuracy of this data. It is important to consider the accuracy of each reason code independently, particularly when we look at the reason codes currently driving the increase in sow mortality. Locomotive challenges (downers and lames) are extremely subjective. Within our managed farm system we see wide variation in classification between downers and lames. We likely underappreciate the total number of downers/lames across farms given the inevitability that some sows will have reason codes assigned following euthanasia and the employee assigning the reason code may not be aware of the inciting root cause nor able to visually identify any locomotive challenges were present prior to euthanasia. Prolapses represent an example of a mortality cause so visually obvious that correct diagnosis can be expected in nearly every situation. As such, statistical analysis that demonstrates a changing trend can be interpreted as real and meaningful without the need for data verification or validation. (Figure 2)

Lameness

While our records tell us that lameness as a cause of death is increasing we do not completely understand whether the prevalence or severity of lameness in sows is actually increasing. Producers and veterinarians are currently more aggressive at targeting sows for euthanasia than at any time in the history of pig production. This combined with the loss of highly effective tools to treat some forms of infectious lameness has inevitably contributed to a larger number of lame animals becoming mortalities. Like downers, we too often spend the majority of our resources evaluating infectious route causes and interventions for lameness. Adults in most species rarely suffer from infectious lameness. Our producers are frustrated with veterinarians as we continue to recommend injectable therapy in the face of very poor response. This suggests our diagnosis of infectious lameness is more often than not incorrect. Before more dollars are spent on an inappropriate intervention we should spend at least a share of those dollars on diagnostics to better characterize the true cause of lameness.

Gilt selection rates have increased across the United States female populations going back to the introduction of PED. Epidemic disease followed by industry expansion has put pressure on our ability to select gilts of appropriate structure. This pressure is further amplified by the lack of skilled herdsman available to complete gilt selection, most evident at the commercial level where farm staff often select their own gilts. Even the most skilled of gilt

Figure 2: Annual percent sow mortality by reason code

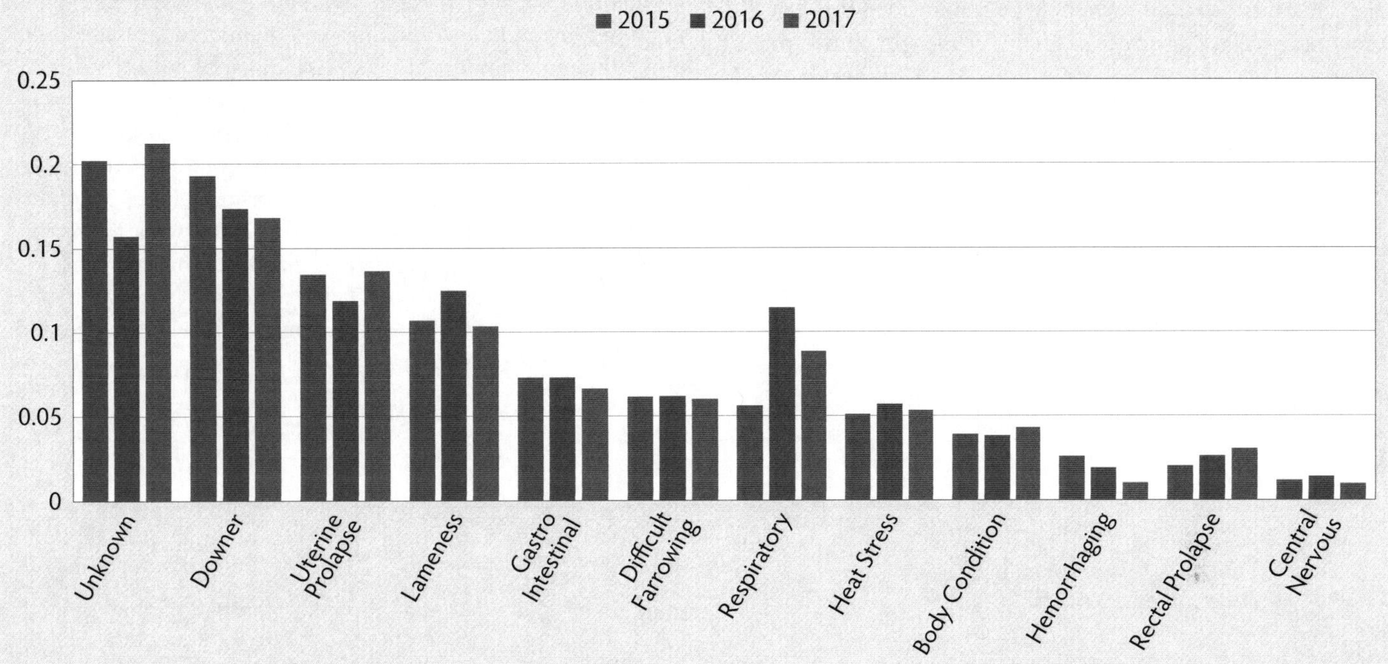

selection officers find it impossible to predict what gilt confirmation will look like when the gilt becomes a sow. Gilts are selected before they reach half of their final bodyweight. The physical changes that accompany skeletal maturity, pregnancy and parturition along with continued growth and weight gain will inevitably modify the likelihood of lameness and therefore limit the value of any conformational assessments made at five months of age.

Downers

This is the most broad categorization of sow death loss which we see increasing. Prior to death these sows are often laterally recumbent and not able to rise. Frustratingly, on postmortem we rarely find the inciting factor and often are unable to even identify secondary lesions associated with whatever route cause may have been present. When a lack of infectious disease is evident on gross necropsy we should challenge ourselves as practitioners to evaluate other disease modalities. Nutrition, Degenerative disease, Auto immune disease, Neoplasia and Traumatic disease should all be evaluated. Unfortunately none of these disease processes lend themselves too easy gross observations. Further diagnostic means of understanding these health challenges are nowhere near as sophisticated or practical as a diagnostic techniques to evaluate infectious disease processes. Industry support is needed to develop improved practical and cost effective diagnostic techniques to evaluate the root cause of Downer sows.

Prolapses

Prolapses represent the most frustrating of all sow mortality reasons to producers and veterinarians. There is no effective treatment other than euthanasia and we have no ability to predict onset prior to occurrence. Numerous production systems and veterinary clinics have attempted to identify specific risk factors associated with increased prevalence with minimal success. Specific risk factors evaluated include:

- Mycotoxins
- Hypocalcemia
- Vitamin deficiency
- Litter size
- Genotype
- Constipation
- Phytase/phosphorus
- Hormonal changes
- Oxidative stress
- Diet type (mash vs. pellet)
- Respiratory disease
- Enteric disease
- Feeding patterns
- Feeder type
- Constipation
- Housing design
- Ambient temperature
- Amino acid levels

Discussion

While our efforts to date have developed more questions than answers many in the industry should be commended for their hard work in collecting evidence on this challenging problem. The National Pork Board has commissioned a Working Group to coordinate investigative efforts with a priority given to Sow Pelvic Organ Prolapse in addition to a review of Lameness and Sudden Death. Current Working Group efforts are focused on gathering epidemiological information to further characterize our collective Problem Statement. This better positions the industry to propose researchable hypotheses which will drive research in 2018 and beyond.

As veterinarians I feel it is our charge to recognize the low to moderate economic impact of sow mortality while at the same time championing change in the spirit of continuous improvement for animal well-being and welfare. We took an oath to use our scientific knowledge and skills for the benefit of animal health and never is health more compromised than when a mortality event occurs. If we don't champion this effort no one will. Don't agonize over the difficult task ahead, influence the resources around you to further improve the quality of our Problem Statement while taking creative approaches toward intervention and diagnostic efforts. We must challenge ourselves to get past flawed diagnoses and intervention efforts. Stay current on the work being done with locomotive issues and prolapses and be a pioneer in attempting novel diagnostic and intervention strategies.

Next generation sequencing and metagenomics: Interpretation for the practitioner

D. Marthaler[1], PhD; T. Knutson[2], PhD
[1]Veterinary Diagnostic Laboratory, Kansas State University, Manhattan, Kansas;
[2]Minnesota Supercomputing Institute, University of Minnesota, Minneapolis, Minnesota

Introduction

Over the past years, sequencing technology has greatly advanced, and the volume of sequence data has greatly increased while the cost associated with sequencing per sample has decreased. The advancement in technology and decrease in the cost leads to new opportunities in pathogen discovery and using the whole genome of viruses to understand their epidemiology and phylogenetic origins. The advancement in Next Generation Sequencing (NGS) technology has opened doors to investigate complex diseases and better understand the virome and microbiome associated with disease complexes. Swine health is multifactorial and includes the interaction between the host, nutrition, environment, and pathogens; therefore, NGS can reveal viral diversity and evolution allowing practitioners to design better health interventions and minimize the impact of infectious diseases in swine health and production.

Sequencing technology

Currently, there are three types of sequencing technology: traditional sequencing (Sanger technology), second generation NGS (short sequences or reads), and third generation NGS (long reads). Most practitioners get their porcine respiratory and reproductive syndrome virus, influenza A virus, and rotavirus sequence results using Sanger technology. Sanger technology involves amplification of the target (bacteria or virus) using PCR. Then, the PCR product is purified, and amplified several times adding colored nucleotides. The PCR product is run through a capillary gel electrophoresis, and the laser reads the colored bases, yielding sequence data (trace file). The length of trace file is between 200-1,000 nucleotides long, and multiple trace files are combined to verify the accuracy and generate a final consensus sequence of the pathogen of interest. Since multiple strains of the same virus can be found on a farm, traditional sequencing is biased due to the amplification process, where the predominant virus strain appears as the consensus sequence and minor strains can go undetected. In contrast, NGS generates millions of reads for all viral strains in the sample allowing for easier detection of both major and minor strains.

The NGS process is completely different than Sanger sequencing. With Sanger sequencing, PCR amplification identifies the specific pathogen of interest while NGS sequences everything from the sample allowing us to better estimate the diversity and distribution of the viral population and/or obtain the complete genome in a shorter period of time. Physical and chemical treatments are used prior to extraction of the nucleic acid to remove the host DNA/RNA, but a majority of the sequence data can still be uninformative host information. Most of the Veterinary Diagnostic Laboratories (VDLs) conduct NGS using the Illumina MiSeq (`www.illumina.com`). With Illumina sequencing, the nucleic acid is broken into millions of fragments of 500-800 nucleotides. Then, adapters and unique bar codes are added to each fragment to differentiate between samples, which allows for multiple samples to be loaded into one reaction of the machine. In the NGS machine, the fragments are briefly amplified and sequenced, creating millions of sequences (reads) with lengths of 50, 150, 250, or 300 nucleotides. Illumina sequencing is highly accurate, but interpretation and creation of the viral genome can be difficult given the amount of uninformative data and length of the read.

The third generation NGS is still being developed, but it is making significant advancements in a short period of time. The most exciting is the hand-held sequencer from Oxford Nanopore (nanoporetech.com). The Nanopore sequencer utilizes a pore that reads the DNA sequence regardless of the length, which yields very long reads of sequence (> 50 kb average). In addition, the Nanopore generates data in real time making very accurate viral strain classifications possible within an hour. While this technology generates long reads making whole genome assembly easier, the error rate is much higher than Illumina. Plus, the uninformative data is still sequenced, requiring bioinformatic tools to find the pathogen of interest.

Metagenomic sequencing

Since everything in the sample is sequenced with NGS, metagenomics becomes a powerful tool to understand health parameters in a farm, system, and swine company. Metagenomics relies on bioinformatic tools to identify the organism associated with each read identified in a sample. The Kraken and Krona programs are used by a few of the VDLs for sequence identification. Kraken utilizes a database of publicly known sequences from GenBank, and Krona provides an interactive taxonomical pie chart of the classified reads in a sample (Figure 1).[1] In this fecal sample, 48% bacterial reads (red), 33% pig reads (orange), and 2% viral reads (blue) were identified while 13% of the reads were not identified (green). These unknown reads contain host or bacterial data not in the Kraken database since the host and bacterial reference strains are used to create the database, or the unknown reads contain a new pathogen. By expanding the portion containing the

Figure 1: Krona chart from a fecal sample

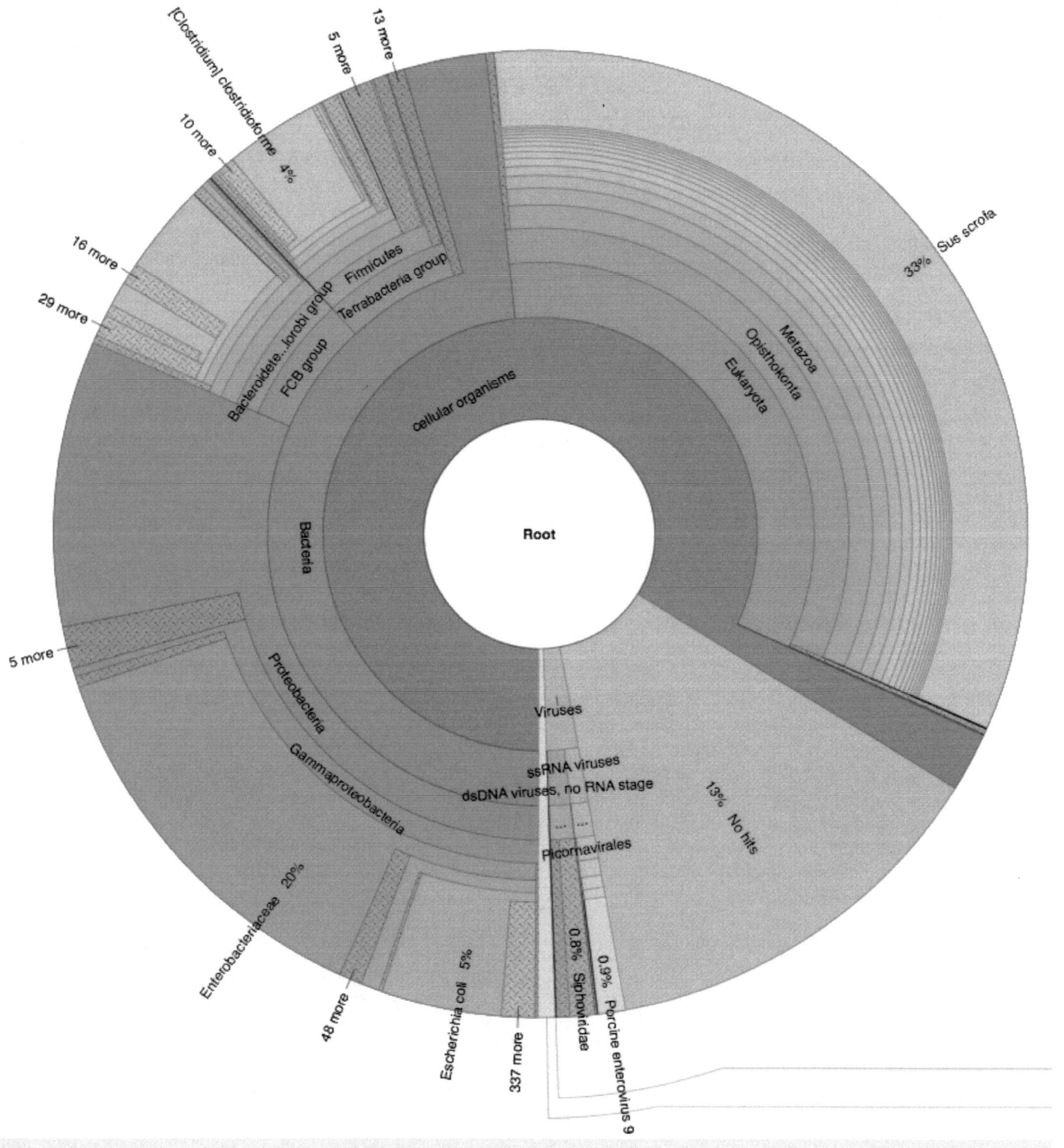

viral reads, 40% of these reads were identified as Enterovirus G (red) while a majority of the remaining viral reads were bacterial phages (yellow) (Figure 2). The Enterovirus G reads generated a whole genome, identifying a new viral strain containing an insertion of a papain-like cysteine protease.[2]

Another benefit of metagenomics is to investigate viral populations within a herd via whole genome sequencing. In animal populations, viral evolution varies due to introductions of new viral strains and varying immune statuses within the population. Hence, different viral subpopulations may emerge or subside over time within a herd, and these viral populations can be investigated using NGS. This is especially useful for characterizing minor

Figure 2: Viral reads from a fecal sample

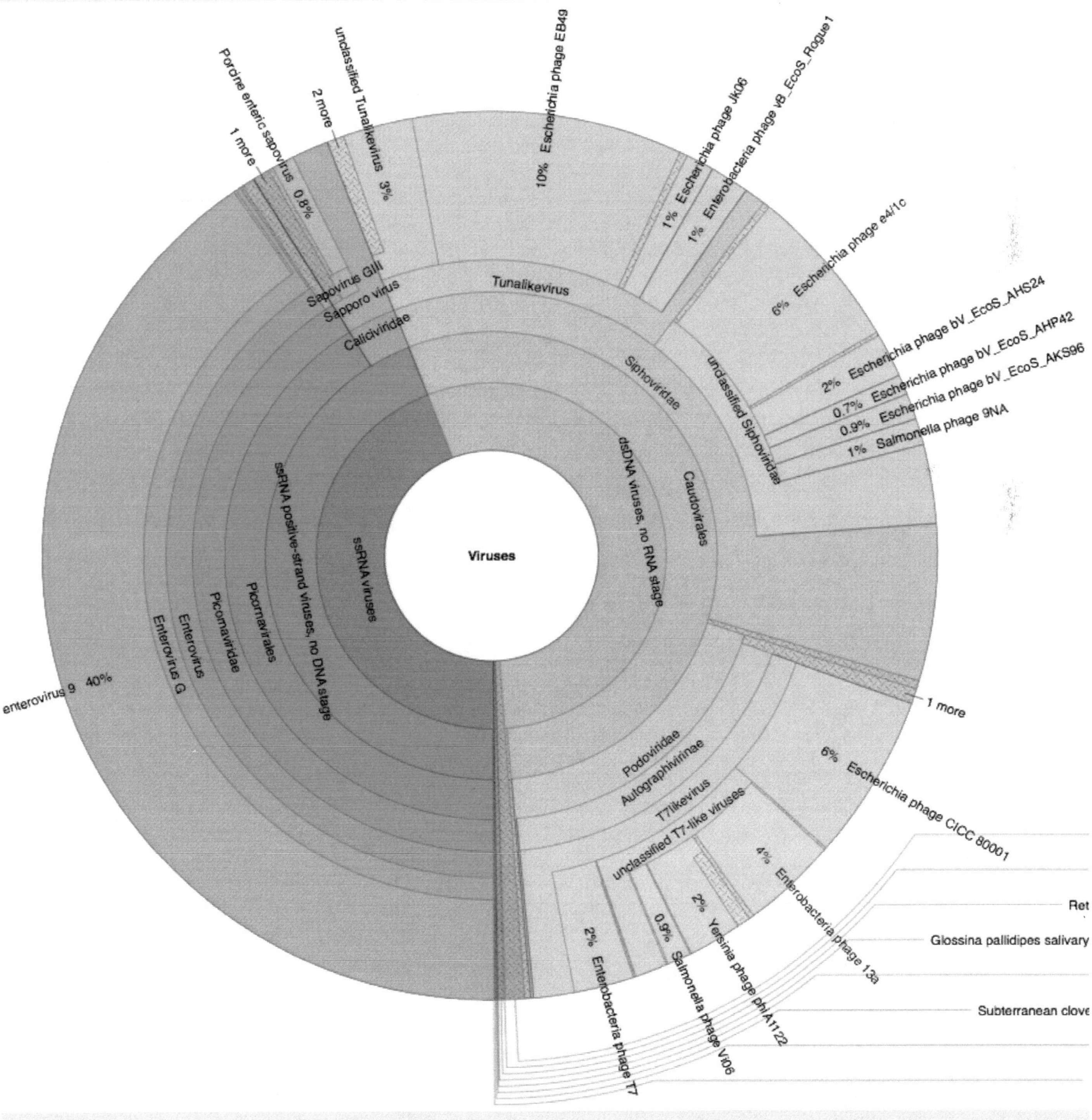

viral strains in a population that have not yet become pathogenic or predominant. A serum sample was sequenced to investigate a clinical case of PRRS from a herd experiencing an increase in clinical signs while ORF5 sequencing identified the PRRS strain that had been historically circulating in the herd. Two different sequences of the ORF2s-ORF4 were generated while no other changes occurred throughout the remaining genome, including ORF5 (Figure 3). The ORF2a-ORF4 proteins elicit neutralizing antibodies, and a viral subpopulation may be emerging within the herd, which was not indicated by Sanger sequencing of ORF5.

Interpretation of the NGS

Since metagenomics is in the early stages for clinical diagnostics, interpretation of metagenomics and the viral population is a team effort between all the individuals associated with the case (swine veterinarian, pathologist, diagnostician, etc.). The infancy of NGS's interpretation is very similar to the early days of PCR by becoming familiar with Ct values and their relationship to clinical disease. Interpretation of metagenomics on serum or tissue is slightly easier. For serum, a limited number of viruses cause viremia, and their identification in serum suggests viral replication. With tissues, the pathologist can confirm the presence of viral replication. Fecal or intestinal samples can be more difficult due to the number of viruses that can be normally present in the samples, making the linking of any particular virus to a clinical disease difficult. While coronaviruses and rotaviruses cause clinical disease, the clinical relevance of astrovirus, enterovirus G, sapelovirus, teschovirus, sapovirus, norovirus, and kobuvirus to swine health are still in question. Individual or co-infections of these viruses are found in many enteric samples. As clinical disease is a multifactorial process, all of these viruses could be reducing animal health. Thus, it's extremely important to correlate the findings with clinical disease seen by the swine veterinarians.

Conclusion

As the NGS technology continues to advance, the volume of sequence data is significantly increasing in a shorter period of time. While the bioinformatic tools are advancing, we can do so much more than simply generate a consensus sequence, but as more in-depth analysis is required, the longer it takes to generate results. However, this analysis can lead to new insights in disease epidemiology within a herd and better inform decisions to minimize the impact of infectious diseases in swine health and production. The future possibilities of understanding disease with NGS are endless.

References

1. Wood DE, et al. Kraken: ultrafast metagenomic sequence classification using exact alignments. Genome Biol. 2014 Mar 3;15(3):R46.

2: Knutson TP, et al. A porcine enterovirus G associated with enteric disease contains a novel papain-like cysteine protease. J Gen Virol. 2017 Jun;98(6):1305-1310.

Figure 3: Two different regions of the ORF2a-ORF4 from a single PRRS sample. The red boxes highlight the changes between the two sequences

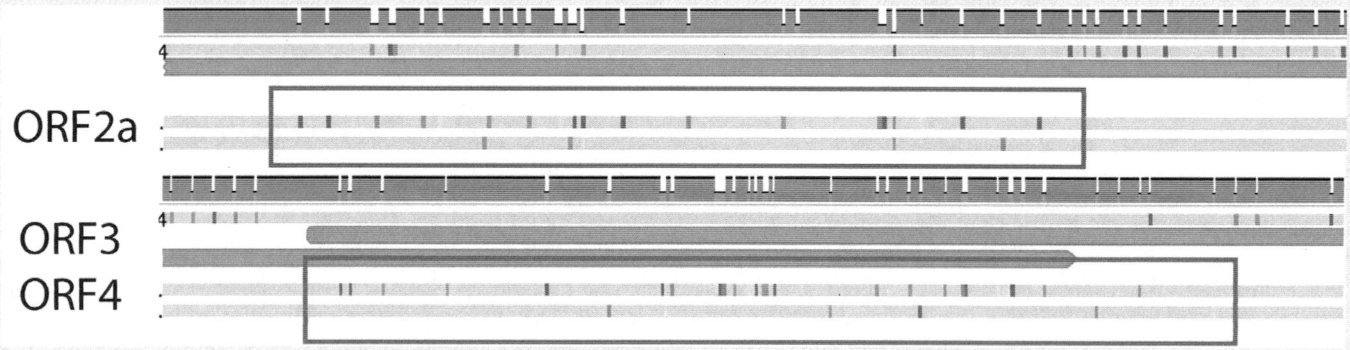

Porcine circovirus type 3: Practitioner perspective

Emily B. Byers, DVM
Prestage Farms, Clinton, North Carolina

Introduction

In 2015, a North Carolina herd experiencing poor reproductive performance and a high sow mortality was diagnosed with porcine circovirus type 3 (PCV3). This is the first known diagnosis of PCV3 in association with reproductive failure, mortality, and porcine dermatitis and nephropathy syndrome (PDNS). Palinski, et al. described the molecular characterization of the PCV3 identified in this clinical case in addition to retrospective analyses identifying PCV3 in historically inconclusive PDNS and post-weaning multi-systemic wasting syndrome (PWMS) cases.[10-11]

Clinical case experience

History

The herd consisted of two commercial breed-to-wean "sister" sow farms (approximately 7,500 total sows), a gilt acclimation site, and a separate large wean-to-finish (W-F) site containing the downstream production. Historically, the sow farms had been considered PRRSV positive-stable-IIa[5] and relatively high-health for the system. Gilts entering the sow farms were routinely vaccinated prior to breeding with commercially available products for porcine reproductive and respiratory syndrome virus (PRRSV), porcine circovirus type 2 (PCV2), *Mycoplasma hyopneumoniae* (MHP), influenza A virus of swine (IAV-S), porcine parvovirus (PPV), leptospirosis, and erysipelas.

In February 2014 the sow farms became infected with porcine epidemic diarrhea virus (PEDV). The sow farms recovered as expected within 14 weeks following a typical intervention strategy, including enteric tissue feedback, piglet depopulation, and enhanced sanitation practices. Immediately following PEDV recovery, productivity was above average for the system, meeting or exceeding target key performance indicators. However, in late Summer 2014 the herd began to perform below expectations by failing to meet targets such as conception rate, farrowing rate, sow mortality, total pigs born, and pigs born alive. It was difficult to identify consistent performance trends, and numerous veterinary visits and diagnostic investigations at the sow farm failed to identify a cause.

Concurrently, mortality began to increase at the downstream W-F site, which was located approximately 30 miles from the sow farms. Initially, diagnostic investigations yielded diverse viral pathogens, including PEDV and multiple strains of PRRSV and IAV-S, which were similar to the endemic W-F site viruses and viruses known to circulate at neighboring farms. As mortality continued to increase over time, monthly mortality surveys were conducted by submitting routine core organ samples from pigs found dead of unknown causes. These diagnostic surveys consistently included features typical of porcine circovirus associated disease (PCVAD)[14] and histopathologic lesions suggestive of PDNS;[14] but PCR and immunohistochemistry were all negative. Myocardial lesions similar to those described by Phan et al.[12] were occasionally reported.

A definitive diagnosis was not achieved through diagnostic investigations at the W-F site, and further investigation ceased in early 2015 when the sow farms were infected with PRRSV following the emergence of a new strain in North Carolina (RFLP 1-7-4). Following interventions with a commercial attenuated PRRSV vaccine, the sow farms were monitored bi-weekly via qPCR of 30 pooled piglet sera and began producing field-virus-negative pigs within approximately 12 weeks. Despite apparent clinical PRRSV recovery and return to the herd's pre-PRRSV baselines, the herd continued to perform below expectations.

Persistent reproductive failure and sow mortality

Consistent patterns in individual key performance indicators were difficult to establish at the sow herd level, but reproductive failure was apparent regardless of gestation stage. Reproductive failure included early embryonic deaths, delayed returns to estrus, abortions, mummified fetuses, and stillbirths. Through further investigation with caretakers there appeared to be a correlation between increases in acute sow deaths, abortions, and increased mummified fetus rates. Caretakers noted that sows dying of acute, unknown causes were found to have skin lesions best described as round to irregular, occasionally coalescing, purple macules and papules distributed from the chest and inner forelimbs (Figure 1), along the abdomen (Figure 2), and extending through the hindquarters to the perineal area. These same lesions were also reported in some sows just prior to abortion. Additionally, caretakers observed that while mummified fetus rates remained high on a per sow basis, it was inconsistent from week to week. Dramatic increases in mummified fetuses were typically related to individual sows giving birth to large litters of predominantly mummified fetuses of varied crown-to-rump lengths (Figure 3), suggesting that fetal death occurred at different stages throughout gestation. The syndromes described were highly suggestive of PCV2;[8, 13-14] however, routine diagnostics performed on sow and piglet serum, fetal tissues (abortions, mummified fetuses, and stillbirths), and placental umbilical cord serum tested negative for PCV2. All samples were also negative for other common pathogens, including, PRRSV and PPV.

Necropsy was performed from sows dying with PDNS-like skin lesions, and tissue samples were submitted to a swine diagnostic laboratory. Histopathologic lesions were consistent for each sow

Figure 1: Inner forelimb and abdominal skin lesions of PCV3-positive Sow #1

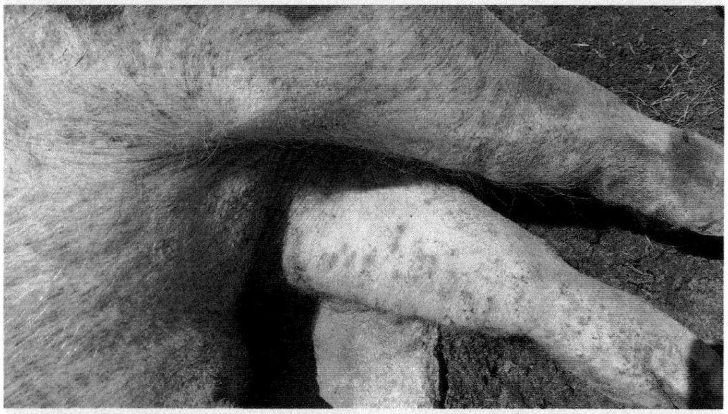

Figure 2: Abdominal skin lesions of PCV3-positive Sow #2

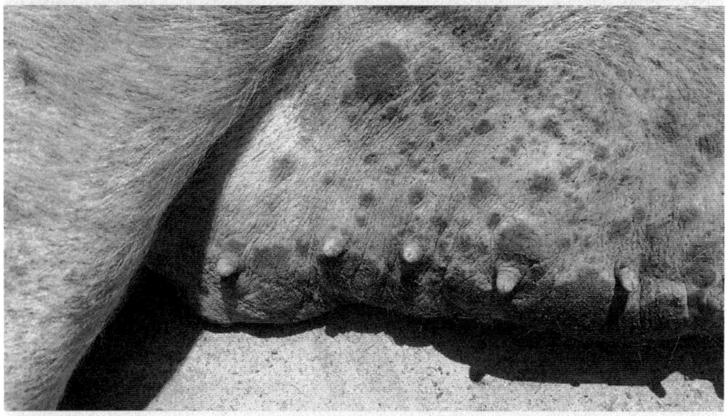

Figure 3: Representative litter of mummified fetuses of varied crown-to-rump lengths

submitted, and typical lesions were suggestive of PCV2 infection and PDNS.[14] Histologic skin lesions included acute necrotic dermatitis and epidermitis associated with lymphoplasmacytic perivascular accumulation. Nephritis was characterized by dilation of cortical tubules with proteinic fluid, attenuation and regeneration of the tubular lining epithelium, and diffuse infiltration of the cortical interstitium and glomeruli with large clusters of lymphocytes and macrophages. Lymphoid depletion was also a feature. Despite a strong suspicion of PCVAD, all sow tissues (kidney, lymph node, lung, skin) and additional mummified fetal tissues remained negative for PCV2 by qPCR and IHC. Samples were also negative for PRRSV, PPV, and IAV-S. Definitive diagnosis remained inconclusive.

In the month following the inconclusive diagnoses, the herd reported a 10.2% increase in sow mortality and 0.6% decrease in conception rate. Subjectively, piglet quality and viability at birth was not affected, however, liveborn pigs remained lower than target with a rate of 1.19 mummified fetuses per litter. Due to the continuity of the syndrome, further diagnostics were pursued.

Identification of a novel porcine circovirus[10-11]

Metagenomic sequencing and genetic analysis performed on both pooled fetal tissue homogenate and pooled sow tissue homogenate identified a, "highly divergent new species of porcine circovirus," designated PCV3. In the samples analyzed PCV3 was the only pathogen identified by metagenomic sequencing. Following establishment of a PCV3 qPCR, fetal tissues consistently tested positive with high levels of detectable virus evidenced by Ct values ranging from 16.7 to 21.3. Analysis of skin, lungs, kidneys, and lymph nodes from sows with PDNS-like lesions detected PCV3 by PCR and IHC. Repeated analysis by PCR, IHC, and metagenomic sequencing failed to identify PCV2 in these samples. It was hypothesized that these results implicated PCV3 as a pathogen responsible for PDNS-like lesions in sows, and vertical transmission of the virus was responsible for the reproductive failures observed within the herd. Subsequently, following observation of similar skin lesions in mature gilts at the gilt acclimation site, PCV3 was detected by PCR in serum and oral fluids.

While the PCV3 diagnosis provided a likely etiology to the syndrome, it created more questions than answers. What is the epidemiology? What is the pathogenesis? What is the significance of the virus to an integrated system? What is the significance of the virus to the industry? How is the virus eliminated from a herd? Is it possible to detect PCV3 in the absence of PDNS-like lesions and/or reproductive failure? Is PCV3 likely to emerge as PCV2 did, with sporadic reports of disease during late 1980s, becoming considered a significant pathogen by the late 1990s, and finally being associated with severe disease epidemics by 2003?

Intervention and outcome

Following the PCV3 diagnosis in late summer, the producer was concerned with eliminating the virus from the herd. This posed several challenges as little was understood about the virus, and no vaccines for PCV3 existed. Doing nothing and letting the virus run its course was not effective, and the underlying clinical signs, including sporadic incidence of PDNS-like disease, continued into the fall of 2015. Herd homogenization via live virus inoculation or tissue feedback was not a favorable option for the system. While it is theorized that little or no cross-protection exists between PCV2 and PCV3, it was unknown if a whole herd vaccination with commercial PCV2 vaccines would be effective.

Having no well-understood options, the producer elected to vaccinate both sow farms with a commercial PCV2 vaccine, as had been previously done in the system to control vertical transmission and reproductive disease associated with PCV2. Both sow farms reacted strongly to this vaccination in October 2015. Sows experienced pyrexia and anorexia in addition to an abortion storm that affected approximately 5% throughout gestation, with some breed groups as high as 30%. The cause of the severe reaction was not determined. Possible causes for the abortion storm include reaction to adjuvants, blood-borne dissemination of PCV3 through the herd via needles, or other unidentified conditions. Repeated diagnostics found no evidence of PRRSV. Following mass vaccination, no additional cases of sow mortality due to PDNS were observed, and reproductive performance improved at the sow farm level. Downstream W-F livability of harvested market groups significantly improved within 9 months of the sow farm vaccinations.

Conclusion

When asked by colleagues if PCV3 is truly a disease, this practitioner's conclusion is currently, "I'm not sure. I think it depends." Multiple reports of PCV3 identification and diagnostics have since been published in the USA,[10-12] Brazil[17], China,[1-2,4,6,15,18-20] Italy,[3] Korea,[7] and Poland.[16] Some implicate PCV3 in clinical disease, but reports of PCV3 identification in the absence of clinical disease also exist. If PCV3 is identified in swine with clinical PCVAD / PDNS, and histopathology consistent with circovirus infection exists in the absence of PCV2 identification, then PCV3 is a likely pathogen. Understanding reproductive failure as it relates to PCV2 is challenging,[8] and understanding the reproductive role of PCV3 seems even more challenging.

Through anecdotal experience, in the absence of PDNS-like lesions, mummified and stillborn fetal tissues have been a reliable sample for PCV3 detection; however, placental umbilical cord serum, piglet serum, and sow serum seem less consistent. PCV3 could likely be detected in processing fluids, but more research is needed to understand significance. It is important to understand that, like PCV2, a diagnosis should not be assigned based on qPCR results alone. One must consider a positive result in context with clinical signs and rule out other known primary pathogens, such as PRRSV, PCV2, and PPV.

In conclusion, PCV3 has the potential to emerge as a significant pathogen similar to PCV2, which is disconcerting. However, it is unknown under what circumstances, if any, this emergence will occur. Further research is needed to close the knowledge gap, particularly since the lack of genetic relatedness to PCV1 and PCV2 implies that there is likely no cross-protection among these viruses.

References

1. Chen GH, Mai K, Zhou L, et al. 2017. Detection and genome sequencing of porcine circovirus 3 in neonatal pigs with congenital tremors in South China. Transbound Emerg Dis. https://doi.org/10.1111/tbed.12702.

2. Deng J, Li X, Zheng D, et al. 2017. Establishment and application of an indirect ELISA for porcine circovirus 3. Arch Virol. https://doi.org/10.1007/s00705-017-3607-7.

3. Faccini S, Barbieri I, Gilioli A, et al. 2017. Detection and genetic characterization of porcine circovirus type 3 in Italy. Transbound Emerg Dis. 2017;64(6):1661-1664.

4. Fan S, Ku X, Chen F, et al. 2017. Complete genome sequence of a novel porcine circovirus type 3 strain, PCV3/CN/Hubei-618/2016, isolated from China. Genome Announc 5:e00100-17. https://doi.org/10.1128/genomeA.00100-17.

5. Holtkamp DJ, Polson DD, Torremorell M, et al. Terminology for classifying swine herds by porcine reproductive and respiratory syndrome virus status. *J Swine Health Prod.* 2011;19(1):44–56.

6. Ku X, Chen F, Li P, et al. 2017. Identification and genetic characterization of porcine circovirus type 3 in China. Transbound Emerg Dis. https://doi.org/10.1111/tbed.12638.

7. Kwon T, Yoo SJ, Park CK, Lyoo YS. 2017. Prevalence of novel porcine circovirus 3 in Korean pig populations. Vet Microbiology. https://doi.org/10.1016/j.vetmic.2017.06.013 .

8. Madson D, Opriessnig T. 2011. PCV2-associated reproductive failure. Proc. 42nd AASV Annual Meeting. 1:427-432.

9. Madson D, Patterson A, Ramamoorthy S, et al. 2009. Effect of porcine circovirus type 2 (PCV2) vaccination of the dam on PCV2 replication in utero. Clin Vaccine Immunol. 2009 Jun;16(6):830-4. https://doi.org/10.1128/CVI.00455-08. Epub 2009 Apr 8.

10. Palinski R. The application of metagenomic sequencing to detect and characterize emerging porcine viruses. Dissertation. Manhattan, Kansas: Kansas State University; 2016. 182 p. http://hdl.handle.net/2097/34468.

11. Palinski R, Piñeyro P, Shang P, et al. 2017. A novel porcine circovirus distantly related to known circoviruses is associated with porcine dermatitis and nephropathy syndrome and reproductive failure. J Virol 91:e01879-16. https://doi.org/10.1128/JVI.01879-16.

12. Phan TG, Giannitti F, Rossow S, et al. 2016. Detection of a novel circovirus PCV3 in pigs with cardiac and muti-systemic inflammation. Virology Journal (2016) 13:184. https://doi.org/10.1186/s12985-016-0642-z.

13. Pittman JS. Reproductive failure associated with porcine circovirus type 2 in gilts. *J Swine Health Prod.* 2008;16(3):144–148.

14. Segales J, Allan GM, Domingo M. Porcine circoviruses. In: Zimmerman JJ, Karriker LA, Ramirez A, Schwartz K, Stevenson GW, eds. Diseases of Swine. 10th ed. Ames, Iowa: Wiley-Blackwell; 2012.

15. Shen H, Liu X, Zhang P, et al. 2017. Genome characterization of a porcine circovirus type 3 in South China. Transbound Emerg Dis. https://doi.org/10.1111/tbed.12639.

16. Stadejek T, Wozniak A, Milek D, Biernacka K. 2017. First detection of porcine circovirus type 3 on commercial pig farms in Poland. Transbound Emerg Dis. https://doi.org/10.1111/tbed.12672.

17. Tochetto C, Lima DA, Varela APM, et al. 2017. Full-genome sequence of porcine circovirus type 3 recovered from serum of sows with stillbirths in Brazil. Transbound Emerg Dis. https://doi.org/10.1111/tbed.12735.

18. Wand J, Zhang Y, Wang J, et al. 2017. Development of a TaqMan-based real-time PCR assay for the specific detection of porcine circovirus 3. Virological Methods. 248: 177-180. https://doi.org/10.1016/j.jviromet.2017.07.007.

19. Zhang J, Liu Z, Zou Y, et al. 2017. First molecular detection of porcine circovirus type 3 in dogs in China. Virus Genes. https://doi.org/10.1007/s11262-017-1509-0.

20. Zheng S, Wu X, Zhang L, et al. 2017. The occurrence of porcine circovirus 3 without clinical infection signs in Shandong Province. Transbound Emerg Dis. 2017;64:1337–1341. https://doi.org/10.1111/tbed.12667.

Porcine circovirus type 3: A pathologist's perspective

S. E. Detmer[1], DVM, PhD; F. Vannucci[2], DVM, MSc, PhD
[1]Western College of Veterinary Medicine, Saskatoon, Saskatchewan;
[2]University of Minnesota Veterinary Diagnostic Laboratory, St. Paul, Minnesota

Introduction

Two diagnostic cases were submitted to Prairie Diagnostic Services by the same veterinarian from two different sites within the same farm system. Necropsies were performed by the submitting veterinarian and portions were received. Routine histopathology was performed for both cases.

Case 1

Tissues were submitted from 3 grower pigs, including a "myco sentinel" that was severely stunted and the veterinarian requested work up for *Mycoplasma hyopneumoniae*. The other two pigs had fibrinous pericarditis and vegetative valvular endocarditis.

The "myco sentinel" had predominantly lymphoplasmacytic inflammation noted in the heart. There were minimal changes observed in the lungs. Bronchiolar and alveolar walls were thickened 2-3 times normal size by predominantly lymphocytes and plasma cells admixed with small numbers of macrophages and neutrophils. In the heart, arteries were frequently surrounded by lymphoplasmacytic inflammation which sometimes infiltrated into the vascular wall. In the other two pigs, foamy macrophages were frequently seen in the alveoli of the lungs and lymphoplasmacytic inflammation was found in the liver and kidneys. The hearts of both pigs had fulminant bacterial infections of the mitral valve accompanied by necropurulent myocarditis.

Case 2

Tissues were submitted from a grower pig with skin lesions. On gross examination, there were scattered grey-white pinpoint foci in the kidneys, the lymph nodes were slightly enlarged and congested, and the skin had variably sized, raised and ulcerated lesions that were frequently are covered with dark crusts. In general, the microscopic lesions were predominantly in and around blood vessels and more lymphoplasmacytic than histiocytic.

More specifically, the skin had acanthosis and hyperkeratosis with serocellular crusts. Most significantly, the blood vessels and adnexa were surrounded by a mixture of lymphocytes, histiocytes, plasma cells, eosinophils and occasionally neutrophils. Limited vasculitis was also found. The highly cellular tissue of the lymph node lacked secondary follicle formation and was densely composed of mononuclear cells, primarily lymphocytes. In the spleen, the densely cellular follicular centers have frequent mitoses and variably sized lymphocytes. The surrounding red pulp contains increased numbers of mononuclear cells along with cellular and nuclear debris. Multifocally, there were lymphoplasmacytic infiltrates in the interstitium of the kidney and occasional glomeruli. There were a few foci containing mild fibrosis. Multifocally, a few tubules were degenerated and contained cellular casts. Within the heart, there was mild interstitial edema with a few lymphohistiocytic infiltrate and scattered degenerated cardiac muscle cells.

Additional diagnostic tests were requested for both cases that included immunohistochemistry (IHC) for porcine circovirus type 2 (PCV2), and in-situ hybridization (ISH) for PCV2 and porcine circovirus type 3 (PCV3).

Results

Both PCV2 and PCV3 were found in Case 1, and only PCV3 was found in case 2 (Table 1).

Discussion

Duplex ISH for PCV2 and PCV3 is a research assay. It is not fully validated and has given negative results for PCV2 on the same tissues where PCV2 was detected by IHC (Detmer, personal experience). The opposite has also occurred where ISH was positive and IHC was negative. Therefore, negative results for either assay should be interpreted with caution, particularly when PCR is positive. However, the consistency of the locations where positive staining for both PCV2 and PCV3 were found between cases does provide corroboration of the results reported here.

PCV3 detection in China, USA, Korea, Brazil, Italy and Poland has been reported from real time PCR assays and sequences have been published,[1-7] but there are no published reports on epidemiology or pathogenesis of infection or co-infection with PCV2. In one case report, non-suppurative myocarditis and arteriolitis were the major microscopic findings for a case of PCV3, similar to the "myco sentinel" presented here.[1] Sows with PDNS-like lesions from North Carolina had similar lesions to the pig in Case 2, but in the adult animals, abundant PCV3 antigen was found in the skin and kidney lesions.[2]

PCV2, on the other hand, has been extensively research for the role it plays in PCVAD, PDNS and PMWS. PCV2 is also associated with abortions, stillbirths and fetal mummification. The virus is widely dispersed and moderately controlled by current vaccine strategies.

While viremia is of variable duration, replication and persistence of PCV2 is primarily associated with macrophages and monocytes of lymphoid organs and the lung. The most common microscopic lesion found in vaccinated animals in lymph nodes is lymphocytic depletion often accompanied by histiocytic infiltration. The characteristic "Botryoid" inclusion bodies are rarely seen in vaccinated animals.

Table 1: Results of immunohistochemistry (IHC) and in-situ hybridization (ISH) for PCV2 and PCV3

Pig identification	Organ	IHC PCV2	ISH PCV2	ISH PCV3
Case 1, "myco"	Heart	Suspect	**Positive**	**Positive**
	Lung	Suspect	**Positive**	Negative
Case 1, pig 2	Heart	Negative	**Positive**	Negative
	Liver, lung, kidney, spleen	Negative	Negative	Negative
Case 1, pig 3	Kidney, spleen	Negative	**Positive**	Negative
	Heart, liver, lung	Negative	Negative	Negative
Case 2	Spleen	Negative	Negative	**Positive**
	Heart, liver, lymph node, skin, kidney, lung	Negative	Negative	Negative

Comments

The major diagnostic finding associated with PCV3 in these cases were lymphoplasmacytic inflammation centered predominantly within and surrounding blood vessels with occasional vasculitis. The overwhelming histocytic response seen with PCV2 has not been seen with PCV3, unless there is a co-infection. The farms involved in this case study were using a commercial PCV2 vaccine, and therefore, the PCV2 lesions were not a significant finding.

In cases where a veterinarian suspects PCV2-related disease but the virus is either not detected or not at level that is expected to cause the amount of disease seen on the farm, then additional testing for PCV3 and other pathogens is recommended.

References

1. Phan, et al. Detection of a novel circovirus PCV3 in pigs with cardiac and multi-systemic inflammation. Virol J. 2016;13:184.
2. Palinski, et al. A novel porcine circovirus distantly related to known circoviruses is associated with porcine dermatitis and nephropathy syndrome and reproductive failure. J Virol 91:e01879-16.
3. Kwon, et al. Prevalence of novel porcine circovirus 3 in Korean pig populations. Vet Microbiol. 2017;207:178-180.
4. Tochetto C, et al. Full-Genome Sequence of Porcine Circovirus type 3 recovered from serum of sows with stillbirths in Brazil. Transbound Emerg Dis. 2017 [Epub ahead of print].
5. Faccini S, et al. Detection and genetic characterization of Porcine circovirus type 3 in Italy. Transbound Emerg Dis. 2017;64(6):1661-1664.
6. Stadejek T, et al. First detection of porcine circovirus type 3 on commercial pig farms in Poland. Transbound Emerg Dis. 2017;64(5):1350-1353.
7. Zheng S, et al. The occurrence of porcine circovirus 3 without clinical infection signs in Shandong Province.Transbound Emerg Dis. 2017;64(5):1337-1341.

Porcine parainfluenza virus – Practitioner perspective

Aaron Lower
Carthage Veterinary Service, Carthage, IL

Clinical history

In 2012 and 2013, a 5,000 head breed to wean herd was experiencing respiratory disease in late suckling and post wean pigs. 20% of the pigs would sneeze and cough around 14 days of age through weaning. Post wean, cough would be evident for 0-2 days and quickly subside. In the first 7 days post wean, pigs were lethargic and anorexic. Fallout pulls were 10-15% however mortality increase would only be 0 – 0.5%. Clinical assessment of individual and mass antimicrobial treatment was generally unrewarding. The clinical signs would wax and wane every few months. This particular herd sold the wean pigs to multiple customers which prompted a more aggressive diagnostic project to determine the health driver.

Diagnostic case description

A series of tissue cases were submitted to diagnostic laboratories. The farm was negative to PRRS and IAV on all samplings. Bacterial culture would be variably positive but not the same bacteria on each case (*S. suis, H. parasuis, M. hyorhinis, P. multocida).* Histopathology was very consistent between the cases suggesting viral driver on all cases:

Lung: Pigs A-D have similar microscopic lesions and characterized by bronchiolitis. Large and medium sized bronchioles are lined by dysplastic, attenuated, or rounded epithelium that is further accentuated by intra-epithelial microabscesses. Cilli is often lost, while lymphocytes variably expand the lamina or adjacent adventitia. In addition, few pigs have mild suppurative bronchopneumonia.

Figure 1: PPIV infected lungs

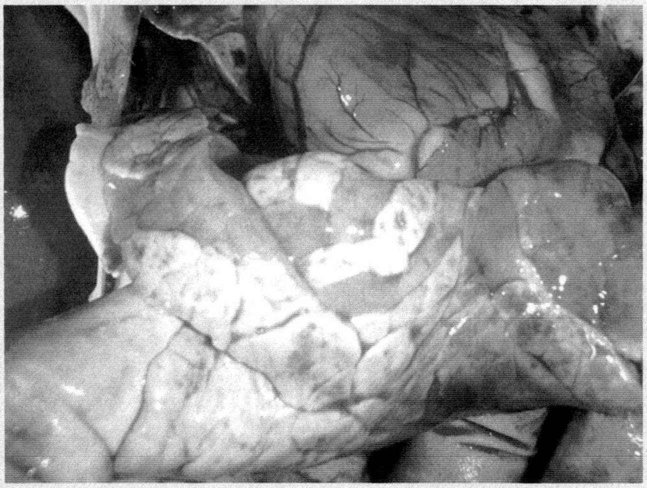

Trachea: Lesions are similar in pigs A-D. Mucosal is ulcerated, attenuated or disrupted by transmigrating inflammatory cells and there is squamous metaplasia of the epithelium. Lymphocytes and plasma cells invade the submucosa.

Diagnosis:
Tracheitis and bronchiolitis, multifocal and severe, lymphocytic (pigs A-D).

Comments:
Lesions in pigs A-D are consistent with infection attaching respiratory epithelium
Due to the consistent histopathological suggestion of a viral driver with a lack of IAV, Iowa State Veterinary Diagnostic Laboratory was asked to look for other viral drivers. Dr. Scanlon Daniels had also been corresponding with the lab on cases with a similar clinical description. Drs. Madson and Yoon determined that the samples were positive on a Pan-paramyxoviridae PCR which sequenced as porcine parainfluenza virus type 1.

Treatment and prevention

After PPIV1 was identified in the herd, options regarding autogenous vaccine development were explored. Samples from this herd were sent to two Universities and three biological companies for virus isolation. It is discussed that cattle PIV3 is very easy to isolate, but porcine PIV has proven to be difficult. None of the companies were able at that time to isolate the virus. PEDV introduction into the United States turned all focus to management and development of vaccine options to that disease, resulting in a lapse in focus on PIV.

In 2015, Harrisvaccines developed a PIV autogenous vaccine based on their replicon technology. The hemagglutinin and fusion proteins are most important for immune recognition and protection so both proteins are included in the vaccine. The original formulation utilized was from an Oklahoma isolate and put into two breed to wean herds. Sows were vaccinated twice prefarrow and suckling pigs assessed. Clinically, suckling pig health was not improved with this strain formulation. Due to lack of response, 100% homologous vaccine was developed for both of the herds and implemented. Clinically, respiratory heath did seem to improve in both of the herds but clinical signs were still present. Nasal swabbing of pigs at weaning show a decreased prevalence but virus can still be identified at weaning and on oral fluids post weaning. To date, both of those herds continue to use replicon based autogenous vaccine.

Discussion

Prevalence of this virus is widespread within the industry and can be found in clinical and non-clinical herds. Additionally, it can be found in nursery and finisher pigs indicating it can be a persistent infection and does not seem to clear like influenza. Due to the mild clinical signs, my perception is that if this virus does cause respiratory clinical signs, it only does so during initial infection in the suckling and post wean phase.

Due to the high prevalence and persistence, it is important that you put together all three components to make a clinical diagnosis:

1. Clinical signs – Cough and sneeze in suckling and post wean pigs. Post wean depression and anorexia for ~4-7 days. Low mortality, high morbidity.

2. PCR positive results for PIV. PCR negative results for IAV and PRRS.

3. Histopathology that describes lesions of an epitheliotropic virus.

I have been working with the virus since 2013 and only have 3 sow herds that meet this definition.

There seems to be some response to the replicon vaccine technology (Merck) but it is difficult to measure due to the waxing and waning of clinical signs in post wean pigs. It is recommended to use 100% homologous isolate at this time. Dr. Phil Gauger (Iowa State VDL) has been successful at virus isolation on a few isolates and will open up the opportunity to look at other autogenous vaccines for alternative formulation and delivery. However, success of virus isolation is still relatively low so movement forward with other autogenous vaccine formulation has not moved forward to date.

The clinical syndrome of mild respiratory disease in suckling and post wean pigs has been frustrating for veterinarians and producers. Research and improvement in sequencing technology will help us determine if PIV is a key driver of this syndrome or if it requires co-infection. It is also very possible a separate virus or bacteria may be driving the clinical syndrome.

Porcine parainfluenza virus type 1 (PPIV-1): Diagnostic laboratory perspective

P. Gauger, DVM, PhD; M. Welch, BS; J.Y. Park, PhD, PhD; K.M. Harmon, PhD; J. Zhang MD, PhD; P. Piñeyro DVM, MVSc, DVSc, PhD; L. Gimenez-Lirola, MS, PhD
Department of Veterinary Diagnostic and Preventive Animal Medicine, College of Veterinary Medicine, Iowa State University, Ames Iowa

Introduction

Respiratory disease is a major health concern to the United States (US) swine industry due to poor growth performance, increased antimicrobial use and added vaccination costs.[1,2] Several viruses play a role in the porcine respiratory disease complex (PRDC) and most often involve primary pathogens that include porcine reproductive and respiratory syndrome virus (PRRSV), influenza A virus (IAV), and porcine circovirus type 2 (PCV2) among others. However, diagnostic investigations have demonstrated that clinical respiratory disease in swine is often polymicrobial involving coinfections with different viruses and bacteria.[3] A recent publication classified swine respiratory pathogens into groups of major or minor relevance to the pork industry.[3] Pathogens of minor significance included the *Paramyxoviridae* family consisting of several genera that may contribute to porcine respiratory disease (PRD) in specific geographic regions but are not common in the US. Among these, porcine rubulavirus and Nipah virus have been documented in affected pig populations as playing a significant role in porcine respiratory disease.[4,5] In addition to Nipah virus, Menangle virus and Newcastle disease virus have also been detected in swine.[6-9] Pigs are the primary reservoir of porcine rubulavirus and cross-species transmission of paramyxoviruses from host to swine have been documented.[6,8] Recently, a new Paramyxovirus detected in swine in Hong Kong, and now the US, has been designated taxonomically *Porcine Respirovirus type* 1 or more commonly by the virus name porcine parainfluenza virus type 1 (PPIV-1). PPIV-1 has been regarded as a potential respiratory pathogen in nursery and grow-finish pigs.[10]

Etiology

Parainfluenza virus is a member of the *Paramyxoviridae* family, subfamily *Paramyxovirinae* and the genus *Respirovirus*. This genus consists of five previously recognized species including bovine parainfluenza virus 3 (BPIV-3), human parainfluenza virus 1 (HPIV-1), human parainfluenza virus 3 (HPIV-3), Sendai Virus (SeV), and Simian virus 10.[11] The sixth species includes PPIV-1 although taxonomically, all species are currently referred to as *Respiroviruses.*[12] PPIV-1 is a negative-sense, single-stranded RNA virus of approximately 15 kb in length.[13] The virus consists of six genes that encode six proteins and two accessory proteins associated with the phosphoprotein that are translated through a process known as RNA editing. Two PPIV-1 proteins important for virus replication in host cells include the fusion (F) protein that induces fusion between the viral envelope and the host cell membrane and the hemagglutinin-neuraminidase (HN) gene that encodes the cell attachment protein and facilitates viral entry into a host cell.

Epidemiology

A report from Hong Kong in 2013 described a parainfluenza virus belonging to the family *Respirovirus* that was detected from nasal swabs collected from swine in an abattoir.[10] A total of 951 samples from 386 pigs were collected from a slaughter house from 2008 to 2012 during a surveillance project. Nasopharyngeal swabs, rectal swabs, blood, lung and liver samples were collected and analyzed for Paramyxoviruses by PCR. Three nasopharyngeal samples were positive and sequencing confirmed a novel paramyxovirus, designated porcine parainfluenza virus type 1 (PPIV-1). Phylogenetic analysis revealed a distinct cluster separate from known *Respiroviruses* although closely related to HPIV-1 and SeV. PPIV-1 is currently detected routinely in veterinary diagnostic labs in the US.[11]

At the time of this submission, the Iowa State University Veterinary Diagnostic Laboratory (ISU VDL) sequenced 33 HN genes and 32 F genes as well as one PPIV-1 whole genome from an isolate. The isolate, MN25890NS/2016, HN gene shared 95.7% - 98.2% nucleotide homology with the 33 HN ISU VDL sequences and 96.0% - 98.3% nucleotide similarity with the PPIV-1 HN genes reported by Palinski et al., 2016 at the Kansas State University Veterinary Diagnostic Laboratory (KSU VDL).[11] In contrast, MN25890NS/2016 shared only 90.6% - 96.6% nucleotide similarity with the Hong Kong HN sequences reported in 2013.[10] The F protein of the 32 ISU VDL sequences shared 94.2% - 98.5% nucleotide homology with MN25890NS/2016 and 97.8% - 98.3% with the KSU F gene sequences and only 89.9% - 95.2% homology with the Hong Kong sequences.[10,11] These data suggest geographic genetic differences in the PPIV-1 circulating in swine.

Pathogenesis

Reports of PPIV-1 detected by real-time, reverse transcription polymerase chain reaction (rRT-PCR) in swine with respiratory illness and evidence of field infections coinciding with lung lesions have supported its role in the PRDC.11,13 A recent publication described a case of recurrent respiratory disease in swine approximately 10-21 days of age with clinical signs of cough, minor sneezing and serous nasal discharge of unknown etiology.[11] Nasal swabs tested at the ISU VDL were positive for PPIV-1 and samples transferred to the KSU VDL. The use of metagenomics sequencing revealed the presence of several viruses including PPIV-1. Additional samples submitted from ten-day-old pigs also contained PPIV-1 in nasal turbinate with concurrent detection of astrovirus 4, kobuvirus, and porcine hemagglutinating encephalomyelitis virus (PHEV). Although PPIV-1 was detected in these samples, PCV2 and PHEV have also been associated with respiratory disease in swine.[3] However, the use of next generation sequencing has also supported an association between PPIV-1 and respiratory disease in swine.

PPIV-1 has been isolated in the laboratory and evaluations of the pathogenesis, clinical signs and lesions associated with PPIV-1 infection are in progress. Current research in naturally infected pigs has been limited with the risk of co-infections that may confound interpretation of results. A 2016 investigation of PPIV-1 in swine conducted at the KSU VDL involved pigs from a herd naturally infected with PPIV-1 based on previous diagnostic submissions.[11] Eleven pigs selected at weaning (18-19-days-old) were transferred to experimental facilities in Nebraska. Pigs were observed for two weeks without evidence of clinical disease. However, at the start of the study, 54% (6/11) of the pigs were PCR positive for PPIV-1 with three additional pigs that became infected during the course of the observation period. Based on the results of this KSU study, the length of time for viral shedding was 2-10 days based on PCR. Histopathology examination on lung, trachea and nasal turbinate for eight of the eleven animals revealed only atelectasis in the lung and a mild decrease in goblet cells and/or cilia of the trachea. *In situ* hybridization (ISH) in nasal turbinate epithelium and trachea of one pig confirmed the presence of PPIV-1.

Serum from two of three pigs collected at the end of the two-week study were positive for PPIV-1 antibody using an immunoprecipitation coupled to PCR detection (ICPD) assay. In contrast, seven of the eleven experimental animals were positive for F protein ELISA antibodies at 0 dpi suggesting the presence of maternal antibody.[11] However, two pigs that were ICPD antibody positive at 14 dpi were negative for F protein ELISA antibody at 0 dpi, but shed PPIV-1 during the trial and proceeded to demonstrate evidence of seroconversion at 14 dpi with the detection of F protein ELISA antibody. Further experimental studies evaluating the pathogenesis of PPIV-1 are in progress.

Clinical signs and lesions

Clinical signs and lesions associated with PPIV-1 infection in swine have not been well characterized. Palinski et al., 2016 described the clinical disease and lesions associated with naturally infected, weaning-age pigs that were closely monitored at an animal research unit after they were moved to the facility from a farm with a prior history of PPIV-1 infection. Clinical signs were not observed in the pigs during the two-week monitoring period and gross lesions were not described. Microscopic lesions in the trachea and nasal turbinate were non-specific and included a loss of goblet cells and cilia in the trachea and the presence of neutrophils in respiratory epithelium and mononuclear cells in the submucosa of the nasal turbinate.[11] A thorough diagnosis of PPIV-1 clinical disease requires the presence of clinical signs that are correlated with gross and microscopic lesions with detection of the agent in affected tissues. Diagnostic reports have detected PPIV-1 by rRT-PCR in the absence of other known respiratory pathogens in nursery and grow/finish swine with coughing or respiratory disease suggesting a presumptive diagnosis.

Diagnostics

Diagnostic assays used to detect PPIV-1 have been established in spite of the difficulty that is associated with interpreting the outcome without specific histopathological lesions. The ease of sequencing viruses and the availability of nucleotide sequences in public databases have allowed diagnostic labs to quickly develop PCR assays to detect novel viruses such as PPIV-1. In addition, published PCR assays can be implemented into diagnostic labs after optimization and validation using primers and probes specific to the sequence associated with the pathogen of interest.

Currently, the most common diagnostic test available for PPIV-1 is PCR. Although it has been suggested PPIV-1 resides in the upper respiratory track, samples appropriate for PCR would be analogous to those used to detect IAV-S including lung, nasal swabs and oral fluids. In addition, bronchoalveolar lavage fluid, trachea and nasal turbinate are also appropriate for PCR and detection of PPIV-1.

Sequencing of PPIV-1 is available and currently conducted on the HN and F genes due to their genetic diversity.[14] Diagnostic assays beyond PCR and sequencing, including immunohistochemistry (IHC) and serology, are in development. Immunohistochemistry or ISH assays would help confirm the presence of PPIV-1 in affected tissues and aid in the diagnosis of PPIV-1 clinical disease.

ISU VDL PPIV-1 Diagnostic Cases

The ISU VDL has received several swine diagnostic submissions experiencing respiratory disease, lethargy, coughing and sneezing with microscopic lesions consistent with bronchointerstitial pneumonia. Lung lesions also included bronchiolitis with bronchiolar epithelial necrosis, a common lesion caused by infection with IAV. Lung tissue, nasal swabs and oral fluids submitted with these accessions were PCR negative for pathogens commonly associated with the PRDC, including influenza, which prompted further investigations. Additional diagnostics using genus-level PCR assays detected a virus with genetic similarity to a porcine parainfluenza virus type 1 reported in Hong Kong suggesting an association between PPIV-1 and respiratory disease in swine.[10]

Results from 931 swine accessions submitted to the ISU VDL since 2014 requesting rRT-PCR were analyzed. Accessions (diagnostic cases) were submitted from swine of different ages with a complaint of PRD and including a request to test for PPIV-1 among other differentials for PRDC. Cases included fresh and formalin fixed tissues from eighteen states (AK, AR, AZ, GA, IA, IL, IN, MI, MN, MO, NC, NE, NV, OH, OK, PA, SD, TX). Approximately 43.1% (401/931) of these cases were PPIV-1 rRT-PCR positive. Positive cases have increased from 27.3% (9/33) in 2014 to 44.4% (251/565) in 2017 (Figure 1). These data suggest the number of PPIV-1 positive accessions have increased over time (Figure 2).

Figure 1: Total number of PPIV-1 diagnostic cases and rRT-PCR positive cases submitted to the ISU VDL 2014 - 2017.

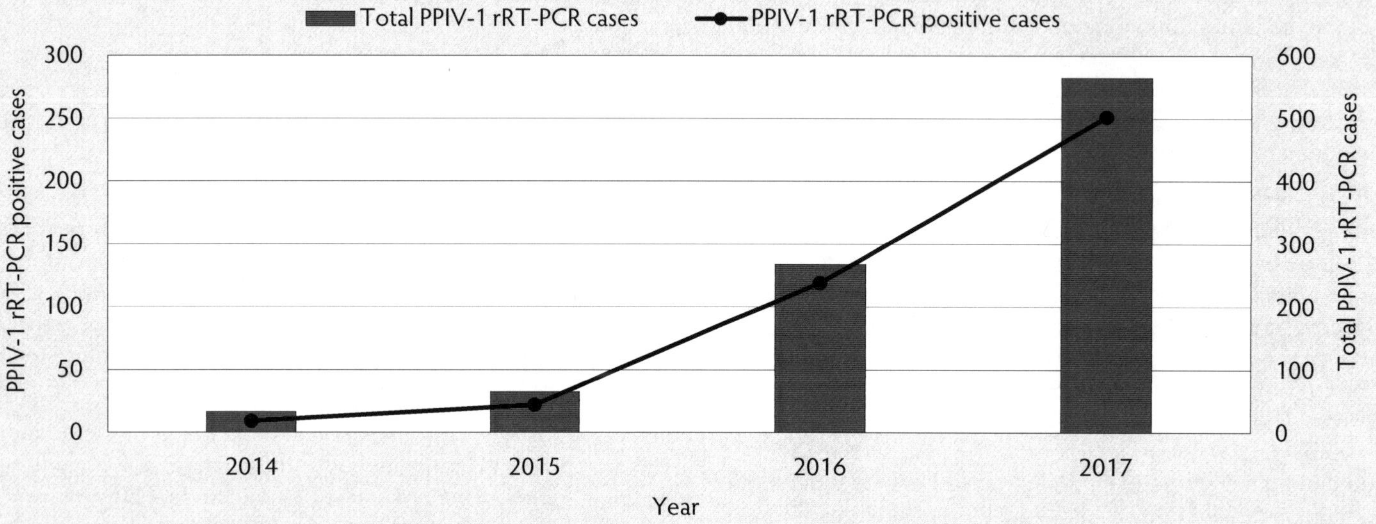

Figure 2: Total number and percent PPIV-1 rRT-PCR positive diagnostic cases submitted to the ISU VDL 2014 - 2017 per month.

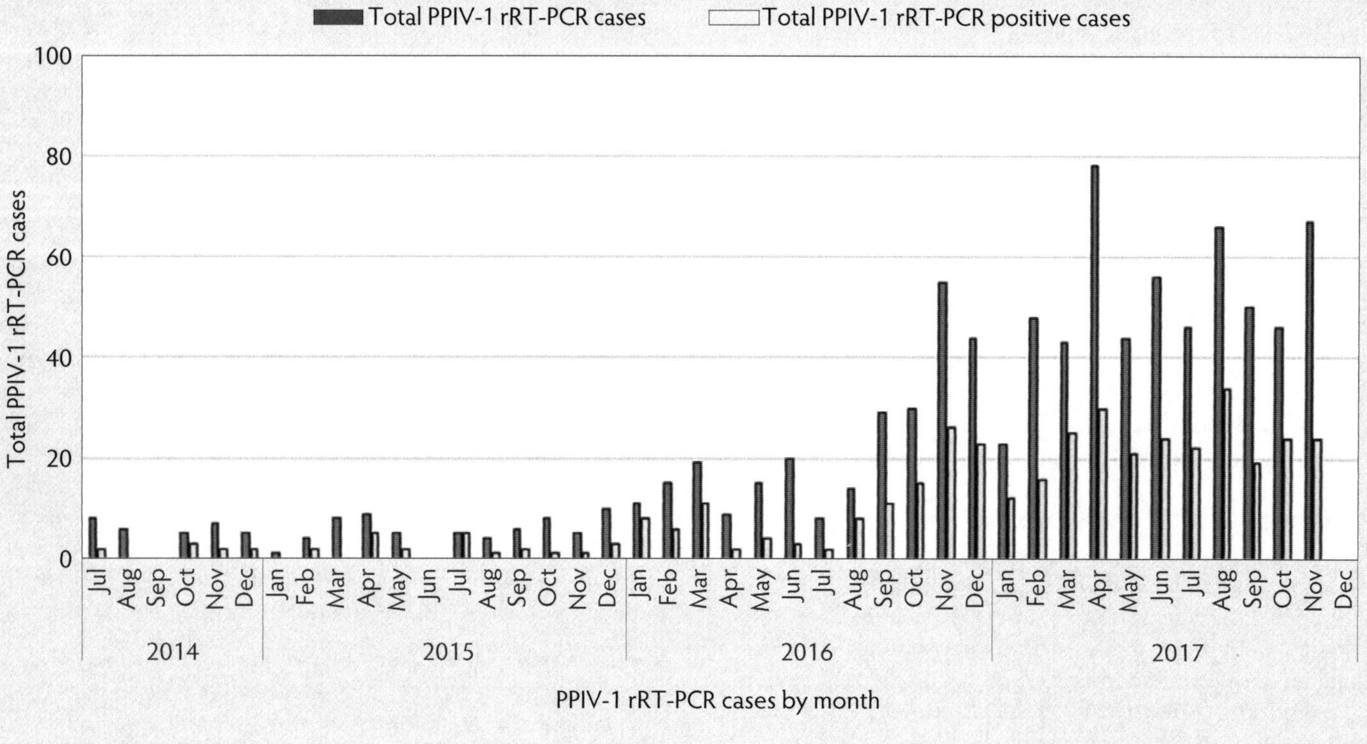

Samples for PPIV-1 rRT-PCR most often included lung, nasal swabs, and oral fluids. Additional sample types appropriate for PPIV-1 rRT-PCR include oropharyngeal swabs, turbinate, laryngeal swabs, trachea, bronchiolar swabs, and bronchoalveolar lavage. Percent positive PPIV-1 rRT-PCR samples by specimen type included 14.7% (59/401) for lung, 42.4% (170/401) for oral fluid, 26.7% (107/401) for nasal swabs and 15.7% (63/401) for all other sample types (Figure 3). Collectively, the percent PPIV-1 rRT-PCR positive specimens from the total number of each sample type tested included 14.8% (59/400) of lungs, 68.3% (170/249) of oral fluids, and 64.5% (107/166) of nasal swabs. Additional diagnostic samples with percent positive above 40% of the samples tested included oropharyngeal swabs, turbinate, laryngeal swab, and oral swabs. However, data is biased due to the restriction of testing samples collected from field cases of PRD, potential quality issues associated with specimens (autolysis, organic material, temperature control), and the collection of samples from the pigs that do not represent the clinical signs of the population.

Pig ages represented in this analysis included 12.7% (118/931) neonate/suckling piglets, 41.4% (385/931) nursery pigs, 12.1% (113/931) grow/finish swine, 2.4% (22/931) adult swine and 31.5% (293/931) of unknown age. Percent of PPIV-1 rRT-PCR positive accessions by age included 18.1% (75/401) Neonate/suckling piglets, 37.2% (149/401) nursery pigs, 10.2% (41/401) grow/finish swine, 1.2% (22/401) adult and 32.7% (131/401) of unknown age (Figure 4). The percent PPIV-1 rRT-PCR positive of the number of specimens submitted per age group included approximately 63.6% (75/118) of the neonate/suckling piglet samples, 38.7% (149/385) nursery pig samples, 36.3% (41/113) grow/finish samples, and 22.7% (5/22) of the adult samples.

The ISU VDL also conducted a 2016 prospective PPIV-1 evaluation of diagnostic samples submitted from swine for PPIV-1 analysis. Samples were tested by the ISU VDL rRT-PCR for the presence of PPIV-1 regardless of clinical history, age or specimen type. Samples included oral fluid, lung, nasal swab, environmental sample, serum, tonsil swab, oropharyngeal swab, bronchoalveolar lavage, fibrin, feed, brain, fetal tissues, tissue composite, urine, undesignated fluids, feces, joint fluid and spfigureinal cord. A total of 1,385 samples were tested and 34.1% (472/1385) were PPIV-1 rRT-PCR positive. The diagnostic samples that tested positive included oral fluid (75.4%), lung (7.2%), nasal swab (13.1%), environmental (2.3%), swabs (unknown source) (1.5%), bronchoalveolar lavage (0.2%), and a feed sample (0.2%) (Figure 5). Samples originated from twenty-seven states with Iowa, North Carolina and Illinois representing the largest number of accessions and largest number PPIV-1 rRT-PCR positive. Some samples were also tested by PCR for IAV and 21.1% (115/544) oral fluid, 4.5% (12/269) of lung, 7.4% (13/176) nasal swabs were PCR positive

Figure 3: Total number of specimen types submitted for PPIV-1 rRT-PCR, total number PPIV-1 rRT-PCR positive specimen types, and percent of total PPIV-1 rRT-PCR positive specimen types submitted to the ISU VDL 2014 - 2017 with each specimen type representing an individual case.

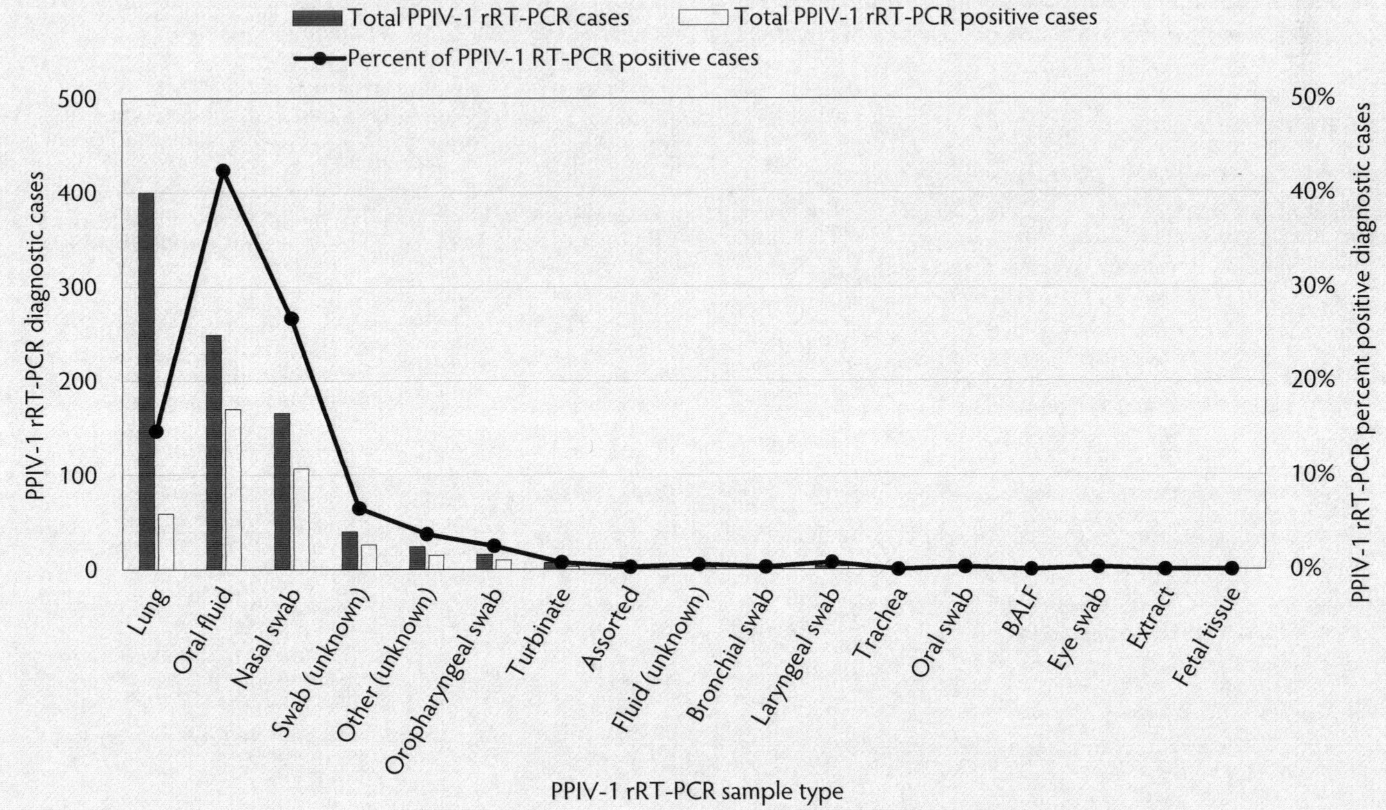

Figure 4: Total number of samples submitted for PPIV-1 rRT-PCR , total number PPIV-1 rRT-PCR positive diagnostic cases, and percent of total PPIV-1 rRT-PCR positive diagnostic cases submitted to the ISU VDL 2014 - 2017 representing different production phases.

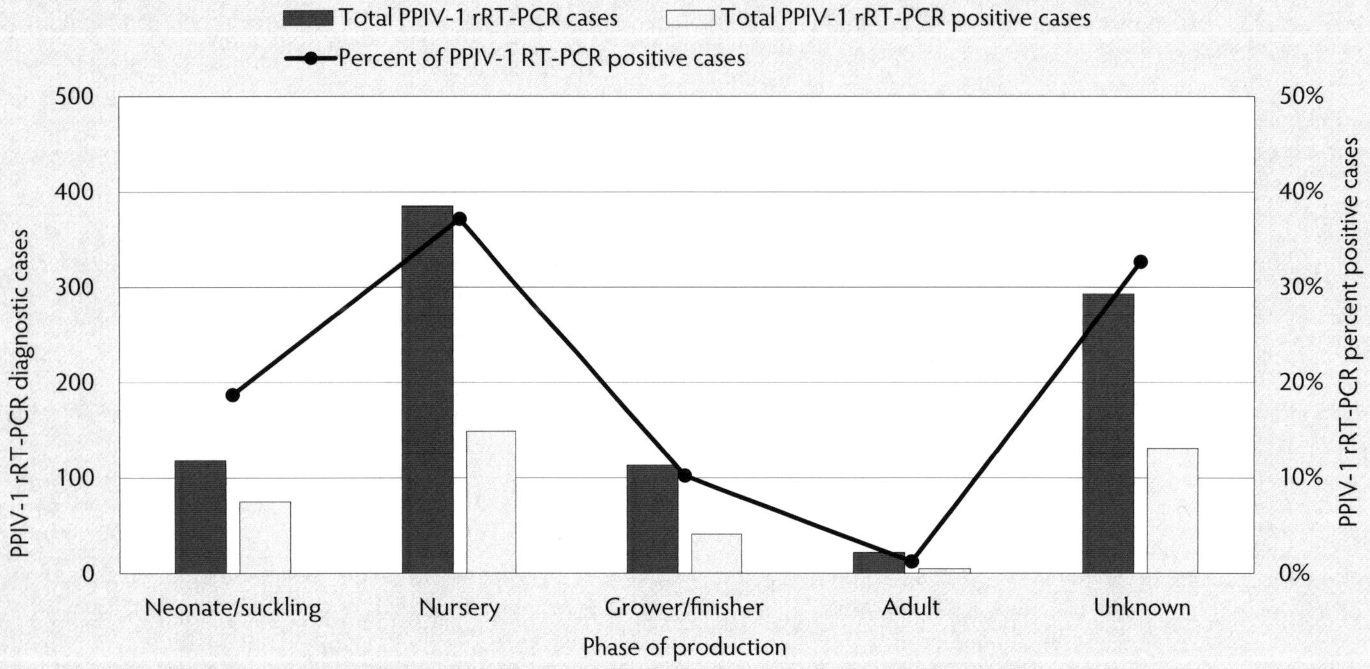

for both pathogens. Based on sample type, oral fluid is the most likely to be PPIV-1 PCR positive compared to lung and nasal swabs. However, in cases submitted with a specific history of respiratory disease, nasal swabs were more likely to be positive for the presence of PPIV-1 compared to other sample types.

Summary

These results suggest PPIV-1 detection, but not necessarily clinical disease, may be widespread in the US pig population. PPIV-1 can be detected in swine of all ages regardless of clinical history. Further evaluation is necessary to understand PPIV-1 clinical significance and effect on production as well as potential clinical disease.

- PPIV-1 is widespread in the US swine population.
- PPIV-1 has been primarily detected in nursery and grow-finish swine.
- PPIV-1 has been detected most often in the upper respiratory tract.
- Appropriate sample types for detecting PPIV-1 by rRT-PCR include oral fluids, lung and nasal swabs among others.
- Appropriate samples for diagnosing PPIV-1 disease include fresh and formalin fixed lung for PCR as well as histopathology evaluation.

References

1. Holtkamp D, Kliebenstein J, Zimmerman J, et al: Economic analysis of PRRS Virus Elimination from a Herd. Iowa State University Animal Industry Report AS-Leaflet R2678, 2012.

2. Chae C: Porcine respiratory disease complex: Interaction of vaccination and porcine circovirus type 2, porcine reproductive and respiratory syndrome virus, and Mycoplasma hyopneumoniae. Vet J 212:1-6, 2016.

3. Opriessnig T, Gimenez-Lirola LG, Halbur PG: Polymicrobial respiratory disease in pigs. Anim Health Res Rev 12:133-148, 2011.

4. Hooper P, Zaki S, Daniels P, et al: Comparative pathology of the diseases caused by Hendra and Nipah viruses. Microbes Infect 3:315-322, 2001.

5. Rodriguez-Ropon A, Hernandez-Jauregui P, Sanchez-Torres L, et al: Apoptosis in lymph nodes and changes in lymphocyte subpopulations in peripheral blood of pigs infected with porcine rubulavirus. J Comp Pathol 128:1-8, 2003.

6. Chua KB, Goh KJ, Wong KT, et al: Fatal encephalitis due to Nipah virus among pig-farmers in Malaysia. Lancet 354:1257-1259, 1999.

7. Ellis JA: Bovine parainfluenza-3 virus. Vet Clin North Am Food Anim Pract 26:575-593, 2010.

8. Philbey AW, Kirkland PD, Ross AD, et al: An apparently new virus (family Paramyxoviridae) infectious for pigs, humans, and fruit bats. Emerg Infect Dis 4:269-271, 1998.

9. Stephan HA, Gay GM, Ramirez TC: Encephalomyelitis, reproductive failure and corneal opacity (blue eye) in pigs, associated with a paramyxovirus infection. Vet Rec 122:6-10, 1988.

10. Lau SK, Woo PC, Wu Y, et al: Identification and characterization of a novel paramyxovirus, porcine parainfluenza virus 1, from deceased pigs. J Gen Virol 94:2184-2190, 2013.

11. Palinski RM, Chen Z, Henningson JN, et al: Widespread detection and characterization of porcine parainfluenza virus 1 in pigs in the USA. J Gen Virol 97:281-286, 2016.

12. Lamb RAP, G.D.: Paramyxoviridae: the viruses and their replication. (ed 5). Philadelphia, Lippincott Williams & Wilkins, 2007.

13. Yoon KJ.; Sun D.;Stevenson G ASHHMDBECVGPPASK: Parainfluenza: Influenza-like syndromes, Proceedings, American Association of Swine Veterinarians, Orlando, FL, February 28-March 3, 2015, 2015 (available from

14. Zeltina A, Bowden TA, Lee B: Emerging Paramyxoviruses: Receptor Tropism and Zoonotic Potential. PLoS Pathog 12:e1005390, 2016.

Figure 5: Prospective ISU VDL PPIV-1 rRT-PCR analysis of various sample types indicating the total number of samples types analyzed, total number PPIV-1 rRT-PCR positive by sample type, and percent PPIV-1 rRT-PCR positive by specimen type conducted in 2016.

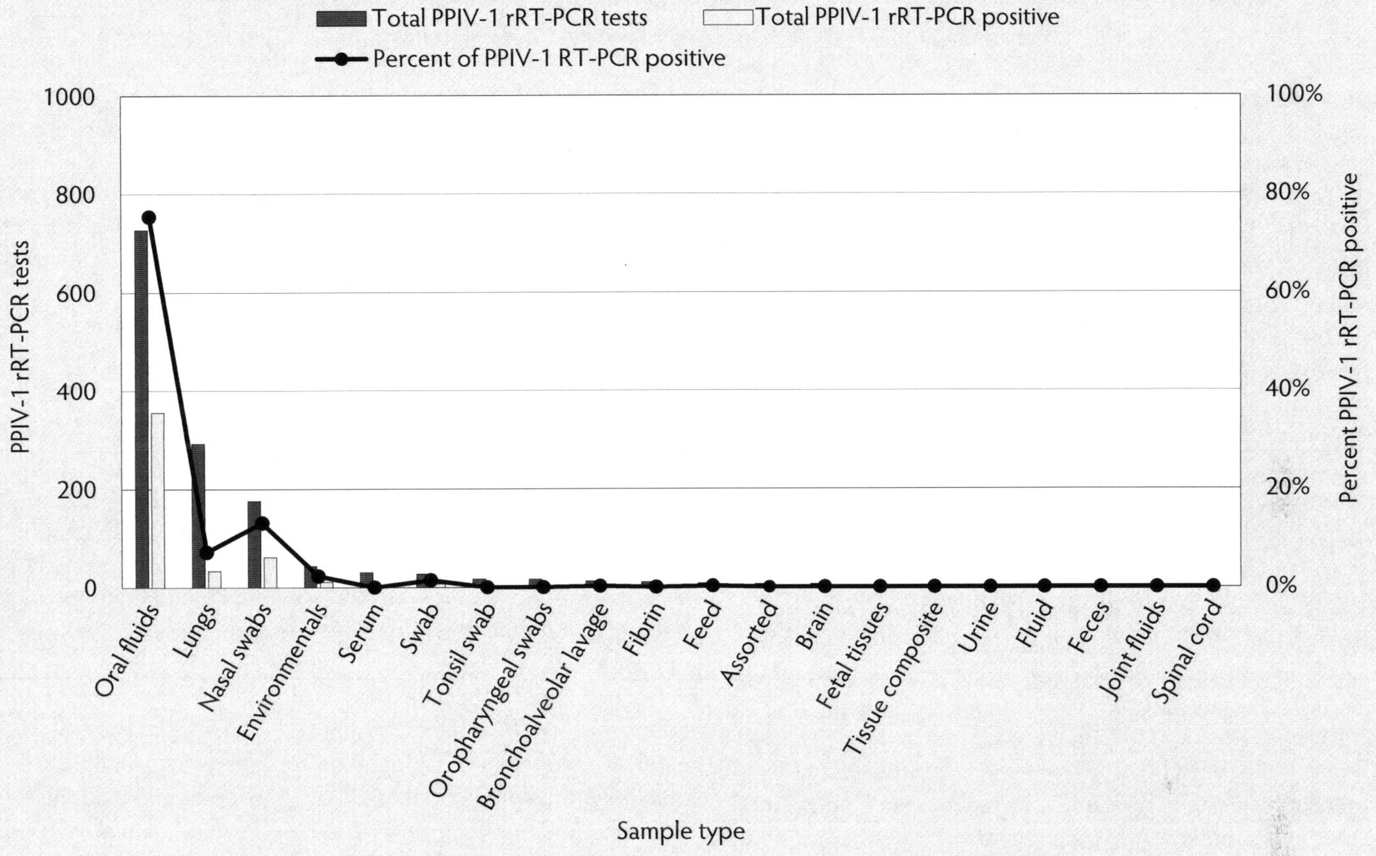

Porcine sapelovirus: Practitioner perspective

Brian Payne[1], DVM; Abby Patterson[2], DVM, PhD; Paulo Arruda[3], DVM, PhD; Franco Matias Ferreyra[4], DVM
[1]Pipestone Veterinary Services, Sycamore, Ilinois; [2]Boehringer Ingelheim Animal Health, Ames, Iowa; [3]VRI, Ames, Iowa; [4]Iowa State University, Ames, Iowa

Farm description

The described case occurred on an 1,100-head farrow-to-finish farm with two sites 2.0 miles apart (Site 1: Sows and nursery; Site 2: Wean-finish, finisher and gilt development). The farm is positive for field infection PRRS, *Mycoplasma hyopneumoniae* and influenza-A. In addition, the farm experienced pestivirus-associated congenital tremors in 2012.

Current case history and diagnostics

The farm has experienced growing pig neurological cases since at least 2008. In 2016, the incidence of cases began being more closely monitored and it was observed that nearly 100% of all weekly 500-head groups had individual neurologic cases in pigs between nine to twelve weeks of age. A typical case description includes: Day 1–Pig is observed as swaying, staggering, leaning on the gate while walking. The pig is pulled into the alley for treatment. Day 2–Neurologic signs progress, pig starts to "bunny-hop" when encouraged to rise and walk (front legs are curled underneath and the back legs extend simultaneously pushing the pig forward). The pig stops and lays down when not prompted to move. Day 3–The pig becomes recumbent, paddling of the legs is not often noticeable, nystagmus may or may not be present. Pig succumbs this day or on day 4 regardless of antibiotic or palliative care; all morbid pigs (100%) die. Neurologic tissues from mortalities were submitted to multiple laboratories. Histopathology results were suggestive of a viral infection due to the severe, non-suppurative, encephalomyelitis with perivascular cuffing observed in multiple sections of spinal cord and brain stem. Further PCR testing confirmed the presence of porcine sapelovirus (PSV); other viral differentials (porcine enterovirus and porcine teschovirus) were not detected. Since 2014, the average total mortality in all groups was 5.3%. Within total mortalities, those animals which showed clinical neurological signs prior to death were considered neurologic mortalities; mortality associated with these cases was 0.8% (Figure 1). On the farm, 95% of neurologic mortalities occurred in the first 90 days post-weaning (Figure 2).

To determine the prevalence of PSV on the farm, a multi-age fecal swab PSV PCR survey was conducted. Results indicated that PSV was ubiquitous in weaned to market age pigs, but less prevalent in gestating sows (33%), farrowing sows (0%) and suckling pigs (16%) (Table 1). In addition, 15 serum samples from two age groups were collected for serological evaluation. Virus neutralizing (VN) antibodies were detected in 20% and 100% of weaned pigs and late finishing pigs, respectively. The average VN titers were approximately 20 × higher in finisher pigs (Figure 3).

Current status and discussion

To further investigate the role of PSV on the farm, an autogenous vaccine program was implemented in the fall of 2017. The protocol included two, 2mL doses of vaccine (Newport Labs, Worthington, MN) administered at three and six weeks of age. Vaccinates (~83%) and non-vaccinates (~17%) were commingled. Clinical signs and mortality were monitored for thirteen groups (~500pigs/group). Brain and fibrin swabs were collected as supportive data. Sapelovirus is highly suspected to be the cause of most of the neurologic diseases in this farm. However, the wide distribution of PSV found on the farm is inconsistent with the low number of clinical cases. Possible reasons include: 1) PSV is present but not the primary etiologic agent, 2) lack of uniform herd immunity leading to endemic circulation of the virus with clinical signs being observed in susceptible animals, or 3) the impact of PSV on pigs ranges from clinical to subclinical presentations. In March 2017, the Swine Health Information Center (SHIC) has reported an increase in central nervous system cases to veterinary diagnostic laboratories.[1] Coordination, nationally, to educate and monitor potentially emerging diseases is important to quickly determine significance, etiology and prevention.

References

1. Swine Health Information Center (SHIC); http://www.swinehealth.org/wp-content/uploads/2017/03/CNS-VDL-Reports-2017.pdf.

Figure 1: Total and neurologic mortality percentage by wean group, since 2014, %

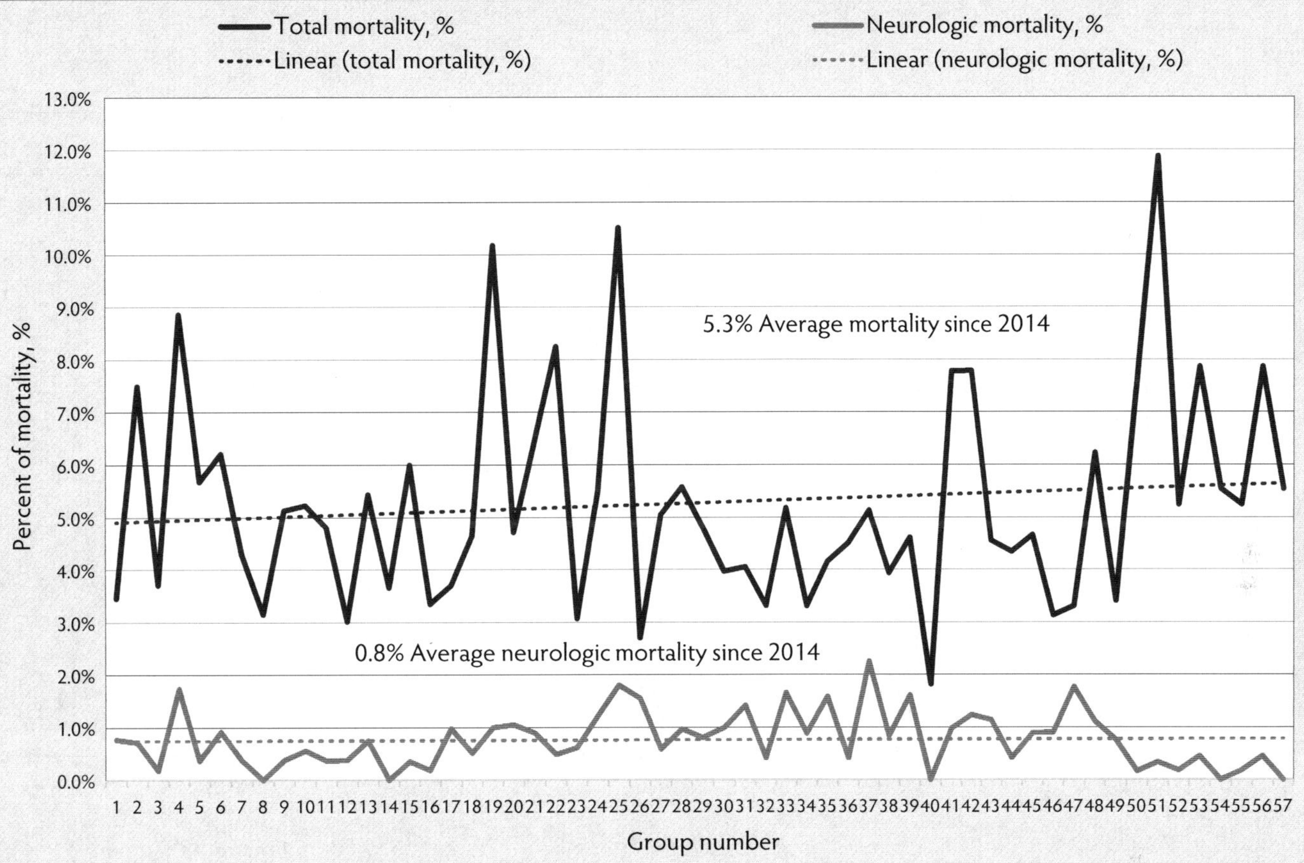

Figure 2: Total neurologic mortality and percent of neurologic mortality, by days post-placement, since 2014

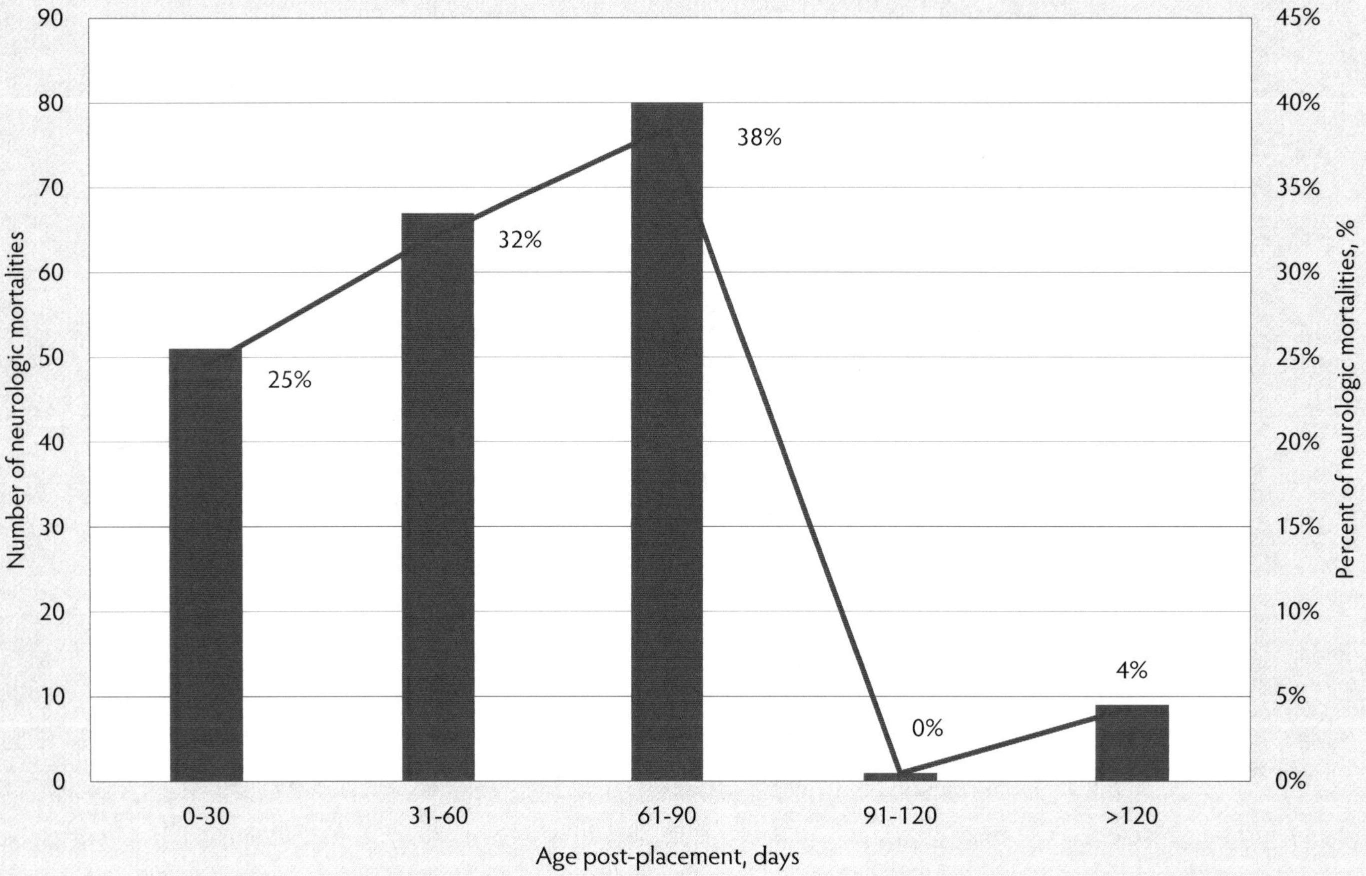

Table 1: Sapelovirus PCR results of fecal swab pools, by age group and location

	Designation		% detected*
Site 2	1	4 wks	6/6 (100%)
	2	7 wks	9/9 (100%)
	3	11 wks	9/9 (100%)
	4	15 wks	9/9 (100%)
	5	19 wks	9/9 (100%)
	6	23 wks	9/9 (100%)
	7	Gilts (finisher stage)	8/9 (89%)
Site 1	8	Gestating sows	3/9 (33%)
	9	Farrowing sows	0/9 (0%)
	10	Suckling piglets	1/6 (17%)
	11	Nursery (4-6 weeks of age)	9/9 (100%)

* Number of positive fecal swab pools / total number of fecal swab pools tested; ~5 swabs/pool were tested

Figure 3: Virus neutralizing (VN) percent positives data in weaned pigs and late finishing pigs

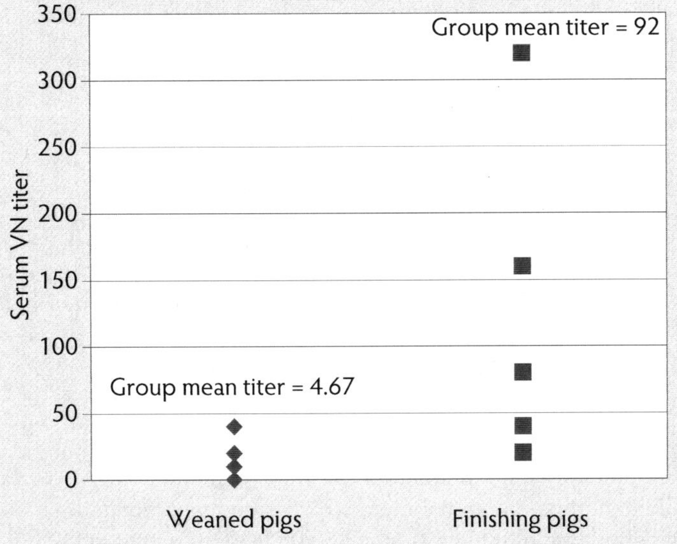

Porcine sapelovirus: Pathologist/research perspective

B. Arruda[1], DVM, PhD; K. Schwartz[1], DVM, MS; P. Arruda[2], DVM, MS, PhD
[1]Iowa State University, Ames, Iowa; [2]VRI, Ames, Iowa

Background

Sapelovirus, formerly classified as an *Enterovirus*, is a newly assigned genus within the family *Picornaviridae*, a family which consists of 29 genera that cause a wide range of diseases in humans and animals.[1,2] *Sapelovirus A*, one of 3 current species of *Sapelovirus*, was formerly known as porcine sapelovirus (PSV) and porcine enterovirus 8.[1] Unlike porcine teschovirus (PTV) or porcine enterovirus (PEV), both of which were also previously assigned to the genus *Enterovirus*, PSV causes unique type II cytopathic effect in porcine kidney cells.[3,4,5] All three viruses (PSV, PTV, PEV) can be found in swine feces.[6,7] Recently, PSV has been linked to non-suppurative poliomyelitis in pigs in the United Kingdom[8] and United States.[9] Due to the endemic nature of these viruses, aseptic collection of central nervous system (CNS) tissue from clinically affected pigs is imperative for an accurate etiologic disease diagnosis.

Anecdotally, cases of non-suppurative poliomyelitis have been rare; however, diagnoses of viral poliomyelitis are increasing in frequency, especially compared to past decades. PSV is detected with some frequency in the CNS tissue of pigs with atypical neurologic signs and non-suppurative poliomyelitis in cases submitted from broad geographical areas to the Iowa State University Veterinary Diagnostic Laboratory (ISU VDL). Information from submission forms indicate that cases often represent a population of pigs varying from 2,400 to 10,500 animals. Herd history suggest the reported case fatality rate approaches 100%, with a herd mortality due to PSV disease reported to range from 1-10% but is commonly 3-4%. Additionally, veterinarians dealing with the disease often indicate a high recurrence rate in subsequent groups of newly placed pigs that can span weeks to years. Affected pigs are usually 4 to 12 weeks of age. Duration of clinical signs in individual pigs is commonly less than 4 days, with many euthanized based on welfare concerns. Clinical disease in a group of pigs commonly spans weeks. Treatments that include anti-inflammatories and antimicrobials are reported as unrewarding.

To assess the frequency of detection of PSV in cases of neurologic disease with viral lesions in CNS tissue, a subset of cases submitted to the ISU VDL was reviewed. Thirty neurologic cases were selected on the basis of histologic lesions attributable to a viral infection and were tested with a nested multiplex PCR for PSV, PTV and PEV. One or more viruses were detected in 19 of those 30 cases. PSV was detected in a majority of cases (14 of the 19), either alone (6 cases) or with PTV or PEV (four and three cases, respectively). All three viruses were detected in one case. No etiologic agent was identified in one-third of the cases in which lesions suggestive of neurotropic virus insult were present. This may be due to sample selection, duration of infection or diagnostic tool(s) applied. Porcine astrovirus type 3 (PAstV-3), not a member of the *Picornaviridae* family, was later detected by PCR and RNAscope® ISH in a portion of those cases that previously lacked an etiologic diagnosis.

Models of PSV-associated disease have been described in the scientific literature; however, limitations in study design, sample collection, reporting of results, lesion descriptions unreconciled with images provided, and the particular assays used have thus far limited the interpretation of results and study conclusions.

Clinical signs

Clinical signs reported in cases submitted to the ISU VDL include mild to moderate ataxia of the hind limbs, paresis (front and/or hind limbs), and lateral recumbency. These signs are similar between different viral causes of poliomyelitis: PSV (Video 1), PTV (Video 2), and PAstV-3 (Video 3). Front limb paresis is a more prevalent observation in pigs with PSV CNS disease compared to PTV CNS disease. Pigs remain centrally aware and paddling or vestibular syndrome do not appear to be a consistent or primary feature.

Sample selection and collection

As with all disease investigations, selection of the appropriate animal(s) to sample is imperative but a challenge. A written, detailed case definition that includes a defined set of clinical signs and expected progression of disease to serve as reference and guide is helpful. Laterally recumbent pigs or pigs later in the course of disease progression may not provide the most appropriate sample for an etiologic diagnosis. Ambulatory pigs with ataxia and early clinical signs are preferred. Aseptic collection of proper CNS tissue is essential. Submission of brainstem and sections of cervical and thoracic spinal cord will markedly increase diagnostic sensitivity by decreasing false negatives as compared to submission of cerebrum alone.

Diagnostics

Gross lesions are not present in the CNS tissue of pigs affected by a viral poliomyelitis. Histologic lesions of a viral poliomyelitis include multifocal gliosis, neuronal degeneration and necrosis, satellitosis, and non-suppurative perivascular cuffs in the grey matter of the spinal cord and brainstem. The presence or absence of lesions in the cerebrum depends on the extent of lesions elsewhere. In cases with more extensive lesions in the spinal cord and brainstem, Virchow-Robin spaces in some sections of evaluated cerebrum are occasionally expanded by inflammation. Similar lesions are seen in PTV and PAstV-3. Currently, the ISU VDL uses a PSV and PTV R&D qRT-PCR to confirm the presence of viral genetic material.

Diagnostic result interpretation

Viral poliomyelitis and concurrent detection of PSV and PTV in CNS samples is likely a result of fecal contamination rather than a dual infection. This is not an uncommon finding in cases submitted to the ISU VDL due to the sample collection process within farm settings. In these cases, an additional submission to the diagnostic laboratory of aseptically collected brainstem and spinal cord or sections of the vertebral column containing cervical and thoracic spinal cord or intact pigs (liver preferred) are necessary for an etiologic diagnosis.

Important next steps

Basic research, field investigations, animal inoculation studies, and diagnostic assay development are needed to investigate the ecology, epidemiology and pathophysiology of PSV and to provide a basic foundational understanding of PSV in swine herds, including:

- prevalence of PSV in feces in the US swine herd on affected and non-affected farms
- seroprevalence of PSV in the US swine herd and seroprevalence of PSV in affected versus non-affected flows
- age distribution of fecal shedding and antibodies and the correlation between these and the development of PSV CNS disease
- antigenic and genomic differences between PSV fecal and CNS strains
- virulence variation and risk factors between PSV fecal versus CNS strains.

To begin to answer some of the unknowns, diagnostic assays and techniques need to be developed, including:

- Development of an ELISA in which the diagnostic sensitivity and specificity and analytical specificity (cross-reactivity) is determined using known-status samples following an animal inoculation study. This would allow for the investigation of seroprevalence and response to vaccination. As well as the impact of immunity on the development of disease.
- Use of an *in situ* assay such as RNAscope® ISH for PSV and PTV would provide a tool to utilize in cases with dual detection of PSV and PTV without the need for additional submissions and would provide a more accurate assessment of the number of atypical neurologic cases due to PSV and PTV.
- Investigation of the genetic and antigenic variability between strains of PSV. This includes development of a reliable virus isolation protocol for case material. Neither the genomic nor the amino acid differences that may impact neural tropism between PSV fecal strains and PSV CNS strains are available in the scientific literature. Understanding the difference, if any, between neurotropic PSV and fecal PSV strains could significantly impact disease diagnosis, prevention and mitigation. Furthermore, it may help explain why only a portion of animals develop viral poliomyelitis despite it being highly prevalent in the feces of pigs from an affected farm.

Despite the obvious clinical and economic impact, very little is known about the ecology, epidemiology and pathophysiology of PSV in the United States swine herd. Swine practitioners and producers have no information about prevention and mitigation protocols or access to prevention and mitigation tools. Once PSV is diagnosed in a herd, producers are left only to incur economic loss and risk a serious welfare issue.

References

1. Adams M et al. Ratification vote on taxonomic proposals to the International Committee on Taxonomy of Viruses. *Arch Virol* 2015;160:1837–1850.
2. Racaniello VR (2013). Picornaviridae. In Fields Virology, 6th edn, vol. 1 pp. 453–489. Edited by Knipe DM, Howley PM. Philadelphia, PA: Lippincott Williams & Wilkins.
3. Knowles NJ et al. Classification of porcine enteroviruses by antigenic analysis and cytopathic effects in tissue culture: Description of 3 new serotypes. *Arch Virol* 1979;62: 201–208.
4. Honda E et al. A serological comparison of 4 Japanese isolates of porcine enteroviruses with the international reference strains. *Nippon Juigaku Zasshi* 190;52:49–54.
5. Palmquist JM et al. Detection of porcine teschovirus and enterovirus type II by reverse transcription-polymerase chain reaction. *J Vet Diagn Invest* 2002;14:476–480.
6. Cano-Gomez C et al. Teschoviruses and sapeloviruses in faecal samples from wild boar in Spain. *Vet Microbiol* 2013;165:115–122.
7. Chen J et al. Complete genome sequence of a novel porcine Sapelovirus strain YC2011 isolated from piglets with diarrhea. *J Virol* 2012;86:10898.
8. Schock A et al. Investigation into an outbreak of encephalomyelitis caused by a neuroinvasive porcine sapelovirus in the United Kingdom. *Vet Microbiol* 2014;172: 381–389.
9. Arruda PHE et al. Detection of a novel sapelovirus in central nervous tissue of pigs with polioencephalomyelitis in the USA. *Transbound Emerg Dis.* 2017;64:311–315.

Porcine deltacoronavirus: Researcher's and practitioner's perspectives

Dick Hesse[1], MS, PhD; Katherine Wedel[2], DVM
[1]Kansas State University, Manhattan, Kansas; [2]Iowa Select Farms, Iowa Falls, Iowa

Researcher's perspective

A novel virus in the *Deltacoronavirus* genus has also been recently associated with swine. Porcine deltacoronavirus (PDCoV) was first detected in 2012 during a retrospective study conducted in Hong Kong using PCR surveillance to examine rectal swabs from animals including domestic pigs. The virus was not found to be associated with clinical disease in pigs until 2014, when it was detected on 5 Ohio farms experiencing outbreaks of diarrhea in sows and piglets. These outbreaks were unique in that the majority of affected animals had detectable PDCoV in the absence of other pathogens known to cause enteric diseases in swine. Co-infections with PEDV and PDCoV were detected in some of these affected herds, but not all. Since this initial report, PDCoV has been detected in many herds in the United States and Canada, with many PDCoV positive premises in the US co-infected with Few studies have reproduced clinical diarrheal disease secondary to experimental PDCoV infection in gnotobiotic and conventional pigs, but these studies have been limited in piglet numbers and study time course. The current study[1] demonstrates the clinical course of PDCoV infection in a rigorous study of sequential piglet sacrifice in which diarrheal disease is correlated with histologic lesions and intra-lesional viral antigen demonstrated using immunohistochemistry (IHC). The presence of virus by qRT-PCR over the time course of infection was also demonstrated in feces, nasal secretions, tissues, blood and serum, and oral fluids. Dams of the piglets were similarly characterized by qRT-PCR analysis of feces, nasal secretions, tissues, blood plasma and serum, and IHC. The results of this study demonstrate that PDCoV causes clinical enteric disease in a variety of ages of swine that is clinically and pathologically similar to other swine enteric coronaviruses.

Materials and methods

This study was conducted using 87 conventional piglets and their 9 dams, including aerosol and contact controls to emulate field conditions. At 2-4 days of age, sows and piglets were administered a PDCoV inoculum generated from a field sample testing negative for other enteric disease agents with assays including metagenomic sequencing. Serial samples of blood, serum, oral fluids, nasal and fecal swabs, and tissues from sequential necropsy were collected throughout the 42-day study for quantitative real time reverse-transcription polymerase chain reaction (qRT-PCR), histopathology, and immunohistochemistry (IHC).

Results and discussion

Diarrhea developed in all inoculated and contact control pigs, including sows, by 2 days post-inoculation (DPI) and aerosol control pigs by 3-4 DPI, with resolution occurring by 12 DPI. Atrophic enteritis was observed in the jejunum and ileum of affected piglets from 2 to 8 DPI, and corresponded to positive antigen detection by IHC. Mesenteric lymph node and small intestine were the primary sites of antigen detection by IHC and tissue qRT-PCR, and most piglets and sows were qRT-PCR positive in these tissue samples out to the end of the study. The inoculated piglets were qRT-PCR positive in fecal and nasal swabs to 21 DPI. This study demonstrates the time course of pathogenicity and virus kinetics of PDCoV in conventional neonatal and adult swine.

Acknowledgements

This study was supported by the National Pork Board (#14-182). This work was supported in part by funds through USDA Animal Health and Disease Research Program under the provisions of Section 1433 of Subtitle E, Title XIV of Public Law 95-113 and the Kansas State Veterinary Diagnostic Laboratory.

References

1. Vitosh-Sillman S, Loy JD, Brodersen B, et al Experimental infection of conventional nursing pigs and their dams with porcine deltacoronavirus. *J Vet Diagn Invest*, 2016; 28:486-497.

Practitioner's perspective

Porcine deltacoronavirus can have a significant impact both in a sow farm and in nursery and grow-finish pigs. The production impact of PDCoV is generally less than that of Porcine Epidemic Diarrhea Virus (PEDV), however, many of the considerations and action plans for PDCoV are very similar to those of PEDV.

Clinical signs and mitigating risk

In a breed-to-wean farm, PDCoV clinical signs may start in sows or piglets. In the case of one recent break within the Iowa Select Farms system, clinical signs began in recently weaned sows. On the morning that clinical signs were reported, an estimated 5% of sows had diarrhea and 20% of sows were off-feed. While diarrhea was first noted in wean sows, clinical signs quickly became scattered throughout all gestation barns. Three days after the initial clinical signs, 15-20% of sows in gestation had diarrhea. Young pigs had an increase in scours, but wean pigs were not scouring at that time. Clinical signs may vary depending on where the infection first occurs, which population of animals are affected first, and the layout of the farm. Fecal samples or fecal swabs are easy to collect and very sensitive for PDCoV testing.

When a PDCoV break is suspected, it is crucial to begin actions to mitigate risks to other farms within a system. The following areas should be considered to create an action plan and investigate possible areas of disease introduction:

1) Transport: Investigate recent gilt introductions and cull loads. Have a plan in place for how trailers will be washed to mitigate risk to other farms. Additionally, ensure that gilt sources have not shown any clinical signs of disease. Expect that wean pigs from a PDCoV farm will show clinical signs of disease, so a plan should be created to care for these pigs.
2) Facilities / maintenance: Examine visitor and supply entry records for possible areas of disease entrance. It is important to alert maintenance staff and put measures in place to ensure that supplies and tools are not taken from the affected farm to another site.
3) Semen delivery: If possible, it may be best to have an employee meet the semen driver on the road to reduce the risk of having PDCoV tracked back to the boar stud.
4) Outside supplies or contractors: Consider any other deliveries or services to the farm that may need to be re-routed. For instance, drinking water deliveries or pit pumpers may need to be changed to reduce risk to other farms.
5) Mortality removal: If rendering is used for mortality removal, it will be important to change the schedule of the rendering truck so as not to move virus from the affected farm to another site.

Exposure and follow-up actions

Upon diagnosis of a sow farm, ensuring exposure of all adult animals is critical in order to clean up the virus as quickly as possible. A herd closure of approximately 15-20 weeks should be part of this plan. Sow or piglet fecal material works well for exposure. Fecal material can be collected or piglet intestines can be harvested; this exposure material can be mixed with distilled water in order to achieve the necessary volume to expose the farm. Straining the exposure homogenate through a strainer and cheese cloth is advised to ensure that there are no large particles. A sample of the exposure homogenate can be saved for testing via PCR to ensure viral presence in the exposure material. Backpack sprayers and pressurized garden sprayers work very well for ensuring that each adult animal is exposed on the mucosal surface of their snout or mouth. Thorough straining of the exposure homogenate is critical if using a sprayer for exposure. Water medicators can also be used for exposure, but may be more successful in a pen scenario where sows are more likely to have fecal-oral exposure to the virus. Reload Pack DT® dye and sodium thiosulfate (Boehringer Ingelheim, St. Joseph, Missouri) is a great tool for visualizing which animals have been exposed or for ensuring that the exposure material in a medicator is reaching the end of the water lines, in addition to its utility in buffering the water.

Homogenized exposure is necessary to allow all adult animals on the farm to build immunity to the virus. Approximately two weeks after the exposure, action items can be put in place to clean up the virus. Once the adult animals develop immunity, piglets become the most likely to become infected and shed the virus, so the majority of focus is placed in the farrowing house.

1) Discontinue piglet and sow scour treatments. The majority of diarrhea will be due to PDCoV during this time, and efforts should be made to reduce litter to litter transmission by treating scours.
2) Discontinue piglet movements, hot-boxing or split-suckling, and backfilling sows. If replacement of a sow becomes necessary, the replacement sow should be moved from a room with younger pigs into a room of older piglets.
3) Needles, gloves, and scalpels should be changed between litters at all major piglet handling timepoints, such as processing. Syringes, processing totes, and other equipment should be kept clean. Processing (iron injections, tail docking, and castration) should be done for an entire room at a time. For Iowa Select Farms, the protocol is to wait until the oldest litter is 7 days of age, then process the entire room to eliminate going into an out of rooms on multiple days for processing.
4) Consider discontinuing any injections or vaccinations at weaning and weaning at a younger age. Moving them to the nursery or wean-to-finish barn decreases the risk of viral transmission at the sow farm and may contribute to a quicker clean-up process.
5) It is important to be conscientious of piglet euthanasia to ensure that pigs without a prospect for recovery are not suffering.
6) Farrowing rooms should be all-in, all-out. Thorough inspection of farrowing rooms after washing is necessary to ensure that all fecal material has been removed from surfaces. White-washing should be considered as an additional method to reduce viral load in a farrowing room. Rooms should have one night of downtime prior to loading new sows.

7) Boot washes should be installed in all farrowing rooms, and people movements should be directed from younger pigs to older pigs.
8) Clothing changes after tasks that require extensive pig handling or manure contamination, such as weaning, processing, or power-washing can reduce the risk of viral transmission.
9) Wash and disinfect hallways immediately after animal movements, including weaning and loading rooms.
10) Wash boots when moving between farrowing and gestation.
All of these measures increase internal biosecurity on a farm and reduce the risk of transmission of the viral to other piglets.

Testing

Testing is an important part of ensuring the farm is cleaning up and that gilt entry will be successful without causing clinical signs in naïve animals. Iowa Select Farms elected to start testing four weeks after the whole-herd PDCoV exposure. Swiffer® cloths (Procter & Gamble, Cincinnati, Ohio) have been demonstrated as an effective method for environmental testing for PDCoV. The Iowa Select Farms testing protocol is as follows: Use a Swiffer moistened with phosphate buffered saline (PBS) solution to wipe about a 1 square foot area in five different farrowing stalls of pigs 10+ days of age that are scouring. Place the Swiffer® cloth into PBS solution, wring out the fluid, and place in a tube. Each Swiffer® cloth can be used for five stalls for a total of 30 different litters sampled (six samples at each timepoint).
Samples were collected every two weeks. After three consecutive negative tests, the farm was considered negative.

Clinical impact

Pre-weaning mortality was elevated at affected farms for approximately three weeks after diagnosis and exposure of PDCoV (Figure 1). Approximate time to negative, characterized as three consecutive negative tests at two week intervals, was approximately 12 weeks for affected farms. PDCoV is generally less severe than PEDV, but can still have a significant production impact.

Figure 1: Impact of PDCoV on pre-wean mortality % in a sow farm. Week 0 represents the week of a positive diagnosis of PDCoV.

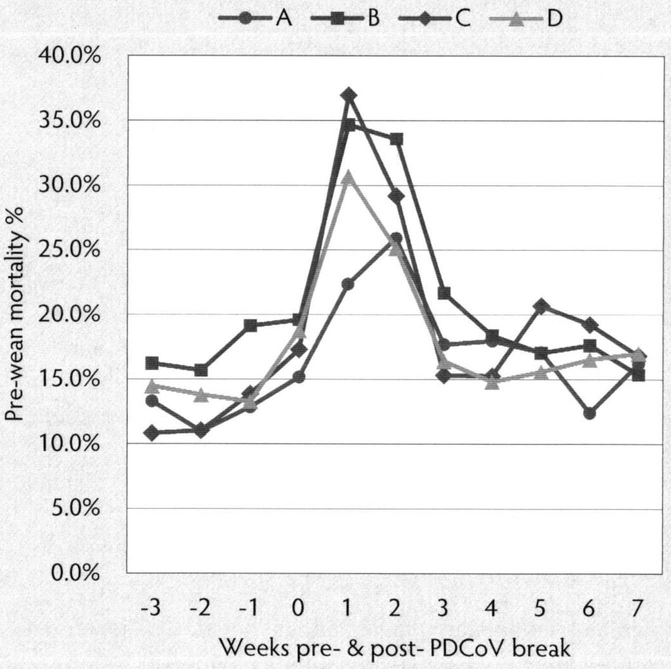

Seneca Valley virus A, a practitioner's perspective

Dr. Laura Bruner
Swine Vet Center

Clinical presentation

Seneca Valley virus (SVV or SVA) is not a new virus to the US. During the summer and early fall of 2015 SVV was causing severe clinical signs in naïve herds across the Midwest. These early outbreaks of SVV caused concern for practitioners and producers, who were inexperienced with disease shedding and eradication procedures.

On August 29, 2015 an operator of a 1000 sow breed to wean farm notified me that gilts had become severely lame. A Minnesota Board of Animal Health visit was set up immediately. On that visit 80% of sows were found to have snout lesions. Many of these animals were slow to rise. Foot lesions were identified primarily on the bottom of pad and were most noticeable when sows laid down and the bottoms of their feet were exposed. Piglet mortality increased by 10% for 2-weeks and quickly came back down to normal (Figure 1). There were no scours noted as reported in other SVV breaks around the same time. Stillborns increased by 14% for one week. Stillborn rate fell back to herd average as gestating sows recovered. (Figure 2).

When deciding on the approach for a SVV break there are two possible options

1. Do nothing and allow virus to move through herd
2. Load/expose/close method.

Do nothing approach

The first option would be to allow natural circulation of virus through the herd without an organized closure and eradication. In this scenario, naïve gilts routinely enter the farm while the current population continue to shed the virus. New gilts entering the farm get exposed to the virus, develop varying degrees of clinical signs, clear virus and generate immunity. The thought being that over time the virus would eventually die out.

Load/close/expose approach

The second option is the load/expose/close approach. This option was chosen by the farm. A system of feedback was set up in order to make sure all animals were exposed. Sows were given feedback consisting of manure from gilts that were recently infected. Every week, manure from animals that were given feedback the prior week were collected and fed back to the next group. This continued for 6 weeks. SVV IFA and SVV ELISA were utilized to look at titers and confirm infection. No new gilts were added to the farm until May 2015, which was 4 months from the last feedback given.

We know through recent research that the sows can shed SVV virus for up to 6 weeks post infection.[1] There are field reports of successful eradication of SVV with 3 to 4 month closures. This conservative approach of a 5 month closure was successful in eradicating SVV from the sow farm. In May of 2015, these SVV exposed animals moved to a naïve farm and no clinical signs were observed in the naïve herd nor were any antibody titers noted post introduction of previously exposed SVV animals.

References

1. Tousignant, Bruner, Schwarz, Vannucci, Rossow, Marthaler, Longitudinal study of Senecavirus A shedding in sows and piglets on a single United States farm during an outbreak of vesicular disease. BMC Veterinary Research (2017) 13:277.

Figure 1: PWM

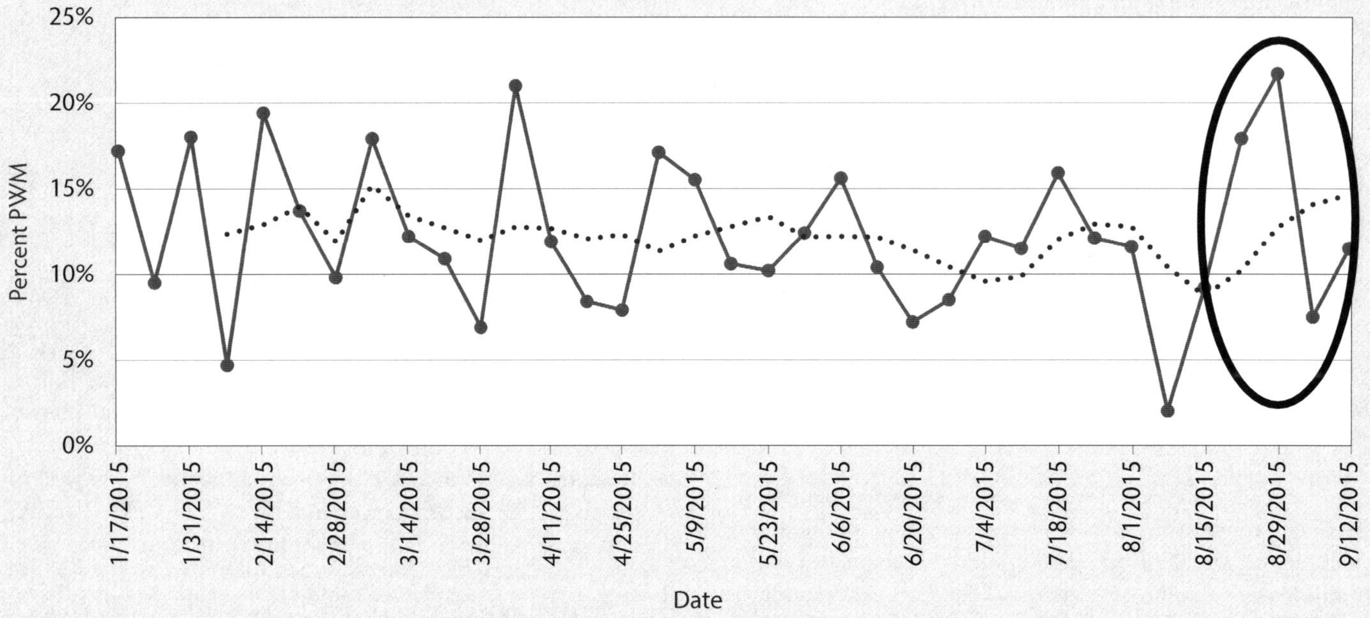

Figure 2: Stillborns and mummies

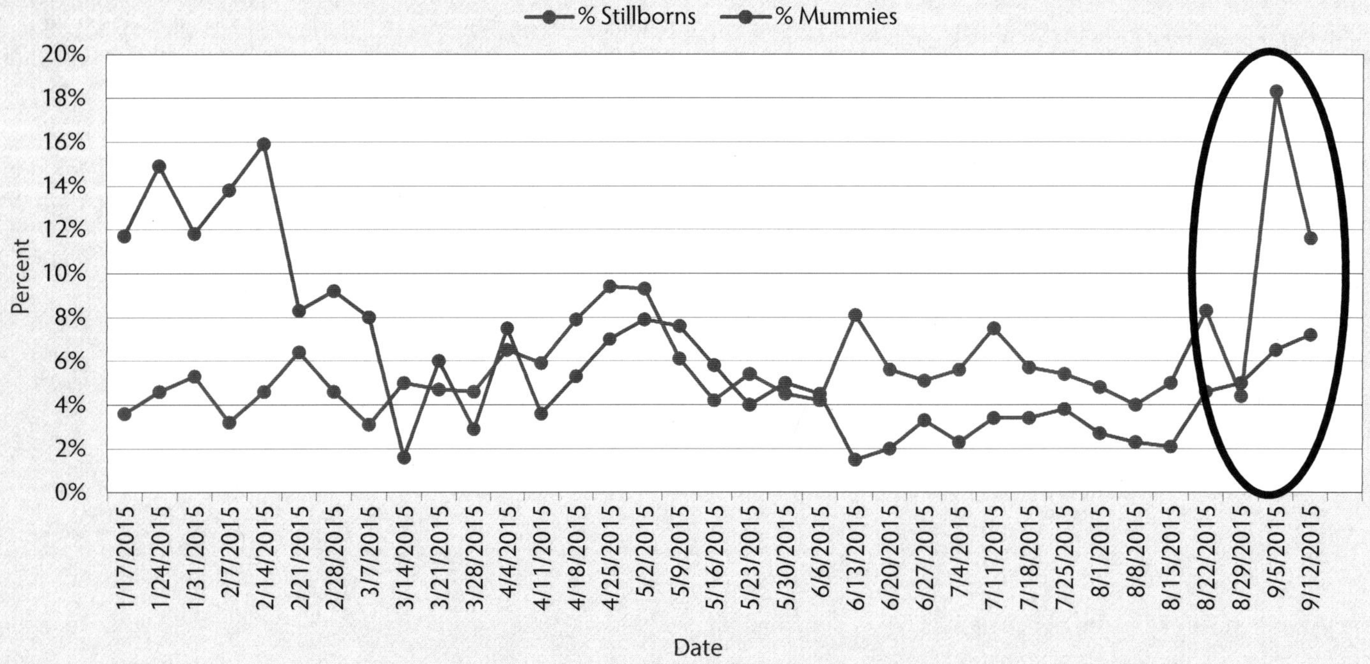

Senecavirus A: Pathologist/researcher perspective

Fabio A. Vannucci, DVM, MSc, PhD
University of Minnesota, St. Paul, Minnesota

Introduction

The swine industry has experienced an emerging concern associated with vesicular disease and transient neonatal mortalities caused by Senecavirus A (SVA). This vesicular disease is clinically indistinguishable from high-consequence foreign animal diseases (FAD), including foot-and-mouth disease virus (FMDV), vesicular stomatitis (VS), vesicular exanthema of swine (VES) and swine vesicular disease (SVD).[1-4]

SVA was originally described in 2002 as a contaminant while cultivating adenovirus-5-based vectors in the cell line PER.C6.[5] It was named Seneca Valley virus 001 (SVV-001) based on the Seneca Creek State Park (Maryland, USA), near the laboratories that first characterized the virus (Neotropix, Inc., Malvern, PA, USA). The virus was speculated to be a contaminant derived from fetal bovine serum or porcine trypsin.

Isolation of SVA in pigs has been reported since 1988.[2] While only sporadic cases had been reported between 1988 and 2014, more than 250 SVA cases have been confirmed in the last two years. Natural reservoirs or potential sources for the emerging epidemic scenario observed since July 2015 remain unknown. Furthermore, basic understanding of the pathogenesis of the SVA infections as well as transmission dynamics is still to be elucidated. The objective of this brief review is to describe and discuss the current impact of SVA on the pork industry, the initial advances on understanding the basic aspects of SVA infections, and the prospective scenarios focusing research priorities to mitigate the economic burden of SVA.

Emerging disruption in the pork supply chain

Approximately in November of 2014, SVA started being detected by PCR in cases of neonatal mortality and vesicular disease in Brazil, and it spread rapidly throughout large swine production systems within the country.[1,6,7] Clinically affected animals developed cutaneous vesicular lesions mainly on the snout and coronary bands.[8] In addition, high mortality in piglets in the first week of life (referred to as Epidemic Transient Neonatal Losses) was associated with high loads of SVA in tissues.[1] This emerging epidemic scenario observed in South America differed from the historical data recorded in the US. The National Veterinary Services Laboratories has documented approximately 20 SVA isolates in 24 years (1988-2012) from different case submissions, including idiopathic vesicular disease, lameness and neonatal mortality. The reports were usually derived from sporadic cases affecting an individual animal.[9] One exception resembling the epidemic scenario observed in South America was reported in June of 2007 in Canadian market hogs that arrived at a harvest facility in Minnesota from Manitoba.[10] Although SVA was identified in affected animals, the case was reported as idiopathic vesicular disease, due to the lack of consistent evidence of SVA as a causative agent of vesicular disease.

In July 2015, commercial swine herds in large integrated systems in the US started experiencing a similar epidemic scenario of vesicular disease and neonatal mortality associated with detection of SVA by PCR.[11] Thereafter, the Morrison Swine Health Monitoring Program (MSHMP) at the University of Minnesota started recording the weekly incidence of cases.[12] SVA cases have spread across several states in the US ranging from major swine producing states in the Midwest to the Hawaiian Islands. From July 2015 through November 2017 more than 240 cases in the US have been recorded (Fig. 1). Cases of SVA have also been reported in Canada, China and Colombia [13-15].

Figure 1 shows a seasonal pattern in 2015 and 2016 with high incidence of cases concentrated in late summer and early fall. However, this seasonal curve has not been clearly reproduced in 2017, when the cases were regularly distributed throughout the year. Interestingly, current data from Minnesota have demonstrated regional spreading to neighbor counties following the two first cases (orange counties) diagnosed in the area (Figure 2). The regional spreading may indicate the role of different viable modes of transmission including mechanical and/or biological vectors.

The similarities among SVA and foreign animal diseases have raised numerous false alarms, and significant disruptions in the pork supply chain, including: (i) temporary closure of pork processing plants until the diagnosis of foreign animal diseases has been ruled out, (ii) reallocation of resources for diagnostic investigation of FMD and other vesicular diseases, (iii) mortality of neonatal piglets and (iv) increased culling rates in affected sows. In addition, this scenario has drawn attention from the official authorities due to potential compliances, especially if the disease becomes endemic in swine herds.

Since the emergence of this epidemic picture associated with SVA infections, foreign animal disease investigations went from an unusual event to a routine procedure, draining limited resources previously allocated to rare events. This scenario has created important challenges where the need to protect the swine industry from catastrophic FAD events intersects with commerce.[16]

Initial research efforts and remaining questions to be resolved

After the characterization of the virus (SVV-001),[17] two independent studies reported failed attempts to experimentally reproduce the disease in pigs.[18,19] In 2006, Knowles et al described the presence of neutralizing SVA antibodies in swine, cattle and mice.[19]

Figure 1: National distribution by states and weekly incidence of SVA cases reported by the SHMP (Jul 2015 - Nov 2017). Sow: sow farm; NGF: finishing farm; UNK: unknown. EWMA: exponentially weighted moving average. Summer: red shaded. Fall: yellow shaded.

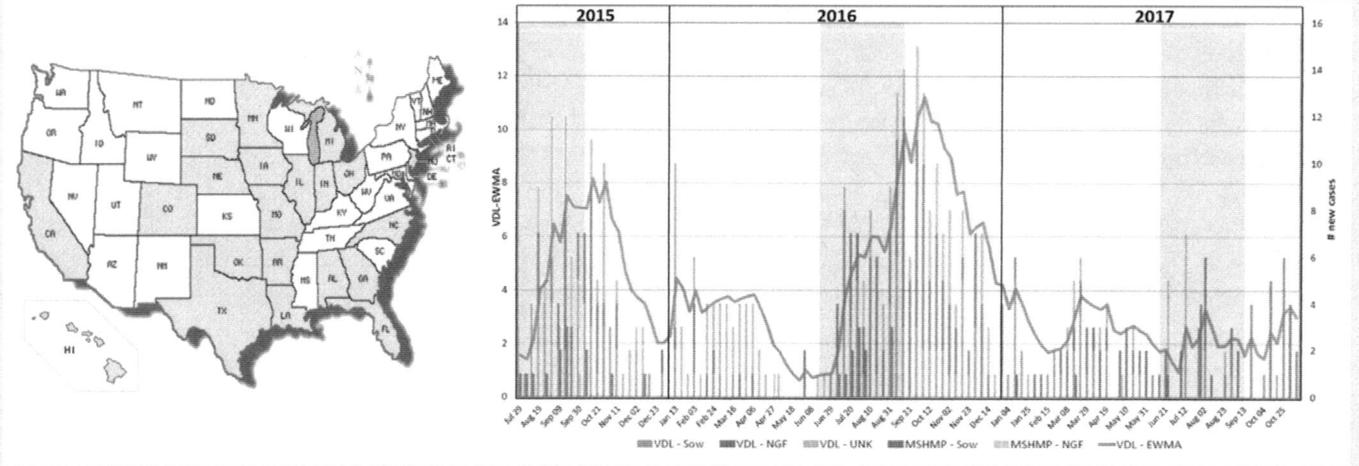

Figure 2: Regional spreading of SVA cases in Minnesota. Orange: First cases (Jul 2016). Yellow: Subsequent cases.

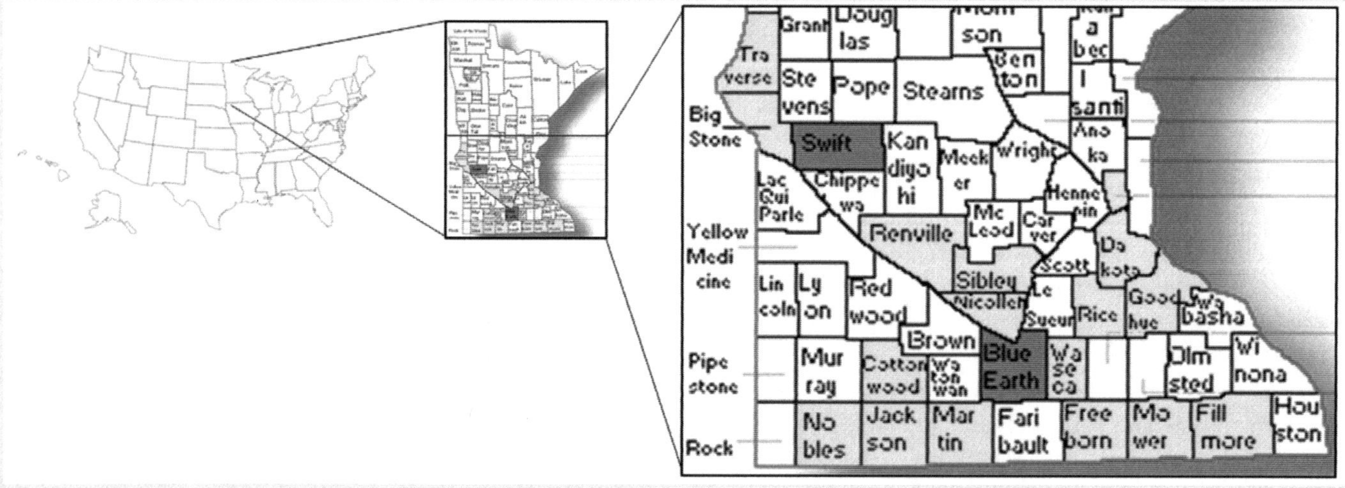

Although these findings highlight the potential ability of SVA to infect a broader range of host species, the clinical relevance of the virus, and its potential association with vesicular disease or any other specific clinical syndrome, was unclear at that time. This scenario led to minimal investments and advances on research in the veterinary field. In contrast, significant progress has been reported in the last decade focusing on the use of the strain SVV-001 for cancer treatment based on its oncolytic properties against small-cell lung cancer, neuroendocrine tumors and others. The published literature in the medical field has suggested a potential selective ability of SVV-001 to infect, replicate and cause cytotoxicity in cancer cells without significantly damaging normal tissues.[17,20]

The recent interest on studying SVA infections in the veterinary field started in 2015 with the sequencing of the complete genome of SVA from vesicular fluid and sera of sows affected with vesicular disease.[1] The association of SVA with vesicular disease became clearer through the demonstration of SVA within the lesions[21] (Figure 3). Notably, high levels of the virus were detected in various tissues without histological lesions from piglets affected by transient acute neonatal mortality (Figure 3). Subsequently, vesicular disease has been experimentally reproduced using a SVA strain isolated from recent outbreaks.[3,4] However, experimental reproduction of the neonatal mortality has not been reported so far.

After recognizing SVA as the causative agent of vesicular disease using contemporary strains, questions arose regarding genotypic differences that would reflect an increase of virulence of recent

Figure 3: A) Vesicular disease. SVA (red) replicating within intraepidermal vesicles (arrows). **B)** Neonatal mortality. SVA (red) replicating within the lamina propria of an intact intestinal villi, indicating the viremic nature of the infection.

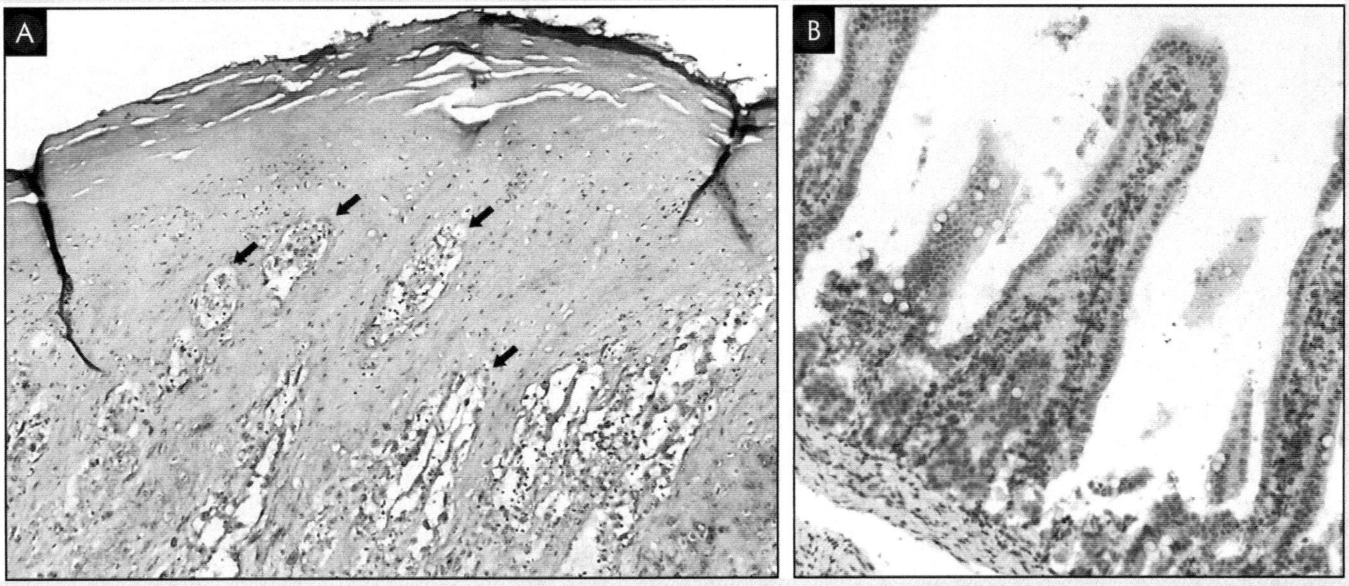

isolates compared with strains historically reported. Genetic divergences were mainly noted from historical isolates obtained prior to 2002. Based on the VP1 gene, 18 polymorphic sites and 7 residue shifts were observed comparing isolates prior to and after 2002 (Table 1). Amino acid changes on structural proteins (VP1, VP2 and VP3) on the surface of the viral capsid have been important mechanisms of immune evasion of picornaviruses in order to succeed as a pathogen over the years.[22] But these specific genomic changes and their potential impact on cross-protection among different SVA strains have not been characterized.

The dynamics of the vesicular disease in experimentally-infected animals is characterized by a short incubation period (2-5 days) followed by an acute and self-limiting clinical disease. Transient and short-term viremia is also observed from 1 to 10 days after exposure. Antibody responses have been detected as early as five days post-infection.[18] Seroprevalence of 80% has been reported in an affected sow farm at 180 days post-clinical signs.[15] Shedding of the virus has been detected for approximately 28 days post-infection in feces, saliva and nasal secretion.[3] Intriguingly, persistence of SVA in tonsil tissues was observed 38 days after experimental inoculation in finishing pigs, and 91 days post-clinical signs in boars naturally-exposed to the virus (Figure 4). A diagnostic investigation in naturally-infected boars also revealed persistent shedding of SVA in the semen associated with the presence of the virus and inflammatory lesions (orchitis) in the testes (Figure 4). Although asymptomatic carrier animals in picornavirus infections were first established for FMDv in 1959 by Van Bekkum, when the virus was recovered from infected animals more than 28 days post infection,[23] the definition of carrier state and its impact on transmission are still debatable.[24] Based on the clinical and relative genetic similarities between SVA and FMDV (family Picornaviridae), potential reservoir species for SVA may represent a constant threat as a source of new outbreaks in swine herds.

Monitoring programs have successfully documented SVA cases at the national level. However, little is known about the epidemiology and the transmission dynamics of SVA within and between swine herds. The regional spreading observed above may indicate the role of different viable modes of transmission including insects, fomites and airborne.

Preliminary research has been conducted focusing on mitigating the potential regional spreading of SVA through the evaluation of disinfectants against SVA under laboratory conditions, and the survivability of the virus in different surfaces, including aluminum, stainless steel, rubber, cement, and plastic surfaces at 25°C and 4°C. The most satisfactory results were observed using household bleach at 1:20 dilution, which was able to inactivate ≥ 99.99% of the virus within 10 to 15 minutes on aluminum, rubber, and plastic.[25] Hydrogen peroxide also was efficacious in inactivating SVA.[26] Phenolic and quaternary ammonium disinfectants at the manufacturer's recommended concentration were not fully efficacious. However, whether these findings would impact the transmission and spread of SVA infection under field conditions is still to be elucidated.

The epidemic scenarios of acute vesicular disease due to SVA infections have been commonly reported in sow farms, finishing sites and packing plants at or after arrival of market pigs. The biosecurity standards among each of these sites are significantly different. While we have observed SVA outbreaks in sow farms that have been negative for PRRSV and *M hyopneumoniae* for several

Figure 4: Persistently-infected boar, 91 days post-clinical signs. **A)** Tonsil. SVA (red) **B)** Testes. Orchitis, lymphocytic inflammatory infiltrate (arrow) and degeneration of seminiferous tubules (arrow heads). **C)** Testes. SVA (red) associated with histological lesions.

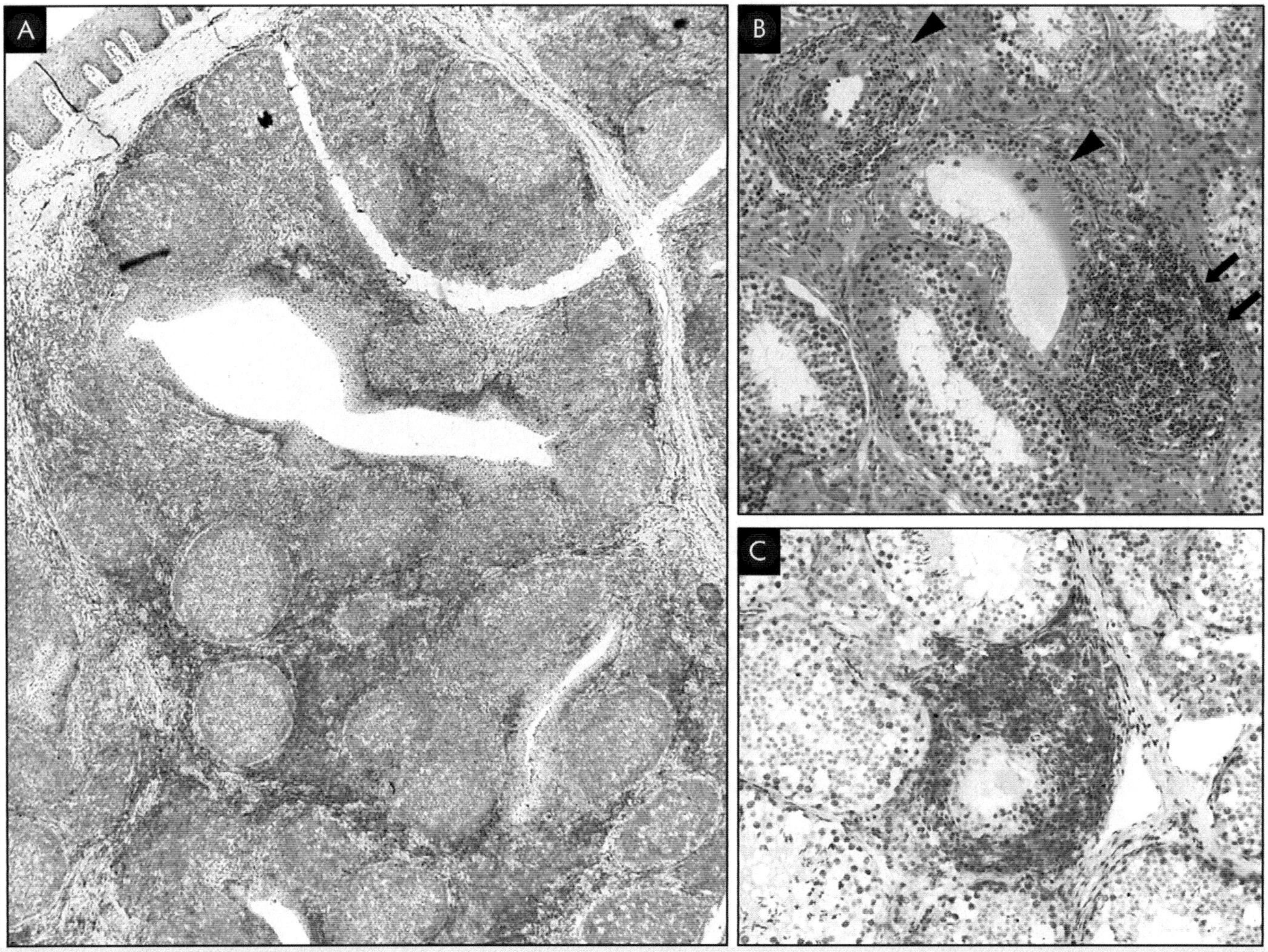

years, and have applied high biosecurity procedures, the disease also has been recurrent on finishing sites and packing plants with variable or minimal biosecurity standards.

Diagnostic investigations in sow farms and packing plants have demonstrated presence and persistence of the virus in the environment. In sow farms, SVA was detected in air samples up to two weeks after first observation of the clinical signs, and viable virus was recovered in environmental samples (farrowing and gestation crates) approximately eight weeks after the initial outbreak. In packing plants, the virus was identified in oral fluids from animals in waiting pens, and also from environmental samples collected from pen floors, railings and unloading docks. An interesting finding was isolation of SVA in environmental samples from pen railings, confirming the presence of viable and potentially infectious virus in the environment.

Final considerations

The current challenges associated with SVA infections have been primarily based on the disruptions in the pork supply chain due to the recurrent foreign animal investigations of vesicular lesions in finishing site and packing plants. Initial research efforts have been focused on defining SVA as the causative agent of vesicular disease. Although basic aspects of the pathogenesis of this infection, including those involved in the neonatal mortality, need to be explored, there is an avenue of opportunity for applied research focusing on identifying potentially viable modes of transmission, and whether biosecurity procedures currently established in the swine industry may help to prevent the spread of SVA among pig populations. Lastly, based on the nature of picornaviruses, and their ability to evade neutralizing antibody response through mutation on the viral capsid structure, it is critical to investigate the evolutionary aspects possibly associated

Table 1: Amino acid changes in 18 polymorphic sites from VP1 gene in historical (red) and contemporary (blue) isolates. (Adapted from Segales et al 2016).

	Amino Acid Location in VP1																	
Strain Name	10	57	62	63	65	93	94	97	133	155	161	172	197	210	221	232	239	250
USA/NorthCarolina/88_23626/ 1988/EU271759	V	H	L	.	V	.	.	.	.	.	.	.	.	V	.	A	.	N
USA/Minnesota/88_36695/1988/ EU271760	V	H	L	.	V	.	.	.	.	.	.	.	.	V	.	A	.	N
USA/Iowa/89_47752/1989/ EU271757	V	H	L	.	.	.	.	.	.	.	.	.	.	.	.	A	.	.
USA/NewJersey/90_10324/ 1989_12/EU271761	V	H	L	.	.	.	.	.	.	.	.	.	.	.	.	A	.	.
USA/Illinois/92_48963/1992_09/ EU271762	V	H	L	.	.	.	.	.	.	.	.	.	.	.	.	A	.	.
USA/Louisiana/1278/1997/ EU271763	.	.	.	.	.	.	.	.	N	.	.	.	.	V	.	.	.	.
CHN/CH-01-2015/KT321458	.	.	A	T	T	V	S	D	.	.	Y	.	.	.	I	.	V	.
CAN/Manitoba/11_55910_3/ 2011/KC667560	.	.	E	T	.	V	.	D	.	.	Y	T	A	.	.	.	V	.
USA/IA40380/2015/KT757280	.	.	A	T	.	V	.	D	.	.	Y	T	.	.	I	.	V	.
USA/IA46008/2015/KT757282	.	.	A	T	.	V	.	D	.	.	Y	T	.	.	I	.	V	.
USA/SD41901/2015/KT757281	.	.	A	T	.	V	.	D	.	.	Y	T	.	.	I	.	V	.
BRA/UEL-SVV-B2/15/KR075678	.	.	A	T	.	V	.	D	.	.	Y	T	–	–	–	–	–	–
BRA/UEL-SVV-A1/15/KR075677	.	.	A	T	.	V	.	D	.	.	Y	T	–	–	–	–	–	–
BRA/GO3/2015/KR063109	.	.	A	T	.	V	.	D	.	.	Y	T	.	.	I	.	V	.
BRA/MG1/2015/KR063107	.	.	A	T	.	V	.	D	.	.	Y	T	.	.	I	A	V	.
BRA/MG2/2015/KR063108	.	.	A	T	.	V	.	D	.	.	Y	T	.	.	I	A	V	.

with the recent emergence of clinical disease, since the virus has been isolated from swine in the US since 1988.

References

1. Vannucci FA, Linhares DC, Barcellos DE, Lam HC, Collins J, Marthaler D. dentification and Complete Genome of Seneca Valley Virus in Vesicular Fluid and Sera of Pigs Affected with Idiopathic Vesicular Disease, Brazil.

2. Segalés J, Barcellos D, Alfieri A, Burrough E, Marthaler D; Senecavirus A: An Emerging Pathogen Causing Vesicular Disease and Mortality in Pigs?; Vet Pathol. (2016) Jul 1.

3. Joshi L, Fernandes M, Clement T, Lawson S, Pillatzki A, Resende T, Vannucci F, Kutish G, Nelson E, Diel D. J. Pathogenesis of Senecavirus A in finishing pigs.Gen. Virol. (2016) 97(12):3267-3279.

4. Montiel N, Buckley A, Guo B, Kulshreshtha V, VanGeelen A, Hoang H, et al. Vesicular Disease in 9- Week-Old Pigs Experimentally Infected with Senecavirus A. Emerg Infect Dis. 2016; 22: 1246–1248.

5. Hales LM, Knowles NJ, Reddy PS, Xu L, Hay C, Hallenbeck PL. Complete genome sequence analysis of Seneca Valley virus-001, a novel oncolytic picornavirus. J Gen Virol. 2008 May;89:1265-75.

6. Leme RA, Zotti E, Alca^ntara BK, et al. Senecavirus A: an emerging vesicular infection in Brazilian pig herds. Transbound Emerg Dis. 2015;62(6):603–611.

7. Leme RA, Zotti E, Alcantara BK, Oliveira MV, Freitas LA, Alfieri AF, et al. Senecavirus A: An Emerging Vesicular Infection in Brazilian Pig Herds. Transbound Emerg Dis. 2015; 62: 603–611.

8. Leme RA, Oliveira TE, Alcantara BK, Headley SA, Alfieri AF, Yang M, et al. Clinical Manifestations of Senecavirus A Infection in Neonatal Pigs, Brazil, 2015. Emerg Infect Dis. 2016; 22: 1238–1241.

9. Singh K, Corner S, Clark SG, Scherba G, Fredrickson R. Seneca Valley virus and vesicular lesions in a pig with idiopathic vesicular disease. J Vet Sci Tech- nol. 2012;3(1):6.

10. Pasma T, Davidson S, Shaw SL. 2008. Idiopathic vesicular disease in swine in Manitoba. Can. Vet. J. 49:84–5.

11. Canning P, Canon A, Bates JL, Gerardy K, Linhares DC, Piñeyro PE, Schwartz KJ, Yoon KJ, Rademacher CJ, Holtkamp D, Karriker L. Neonatal Mortality, Vesicular Lesions and Lameness Associated with Senecavirus A in a US Sow Farm. Transbound Emerg Dis. 2016 Aug;63(4):373-8.

12. Morrison, B. Swine Health Monitoring Program, College of Veterinary Medicine, University of Minnesota, 2017.

13. Wu Q, Zhao X, Bai Y, Sun B, Xie Q, Ma J. The First Identification and Complete Genome of Senecavirus A Affecting Pig with Idiopathic Vesicular Disease in China. Transbound Emerg Dis. (2016) Aug 18.

14. Sun, D., Vannucci F., Corzo C., Marthaler, D. Emergence and whole genome sequence of Senecavirus A in Colombia. Transbound Emerg Dis 2017 (in press).

15. Goolia M, Vannucci F, Yang M, Patnayak D, Babink S, Nfon CK. (2016). Validation of a competitive ELISA and a virus neutralization test for the detection and confirmation of antibodies to Senecavirus A in swine sera. J Vet Diag Invest vol 29, Issue 2, pp. 250 – 253.

16. Recommendations from the Senecavirus A Working Group To Reduce Disruptions to Commerce in the Pork Industry, AASV,2017. https://www.aasv.org/documents/SVA_WG_FINAL_171012.pdf

17. Reddy PS, Burroughs KD, Hales LM, et al. Seneca Valley virus, a systemically deliverable oncolytic picornavirus, and the treatment of neuroendocrine can- cers. J Natl Cancer Inst. 2007;99(21):1623–1633.

18. Yang M, van Bruggen R, Xu W. Generation and diagnostic application of monoclonal antibodies against Seneca Valley virus. J Vet Diagn Invest. 2012 Jan;24(1):42-50.

19. Knowles NJ, Hales LM, Jones BH, et al. Epidemiology of Seneca Valley virus: identification and characterization of isolates from pigs in the United States. EUROPIC 2006: 14th Meet Euro Study Group Mol Biol Picornaviruses 2006; Abstract G2.

20. Burke MJ. Oncolytic Seneca Valley Virus: past perspectives and future directions. Oncolytic Virotherapy. 2016;5:81-89. doi:10.2147/OV.S96915.

21. Resende TP, Marthaler DG, Vannucci FA (2017) A novel RNA-based in situ hybridization to detect Seneca Valley virus in neonatal piglets and sows affected with vesicular disease. PLOS ONE 12(4): e0173190.

22. Mateu MG. Antibody recognition of picornaviruses and escape from neutralization: a structural view. Virus Res. 1995 Sep;38(1):1-24.

23. Van Bekkum, H.S. Frenkel, H.H.J. Frederiks, S. Frenkel, Observations on the carrier state of cattle exposed to foot-and-mouth disease virus, Bull. Int. Epizoot. 51 (1959) 917–922.

24. Bronsvoort BM, Handel IG, Nfon CK, Sørensen KJ, Malirat V, Bergmann I, TanyaVN, Morgan KL. Redefining the "carrier" state for foot-and-mouth disease from the dynamics of virus persistence in endemically affected cattle populations. Sci Rep. 2016 Jul 6;6:29059.

25.Singh A, Mor SK, Aboubakr H, et al. Efficacy of three disinfectants against Senecavirus A on five surfaces and at two temperatures. J Swine Health Prod. 2017;25(2):64–68

26.Hole K, Ahmadpour F, Krishnan J, Stansfield C, Copps J, Nfon C. Efficacy of accelerated hydrogen peroxide(®) disinfectant on foot-and-mouth disease virus, swine vesicular disease virus and Senecavirus A. J Appl Microbiol. 2017 Mar;122(3):634-639.

Regional IAV-S vaccination strategies for breeding herds

Clayton Johnson, DVM
Carthage Veterinary Service, Carthage, Illinois

Introduction

Influenza A viruses in swine (IAV-S) circulate widely in the United States annually, peaking in the late fall through the early spring. Although most breeding swine with IAV will recover without performance consequences, IAV can cause severe disease and in some cases death, particularly in immunocompromised and/or co-infected animals. Increased abortions, decreased conception rate and increased sow mortality are common sequelae to a novel IAV-S introduction. It can be difficult to started weaned pigs on sow farms that are IAV-S positive. In one study, the cost per head due to losses such as mortality, culls and tail-enders was $3.23.[1]

Recent surveillance in the Midwest demonstrated that 90.6% of rural farms were positive for IAV-S at least once, based on monthly testing for 12 to 24 consecutive months.[2] Introduction of replacement gilts and infection in suckling pigs played a key role in transmission of the virus, allowing it to frequently become endemic with only minimal clinical signs. Other studies show a cyclical herd-level prevalence pattern with levels increasing during fall, peaking in both early winter and late spring and decreasing in the summer.[3]

Intervention strategies to control influenza often focus on the use of pre-farrow sow vaccines[4-6] with the goal to piglet immunity through colostrum and reduce piglet clinical disease presentation.[7-12] However, an additional challenge to control influenza is the genetic diversity of strains circulating in pig populations. Given the performance and economic impacts of IAV-S, effective control measures are highly desirable to breed-to-wean producers. While IAV-S vaccines sometimes have disappointing efficacy in the field, under the right conditions vaccines can reduce or eliminate transmission of IAV-S in herds.[13, 14] Active surveillance of regional IAV-S viruses is critical to ensuring vaccination efficacy. Successful IAV-S vaccination can improve herd health and lower the risk of transmission to other species, including humans.

While there is a tremendous amount of skepticism toward the effectiveness of influenza A virus (IAV) vaccination of swine breeding herds, with appropriate evidence veterinarians are capable of improving immunocompetence against regionally dominant IAV-S strains. Selection of appropriate vaccine strains is critical and veterinarians have a growing set of tools at their disposal to support strain selection. This paper will review the tools our team uses to characterize regional IAV-S strains and determine which strains are most appropriate to include in a multivalent killed autogenous vaccination program.

Practice/system specific tools for characterizing regional IAV-S strains

Many diagnostic samples are acceptable for laboratories to detect influenza infection in pigs. Samples collected by producers and veterinarians are generally submitted to NAHLN or private laboratories.

Polymerase chain reaction

Polymerase Chain Reaction (PCR) can identify the presence of influenza viral RNA in respiratory samples. This assay provides results quickly, generally within 24 hours of submission. It is important to remember the detection of influenza viral RNA by PCR does not necessarily indicate detection of viable virus or on-going influenza viral replication. PCR is required to verify the sample contains IAV-S and is capable of contributing evidence toward a regional vaccine strategy. Lung tissue, nasal swabs, tissue swabs (small airways), oral fluids snout wipes and udder wipes have all been demonstrated as effective samples for IAV-S PCR testing. Once a sample has been verified as PCR positive subtyping the hemagglutinin and neuraminidase genes via PCR is commonly completed. Sequencing and virus isolation allow further characterization once subtype has been determined.

Sequencing

Influenza viruses are constantly evolving. Sequencing allows year-round surveillance of circulating IAV-S viruses to monitor changes to specific parts of the IAV-S genome that are antigenically and clinically relevant. Comparing the nucleotide sequence from one virus gene to a different virus gene gives us a picture of variation between the two viruses. Genes that affect the IAV-S surface protein structure are particularly important. Changes in nucleotides often result in changes of amino acids which often result in changes in the surface proteins. This can affect virus transmission, virulence and immunogenicity. IAV-S viruses have eight gene segments. Of the eight genes that make up an IAV-S virus, veterinarians focus sequencing efforts on one gene segments: the hemagglutinin (HA) gene. The HA genes determine the structure of a primary surface protein and has tremendous impacts to the immune system's ability to detect and neutralize IAV-S.

High throughput serum neutralization

In vivo, neutralizing antibodies are critical for viral clearance. High-throughput serum neutralization (HTSN) assays allow us to antigenically categorize IAV-S isolates. Uncategorized viruses are tested using a panel of antisera representing the H1 or H3

IAV-S subtypes and the results expressed as a serum neutralization ratio. Antisera should be generated against contemporary isolates representing circulating H1 & H2 SIV subtypes. When compared to traditional HA inhibition (HI) HTSN assays show good correlation. This tool allow the practicing veterinarian assess a strains ability to provide cross protection against other viruses within and across commonly circulating IAV-S clades.

Key amino acid sites

Single amino acid substitutions within the HA gene are likely responsible for antigenic differences between antigenic clusters.[15] This is critical information as the HA gene is the primary target of protective immune responses and the major target of vaccines. Identifying changes in these amino acids means a change in the HA protein is likely and as such, vaccines likely need to be updated. Specifically, H3 amino acid sites 145, 155, 156, 158, 159 and 189 should be monitored. H1 HA key amino acid sites are currently being identified and validated.

Industry tools for characterizing regional IAV-S strains

USDA Swine Health Monitoring Surveillance

USDA divides the Unites States into 5 Regions providing IAV-S subtyping and sequence identification by Clade for each distinct geographical area. The number of swine isolates in each clade is published for each region each quarter. Remarkable regional differences can be identified allowing the veterinarian to assess whether the isolates seen within the herds they service are common or unique within their geographical area.

University of Minnesota reference trees

Reference trees including isolates in currently available commercial IAV-S vaccines are available via the University of Minnesota. These reference documents allow comparison of your isolates to "shelf ready" vaccine opportunities. These reference trees are free to University customers who obtain their sequences via the University of Minnesota or for a fee for outside generated sequences.

Iowa State University FLUture

This interactive web-based tool was developed to provide diagnostic information from an Influenza A Virus database of test results, metadata, and sequences collected at the Iowa State University Veterinary Diagnostic Laboratory. Data is derived from diagnostic samples submitted from farms throughout the United States and North America. Regional incidence of IAV-S positive cases processed by the USDA/ISU-VDL can be viewed. Heat maps can be created that demonstrate the distribution of HA and neuraminidase subtype combinations over time. Pre-created, state-specific heat maps are available demonstrating the number of positive IAV-S submissions for each state, with the user being able to define the date range observed.

Discussion

Veterinarians have a large set of IAV-S characterization tools to support a regional IAV-S breeding herd vaccination strategy. Active surveillance of your breeding herds allows you to identify which herds are infected and what isolates are present. Sequencing of these isolates allows comparison across farms and to commercially available vaccine strains. Key Amino Acid comparisons provide a more in depth evaluation of H3 IAV-S subtypes to ensure that critical antigenic sites match between vaccine candidates and currently circulating field isolates. HTSN allows for an estimation of cross-protection across clades, helping veterinarians select isolates for autogenous vaccine production which will provide the broadest protection possible. Industry tools allow the veterinarian to observe national, regional and state IAV-S trends. These trends provide perspective on whether the IAV-S field isolates and/or regional vaccine strains are congruent with trends observed in other farms and may provide a preemptive warning signal that changes to regional vaccination strategies are required.

Active surveillance systems should remain in place after implementation of regional IAV-S breeding herd vaccination strategies to allow intervention evaluation. At a minimum, farms with IAV-S clinical signs should be evaluated via the characterization processes listed above. Efficacy can be objectively evaluated by monitoring Sow Mortality, Conception Rate, Pre-Wean Mortality and Abortion rates in and around the time of IAV-S activity. Updates should be made to regional IAV-S breeding herd vaccination strategies based on routine program evaluation.

References

1. Haden C, et al. Assessing production parameters and economic impact of swine influenza, PRRS and Mycoplasma hyopneumoniae on finishing pigs in a large production system. . In: The 2012 proceedings of the annual meeting of the American Association of Swine Veterinarians. 75-76.

2. Connor J, et al. Management and challenges of dealing with swine influenza. In: The 2013 proceedings of the Allen D. Leman Swine Conference.

3. Orlando Chamba Pardo, Fabian et al. Influenza Herd-Level Prevalence and Seasonality in Breed-to-Wean Pig Farms in the Midwestern United States. Frontiers in Veterinary Science. Oct 2017. Volume 4. Article 167.

4. White LA, et al. Influenza A virus in swine breeding herds: combination of vaccination and biosecurity practices can reduce likelihood of endemic piglet reservoir. Prev Vet Med (2017) 138:55–69.

5. Reynolds JJ, et al. Mathematical modeling of influenzaA virus dynamics within swine farms and the effects of vaccination. PLoSOne (2014) 9(8):e106177.

6. USDA. Swine 2012. Part II: Reference of Swine Health and Health Management in the United States, 2012. Fort Collins, CO: USDA–APHIS–VS–CEAH–NAHMS (2017). Available from: https://www.aphis.usda.gov/aphis/ourfocus/animalhealth/monitoring-and-surveillance/nahms/nahms_swine_studies.

7. Cador C, et al. Maternally-derived antibodies do not prevent transmission of swine influenza A virus between pigs. Vet Res (2016) 47(1):86.

8. Choi YK, et al. Evaluation of transmission of swine influenzatype A subtype H1N2 virus in seropositive pigs. Am J Vet Res (2004) 65(3):303–6.

9. Loeffen WL, et al. Effect of maternally derived antibodies on the clinical signs and immune response in pigs after primary and secondary infection with an influenza H1N1 virus. Vet Immunol Immunopathol (2003) 92(1–2):23–35.

10. Sandbulte MR, et al. Divergent immune responses and disease outcomes in piglets immunized with inactivated and attenuated H3N2 swine influenza vaccines in the presence of maternally-derived antibodies. Virology (2014) 46(4–465):45–54.

11. Renshaw HW. Influence of antibody-mediated immune suppression on clinical, viral, and immune responses to swine influenza infection. Am J Vet Res (1975) 36(1):5–13.

12. Kitikoon P, et al. The immune response and maternal antibody interference to a heterologous H1N1 swine influenza virus infection following vaccination. Vet Immunol Immunopathol (2006) 112(3–4):117–28.

13. Corzo, C.A., et al. Observations Regarding Influenza A Virus Shedding in a Swine Breeding Farm After Mass Vaccination. J. Swine Health Prod. 2012, 20, 283–289.

14. Romagosa, A., et al. Vaccination of influenza a virus decreases transmission rates in pigs. Vet. Res. 2011, doi:10.1186/1297-9716-42-120.

15. Lewis NS, et al. Substitutions near the hemagglutinin Receptor-Binding Site Determine the Antigenic Evolution of Influenza A H3N2 Viruses in U.S. Swine. Journal of Virology. 2014 May, 88(9): 4752-4763.

Systems approach to PRRSV management

Hans Rotto, DVM, MS; Peter Schneider, DVM, MS; Jessica Arnold, DVM; Caleb Robb, DVM
Innovative Agricultural Solutions LLC, Ames, Iowa

Introduction

Porcine reproductive and respiratory syndrome virus (PRRSV) continues to plague the swine industry and cause swine producer's financial loss. Major challenges to PRRSV management include the increasing difficulty in maintaining PRRSV control in areas of high pig density and the growing diversity of PRRSV within, and between, production systems. One approach to managing a herd challenged by PRRSV instability has six steps: 1) Uniform immunity of the gilts and sows, 2) Gilt acclimation and immunity management, 3) Delayed herd closure, 4) Farrowing house management to prevent virus propagation, 5) Downstream flow monitoring, and 6) Wean to finish production strategies. The following discussion is a brief outline of practice strategies utilized to minimize production and financial losses related to PRRSV in swine practice.

Discussion

Uniform immunity of the gilts and sows

The main goal of PRRS control is to eliminate wild-type PRRSV replication in the sow herd. Tools for stabilization may include live virus inoculation (LVI), modified live PRRSV vaccine (MLV), and autogenous killed PRRSV vaccine. If the herd has had a recent introduction of a new PRRSV, LVI is used to create uniform immunity. Virus is isolated and the process of getting it into a killed autogenous vaccine is initiated. Once stabilized, farms may use the multivalent autogenous PRRSV vaccine as a whole farm vaccination two to four times a year. Some herds will use a rotation of MLV vaccination with killed vaccination. The ongoing use of vaccine in breeding females in the high-risk areas has proven to provide a more predictable pig flow through marketing. The use of both MLV and multivalent killed vaccines have been beneficial to herds by promoting a high level of immunity across the sow herd while providing specific virus protection. It is unfortunate that some producers believe that the use of MLV or killed autogenous PRRSV vaccines are mutually exclusive decisions, when in combination they have been useful.

Gilt acclimation and immunity management

The next step is the process of preparing gilts to move into the farms with as much homologous immunity as possible. It is common to assume that live virus inoculation (LVI) is the only method to achieve this goal. LVI can be used effectively if controlled and health assurance diagnostic testing is used to monitor the process. Strict access control and inventory management of any virus used for LVI is essential. Testing of each group of gilts to confirm they are positive with the expected virus is necessary. This is done by sequencing viruses, managing a dendrogram for each sow farm and comparing sequences. The virus used for LVI is often the last virus that was laterally introduced to the herd.

Some producers are not comfortable with the process of LVI. To achieve homologous protection, killed PRRSV vaccine containing herd specific viruses has been used. Prior to administering this multivalent autogenous killed PRRSV vaccine, it is common to use MLV to activate the immune system of the gilts. Two doses of killed autogenous vaccine are then given as part of the gilt acclimation vaccination schedule.

The most successful process in gilt acclimation has been the use of both LVI and the multivalent autogenous killed vaccine. The use of all tools in gilt acclimation has proven helpful to minimize the change of resident viruses into immunologically different viruses causing instability or re-breaks.

Delayed herd closure

If the herd has had a lateral virus introduction and has recently used LVI to stabilize the breeding and gestation herd, a herd closure often helps complete the process. In the first twelve weeks after LVI, there is a lot of virus in the sows, gilts and in the farrowing house. Prior to arriving to the sow herd, all gilts should undergo the acclimation and immune management strategy for PRRSV and other resident pathogens. In the first twelve weeks after LVI, these gilts are relatively close in immune status to the sow herd so can be added on normal schedule with little expected impact on farm stability. For some herds, as breeding and gestation has returned to normal production, closing or delaying inputs of gilts has been an effective step in stabilizing the herd. Eight to ten weeks of closure, beginning approximately twelve weeks post LVI, can help manage the completion of sow herd stability by decreasing pathogens that get stirred up as new arrivals come into the herd.

Farrowing house management to prevent virus propagation

It has been remarkable how many herds have reached normal production but are still leaking positive pigs in the early grow out phase. A complete review of farrowing house processes, like piglet movement, employee movements, work patterns (ex. from weaning to processing pigs with no clean up) and sanitation of supplies and equipment, is important to close the deal on getting PRRSV off the farm and providing more stable downstream production. There have been many herds where the goal of reaching negative production was delayed by the unintentional recirculation of virus in the farrowing house. Educating employees and auditing this process are important to achieving stable pig flow.

Downstream flow monitoring for stability

New and innovative tools help veterinarians and producers answer questions about virus circulation in sow herds. Diagnostic monitoring of processing fluids is helpful to determine the sow herd stability, and family oral fluids increase diagnostics sensitivity in late lactation litters. The weaned pig and downstream production remains the report card of gilt acclimation and the ongoing sow farm stabilization. Monitoring pigs post weaning is useful to see if the practices in the farrowing house have pushed virus off the farm. Diagnosing timing of seroconversion and sequencing the virus(es) provides a better understanding of what viruses are active in the herd and influences the decision on what steps toward stabilization may be needed.

Wean to finish production strategies

Once weaned pigs remain negative for PRRSV through the first 7 weeks post weaning, there is a discussion on the potential value of MLV vaccination to growing pigs. The pig density and diversity of the location of the growing pigs is often the biggest part of this decision. Today we still place many non-vaccinated pigs throughout Iowa with success. The decision to vaccinate or not is driven by the frequency and severity of lateral PRRSV infection in the area. "Does the benefit from vaccination of groups which are subsequently infected by PRRSV justify vaccination of all groups placed in the area?"

Prevention is worth it

Biosecurity to prevent the introduction of new viruses into farms is still the best management of PRRSV. Filtration of farms has highlighted this importance by reducing aerosol transmission. All other potential routes of virus introduction must be mitigated or the efforts are for naught. We have the tools and knowledge to stabilize and eliminate viruses in herds. Preventing new viruses from getting into farms simplifies and sustains the whole herd productivity.

Summary

Some herds experience more risk of new introductions of PRRSV due to pig density and diversity in their region. The six points outlined are guidelines that have been used to minimize the production and financial losses associated with PRRSV. Many of these points need to be catered to the specific flow and facilities and there can be constraints that guide which tools are best for that farm.

Acknowledgments

Thank you to the producers who allow us to help. Thanks to my colleagues in practice.

Perennial herd closure

E. Spronk, DVM
Eichelberger Farms, Wayland, Iowa

Introduction

Swine diseases, such as porcine reproductive and respiratory syndrome virus (PRRSV), porcine epidemic diarrhea virus (PEDV), and porcine delta corona virus (PDCoV), are more prevalent during the fall and winter months.[1] These diseases have great economic impact not only to the individual sow farm but also to downstream production.[2] Due to the financial implications of these disease challenges, many companies have increased biosecurity practices during the higher risk months. These practices range from adding simple biosecurity practices like wearing boot covers to higher cost practices like seasonal filtration of air entering the farm. Eichelberger Farms has taken a different approach to controlling the seasonal challenges by implementing perennial herd closures on our Iowa farms in swine dense areas.

Material and methods

Eichelberger Farms has several sow farms located the swine dense region of Southeast Iowa, which has historically resulted in high incidence of disease challenges. Due to increased focus over the last several years, the system has been able to improve several aspects of biosecurity. However, despite increased focus on disease prevention, their history of wild-type PRRSV strains infecting several sow farms during the winter was still a frequent occurrence. Management strived to further reduce frequency of these breaks and control internal disease movement by implementing strategies used during PEDV and PRRSV eliminations. This included several different items: 1) seasonal closures during peak disease challenge months 2) strict McREBEL year round.

Timing of initial PRRSV challenges, sourced from historical system data, were used to determine the timeline of the seasonal closure. This data indicated that PRRSV pressure in the area increases shortly after crops were harvested and manure hauling season began. This means the closures usually begin around the middle of October. The duration of the closures vary due to capacity of each specific farm, farrowing rate and herd parity distribution. To maximize the length of time farms are able to close we modify our culling protocols starting in late summer to allow farms to cull older parity animals that would farrow during the closure and potentially require replacement with a gilt. Encouraging an increased replacement rate early allows the farms to decrease the replacement rate during the closure, allowing farms to cut gilt breeding target by 30-50%. A majority of our farms are able to remain closed until the middle of March utilizing this modified culling strategy. Gilts are loaded into farms based on current age and targeted breeding dates, and are limited by available pen space. Whole herd vaccinations with PRRSV and Influenza Type A virus vaccine are implemented soon after the closure is initiated to establish a similar baseline of immunity.

The implementation of strict McREBEL on the sow farms includes not moving any pigs unless the entire litter is moved to a nurse sow, changing gloves between any piglet contact from different litters, rotating processing sets of tools between litters and not stepping in any crates until weaning. By utilizing the needless technology for piglet injections we were able to eliminate the need for changing needles between each litter.

Results and discussion

Since implementing these strategies over the last three years the system has seen a drastic decrease in enteric challenges in the farrowing house, a reduction in nonresident PRRSV strains found in wean pigs and a decreased duration of PRRSV positive pigs weaned. The farm with the history of most PRRSV activity had 8 strains sequenced during the 6 years prior to implementation of our closure protocol and has only had 1 strain since the closure - the strain that most recently infected the farm, prior to this closure protocol. In addition, since implementing this protocol in our system, only 1 farm has broken with a wild type PRRSV strain that the farm had not been exposed to previously and this strain matched the PRRSV from the other sow farms in the system.

When comparing the production records for 2 years prior to the closure protocol and the 2 years post implementation, the sow farms have shown a steady decrease in prewean mortality, and an increase in farrowing rate and live born. These have contributed to an average increase in pigs weaned per female of 2.65 with a range of 0.8 through 5.58.

In addition to the benefits in PRRSV control, scour control and production benefits, there has been an influence on downstream health. The farms are currently *Mycoplasma hyopneumoniae* (M hyo) positive but the downstream wean to finish data and diagnostics have shown a decrease in the prevalence of clinically infected groups. The system has taken steps to further the control of M hyo in the system by eliminating M hyo in the gilt source. This has allowed more strategies like herd vaccinations and medication programs to help reduce the M hyo shedding from dam to pig, and opened up the possibility of M hyo elimination from the commercial sow farms.

Eichelberger farms feel that several factors have contributed to the increased health and performance of the sow farms over the last 2 years but we feel that the perennial closures, strict McREBEL practices and *Mycoplasma hyopneumoniae* control have all impacted the results.

References

1. Dr Bob Morrison's Swine Health Monitoring Project.

2.Holtkamp D et al. Assessment of the economic impact of porcine reproductive and respiratory syndrome virus of the United States pork producers. *J Swine Health and Prod.* 2013; 72:84.

Tech tools for disease management

Maryn Ptaschinski
National Pork Board, Clive, Iowa
Institute for Infectious Animal Disease, Texas A&M University, College Station, Texas

As veterinarians we often find ourselves overwhelmed with work and required to make decisions about animal health without all of the evidence we need or the time to collect it ourselves. The easy decisions are often easy because we have assimilated enough experience and evidence to know what we are dealing with and how to fix it. The more complicated decisions are hard because we lack the evidence or experience to fully understand or tease out all of the facets of the problem and how they are linked.

Veterinarians have access to many different types of data that we can use to make decisions including observational field data (clinical pictures, field diagnoses and conditions, interventions, etc), laboratory data, and production data (mortality, epidemiological information, efficiency metrics, costs, etc). In order to make the best possible decisions, we need to bring many of these pieces together in the same place. This allows better characterization of health issues and the effectiveness of interventions, helping to better evaluate our actions and provide more evidence-based justification for them.

Having a solid understanding of disease recognition, tracking, risk factors, and mitigations is critical to managing endemic disease. Disease expression and severity can be greatly influenced not only by the disease-causing organism, but also environmental factors, coinfections, previous exposure, vaccination, medication or other interventions, and concurrent stressors. Collecting better, more accurate observational data about the animals and the conditions they are in in a useable form can help us recognize disease earlier, monitor disease progression and characterize associated risk factors. This field data should be recorded near real-time, preferably on site, and as much as possible quantifiable so it can be compared, summarized, and searched.

Another major factor in understanding and managing endemic disease is having accurate laboratory results that are useable. Analysis of laboratory results can often be difficult as results are frequently in different formats if they aren't coming from the same laboratories. They may also be in formats that don't lend themselves well to analysis. Translating these laboratory results into a uniform and useable format is a must if a complete and accurate understanding of the disease picture is desired.

Combining quality production data, including movements, sources, flows, mortality, feed conversion, and other important production parameters with useable laboratory and field data gives us a robust ability to ask a wider variety of questions of the combined data that in-turn gives a much more complete picture of what is actually happening in the barn and/or system.

A proof-of-concept information technology system that facilitates the consolidation of data from these multiple sources has been successfully piloted. This technology allows in-depth analysis of this consolidated data, allowing practitioners to make more evidence-based decisions and gain a better understanding of the influence of interrelated cofactors, effectiveness of interventions, and cost of disease.

Mycoplasma hyopneumoniae: Lateral transmission and gilt exposure methods

P. Yeske
Swine Vet Center St. Peter, Minnesota

Introduction

Mycoplasma hyopneumoniae is the causative agent of Enzootic Pneumonia,[1,2] a highly prevalent clinical condition that influences the production outcome of swine operations.[3] *Mycoplasma hyopneumoniae* is a pathogen that has been in the swine industry for a long time and continues to be a significant problem for the swine industry. The major effects are in the grow finish phase of production with effects of: decreased average daily gain, increased mortality and lower percent marketed pigs to the primary market, poorer feed conversion and increase medication cost of treatment in feed, water and injectables.

These costs are significant when using actual production numbers from records accumulated from 2007-2015, and plugged into an economic model the resulting cost was $4.99 per pig.[4] As well as other data reported from other farm systems at $2.85.[5]

Research data suggests that disease status in the sow farm have an important effect on downstream flows.[6] One of the biggest challenges is proper acclimatization of the gilts going into the sow herd to control the level of shedding to piglets in litter and subsequent disease problems. Reducing the amount shedding is important to reduce the amount of downstream disease in the pigs.[6] Historically this was not a problem because most of the replacement gilts were born in positive herds or raised internally in the herd and were infected early in life with plenty of time to reduce shedding by the time of farrowing. Today most all of the replacement gilts are *Mycoplasma. hyopneumoniae* negative requiring that they get acclimatized once they get to the sow farm. One of the challenges is getting the gilts infected. Work has been done with seeder gilts to expose negative gilts which took (6 infected): (4 negative) gilts to get this done in 30 days.[7] Allowing these gilts, the necessary time following infection to reduce shedding at farrowing.[8] This is critical weather you want to stabilize a positive sow farm to reduce the impact of clinical disease in the finishing phase or if you want to do a herd closure for elimination protocol.

Mycoplasma hyopneumoniae elimination programs have been overall highly successful with a fair number of herds doing this.[9] Herds that look at doing elimination are primarily herds that are struggling to get gilts exposed on a consistent basis and herds that are doing a filter project as well as farms that are just tired of dealing with clinical problems of *Mycoplasma hyopneumoniae*. One of the first questions is will I really be able to get all the benefits since my pigs are predominantly grown in pig dense areas, won't my pigs just get re-infected in the area anyway?

Objectives of gilt exposure

- The main objective is to reduce the numbers of *Mycoplasma hyopneumoniae* shed to piglets and subsequent disease problems in the finishing phase.
  - This is accomplished is by getting the gilts to be colonized and stop shedding by the time they farrow their first litter.
- Establish the day zero for elimination of *Mycoplasma hyopneumoniae* program.

Methods of exposure

The first step is to identify gilts that are *Mycoplasma hyopneumoniae* positive to use to infect the negative incoming gilts at an early enough age to allow them to not be shedding by the time of farrowing. These negative gilts would ideally be infected by 84 days of age. Figure 1 demonstrates the time needed to reduce shedding in farrowing gilts and their piglets.

There are several methods of exposing these negative gilts.

- Using seeder animals has been done for a long time by the industry and works well if the right animals are used and there is plenty of time. Depending on the age of replacement gilts this may be difficult. To achieve a shorter time such as 30 days the work shows that a large number 6:4 of seeders is needed to be 100% successful.[7] One of the challenges with this system if the infection dies out in the seeders it can be difficult to get restarted resulting in problems in finishing population and it may take some time to get the herd stability reestablished costing a lot in performance during the process.
- Intra tracheal inoculation has been done in research work and challenge models. This can be done using a lung homogenate inoculum from positive animals from the farm to avoid cross contamination of other Mycoplasmas or disease.[10,11] This method although effective is very labor intense and more dangerous to staff due to restraint methods that have to be used.
- Aerosol inoculation has been done for other diseases and by other species for vaccination so is a possibility. This can be done using the same lung homogenate inoculum from positive animals from the farm. This method has the advantage of less labor and animals don't have to be restrained to get this job done. There are some technical steps required including: having a smaller air space to work with animals, shutting down ventilation for 30 minutes and watching barn temperatures to get post exposure time as long as possible as well.[12]

Figure 1: Gilt *Mycoplasma hyopneumoniae* exposure timeline

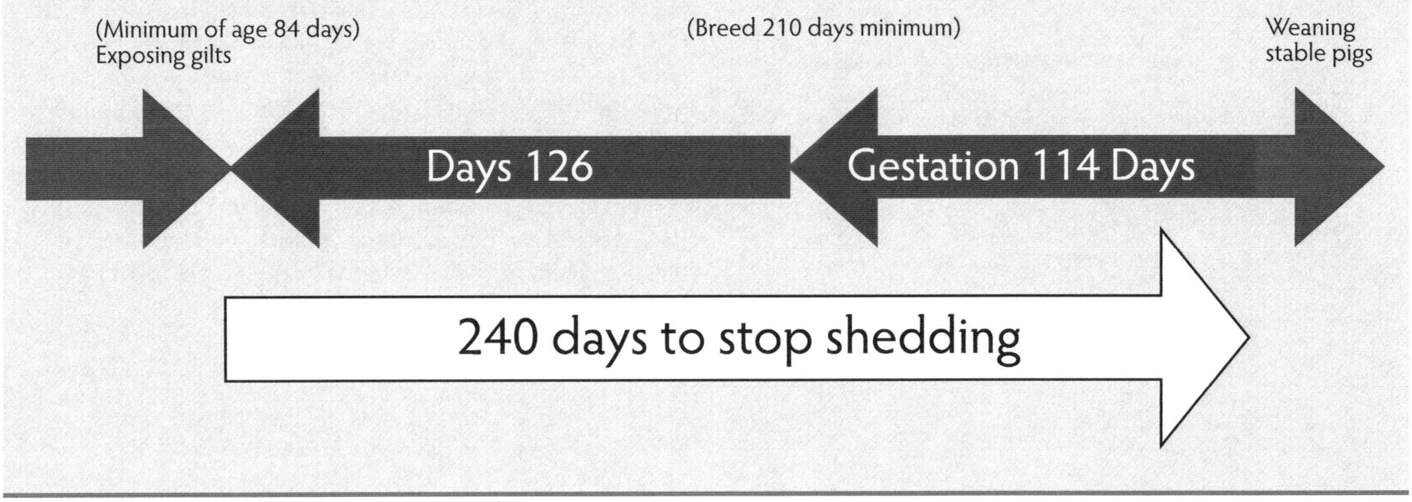

For the intratracheal and aerosol inoculations will need a source of inoculum. The best is from the herd itself since there is no possibility of introducing a different pathogen to the farm. This can be done by identifying animals with clinical signs and testing with either tracheal swabs or laryngeal swabs to be sure animals are positive for *M hyopneumoniae* by PCR. Some systems are treating the potential donor animals with Excede to reduce other potential pathogens from being introduced with the inoculum. Once positive animals are identified they can be humanely euthanized to harvest the lungs and the tissue can be put into a (Ninja) blender with equal volumes of Friis media to make a lung homogenate. Once this is completed then can filter (panty hose has worked the best) the homogenate to be diluted with additional Friis media and used for exposure by either method.

One of the keys to this process is having a good diagnostic protocol to confirm that gilts have been properly exposed. Testing every group to be sure that this has been accomplished is key to the success of these programs whether for herd stabilization or for elimination protocols.

Objectives for lateral transmission study

- Determine the incidence rate of lateral infections in finishing phase.
- Risk factors that increased their risk of site being laterally infected with *Mycoplasma hyopneumoniae.*

Methods for determining lateral exposure

First step was to identify finishing sites that were sourced from *Mycoplasma hyopneumoniae* negative farms that were in pig dense areas. Source sow farms were tested to validate the status of the farm to be sure that it was truly negative by testing 30 sows with IDEXX ELISA and any suspects to be followed up with laryngeal swabs. Sites were run on an all in all out by site basis. A total of 50 sites were tested in summer / fall season and 50 herds that were tested in the Winter / Spring season to look for potential seasonality. A total of 100 sites were done.

Site status was determined by testing a 95/10 statistical sample of 30 head serologically using *Mycoplasma hyopneumoniae* IDEXX ELISA test. The samples were taken just prior to marketing so that maximal exposure time could be used. If S/P values were over 1.5 on IDEXX ELISA sites were considered positive if more than 3 positives below this level, then the herd was retested using 50 laryngeal swabs.

A questionnaire was filled out for each site to determine possible risk factors for sites that turned positive. Closeout information was collect from all of the sites to compare performance differences.

Results of lateral transmission study

100 sites were monitored across both seasonal periods from 5 different production systems with 10 sites per system per season in order to keep things as consistent as possible but not just one production system. A total of 6% of the sites were positive for Mycoplasma hyopneumoniae from lateral source introduction. All positive sites had some clinical signs when samples were collected with clinical signs of coughing. 1 site was confirmed with serology and the other 5 sites with retesting and laryngeal swabs. Figure 2 shows the results for each season and cumulative total. Because of the low number of positive sites there were no clear risk factors other than area density that could be evaluated, and due to the variance between production systems, there were no significant production parameters identified.

Discussion

The goal of intentional *Mycoplasma hyopneumoniae* exposure is to establish exposure, infection and clearance of the organism before the gilts farrow. This can be useful in both a herd stabilization plan that is being used to make pigs with a relatively low load of *Mycoplasma hyopneumoniae* at the time of weaning or to establish time zero for an *Mycoplasma hyopneumoniae* elimination programs. It will depend on the herd and the long-term control

Figure 2: Results of lateral transmission study by season and total

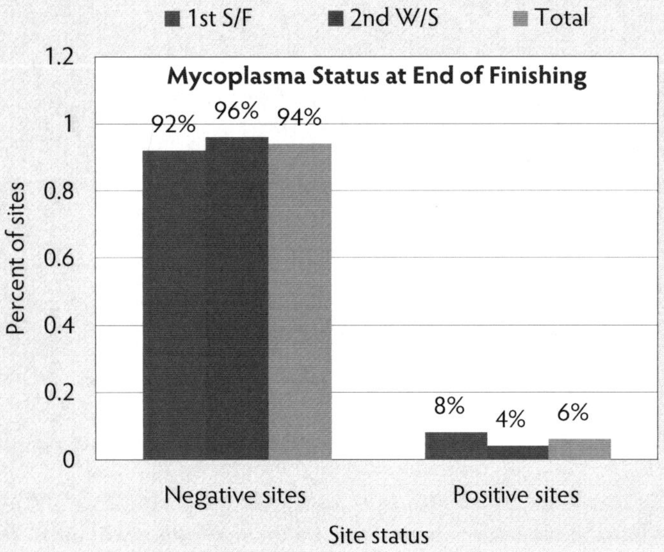

strategy. Certainly not all herds will follow the same plan but to reduce the economic impact of *Mycoplasma hyopneumoniae*. However, regardless of the plan for stability they will first need to stabilize the herd and controlled exposure (intratracheal or aerosol) are the most reliable ways to get this done. No matter what the protocol for acclimatization, a good diagnostic plan is needing to follow every group of gilts that enters the herd to make sure that this is successfully completed. One of the big challenges is there is just not a lot of time to get this done. One of the challenges is that there needs to be a way to get this done repeatability. With nearly all replacement gilts coming from negative herds this is more of a challenge than it was historically when replacement gilts were positive and has resulted in *Mycoplasma hyopneumoniae* surfacing as a clinical problem again.

Lateral transmission of *Mycoplasma hyopneumoniae* occurs in negative pigs placed in high pig density regions. It just doesn't happen very often, only 6% of the time there was not much of a difference in seasonal affect, but more were identified in summer / fall season tested. If a system is having difficulty with *Mycoplasma hyopneumoniae* control, it is likely due to the instability of the sow farm and not getting good gilt exposure rather than sites becoming infected downstream. This should no longer be a hurdle to over come for farms looking at elimination as a possibility and the numbers generated here can be used in modeling the impact of lateral transmission in the partial budgets for these types of projects.

References

1. Mare, C.J., Switzer, W.P., 1965. New Species: *Mycoplasma hyopneumoniae*; a causative agent of virus pig pneumonia. *Vet. Med. Small Anim. Clin.* 60, 841– 846.

2. Goodwin, R.F.W., Pomeroy, A.P., Whittlestone, P., 1965. Production of enzootic pneumonia in pigs with a mycoplasma. *Vet. Rec.* 77, 1247–1249.

3. Thacker, E.L., Minion, F.C., 2012. Mycoplasmosis. In: Zimmerman, J.J., Karriker, L.A., Ramirez, A., Schwartz, K.J., Stevenson, G.W. (Eds.), *Diseases of Swine*. Iowa State University, pp. 779–797.

4. Linhares, D., 2017: A field study on economics of *Mycoplasma hyopneumoniae* elmiantion. Allen D Leman conference.

5. Yeske, P., 2016, 2016: *Mycoplasma hyopneumoniae* Elimination. Proceedings from AASV annual meeting p 376-381.

6. Schwartz, M., 2015: Cost of *M Hyopneumoniea* in growing pigs. Allen D Leman conference.

7. Fano, E., Pijoan, C., Dee, S., Deen, J., 2007. Effect of *Mycoplasma hyopneumoniae* colonization at weaning on disease severity in growing pigs. *Can. J. Vet. Res.* 71, 195–200.

8. Roos, L., et. al. 2016. A model to investigate the optimal seeder-to-naïve ratio for successful natural *Mycoplasma hyopneumoniae* gilt exposure prior to entering the breeding herd. *Veterinary Microbiology* 194 p 51-58.

9. Holst S. 2015. Elimination of *Mycoplasma hyopneumoniae* from breed-to-wean farms: A review of current protocols with emphasis on herd closure and medication. *J Swine Health Prod.* 2015;23(6):321–330.

10. Pieters, M. 2016. Current efforts on *Mycoplasma hyopneumoniae* disease control. Allen Leman Swine Conference. St. Paul, MN.

11. Sponheim, A. 2017. A Diagnostic Approach to Confirm Day Zero. Allen Leman Swine Conference. St. Paul, MN.

12. Toohill, E., 2017. Achieving Day Zero in Large Swine Operations. Allen Leman Swine Conference. St. Paul, MN.

Experiences managing wean-to-finish PEDV in a production system

Lynn Pavlovic, VMD
Iowa Select Farms, Iowa Falls, Iowa

Introduction

Porcine epidemic diarrhea virus (PEDV) has placed an economic burden on the US swine industry since 2013 and even longer worldwide. Field experiences consistently show that disease caused by PEDV increases mortality and reduces growth and development.[1] A large focus is placed on preventing disease in the breeding herd since mortality in suckling pigs often reaches 50-80%.[2] This impacts future performance due to extensive dehydration and electrolyte imbalance. It can cause rapid spread of diarrhea in growing pigs, and nursery-age mortality may reach 1-3%.[2] In older hogs, producers may be unaware that pigs are clinically ill as the disease is often self-limiting within 1-2 weeks of onset.[3] Signs of fever, diarrhea, decreased feed consumption, and lethargy may be overlooked; however, the negative consequences can be significant. This poses a challenge when marketing hogs in areas of high disease pressure, especially during colder weather when exposure and survivability of the virus is at its peak.

Impact

It is well documented that PEDV negatively impacts average daily gain (ADG), average daily feed intake (ADFI), and reduces growth and digestibility, especially when combined with other health challenges such as porcine reproductive and respiratory syndrome virus (PRRSV) or bacterial co-infections.[1,4] This impairs performance of the pigs that survive a PEDV challenge at birth, as well as pigs infected throughout the wean-to-finish phase. It can affect a system's overall performance loss depending on the number of farms infected. Last winter, our company experienced an increase in disease challenge in late finishing, specifically during marketing. From November 2016 through February 2017, 18% of sites marketing broke with widespread diarrhea. Of these farms, 47% were confirmed PEDV positive via PCR, and the remainder were presumed to be positive with either PEDV or porcine delta coronavirus (PDCoV) due to clinical signs. These symptoms were typically noted 1-3 days after a transportation event, and the sites were added to our company quarantine list. We witnessed the seasonality of the disease as the percentage of farms that broke during marketing dropped to 1.3% between March 2017 and November 2017.

Based on records from company quarantine sites, lateral PEDV breaks affect our ADG and time to market. Harvest data confirmed a 1.98 kg difference in average carcass weight between farms that broke during interim cuts vs non-PEDV farms. Introduction of this disease during late finishing reduced our overall ADG by 21.4%. Prior studies have shown a 14.8%[5] and a 19%[4] reduction during the growing phase in the weeks following exposure. During the winter of 2016/17, we experienced an average increase of 4.6 days to close out a site when comparing PEDV positive farms to PEDV negative farms.

Outcomes described above do not factor in the downstream impact of sow farm PEDV breaks or the additional setback of previously positive sites breaking during the following turn with newly placed pigs. Increased wean-to-finish breaks also pose a greater risk of sow farm exposure. Thus, the financial impact in finishing hogs can be significant. It is important to identify the area of disease introduction, develop a biosecurity pyramid, and focus on methods of preventing and eliminating the virus. One must also account for both losses and expenses and adjust prevention methods accordingly.

Surveillance

There are varying levels of surveillance across producers and systems, especially since they experience different risks of transmission. We know that PEDV is prevalent during certain times of the year so it is all about mitigating the risk. According to MSHMP data looking at coronaviruses in sow farms across several producers, the cumulative incidence for the past 3 years has been around 6-7%.[6] Through PEDV exposure of multiplication herds and seasonal sow farm vaccination, we provide protection to our breeding animals; however, warm weather months are crucial times to eliminate virus that is present across all levels. This is especially important prior to the fall pumping season, when our routine testing costs increase and potential exposure begins to rise. Knowing overall prevalence within our system provides a more accessible method to identify associated threats and areas of disease introduction.

Throughout the year, our wean-to-finish sites are tested for PEDV via PCR on oral fluids any time the virus is suspected by clinical signs. If a site breaks early in the growing phase, it is added to a quarantine list, moved to the end of our biosecurity tier, and re-tested 12 weeks later. Late finishing breaks are either confirmed or suspected positive, and they are added to the list. To ensure that the virus is eliminated from these previously positive farms, new animals are tested by PCR at 2 weeks post-placement. The transportation team, rendering service, load crews, sort crews, and feed mills are notified of any quarantine sites. This affects vehicle movements, the flow of daily activity, and clean-up procedures following the closeout.

In order to ensure all equipment is clean prior to pumping manure, we have implemented routine testing of agitator pumps, hoses, tractors, and trucks contingent on an internal tier system. Equipment moving into a high health tier will require a PEDV negative swab result, downtime, and notification of a Health Services member prior to entering the facility. During times of particularly high viral challenge, we have also added testing of truck wash bays and regular inspections.

Interventions and prevention

Reducing PEDV prevalence and impact requires a combination of immunity and biosecurity. It is vital to develop a biosecurity pyramid and determine downtime requirements for entry into each farm. Assessing areas of shared labor, vehicle and supply contamination, and other high-risk factors will help regulate potential disease spread. All of our finishers maintain all-in-all-out status. Due to challenges with re-breaks and compounding co-infections, we have placed a greater focus on the biosecurity surrounding mortality removal, office sanitation, and respecting the clean/dirty line. Load crews keep handling equipment on site, and all caretakers wear farm-specific clothing, shower into and out of their first site of the day, and shower out of all remaining sites. Caretakers are also provided separate boots for mortality removal that are disinfected daily. Once a site is infected with PEDV, they stop movement of common supplies such as medications, syringes, and paperwork. Clothing is machine-washed and dried after chores.

In general, all farms are washed and disinfected with a broad-spectrum virucidal product following every turn. All nurseries, quarantine sites, and farms within 3 miles of sow/gilt herds are whitewashed to ensure confidence in eliminating pathogens. Common supplies that remain at the farm, such as sort boards, rattle paddles, brooders, pig mats, boots, and portable load-chutes, are also washed. Caretakers are in charge of cleaning offices and disinfecting floors and furniture. If a site manager or supervisor determines that the wash is inadequate, it is re-washed. Ideally, these sites would also be dry prior to movements.

Since fecal contamination of fomites plays a large role in spreading the virus, focus is placed on transport biosecurity, especially since the majority of our PEDV breaks occur within a few days of a transportation event. In order to prevent the risk of disease, all multiplication, sow, wean, and feeder pig trucks go through inspection at our internal truck washes. This follows a rigorous wash, disinfection, and baking process. Rendering trucks dedicated for sow farms and GDUs follow a similar protocol. In addition, we have implemented routine internal audits of goosenecks and semi-trailers. The audit checklist includes categories for wood chips, unwashed fecal material, and trailer damage, with a focus on handling equipment that will be in direct animal contact.

The marketing side follows a different protocol, as the risk and associated setbacks of late finishing disease are often thought to be less than the cost of cleaning these vehicles. It has been shown in previous studies that packing plants pose a significant risk, with a documented proportion of PEDV contaminated trailers increasing from 6.6% to 9.2% after unloading.[7] Therefore, we recognize the importance of washing and disinfecting these trailers after every load when going into protected finishers in sow farm dense regions. In 2016, we moved to washing market trailers in non-protected areas following every fourth load regardless of cut. This reduced total washes by 64%, which resulted in $230,000 average annual savings to offset the aforementioned cost in gain and time to market. It has the added benefit of reducing mileage, increasing efficiency, and expanding room at the truck washes. During times of high disease introduction, we have implemented routine swabbing and testing at the truck wash, as well as whitewashing the bays. In addition, we have increased crew and driver biosecurity through training and load-out audits, focusing on clean/dirty lines, wearing booties, and maintaining safe and undamaged equipment.

There are additional risks during pumping seasons when we utilize external equipment and vehicles. A biosecurity tier specific to nutrient management decides which farms take priority, and we keep dedicated pumpers for high health status sites early in the season prior to moving into commercial finishing. Besides routine testing and downtime, all applicator equipment is inspected prior to entering the facility. Farm supervisors fill out a checklist, which ensures that all fecal material has been removed and that equipment is disinfected and dry. They are also in charge of disposing of any debris following the procedure. All pumping personnel are trained or supervised by a qualified individual. Protocols undergo continuous review so that certain sites maintain priority to ensure the best prevention methods.

Challenges and limitations

Many opportunities exist to enhance the way we manage PEDV cases and improve biosecurity. Considerations of action vs cost may change depending on the incidence and severity of disease, as well as compounding issues with other health challenges. We have experienced setbacks from creating PEDV positive environments for future turns and have had to prioritize and re-examine our washing/whitewashing so that it is feasible to accomplish on a large-scale. Inspecting quarantine sites after washing is not always possible with quick turnaround, and ultimately, any increased viral load in finishing sites poses a greater risk to the entire system. During weeks of high disease introduction last winter, the losses from decreased carcass weights equaled our savings from reduced trailer washes. Therefore, cost vs risk should be evaluated regularly to determine if we are making the right management decisions. Due to an increased incidence, we have expanded our testing. More importantly, case documentation improved with the use of a spreadsheet that records the site, time of break, and testing information so that we can evaluate our methods. Every year, we learn new areas of focus through both our successes and challenges. Though we cannot predict the impact of PEDV in the coming years, it is important that we are prepared.

References

1. Alvarez, J., J. Sarradell, R. Morrison, A. Perez. 2015. Impact of Porcine Epidemic Diarrhea on Performance of Growing Pigs. *PLoS ONE* 10(3): e0120532.

2. Pospischil, A., A. Stuedli, M. Kiupel, et al. 2002. Update on porcine epidemic diarrhea. *J Swine Health Prod*. 10(2): 81-85.

3. Lee, C. 2015. Porcine epidemic diarrhea virus: An emerging and re-emerging epizootic swine virus. *Virol J*. 12: 193.

4. Schweer, W.P., K. Schwartz, E.R. Burrough, K.J. Yoon, J.C. Sparks, N.K.Gabler. 2015. The effect of porcine reproductive and respiratory syndrome virus and porcine epidemic diarrhea virus challenge on growing pigs I: Growth performance and digestibility. *J Anim Sci*. 94(2): 514-522.

5. Curry, S.M., K.A. Gibson, E.R. Burrough, K. Schwartz, K.J. Yoon, N.K. Gabler. 2016. Pigs inoculated with porcine epidemic diarrhea virus have decreased growth performance and tissue accretion compared with controls after 42 d. *J Anim Sci*. 94(2): 120-121.

6. Vilalta, C. 2017. Dr. Morrison's Swine Health Monitoring Project (MSHMP) Annual Summary. https://www.vetmed.umn.edu/sites/vetmed.umn.edu/files/shmp_2016l18.2_mshmp_annual_summary_-_science_page.pdf.

7. Lowe, J., P. Gauger, K. Harmon, J. Zhang, J. Connor, P. Yeske, T. Loula, I. Levis, L. Dufresne, R. Main. 2014. Role of Transportation in Spread of Porcine Epidemic Diarrhea Virus Infection, United States. *Emerg Infect Dis*. 20(5): 872-874.

Subclinical ileitis: Diagnostic monitoring, R^2, and economics

Nathan Winkelman, DVM
Swine Services Unlimited Inc. Rice, Minnesota

Introduction

Ileitis (*Lawsonia intracellularis, LI* or PPE) is considered the most common grow-finish enteric disease in North America. Of the 3 forms of this disease: PHE (Porcine Hemorrhagic Enteropathy), PIA (Porcine Intestinal Adenomatosis) and Subclinical ileitis, the latter is considered by far the most ubiquitous and economically important. A primary challenge for today's swine practitioner is to continually monitor their clients' ileitis control programs (vaccines and/or antibiotics) with anti-mortem diagnostics (fecal PCR and IPMA serology) to determine if subclinical ileitis is present and if it is of economical significance. It can be difficult to interpret *Lawsonia intracellularis* PCR diagnostics due to confounding vaccine shedding with field infection shedding. It can also be difficult to differentiate vaccine immune response vs. field infection LI titers. However, due to more advanced quantitative Fecal PCR assays with ct values and recent advances in serological IPMA (immunoperoxidase monolayer assay) antibody assays with titers, one can better interprete ante-mortem diagnostics to diagnose both clinical and subclinical ileitis to help evaluate ileitis control programs. Proper interpretation begins with an understanding of the control program in place and an understanding of how ileitis progresses through a population of pigs.

Disease progression

In general, pigs will start to shed *LI* bacteria 7-14 days post-exposure and continue shedding for at least 6 weeks, and up to 10 weeks post-exposure. Seroconversion begins about 2-3 weeks post-exposure, depending on the exposure dose, and may last for 6-14 weeks, or longer, depending on how high the titer peaks and the duration of fecal shedding.

The onset and degree of ileitis clinical signs (diarrhea, gauntness), fecal shedding, seroconversion, and intestinal lesions is largely dependent on the (1) age of the pigs infected and (2) the *Lawsonia* exposure dose.[1] In research trials, younger nursery pigs may show clinical diarrhea and gauntness in about 80% of challenged pigs with 20% being subclinical. While pigs 10 weeks of age or older may show only 20-40% with clinical signs, and 60-80% being subclinically infected. Mortality in *LI* challenged pigs at 4-6 weeks of age will be about 5-20%, whereas in LI challenged grow-finish pigs mortality will be only 0-5%. Older pigs, while still infected, exhibit fewer clinical signs and are more likely to be subclinically infected. They have an amazing ability to hide clinical diarrhea while still having significant levels of disease.

The *Lawsonia intracellularis* challenge dose (or natural exposure dose) also changes the onset and severity of disease parameters. Table 1 shows significant differences in all performance, clinical, and pathological parameters in 5 week old pigs challenged with a high (10^{10}), medium (10^9), or a low (10^8)*LI* dose per pig.

Ante-Mortem Diagnostics

Fecal PCR is the Gold Standard Ante-Mortem Diagnostic: This assay has a 90% sensitivity and 100% specificity and can detect *L. intracellularis* quantitatively down to 10^2 per gram of feces (Beckler[2]). Diagnostic labs report ct values which correspond to qPCR tables provided by each lab. However, the PCR assay is not standardized between D-Labs and can differ by one or two logs for the same ct value between laboratories. It is up to the veterinarian to determine if a positive PCR results reflects significant ileitis disease or simply the presence of underlying *Lawsonia* in the environment or vaccine. This determination of the level of *Lawsonia* shed per gram of feces can then correlate to IHC lesion scores, ADG, and herd profitability.

IPMA Serology[3]: There are at least two commonly used *L. intracellularis* serologic tests: the IPMA and ELISA. A comparison of them is reported at this conference by Vasquez.[4] For this report, the IPMA serology assay will only be discussed. The IPMA test (U of MN VDL) has recently improved by reporting titers rather than just positive or negative results. The temporal relationship between identification of positive fecal PCR and IPMA titers in the field is 2 to 8 weeks from the time of initial fecal shedding to initial IgG titers.[5] In general, the higher the Li exposure dose, the quicker the IPMA immune response. For example, in pigs chal-

Table 1: Performance, clinical, and pathology parameters

	High 10^{10}	Medium 10^9	Low 10^8	Control
% Mortality	25	6.3	0	0
ADG	.20	.41	.57	.97
ADFI	.98	1.2	1.4	1.7
FE (feed/gain)	5.0	2.9	2.4	1.7
% ADD	63	50	39	16
% ACISD	18	11	5	0
Lesion length	171	85	73	0
Lesion score	3.1	2.0	1.8	0
IHC score	3.4	2.9	2.2	0

ADD- Abnormal diarrhea days
ACISD- Abnormal clinical impression score days

lenged (exposed) with a high 10^{11} dose initially begin sero-conversion in 2-3 weeks and those challenged with a low 10^3 dose initially sero-converted at 4-5 weeks post-challenge.

IPMA titers are usually measured from a low titer of 1:30 to a high titer of 1:960. However, for an additional cost, serum can be plated out to determine the titer endpoint of the assay. Higher shedding levels or longer duration of shedding is associated with longer duration of serum positives for IgG antibodies against *L. intracellularis*. Pigs exposed to more severe disease such as PHE mortality and high disease challenges in research trials will have higher titers (1:480 to > 1:7680) in general. In an extreme example, a "super shedder" chronically infected experimental pig was still shedding high fecal ct levels (23.2 of 40ct) with an IPMA titer of 66,120 at 10 weeks following challenge.[6]

The approximate ½ - life of *Lawsonia* IgG is 21 days, therefore some pigs can maintain titers up to 12 weeks following a PHE infection. Pigs with subclinical or mildly clinical infection have lower titers which may last for only a few weeks.

Clinical and subclinical ileitis

Clinical ileitis: The first goal is to not miss clinical ileitis! (The second goal is not to miss subclinical ileitis!) – Swine veterinarians and producers need to first identify and diagnose clinical ileitis (PIA and PHE). Too many of us ignore low levels of clinical ileitis, assuming it is not much of a problem, when it can be and is a very significant problem causing decreased performance. Whenever clinical diarrhea is present at all, it can be the "tip of the iceberg", with the majority of the pigs in the barn already subclinically infected.

Diarrhea caused by *Lawsonia intracellularis* can be of almost any consistency and a variety of colors.. Fecal consistency and fecal color matter: diarrhea consistency is scored mild (soft, "cow-pie"), moderate (< 50% liquid), and severe (very watery). Diarrhea fecal color is scored mild (yellowish: bile-stained from anorexia), moderate (some blood -orange, pink, dark brown, or brown red) or severe (melena: black-red, red-back). Fecal consistency and color are scored separately because you can have one without the other. For example, moderate brown-red bloody color can be found in normal formed feces without diarrhea. Any and all of these scores can be caused by LI with high levels of bacteria shedding, significant disease and PPE pathology. Test any suspect feces with a PCR for *Lawsonia* and appropriate diagnostics for other pathogens as you deem necessary.

Consider carrying a "fecal kit" during vet visits to the farm. This kit contains swabs and fecal collection tubes/cups in a bag to allow one to collect fecal samples for PCR analysis right away. Any abnormal fecal color or consistency, especially in pigs > 100 lbs, is likely ileitis and is likely significant. Positive samples may indicate it's time to re-evaluate the current control program. **Do not walk by clinical diarrhea without confirming a diagnosis.**

Subclinical ileitis: The second goal is to not miss subclinical ileitis! Subclinical Ileitis is defined as having a confirmed ileitis diagnosis (fecal or intestinal PCR, seroconversion, PPE gross lesions, and/or IHC on necropsy) without having clinical signs of diarrhea, fecal color, or gauntness. The prevalence of subclinical ileitis in US swine herds is considered to be high. NAHMS surveys have shown the clinical prevalence to be 35-42 %, with 96% of US swine herds with subclinical or clinical ileitis.

The classic subclinical *Lawsonia* dose titration trial[7] showed with decreasing *Lawsonia* challenge doses from 10^8 to 10^4 per pig, that pigs without any clinical signs still had a significant 39% decrease in ADG(.55 vs .88 lbs/day) and a 27% poorer FCR (1.63 vs 2.07). This significant performance loss with subclinical ileitis in the presence of gross and microscopic IHC lesions and fecal shedding has been repeated in other trials.[8] Beckler[9] has shown that of 271 time points when a pig was shedding *L. intracellularis*, 74.4% (201/271) of those time points, the pig had normal feces and 93% (252/271) of the times when a pig was shedding *Lawsonia*, the fecal score was 0 or 1. Again, indicated a high prevalence of subclinical pigs with positive shedding.

Diarrhea is not easy to reproduce in challenge trials in pigs older than 10 weeks of age. This "seeder pig" trial model (Table 2) had two, 10 week old *Lawsonia* challenged seeder pigs per pen co-mingled with 23 "contact pigs" to mimic a natural infection in grow-finish pigs. At the peak of infection, 27 days after seeder pig challenge, 14 of 23 (61%) still did not have any clinical signs of diarrhea, gauntness, or fecal color scores, but 100% were shedding high levels of *Lawsonia* (ct values 18.57 to 31.86) and all but one of them had moderate IHC scores and PPE Gross Severity Scores, indicating significant disease in subclinically infected pigs. Figure 1 shows the % IHC Positive and Mean IHC Scores from those naturally infected pigs euthanized later on Days 42, 56, and 70 after seeder pig exposure. The % positive IHC PPE lesions decrease over time as the pig's immune system reacts to the infection.

Natural shedding patterns show 90% of contact pigs shed high levels of LI for 6 weeks, 30% at 8 weeks and 0 at 10 weeks after seeder pig exposure.[6] Therefore, there is only a 6 to 8 week time period in which ileitis shedding, PPE lesions and performance loss occur in a barn before the disease naturally heals itself. The veterinarian must identify this period with the aid of routine ante-mortem diagnostics when monitoring control programs.

Correlation coefficients (R^2):

What is the association between the fecal PCR ct value and ileitis lesions and performance?

Fecal PCR to IHC lesions – Fecal PCR ct values are highly correlated to IHC scores of intestinal PPE lesions. IHC is considered the gold-standard post-mortem diagnosis for ileitis. Burrough[10] reported that of 63 ISU suspect ileitis cases of which both fecal and intestinal samples were submitted, ct values ranged from 16.94 to 37.66. Thirty-three fixed intestines were negative

Table 2: Euthanized "contact" pigs 27 days after "seeder" pig exposure

	ID	BW day 0	BW day 27	ADG	IPMA result		Fecal PCR result		IHC	PPE gross severity score	Clin obs* day 27
1	4	61.4	111.7	0.72	POS	480	POS	18.57	2	870	0-2-2
1	24	74.1	130.0	0.80	POS	30	POS	31.86	2	51	2-2-0
1	38	70.5	114.9	0.63	POS	240	POS	20.41	4	550	0-2-0
1	45	67.4	83.8	0.23	POS	30	POS	26.58	2	0	0-0-0
1	54	70.7	128.7	0.83	POS	240	POS	19.04	2	11	0-0-0
1	57	59.3	130.7	1.02	POS	240	POS	21.52	4	63	0-0-0
2	61	72.3	110.7	0.55	POS	60	POS	21.11	2	22	0-0-0
2	71	63.6	130.3	0.95	POS	60	POS	26.56	2	0	0-0-0
2	77	57.2	116.9	0.85	POS	120	POS	20.12	3	24	0-2-0
2	84	60.6	112.2	0.74	POS	30	POS	20.78	3	7	0-1-0
2	95	62.6	120.7	0.83	POS	30	POS	22.70	3	18	0-0-0
3	6	59.0	124.2	0.93	POS	30	POS	26.38	2	6	0-1-0
3	10	67.4	125.1	0.82	POS	30	POS	25.22	3	15	0-3-0
3	14	65.8	117.6	0.74	NEG	15	POS	26.45	2	0	0-0-0
3	21	64.6	109.3	0.64	POS	480	POS	23.38	2	0	0-0-0
3	47	61.0	123.9	0.90	POS	30	POS	22.81	4	13	0-0-0
3	49	61.8	111.8	0.71	POS	120	POS	26.65	2	10	0-0-0
4	7	62.0	116.5	0.78	POS	1920	POS	24.61	3	17	0-0-0
4	11	65.7	112.9	0.67	POS	240	POS	18.19	4	930	0-3-2
4	25	59.9	110.7	0.73	POS	120	POS	21.83	2	18	0-0-0
4	50	59.9	119.9	0.86	POS	60	POS	22.88	0	20	0-0-0
4	82	60.4	108.4	0.69	POS	120	POS	22.22	3	31	0-0-0
4	96	58.3	92.5	0.49	POS	7680	POS	23.54	3	300	1-2-0
SUMMARY		63.72	115.80	0.74	21/21	113.0	5/21	23.19	2.57 22/23	129.4 19/23	2/23-9/ 23-2/23

* Clinical Observations: AAS-FS-FC

and thirty were positive on IHC. There was a strong negative correlation between the ct value of a positive PCR and the quantity of L. intracellularis antigen detected by IHC. On the basis of these results at ISU VDL PCR ct values ≤ 20.0 had a positive predicative value of 100% for the presence of IHC positive PPE lesions, and PCR ct values ≥ 30 were associated with a negative predictive value of ≥ 95% for these values ($P \leq 0.05$). Beckler[9] also reported a positive association between intestinal PPE lesion length (cm), PPE Gross Lesion Scores, and IHC Scores, vs. fecal shedding on the day of necropsy ($P \leq .0001$ for all three comparisons).

At the U of MN VDL, fecal PCR's with ct values < 32 of 40 ct are considered significant and highly associated with positive IHC scores and PPE lesions. Ct values from > 33 to < 35 may or may not be significant. Those ct values > 35 to 40 are considered "suspect". Of 3,577 LI PCR fecal samples submitted to the lab from 2013-2016, there were 41%, 25%, and 34% in each of the above three categories, respectively.[11]

PPE lesions to average daily gain (ADG): Winkelman[12] reported the length of PPE lesion at necropsy was found to be a useful quantitative measure of disease severity. In that study, PPE lesion length was inversely correlated with performance as shown where lesion length is plotted against ADG. The relationship between PPE lesion length and ADG was linear ($R^2 = 0.97$, correlation coefficient = 94 %) as shown in Figure 2. It should be noted that this was a severe disease challenge in nursery pigs. Although the association of PPE pathology to ADG may not always be this strong, the presence of lesions will always explain some of the performance variation that exists.

Figure 1: Contact pigs % IHC positive and mean IHC score over time

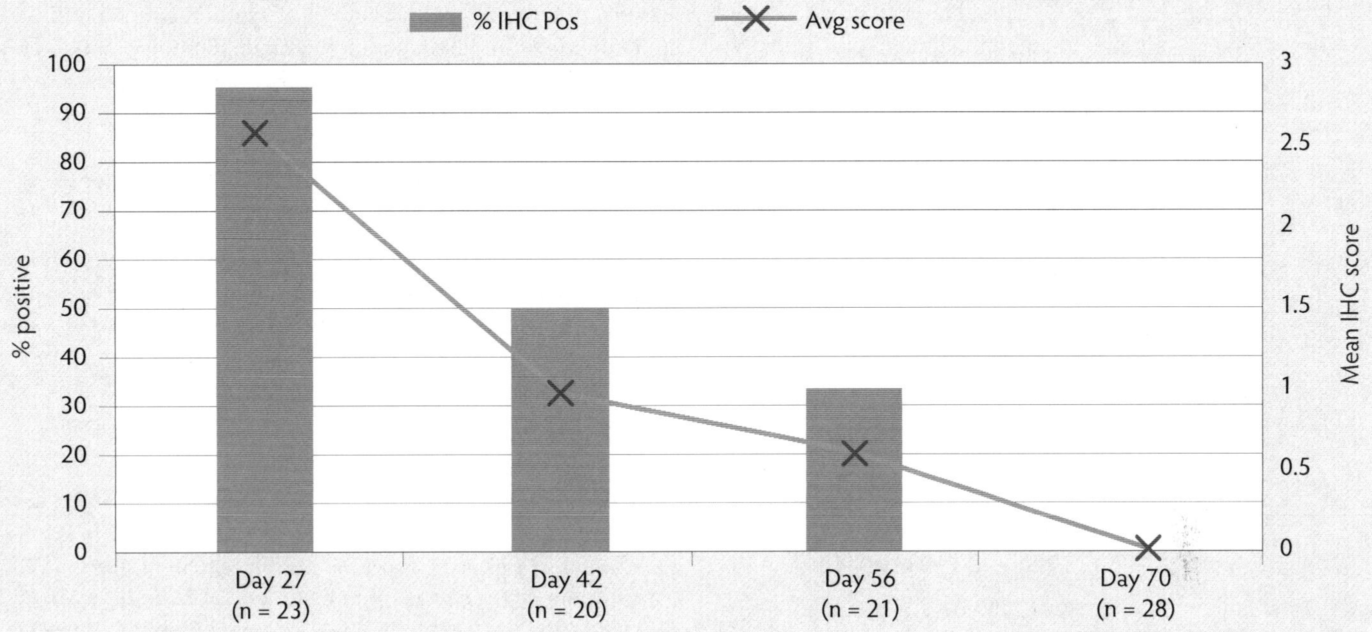

Therefore, intuitively it can be assumed that the more *L. intracellularis* shed per gram of feces (i.e. the lower the ct value) the more PPE intestinal lesion length, the higher the severity score and higher IHC score in that pig. Also intuitively, the more intestinal pathology, the poorer the performance (ADG, ADFI, and FCR) of that pig or population of pigs. Economic calculations on this performance loss, even in subclinical pigs, can cost about $3.50 to $5.00 per pig.

Monitoring the system. Know what to expect

Although commercial vaccines and feed antibiotics are very cost-effective, they show only moderate efficacy in controlling clinical and subclinical ileitis in many cases. They do not prevent infection nor do they eliminate the infected carrier pig. (except carbadox).

In order to interpret ante-mortem diagnostics, veterinarians must understand if fecal shedding and/or an immune response is anticipated from the vaccines or the DLI® (diluted *Lawsonia intracellularis*) Immunity program. (DLI is a patented method of providing immunity to control and prevent both clinical and subclinical ileitis by oral inoculation of pigs with an autogenous subclinical LI dose followed by an antibiotic pulse two weeks later).[13] In herds with only antibiotic control programs, fecal PCR or *Lawsonia* serologic analysis is not confounded, making diagnostic interpretation easier.

Comments on serology

Mucosal immunity and cell-mediated immunity at the gut level is what affords protection against LI. However, we don't have a convenient way to measure it. Therefore, we use IgG immunity as our best tool currently. Positive titers only indicate prior infection or vaccine exposure. However, high titers usually indicate field infection and not vaccine titers.

- **Enterisol**®–Expect very little if any IPMA immune response from the vaccine. There was no immune response after 42 days post-vaccination.[14]
- **DLI**® – Expect low to moderate titers in 50% of pigs by 4 weeks post administration, with increase percentage of positive pigs over time with titers ranging from 1:30 to 1:480.
- **Porcilis**® – Expect at least 50% of pigs positive within 2 weeks and 90% positive by 40 days post-vaccination. IPMA vaccine titers range from 1:30 to 1:240 typically.[42]

As previously stated, many pigs infected with ileitis infections will have titers ranging from 1:30 to ≥ 7,680; but if high titers are found, they are likely due to field infections and not vaccines.

Comments on fecal shedding[14]

- **Enterisol**® – Vaccinated pigs will shed low levels of vaccine *Lawsonia intracellularis*. For example, at 2 weeks post-vaccination 10% were positive (ct 34), 45% were suspect and 45% were negative. In general, the vaccine ct value will be low positive and not associated with PPE lesions.
- **DLI**® administered pigs had 66% positive (ct 28.7-35.0 of ct 40) and 33% suspect levels 14 days post-DLI. This is useful for compliance monitoring to ensure pigs have received the subclinical dose prior to the recommended antibiotic pulse program. Depending on the antibiotic pulse, shedding will stop completely within 2-6 weeks after DLI administration.

Figure 2: Relationship of lesion length to daily gain with plot of treatment results

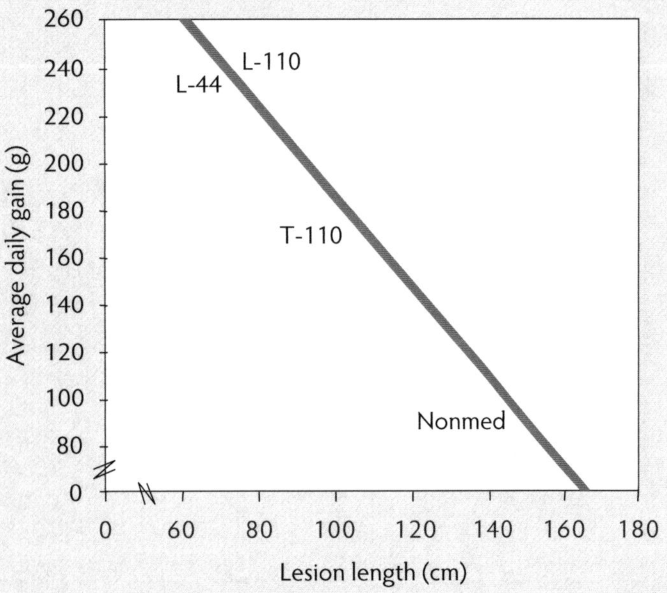

• In **Porcilis**® or antibiotic control programs, *Lawsonia* shedding is obviously not confounded with the vaccine.

Comments on antibiotics and ileitis

Antibiotics are still imperative for the treatment, control and prevention of ileitis. Those classes of antibiotics which are absorbed systemically and actively concentrate in the enterocyte cytoplasm where *Lawsonia intracellularis* replicates are efficacious: macrolides, lincosamides, tetracyclines, pleuromutilins, and quinoxalin-like (Mecadox®). Other classes of antibiotics that are not very effective are amino-glycosides (neomycin, gentamicin); B-lactams (penicillin, amoxicillin, and ceftiofur) sulfonamides, and bacitracins (BMD). Tulathromycin (Draxxin®), a macrolide injectable, has ironically been shown to be in-effective in treating *Lawsonia intracellularis*-likely because it is primarily renally excreted

Antibiotics and subclinical ileitis: Mecadox® (carbadox) is unique – Randomized controlled LI challenge trials have shown that many of these antibiotics control clinical signs, reduce fecal shedding, PPE intestinal lesions, and improve performance. In 2 × 2 controlled studies, the combination of vaccines and antibiotics together are statistically better at controlling Lawsonia infection than either one alone.[15]

However, only carbadox at 50 g/ton for 14 days eliminates the carrier state completely, and completely prevents colonization or infection while being fed to pigs challenged with *Lawsonia intracellularis*.[16] It is the only antimicrobial that prevents or eliminates both clinical and subclinical ileitis.

Sample frequency to monitor

To detect subclinical ileitis in a pork production system at a 90 % confidence level and a 10% prevalence would require sampling 30 fecal and serums about every 4 weeks beginning at 6-10 weeks old. However, periodic sampling of 2 or 3 age groups in finishing via fecal PCRs, oral fluid PCRs, and/or serology will often be enough to monitor the program and detect the presence of clinical or subclinical ileitis. If ileitis is present, then consider changing the vaccination and/or feed antibiotic program.

Also consider sampling replacement gilts to catch subclinical ileitis before it gets into the sow farm. Commercial vaccines do not prevent subcinical infection and finding ileitis early may prevent a PHE outbreak later.

Five take home messages

1) **Clinical Ileitis** – Do not" walk by" clinical ileitis. Analyze any form of diarrhea or abnormal fecal color with a fecal PCR assay for *Lawsonia intracellularis* to evaluate its significance.

2) **Subclinical Ileitis** – Do not be complacent and assume ileitis is under control because no obvious clinical diarrhea is present. Realize subclinical ileitis is ubiquitous and potentially costly. Look for it with fecal PCR, serology or oral fluid PCR's analysis.

3) **Ante-mortem diagnostics**: PCRs—Current knowledge indicates when U of MN VDL fecal PCR ct values are ≤ 32 of 40ct that significant PPE lesions and poor pig performance exists. At ISU VDL fecal PCR ct values of ≤ 20 of 35ct had a 100% Predictive Value for the presence of IHC positive PPE lesions. **Serology**—High IPMA titers ≥ 960 also likely indicate past or current exposure to significant ileitis disease pressure. Request titers for herd monitoring.

4) **Understand your ileitis control program** – Know what to expect for *Lawsonia* titers or fecal PCR's with ileitis vaccines, DLI® immunity, and/or feed medications. This will allow you to better interpret field infections.

5) **Re-evaluate the program** - If significant levels of clinical or subclinical ileitis are present, then re-evaluate your ileitis control program and make changes to maintain herd profitability.

References

1. Guedes R, Winkelman N, Gebhart C, Relationship Between the Severity of Porcine Proliferative Enteropathy and the Infectious Dose of *Lawsonia intracellularis*; Veterinary Record (2003) p153, 432-433.

2. Beckler D, Armbruster G, Gebhart C, Development and Quantitative PCR assay for the Detection and Enumeration of *Lawsonia intracellularis*; AASV 2010 Poster Session.

3. Guedes R, Gebhart C, Deen J, Winkelman N, Validation of an Immunopero idose Monolayer Assay as a Serologic Test for Porcine Proliferative Enteropathy; Ref. Diagnosis Invest 14: 518-530 2002.

4. Vasquez E, Comparison of IPMA and Commercial Blocking ELISA for *Lawsonia intracellularis*; AASV 2018.

5. Hammer JM, The Temporal Relationship of Fecal Shedding of *Lawsonia intracellularis* and Seroconversion in Field Cases; JSAAP Vol. 12 No 1 pp29-33.

6. Winkelman N, Mueller A, Winscher R, Validation of an Ileitis Mucosal Homogenate Seeder Pig Model to Study the Natural Transmission of *Lawsonia intracellularis* in Grow-Finish Pigs; 48th AASV (2017) pp 170-174.

7. Paradis M, Gebhart CJ, Tool D, et al, Subclinical Ileitis: Diagnostic and Performance Parameters in a Multi-Dose Mucosal Homogenate Challenge Model; J of Swine Health and Prod., 2012, 20(3): 137-141.

8. Winkelman N, et al Use of a Challenge Model to Measure the Impact of Subclinical Porcine Proliferative Enteritis on Growth Performance in Pigs; AASP 1998 pp 209-211.

9. Beckler D, Armbruster G, Rutten-Ramos S, Evaluation of Fecal Shedding by a High-Throughput qPCR Assay in a *Lawsonia intracellularis* Challenge; Proc. AASV (2012) pp 149-153.

10. Burrough ER, Rotolo ML, Gauger PC, et al, Correlation of *Lawsonia intracellularis* Semi-Quantitive Fecal Polymerace Chain Reaction Assay Results with the Presence of Histologic Lesions of Proliferative Enteropathy and Positive Immunohistochemical Staining; J. Swine Health Prod. 2015; 23(4); 204-207.

11. Vannucci F (personal communication).

12. Winkelman NL, Measurement of the Correlation Between Lesion Length and Reduction in Average Daily Gain Using A *Lawsonia intracellularis* Challenge Model; Proc AASP 1999, 241-2.

13. Winkelman NL, Mueller A, Domangue R, Effectiveness of DLI[R] (diluted *Lawsonia intracellularis*) "Vaccine" followed by a Denegard[R] LC, Aivlosin, or Mecadox in Finishing Pigs Challenged with a High Dose LI mucosal homogenate. AASV 2014 pp 273-278.

14. Winkelman N, Mueller A, Domangue R, Comparison of DLI[R] Controlled Immunity and Two Commercial Ileitis Vaccines (Porcilis[R] and Enterisol[R]) in Grower Pigs Challenged with *Lawsonia intrcellularis* AASV 2018.

15. Kinsley K, Winkelman N, Gebhart C; Elimination of *Lawsonia intracellularis* from Pigs Fed Carbadox (Mecadox[R]). AASV 2006 pp 157-161.

Crucial to criminal: The range of perspectives on antimicrobial use in pork production

L. Karriker, DVM, MS, DACVPM
Iowa State University College of Veterinary Medicine, Ames, Iowa

Introduction

There are things known and there are things unknown, and in between are the doors of perception.

– Aldous Huxley

Most misunderstandings in the world could be avoided if people would simply take the time to ask, "What else could this mean?

– Shannon L. Alder

As an example of how hard it to understand this issue, I got even the title wrong when I submitted it for the program. I used "perspective" which is a physical, objective, measurable quality. It generally refers to alternative views of the same physical object when height, angle or distance is changed. It probably would have been more correct to use the term perception in the title. Wikipedia describes perception as "the organization, identification, and interpretation of sensory information in order to represent and understand the presented information, or the environment" and goes on to say that perception "is not only the passive receipt of these signals, but it's also shaped by the recipient's learning, memory, expectation, and attention" and that most of "this processing happens outside conscious awareness."

With that in mind, remember what our mission statement says: "It is the mission of the American Association of Swine Veterinarians to:

- increase the knowledge of swine veterinarians
- protect and promote the health and well-being of pigs
- advocate science-based approaches to veterinary, industry, and public health issues
- promote the development and availability of resources that enhance the effectiveness of professional activities
- create opportunities that inspire personal and professional growth and interaction
- mentor students, encouraging life-long careers as swine veterinarians"

Much of the AASV annual meeting is focused on increasing knowledge using objective observations and analysis. We advocate science-based approaches. Science based-approaches often assume that there is one truth that can be repeatedly distinguished from the variation and noise in measurement if we just get a big enough sample size, measure accurately, and apply objective statistical analysis. If fact, we work hard to remove the effects of perception ("bias?") from the knowledge generating process. Advanced degrees and entire statistical departments exist to reduce the effects of variation and bias on our investigation.

In short, our most valued swine veterinarian skill sets are terrible impediments to understanding the perceptions of others about what we do. We are trying to accurately measure, quantify and respond to a complex process that occurs to us and others without our conscious awareness. An accurate point-in-time understanding of the perceptions of a limited population about the use of antimicrobials in swine medicine under a single defined circumstance is a tough challenge and understanding the scope and breadth of the landscape of perceptions may be impossible.

In a study of Belgian, French, German, Swedish and Swiss pig farmers, Visschers et al reported that a convenience sample of farmers (n-281) generally perceived their own antibiotic use to be lower than peers within the same country and lower than that of farmers in other countries.[12] Although in this study, perceptions were not compared to actual use, the consistent perception that farmers used less than peers suggest that either their own relative use is underestimated or the use of peers is overestimated (assuming no bias in the study sample that would have selected for farmers actually using less). Further, farmers were significantly more worried about financial and legal issues in this cohort than emergence of antimicrobial resistance (AMR). However, C. Garforth warns us that "Their decisions cannot be predicted on the basis of simplistic notions of 'economic rationality'. While costs and returns are clearly important in weighing up choices (and farm management economics has given us several useful tools and methodologies for analysing these), farmers operate within a social context that both constrains and facilitates their behavioural choices. They have complex sets of core values, just like anyone else..."[3]

Discussion

How powerful are perceptions? In several human studies the effect of expectations and perceptions on antimicrobial prescribing practices reveals that perceptions, often erroneously, lead to changes in prescribing practices. Karras et al evaluated antibiotic use in academic emergency department cases of acute diarrhea.[4] They observed that physicians were more likely to prescribe antibiotics when they believed that patients expected them. However, they were only correct in identifying which patients expected antibiotics in approximately 33% of cases. Prescription practices influenced patient's reports of satisfaction with 100% of patients receiving antibiotics reporting satisfaction with care versus 90% reporting satisfaction among the patients that did not receive antibiotics.

A similar study evaluated respiratory infections treated in the emergency departments of 10 hospitals associated with academic units.[8] In that study, physicians were also more likely to prescribe antibiotics when they perceived patients expected them. However, as was the case with diarrhea patients, physicians in this study correctly identified only 27% of the patients that expected antibiotics. In this study, satisfaction was not influenced by whether patients received antibiotics. Satisfaction was reported by more patients that believed they had a better understanding of their illness (92%) versus those that believed they had no improved understanding (72%).

Klein et al. reported that for upper respiratory infection (URI)/pneumonia visits, 64.2% of patients were given antibiotics but that the prescribing rates for the 69 clinicians surveyed ranged from 7 - 91%.[5] Not surprisingly, clinicians that perceived antimicrobial therapy to be essentially harmless had higher rates of prescribing than those that believed "antibiotics may be harmful."

The association between satisfaction and prescribing practices has been confirmed more recently by the work of Cordelia R. Stearns et al.[10] In acute upper respiratory infections (URIs) antibiotic treatment was significantly associated with overall visit satisfaction and patients receiving antibiotics were more likely to report satisfaction with their treatment by the provider and the explanation of treatment they were given.

A. R. McCullough et al. conducted a systematic review of the public's knowledge and beliefs about antibiotic resistance.[6] They screened 3537 articles and extracted data from 54 studies representing 55,225 participants. Conclusions from the analysis included: "...the public have an incomplete understanding of antibiotic resistance, misperceptions about it and its causes, and do not believe they contribute to its development." They reported that "some participants had heard of antibiotic resistance, but most believed it referred to changes in the human body." McCullough conducted a similar review of clinicians (physicians) that included 57 studies and a total of 11,593 clinicians (physicians).[7] The proportion who believed AMR was a problem for their practice was smaller than the proportion who believed it was a problem nationally and they attributed responsibility for antibiotic resistance to patients, other countries and healthcare settings. Resistance was considered a low priority and a distant consequence of antibiotic prescribing. The report was titled "Not in my backyard: a systematic review of clinicians' knowledge and beliefs about antibiotic resistance."

The point here is not to impugn the human medical community or assign blame for AMR. Rather, the important message from these studies is that very well educated people struggle to: a) correctly identify the perceptions and expectations of others, and b) very well educated people do modify critical actions based on their own beliefs even in clinical and evidence based environments.

Several studies indicate that predictors of AMR and appropriate use of antibiotics include ethnicity, education and age but that the degree of association of these predictors varies by population.[1,2]

While the synthesis of second-hand information about AMR and appropriate use into perceptions clearly varies, it could be expected that first hand experiences might improve consistency of perceptions. However, Ritchie et al showed that patients that previously experienced an antibiotic related adverse drug reaction (aADR) did not differ in their perception of antibiotic safety or expectation of antibiotic treatment for respiratory infections from patients that had no previous aADR.[9]

Conclusions and actions

It is difficult to find reports that investigate the perceptions of people about antimicrobial use in animals. We might be well served to extrapolate a starting point from the human medicine studies described here combined with what we know about populations and organizations who might seek to influence our practice of swine medicine. Even a cursory attempt of this quickly illustrates that perceptions are likely to be less accurate and more varied than even suggested by the human perception of human data. This is, in part, due to: decreased first-hand knowledge of swine production and medication, the complexity of production systems, the persistent selection of messaging points to effect particular agendas, the persistence of sensational but inaccurate information about swine treatment with antimicrobials, societal shifts in how "information" or sensory input is acquired, ethnic and religious influence on the acceptability of pigs and pork, messaging that deliberately segregates antimicrobial use from welfare states, and presence or absence of competition for people's attention.

Understanding and influencing perceptions will require a willingness to spend substantial time and effort to listen to viewpoints that we may find difficult to hear and prefer to dismiss. In the last fifteen years, there has been a societal shift to peer-to-peer transfer of information and away from central one-to-many distributions. Further, a number of studies suggest that proximity of the source and the frequency with which a message is heard influences perceptions of credibility. The implications for swine veterinarians is that annual dissemination of research from a central national source may have far less impact than constant, continuous, local discussion of the issue. Taking what you learn at this meeting back and applying it immediately in practice has always been a desirable standard for new swine medicine technology but if we seek to influence perceptions about antimicrobial use and the contribution of pork production to AMR, we will each need to return to practice with what we've learned and share it in hundreds or thousands of local venues from dialog at the barbershop to the comments section on Facebook.

References

1. Cespedes, A., et al. Knowledge, attitudes, and practices regarding antibiotic use among Latinos in the United States: Review and recommendations, AJIC. 2006; 34(8):495-502.

2. Corbett, Kitty K., et al. Appropriate antibiotic use: variation in knowledge and awareness by Hispanic ethnicity and language, Preventive Medicine 40 (2005) 162–169.

3. Garforth, C. Livestock Keepers' Reasons for Doing and Not Doing Things Which Governments, Vets and Scientists Would Like Them to Do, Zoonoses and Public Health 62 (suppl. 1) (2015: 29–38).

4. Karras, D., et al. Antibiotic Use for Emergency Department Patients With Acute Diarrhea: Prescribing Practices, Patient Expectations, and Patient Satisfaction, Ann Emerg Med. 2003; 42:835-842.

5. Klein, E.Y. et al. Categorical Risk Perception Drives Variability in Antibiotic Prescribing in the Emergency Department: A Mixed Methods Observational Study, SGIM. 2017; 32(10):1083–9.

6. McCullough, A.R., et al. Not in my backyard: a systematic review of clinicians' knowledge and beliefs about antibiotic resistance, J Antimicrob Chemother 2015; 70: 2465 – 2473.

7. McCullough, A.R., et al. A systematic review of the public's knowledge and beliefs about antibiotic resistance, J Antimicrob Chemother 2016; 71:27–33

8. Ong, S., et al. Antibiotic Use for Emergency Department Patients With Upper Respiratory Infections: Prescribing Practices, Patient Expectations, and Patient Satisfaction, Ann Emerg Med. 2007; 50:213-220.

9. Ritchie, S. R., et al. Previous antibiotic-related adverse drug reactions do not reduce expectations for antibiotic treatment of upper respiratory tract infections, Journal of Global Antimicrobial Resistance 10 (2017) 256–260.

10. Stearns, C., et al. Antibiotic Prescriptions Are Associated with Increased Patient Satisfaction With Emergency Department Visits for Acute Respiratory Tract Infections, Acad Emerg Med. 2009; 16(10):934-41.

11. Visschersa, V.H.M, et al. A Comparison of Pig Farmers' and Veterinarians' Perceptions and Intentions to Reduce Antimicrobial Usage in Six European Countries, Zoonoses and Public Health 2016, 63, 534–544.

12. Visschersa, V.H.M, et al. Perceptions of antimicrobial usage, antimicrobial resistance and policy measures to reduce antimicrobial usage in convenient samples of Belgian, French, German, Swedish and Swiss pig farmers, Preventive Veterinary Medicine 119 (2015) 10–20.

Antimicrobial use: Current EU perspective

MEC White, BVSc, LLB, DPM, FRCVS
Independent Pig Consultant and President of the Pig Veterinary Society

* * *

Post antibiotic apocalypse [1]
... emergence of "nightmare bacteria" [2]

Most common infections will no longer have a cure and once again could kill unabated [3]

Antimicrobial Resistance is the major global health challenge of our time [4]

* * *

These quotes are as significant for who said them as much as the disturbing messages they carry. The primary concern of Antimicrobial Resistance (AMR) relates to human health and therapy and the perceived risk of cross-contamination from animals of bacteria selected for resistance by veterinary use. The evidence for this is often overstated, and there is now a general acceptance in Europe that the major cause of AMR in human medicine results from human usage. Nevertheless, the concept of "One Health", in which humans, animals and the environment all interact, cannot be ignored, but the drive to reduce and refine use of antibiotics in animals is more a socio-political issue than a scientific one. This was re-emphasised in September 2016 when 193 nation states sitting within the UN general Assembly made secure commitments to address AMR[5] which complemented AMR declarations from both the G7[6] and the G20.[7]

Such a state of affairs leaves the subject open to capture by lobby groups and marketing executives and this has helped to drive policy in Europe over the last seven to ten years. In Europe at least, arguments relating to scientific evidence of AMR transfer from animals to man have largely been side-lined in pursuit of the political imperative. As a veterinary surgeon this approach does not sit comfortably but one has to face reality whilst retaining the primary duty to ensure health and welfare.

A brief history in the UK and Europe

In the early 1980's antibiotic usage in pig keeping was widespread and extensive. Antibiotic growth promoters were incorporated in feed without veterinary prescription and, whilst therapeutic antibiotics were veterinary prescription only (at least in some countries), medication of feed to control widespread enzootic infectious diseases such as *Mycoplasma hyopneumoniae* and its complications was extensive, often covering pigs from weaning to slaughter.

The introduction of maximum residue levels and withdrawal periods led to considerable scaling back of use of in feed antibiotics in finishing pigs; the availability of *Mycoplasma* vaccines from the mid 1990's has further helped reduce use, and in 2006 a complete ban on antibiotic growth promoters was introduced across the EU. As such, in theory in the EU antibiotics are now only available under a veterinary prescription.

However, the structure of most western European pig farming – with the exception of for example modern Spanish production – is single or twin site continuous flow operations. With type 1 porcine reproductive and respiratory syndrome (PRRS) and porcine circovirus and associated diseases (PCVAD) – both immune modulating diseases – superimposed, chronic enzootic herd bacterial challenge is commonly persistent particularly post-weaning, meaning that antibiotic treatment remains essential.

Due to these challenges, programmes have evolved whereby antibiotics are widely used for prophylactic purposes in the long-term, particularly in pigs from weaning (3 to 4 weeks) to 10 to 12 weeks of age.

In addition to pressure to reduce overall use of antibiotics, focus of attention has been placed on the use of actives viewed as critically important for human use with different countries taking different views. Consistency at present is lacking.

EU pig herd

The largest sow populations in Europe are in Spain (2.5 million), Germany (2 million), Denmark (1.2 million), France (1 million), Netherlands (1 million) and Poland (< 900,000), By contrast UK (400,000), Ireland (140,000) and Sweden (140,000) are very small in production terms.[8]

The total EU sow population of approximately 12 million compares to 6 million in the United States and approximately 1.25 million in Canada. However, as in North America, there is considerable cross border movement of weaners for finishing, with Denmark exporting 40 percent of its weaners and the Netherlands 15 percent. (Figure 1) [9]

The largest producers of pig meat are Germany and Spain (24% and 17% of the EU total production of 23.4 million Tonnes) – unsurprising given their sow populations augmented with imported weaners along with slaughter pigs derived from neighboring countries.

How is antibiotic use measured?

Unfortunately, there is no single agreed system of measuring antibiotic use, and in many countries usage is not broken down by sector or species. The European Medicines Agency (EMA) collates figures, albeit delayed, for overall agriculture sales of antibiotics and show that the highest pro rata consumers of antibiotics for food producing animals were Spain, Italy and Cyprus. (Figure2)[10]

Figure 1: Net movement of weaners within EU[9]

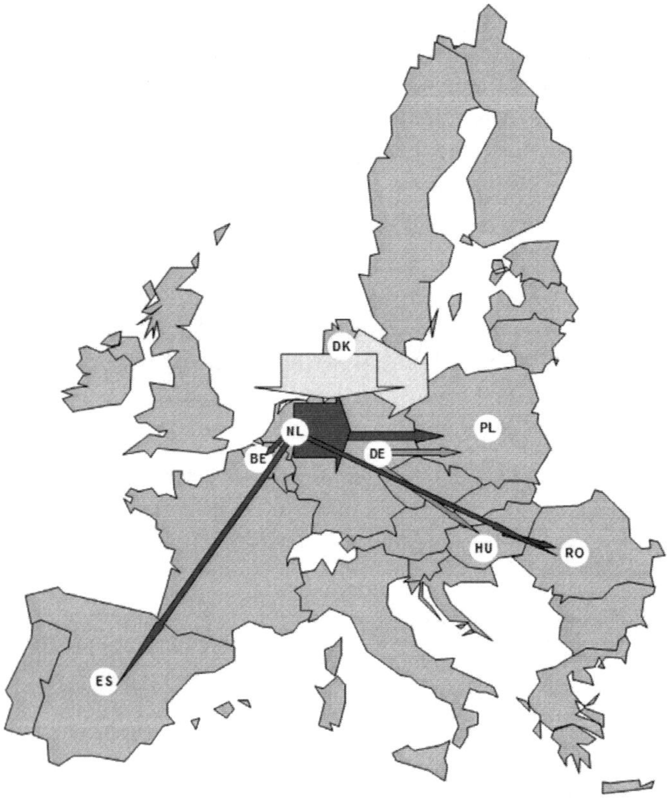

However, these figures hide the breakdown of livestock and the populations within each country.

The standardised method adopted by the European Medicine Agency is to express use in terms of milligrams of active antibiotic salt per kilogram population correction unit (PCU). In this a weight is assigned to each species (and subclass within each species) at which animals are most likely to be treated. For pigs this assumes weaners are expected to weigh 25kg and finishing pigs 65kg. (Sows are assigned a nominal weight of 240kg.) [11,25]

This methodology is useful for whole population comparisons but is criticised as being inappropriate for on farm measurement. However, no specific agreed alternative system has been adopted with alternative approaches employed, including:

- Defined average daily dose (DaDD)
- Defined daily dose (DDD)
- Defined course
- Mg/kg liveweight produced
- Mg/kg deadweight produced.
- Treatment incidence (TI200d)

The Dutch and the Danes have been at the forefront of antibiotic reductions in Europe, and both favor a DDD system, although they use different calculation methodology, making comparison using this system impossible (see below).

Drivers for change

Whilst the subject of AMR has been discussed for many years, major turning points were the Dutch needing to address LA-MRSA in pigs that caused human disease in 2004 and later in 2010 with Denmark beginning to address use.

This was followed by the Defra UK Five Year Strategy 2013,[12] aiming to reduce and refine use of antibiotics in UK agriculture, the WHO Global Action Plan on AMR 2015,[13] and a series of reports compiled by Lord (Jim) O'Neill for the UK Government in 2015-6. In the final O'Neill report a recommendation was made to reduce antibiotic use on a multinational level in all food producing animals to 50mg/kg PCU by 2020[14] – a target that was accepted by the UK Government to be achieved by 2018. As can be seen from Figure 2 some countries (particularly Italy, Spain and Cyprus) have a very long way to go whilst the Nordic countries (only Sweden and Finland being in the EU) are held up as paragons of virtue.

Sweden in particular and its pig industry are often cited by pressure groups as a model for the UK and the rest of Europe. However, their pig industry is small (140,000 sows) and almost totally internally focused to feed a small national population of 10 million people who pay the highest price for food in the EU. Cost of pig production consistently exceeds the EU average by 10 to 15 percent. Additionally, Sweden remains PRRS free, having eradicated the disease by depopulation within months following its recognition in 2007 and have maintained import restrictions to keep it out.[15] It therefore cannot be viewed as typical of pig production and consequent antibiotic use.

In 2016, following the O'Neill report, the Responsible Use of Medicines in Agriculture Alliance (RUMA) set up a Targets Task Force (TTF) to set antibiotic reduction targets for the UK for all agricultural sectors. These targets were both quantitative – looking for absolute reductions – and qualitative in terms of curtailing use of critically important antibiotics. The Dutch and the Danish activity to reduce use has however preceded this.

Why is pig use of antibiotics high?

There are a number of reasons why use of antibiotics in pigs exceeds that of most other sectors. The UK poultry meat industry began to address their usage in 2012 and have seen a 71 percent drop in five years.[16] This has been achieved by a range of actions that have incorporated husbandry changes, corporate investment, vaccine application, and attitudinal change. The challenge is for the pig industry to follow.

The TTF report[17] highlighted many factors driving antibiotic use on UK pig farms:

a. A wide range of persistent enzootic bacterial disease challenges for which vaccines are not available, reliable or economically attractive (e.g. *Streptococcus species*, *Haemophilus parasuis*, *Brachyspira species*, *Lawsonia* etc).

Figure 2: Antibiotic sales for all livestock species in the EU 2014[10]

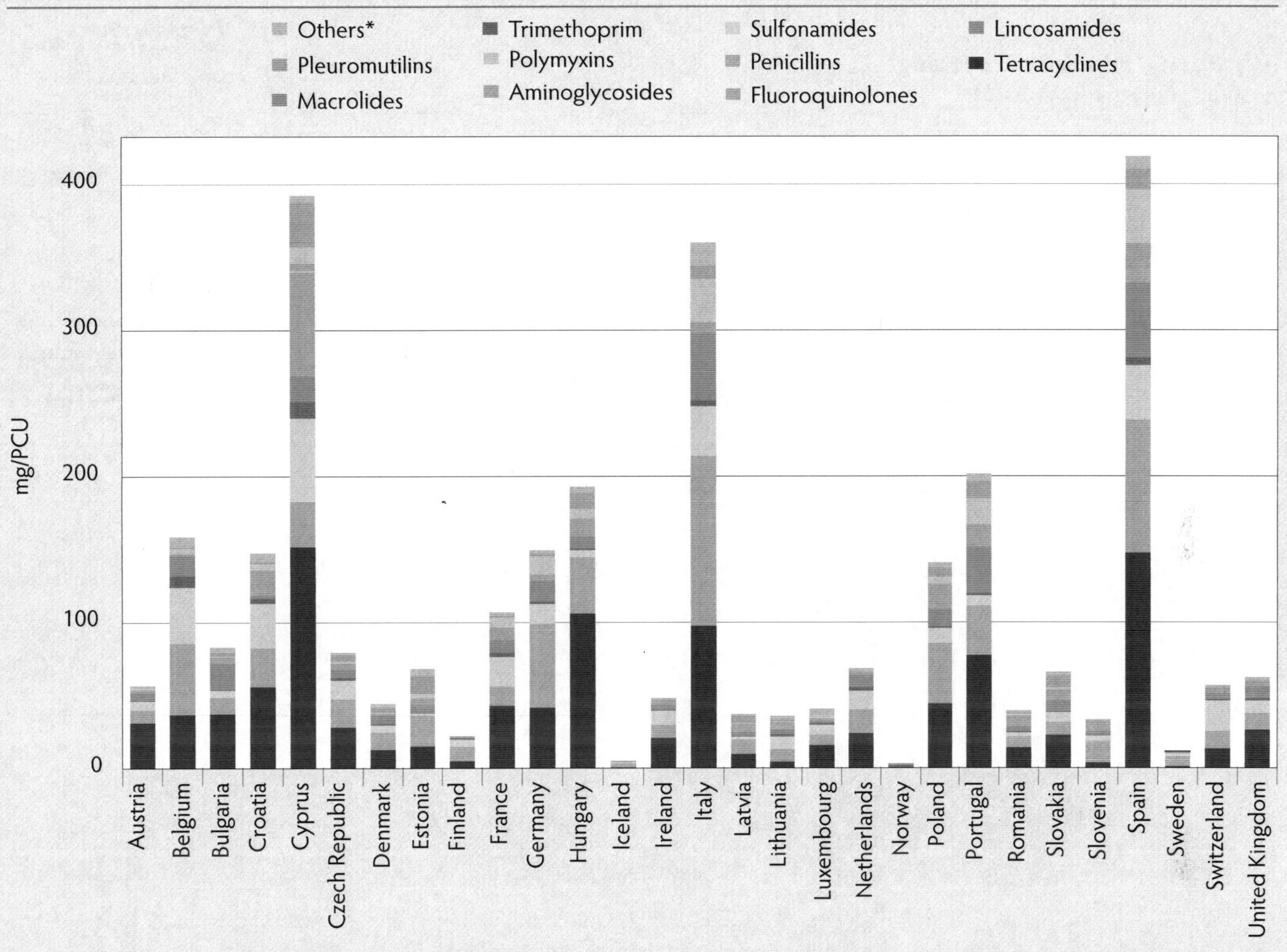

b. Continuous production systems at different points within the production cycle that allow enzootic disease to continuously recycle in a dynamic farm population.
c. Presence of major enzootic non-bacterial diseases which have immune modulating effects on pigs, "opening the door" to secondary bacterial disease (PRRS, *Mycoplasma hyopneumoniae* and PCVAD).
d. Variable or inadequate vaccinal control of these and other agents.
e. High proportion of UK pigs kept on solid floor (straw bedded) systems, creating hygiene challenges and increased faecal/oral recycling.
f. Old buildings compromising environmental control and husbandry, (e.g. pig flow, hygiene measures)
g. Poor understanding of environmental needs of pigs and poor control systems.
h. Inadequate internal and external biosecurity measures.
i. Geographical clustering of pig farms, leading to disease exchange and spread.
j. Ease of medication via feed, encouraging prolonged treatments to fit with feeding regimes. Conversely limitations on water based medication application.
k. 'Cheap' medication, especially via in feed route producing low cost insurance against disease threats.
l. Habit.
m. Fear of consequences of withdrawal following historical bad experiences.
n. Pressures on farm margins requiring avoidance of production challenges and limiting investment in infrastructure, husbandry techniques etc.
o. Planning constraints limiting new buildings and innovative changes.

Many of these features are reflected across Europe particularly in countries with long established industries that have not enjoyed major reinvestment.

Figure 3: CIA classification by different bodies[18]

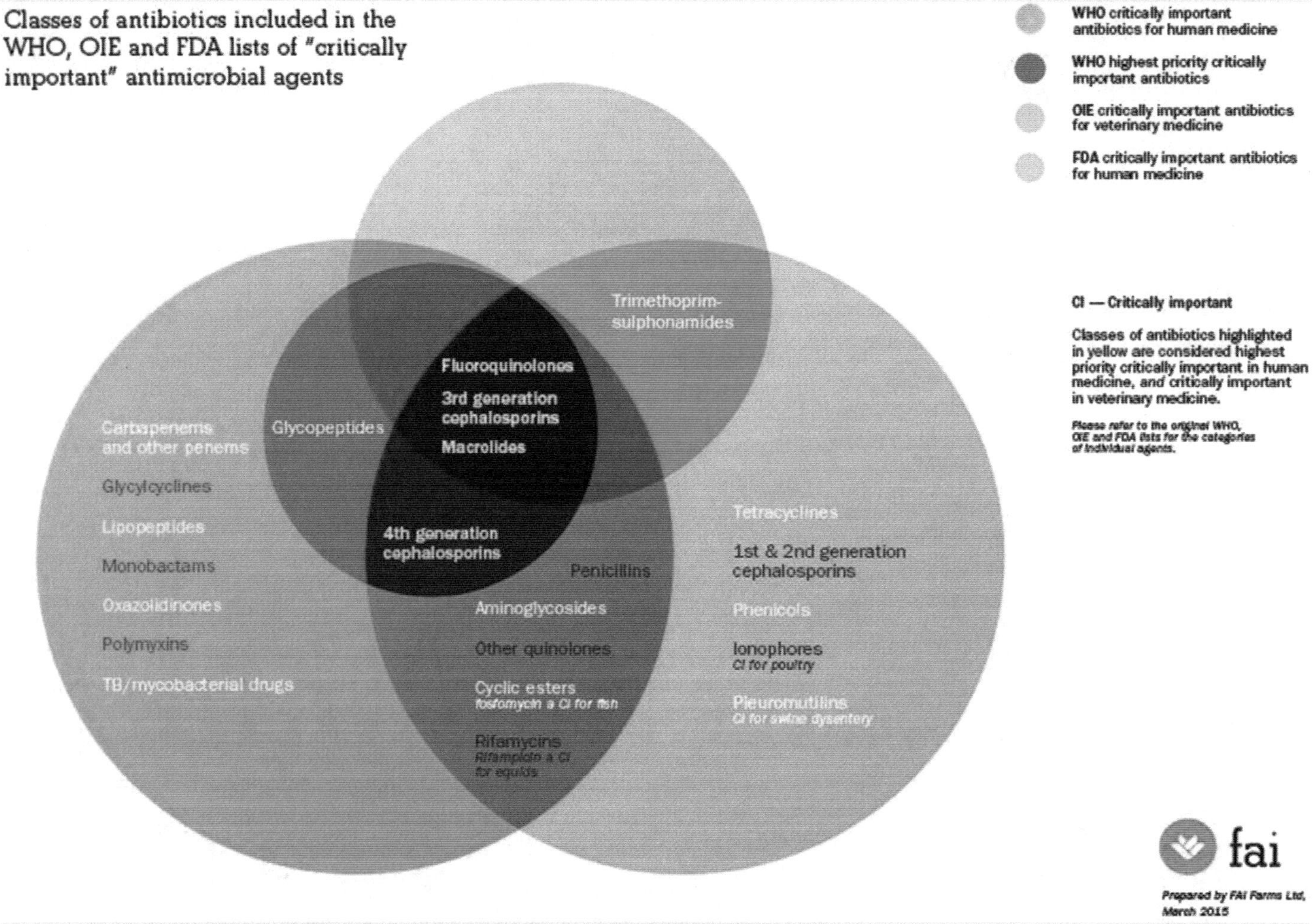

Critically important antibiotics

Certain active compounds have become of critical importance for use in human medicine. However, there is no universal agreement as to which groups are included; the most widely embracing list is produced by the WHO but the European Medicines Agency have reviewed this and currently view Fluoroquinolones, third and fourth generation Cephalosporins, and Colistin as the highest priority, critically important antibiotics (HP-CIA's) and thus have been adopted in the UK. Individual countries and sectors have gold plated this by imposing voluntary or legislative bans on use of such CIA's. (The UK poultry industry have added Macrolides to the list due to issues relating to Campylobacter jejuni food poisoning in humans.

In the US, the FDA classify Fluoroquinolones, third and fourth generation Cephalosporins, Macrolides and Trimethroprim/Sulpha as CIA's for human medicine. (Figure 3)[18]

Who is doing what in Europe?

The Dutch and the Danes have largely led the way to reduced antibiotic use in livestock over the last 10 years with a structured, rule-based approach.

In Denmark, legislation was passed in 2010, which required submission of data to a central database, and average daily dose per 100 animals over 9 months was used to benchmark production. Yellow and red card systems were introduced to control and penalise producers using high levels above pre-set thresholds. Routine prophylactic use is banned (but "treatment" during incubation is permitted).

Latest figures from DANMAP suggest rates of reduction in use has slowed with pig use stated as 44mg/kg PCU.[19] Figure 4

In the Netherlands, following outbreaks of LA-MRSA in people linked to pigs, the Minister of Agriculture, encouraged by the Ministry of Health, "requested" reduction in use of 20 percent in 2011, a further 50 percent in 2013 and a further 20 percent by

Figure 4: Antimicrobial consumption in the pig production, and the distribution on age groups, Denmark (DANMAP 2017)[19]

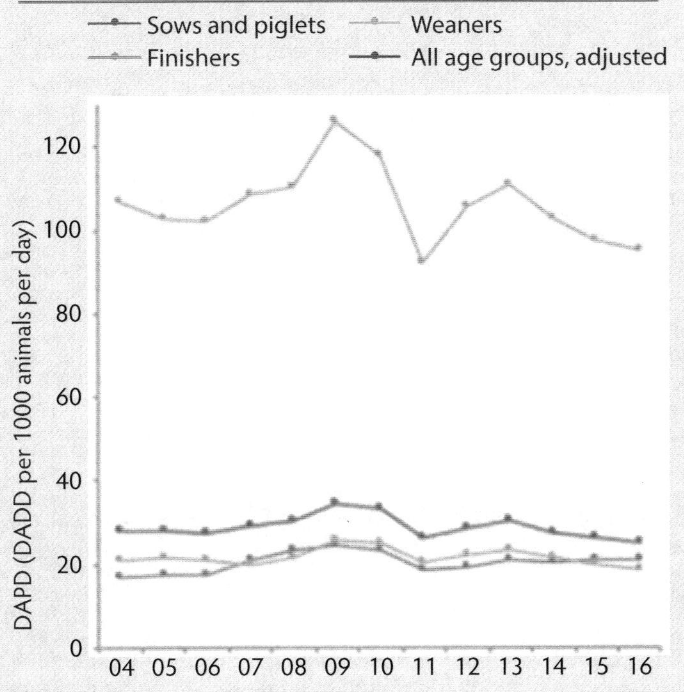

2015. Threshold use on farm was set by the National Veterinary Medication Association at ADDD/year of 10, with enforcement action on producers and vets if greater than 22. (ADDD/year is based on average weight percent on farm during the year, not on production.)[20] In addition, in 2013, a 'drug law' was passed to impose a treatment formulary on vets and producers with the choice of antibiotics limited and driven by MRSA problems which has not declined. Dutch pig use is now 50mg/kg PCU, although anecdotally pig health has suffered as a result of cutting use too far too fast.

As a veterinarian it is interesting to note that despite these reductions, the prevalence of LA-MRSA CC 398 in pigs in Denmark has risen dramatically and has not declined significantly in the Netherlands.[21,22]

In Belgium an advisory approach has been taken, with farms provided with "coaching" to improve external and internal biosecurity and husbandry which has led to a 50 percent reduction in antibiotic use in nursery/finishers.[23] Overall use has declined 20 percent in 5 years with a further aim to halve pig industry use from 2011 to 2020.[24]

Recent data within the UK reveals that overall antibiotic sales in livestock is down 27 percent 2014 to 2016 and pig use has declined 34 percent in twelve months 2015 to 2016.[25] (Tables 1, 2). In particular there has been a massive shift away from use of in feed medication,[25] (Table 3) and a reduction in most classes of antibiotics. Figure 5.[25] Target reductions set for the UK pig industry are to reach 99mg/kg by 2020 from a start point of 264 in 2015 - a planned total reduction of 63 percent over a 5-year programme (2016-2020).[17] Compulsory recording via an electronic medicine book has recently been introduced to enable measurement of use at individual and industry levels.

Use of some CIA's (particularly Fluoroquinolones and 3rd/4th generation Cephalosporins) in the main EU pig producing countries is low and is targeted by most to remain so. Colistin use is still high in some countries with no evidence as yet that those countries with high in-feed use (Germany, France, Spain) are planning reduction. CIA usage in UK pigs has reduced dramatically from a low base in 2015 to 2016. (Figure 6)[25]

What about the USA?

Limited data is published in comparable form for US production. Smithfields have openly published their use for 2015 to 2016 for both their EU and US production areas. When converted to mg/PCU as per the EMA method declared usage is as shown in Table 4.[26]

How recent changes to legislation will affect the US use remains to be seen, but whilst the trends occurring within the EU are reflected here the absolute usage levels are high.

Where now?

The momentum for reduced reliance on antibiotics in pig production has gathered considerable pace over the last few years, and in some countries reductions have been sizeable albeit with some concerns over the effects on pig health.

The drive within Denmark historically and in the UK now is to improve pig health and therefore reduce antibiotic use rather than reduction for the sake of it. Such priorities have not necessarily been followed elsewhere. Time will tell how successful reduction programmes will be in the UK.

In some European countries such as the UK and more recently Ireland[27] the drive to reduce use has come from industry and veterinary channels (with encouragement from regulators) whereas the more rule-based mentality in some Northern European countries has led to legislation and enforcement. The challenge remains for the rest of Europe and the rest of the world to act.

Marketing groups should be restrained from making antibiotic use a competitive affair and, whilst in the US "no antibiotics ever" and "antibiotic free" labelling is available, it has been resisted in Western Europe to date.

References

1. Dame Sally Davies, CMO England. 2015. Available at `http://www.bbc.co.uk/news/health-21702647`. Accessed 11 December 2017.

2. CDC Director. 2013. Available at `https://www.healio.com/infectious-disease/news/print/infectious-disease-news/%7B24735a61-84f9-49c3-8083-cc981828e391%7D/cre-deemed-nightmare-bacteria-by-cdc`. Accessed 11 December 2017.

Table 1: Tonnes and (% of total sales) of active ingredient of antibiotic sold for species category indicated, 2012 - 2016[25]

		2012	2013	2014	2015	2016
Indicated for food producing animals only	Pigs and poultry only	235	217	235	214	127
	Pigs only	66	63	66	50	39
	Poultry only	47	43	43	38	26
	Cattle only	14	14	13	14	18
	Fish only	2.1	0.8	2.4	0.7	1.6
	Sheep only	0.2	0.2	0.1	0.2	0.2
	Multiple food producing species	32	30	24	30	33
	Total	**396**	**368**	**383**	**347**	**244**

Table 2: Tonnes of active ingredient of antibiotic sold for all species by class, 2012 - 2016[25]

	2012	2013	2014	2015	2016
Tetracyclines	**201**	**194**	**181**	**166**	**116**
Trimethoprims/ Sulphonamides	**80**	**61**	**71**	**68**	**68**
Trimethoprims	13	10	12	11	11
Sulphonamides	66	51	59	57	57
ß-lactams	**94**	**94**	**95**	**81**	**80**
1st/2nd Generation Cephalosporins	5	5	5	5	6
3rd/4th Generation Cephalosporins (kg)	(1328)	(1192)	(1332)	(1202)	(1071)
Penicillins	19	20	12	9	17
Other Penicillins	69	68	77	66	57
Aminoglycosides	**22**	**24**	**24**	**23**	**22**
Streptomycins	10	11	9	10	16
Neomycin and framycetin	1	1	1	1	1
Other aminoglycosides	12	9	14	13	5
Macrolides	**41**	**40**	**48**	**38**	**29**
Fluoroquinolones (kg)	**(2381)**	**(2562)**	**(2590)**	**(2532)**	**(1796)**
Other	**24**	**21**	**24**	**28**	**20**
Colistin (kg)	(606)	(728)	(837)	(870)	(128)
Total	**464**	**436**	**445**	**408**	**337**

3. World Health Organisation: World Health Day 2011. Available at http://www.who.int/mediacentre/news /statements/2011/whd_20110407/en/. Accessed 11 December 2017.

4. Lord Gardiner of Kimble Parliamentary Under Secretary of State @ RUMA/ VMD conference day 29/10/17 London

5. United Nations General Assembly. 2016. Available at https://www.un.org/pga/71/wp-content/uploads/ sites/ 40/2016/09/ DGACM_GAEAD_ESCAB-AMR-Draft-Political-Declaration-1616108E.pdf. Accessed 11 December 2017.

6. Declaration of G7 Health Ministers. 8-9 October 2015. Available at https://www.bmg.bund.de/fileadmin/ Dateien/3_Downloads/G/G7/G7_Health_Ministers_Declaration_AMR_and_EBOLA.pdf. Accessed 11 December 2017.

7. G20 declaration and EU support. 2016. Available at http://europa.eu/rapid/press-release_STATEMENT-16-2967_en.htm. Accessed 11 December 2017.

8. Eurostat. 2016. Available at http://ec.europa.eu/eurostat/statistics-explained/index. php /Agricultural_production_-_animals. Accessed 11 December 2017.

9. Eurostat. 2013. Available at http://ec.europa.eu/eurostat/statistics-explained/index.php /File: Net_ exchanges_of_piglets,_2013_.png. Accessed 11 December 2017.

10. ESVAC report. 2016. Available at http://www.ema.europa.eu/docs/en_GB/document_library/Report /2016/10/WC500214217.pdf. Accessed 11 December 2017.

11. ESVAC report.. 2011. Available at http://www.ema.europa.eu/docs/en_GB/document_library/Report /2011/09/WC500112309.pdf. Accessed 11 December 2017.

Table 3: Tonnes and (% of total sales) of active ingredient of antibiotic sold for all species by route of administration, 2012 - 2016[25]

	2012	2013	2014	2015	2016
Premix	287 (62%)	263 (60%)	281 (63%)	233 (57%)	148 (44%)
Oral/Water*	108 (23%)	109 (25%)	100 (22%)	109 (27%)	97 (29%)
Injectable	49 (11%)	47 (11%)	45 (10%)	50 (12%)	72 (21%)
Tablets	16 (3%)	14 (3%)	16 (4%)	13 (3%)	16 (5%)
Intramammary	4 (< 1%)	3 (< 1%)	3 (< 1%)	3 (1%)	4 (1%)
Total	**464**	**436**	**445**	**408**	**337**

* Excluding tablets, including bolus preparations.

Figure 5: Change in antibiotic usage (%) recorded in eMB between 2015 and 2016 by class[25]

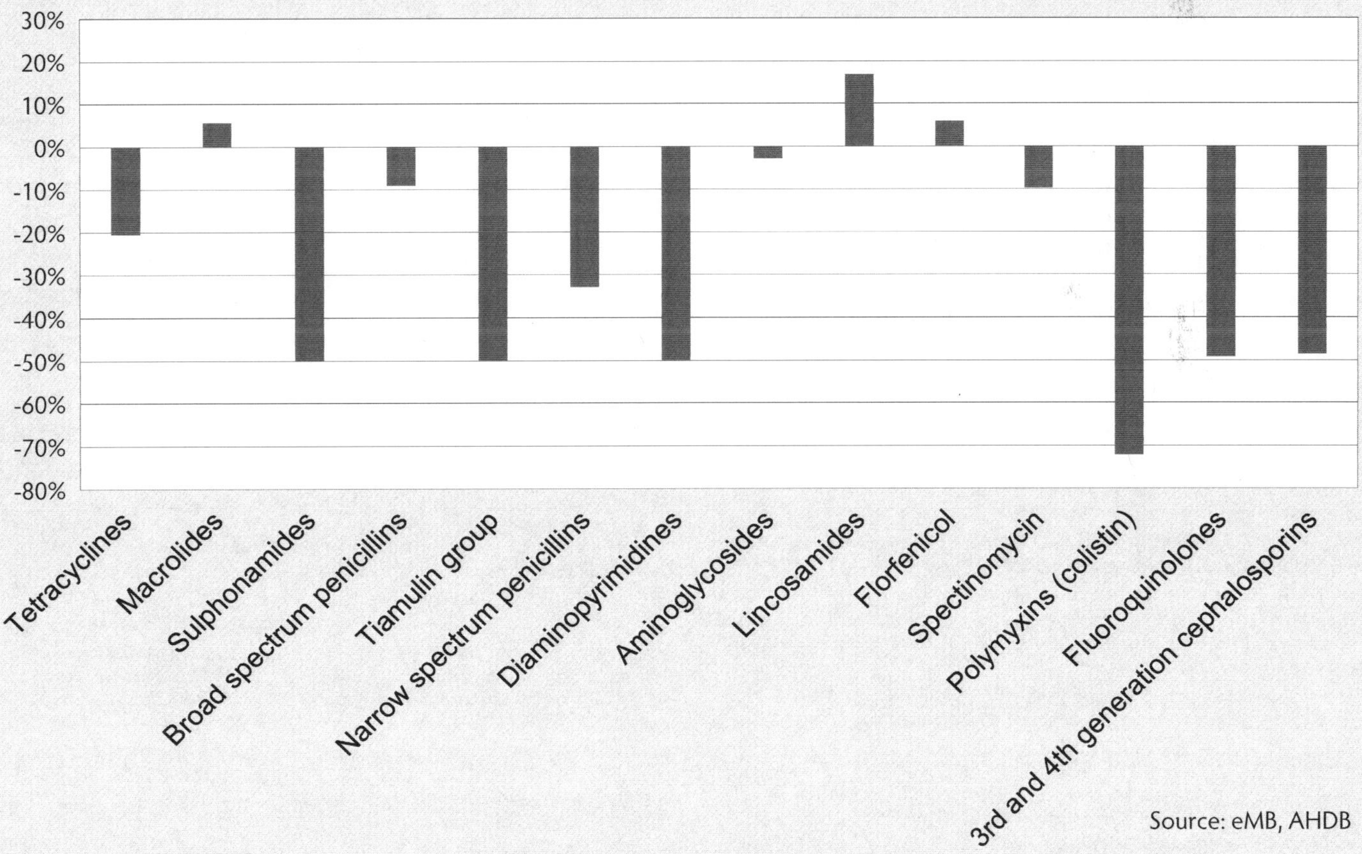

Source: eMB, AHDB

Figure 6: HP-CIA usage in pigs recorded in eMB by class, 2015-2016[25]

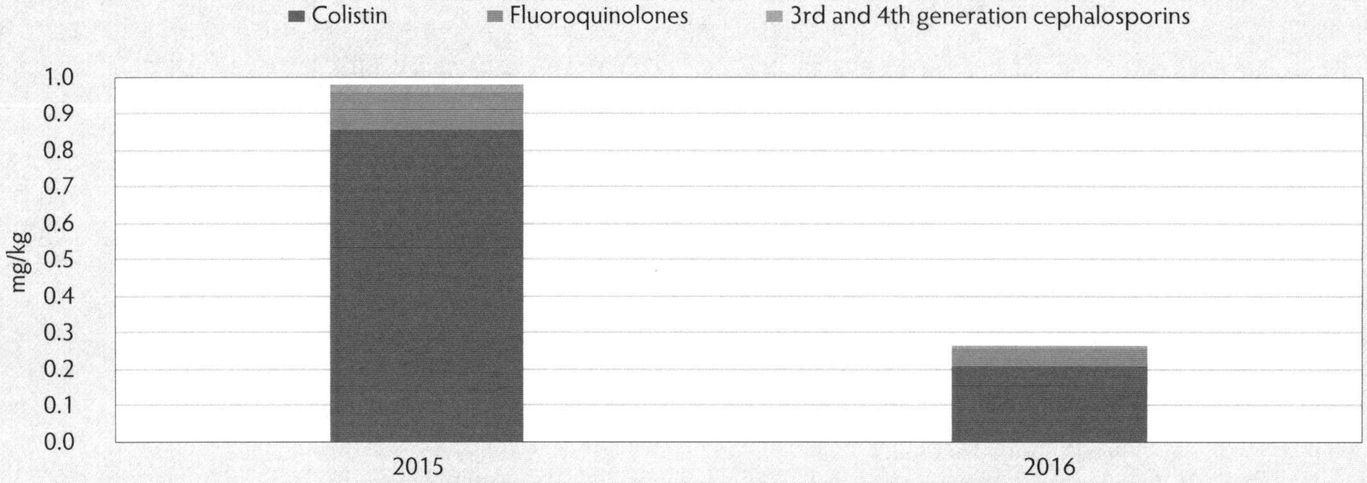

Source: eMB, AHDB

Table 4: Declared and calculated antibiotic use in pigs by Smithfields for EU and US production areas, 2015 and 2016[26]

		mg/lb	mg/kg	mg/PCU
US	2015	152	334	568
	2016	118	260	441
EU	2015	142	312	530
	2016	129	283	481

12. UK Five Year Antimicrobial Resistance Strategy 2013 to 2018. 2013. Available at https://www.gov.uk/ government/uploads/system/uploads/attachment_data/file/244058/20130902_UK_5_year_AMR_strategy.pdf. Accessed 11 December 2017.

13. WHO global Action Plan on Antimicrobial Resistance. 2015. Available at http://www.who.int/anti microbial -resistance/global-action-plan/en/. Accessed 11 December 2017.

14. The Review on Antimicrobial Resistance. J O'Neill. 2016. Available at https:// amr-review.org/ sites /default/files/160525_Final%20paper_with%20 cover.pdf. Accessed 11 December 2017.

15. Frössling J, Agren EC, Eliasson-Selling L, Lewerin SS. 2009. Probability of freedom from disease after the first detection and eradication of PRRS in Sweden: scenario-tree modelling of the surveillance system. Prev Vet Med. 2009 Oct 1;91(2-4):137-45.

16. British Poultry Council Antibiotic Stewardship Report.2017. Available at http://www.britishpoultry.org. uk/download/bpc-antibiotic-stewardship-report-2017/. Accessed 11 December 2017.

17. RUMA Targets Task Force Report. 2017. Available at http://www.ruma.org. uk/wp-content/ uploads /2017/10/RUMA-Targets-Task-Force-Report-2017-FINAL.pdf. Accessed 11 December 2017.

18. Figure 2. 2016. Available at http://www.faifarms.com/portfolio-item/what-constitutes-critically-important-antimicrobials/. Accessed 11 December 2017.

19. DANMAP. 2017. Available at https://www.danmap.org/~/media/Projekt%20 sites/Danmap /DANMAP %20reports/DANMAP%202016/DANMAP_2016_web.ashx. Accessed 11 December 2017.

20. H Van Beers. 2013. Presentation to PVS Antibiotics meeting. AMR in Livestock; A Dutch Approach. 13 December 2013 Birmingham, UK.

21. Larsen J, Petersen A, Larsen AR, Sieber RN, Stegger M, Koch A, Aarestrup FM, Price LB, Skov RL. Danish MRSA Study Group. 2017. Emergence of livestock-associated methicillin-resistant Staphylococcus aureus bloodstream infections in Denmark. Clin Infect Dis. 2017 May 30. doi: 10.1093/cid/cix504. Available at https://www.ncbi.nlm.nih.gov/pubmed/28575216. Accessed 11 December 2017.

22. Dierikx C M et al. 2016. 10 years later: still high prevalence of MRSA in slaughter pigs despite significant reduction in antimicrobial use in pigs in NL *J Antimicrobial Chemotherapy* 71 2414-2418

23. Dewulf J. 2017. Pigs 2020 AHDB Pork Conference Solihull UK 7 June 2017.

24. Sjolund et al. 2016. Quantitative and qualitative antimicrobial usage patterns in farrow-to finish Pig herds in Belgium, France, Germany and Sweden, *Preventive Veterinary Medicine*, 13:41-50. Available at http://www.ncbi.nlm.nih.gov/ pubmed/27435645. Accessed 11 December 2017.

25. VARSS Report. 2017. Available at https://www.gov.uk/government/ publications/veterinary-antimicrobial-resistance-and-sales-surveillance-2016. Accessed 11 December 2017.

26. Smithfield Food's Sustainability Report. 2016. Available at https://www. smithfieldfoods.com/integrated-report/2016/introduction. Accessed 11 December 2017.

27. Ireland's National Action Plan on Antimicrobial Resistance 2017-2020. 2017. Available at http://health.gov.ie/blog/publications/irelands-national-action-plan-on-antimicrobial-resistance-2017-2020/. Accessed 11 December 2017.

What to expect when you're not expecting: A veterinarian, producer and feed mill perspective of an educational FDA audit

Peter Schneider, DVM, MS

On January 1, 2017 new regulations regarding the usage of medically important drugs in livestock feeds took effect for livestock producers in the United States. Over the decade proceeding 2017 the U.S. Food and Drug Administration (FDA) had taken numerous steps to place more control of antibiotic usage under the guidance of veterinarians. The most critical change required a veterinary feed directive (VFD) to be issued for medically important antibiotics used in livestock feed. Initial recognition that changes were to be made in the oversight of antibiotics used in livestock production was first noted in 2010, and the final ruling was codified in the Code of Federal Regulations in June 2015. None the less details regarding the implementation of VFDs and guidance to the parties involved with the generation and usage of VFDs remained vague until mere months prior to January 1, 2017.

On March 21, 2017 I was involved in an FDA audit to evaluate the implementation of the new VFD rules. The goal of the FDA VFD audit was to evaluate compliance among all levels of the livestock industry involved in the VFD process, namely feed mills, producers and veterinarians. To do this the FDA developed a vertical pathway to streamline the evaluation process and to simplify auditing. The Center for Veterinary Medicine (CVM) at the FDA would start the audit process by selecting a feed mill for evaluation by FDA inspectors. There was no advanced notice of the audits for feed mills; the FDA investigators showed up unannounced, introduced themselves and the audit process and asked to see VFD records from a variety of producers who purchased medicated feed from the mill. Then, investigators would select a handful of the VFDs evaluated and follow through to the veterinarian who produced the VFD and to the producer who fed the feed to determine their level of compliance.

At all levels the completeness of VFDs was evaluated. An inspection checklist was generated by the FDA to standardize the process for VFD inspection. Criteria for VFD evaluation included the name and contact information for the issuing veterinarian and producer, premise information where animals were located, name of the drug(s) included in the medicated feed, species, production class, number of head fed, level and duration of drug(s) to be used, indication for drug usage, withdrawal time, special instructions or cautionary statements, number of refills (if applicable), a statement indicating extralabel use of medicated feeds is prohibited, affirmation of the intent for combination of VFD drugs, and the signature of the veterinarian. More specific audit criteria were also laid out for the individual stakeholders involved in the VFD process. Sections were dedicated to evaluation of distributors (feed mills), veterinarians and producers.

All parties were audited on their VFD records storage. Specifically, inspectors checked that VFDs are maintained for up to two years. Distributors needed to show mill receipts and distribution records for medicated feed. In addition, distributor records needed to demonstrate that the correct amount of drugs were utilized for each batch of feed, and that feed formulas and labels correctly matched the VFD orders. Feed labels must also have had the VFD caution statement included. Veterinarians must have shown proof of licensure in the state where the VFD authorized feed is being fed, and must have understood the state's VCPR rules. Most of the producer specific questions focused on producer compliance with the instructions stated on the VFD.

New regulations required on-farm mixers who may distribute medicated feeds to contracted production sites to notify the FDA of their intent to distribute medicated feed. Due to this regulation, it was necessary to register my on-farm mill with the FDA; this is likely how my mill was selected for the VFD audit process by CVM. The FDA inspector was joined by a Livestock Compliance Officer from the Iowa Department of Agriculture and Land Stewardship. Together they completed the audit of my VFD records.

The audit process was simplified due to my status as the feed distributor, veterinarian and producer. I produced one copy of VFDs to suffice for all three aspects of the audit process and shared copies of my feed records from the farm mill. The process was quick, lasting about an hour, and was educational. The inspectors made it clear that the process was about identifying flaws with the early implementation process of the new VFD regulations only; there was no inspection of the feed mill or questions regarding non-VFD topics in the veterinary or producer portions of the audit. Though the process can result in corrective actions being assigned, indications were that the FDA feels an approach of "Educate Before Regulate" is most appropriate at this time. This line of thought is likely to continue through at least 2018. It is important to note that the FDA sees the current VFD audit program as different from traditional mill audits that have been done previously and will continue into the future. Additionally, the inspector made it clear that if the FDA suspects there are abuses in the VFD program they will perform a regulatory audit of the suspected offenders.

I did have a few simple flaws in my VFDs. As a clinic we decided to generate our own VFD program instead of using a commercially available VFD management tool. Due to formatting errors in the program a signature was left off one VFD. In another case wording caused the duration of medicated feed usage to be too ambiguous; the investigators felt it did not specify the exact duration of usage and left this up to the producer to decide. In a regulatory audit these errors would likely result in minor citations. The auditors were quick to point out that the extent of a corrective action remains in the purview of the investigator completing an audit, and that not all investigators may evaluate the severity of an infraction the same way.

During a follow up conversation with the FDA inspector who completed my audit, I asked about common mistakes that have been identified in other VFD audits. The most common findings for mills and veterinarians are incorrect addresses assigned to production sites. Inspectors are finding VFDs assigned to homes or company offices, rather than production sites. For producers, the most common finding is a lack of VFD records. Overall compliance has been good; the use of VFD management tools, like Global Vet Link, have kept the process consistent for most VFD stakeholders. The FDA will likely have a positive outlook on the current steps that have been in place if systemic failures are not detected during FDA VFD audit processes.

Measuring antibiotic use in pork production: Why, how, and for whom?

Peter Davies
College of Veterinary Medicine, University of Minnesota

Today, antibiotic resistance (AMR) is the hottest topic in medicine. After 50 years of debate and discussion, antibiotic use (ABU) in animals, and specifically the role it may play in AMR in humans, remains controversial. Although the contribution made by ABU in animals to the clinical burden of AMR in humans is uncertain, the popularity of the 'One Health' paradigm has cemented the need to incorporate animal use of antibiotics in programs to address AMR. Better measurement of ABU in animals is now a ubiquitous goal of such programs, and it is hard to argue against the idea that measurement of any item will assist efforts to manage it more effectively.

Systematic efforts to quantify antibiotic use in animals have mostly evolved over the last 20 years. Due to availability of data (albeit of variable quality among countries), analysis of national sales data has been the predominant approach. However, national sales data, even of the best quality, arguably give little if any insight into actual patterns of use, and particularly at the level of individual species. Some entities advocating for more detailed data on antibiotic use in food animals include:

- World Health Organization policy to standardize protocols to facilitate global harmonization in surveillance of ABU in humans and animals, and of AMR
- European Union directive to harmonize ABU surveillance in food animals across countries
- US National Action Plan for Combatting Antibiotic Resistant Bacteria goal of better surveillance of ABU in food animals
- US Government agencies seeking to gauge the impact on ABU of FDA GFI 209/213 and the VFD regulations
- Consumer groups/politicians advocating stricter regulatory oversight of ABU in food animals (e.g., the Delivering Antibiotic Transparency in Animals Act)
- Initiatives in some EU countries (Denmark, Netherlands, Belgium) to benchmark AMU at the individual user (farm, veterinarian) level
- Demands from downstream customers for assurances related to ABU in the supply chain
- Need for reliable ABU data to enable meaningful risk assessment of the impact of ABU in food animals on human health (McEwen and Singer, 2006)

While the consensus that better measurement of ABU in food animals is desirable is easy to accept, the devil always lies in the details. The objective of measurement cannot just be measurement! Designing an efficient and effective measurement system (for anything) is contingent upon clearly articulating the purpose. The 'why' of measuring ABU, is what drives the 'how' (Figure 1). The measurement process needs to be tailored to enable useful inferences in a context that will support effective communication and actions to achieve the desired purpose. The ability to optimize the scope (e.g., census vs. sample), scale, and specific metrics to be adopted is also dependent on the clarity of the end goal.

Figure 1: Schematic for constructing a system to measure antibiotic use in food animals

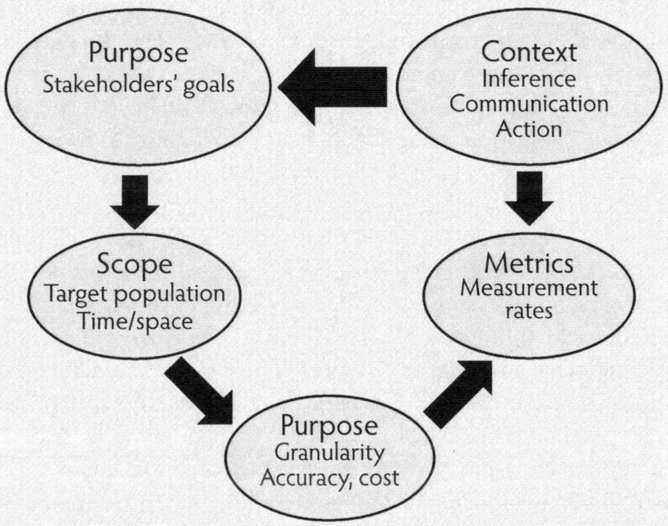

Defining the purpose – why?

1.Reduce the impact of ABU in animals on clinical resistance in human (and/or veterinary) medicine

The current momentum surrounding AMR is derived from the accepted crisis of clinical resistance in human medicine, which is generally attributed to ABU in any or all sectors, with ABU in human medicine most directly implicated. Although some concerns of clinical resistance are emerging in veterinary medicine (e.g., *Mannheimia* in cattle, *S. pseudintermedius* in dogs), they fall far short of a 'crisis' but do constitute warning bells and a sound argument for improving stewardship of ABU in veterinary medicine.

If the purpose of measuring ABU in animals is to reduce the impact of clinical AMR in human medicine, several issues stand in the way of optimizing the process of measurement. Firstly, the problem of how to quantify AMR in human medicine is not trivial. Data on the incidence of treatment failures (not to mention how this should be defined) are sparse. Should AMR be quantified at the level of the organism, phenotypic or genotypic of clinical resistance, or encompass the broader microbiome and resistome. The CDC has listed priority problems (organism/AMR profiles) of greatest concern, but few of these (e.g., *Salmonella, Campylobacter*) are known to have significant reservoirs in animals, and many (e.g., multiresistant gonorrhea) have no link to animals. Even if a system were available to quantify AMR in human medicine at the collective (i.e. all organisms vs. specific pathogens), the next issue would relate to attribution of AMR in relation to human vs. animal vs. other antibiotic uses. This would require more understanding of the mechanisms and pathways of selection and transmission of resistant organisms within and among species, and of how these could be influenced by specific patterns of usage in AMR. The concept of 'reducing selection pressure' is often thrown about, but without any shared understanding of what this really means. In a virtual vacuum of knowledge of how specific ABU practices in food animals impact any specific AMR problems in human medicine, it is arguably impossible to design systems to quantify ABU in animals that provide meaningful and actionable information to impact that outcome. This is presumably the reason why, to date, no effort to quantify ABU in food animals has been linked with specific targets for improving AMR problems in human medicine. After hearing presentations advocating the implementation of programs like those in Denmark and the Netherlands, I have posed the question of whether these programs have delivered, or even anticipate, any measurable benefits for AMR in human medicine. Responses have been along the lines of 'maybe in 5 years we will see something' or 'I believe this is the right approach, even if we cannot measure any benefit'.

2. Reduce ABU – less is better!

Given the inability to link ABU practices in animals quantitatively with AMR in human pathogens, simple reduction in ABU *per se* has emerged as the goal of current European initiatives. Generally these involve either mandated targets with no biological of scientific basis (e.g., in the Netherlands), and/or statistically based methods that target 'high users'. There is some logic underpinning these efforts (more use must be worse!), but they are arguably more 'faith based' than 'evidence based', and use completely arbitrary targets. To date, there has been no prediction of how any stated reduction in ABU would impact AMR in human medicine generally, or with respect to any specific pathogen/resistance pairs.

The most tangible benefit of this approach is arguably educational, in that mandatory reduction forces scrutiny of ABU practices in the field, which ultimately should lead to reduction in 'inappropriate' use (however defined), or at least less necessary uses. However, this approach raises several questions, particularly related to defining what specific metric(s) should be used for ABU. I would argue that the diversity of approaches that have been used to quantify ABU in animals in part results from the fact that none of them has been quantitatively or mechanistically linked to any specific AMR outcome in human or animal pathogens. Even the question of what antibiotics to include (medically important vs. not - what about ionophores?) is not straightforward. Conceptually, changes in ABU practices that replace medically important antibiotics with non (or less) medically important antibiotics would be deemed to be successful. However, they could also result in an increase, or no change, in ABU, depending on the metric employed (e.g., total weight). Industry-wide initiatives that target high users require substantial investment and compliance, yet to date provide relatively crude information that excludes the dose, timing and duration of antibiotic administration that are essential to understanding actual ABU practices.

3. Inform and motivate antibiotic stewardship

Increasingly there are reports that a considerable proportion of ABU in human medicine is 'inappropriate'. It is unlikely that the situation in veterinary medicine is very different, although in most countries the additional issue that veterinarians, unlike doctors, can sell what they prescribe is often an added criticism. How to define what is appropriate, and who is best qualified to decide, are not trivial questions. However, what is not in dispute is that to improve ABU practices, it is essential to have both measurement and discussion of what those practices entail. Inevitably, in some instances values will have more influence than data. For example, there is little evidence that growth promotion uses of antibiotics were more detrimental with respect to emergence of AMR than therapeutic uses. However, the perceived benefits for animal health and well-being were much less, and growth promotion uses became philosophically untenable. The current discussions targeting 'prevention' uses are an obvious next step. I would argue that prevention as currently practiced in the US industry involves both appropriate and inappropriate ABU. It is a challenge to our profession to understand and communicate the difference, and that there likely is no 'one size fits all' answer. Animal health and welfare impacts need to be a part of the discussion, as in many situations tradeoffs will be real, and veterinarians have to address the problems they confront 'here and now'.

Measurement systems for ABU designed to improve stewardship need to be customized for each species and even sector within species (e.g. dairy and beer), and include details on both ABU practices and understanding of prescribing behaviors. This is a relatively new area of research in human medicine that needs to be advanced in veterinary settings.

Resisting the urge to compare

Many 'external' voices (i.e. not involved in the industry or veterinary profession) for more granular measurement of ABU in food animals, and some internal voices, appear to be captivated by the urge to compare ABU among industries or among countries in complete isolation from any biological or medical context. I would argue that comparison of ABU between species, although an understandable curiosity, is biologically meaningless.

Such comparisons may be seized upon as ammunition for scolding, but serve no other purpose.

Comparisons across countries, but within species, have the potential to provide some meaning, but will inevitably by over simplistic and in the absence of comparative information about health challenges are likely to be misleading. For instance, Uganda uses substantially more anti-malarials than Finland – but this tells us nothing about whether this is more or less appropriate. All diseases are influenced by geoclimatic factors, and major pathogens of swine (e.g., PRRS, PCV2) vary greatly in their temporospatial incidence and impact. Comparing collective ABU while ignoring the diversity of disease challenges among locations.

Measurement of ABU in food animals is an essential component of efforts to optimize ABU. Interpretation will be meaningful if focused on individual species, or even sectors within a species (beef, dairy and veal production are very different industries). Limited understanding of how specific ABU practices influence the emergence of priority AMR concerns is perhaps the biggest impediment to making meaningful inferences about how practices should be improved. Identification of ABU practices that provide limited benefit to animal health and well being is probably the lowest hanging fruit for effective reduction in ABU, but requires effective assessment of outcomes in addition to measurement of use.

Antibiotic use monitoring in cattle

Michael D. Apley, DVM, PhD, DACVCP
Kansas State University, Manhattan, Kansas

Introduction

Antibiotic use monitoring has the potential to be both a useful benchmarking tool which can promote and support antibiotic stewardship in veterinary medicine, and also the potential to be a merciless, untargeted policy hammer which sets back animal welfare a couple of decades. There are several steps which can help us achieve the first, and avoid the latter.

Attributes of a useful antibiotic use monitoring system

Table 1 illustrates what we would like to have, and like not to have in an antibiotic use monitoring system. To change this from a wish list to a reality list, switch "easy" and "resource intensive"; Table 1 now becomes two lists of associated characteristics of monitoring systems and programs. You can have easy, or you can have useful, but it is very difficult to have both. This is a problem in communication regarding antimicrobial use programs. If all that is wanted is justification for policy, then nothing on the left side of Table 1 matters; the left side is only of interest to those wishing to move antibiotic stewardship forward.

First, defining stewardship

Stewardship has a different meaning between physicians and veterinarians. The Infectious Disease Society of America has a stewardship definition which does not include infection prevention.[1] In human medicine, infection prevention and antibiotic stewardship are considered two separate but overlapping processes. However, in veterinary medicine, it is rare that two separate people or groups address these two issues, they are addressed together. Therefore, the concept of antibiotic stewardship in veterinary medicine is all encompassing, including disease prevention and the judicious use of antibiotics when they are needed (Figure 1).

Within this stewardship cycle, there are aspects of antibiotic use which may be benchmarked to make the veterinarian and their clients aware of the primary disease challenges being experienced by others, and how they are being addressed. The nature of different production systems creates a situation in which the optimal antibiotic stewardship benchmarking metrics are often specific to the species, and even the production system type within species.

Table 1: The desired attributes of an antibiotic use monitoring system

What we would like...	We would not like it to be....
Coupled to cause	Uncoupled to cause for use
Accurate	Approximate
Granular	Aggregate
Current	Significant lag
Easy	Resource intensive
Enables benchmarking	Policy driver only

Appropriate numerators for beef cattle

The implication of all discussion of numerators and denominators in this paper is that they are associated with cause for use. With that in mind, there are three basic components of numerators which allow for evaluation of use from different perspectives.

The amount of drug used is dependent on both the number of animals receiving the dose and the regimen (primarily dose, frequency, and duration here) required to achieve a clinical effect. This clinical (in-vivo) potency against target pathogens is a driver of the required treatment, control, or prevention regimen, but so are factors such as bioavailability from any route other than intravenous administration, protein binding, metabolism to metabolites with varying degrees of activity, and elimination characteristics. These clinical realities make the comparison of relative kgs of use for different antibiotics a useless exercise, because the same factors affecting the regimen for clinical effect also affect the potential for selection of resistant organisms. The reality of resistance selection is that the outcome is dependent on both the antibiotic and the bacterial population which is exposed to the antibiotic.

Two things more useless than using kg of actual use to compare antibiotic groups are using sales data to do it, or combining all drug classes together to compare between species or countries. Neither of these approaches provides useful tools to the person at the point of the use decision.

The number of regimens are more straight forward to quantify from use records, although a regimen may result in only a day or so of actual exposure up to more than 200 days if used for an entire feeding period. The regimens are indicative of the number of animals exposed to a drug, without consideration of the nature of the regimen (dose, duration, frequency).

The days of exposure give an idea of how long the antibiotic pressure for potential resistance selection is present. The next logical question would then be "how long is too much"? Herein lies the problem; other than thinking that the shorter the better, we really don't have concrete evidence for how the combination of magnitude and duration of exposure serves to minimize or maximize the selection for resistant bacteria.

And the next challenge is how many days of exposure to assign to single injection treatments, which are common in cattle.

Figure 1: Antibiotic stewardship

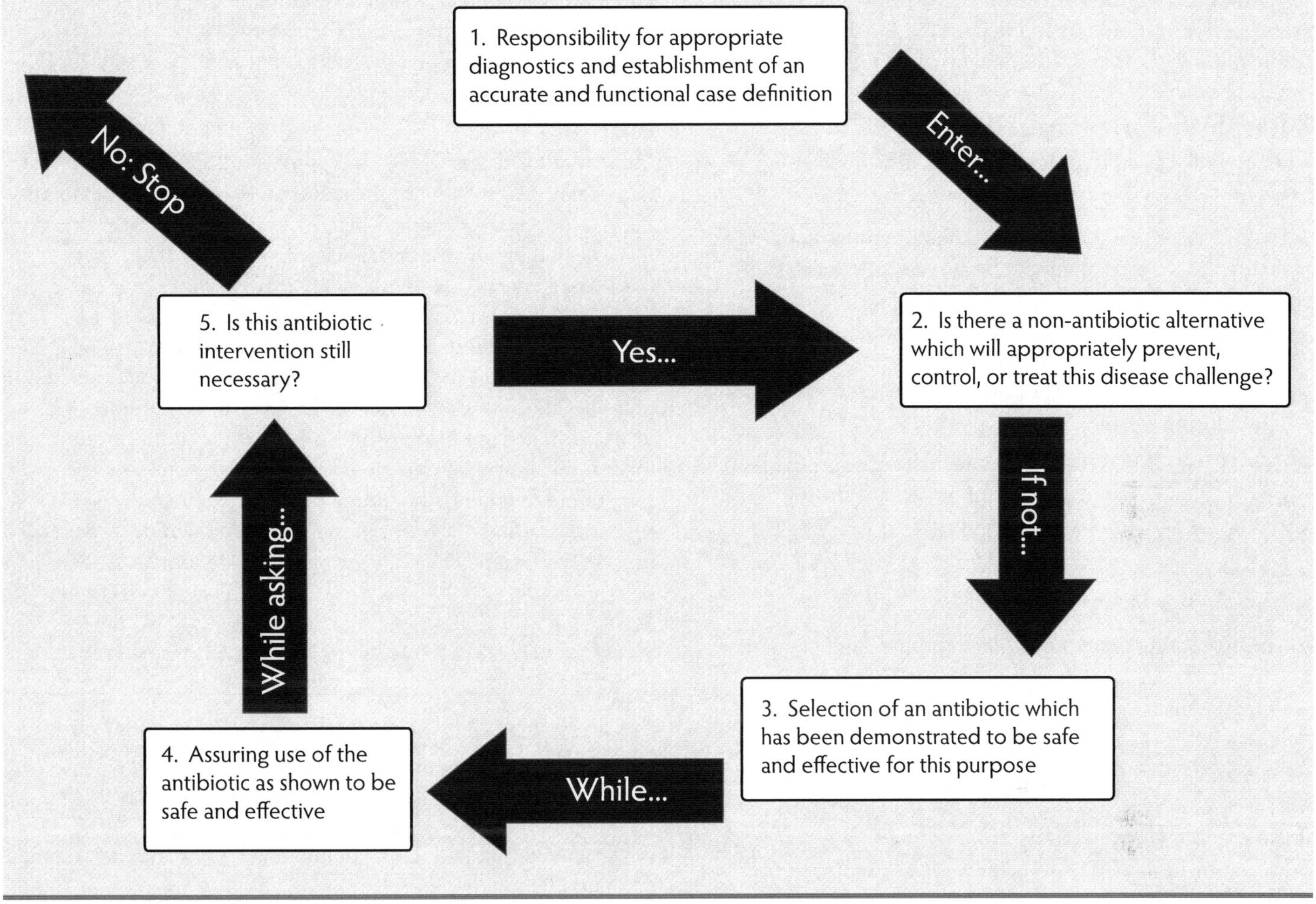

Appropriate denominators for beef cattle

After looking for a denominator for kg of use, one is tempted to completely discard the kg numerator. But, the kg metric has become a staple of antibiotic use monitoring and is likely here to stay. One proposed denominator for kg of drug is an estimate of the mass of animals sold during the monitoring period, or the estimated mass of the population as some point during the monitoring period. The problem with the population mass is that it requires estimating some type of average mass, which may not reflect the size of the animals at the time of exposure. When looking at a policy goal of comparing species, then this is one of the few options, but don't confuse it with something that can meaningfully drive stewardship through management decisions at the production level. Biomass as a denominator has been proposed in the United States, with comments on this approach recently being requested by the Food and Drug Administration Center for Veterinary Medicine (FDA/CVM).[2]

Another option for a denominator to pair with the kg numerator is the number of animals rather than mass, whether sold or determined at a consistent time point during the monitoring period. This would make the most sense for comparing antibiotic use within specific segments for a species; however, this denominator loses clarity across species because it does not take mg/kg dosing realities into consideration.

When the numerator is number of animals exposed (regimens) or duration of exposure, then a time component makes the most sense. In feedlot, the term "days on feed", or DOF, describes the period an animal has been in the feedlot. The DOF values can be added up and divided by 365 to derive a fairly intuitive "animal year". The results are values for number of regimens and number of days exposure per animal year.

What sources of data are available?

Data collection sources in feedlots vary across the spectrum of production systems and sites. In some cases, there are available sources of batch-specific ration antibiotic inclusions coupled with individual animal treatment records which contain weight, drug, dose, indication, and duration. At the opposite end, some records may consist only of purchase receipts coupled with verbal protocols. In the authors experience with evaluating antibiotic use records across multiple species and segments within these species, one of the major lessons of the antibiotic use monitoring programs is that perhaps broader incorporation of a consistent antibiotic record format would also lead to advancing understanding of the disease pressures being encountered.

What do the metrics tell us?

Looking at all 3 components of antibiotic use monitoring, we have...

- an estimate of total amount of different antibiotics used as expressed by kg in relation to some type of animal mass or population number,
- an estimate of the total number of animals exposed in relation to a time period (animal year), and,
- an estimate of the total duration of exposure for these animals during an animal year.

The big concept is that these components of measuring use are only truly valuable for advancing antibiotic stewardship at the treatment decision level when they are...

- associated with cause for this use,
- as determined by treatment records.

When these components are in place, the antibiotic use metrics are in reality a proxy for disease pressure on a production system. Looking again at Figure 1, that makes sense as our first obligations are case definitions, diagnostics, and then non-antibiotic alternatives for disease prevention, control, and treatment.

What are others doing?

There is a wealth of literature, both reporting antibiotic use, and providing input into the most appropriate metrics. A comprehensive review is beyond the scope of these proceedings; however, there are some key papers which highlight thoughts on collecting and applying antibiotic use data. For the purposes of beginning to explore the subject of antibiotic use metrics, the paper by Collineau, et al., is an excellent start with summaries of different metrics and their characteristics.[3] There is one especially telling quote from the abstract for this paper; "... there is a clear lack of standardization, resulting in poor transparency and comparability".

Summary

The choosing of reasonable metrics for monitoring antibiotic use in food animals starts with the end goal. When there is a commitment to providing the resources to drive antibiotic use at the level of those making the use decisions, then it is clear that use record-based systems are worth their higher resource requirements. This is especially clear when it is recognized that successful, antibiotic use monitoring becomes a proxy for disease pressure due to being closely correlated with cause for the use. Other, less granular, approaches such as sales-based estimates coupled with an extrapolated animal population estimate are sufficient only for macro-monitoring of use trends. The basic question is whether the desired outcome is providing a tool for advancing stewardship at the end-user level, or rather to provide a crude instrument for policy pressure where the only goal is reduction.

References

1. IDSA. Promoting Antimicrobial Stewardship in Human Medicine. http://www.idsociety.org/Stewardship_Policy/: Infectious Disease Society of America, 2017.

2. FDA/CVM. FDA's Proposed Methoid for Adjusting Data on Antimicrobials Sold or Distributed for Use in Food-Producing Animals, Using a Biomass Denominator. https://www.fda.gov/downloads/AnimalVeterinary/SafetyHealth/AntimicrobialResistance/UCM571099.pdf: Food and Drug Administration Center for Veterinary Medicine 2017.

3. Collineau L, Belloc C, Stark KD, et al. Guidance on the Selection of Appropriate Indicators for Quantification of Antimicrobial Usage in Humans and Animals. *Zoonoses Public Health* 2017;64:165-184.

Swine Health Information Center 2017 progress

P. Sundberg, DVM, PhD, DACVPM
Swine Health Information Center, Ames, Iowa

Introduction

The Swine Health Information Center was funded by a National Pork Board grant in July of 2015. Its mission is to protect and enhance the health of the United States swine herd through coordinated global disease monitoring, targeted research investments that minimize the impact of future disease threats, and analysis of swine health data.

To give the Center strategic direction and set policy, a Board of Directors was named for the newly formed, independent corporation. And, to help guide and give oversight to its programs and research, two working groups were formed with volunteer members from pork industry associations, pork production, veterinary medicine, academia, USDA and allied industry. Because of our linked production, SHIC also has engaged Canadian counterparts to help coordinate disease information and response with their domestic industry.

The Monitoring and Analysis Working Group is charged with assessing foreign, transboundary production disease risk using information from a variety of sources. The outcome of this assessment is the on-going prioritization of the Swine Disease Matrix (Table 1.). The Working Group is also responsible for improving the health of the nation's swine herd through new project development and oversight of on-going projects. These include monitoring for domestic diseases affecting swine health and analyzing health and other data to support on-farm and prospective producer decision making.

The Preparedness and Response Working Group is charged with oversight of the Swine Disease Matrix research and funding decisions to fulfill the Matrix research and other science-based objectives. It is also responsible for advising and oversight of SHIC's role in the emerging swine diseases response planning. That includes the appropriate SHIC response to an emerging swine disease and for the information and analysis necessary to support the proportional pork producer and pork industry response to these emerging diseases.

2017 Progress

Swine Disease Matrix

In collaboration with the American Association of Swine Veterinarians, the National Pork Board and the National Pork Producers Council, the Swine Health Information Center continues to review and prioritize, based on changing conditions or new information, a list of endemic and foreign swine pathogens, known as the Swine Disease Matrix. The SHIC Monitoring and Analysis Working Group reviewed the list and has prioritized it to help to focus SHIC activity on the highest risks to the US pork industry. Prioritization was done by assigning a 1 to 10 score to each of three criteria and averaging the scores. The likelihood of introduction of the transboundary pathogen into the United States or of the endemic pathogen changing and causing a newly emerging disease or syndrome was scored. Scoring the likely effect on production economics of disease from the pathogens was also done. In addition, the probable effect on domestic and/or international markets from an outbreak due to the pathogen was scored.

Organizing and funding the research needed for diagnostic preparedness for the prioritized diseases of the Swine Disease Matrix is a core activity of the Center. In 2016, SHIC Matrix research focused on the ability to detect the Matrix pathogens via nucleic acid detection, using platforms that are commonly available in the US major veterinary diagnostic laboratories, for example PCR testing. Given that some of these assays may already be produced and available in other countries, the research effort included an assessment of the fitness for purpose of these tests.

The SHIC 2017 Swine Disease Matrix research focused on the development and validation (analytic and diagnostic) of antibody detection assays for antibodies important for monitoring for emerging diseases, determining freedom from disease (after an outbreak) or defining the extent of disease spread. The outcome of the research will be assays with a consistent and reliable means to screen populations using high throughput capacity (high volume of samples and quick turnaround time), the ability to detect antibody in sera and oral fluids and DIVA capability if vaccine availability justifies the need for a DIVA assay.

The current Swine Disease Matrix priority list and the outcomes of these research programs are presented in Table 1.

Investigating swine health risks from feed and common inputs to production

In preliminary findings, a study conducted by Pipestone Applied Research and South Dakota State University shows the potential for foreign and endemic viruses to contaminate and survive in feed ingredients, including soybean meal (SBM) and dried distillers grains (DDGs). The study is based on a model simulating the transboundary movement of contaminated ingredients that was developed to identify "high risk combinations" of viruses and feed ingredients.

This research examined multiple viruses for their ability to survive under shipping conditions coming into the U.S. Preliminary data indicate the survival of viruses and surrogate viruses, including the Seneca Virus A (surrogate for FMDV and of interest itself), Bovine Herpesvirus-1 (surrogate for Pseudo-rabies virus) and PRRSV (using PRRSV 174), during the 37-day study period.

Table 1: Swine Disease Matrix

SWINE DISEASE MATRIX

Representative virus affecting swine	Production impact	Domestic/ foreign market impacts	Likelihood of introduction or emergence	Numerical average	PCR capability	Oral fluid ELISA (2017)
Foot and mouth disease virus	9	9	9	9.0	X	X
Classical swine fever virus	9	9	5	7.7	X	X
African swine fever virus	9	9	5	7.7	X	X
Pseudorabies virus*	8	8	5	7.0	X	X
Influenza A virus	4	8	8	6.7	X	
Nipah virus*	8	9	2	6.3	X	
Ebola-Restin*	8	9	2	6.3	X	X
Porcine epidemic diarrhea virus	6	4	7	5.7	X	
Chinese high path PRRS virus *	6	5	5	5.3	X	
PRRS virus	6	3	6	5.0	X	
Porcine teschovirus (Teschen/PTV1)*	5	5	5	5.0	X	X
Japanese enchephalitis virus*	5	5	4	4.7	X	
Getah virus*	5	5	4	4.7		
Transmissible gastroenteritis virus	5	4	4	4.3	X	
Menangle virus*	4	4	4	4.0	X	
Porcine circovirus	4	2	5	3.7	X	X
Circovirus 3*	4	2	5	3.7	X	X
Porcine rotavirus	4	1	5	3.3	X	
Swine vesicular disease virus	4	3	2	3.0	X	
Vesicular exanthema of swine virus*	3	4	2	3.0		
Porcine rubulavirus*	5	2	2	3.0	X	
Seneca Valley virus*	3	3	3	3.0	X	
Porcine parvovirus	3	1	5	3.0	X	
Porcine deltacoronavirus	3	2	3	2.7	X	
Porcine parainfluenza 1 virus*	3	2	3	2.7	X	X
Atypical swine pestivirus*	3	2	3	2.7	X	X
Influenza C virus*	2	2	2	2.0	X	X
Porcine respiratory coronavirus*	2	2	2	2.0	X	
Hemagglutinating encephalomyelitis virus*	2	2	2	2.0	X	
Encephalomyocarditis virus*	3	1	2	2.0	X	
Hepatitis E virus*	1	3	1	1.7	X	X
Porcine adenovirus*	2	1	2	1.7		
Porcine kobuvirus*	2	1	2	1.7	X	
Porcine sapovirus*	2	1	2	1.7		

Table 1, continued: Swine Disease Matrix

Representative virus affecting swine	Production limpact	Domestic/ foreign market impacts	Likelihood of introduction or emergence	Numerical average	PCR capability	Oral fluid ELISA (2017)
Orthoreovirus*	2	1	1	1.3		
Sendai virus*	1	1	2	1.3		
Porcine cytomegalovirus*	2	1	1	1.3	X	
Vesicular stomatitis virus*	1	1	1	1.0		
Chikungunya virus*	1	1	1	1.0		
Rabies virus	1	1	1	1.0		
Porcine bocavirus*	1	1	1	1.0		
Porcine astrovirus*	1	1	1	1.0		
Swine pox virus*	1	1	1	1.0		
Porcine sapelovirus*	1	1	1	1.0	X	X
Porcine torovirus*	1	1	1	1.0	X	
Swine papillomavirus*	1	1	1	1.0		

* Fact sheet found on **www.swinehealth.org**.

Ingredients frequently supporting virus survival include SBM, lysine, choline, and Vitamin D, and, in some cases, DDGs. None of the viruses survived the 37-day incubation period in the absence of a feed component matrix.

For the purpose of African Swine Fever Virus evaluation, an Eastern European model (Warsaw, Poland to Des Moines) was developed. Using the appropriate biosecure laboratories at Kansas State University, at designated intervals post-inoculation, samples were tested by PCR for the presence of viral nucleic acid and viability by virus isolation and swine bioassay (as needed).

These results suggest a subset of contaminated feed ingredients could serve as vehicles for foreign animal disease, other transboundary introduction in the US, and possibly circulation of viruses within the US. In particular, the PRRSV data may provide new insights and areas of further study on the role of area spread. Heat treatment during corn processing into DDGs and soybean conversion into SBM should neutralize pathogens present on the corn kernel or bean prior to processing. However, research at Kansas State University has shown the potential for porcine epidemic diarrhea virus (PEDV) contamination of feed during the milling process if PEDV is present within the feed mill emphasizing the need for feedmill biosecurity plans (**https://www.aasv.org/shap/issues/v24n3/v24n3p154.pdf**). And, more information is needed about oral viral infective doses to accurately assess risk.

But further mitigation research has already begun. This project includes testing of a variety of feed additives or other processes that might be able to neutralize these pathogens to mitigate risk when added to feed components prior to shipping to the US, on arrival to the US or during milling in the US.

Global and domestic swine disease monitoring

A near real-time monitoring system of swine diseases around the world includes identification of potential hazards due to new diseases or changes in current diseases' status, screening steps to evaluate the information collected and informing the US pork industry through regular, timely reporting. Multiple sources of information may contribute to the development of this global monitoring system for swine diseases. Sources of information may be classified as "soft or unofficial" (rumors or data that may or may not have been corroborated) and "hard or official" (confirmed by national or international agencies). The global disease monitoring system has been developed at the University of Minnesota, with collaboration with the USDA/APHIS Center for Epidemiology and Animal Health, to result in a strong private-public-academic partnership.

On a regular basis, data are evaluated by a group of swine health experts, including input from the USDA/APHIS Center for Epidemiology and Animal Health, and a report generated. The summary includes interpretation from the experts that add their impression about the event – does it sound accurate, should I care about this, why or why not. Follow-ups with local contacts is also done, when possible. The information is graded to reflect a consensus of risk to the US pork industry and the report released. The system is tested and is expected to be fully operational by early in 2018.

SHIC has also funded an Iowa State University and University of Minnesota collaborative project to develop a system for taking advantage of the SHIC and USDA funded project that standardizes VDL swine test results reporting. The veterinary diagnostic laboratories of Iowa State University, Kansas State University, University of Minnesota and South Dakota State University have worked with Clemson University on this standardization, with the capability of implementation by additional veterinary diagnostic labs. This project has improved the ability of the diagnostic labs to communicate veterinary diagnostic lab test results and improved the ability for analysis of these data to offer real-time status of US disease trends and outbreaks. These data from VDLs, and other sources, are compiled while respecting producer-specific confidentiality, analyzed, reviewed to provide context and interpretation of the information and reported. The first report of this collaborative project is expected by early 2018.

Farm disease risk and outbreak prediction

SHIC is funding a project to generate farm-level risk scores and create unique dynamic maps of PRRS/PED risk that account for recent animal movements, present disease distribution, and environmental factors. The algorithms developed will then be used to test against other pathogens. The development of the algorithms, with their validation, is scheduled for the first quarter of 2018.

Because of its large spatial and temporal scope, incorporation of animal movements, and the use of innovative data science methods, this project has the potential to generate new knowledge about the interacting processes that contribute to viral spread. In addition, these farm-level forecasts of risk are unique from similar efforts in that these scores change dynamically through time as they account for pig movements and infection status of neighboring farms. Producers will have access to their own farm-level predictions that will enable near real-time, appropriate actions. Ultimately, this project will provide tools available for coordinated actions of producers and practitioners that could also control disease outbreaks.

Rapid Response Program

A SHIC funded Rapid Response Program enables epidemiological investigations of emerging transboundary or endemic swine disease outbreaks. A Rapid Response Corps of practicing veterinarians, state and federal animal health officials or their representatives, and veterinary epidemiologists are being trained and ready for deployment in different regions of the US. (Figure 1).

Figure 1: Rapid Response Program regions

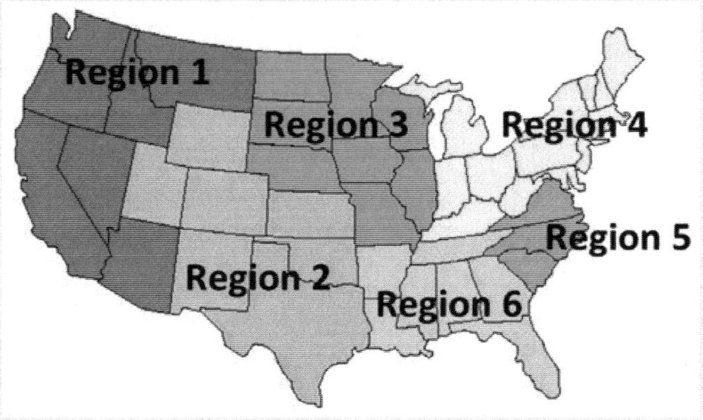

While Corps members will be required to participate in the online training available at **www.swinehealth.org**, it is open to all who are interested in the Rapid Response Program or wish to grow their knowledge of epidemiological investigation and processes. The objective of the training program is to standardize the conduct of epidemiological investigations conducted by the Rapid Response Corps in the event of a swine disease outbreak. There are four training modules and a short quiz at the end of each module assesses content understanding by the participant.

When a disease investigation is requested by a producer or veterinarian, SHIC will coordinate engaging Rapid Response Corps members in the appropriate region. Within 72 hours of notification, a Rapid Response Team will be able to be on site to conduct the epidemiological investigation in cooperation with the producer and veterinarian of record. SHIC has a cooperative agreement with USDA to help fund the Rapid Response Corps investigations.

Diagnostic fee support

To help more producers and their veterinarians solve outbreaks from unknown causes, SHIC developed a system to help offset diagnostic fees for further investigation after the initial diagnostics are completed. To request support of diagnostic fees the diagnostician of the case must contact SHIC.

In cases of unresolved high or ongoing morbidity or mortality a panel of diagnosticians will review the case and must concur that the most likely differential diagnoses have been addressed. To ensure that a foreign animal disease is not being missed, the SAHO of the premises' state must be aware of the lack of a satisfactory diagnosis for the case and has considered if a foreign animal disease investigation should be initiated.

More information and the forms needed to instigate the process can be found at www.swinehealth.org. A novel Sapelovirus, causing morbidity and mortality via CSN infection, has been identified through this process.

SHIC monthly e-Newsletter

The monthly SHIC e-Newsletter contains updates and information on current and new projects and other up-to-date swine health information. If you are interested in receiving it, please send your email address to **psundberg@swinehealth.org** and you will be added to the distribution list.